1 MONTH OF
FREE
READING

at

www.ForgottenBooks.com

By purchasing this book you are eligible for one month membership to ForgottenBooks.com, giving you unlimited access to our entire collection of over 700,000 titles via our web site and mobile apps.

To claim your free month visit: www.forgottenbooks.com/free539771

English
Français
Deutsche
Italiano
Español
Português

www.forgottenbooks.com

Mythology Photography **Fiction**
Fishing Christianity **Art** Cooking
Essays Buddhism Freemasonry
Medicine **Biology** Music **Ancient
Egypt** Evolution Carpentry Physics
Dance Geology **Mathematics** Fitness
Shakespeare **Folklore** Yoga Marketing
Confidence Immortality Biographies
Poetry **Psychology** Witchcraft
Electronics Chemistry History **Law**
Accounting **Philosophy** Anthropology
Alchemy Drama Quantum Mechanics
Atheism Sexual Health **Ancient History**
Entrepreneurship Languages Sport
Paleontology Needlework Islam
Metaphysics Investment Archaeology
Parenting Statistics Criminology
Motivational

A

N E W   S Y S T E M

O F

G E O G R A P H Y:

IN WHICH IS GIVEN,

A General Account of the SITUATION and LIMITS, the
MANNERS, HISTORY, and CONSTITUTION, of the
feveral KINGDOMS and STATES in the known World;

And a very particular Defcription of their *Subdivifions* and *Dependencies*; their
Cities and *Towns, Forts, Sea-ports, Produce, Manufactures* and *Commerce.*

By A. F. BUSCHING, D. D.

Profeffor of Philofophy in the Univerfity of GOTTINGEN, and Member of the
Learned Society at DUISBURG.

Carefully Tranflated from the laft Edition of the GERMAN Original.

To the Author's Introductory Difcourfe are added three *Effays* relative to the Subject.

Illuftrated with Thirty-fix *Maps*, accurately projected on a new Plan.

IN SIX VOLUMES.

VOLUME the SECOND.

CONTAINING,

HUNGARY, TRANSYLVANIA, SCLAVONIA, DALMATIA,
TURKEY in EUROPE, PORTUGAL, SPAIN and FRANCE.

L O N D O N:
Printed for A. MILLAR in the Strand.
M DCC LXII.

CONTENTS

TO THE

SECOND VOLUME.

The KINGDOM of HUNGARY.

INTRODUCTION	page 3	Borſod	page 64
Proper Hungary	17	Jaſi	65
Lower Hungary	18	Cſongrad	66
Upper Hungary	54	Maramarus	67
Greater Sitz	56	Szathmar	68
Saros	59	Szabolts	69
Zemplin	60	Zaraud	70
Abaujvar	61	Cſanad	71
Torner	62	Bannate	72
Gomor	63		

TRANSYLVANIA.

Szolnock	79	Orbai Szek	83
Kolos	80	Aranyas	84
Weiſſenburg	81	Altland	85
Hunyad	82	Borzeland	87

CONTENTS.

HUNGARIAN ILLYRICUM,

Or the KINGDOMS of

SCLAVONIA, CROATIA, and DALMATIA.

SCLAVONIA	page 89	Zeng	page 96
Verotz	ibid.	Zwonigrod	97
Sirmi	90	Zara	98
Save	91	Macarfca	100
CROATIA	92	Saba	101
Krentz	ibid.	Ocero	102
Warafdin	93	Ragufa	103
Banat Croatia	94	Turkifh	105
DALMATIA	95	Zenta	106

TURKEY in EUROPE.

Introduction	109	Greece	147
Turkifh Illyricum	121	Candia	155
Bulgaria	125	Santorin	157
Romania	128	Cerigo	158
Macedonia	135	Cephalonia	159
Albania	139	Corfu	160
Theffaly	141	Walachia	161
Livadia	142	Moldavia	165
Morea	· 144	Tartary	167

PORTUGAL

C O N T E N T S.

P O R T U G A L.

Introduction	page 175	Algarve	page 230
Entre Douro e Minho	191	Madeira	234
Traz-os Montes	197	Miguel	235
Beira	202	Tercera	236
Eftremadura	212	Praya	237
Alentejo	223	Flores	238

S P A I N.

Introduction	241	Granada	291
Galicia	264	Murcia	295
Afturias	268	Valencia	296
Bizcaya	270	Caftile	300
Giupuzcoa	271	Old Caftile	308
Alaba	273	Navarre	312
Leon	274	Aragon	314
Eftremadura	277	Catalonia	319
Andaluzia	280	Majorca	325
Seville	281	Ivica	329
Cordova	289	Formentera	330
Jaen	290		

F R A N C E.

Introduction	333	Roufzillon	481
Paris	363	Navarre and Bearn	485
Ifle de France	383	Guyenna	486
Picardy	396	Gafcogne	500
Artois	405	Saintonge	507
Champagne	409	Angonoism	509
Burgundy	420	Aunis	510
Dombes	430	Poitou	512
Dauphine	432	Bretagne	516
Provence	443	Normandy	523
Languedoc	461	Havre de Grace	533
Foix	484	Maine	534
		Perche	

CONTENTS.

Perche	page 536	Saumur	page 566
Orleanois	537	Flandres	ibid.
Nivernois	543	Berry	569
Bourbonnois	546	Dunkirk	571
Lionnois	547	Metz	574
Auvergne	550	Lorrain and Bar	578
Limofin	555	Verdun and the Verdunois	589
La Marche	557	Toul and Toulois	590
Berry	558	Alface	591
Touraine	561	Franche Comte	607
Anjou	564		

THE

T H E

N G D O M

O F

N G A R Y,

Hungarian language, Magyar Orszag,

A N D   I T S

PORATED COUNTRIES.

———————————

B

CONTENTS

Perche   page 536
Orleanois   537
Nivernois   543
Bourbonnois   546
Lionnois   547
Auvergne   550
Limofin   555
La Marche   557
Berry   558
Touraine   561
Anjou   564

Saumur
Flandres
Berry
Dunkirk
Metz
Lorrain and Bar
Verdun and the ꞌ
Toul and Touloiꞅ
Alſace
Franche Comte

THE

KINGDOM

OF

HUNGARY,

In the *Hungarian* language, Magyar Orszag,

AND ITS

INCORPORATED COUNTRIES.

THE
KINGDOM
OF
HUNGARY

In the Hungarian Language, Magyar Ország,

AND ITS

INCORPORATED COUNTRIES.

A N

INTRODUCTION

TO THE

KINGDOM OF HUNGARY,

In the *Hungarian* language, MAGYAR ORSZAG,

AND ITS

INCORPORATED COUNTRIES.

§. 1.. AMONG the numerous maps of *Hungary*,. which are owing to the frequent wars with *Turkey* during the two laſt centuries, and of which this kingdom ·was the theatre, one of the beſt is that of captain *Mullen*, performed by order of the Revenue. chamber of *Hungary*, and made public in 1709, with a Dedication to the Emperor *Joſeph*, and reprinted afterwards by *Homann* on four large ſhèets. *Haſen's tabula Hungariæ ampliori ſignificatu—ex recentiſſimis pariter et an. tiquiſſimis·relationibus et monumentis concinnata*, &c. publiſhed·by *Homann's* heirs in 1744, is a work of the greateſt attention and moſt critical exact. neſs; and exhibits both the modern and ancient ſtate of this kingdom : But in the latter it is more perfect and unexceptionable than in what relates to the former. *Homann's Danube* repreſents not only *Hungary*, but in. cludes alſo *Turkey* in *Greece*.

§. 2. The ancient name of *Hungary* was *Pannonia*, ·which it received from the *Pannonians*, ·who are deſcendants of the *Slavi*, and arrogantly ſtiled themſelves *Panovæ*, *i. e.* Lords; from the *Sclavonian* word *Pan*, a Lord; whence foreigners have accordingly called them *Pannonians*. But the ancient *Pannonia* did not.include all the preſent kingdom of *Hungary*,

nor

nor does modern *Hungary* comprehend all the ancient *Pannonia*. That part of modern *Hungary* betwixt the *Carpathian* mountains, the *Danube* and the *Theifs*, was by the Ancients termed *Jazygum Metanaftarum regio*. They were called *Metanaftæ* as being emigrants hither from other places: But the origin of the word *Jazyges* is uncertain. The appellation of *Hungary* is derived from the *Hungarians*, a race of the *Huns* who, in the ninth century, as fhall be fhewn hereafter, took poffeffion of thefe countries.

§. 3. The word *Hungary* is taken both in a limited and extenfive fenfe. In the former it is bounded on the fouth by *Servia* and the river *Drave*, which feparates it from *Sclavonia*; to the caft by *Walachia* and *Tranfylvania*; to the north by the *Carpathian* mountains, which feparate it from *Poland*; and weftward by *Moravia*, *Auftria* and *Stiria*. But in its more extenfive fenfe, it comprehends *Sclavonia, Dalmatia, Bofnia, Servia* and *Tranfylvania*; and even *Moldavia* and *Walachia* are included in it.

§. 4. *Hungary* lies in the northern temperate zone. Its northern parts indeed are mountainous and barren, but healthy: the fouthern, on the contrary, are level, warm and fruitful; but, by reafon of the many fwamps and moraffes not fo wholefome as the northern. The moft common diftempers in *Hungary* are the gout in its feveral fpecies, and the fever; the laft of which is even called the *Hungarian* ficknefs, and is fo deftructive to foreigners as to have given rife to the common faying, *Hungary* is the *German* foldiers grave. The firft fymptoms of this diftemper are nodes or tubercles on the hands and arms, which, if rubbed in time with vinegar, falt and garlick till they difappear, a ftop is thereby put to the progrefs of the diftemper. The chief caufes of thefe difeafes are intemperance in eating and drinking, together with the alternative of very hot days and as cold nights, by which, without more than ordinary care, colds are foon contracted. The plague comes from *Turkey*, and fpreads here by contagion. *Hungary* has fuch a fuperfluity of the neceffaries and enjoyments of life, that a traveller has faid of it, *Extra Hungariam non eft vita, et fi eft, non eft ita*: i. e. Out of *Hungary* there is no living, or if there be living it is not life. The level country produces the following vegetables; grafs, efculent plants, tobacco, faffron, afparagus, melons, hops, corn, pulfe, millet, buckwheat, delicious wine, fruits of various kinds, peaches, mulberry-trees, cheftnuts and wood. Of minerals, here are gold, filver, copper, iron, lead, quickfilver, cinnabar, antimony, auripigmentum or yellow orpiment, fulphur, vitriol, marcafite, falt native and factitious, faltpetre, magnets, afbeftos, or earth-flax, marble of feveral colours, alabafter and gems, though the latter indeed are very different from the oriental. Of animals, fine horfes, moftly moufe-coloured, and of which incredible numbers are exported, buffaloes, horned cattle, affes, mules, fheep, goats, fwine, and many fpecies of wild beafts, birds and fifhes, ftags, deer, chamois, bears, lynxes, partridges, wood-cocks, moor-fowl, pheafants, bees, &c. The

mountains,

·mountains, with which *Hungary* is environed on all fides, have eminences producing excellent wine, and others contain valuable minerals.

§. 5. *Hungary* has alfo divers kinds of fprings : namely, warm baths ; as thofe of *Erlau, Buda, Beimotz, Eifenbach, Peſt, Ribat, Raitz, Zips, Sklen, Stubna, Gran, Trentfchin* and *Waradin:* acid fprings; as thofe of *Novigrad, Saroiz, Szulad, Trentfchi, Altheufel,* &c. Mineral fprings; as the *Neuheufel* in *Herren-grund,* and the *Schmolnitz,* which contains *àqua fortis :* likewife lethiferous fprings in the Gefpanfchaft or Palatinate of *Altheufel:* petrifying fprings in the Gefpanfchaft of *Liptauer ;* and freezing fprings in that of *Torn.*

§..6. The chief mountains in *Hungary* are the *Carpathian (Tatra)* the weſtern boundaries, and running northward. In the county of *Zipſer* they are higheſt, and in clear weather command a·view both· of *Erlau* in *Hungary* and *Cracow* in *Poland.* At the root they are overgrown with common trees, higher up with large ones, and at a greater inteival, which makes as it were the third region, with bruſhwood. The fummit is a chaos of frightful crags and precipices continually covered with fnow, and lakes of very tranfparent water lie betwixt them. *Fatra* in the Gefpanfchaft of *Liptau, Matra* in that of *Heve,* and *Bacony* foreſt.

§. 7. The chief rivers are

The *Danube, Danubins,* the fource of which lies in the principality of *Furſtenberg,* near *Don Efchingen* in *Swabia,* runs eaſtward through *Germany, Hungary* and ·*Turkey* ; and after·receiving fixty navigable rivers, and, including the fmaller, above one hundred and twenty in the whole, difcharges itfelf, by feveral outlets, with fuch violence ·into the *Black-fea,* that both the ſtream and water are perceptible in it for feveral miles diſtance. Concerning· the difputes about the origin of this large and celebrated ſtream, I refer the reader to Mr. *Hauber's* hiſtoiical account of the maps of the circle of *Swabia.* The courfe of the whole river, with all the adjacent countries and 'towns, are delineated on twenty-eight fmall maps, with an account of them in a little work entitled, *Der Wegen des hochſt-tapferen Kaiſer-adlers heldenthaten Siegberuhmte Donau fluſs,* i. e. *The Danube, famous for the heroic atchievements and fignal victories of the intrepid Imperial eagle, Nurnberg,* 1687, fmall quarto : alfo to *Sigmund von Birken neu vermehrten Donau ſtrand-mit 'landcharten and vielerſtadte Kupern, in* 12*mo.* i. e. *Sigmund von Birken's Danube ſhore·enlarged and ·embelliſhed with Maps and Plans of feveral towns, duodecimo.* This river, after paſſing the frontiers of *Germany,* was anciently ·called the *Iſter:*

The *Morau, Marus,* or *Marcus,* feparates *Auſtria* and *Moravia* from *Hungary,* falling into the *Danube* fome miles from *Preſburg.* But we referve a fuller defcription of this river for the article *Moravia.*

، The *Wag, Vaguſs, Cuſus,* iſſues from·the *Carpathian* mountains and runs into the *Danube.*

The

The *Gran, Granna, Granus,* alfo has its fource in the *Carpathian* mountains and joins the *Danube.*

The *Theifs* or *Teys, Tibifcus, Patifcus, Tiffus,* rifes in the *Carpathian* mountains, and whilft among them has a rapid and clear ftream, but afterwards becomes flow and turbid. It receives the fmaller rivers of *Bodroch, Hornat, Szamos (Samofius) Koros (Chryfius)* and *Maros (Marufius)* and four miles above *Belgrade* falls into the *Danube.* No river in *Europe* equals it in plenty of fifh, the *Theifs* fwarming in a manner with the multitude of them: But both in this and other rivers of *Hungary* this abundance varies, the fifh often removing from one river to another.

The *Temes, Temeffus,* rifes in the *Trongate* mountains of *Tranfylvania,* waters *Karanfebes, Lugos* and *Temefwar,* afterwards divides itfelf, and having produced feveral marfhes falls into the *Danube* not far from *Panezorfa.*

The *Drave, Dravus,* iffues out of *Stiria,* feparates *Hungary* and *Sclavonia,* and at laft falls into the *Danube* near the town of *Darda.*

Araba, commonly called the *Raab,* iffues in *Stiria,* and joins the *Danube* near *Raab.*

Leitha, Litaha, a fmall river betwixt *Auftria* and *Hungary,* falls into the *Danube* near *Altenburg.*

Here are alfo two remarkable lakes, both abounding with fifh; namely, *Balaton* or *Plattenfee,* in the Palatinate of *Simigen,* eight miles long and in fome places two broad; and *Neufiedlerfee, lacus Peifonius,* lying betwixt the Palatinates of *Oldenburg* and *Wieffelburg.* By this river the *Oedenburghers* pretend to foretel whether it will be a good year for wine or not, as when full of water it is faid to portend a bad vintage, whereas a fmall quantity betokens plenty and excellence of wine. Among the *Carpathian* mountains are four lakes.

§. 8. The inhabitants are of different origins. The true *Hungarians* are defcendents of that fierce people called by the fame name, who by dint of arms feated themfelves here in the year 888 ; and, though much more civilized than their anceftors, yet feveral traces of a *Scythian* extraction are ftill vifible among them. The *Jafigi* and *Cumani* are alfo included in their number. Of the former mention has been made above. The latter were driven out of their fettlements by the *Tartars,* and in 1234 entered *Hungary,* in the reign of King *Bela* the fourth, who beftowed on them the fertile country betwixt the *Teys* and *Temefwar.* They formerly enjoyed great privileges, but in 1638 thefe were revoked. Among the titles of the Governor of the Kingdom is that of *Judex Cumanorum.* Another part of the inhabitants are of *Sclavonian* extraction, and this includes the *Bohemians, Croats, Servians* or *Ratzians, Ruffians* and *Vandals.* Thefe inhabit the caft and northern parts of *Hungary,* as the Gefpanfchafts of *Prefburg, Neutra, Trenfi, Arva, Lipto, Thuro, Altheufel, Barfe, Novegrad, Han, Gomo* and *Oedenburg.*

Oedenburg, and are likewife found difperfed all over *Hungary*. They feem to have been fettled here from remoteft antiquity. The *German* nations are, the *Auftrians*, *Stirians*, *Bavarians*, *Franks*, *Swabians* and *Saxons*. Thefe feem to have entered *Hungary* much about the time when the *Saxons* feated themfelves in *Tranfylvania*, namely under the Kings *Geyfa* the fecond and *Andrew* the fecond. But in procefs of years war, commerce, and the fruitfulnefs of the country drew hither other *Germans*, who are alfo confiderably increafed fince the fovereignty of the houfe of *Auftria*. The *Walachian* or *Olachi*, who inhabit the country next to *Tranfylvania* and *Walachia* are to all appearance defcendants of the *Romans* that fettled in *Dacia*. In the defcription of *Tranfylvania* we fhall give a further account of them. Among the foreigners in *Hungary* are the *Greeks*, who removed hither for the fake of a more advantageous commerce; the *Jews*, the number of whom was formerly much greater there than at prefent; the *Turks* and *Zigduns*, *Zingari* or *Czigani*, a wandering people and of very uncertain origin. Many of thefe are Smiths and Muficians. The *Hungarians* (among whom are included the *Slàvi*) and the *Germans*, who in the public acts of the Kingdom were alone ftiled *Regnicolæ* and *States* of *Hungary*; whereas all the reft were called *Externi*, *Extranei* and *Forenfes*. From what has been faid, it appears, that the inhabitants muft neceffarily have been formed at firft of different difpofitions, though by frequent intercourfe they now refemble one another greatly. They are for the moft part of a fanguine choleric temper; the nobility are numerous, and both in their drefs and tables affect great pomp and magnificence, yet apply themfelves notwithftanding to learning and rural improvements, but more to war, hunting and martial exercifes. The *Hungarian* Habit is quite different from that of all other *European* nations, and may be faid to be both convenient and graceful.

§. 9. In *Hungary* are four common languages; the *Hungarian*, which is that of the proper *Hungarians*, is of *Scythian* origin, without the leaft affinity to any of the *European* tongues, and of one unvaried dialect. In writing, the *Hungarians* ufe the *Roman* characters; the *German* has its different dialects according to the different nations of the *Germans* fettled here. The *Sclavonian*, which derives its origin from the *Sarmatian*, is divided into the *Bohemian*, *Croatian*, *Vandalian*, *Ratzian* and *Ruffian* dialects. The *Walachian* tongue is allied to the *Italian*, and formed by a mixture of *Latin* and *Sclavonic*: Laftly, the *Latin* is not only fpoken by the literati and gentry, but alfo by the commonality, though thefe often break *Prifcian's* head. The *Czigianians* have a fpeech compofed of a corruption of the *Walachian*, *Sclavonic*, *Hungarian* and other languages.

§. 10. This country, it is highly probable, was bleffed with the knowledge of Chriftianity foon after our Saviour's time, the Apoftle *Paul*, in his Epiftle to the *Romans*, ch. xv. v. 19, faying that *he had preached the Gofpel from Jerufalem to Illyricum*. Now *Illyricum* containing a great part of the

modern

modern *Dalmatia, Croatia* and *Sclavonia*, doubtlefs fome account thereof, muft have reached *Pannonia*, as it bordered on the confines of the former. In the fourteenth century *Sermia* had bifhops, as likewife the *Goths* in *Dacia*. But it was the tenth century before chriftianity was thoroughly eftablifhed in *Hungary*, when, in the year 969 or 975, *Deyfa*, Prince of the country, was baptized; and it ftill made a farther progrefs under his fon and fucceffor *Stephen*, whofe zealous endeavours gained him the title of Apoftle and Saint. In 1523 the Reformation began here by *Martin Cyriacus*, a native of *Leutfchau*, who firft preached the pure doctrine of the Gofpel in this country; but it had made its way before, into *Tranfylvania*, in 1521, by means of fome *Lutheran* writings. From that time great numbers of *Hungarians* went to ftudy in *Germany*, and, at their return into their country, were looked upon as difciples of *Luther*. *Zwinglius*'s doctrine concerning the Sacrament had been made known in *Hungary* a little before or immediately after 1530 by *Matthias Devay*; and before the year 1557 *Calvin*'s doctrine had alfo been heard of, and embraced by great numbers. But from the time the Jefuits got footing in *Tranfylvania*, and afterwards in *Hungary*, the Proteftants underwent many fevere trials, efpecially in the beginning of the feventeenth century. Under *Ferdinand* the third the Proteftants were deprived of feveral churches, and of more under the Emperor *Leopold*, in whofe time, at the diet of *Sopron* or *Oldenburg* it was enacted that the Reformed fhould not poffefs above two churches in each Palatinate. Nor could their enemies reft here till they had driven them out of all the churches which had not been exprefly mentioned in the twenty-fixth article of the diet of *Sopron*; and accordingly above three hundred were actually taken from them and fhut up. Of all thefe Palatinates the greateft fufferer was that of *Eifenburg*, which in many parts has not a place of public worfhip within ten miles or more. The *Vandals* too had fix churches, and now are without fo much as one place where worfhip is performed in their own language. Travelling for improvement in knowledge is, in a great meafure, prohibited the Proteftants: Their fchools are alfo confined to the Syntax, and no teachers of the Sciences allowed among them, excepting that fince 1751 they have been indulged with a college at *Oedenburg*, and a gymnafium at *Eperies*. A fuccinct hiftory of the Proteftant church in *Hungary* may be read in the following piece, *Kurze und zuverlaffige Nachricht von dem zuftande der proteftantifchen in dem Konigreiche Ungaren, befonders von den gegenwartigen gefahrlichen Umftanden derfelben*, 1743, in 8vo. 5 *Bogen*: i. e. *A fhort and authentic account of the ftate of the proteftant churches in the Kingdom of Hungary, and particularly of their prefent dangerous fituation*, in octavo, five fheets. Though the Roman Catholics fcarce make a fourth part of all the inhabitants of *Hungary*, and do not pay above a fixth to the public contributions, yet their Religion has notwithftanding the afcendant at prefent. At the head of it are two arch-

bifhops

bifhops and nine bifhops, who, as well as the Probfts and aichdeacons, are nominated by the King, but confirmed by the Pope. The papal power, however, in *Hungary* is not at that height it is in other kingdoms. No appeals are to be made to the Pope. He can alfo only confirm, but not dif-pofe of, any ecclefiaftical poffeffions: All that the proteftants enjoy, how-ever, is a toleration under their fuperintendants and elders. In *Dalmatia*, *Croatia* and *Sclavonia*, the Roman-catholics only are qualified to hold lands. The *Ratzhians*, *Ruffians* and *Walachians* profefs the *Greek* church, which has been privileged fince the year 1690 by the Emperor *Leopold* and other Kings: Part of them have entered into a coalition with the *Romifh* church, and are therefore called *Ecclefia redunita*; but the reft, who ftill perfevere in the feparation, are diftinguifhed by the appellation of *Ecclefia diffentiens:* The former have two bifhops, namely, of *Buda* and *Muncats*; the latter three, *viz.* of *Buda*, *Neu-azad* and *Peterwaradin*, who are fuffragans to the archbifhop of *Carlowitz*. The Anabaptifts and Memonifts are chiefly fettled in the neighbourhood of *Prefburg*; but the *Jews* are difperfed in moft of the confiderable towns, though under the burthen of double taxes of all kinds, which were firft impofed upon them by *Rodolph* the fecond.

§. 11. Learning among the Roman-catholics is principally cultivated by the Jefuits, who in the univerfities of *Tirnau*, *Buda*, *Raab* and *Cafchau* are the profeffors of divinity, philofophy, mathematics, rhetoric and other fci-ences; as alfo in feveral colleges. But the *patres piarum fcholarum* only teach the belles lettres. The Benedictines, Paulines, and other orders, apply themfelves after their manner to learning in their feveral convents. The Lutherans and Calvinifts, having laid the foundation of the fciences in their fchools and gymnafiums, go, if licence can be obtained, to the univerfities in *Germany*, *Holland* and *Switzerland*. But it has been obferved, in §. 10. that very narrow bounds have lately been prefcribed to their ftudies. The oriental chriftians of the *Greek* church alfo begin to fhow a difpofition for ftudy beyond what they were wont. The law was formerly taught only in private, but at prefent there is a public profeffor appointed for it in the univerfity of *Tirnau*, and even a particular college erected for that purpofe at *Erlau*.

§. 12. The burghers follow arts, handicrafts and trade, but the laft is almoft wholly engroffed by the *Greeks*. From *Hungary* is exported wine, faffron, oil, metals, minerals, cattle, leather, wool, tallow and wax; and its imports are fpices, tin, filk and other foreign goods. The pieces of coin current in *Hungary* are, a heller * (babka) an Ungrifch (penz Kralovfzky) which in *Upper Hungary* goes for the fixth part of a grofch †, and in *Lower*

---

* The Heller, according to *Paraire*'s table, is an imaginary coin ufed in the dutchy of *Holftein* and electorate of *Saxony*, worth about ⁵⁄₈ of our money.

† The Grofch, or, as it is pronounced, Grofh in *Hungary*, is worth, according to the above cited author, ⁷⁄₁₀ *Englifh*.

Hungary for the fifth, and the hundredth part of a *Rhenish* gulden; a grofchel (pataz) the fourth part of a grofch; a Kreutzer (kreytzar) a *German* coin, the third part of a grofch and the fixtieth of a *Rhenish* gulden; the pulgrotz (paltura) half a grofch and the fortieth part of a gulden; a grofch (garas great) in *Upper Hungary* worth fix and in *Lower Hungary* five ungrifch and the twentieth part of a gulden; fiebner (hetes fzedmak) worth feven kreutzers; a fiebentehner (fcfztak marias) worth feventeen kreutzers; a halbergulden (pul zlaty) the half of a * guld; an ungrifch gulden (wherfzky zlaty) feventeen grofchen and an half; a *Rhenish* gulden (nemeczky zlaty) twenty grofchen, or in *Upper Hungary* is worth one hundred and twenty, and in *Lower Hungary* one hundred, ungrifch, a thaler is equal to two *Rhenish* guldens. The ducats are of two forts: The *Kremnitz* ducat is worth four †guilders four grofchen, but a *Tranfylvanian* goes only for three guilders.

§. 13. With refpect to the hiftory of the country, it appears, from the records of paft ages, that the ancient weftern inhabitants of *Hungary* were called *Pannonians*; the northern, *Jazygians*. (§. 2.) The *Romans* reduced *Pannonia* and kept it almoft four hundred years, till in the fourth century the *Vandals* drove them out of it, and held it forty years; but in the year 395, when they advanced towards *Gaul*, the *Goths* took poffeffion of their fettlement; but thefe were alfo, in their turn, obliged to refign their new poffeffions to the *Huns*, who had likewife driven them from their ancient habitations. The *Huns* appear from all circumftances to have been a people of *Scythia*, but whether the fame with the *Avareri* or *Abareri*, or diftinct from them, hiftorians are not agreed, fome afferting the former, and others being as pofitive with regard to the latter; thefe alledge the *Huns* invaded *Hungary* but twice, whilft the others labour to prove three feveral incurfions. The firft was about the year 376, when they paffed the *Tanais* and drove out the *Oftrogoths* and other nations. But in the beginning of the fifth century, when headed by *Attilah* the *Dacian*, not to mention the more extenfive progrefs of their arms, they fubdued all *Pannonia*, the two *Thracias*, part of *Nifia*, and the country of the *Metanaftic Jazygians*. After the death of *Attila*, the *Gepideri* and *Goths*, with the affiftance of the *Romans*, among other advantages, wrefted *Dacia* and *Pannonia* out of the hands of the *Huns*. About the year 564, the *Avari*, a people of the eaftern parts of *Scythia*, having fhaken off the *Turkish* yoke, entered *Pannonia* under the command of *Shagun*, difpoffeffed the *Goths*, and over-run the country. Their firft fettlement was at *Sirmium*, but they were afterwards frequently defeated by the *Bulgarians* and the Emperor *Charles* the great, who made ufe alfo of many endeavours for the propagation of chriftianity among them. Thofe who hold the *Huns* and *Avari* to be one people, look on this as the

---

* Two fhillings and four pence *English*. See *Pmaire*'s table.
† The *German* guilder is two fhillings ard fcui pence *English*.

fecond invasion of *Pannonia* by them. In the year 888, the *Huns*, under
the name of *Hungarians*, made a fecond irruption into *Pannonia*, as auxiliaries to *Arnulph*, Emperor of the weft, and *Leo*, Emperor of the caft,
againft the *Bulgarians* and *Sclavonians*, whom they reduced. They had
feven commanders ; and *Germany* and *Italy* afterwards felt the terrible effect
of their ferocity. By degrees their manners took a more civilized turn,
and efpecially when in the latter part of the tenth century their Prince,
Geyfa, embraced the chriftian religion. His fon *Stephen*, in 997, became the
firft native King of *Hungary*, completed the eftablifhment of the chriftian
religion, erected bifhoprics, abbeys and churches, annexed *Tranfylvania*
as a province to *Hungary*, and at his death was canonized. After him followed a fucceffion of twenty Kings, natives of the country, of whom the
fecond, by name *Peter*, put himfelf and kingdom under the protection of
the Emperor *Henry* the third. The eighth, *Ladiflaus* the pious, enlarged
his dominions with *Sclavonia*, *Croatia* and *Dalmatia*, and was greatly
efteemed by his people. The tenth, named *Stephen* the third, by his
marriage with a *Polifh* Princefs, added the diftrict of *Zip*. The twelfth,
Geyfa the fecond, in 1154, invited the *Saxons* into *Tranfylvania*. The
feventeenth, *Andrew* the fecond, conferred great privileges on the nobility,
and even empowered them to oppofe the King, if he fhould attempt any
thing againft the laws of the Kingdom. And this was not repealed till the
year 1688. The nineteenth, *Stephen*, compelled the *Bulgarians* to pay him
tribute : And the laft, *Andrew* the third, died in the year 1301. On
this followed a fucceffion of twelve foreign Kings, of whom *Lewis* the firft,
in the year 1356, reunited to the Kingdom the whole Province of *Dalmatia*, which had been fo often attacked by the *Venetians*. In 1390 *Sigifmund*
rendered *Walachia* and *Moldavia* tributary to him, but mortgaged to *Poland*
thirteen towns of the Gefpanchaft of *Zip*. *Matthias* conquered *Silefia* and
Moravia from the *Bohemians*. Under *Uladiflaus* the fecond, the *jus confue-
tudinarium*, called *tripartitum*, took place ; and *Lewis* the fecond, the laft
of thefe Kings, in 1526, fell in an unfuccefsful battle againft the *Turks*
near *Mohats*. The Kingdom devolved next to the houfe of *Auftria*, in
which it continues to this day. The firft King of this family, *Ferdinand*
the firft, brother to *Charles* the fifth, had fo powerful a competitor in *John
Zapolya*, that after a long ftruggle he was obliged to yield up *Tranfylvania*
and fome lands in *Hungary* to him, which ceffion was alfo confirmed by
his fon and fucceffor *Maximilian* the fecond. *Radolphus* the fecond, during
his life, was compelled, by the concurrence of the *Hungarians*, to refign
the crown to his brother *Matthias*. He was fucceeded by *Ferdinand* the
fecond, grandfon to *Ferdinand* the firft, who, in the year 1620, was difpoffeffed of the Kingdom by *Bethlengabor*, Prince of *Tranfylvania*; but in
the following year he too was obliged to quit it. *Ferdinand* the third was

C 2

alfo

alſo engaged in a war with *George Rakotzy*, Prince of *Tranſylvania :* His ſon, *Ferdinand* the fourth, had been elected and crowned King of *Hungary*, but died before his father ; and *Leopold* his brother, in 1654, ſucceeded to the crown. In this reign it was that the religious feuds broke out into a very bloody inteſtine war ; and though *Count Teckeley* brought the *Turks* into the country, it was attended with little advantage to him. *Tran-ſylvania* was again united to the Kingdom. The *Hungarian* malecontents afterwards found a proper head in *Francis Rakotzy*, under whom, after the death of the Emperor *Leopold*, they continued the war againſt his ſucceſſor *Joſeph*, till they were reduced to obedience in 1712. The ſame year died *Joſeph*, and to him ſucceeded his brother *Charles* the ſixth, who at the peace of *Paſſarowitz* acquired the whole Bannat of *Temeſwar*, part of *Walachia*, the moſt conſiderable portion of the Kingdom of *Servia*, with *Belgrade* its capital, and a part alſo of *Croatia* and *Boſnia*. But in 1739 the *Turks* re-covered *Belgrade* with all *Servia*, *Auſtria* and *Walachia*, the iſland and caſtle of *Orſava*, Fort *St. Elizabeth* and the fortreſs of *Sabatz*. In 1722, in a Diet held at *Presburg*, the hereditary ſucceſſion in *Hungary* was ſecured to the houſe of *Auſtria*, that in caſe of failure of male heirs females ſhould be capable of the crown. Accordingly, the Emperor *Charles* the ſixth dying in the year 1740, his eldeſt daughter, *Maria Tereſa*, aſcended the throne and was crowned in 1741. Her Majeſty is conſort to the preſent glorious Emperor *Francis Stephen*, whom the States of the Kingdom, in 1741, inveſted alſo with the joint ſovereignty.

§. 14. The King of *Hungary*, by the laws is to be ſtiled Catholic, to which is added the title of Apoſtolic, on account of the zeal which *Stephen* the firſt ſhewed in the converſion of the *Hungarians*. The regalia, namely, the golden crown made in the eleventh century, the ſceptre, King *Stephen's* ſword and mantle, gloves and ſhoes, with the ſilver croſs, the mark of the Apoſtolic function, are kept in the caſtle of *Presburg ;* and in this city is performed the coronation of the King by the archbiſhop of *Gran.* The arms of the Kingdom are a ſhield longitudinally divided ; the right field gules, divided by four bars argent : the left quarter is alſo gules with an archiepiſcopal croſs argent, ſtanding on a triple hill vert.

§. 15. Since the year 1687, *Hungary* is become an hereditary Kingdom to the Archducal houſe of *Auſtria* ; and by an act of the Diet of *Presburg*, in 1723, the Princeſſes alſo ſucceed ; by which means, on the extinction of the *Caroline* ſucceſſion, the crown falls to the *Joſephina* branch, and, on failure of them, to the offspring of the Princeſs *Leopoldina* in *Portugal.* The firſt hereditary Prince was ſtiled Duke of *Hungary*, but at preſent his title is that of Archduke of *Auſtria.*

§. 16. The States of *Hungary* are divided into four claſſes, which in the laws are comprehended under the collective word *Populus.*

To

To the firſt belong the prelates, who direct all religious matters, and have precedence of all the others, except that the governor of the Kingdom gives place only to the Archbiſhop of *Gran*. Theſe are

The Archbiſhops of *Gran* and *Kolocza*. The former is primate of *Hungary*, chief Secretary and Chancellor, *Legatus natus* of the papal ſee, and Prince of the holy *Roman* Empire. He alone crowns the King, is perpetual Count of the Geſpanchaft of *Gran*, creates even noblemen, and never takes an oath himſelf, but his official ſwears in his ſtead, *&c.* Under him are the ſix biſhops of *Erlau, Nitra, Raab, Vatz, Fünfkirchen*, or *Five-churches*, and *Veſzprim*; to whom alſo may be added the *Greek* biſhops of *Buda* and *Muncatz*, who are united to the *Roman* church. Next to him is the Archbiſhop of *Kolocza*, with the following ſuffragans, *viz.* the biſhop of *Bats*, which is annexed however to the archbiſhopric of Great *Waradin, Cſanad, Zagrab, Sirmia, Boſnia, Tranſylvania* and *Bakow* in *Walachia*.

The biſhops beforementioned: Theſe are of a double character, temporal and ſpiritual, being, *Bats* only excepted, perpetual Counts of the Geſpanchafts, in which they reſide, and having alſo a ſeat in the council of Diet. The foregoing account ſhews them to be nine in and four out of *Hungary*.

The Abbots, who are ten in number; and of theſe the chief is the Abbot of *St. Martin's*, that abbey having *St. Stephen* for its founder. This abbot depends immediately on the Pope, without any ſubjection whatſoever either to biſhop or archbiſhop. He has alſo his ſuffragans, the abbot of the foreſt of *Bacony* and the abbot of *Tihany*.

The nine Probſts, *præpoſiti*; as the probſt of the chapter of *St. Martin* on the hill near the caſtle of *Scepus*; the probſt of *Presburg*; the probſt of the order of the *præmonſtratenſes* at *Lelefiz*; the probſt of Great *Waradin*, &c. Theſe, in conjunction with the chapters, have a vote in the Diet. But the probſt of *Stuhlweiſſenburg*, or *Alba regalis*, is on a level with the *Hungarian* biſhops.

Obſ. The Jeſuits *Paulini* and *præmonſtratenſes* alſo make a part of the ſtates, and ſit and vote in the diet,

To the ſecond claſs belong the Magnates, or Barons, of the Kingdom, who are

The principal Barons, and, by way of eminence, are ſtiled the Great Barons of the Kingdom, of which they alſo held the chief offices: namely, the Statholder, or *Palatinus regni*, who is the principal and in the moſt weighty concerns of the Kingdom acts as ſovereign; the Court Judge, *judex curiæ regiæ*; the Ban, or Viceroy, *prorex*, of *Dalmatia, Croatia* and *Sclavonia*; the Woywode of *Tranſylvania*, which laſt poſt is now aboliſhed, *Tranſylvania* being governed at preſent by a Statholder; the Treaſurer, *magiſter Tavernicorum regalium*; the great cup-bearer, *magiſter pincernarum*; the Steward of the houſehold; *magiſter dapiferorum*; the Maſter

of

of the horfe, *magifter agafonum* ; the Lord Chamberlain, *magifter tubicu-lariorum* ; the Captain of the yeomen of the guards ; *magifter janitorum* ; and the Grand Marfhai of the court, *magifter curiæ*. See *Car. Andr. Bely commentatio de archiofficiis regni Hungariæ: Hungari Baronatus vocant Lipfiæ,* 1749, 4to. i. e. *Char. Andr. Belius's Account of the chief offices of the Kingdom of Hungary, Lepfic,* 1749, in quarto ; which the *Hungarians* call Baronies.

The inferior Bans, or Counts and Barons.

To the third clafs belong the Gentry, fome of whom have noble manors, and others only the privileges of nobles : The former are ftiled *Nobiles pof-feffionati,* the latter *Armalifæ.*

To the fourth clafs belong the royal free cities, *civitates liberæ atque regiæ;* which are fummoned to the Diet, and are not fubject to the Counts, but hold immediately of the King, and have a council of their own, in which a city judge and burgomafter ufually prefide. Thefe are of two kinds.

Firft, fuch as are fubordinate to the Lord-treafurer, and meet by themfelves : all proceffes againft them muft be brought before him. Of this kind are *Bartfu, Karpona, Cafchau, Comorra, Debretz, Eifenftrdt, Eperies, Gunz, Leutfchau, Modra, Oedenburg, Buda, Peft, Prefburg, Raab, Szathmar-Nemethi, Szakoltza, Szeged, Tirnau* and *Zagrab* in *Croatia.* Second, fuch as are fubject to the *perfonalis præfentiæ regiæ;* and thefe are, *Altfohl, Baka-banya,* or *Neuftadt, Gran, Refmark, Konigfberg, Leutfchau, Libeth, Nagy-banya, Neufohl, Ruft, S. George, Schemnitz, Stuhlweiffenburg, Trentfchin* and *Zeben* in *Hungary* : in *Croatia* and *Dalmatia, Caproneza, Kreutz, Varafdin, Zength.* Some among thefe royal free cities are mine-towns, and under the *Auftrian* government, having formerly been mortgaged to the houfe of *Auftria* by the Kings of *Hungary* ; as *Cremnitz, Schemnitz, Neufohl, Libeth, Bela-banya, Baka-banya* and *Konigsberg* ; feparate from which are *Konigsberg* and *Felfo-banya,* in the Gefpanfchaft of *Szatinmer.*

The next are the fmall free towns, *oppida libera,* among which are fome under *Polifh* jurifdiction ; as, befides the caftle of *Lublyo* and the little town and caftle of *Podolin,* which are mortgaged to the *Poles,* and committed to the government of a Count, there are *Bela, Laibitz, Menhardfdorf, Deutfchendorf, Michelfdorf, Neudorf, Riftdorf, Vallendorf, Fulk, Varollia, Matzdorf, Georgenberg, Durandfdorf,* to which the little town of *Kniefen* has lately been added. The *Heyduck* towns, *oppida Haidonicalia,* which have particular privileges ; namely, *Vamac, Pertz, Hathuz, Bofzormeny, Dorog, Nanas, Szobofzlo, Polgard.* The metal towns, *oppida metallica,* as, *Schmolznitz, Schwedler,* &c. which depend on the Treafury. The *Huffar,* or military towns, *oppida militaria,* of the *Servians* or *Ratzians,* in the Gefpanfchafts of *Batfch, Podrog* and *Temefwar.* Thefe are under the board of war.

The archiepifcopal and epifcopal gentry, who are ftiled *prædialifts,* enjoy the fame privileges as the *Hungarian* nobility. Among thefe the Gentlemen

men

men Vaſſals of the Archbiſhop of *Gran* are divided into two ſeats of juſtice, that of *Vaik* and *Verebel*. They are independent of the juriſdiction of the Geſpanſchaft, and with reſpect to the governor, and likewiſe to the viſcount and judge of the nobility, have their diſtinct magiſtrates.

§. 17. *Hungary* is governed by the King and States through the channels of the Diet, the *Hungary*-office, the Royal Council, the Exchequer, the Geſpanſchafts and Senates of the royal towns. The Diet, *comitia regni, diæta*, are ſummoned by writ from the King, every three years, to meet whenever the ſovereign's ſervice or the public welfare requires it. Accordingly, on the day appointed, the Lords ſpiritual and temporal perſonally appear in the Chamber of the Magiſtrates ; but the towns and gentry ſend two deputies, who meet in the States Chamber. The States lay their repreſentations before the King, who alſo refers to them ſuch articles of publick concern as require their aſſent.

The *Hungary*-office is at *Vienna*, with a Secretary of ſtate at the head. It expedites the royal edicts in municipal, religious and juridical affairs for *Hungary* and the incorporated Kingdoms of *Croatia, Dalmatia* and *Sclavonia*. To this office belong all matters relating to the King, and wholly dependent on his pleaſure : all who ſtand in need of a perſonal audience of the King muſt firſt acquaint this office with it. Excluſive of this, it has very little connexion with the Kingdom, its principal buſineſs being to execute the King's pleaſure.

The Statholder's council, *conſilium regium locumtenentiale*, reſides at Presburg, and, beſides the Statholder, who is preſident, conſiſts of twenty-two counſellors, whom the King chuſes at pleaſure from among the prelates, nobility and gentry. The Emperor *Charles* the ſixth firſt erected it in 1723. In *Hungary* and the incorporated countries it ſuperintends the civil concerns of the towns as regulated by the laws of the land, or conformably to them. It is ſubject to no court-office, but when it has any thing to lay before the King applies immediately to him.

The royal Schatzkammer, or Exchequer, is divided into the *Hungarian* and mine chambers, and takes care of the royal eſtates, incomes and dues. The chamber of *Hungary* ſits at *Presburg*, has a preſident and eighteen counſellors, and looks to the royal demeſnes, regalia, eſcheats, cuſtoms and ſalt-duties. Under it is the office at *Caſchaw* beſides eight provincial commiſſaries for collecting the contributions. The mine-chamber is held at *Cremnitz*, and manages all affairs in the mine-towns appertaining to mines and coinage. It receives orders from the Treaſury at *Vienna*, and under it are the mine-chambers of *Schemnitz, Neuſohl*, in the Geſpanſchaft of *Zip* and at *Konigsberg*. The *Hungarian* Geſpanſchafts, *comitatus*, are ſmall provinces ſubdivided into diſtricts, with their Counts, Viſcounts and Aſſeſſors, who in the King's name hold provincial aſſemblies, or diets, for the conducting, in the beſt manner, for the King and Kingdom, matters civil and œconomical. All theſe Geſpanſchafts receive their names from their ſeveral caſtles : the *Hungarian* name of them, *Varmegye*, ſignifying properly the

jurifdiction or diftrict of a caftle. But in *Hungary* a difference is made betwixt a fort, *arx*, a fortification, *caftrum*, and a caftle, *caftellum*: The two
laft properly denoting the place of a nobleman's refidence.

§. 18. The public revenues confift of contributions, cuftoms, mines and minerals, falt-works, which belong to the crown, royal demefnes and efcheats:
Thefe abundantly anfwer the expences of the court and defence of the frontiers.

§. 19. The Kingdom of *Hungary* can eafily raife an army of one hundred
thoufand men, of which fifty thoufand are in pay, and a like number is
furnifhed by the provinces. The infantry are called *Heyducks* or foot Huffars, and the cavalry Huffars.

§. 20. Juftice is adminiftered in civil affairs in the King's name, after the
manner prefcribed by the laws and cuftoms of the Kingdom. Suits are
carried from the courts of the fmaller towns either to the court of the Gefpanfchaft when a free town, or to the lord of the manor when belonging to
any particular lordfhip. In the royal free cities the firft hearing is before the
judge of the town, the fecond before the council, from which there is an
appeal either to the treafurer or *Perfonalis prefentiæ regiæ* as he is ftiled, who
has alfo the title of prefident of the *tabula regia*. The Mine-court in the
free mine-towns is diftinct from the Town-court, and takes cognizance only
of matters relative to the mines: The mine-judge prefides, but an appeal
lies from him to the Commiffion-court of the mine-towns. The inferior
noble courts are held in each Gefpanfchaft by the lord of the manor for determining caufes relating to the commonality; and where noblemen are
concerned by the court-judges, or judge of the nobles, and the Vice-gefpan;
but a caufe may be carried from thefe to the Gefpanfchaft-court, and afterwards to the *Tabula regia & feptemviralis*. The Middle noble Court, *forum
Nobilium fubalternum*, meets at *Tirnau*, *Guntz*, *Eperies* and *Depretzen*, and
has the trial of all caufes wherein two or more Gefpanfchafts are concerned,
which from thence may be removed to the *Tabula regia & feptemviralis*. The
Upper noble Court, which refides at *Peft*, is divided into the *tabula regia & feptemviralis*, and not only decides caufes brought thither by appeal, but alfo
other important fuits relating to the Nobility. In the firft prefides the *perfonalis
præfentiæ regiæ*, or the King's reprefentative; in the fecond, the Count
Palatine, or, in his abfence, the Court-judge or the Treafurer. The
Tabula femptemviralis is fo called becaufe it formerly confifted of feven perfons, but *Charles* the fixth has fince made an addition of eight more. It
takes cognizance of fuch caufes only as are referred to it from the *Tabula
regia*, examining them and making the neceffary amendments. The ecclefiaftical courts appointed for the difcuffion of ecclefiaftical affairs, are held in
every diocefe and chapter, but with a power of appealing to the Archbifhop, and afterwards to the Pope's Legate; and may be carried at laft to
the Papal court at *Rome**.

We now proceed to give an account of

---

* The author is a little inconfiftent here with what he fays in §. 10, page 9; but in fo long
a work the reader, we doubt not, will have the goodnefs to excufe a flip of fo trifling a nature.
— *Quandoque bonus dormitat Homerus.*   PROPER

PROPER HUNGARY.

WHICH all agree in dividing into *Upper* and *Lower Hungary*, though in a different manner. Some give the name of *Upper Hungary* to that part lying beyond the *Danube* towards *Poland*; calling that, which lies below the *Danube, Lower Hungary*. But others drawing a meridional line from the Gefpanfchaft of *Zip* to the junction of the Banat of *Temefwar* and the Gefpanfchaft of *Sirmi*, call that part towards the weft *Lower Hungary*, making the country which lies eaft of the Line *Upper Hungary*: And thefe laft it is whom we fhall here follow. Secondly, *Hungary* in the *Corp. jur. Hung.* of 1723, artic. 31, conformably to its government, by the great juridical Courts, is divided into four large circles, and fifty-two Gefpanfchafts, *comitatus,* by the *Hungarians* called *Varmegye*.

LOWER HUNGARY;

HUNGARIA INFERIOR,

CONSISTS of the two following circles,
1. The circle beyond the *Danube, circulus Cisdanubianus,* stretching through the upper regions of the western parts. . *Hungary,* from the *Danube* to the *Carpathian* mountains, contains fourteen Gespanschafts, and is inhabited by *Hungarians, Bohemian-Sclavonians, Germans,* and in one part also by *Servians.*

1. The Gespanschaft of *Presburg,* or *Posony Varmegye, comitatus Posoniensis,* lies on the borders of *Austria* betwixt the *Danube* and *Moravia* ; . is twelve miles in length and eight broad. The mountains in it begin the *Carpathian* chain. The country about *Tirnau* is the best and most fruitful. The soil does not indeed want fertility, but scarce a year passes in which the grain is not damaged by mildew. The large rivers in this Gespanschaft are the *Danube, Morau* and *Wag* ; the smaller are the *Dudvag, Blawa, Tirna, Padla-woda, Parna, Gidra, Rudawa,* &c. The air is healthy, particularly on the mountains, but unwholesome among the swamps on the *Danube.* The inhabitants are *Hungarians, Germans, Bohemian-Sclavonians,* with some *Croats,* and great numbers of *Jews.* The dignity of *Palatine,* since the year 1599, has been made hereditary in the *Palfi* family. The whole province is divided into five Districts, by the *Hungarians* called *processus,* and each has " a noble judge." These are,

1. The Upper Outward District, *processus extraneus superior,* called Outward with relation to the Schutt in which it lies : Besides fifty-eight market-towns, it contains five royal free cities; namely,

Presburg, Posony, Presporen, Posonium, Pisonium, a royal free city and now the capital of the Kingdom, lies on the *Danube* at the foot of a mountain on which stands the castle. It is pleasantly situated, and enjoys a better air than most other towns in *Hungary.* The cathedral, dedicated to *St. Martin,* is the place where, from the time of *Ferdinand* the first, the Kings of *Hungary* have been crowned. Here, ever since the year 1723, are held the Diets, the first of which was assembled by King *Sigismund* in 1411. It has also been the residence of the *consilium regium locumtenentiale* ; and, from the days of *Ferdinand* the first, of the Treasury-office for *Hungary* ; and likewise, since the same Emperor's time, of the archbishop of *Gran.* Here

are

are likewife a chapter of fourteen regular canons, whofe Probft is the arch-
deacon of the Gefpanfchaft; a Jefuits college with a gymnafum and church,
befides three other convents and churches, a Proteftant fchool and *Lutheran* church. The city itfelf is ill built; the houfes of the town, properly fo called, not much exceeding two hundred, and the whole ftrength
of its fortifications confifts in a double wall and moat: But the fuburbs
are large and handfome, and contain four convents with churches and two
hofpitals. The fuburb on the fouth fide of the city makes a fine appearance, and is famous for the *Koingfhugel*, or *King's-hill*, a place in itfelf but
inconfiderable, on which the new elefted King, being mounted on horfeback, brandifhes St. *Stephen*'s fword towards the four cardinal points, as a
token that he will defend the country againft all its enemies. It lies in
latitude forty-eight degrees eight minutes. This city is of great antiquity,
and is faid to have been founded by the *Jazygians*. Befides its frequent
fieges it has often fuffered by fire, as in 1515, 63, 90, 1642.

Tirnau, Tinawa, Nagy-Szombat, Tirnavia, a handfome royal free town
on the river *Tirna*, in a pleafant and fertile country, contains nine churches
and as many convents. Here alfo is the archiepifcopal chapter of *Gran*,
which was removed hither when that city was taken by the *Turks* in 1543;
an academy of Jefuits founded in 1635, and the high court of juftice for
the circle beyond the *Danube*. This city has likewife a ftaple-right. It
was built betwixt the years 1230 and 1240, but in 1683 laid in afhes by
Count *Tekely*'s men; and, in 1704, Prince *Rakotzi*'s troops were defeated
near this city.

Modra, a royal free city, containing about three hundred and fifty houfes,
lies in a valley bordering on the *Carpathian* mountains, and is furrounded
on one fide with vineyards. In 1619, 20, 63, 82 and 1705 it fuffered
much by war, and in 1729 the greateft part of it was burnt down.

Bozin, Bofing, Befineck, Bazinga, Bazinium, a fmall, but pretty, royal
free city, on a rifing ground. The inhabitants apply themfelves with great
induftry to the cultivation of vineyards, trade and handicrafts. In 1605,
20 and 55 this place was laid in afhes.

Sanft George, Szent Gyorgy, fanum S. Georgii, a royal city, but fmall and
greatly dwindled. It has notwithftanding a college, *collegium P P. piarum
fcholarum*, and its neighbourhood produces excellent wine. It fuffered extremely in the wars of *Gaberbethlen* and Prince *Rakotzi*. In 1633 it was
facked by the *Turks*, and in 1728 deftroyed by fire.

Five caftles, of which fome belong to the abovementioned free towns.

Presburg caftle ftands on a pleafant eminence two hundred paces weft
of the city of *Presburg*, is the refidence of the Sovereign when here;
and in one of its four towers are kept the regalia, among which the ancient
golden crown is particularly remarkable, but never fhewn to any. Below
it is a fmall town, by the *Hungarians* called *Varallja*, by the *Germans*

Schlofsberg;

Schlofsberg; in which are a great many *Jews*, who make alfo the moſt confiderable part of the inhabitants of *Czuckermandel* a fmall place in' its neighbourhood. To the jurifdiction of this caſtle appertain alfo the two towns of
Samaria and *Szerdakely* on *Schutt* iſland, and thirteen fmaller towns.

St. George's caſtle, which ſtood on a hill near the city of that name is now
a confuſed heap of ruins, yet has a jurifdiction containing *Menſtift*, a market-
town lying below it, the feats of *Kyrali Falva* and *Nemet Gurab*; the fmall
towns of *Gruman* and *Crotortok*, and five market-towns or large villages.

Bozin caſtle lies on the north fide of the town of that name, belongs to
Count *Palfy*, and has feven other caſtles under its jurifdiction.

Bibersburg caſtle, *Vorofko, Cervenny Kamen*, is fituated on an eminence of
the *Carpathian* mountains, and belongs to the *Palfy* family. Under its
jurifdiction are the caſtles of *Szuha*, the towns of *Szuha*; alfo *Dios Cfefzte,
Ompkithal*, · and thirteen large villages.

Szomolon Smolenis caſtle ſtands on a high hill in a woody country, belongs
to Count *Erdody*, having jurifdiction over the little town of *Szomolon* and
Felfo Dios, or *Obernuedorf*, and eight large villages.

Some feats, viz.

Cfekles, belonging to Count *Eſterhaſi.*
Magyar-Bal, to the *Cfaky* family.
Nemet-Gurab, to Count *Palfy.*
Szuha, to the Count of the fame name.
Vodrod, to Count *Zichy.*
Briſſer, the name of two caſtles near each other.

Privileged towns, viz.

Pofoni Varallja, mentioned at *Presburg* caſtle.

Ratuhdorf, or *Rathfcherfdorf*, *Retza*, a fmall town at the foot of a hill, a
mile from *Presburg*, belongs to Count *Palfy*, and is noted for its good wine.
The greateſt part of this town was deſtroyed by fire in 1732.

Landfitz, Cfekles, a fmall town on an eminence near the abovementioned
feat of the fame name.

Wartberg Semptz, Sentz, an old but handfome and populous town belonging to Count *Eſterhaſi.*

Grun-an, Grinava, a fmall town in a pleafant country remarkable for its
excellent wine.

Cziffen, in a large plain, and formerly a place of note, has feveral
owners.

Szuha, a fmall town not far from the feat of that name. The inhabitants apply themfelves to tillage and the cultivation of vineyards.

Cfefzte, a fmall town, near *Bibersburg* caſtle, on a pleafant eminence.

Ompital, a fmall well built town at the foot of a mountain.

Windifch, or *Unter-Nurfdorf*, *Alfo-dios*, *Orefany*, a town belonging to
Count *Palfy.*

Obernurſdorf,

Obernurſdorf, *Felſo-dios,* *Horne-oreſſany,* a ſmall town, famous for a generous and wholeſome kind of wine.

Szomolan, a ſmall town under the caſtle of that name.

Spatza, *Diete* and *Boleraz,* ſmall towns reſembling each other.

2. The Lower Outward Diſtrict, *Proceſſus inferior extraneus,* which, beſides forty-ſix large villages, contains

The two ſeats of *Galantha,* both which belong to Count *Eſterhaſi, Koſſuth,* alſo the property of the ſame nobleman ; *Vizkelith,* which is alſo the name of two other ſeats, one belonging to Count *Eſterhaſi,* the other to the family of *Samogy* ; *Hodi,* belonging to the *Andraſi* family, as *Neboyſza* does to the *Vitzai,* and *Gany* to that of *Farkaſi.*

The town of *Galantha,* a moſt charming delightful place in a fruitful country.

Szered, a ſmall town on the *Wag* famous for its annual fair, and particularly its beaſt market. It lies near *Szempeh,* which belongs to the Geſpanſchaft of *Nutra.*

3. The Upper Diſtrict of the *Schutt, proceſſus inſulanus ſuperior.* The *Schutt,* which ſignifies an iſland in a river, *inſula cituatum,* or *cituorum,* by the *Hungarians* called *Cſallokz,* is formed by the *Danube,* which divides it alſo into three parts; one of which, called *Vitzkoz,* belongs to the former Diſtrict ; the ſecond lies betwixt the two large arms of the *Danube,* is twelve *Hungarian** miles in length, and goes by the name of *Nagy Sziget,* or Great Iſland ; and the third, containing a ſpace of about ſix miles, is called *Sziget Koſs.* This iſland abounds greatly in fruits and herbage, but the grain is frequently deſtroyed by mildew. It has plenty of game, wood and fiſh of all kinds. The *Huns* were driven out of it by *Charles* the Great. Its Upper Diſtrict, beſides ſixty-eight large villages, contains

Eberhard Fort, an old building lying in a plain and moated.

The ſeats of *Puſpoky, Cſakany, Gomba, Illyeſhaza, Nagy-Szrva, Uſzor, Kiralyfalva.*

The privileged towns.

Puſpoky, Biſchdorf, or Biſhop's Village, well built and belonging to the Archbiſhop of *Gran.*

Samaria, Somorja, Sumarein, Schomerin, an old place, the capital of the iſland, and ſeat of the provincial court ; enjoying alſo a great trade, but frequently deſtroyed by fire, belongs to the juriſdiction of *Preſburg* Caſtle.

Cſotortok, Leoper's dorf, in the records *Leopold's dorf.*

Nagy-magyr, large and populous.

4. The Lower Diſtrict of the *Schutt,* beſides ſixty-five villages, contains

The ſeats of *Bar* and *Boos* belonging to Baron *Amade* ; *Suly,* Baron *Maholany's* ; *Beket Falva,* belonging to one of the Counts of *Eſterhaſi* ; and *Sipos,* to the family of *Somogy.*

* The reader will remember that 13⅓ *Hungarian* miles make one degree on the equator.

The

The town of *Szerdahely*, fo called becaufe fituated in the middle of the ifland, is well built, and under the jurifdiction of *Presburg* Caftle.

5. The Diftrict beyond the beginning of the *Carpathian* mountains is about feven miles in length, but of an unequal breadth, and befides thirty-fix large villages contains

The caftles of

Deven, Deben, fituate on a hill at the conflux of the *Morau* and *Danube,* and belonging to the family of the *Palfy.*

Boroftyanko, Ballenftein, ftanding on a fteep rock of the *Carpathian* chain, and belonging to the fame family.

Detreko Blafenftein, Plaveer, delightfully fituated, the property alfo of Count *Palfy.*

Elefco Scharfenftein, having under its jurifdiction the feat and fmall town of *St. John* with feven large villages.

The famous Eremite convent of *Marienthal vallis Divæ Mariæ, Cenobium B. Virginis Mariæ,* in great repute and much reforted to by pilgrims on account of an image there of the Virgin *Mary,* ftands in a healthy and pleafant fituation.

The feats of *Stompha, Detreko, Malatzka,* all three belonging to the *Palfy* family ; *Levard,* in the town of *Nagy-Levar,* appertaining to that of the *Kollonitz* ; St. *John, Deven-Ujfaln* and *Deven,* both belonging to the *Palfy.*

The privileged towns.

Deven, below the feat of the fame name.

Stompha, Stamphen, Stupawa, belonging to Count *Palfy,* and containing the abovementioned feat.

Malatzka, a pretty populous town with a *Francifcan* convent, to which numerous pilgrimages are made four times a year, lies near the caftle of the fame name.

Gajar, in a level country at the conflux of the *Rudau* and *Morau.*

Nagy-Levard, Levary, Grofsfchutzen, peopled by a colony of Anabaptifts and belonging to the *Colonitz* family.

Sanct Johannes, a handfome well built place with two feats.

The *Vaik Tribunal* conftitutes a particular republic with the Archbifhop of *Gran* at the head of it; has alfo its own Statholder, vice-counts, noble judges, prothonotary and treafurer. The gentry are termed predialifts, as has been mentioned in §. 16 of the Introduction to this divifion. This territory lies on the *Schutt,* is divided into the Higher and Lower Diftrict, and contains

Vajka, a large town, concerning the inhabitants of which feveral ridiculous ftories are told.

Doborga, a large village.

Bacsfalva, a large village celebrated for its convent, which is much reforted to by pilgrims.

Five other large villages.

2. The Gefpanfchaft of *Neutra, comitatus Nitrienfis*, about twelve miles in length, and in fome places above fix in breadth, in others fcarce two, and in others ftill lefs; very mountainous but plentifully watered by the *Wag, Nitra, Zitwa, Livina, Dudvag, Blawa, Holefchka, Mijawa, Chwoynitzre,* &c. It has alfo a very cold fpring, the water of which is famous for creating an appetite; a fulphureous and mineral fpring, fome warm baths, good vineyards, and in feveral parts alfo fine corn-land and other advantages. Its inhabitants confift of *Hungarians, Bohemians, Sclavonians* and *Germans.* The palatine, or fuperior court, of this Gefpanfchaft is the bifhop of *Nitra.* It is divided into the five following Diftricts, *viz.*

1. The Diftrict of *Nitra, Proceffus Nitrienfis*, in which are

A few caftles; namely,

Nitra, Nitria, a fortification of confiderable ftrength on a river of the fame name. Mention is made of it fo early as the ninth century. In it is the epifcopal palace, the cathedral, and a chapter houfe for twelve canons. The town, which lies under it, is divided by the river into two parts; is well peopled but meanly built, and, which is worfe, the water is not wholefome; yet the *P P. piarum fcholarum* have a college and gymnafium here. It was erected into a bifhoprick many years before the coming in of the *Hungarians.* St. *Stephen* indeed enlarged it, and *Geyfa,* in the year 1150, endowed it with revenues and annexed a chapter to it. Its Diocefe comprehends not only a confiderable part of this Gefpanfchaft, but the whole of that of *Trentfchi.* Both town and caftle were taken in 1619 by *Bethlengabor,* and in 1623 given up to the *Turks,* but recovered again the following year.

Neuhaufel, Ujar, Nowe-zamki, Ujvarinum on the river *Nitra,* firft fortified in the year 1573, and famous for ten fieges, but in 1724 and 1725 demolifhed by the Emperor's order, infomuch that the town is now an open place, and the inhabitants of it fubfift by tillage and graziery.

The caftle of *Ghymes,* belonging to Count *Forgatfch,* ftands on a high and rocky mountain.

Schempte, Schintawa, Schintavia, a caftle belonging to the *Efterhafi* family near the town of *Szered,* which makes part of the Gefpanfchaft of *Prefburg.*

The towns.

Nitra and *Neuhaufel,* already taken notice of among the caftles.

Suran, formerly a famous ftrong town, but now defencelefs.

Sclye, an old town on the *Wag.* The Jefuits have a college and fchool here. It is fubordinate to the Probfty of *Thurotz,* which belongs to the Jefuits of *Tirnau,* and was formerly fortified.

Urmeny

Urmeny ftands very pleafantly in a high country, and is one of the beft towns in this Diftrict.

Schempte, an open town, mentioned in hiftory, and ftanding on the river *Wag*. The inhabitants apply themfelves with great induftry to tillage and the cultivation of vineyards. This place belongs to thé *Efterbafi* family.

Mocfonok, a mean town fubfifting by tillage.

Komjathy, formerly ftrong and confiderable, now fmall, and indebted for all its ornament to two feats belonging to Count *Forgatfch*, and under the jurifdiction of the caftle of *Ghymes*.

Ujlak, a mean town, the inhabitants whereof fubfift by tillage.

Ghymes, belonging to Count *Forgatfch*, a fmall place but enjoying good corn-land.

About one hundred large villages.

2. The *Baimotz* Diftrict, *proceffus Baimotzenfis*, which, befides ninety-one large villages, contains

Two caftles ; namely,

Baimotz, pleafantly fituated on a hill and pertaining to the *Palfy* family. Near it, on the river *Nitra*, lies alfo a middling town of the fame name, in which is a probfty of St. *Antony*. It has alfo a warm bath of a very good contrivance.

Kefrello-ko lies on a rocky mountain, and belongs to the family of *Maytheny*.

The towns of

Baymotz, mentioned with the caftle.

Privigye, *Priwitz*, *Priwiza*, *Prividia*, a pretty good town, chiefly inhabited by clothiers and fhoemakers. It has two parifh churches, a college, *P P. piarum fcholarum*, and belongs to the lordfhip of *Baymotz*.

Nemet prona, *Nemetzke-prawno*, an old town, but the country around it fruitful.

Sambokret, formerly a confiderable place, but at prefent in a very low condition.

Szacfany, a wretched place belonging likewife to the bifhop of *Nitra*.

3. The *Bodok* Diftrict, *proceffus Bodoikenfis*, in which, befides a hundred and one large villages, are

Bodok, a caftle, belonging to the counts of *Vereny*, but ill fortified.

Nagy-Tapolcfan, *Great Tapoltfchan*, *Welikae*, *Topolcfani*, a large well-built town. Tillage the chief occupation of its inhabitants. Near it ftands *Tovarnok*, a feat of Count *Bereny*, with a lordfhip belonging to it.

Bayna, a town, the inhabitants of which are principally farmers and hufbandmen.

Radofchyn, *Radofnya*, belonging to the bifhop of *Nitri*.

4. The *Ujhely* Diftrict, *Proceffus Ujhelienfis*, in which, befides eighty-two large villages, are the following towns and caftles, *viz*,

Neuftadle

Neuſtadle on the *Wag-Ujhely*, *Nowe Meſto Ujhelinum*, a place which would have made a good figure had not its houſes, which were handſome, been ſo often burnt down. It lies indeed in the juriſdiction of *Bezko*, but is privileged by the King and has an opulent probſty. The red wine, which it produces, is held in great repute. This unfortunate place was deſtroyed by the *Turks* in 1550, 1599, 1620, 1624, 1663.

Cſiete Czachtitze, an old caſtle on a hill, having the town of *Cormano* lying under it, the inhabitants of which are all wine-dreſſers and buſbandmen.

Wrdau, *Wrbowo*, *Verbovia*, a handſome populous town in a pleaſant and fruitful valley. Thoſe of the inhabitants who do not follow tillage and dreſſing of vineyards are handicraftſmen. Here are, in particular, great numbers of taylors who make coarſe winter cloathing for the common people.

Poſteny, *Pieſtyani*, a ſmall town belonging to the juriſdiction of the *Galgots*, is divided between the Counts *Erdody* and *Forgatſch*. By the *Wag*'s ſide are famous warm baths without any regular ſprings, being only pits dug along the ſhore, nearer or farther, according as the river increaſes or decreaſes : and theſe ſerve for occaſional baths. They caſt indeed a great heat, which may be perceived even in the ſoil of the river when ſtirred with the feet.

Vittenz, *Chtelnicza*, a pretty large town, ſubſiſting by grain and wine, and belonging to the juriſdiction of *Joko* and the Counts *Ardody* and *Czobo*.

Joko, a caſtle ſituate on a rocky hill.

Leopolſtadt, *Leopodopolis*, a regular fortification on the *Wag*, ſtanding in a level marſhy place, and built in 1663 for the defence of the country inſtead of that of *Neuhauſel*, which had been demoliſhed. This is the only place in the country capable of making head againſt an enemy.

Galgotz, *Freyſtud*, a caſtle on the *Wag* belonging to Count *Erdody*; was taken in 1663 by the *Turks*, who quitted it the following year. Below it, on the river, lies a town which, beſides a church, has alſo a *Franciſcan* convent. The chief employment of its inhabitants is tillage and vine-dreſſing.

Temetveny, a caſtle on a very ſteep hill belonging to the family of *Cſaky* and *Sando*.

Ujvaros, *Neuſtadl*, *Meſtecſko*, a ſmall town cloſe by *Leopoldſtadt*, in the Diſtrict of the caſtle of *Galgotz*.

Koſztolan, a good town, wholly ſupported by agriculture, which turns to very good account.

O Tura, i.e. *Old Tura*, a ſmall town whoſe inhabitants ſubſiſt by graſing.

Brezowa, a very populous town, the inhabitants of which follow agriculture and handicrafts.

5. The *Szakoltz* Diſtrict, *Proceſſus Szakolſenſis*, in which, beſides forty-eight large villages, are

Skakoltz, the only royal free city of this Gefpanfchaft, lying on the borders of *Moravia,* and having five churches and four convents with a college of Jefuits. This place has fuffered too much by war to be able to make any confiderable figure.

Holitfch, a town, caftle and lordfhip on the *Morau,* belonging at prefent to the Emperor *Francis,* who has alfo by purchafe annexed to it the lordfhip of *Cogniotzo.*

Eikbel, a fmall town with a fulphureous fpring.

Schaffin, Sas-Var, a town and caftle on the river *Mijawa,* remarkable for an image of the Virgin *Mary,* which draws hither great numbers of pilgrims.

Radofchatz, Radofchawtza, a fmall town fituate in a pleafant fruitful plain.

Szenitza, inhabited by a great many gentry, and lying on the river *Chwoyintze* in the jurifdiction of *Berintfch.*

Szabatiftye, a town on the *Chwoyintze* with a feat and colony of Anabaptifts.

Berentfch, an ancient caftle now in ruins.

Karlaiko, an old inconfiderable caftle.

Sandorf, a fmall town in the jurifdiction of *Karlatko.*

Elefko Caftle, a heap of ruins.

3. The Gefpanfchaft of *Trentfchin, Trentna, Varmegye, Trenetriniencis* lies on the river *Wag,* and extends to the frontiers of *Moravia* as far as *Silefia.* The inhabitants *Bohemian-Sclavonians* with fome *Hungarian* gentry : contains

Trentfchin, a royal free city on the *Wag,* having a Jefuits college and fchool. Its caftle, which ftands on a fteep rock, belongs to the jurifdiction of *Illyeshaty.*

Fetfkaw, a walled and populous town on the *Wag* under feveral proprietors. Its caftle was anciently called *Bolondotz.*

Itava, a middling town with a caftle on the *Wag.*

Ugrotz, Zai-Ugrotz, an ancient caftle with a fmall town, belonging to the *Zajani* family.

Rajetz, a middling place, fupported by handicrafts and tillage.

Letava and *Lednitze,* two towns and caftles ftanding high and having lordfhips.

Puchau, a town noted for its cloth.

Vag-Beftertza, Beftritza, Biftricra, a large town on the *Wag,* with a caftle facing it which ftands on a high rock on the oppofite fide of the river.

Silain Zilina Solna, a fpacious town, with fruitful environs, has a college and fchool of Jefuits.

Kifzutza-

Kifzutza-Vihely, Nowe-Meſto Ujtulinum, a conſiderable town on the river *Kifzutza,* from which it receives its name by way of diſtinction from *Neuſtadl* on the *Wag,*

4. The Geſpanſchaft of *Arva, Arva Vatmegye, Arvenſis comitatus,* lies betwixt *Sileſia* and the *Carpathian* mountains, and extends as far as *Poland.* The inhabitants *Sclavonians* and *Poles.* Here are

Welicona, a large town but in a barren country.

Reubin, called alſo *Zubin,* ſmall, and the country around it unfruitful.

Arva Drawa, a fort on a high hill and under ſeveral lordſhips, derived from the family of the counts of *Thurzo,* now extinct.

Twrdoſſin, a mean town in a barren neighbourhood which, together with the other ſmall places thereabouts, belong to the Diſtrict of *Arva.*

5. The Geſpanſchaft of *Lipto, Lipto Varmegye Liptovienſis Comitatus,* is about ſeven *Hungarian* miles in length, one or two broad, and every where full of mountains, which are not only ſaid to be higher than the *Alps,* thoſe of *Switzerland,* and the *Tyraleſa,* but alſo to ſurpaſs them in curioſities and ſubjects of admiration. Its rocks are indeed aſtoniſhing, particularly thoſe of *Deminfalo;* of which one called *Benikova* is perpendicular, and the height three thouſand paces. In theſe rocks are ſeveral vaſt natural caverns, with multitudes of ſtrange figures formed by the petrified water. Bones of an uncommon ſize are alſo found here differently ſhaped and even larger than thoſe of horned cattle or horſes. The chief rivers are the *Wag* and *Biela.* There are a multitude of freſh ſprings in this Diſtrict and ſome mineral and medicinal, among which that of *Boſſi* is particularly worth notice, together with other remarkable waters of different kinds. The exhalations from thoſe of *Szentivan* and *Stamkowan* ſuffocate birds in their flight over it; and in the cavities of the latter is ſometimes found a reſin reſembling amber. The nature of the country admits of very little tillage, and the agriculture is alſo inconſiderable, though its cheeſes are very famous. On the other hand, this Geſpanchaft abounds greatly in metals, particularly the *Botſa* mountains, containing both gold and ſilver mines; the former of which are eſteemed the beſt in *Hungary* and not inferior to the *Arabian.* It affords alſo ſome iron, with antimony, nitre and other minerals. The inhabitants are *Bohemian-Sclavonians* intermixed with *Hungarian* gentry. This territory conſiſts of four Diſtricts.

1. The Eaſt Diſtrict, *Proceſſus Orientalis,* containing

Lipto Ujvar Hradek, a mean caſtle, and the only one in this Diſtrict. It belongs to the exchequer.

St. Nicholas, Scent-Niklos, Swaty Nikulaſs, Nicopalium, a ſmall trading town belonging to the *Pongratz* family, and part of it within the *Hradeck* juriſdiction. It is here that the leſſer county meetings are held. It has alſo a Jeſuits ſchool. This place has frequently ſuffered by fire, eſpecially in 1713, 1719, 1724, 1732.

Wrbetza,

Wrbetza, Verbetza, a poor place near the former, and under the jurifdiction of *Hradeck.*

Gibbae, Geibe, Hybe, a fmall place within the aforementioned jurifdiction, which, befides the Catholic, has a Proteftant *article* church as it is termed.

Botza, a mine-town in a deep vale, confifts of three parts, Upper and Lower *Botza,* or *Joachim's* Vale, and *Bobrow.* The neighbourhood belongs partly to the King, partly to the nobility, and has good mineral waters. The gold of its mines is remarkable for its finenefs, but the works themfelves are in a very bad condition, and the adventurers, as they are called, at whofe charge it is worked, pay only a duty to the King and the gentry.

Thirty-one large villages, among which is *Szent Ivan,* whofe noxious fpring has been mentioned above. In the church-yards hereabouts corpfes are known to lie above one hundred years without decaying.

2. The South Diftrict, *Proceffus meridionalis.*

German Liptfch, Memet-Liptfcha, Nemetfka-Liptfcha, by fome conjectured to be a *Leipfic* colony, pretty large, with two churches, and had formerly mines.. It is the meeting place of the nobility for changing of the provincial officers, and belongs to the *Likau* jurifdiction.

Twenty-eight large villages, among which is *Dermianfalva,* which gives name to the famous caverns.

3. The Weft Diftrict, *Proceffus Occidentalis,* in which are

Lykawa, a caftle on a rock, formerly the only defence of almoft the whole Diftrict, and which was demolifhed in 1707 in Prince *Rakotzy's* wars. Both this caftle and the whole jurifdiction belong to the exchequer.

Rofenberg, Ruzomberg, a populous town which carries on a great trade in falt brought hither on the *Wag;* has a fchool of the *P. P. piarum fcholarum,* a feat, and belongs to the jurifdiction of *Lykawa.* In this place was held a Proteftant Synod in 1607, when four fuperintendents were ordained.

Trium Szlakes, Try-Slyatze, a fmall town, dependent on the convent of *Thurotz,* and fo called becaufe compofed of three villages.

Thirty-eight large villages.

4. The North Diftrict, *Proceffus Septentrionalis,* containing

Szelnitze Selnitza, in the jurifdiction of *Lykaw.*

Trnowetz, Tarnotz, in the fame jurifdiction.

Bobrowetz, Bobrotz, fmall towns.

Thirty large villages.

6. The Gefpanfchaft of *Thurotzi, Thurotz Varmegye, Thurotfienfis Comitatus,* about five miles in length, and in a few places only three broad, but very pleafant, being a fine level environed with high hills; the foil fo fertile, that, in good years, the inhabitants, befides a plenty for home confumption, fell grain to other parts; but, this being feldom the cafe, if the harveft fail, they are under a neceffity of buying. The chief rivers are the

Wag

Wag and the *Thurotz*. *Stubora*, in the jurifdiction of *Hay*, has warm baths; and at *Budicti*, *Dubow*, and other places, are mineral fprings. The inhabitants are a mixture of *Bohemian-Sclavonians*, *Hungarian* gentry, and *Germans*. It confifts of thefe Diftricts:

1. The *Szlabinya* Diftrict, *Proceffus Szlabinyenfis*, and in it are
Szlabinya, a large caftle on a high mountain, divided into the Upper, Middle and Lower, and belonging to the *Revaji* family.

Szent, *Marton*, *Swiaty*, *Martin*, *Martinopolis*, the county town, fituate on the *Thurotz*, pretty large and populous. The public meetings of the nobility are held here, as alfo the affize for the province. It is within the jurifdiction of the caftle of *Szklabinya*.

Szutan, a fcambling town on the *Wag*, fcarce the fhadow of what it was formerly.

Turan, a town pleafantly fituated in a fruitful country among lofty mountains, belongs to the *Nevagi* family and jurifdiction of *Szklabinya*.

Twenty-five large villages.

2. The *Blatnitza* Diftrict, *Proceffus Blatnicenfis*, containing
Blatnitza, a caftle on a high rock, which belongs to the *Revaji* family.

Twenty-three large villages; among thefe *Netzpal*, which is populous and has three feats with an Article-place, as it is called, of the community of the *Augsburg* confeffion. *Bella* is one of the largeft in the whole Gefpanfchaft.

3. The *Mofchotz* Diftrict, *Proceffus Mofcovienfis*, containing
Mofchotz, *Mofchowtze*, *Mofchovia*, much declined from its former wealth and bignefs, though yet in fome repute for its annual fair. Its inhabitants follow tillage and handicrafts. It is under the jurifdiction of *Blatnitza* Caftle.

Thirty large villages.

4. The Diftrict of *Znio* contains
Znio, a caftle antiently called *Thurotz*, which, in 1252, became the refidence of a probfty of the Bleffed Virgin. *Ferdinand* II. and III. granted this place to the Jefuits. Below it lies the mean town of *Znio Varalja*; by the *Sclavonians* called *Klafhtor* on account of its Jefuits college and convent or cloyfter.

Tot Prona, *Slowenfke-Prawno*, *Windifa-Proben*, a fmall place entirely fupported by tillage.

Nineteen large villages; one of them, *Budifch*, is noted for its mineral waters.

7. The Gefpanfchaft of *Altfohl*, *Zolyom Varmegye*, *Comitatus Zolienfis*, is about ten *Hungarian* miles in length, and in fome places four or five broad, but in others much narrower; the country every where rifes into mountains; but feveral of them abound in valuable ores and minerals; fome even in gold and filver and plenty of excellent copper, iron, orpiment, quickfilver,
cryfocolla

cryfocolla and terra figillata. ‗The rivers are the *Gran, Biſtritza, Szalatna* and *Carponia* or *Krupenitza.* Among its mineral fprings the moſt celebrated are thoſe of *Oſztroſk,* half a mile‑from *Vigles* Caſtle, with the fprings of *Altfohl‑* and thoſe near the river *Gran,* with the *Ribar, Satſin,* and *Batzuch* waters. But the warm baths are at *Neuſohl* and *Ribar,* and *Herrengrund* yields vitriol. Agriculture is little followed here, but the country abounds in good breeds of cattle, efpecially multitudes of ſheep. The inhabitants are *Bohemian-Sclavonians.* The gentry are chiefly compoſed of *Hungarians,* and in the towns are great numbers of *Germans.* This province confiſts of two Diſtriċts.

1. The Upper Diſtriċt, *Proceſſus ſuperior,* in which are
: Three royal free towns.

Neuſohl, Boſzertze, Banya, Banſka, Biſtritza, Neoſolium, a royal‑ free town, and the beſt of the mine-towns, ſtands on the river *Gran.* It is built pretty much in the *Saxon* manner, has ſix churches, a Jefuit college and fchool ; is noted for its weekly market and the copper ore in the adjacent mountains. All things are very good and cheap here, except that the ſmelting works give the air an unhealthy taint. In 1500, 1555, 1591 and 1653 it ſuffered greatly by fire.

Libeth, Libetha, a royal free mine-town, but ſmall and mean. It had formerly good copper and iron works, but at prefent its principal dependence is agriculture.

Brizna, Banya, Brezno, the *Bries, Britzna,* a royal free town on the river *Gran,* which, by frequent fires, is now reduced to a ſmall inconfiderable place, whofe principal employment is the breeding of ſheep.

Two caſtles ; namely,

Neuſohl Caſtle, a ſtrong place with two churches, lies on a hill on the north fide of the town of *Neuſohl.*

Liptſche ſtands on a ſteep rock, having a very deep well hewn in it.

Three feats ; namely, thofe of *Radbany, Meſin* and *Deubravitz* ; the laſt in ruins.

Three ſmall towns.

Radbani, on the river *Gran,* near *Neuſohl,* fupported by handicrafts and a ſmall trade, and belongs to the *Radbanſky* family.

Tot-liptſche, a royal privileged but mean town, lying below *Liptſche* Caſtle, fubfifting chiefly by tillage.

Ponek, formerly a copper-town with great privileges, but at prefent produces only iron-ore. The adjacent plain terminates in a ſtately amphitheatre of hills.

Twenty-five large villages ; among which

Herrengrund, Schtania Dolina, Vallis dominorum, having the appearance of a town, lies among the tops of the mines, and all its inhabitants are miners. It is famous for its vaſt copper works, and its vitriol, by which iron is apparently turned into copper. This vitriol is faid to have been difcovered

in

in 1605. Here are twenty odd churches, in some of which it exudes from the sides, and in others oozes out of the earth. Several hundred weight of iron are thus changed every year into copper; but formerly the quantity was greater than at present. The vitriol indeed does not properly change the iron, but insinuates into it the copper particles with which it is saturated. This transmutation requires a fortnight or three weeks, but if the iron be suffered to lie too long in this vitriolic liquor, it becomes at last reduced to a cupreous powder. Here is a society of Jesuits.

2. The Lower District, *Processus inferior,* in which are

Two royal free cities; namely,

Altsohl, Zolyom, Zwolm, Vetusohlium, the chief town of this District, yet in itself but an inconsiderable place, stands high and pleasant, with the rivers *Gran* and *Szalatna* running nigh it. In 1708 it was set on fire by Prince *Rakotzy's* troops.

Corpona, Krupina, Karpfen, of small account, but delightfully situated among vineyards. Has itself a great many fine fruit gardens and a gymnasium *P P. piarum scholarum.* In 1407 it was destroyed by the *Hussites,* afterwards by the *Bohemians,* and in 1708 laid in ashes by Prince *Rakotzy's* men.

Three castles.

Zolyum, joined to the town of *Altsohl,* but standing higher and fortified, belongs to the *Esterhasi* family.

Vigles was formerly a frontier fortress stands on a high rock, and belongs to the same family.

Dobroniwa, Dobrona, stands on a rock, but in a bad condition.

Four seats; namely,

Kiraly-Falva, Garam-Szog, Hajnik and *Ostroluka.*

Six small towns; namely,

Nagy-Szalatna, a small place on the river *Szalatna,* within the jurisdiction of the castle of *Vigles.*

Otschowa, somewhat more populous than the former, but rather meaner, tho' situate in a good corn soil. The native place of Mr. *Matthias Bel,* whose celebrated work of *Hungary* is of great use to us.

Dobrona, Dobring, Dobroniwa, populous and lying near the castle of the same name. It enjoyed indeed formerly royal privileges, but is now under the jurisdiction of the castle of *Altsohl.*

Babasek, Babina, stands in as good a soil as the former, and belongs to . Count *Bereny.*

Szasz, or *Nemet-Pelsotz,* situate in a fruitful country, and belongs to the jurisdiction of the castle of *Zolyom.*

Pelsotz, Pliessowitz, Tot-pelsotz, as well as the preceding, pays no tithes.

This

This town, together with *Szafz Dobrona* and *Babajek;* have the *jus gla-dii immediatum*, which empowers them to appeal to the *Perfonalis prefentiæ regiæ.*

Thirty-fix large villages, among which that of *Ribar* is celebrated for the warm baths in a hill near it ; and their external appearance is no lefs extra-ordinary than their virtues. About fix hundred paces from it towards the fouth, in a fine meadow, which makes part of a fmall delightful valley, is an aperture long noted for its noxious effluvia, which feem to be fulphu-reous, and kill both beafts and birds. A ftream gufhes out here with great impetuofity; which is afterwards notwithftanding abforbed in the aperture. Not far from it is a mineral fpring. Thofe effluvia, however lethiferous, are not poifonous ; for the water may be drank, and the dead beafts and fowls killed by it fafely eaten. See Mr. *Bel's* account in the *Hamburgh* Magazine, Vol. iv. P. 69.

8. The Gefpanfchaft of *Bars, Bars Vermegye, Comitatus Barfchienfis;* is about feven miles long, and from two to three broad. Among feveral others here is the famous *Cremnitz* gold mine. Its chief rivers are the *Gran; Nitra* and *Sitva* or *Zitawa.* Here are not only the mineral fprings of *Bukow* and *Ebedertz,* but alfo the hot baths of *Glafs-hutner* or *Sklenne,* together with thofe of *Eifenbach* or *Wihna,* efteemed fome of the moft famous in all *Hungary.* This country affords a good wine, and the level parts alfo good grain. The *Cremnitz* and *Konigsborg* gold mines are far from being fo rich as formerly. There is little cattle bred in the mountains, fheep excepted ; nor are the vallies more remarkable for graziery. The inhabitants confift of *Hungarians, Bohemian-Sclavonians* and *Germans.* The Province contains the following Diftricts.

1. The Diftrict of *Ofzlan, Proceffus Ofzlanenfis,* in which are
Two royal free towns, namely,

Cremnitz, Kremnitza, Kormotz, the principal mine-town, lying in a deep valley betwixt high mountains, fo as not to be feen, till one is juft upon it. The town of itfelf is but an inconfiderable place, containing two churches, one *Francifcan* convent, not above thirty houfes and a caftle ; but the fuburbs are much larger, and thefe confift of nine ftreets with an alms-houfe and a church. In this town the King has a revenue-office, with a director, receiver, book-keeper and other officers. Here is alfo the mint to which all the other mine-towns bring their gold and filver. It is computed to coin every year about one hundred thoufand ducats, and every ducat to ftand the crown in twelve guilders : which great deficiency muft be compenfated by the richnefs of the other ores. The gold mine country hereabouts is divided into the after and fore mint. The former contains feven mines, and the latter two. But all the nine belong to the King. There are alfo mines appertaining to the town in

general,

general, and to private perfons; but at prefent, thefe, like the others, are
fo exhaufted, that the gold got from them fcarce anfwers the expence.
On a hill near the town is a caftle with a church. The unwholefomnefs
of the air and water occafions a great deal of illnefs among the common
people. Both the mines and towns fuffered extremely in the dif-
turbances raifed by Prince *Rakotzy* and Count *Tekely*; but in 1751 it was
honoured with the prefence of the Emperor *Francis* I. who, in the garb
of a miner, went down a very deep fhaft, and is the only inftance of
fuch venterous curiofity in a Prince.

Konigsberg Uj-banya, Nowa Banya, Nova Todina, i. e. ' the new mine,'
is a mine-town, to which the mountains furrounding it ferve for a wall.
It has but few houfes, and thofe as mean as the fituation is bleak and
difmal. It is however not without two churches, one of which belongs
to the hofpital. The ancient gold mine work of this place, which was
once of fuch confiderable value, is now at an end, in fo much that
the whole fubfiftence of the place depends upon tillage, glafs-works and
brewing of beer.

Small towns and caftles, namely, .

Ofzlany, a fmall town pleafantly fituated; but the foil not the moft fa-
vourable for tillage. In 1662 and 83 it was laid in afhes by the *Turks.*

. .*Holy-crofs, Szent-kerefzt, Swaty-kritz, Fanum fcantæ crucis,* a fmall town
on the *Gran,* in a high and pleafant country, with good corn lands, paftures, and a feat. This place has been feveral times burnt, particularly
in 1726.

Zarnotz Zernovitza, a town on the *Gran,* the neighbourhood pleafant
and fruitful; is noted for good bread, which is carried from thence to
the neighbouring towns, and belongs to the revenue-chamber at *Cremnitz.*

Welka polya, Highmetad, a fmall well-peopled town in a pleafant valley,
belongs to the *PP. Sancti Pauli* of the Eremites.

Sas-ko, Saxon-Stone, a caftle on a high and fteep rock, bolongs, together
with its jurifdiction, to the revenue-office at *Schemnitz.*

. *Rewifchtye Reve,* a caftle on a high mountain on the banks of the *Gran,*
belongs to the revenue-chamber at *Schemnitz.*

A great number of towns; fourteen belonging to the *Sas-ko,* or *Hladomer*
jurifdiction; twelve to that of *Rewifcht;* above eighteen to that of *Holy-*
crofs in the archbifhopric of *Gran;* twenty-four to feveral owners, and
fix to the town of *Cremnitz.* Of all thefe the moft remarkable are

Skleno or *Glafshuts,* famous for its excellent warm baths, of which that
called the Sudatory is particularly remarkable. On the hill over them
ftands the church, the earth of whofe burial-place is fo hot, that in
half a years time the bodies are reduced to duft. This place lies in the
jurifdiction of *Sas-ko.*

. *Alfo-dany,* in the *Rewifchtye* jurifdiction, has a mineral-water.

Nemes-Kofztolany, large, and an article-place of the communities of the *Augsburg* confeffion, where they ¡ have alfo public fervice ; belongs to the family of *Kofztolan*.

Eifenbach, *Wihnye*, a populous town, belongs to *Schemnitz*, and is fa- mous for its warm baths. The water, which is of a furprifing heat at the fprings, is conveyed three hundred and fifty paces through pipes into the bathing rooms, where it becomes of a proper warmth.

2. The *Leva* Diftrict, *proceffus Levenfis*, in which are
The following towns and caftles.

Leva, *Lewitze*, *Lewentz*, a little town belonging to Prince *Efterhafi*, with a caftle near it, noted for being frequently befieged and taken.

Berfenburg, *Tekou*, a very inconfiderable place on the *Gran*, which often overflows the neighbouring plain, gives name to the Gefpanfchaft.

Nagy-Sarlo, *Velike Scharluby*, a fmall town with good corn-land, under the jurifdiction of the Archbifhop of *Gran*.

. Fifty-two large villages, among them *Cfeiko* and *Szolos* famous for very good wine.

3. The *Kis-topoltfan* Diftrict, *Proceffus Kis-topoltfanenfis*, in which are
The following caftles and towns,

Kis-topoltfan, *little Topoltfan*, *Male Topoltfchany*, a fmall town with a feat, has good corn-land, and belongs to the *Zchichi* family, and is the ufual meeting place of the fmall provincial Diets.

Hrufow, a caftle now lying in ruins, in which the records of the county were formerly kept.

Szent Benedick, *Swaty Benedec*, *Fanum S. Benedicti*, formerly a *Bene- dictine* Convent, belonging to the chapter of *Gran*, lies on the high fhore of the river *Gran*, and is fortified, which gives it the appearance of a caftle: near it is a mean town.

Aranyos, *Maroth*, *Morawetz*, *Morawetze*, a fmall mean town, but en- joying good corn-land, belongs to the *Palufky* family.

Fifty fmall villages.

4. The *Verebely* Diftrict and feat of juftice, *proceffus & Sedes Werebelyenfis*, where the gentry, who are called Predialifts, are folely under the jurifdiction of the Archbifhop of *Gran* as their fupreme Lord, they have alfo particular palatines and magiftrates of their own; but the other gentry and peafants are obliged to apply for juftice to the county court. The remarkable places in this Diftrict are

Verebely, *Wrable*, a mean town, on the river *Zfitva*, formerly fortified. To its jurifdiction belong fix large villages.

Twenty three large villages under feveral proprietors.

9. The Gefpanfchaft of *Hont*, *Comitatus Hontenfis*, confifts of two parts, betwixt which lies a part of the Gefpanfchaft of *Novogrod* and *Altfohl*.

1. *Nagy-*

1. *Nagy-hont*, *Great Hont*, reaching from *Bela Banya* to *Maros*, is about nine *Hungarian* miles in length; and in the broadeſt parts, betwixt *Varſaney* and *Kovar*, five in breadth; but towards the ſouth terminates as it were in a point. The country is every where mountainous, but the mountains both in the middle of the country and on the frontiers, eſpecially the latter, abound in gold, ſilver and lead. The rivers are the *Danube, Gran, Ipola*, and ſome other ſmall ones. Near the villages of *Gyogy* and *Szanto* are warm baths; the laſt has alſo a celebrated mineral ſpring, the like of which is found alſo at the villages of *Szalatnya Gyogy* and *Felſo-Palojta*. On the mountain of *Szitna*, the higheſt in the whole Geſpanſchaft, is a ſpring quite cold in ſummer and warm in autumn. Here are alſo ſome medicinal ſprings. The ſouth part of the Geſpanſchaft produces a little grain and a good wine; but the graziery here is inconſiderable, and the air about the mine mountains cannot be ſuppoſed to be very healthy. The inhabitants are *Hungarians, Bohemians, Sclavonians* and *Germans*. In this province is

1. *Schemnitz* Diſtrict containing

Two royal free and fine towns.

Schemnitz, Schemnicium, formerly *Banya* and *Sebnitz* or *Sebenitz*, a pretty large and very populous town in a long valley, the houſes of which ſtand ſcattered a conſiderable way up the acclivity on both ſides. The proteſtant inhabitants, who make above two thirds of the town, are between ſix and ſeven thouſand. In largeneſs and number of mine-works it ſurpaſſes all the other mine-towns in *Hungary*. It has two caſtles, two churches, two chapels, and a college of Jeſuits, with a royal mine-ofhce; and is the reſidence of a chief commiſſioner. The gold and ſilver mines here ſtill produce a conſiderable quantity of ore, and the workmen amount to above five thouſand. But how greatly they are declined appears from this ſingle circumſtance, that at the cloſe of the laſt century, three or four thouſand marks of ſilver were made here every week; whereas now the quantity ſcarce reaches one hundred. The ore here contains more and better gold than that of *Cremnitz*, but the gain attending it is not large. The yearly charges to the crown of the mine-works here amount to above five hundred thouſand guilders.

Bela-Banya, *i. e.* the white mine, *Diln*, formerly *Fejor-Banya*, which name has ſome affinity to the firſt, a mean little town, whoſe mines being exhauſted, the inhabitants apply themſelves to tillage, in which the profit recompences the labour.

The mine-caſtle of *Szitna*, together with its juriſdiction, belongs to the family of *Kohar*.

Three ſmall towns; *viz.*

Szebeklib, founded by a colony of *Saxons*, and, at preſent, the rendezvous of the Diet and the repoſitory of the records of the Geſpanſchaft. This place belongs to the cathedral of *Gran*.

Nemethy,

Nemethy, Nemetz, in 1731 made a privileged town; follows tillage. The name fpeaks it a *German* colony.

Saagh, a mean place, was formerly famous as a probfty of the Bleffed Virgin of the *Præmonftratenfes,* but at prefent belongs to the Jefuits college at *Neufchl.*

Fifty-four large villages, among which is *Szud,* whofe wine is the beft in the whole Gefpanfchaft.

2. The *Bath* Diftrict, *Proceffus Bathenfis,* contains

Baka-Banya, Bugganz, Bukanetz, a royal free and mine-town; but in-ftead of the gold and filver mines, which it was formerly famous for, now fubfifts by tillage. In 1664 this town was facked and burnt by the *Turks.*

Bath, Batowtze, Fraumark, a fmall town, but having good vineyards and corn-land, has alfo a good weekly market, and a very confiderable fair.

Barfony Pilfen, an old mine-town and colony of *Saxons* who formerly were employed in the gold mines, but now fubfift by agriculture, belongs to the jurifdiction of the Archbifhop of *Gran.*

Fifty-four villages, and among them

Maros, fituated on the banks of the *Danube,* and formerly a town.

Szanto has a mineral water faid to be like that of *Seltz.*

Magyarad, noted for its warm baths.

3. The Diftrict of *Bozok, Proceffus Bozokienfis;* in which are

Bozock, Bzowik, a fmall town with a caftle, had formerly a probfty of *Præmonftratenfes,* which is now no more, belonging to the *Tirnau* Jefuits.

Cfabrag, a mine-caftle belonging to the *Kohar* family.

Dregely, a ftrong caftle on a fteep mountain; was taken by the *Turks* in 1552, from whom it was not recovered again till 1593. In 1649 they took it a fecond time. At prefent it belongs to the Archbifhop of *Gran.*

Fifty-feven large villages; among which is *Honth,* formerly the capital of the Gefpanfchaft which bears its name.

Kis-hont, Klein-hont, is from *Teifholtz* to *Rimafzombath,* about three *Hungarian* miles in length; and from *Szuha* to *Tothegymek* at moft about one mile in breadth. The rivers of this territory, which is wholly moun-tainous, are *Rima* and *Szuha.* It has but little agriculture and graziery. The air, however, is healthy, and the mountains yield a good iron. It has alfo feveral mineral fprings. The inhabitants are *Bohemian-Sclavonians,* with a mixture of *Hungarians.* It contains

Rima Szombath, Rimaufka-Szobota, a fmall town on the river *Rima,* but well-built and populous. The neighbourhood is level and turns to good account in tillage. This is the place of rendezvous for the county meetings, is moated and walled. The inhabitants, befides tillage, apply themfelves to trade and handicrafts.

Tifzoltz,

Tifzoltz, Teifzholz, Tifzownik, Taxovia, a fmall town, near which are found iron and load-ftones; it has alfo a mineral fpring.

Raho and *Ofgyan,* feats with villages.

Thirty-two large villages.

10. The Gefpanfchaft of *Neograd, comitatus Neogradienfis, Nograd Var-megye,* twelve *Hungarian* miles long and from five to fix in breadth; is mountainous and woody, though with fome fruitful plains towards the South. Its chief rivers are the *Ipoly (Eipel, Ipel)* and the *Zagyva,* and its chief medicinal fprings are thofe of *Garab, Poltar, Fileke, Efzurger, Kurto, Zfalatna, Tifzowna, &c.* Here are alfo fome fprings, which are cold in fummer and warm in winter. The fouthern parts find their account in til-lage and vineyards, particularly thofe at *Kofda, Rada, Etfeg,* and *Jobbag* which produce an excellent fort of wine; and in fome places great advan-tages are derived from grazing. The inhabitants are *Hungarians,* and *Bohe-mian-Sclavonians.* This Gefpanfchaft is divided according to the epifcopal Diocefes into *Nagy* and *Kis Nograd,* i. e. *Great and Small Nograd,* the for-mer belonging to the Archbifhop of *Gran,* the latter to the Bifhop of *Vatzi.* By another divifion it is reduced into four Diftricts, which are

1. The Diftrict of *Lofoutz, proceffus Lofoutzienfis,* all over mountainous and woody, and contains

Two caftles, namely,

Gats Halitfch, a mine-caftle fo called from the *Halitfchi* who formerly dwelt in *Red Ruffia;* ferves for the county jail, and belongs to Count *Forgatfch:*

Divin, a caftle on a very fteep rock, taken in 1576 by the *Turks,* who held it till 1593. In 1674 it was taken from the rebel *Balaffa,* and de-ftroyed, by which means it is now reduced to a heap of ruins. This caftle belongs to Count *Zichy.*

Three fmall towns, namely,

Lofouz, Lucfenetz, fituate in a plain furrounded by mountains; and, on account of its dirtinefs in rainy weather, jocularly called *Lutetia Hungaro-rum.* Here all the meetings of the Gefpanfchaft are held. With refpect to the great numbers of gentlemen who refide in this town, it may be looked upon as the capital. It has alfo very confiderable annual fairs.

Gats, a mean place belonging to the jurifdiction of the adjoining caftle of the fame name.

Divin has its name from the above-mentioned ruined caftle, to the ju-rifdiction of which it belongs. The inhabitants apply themfelves to tillage and hunting.

Fifty-one large villages.

2. The *Filek* Diftrict, *proceffus Filekienfis,* in which are

Five caftles, namely,

Filek,

Filek, formerly fortified, on a steep rock whereon it was built, and divided into three parts; but having been frequently besieged and taken, it was at last blown up by Count *Tekeley*. Its jurisdiction belongs to Count *Kohar*.

Salgo, a ruined castle on a steep rock, which the *Turks* took by stratagem in 1551, and kept till 1593. In 1726 it became the property of Baron *Szluha* of *Iklad*.

Somsko, a castle on a high rock, which the *Turks* held from 1576 till 93.

Zagyva and *Baglyas Vara*, castles in ruins.

One town, namely *Filek*, under the jurisdiction of the castle of the same name, formerly a place of some consideration, particularly for being the residence of many noble families: But at present it is deserted, and in a very mean and wretched condition. About a hundred paces from hence is an excellent medicinal spring.

Sixty-four large villages.

3. The *Szetseny* District, *processus Szetsenienfis*, containing Six castles, namely,

Szetseny, standing on a rising ground, and very pleasantly situated. The *Turks* were once in possession of it for forty-two years, and afterwards twenty. At present both the castle and its jurisdiction belong to Count *Forgatsch*.

Holloko, a castle on a high, craggy and steep rock, taken in 1552, by the *Turks*, who held it forty-one years. Falling again, in 1663, into their hands, they kept it twenty years. It belongs to the families of *Forgatsch* and *Szemer*.

Bujak, a castle standing betwixt mountains on a steep rock, and belonging to Prince *Esterhasi*.

Ecsegh and *Samson Vara*, or *Fejerko* and *Szanda*, two castles lying in ruins.

The little town of *Szetseny* underwent the same fate with the castle of the same name, under which it lies; and was formerly fortified. In 1719, it was destroyed by fire: In it is a *Franciscan* convent, and the place itself belongs to Count *Fortgatsch*.

Sixty-three large villages.

4. The *Kekko* District, *processus Kekköienfis*, in which are Four castles, namely,

Kekko, Modry Kamen, Plauenstein, formerly a well fortified castle situate on a rock, and taken in 1576, by the *Turks*, who lost it again in 1603. Since its being burnt by *Rakotzy*'s men, a part of it only has been rebuilt.

Balassa-Gyarmath, Nograd, and *Kamor*, castles lying in ruins.

Five small towns, namely,

Rekko, lying below the castle of the same name, irregularly built, but well inhabited. It is under the jurisdiction of *Balassa*.

Balassa-

Balaſſa-Gyarmath, a mean place.

Nagy-Oroſzy, built by a colony of *Ruſſians,* as appears from the name, the *Hungarians* calling it a *Ruſſian Oroſz.* It belongs to the *Starenberg* juriſdiction.

Nograd, formerly a fine town defended by a ſtrong caſtle, but at preſent only a large village under the juriſdiction of the Biſhop of *Natz.*

Vadkert, belonging to the Archbiſhop of *Gran,* and erected into a town by *Charles* VI.

Fifty-five large villages.

11. The united Geſpanſchafts of *Peſth, Pilis,* and *Solth.*

The Geſpanſchaft of *Pilis* was incorporated with that of *Peſth* upon the removal of the King's reſidence to *Buda,* and to that of *Solth,* after the taking of *Alba Regalis* by the *Turks.* The length and breadth of this county is not to be ſtated with any preciſion, the *Cumani* and *Jaſi* having broke off their connection, and the ſeveral frontiers not being yet determined. Amidſt its mountains and woods, however, is a large level lying betwixt the *Danube* and *Theis,* but for the moſt part ſandy and barren. The capital river is the *Danube,* the ſmaller are the *Vajas, Theis, Zagyva, Galga, Kakos,* and *Tapjo.* In the mountainous parts the winters are cold, and the ſummers temperate. On the contrary in the large plains the winters are ſevere, and the ſummers intolerably hot, the warm days being alſo ſucceeded by very cold nights; the muſketto's too here are very troubleſome both to man and beaſt, and good water is alſo very ſcarce. In the mountainous parts is produced an excellent kind of red and white wine. In the ſandy ſoil, which every where occurs, no corn can be raiſed; but in other places tillage ſucceeds ſomething better, though not without a great deal of trouble. On the deſert plains is good paſturage for cattle, where they wander at large. The inhabitants are *Hungarians, Bohemian-Sclavonians,* and *Germans,* with ſome *Dalmatian* and *Thracian* colonies. The whole Province is divided into four Diſtricts.

1. The *Vatzy* Diſtrict, *proceſſus Vatzienſis,* in which are

Peſth, Peſtum, a royal free town in a plain on the *Danube,* over againſt *Buda,* to which in ſummer we go by means of a bridge of boats. Here is the Supreme Court of Appeal, which is divided into the *tabula regia,* or royal board, and the *tabula ſeptemviralis,* or board of ſeven; alſo a large military hoſpital, all of ſtone about two hundred paces ſquare, and three ſtories high; ſix convents, among which is likewiſe a college *P P. piarum ſcholarum,* and ſeveral churches. The town is ſurrounded with a wall and moat. The field *Kakos* near it, on the river of the ſame name, which takes its title from the *ecreviſſes (Kak),* is famous for the Diets held, and Kings choſen, there. In 1526, and 41, it was taken by the *Turks,* who held it till 1602, when it was retaken from them by the *Hungarians,* and defended by them againſt a freſh attack of the *Turks.* But in the following

4 year,

year, being abandoned through fear, it fell once more into the hands of the *Turks*, who in 1684, set fire to it and marched to *Buda*. The *Imperialists*, upon this, indeed, took possession of the town, but abandoned it again, not returning till two years after, when it was reduced to a very poor condition, but was raised by the liberality of the Emperor *Leopald*. In 1721, a commission met here to enquire into the religious grievances of the Protestants, and was closed the following year at *Presburg*.

Vaitz, *Waitzen*, *Vatzia*, *Vatzovia*, a well peopled episcopal city on the *Danube*, pleasantly situated, with fruitful corn-lands. The above bishop-rick was founded by *Geyfa* the Great in 1074, or 75, and its annual income is 50,000 guilders. Besides the boufes belonging to the Bishop, here are three convents, among which is one also belonging to the *P P. piarum scholarum*, together with a gymnasium. The city chiefly owes its prosperous condition to its great annual fair, and beast-market. It has been frequently destroyed by fires, and several times taken. In 1535, a peace was concluded here betwixt King *John* and King *Ferdinand*. In 1543, it was taken by the *Turks*, who held it till 1595. In 1597, it was set on fire by the *Imperialists*; but these afterwards settled in it again, and fortified it. Not to mention other vicissitudes of the like nature, in 1619, it fell again into the hands of the *Turks*, and though they were obliged in 1684, to deliver it up to the *Imperialists*, these relinquished it of themselves; but in 1685, took possession of it again, after it had been burnt by the *Turks*.

· Fifty-three large villages, among which *Aſzod* is the most remarkable; it was called formerly *Oſthmuch*, and looks like a town. It belongs to a person of the name of *Podmanickky*.

2. The *Ketſkemet* District, *proceſſus Ketſkemetenſis*, containing

Ketſkemet, a large and very populous town : it is much resorted to on account of its annual fair, and carries on a lucrative tillage. Besides a *Franciſcan* convent with a church, and a house and church belonging to the *P P. piarum scholarum*, it has likewise a *Lutheran* church.

Koros, a well inhabited town belonging to Count *Reglewitſch* and some other noblemen. The neighbouring country is better adapted to vineyards and pasture than agriculture.

Czigled, a pretty large town in a fruitful country, belonging to the nuns of St. *Clare* at *Buda*. After the battle of *Sicambria* it was the residence of the *Zukler* of *Tranſylvania*.

Twenty-seven large villages, among which is *Janos-Hidja*, or *Hida*, belonging to the *Præmonſtratenſes*, who call it the probſty of *John's-bridge*.

3. The *Pilis* District, *proceſſus Piliſienſis*, is but a barren soil, and though properly belonging to the Circle below the *Danube*, must be treated of here.

Ofen Buda is the name both of an old and new city. *Old Buda* stands in a plain extending itself from the suburbs of *New Buda* to between the

Pilis mountains and the *Danube*, and was formerly called *Sicambria*. At prefent it is a poor mean place belonging to the *Zichy* family. In and about it are to be feen the defolate ruins and the remains of the old city, with feveral *Roman* monuments. *New Buda* was built by *Bela* IV. and is a royal free town. It lies on a mountain on the *Danube*, was formerly the capital of the Kingdom, the refidence of the King, and the largeft and fineft of all the *Hungarian* towns, but has been very much reduced by being often befieged, taken, and deftroyed. Omitting all mention of its ruined church, that of the Afcenfion of the Virgin *Mary* is the principal, near which the *Jefuits* have an academical college and feminary. The *Carmelite* nuns of St. *Claire* have convents here, and the *Francifcans* churches. In the town-houfe is a chapel wherein is kept a confiderable church-treafury. The city is furrounded with walls and moats, and well fortified. Near it ftands a ftrong caftle. It has three fuburbs. In the *Wafferftadt*, which lies on the *Danube*, are two churches and one convent; in the *Rayenftadt*, *Thaban*, is a catholic and *Ruffian* Greek church, and in *Neuftift* is only one church, together with a pillar fifty-two feet high, which in 1690, was dedicated to the Holy Trinity as a thankfgiving for deliverance from a peftilence which ceafed in 1715. Here are fome admirable warm baths called *Gerhardsbad, Burgurbad, Raitzenbad*, the hofpital-bath and the Emperor's, together with a pond of mineral water, having this furprifing property, that when the water is wholly turned off, the warm fprings ceafe flowing, but when the pond is filled a little above half full they return again. The wine produced on the neighbouring hills is red and of a very good fort. Here are alfo excellent melons. This famous city was in the hands of the *Turks* from 1529, till 1686, and notwithftanding all attempts could not be taken from them before the laft mentioned year, when it was in a very deplorable condition. In 1723, it was almoft totally deftroyed by fire. Between this place and *Pefth* is a bridge of boats.

Viffegrad, Vicegrad, Plindenburg, Viffegradum, altum caftrum, was formerly a ftrong caftle on a high hill, in which the *Hungarian* crown was kept; and below it, on the *Danube*, was a town containing upwards of three hundred and fifty boufes, where the Kings ufed to refide on account of its wholfome air, fine gardens, and other inducements. But the good fortune of the place lafted no longer than till the death of *Matthias*; whence the caftle is at prefent become a heap of ruins, and the town a poor village.

Sambek, a caftle fituate on a plain amidft pleafant mountains, with a fmall town, belonging to Count *Zichy*.

The ifland of St. *Andrew, Ros infula*, in the *Danube* is about three *Hungarian* miles and a quarter long, and more or lefs in breadth. It belongs, with its villages, to the *Hungarian* crown-lands.

St. *Andrew, Szent Endre, Fanum S. Andreæ*, on the *Danube*, makes' a better figure and is a more populous town than *Buda*; it is a *Rafcian* co- lony, and belongs to Count *Zichy*.

Twenty-three large villages.

Eugene promontory, or *Eugenius Hyge*, a pleafant mountain in the *Da- nube* covered with vines and woods, at the place where an arm of that river forms the ifland *Efepel*. In this neighbourhood is a pleafant and fruitful plain about one *Hungarian* mile in circumference, containing befides Prince *Eugene*'s feat feveral farm-houfes. Prince *Eugene*, who often made it his place of refidence, had a breed of *Arabian* fheep here.

The ifland of *Efepel* lies a quarter of an *Hungarian* mile below *Buda*, in the *Danube*, and is furrounded on both fides with fmall iflands, ten of which are on the left; but thofe on the right are richer, among which the pheafant ifland is particularly remarkable, being a thoufand paces long and covered with wood, as alfo *Ujvalvififche* ifland, in which is a ruined church, faid to have been built by St. *Margaret*; hence it is that fome geographers have called *Efepel* ifland that of St. *Margaret*. This ifland is five *Hungarian* miles in length, fandy, and not very fruitful, abounding, befides other game, in great plenty of hares. Formerly it was one of the Queen's dower-houfes, and a menagery. In latter times it belonged to Prince *Eugene*, and after- wards to the Emprefs dowager *Elizabeth*. Of the four towns which for- merly ftood upon it, not one now remains.

Katzkeve, a poor little town which takes its name from a *Rafcian* colony, and was formerly a confiderable place. In 1698 it fell into the poffeffion of Prince *Eugene*, who built a ftately feat near it.

. The nine large villages of this ifland were formerly fine, but at prefent are very mean. Of thefe we fhall take notice of *Efepel* only from which the ifland derives its appellation, and *Tekely* the original place belonging to the family of that name.

4. The *Solth* Diftrict, *Proceffus Solthenfis*, containing,

Colocfa, an archiepifcopal city, which formerly flourifhed greatly but de- clined after the death of the Emperor *Lewis*, and together with *Buda* was taken by the *Turks*, in whofe hands it alfo continued till in 1602 it was deftroyed by the *Hungarians*, and afterwards further ruined by the *Turks*. In 1686 it became again an Imperial and archiepifcopal fee, and now begins to thrive. St. *Stephen* founded a bifhoprick here, and *Geyfa* II. an arch- bifhoprick.

Solth, a fmall town not far from the *Danube*, fupported by tillage and graziery. The ancient caftle near it lies in ruins.

Pathay, a large and populous town belonging to the family of *Szarazy*. 4

Twenty-feven villages.

12. *Little*

12. *Little Cumany, Kis-kunok, Cumania minor,* is united with Great *Cumany,* which shall be spoken of below, and takes its name from the *Cumani,* who probably dwelt betwixt the *Dnieper* and *Black-sea* in the country now inhabited by *Precopian Tartars;* but being expelled in 1237 by the *Tartars,* came under their King, *Cuthen,* to *Hungary,* where they were favourably received by King *Bela* the fourth, who, out of a desire of propagating the Christian religion, gave them the best places on the *Theis,* and stiled himself King of the *Cumani.* The province contains three well inhabited towns.

Szent-Miklos, Kun-Szent-Miklos, the first town of the *Cumani,* situate in a level, fertile soil.

Szabad-Szalas, Libera mansio, the second town of the *Cumani,* populous and lying in an open plain, being surrounded with corn-lands and pastures.

Phulep-Szalas, Fulep-Szalas, Philippi Mansio, the third town, situate in a level and fruitful country.

13. The Gefpanfchaft of *Bats, Bats Varmegye, Bathienfis Comitatus,* was formerly united with the Gefpanfchaft of *Badrog,* and is subject, for the most part, to the royal chamber. The inhabitants are *Hungarians,* and among them great numbers of *Servians* and *Rafcians.*

Pandur, a mean *Rafcian* village, but remarkable for the activity of the *Heyducks* against robbers infesting the roads, and from which the *Rafcian* foot in the last war took their name. These *Pandours* were first known in *Germany* in 1741, when Baron *Trenk* marched a body of these irregulars, of about one thousand men, into *Silefia* against the *Pruffians.*

Baja, a well peopled town on the *Danube,* in a fruitful country, belongs to the barons of *Vajay.*

Bachmonoftra, Bathmunfter, Batfmonoftra, formerly a famous probsty, of which some remains are to be seen in the town of the same name.

Zombor, a large and populous town in a fruitful country.

Bats, Bachia, an Episcopal city from which the county takes its name, and formerly famous both as a bishop's see and for the number of its inhabitants, but since much dwindled, and with the bishopric united to *Colocza.*

Szabadka, or *Szent Maria, Sancta Maria,* a large and well peopled military town, and *Rafcian* colony, situate in a pleasant country lying on the *Theis.*

Zeuth, a military village on the *Theis,* but remarkable for the important victory which Prince *Eugene* gained near this place in 1697 over the *Turks,* of whom about thirty thousand were either cut to pieces or drowned.

14. The Gefpanfchaft of *Bodrog, Bodrogh Varmegye, Bodrogienfis Comitatus,* was in 1747 feparated from the Gefpanfchaft of *Bats,* and is, for the most part, added to the exchequer. The inhabitants are *Rafcians* with a few *Hungarians.*

Palanka, a village, formerly surrounded by the *Turks* with a moat and wall.

G 2 *Futtak,*

Futtak, *Futtakum*, a town in a large plain with good pastures, and very commodious for a camp, as the *Imperialists* experienced in 1736.

Romerschantz, *Romanus S. Abarius Ager*, extends from the *Danube* to the *Theis* along a strip of land. Some remains of its large works are still to be seen.

Peter-Vardein-Schanz, *Petri Varadini fossatum*, a large *Rascian* town, surrounded with a wall, and the see of a *Greek* bishop. It lies on the *Danube*, over against *Peter Vardein* in *Sclavonia*.

Kobila is well known for a *Greek* convent of the order of St. *Basil*.

Titul, *Titel*, a town situate at the conflux of the *Danube* and *Theis*, formerly very rich and considerable for its probity, but at present only the shadow of its former magnificence.

Betse, a military town on the *Theis*, near its influx into the *Danube*, and inhabited by *Rascians*.

Martonos, a *Rascian* military town on the *Theis*, near the *Roman Vallum*.

II. The circle below the *Danube*, *circulus Transdanubianus*. The inhabitants *Hungarians*, intermixed with *Croats*, *Rascians* and some *Sclavonians*. It consists of twelve 'Gespanschafts.'

1. The Gespanschaft of *Oedenburg*, *Sopron Varmegye*, *Sopronienfis Comitatus*, borders upon *Austria*. Its inhabitants are *Hungarians* and *Germans*, with some *Bohemian-Sclavonians* and *Croats*.

Eisenstadt, *Kis-Marton*, a royal free-town, in which Prince *Esterhasy* has a magnificent palace. It lies on the frontiers of *Austria*, and formerly also belonged to it for some time, till, in 1625, at the Diet of *Oedenburg*, and in 1637 and 1638, at that of *Presburg*, the states of *Hungary* sollicited the redemption of it.

Oedenburg, or *Edenburg*, *Sopron*, *Sopronuim*, or *Sempronium*, the best of all the royal free towns in this District. It is not indeed large, but is nevertheless well built and populous and has large suburbs. The *Hungarian* chronicles derive the *Latin* name from *Sempronius*, but the *Germans* are said to take it from the destruction of it by the Emperor *Charles* the Great, or *Henry* the third, after which it was rebuilt and peopled with *Germans*. Here is a Jesuits college and a *Lutheran* gymnasium. The inhabitants apply themselves with great industry to the cultivation of vineyards, and the wine is accordingly remarkable for its goodness. In 1605 the town held out a siege, but in 1619 was taken by *Bethlen Gabor*, and in 1676 the greatest part of it was destroyed by fire. Among the Diets which have been held here, that of 1681 is particularly remarkable for the free exercise of religion granted to the Protestants by the Emperor *Leopold*.

Nyek, *Neckenmark*, a middling town in a fruitful country.

Lutsmannsburg, *Luitzburg*, *Lotsmany*, a handsome town conveniently situated.

Forchenstein,

Forchenſtein, Frakuo, a caſtle on a high hill, with a town at the foot of it; which, with the juriſdiction annexed to it, belongs to Prince *Eſterhaſy,* and was formerly for a long time mortgaged to the houſe of *Auſtria,* till, in 1625, 1637 and 1638, the States of *Hungary* petitioned for the redemption of it.

Lackenbach, Landſek and *Roberſdorf* are caſtles and lordſhips.

Hornſtein, Szaruko, a caſtle with a lordſhip ſituate on the borders of *Auſtria.*

Kereſztur, Krentz, a town ſituated in a pleaſant plain, with a fortified caſtle.

Cſepreg, formerly a populous town, but at preſent a poor place, the inhabitants whereof apply themſelves to tillage.

Rabakoz, Rabæ inſula, an iſland formed by the river *Raab,* which, beſides ſpacious and fruitful fields, contains the town of *Cſorna, Chorna,* where is alſo a probſty of the order of *Præmonſtratenſes.*

Kapuvar, a caſtle, fortified with a double wall, not far from the river *Raab,* betwixt moraſſes; belongs to Prince *Eſterhaſy.*

Kakos, Kroiſbach, a town with a caſtle belonging to the biſhop of *Raab.*

Ruſt, a poor place, but one of the royal free towns, lies on the lake of *Neuſiedler,* and is ſupported by fiſhing, tillage and vineyards.

St. Margaret, Fanum S. Margarethæ, a walled town in a fine wine and corn country.

Purbach, Fekete, Varos, a middling town ſurrounded with an old wall and belonging to Prince *Eſterhaſy.*

Bratenbrunn, Szeles-kut, a handſome, well inhabited, walled town, in the juriſdiction of the *Eſterhaſy* family.

2. The Geſpanſchaft of *Wieſelburg, Moſony Varmegye, Moſonienſis Comitatus.* The inhabitants *Hungarians, Germans* and a few *Croats.*

Kitze, Kopſin, a large town and caſtle ſituate in a ſpacious plain, and belonging to Prince *Eſterhaſy.*

Carlburg, Oroſvar, a town and caſtle belonging to Count *Zichy.*

Rajka, Rackendorf, a large village having ſeveral noblemens ſeats about it.

Altenburg O-Var, a beautiful town with a caſtle ſtanding on a ſmall branch of the *Danube* and the *Leitha,* and further ſecured by deep and wide moats. This is the chief crown eſtate in this province, and excluſive of its annual fair, which laſts a week, has a ſchool of the *P P. piarum ſcholarum.* The only way through this country from *Hungary* to *Germany,* lies cloſe by its caſtle. In 1529 it was taken by the *Turks,* and ſet on fire in 1605. In 1619 *Bethlen Gabor* became maſter of it; and in 1621 the *Imperialiſts.* In 1663 this was the place of rendezvous for the Emperor's troops.

Wieſelburg, Moſony, a ſpacious town in a fruitful country, belonging to the Geſpanſchaft of *Altenburg.*

Halb-

Halb-Thurn, Fel-Torony, Hemipigrum, a royal feat, fituate in a level country amidft pheafant fields.

Galos, Galofch, Gols, a large fightly town, in a good corn country, and almoft in the centre of the province.

Neufiedel, Nezider, a fine town, in a good corn and wine country, lies on the lake of *Neufiedler,* and belongs to the lordfhip of *Altenburg.*

Szent Niklos, Fanum S. Nicolai, a moated feat near a village of the fame name, lies in an open plain and belongs to Count *Zichy.*

Leben, Lyben, a famous *Jacobine* abbey.

3. The Gefpanfchaft of *Raab.* The inhabitants *Hungarians* and *Germans* with fome *Ratzians* and *Croats.*

Razo, a village with a feat lying betwixt fome arms of the *Danube,* formerly belonged to Count *Caunitz.*

Hedervar, an old caftle of the family of *Hedervar,* on the ifle of *Schutt,* defcended by inheritance to Count *Vitzay.*

Raab, Gyor, Jaurinum, was not a royal free city till 1742, but is an ancient and ftrong fortrefs, in a pleafant level country, at the conflux of the *Danube,* the *Raab* and *Rabnitz,* by which it is furrounded. Its houfes are all of ftone, the ftreets large and ftrait. It has a bifhop, chapter and univerfity, the profeffors in which are Jefuits. It is fortified with feven baftions, and has always a ftrong garrifon well provided with military ftores. Here are yet to be feen fome *Roman* antiquities. The fortifications of the city and caftle are chiefly the works of the Emperors *Ferdinand* the firft and *Maximilian* the fecond. In 1529, the garrifon abandoned the city out of fear of the *Turks,* having firft fet fire to the caftle. In 1566 it was burnt down. In 1594 it was by agreement given up to the *Turks,* from whom, in 1698, Count *Adolphus* of *Swartzenburg* recovered it by ftratagem. In 1749 the churches and fchools of the *Lutherans* and Reformed were fuppreffed.

Martinsberg, Szent Martin, Fanum S. Martini, the principal Benedictine abbey in *Hungary,* built by St. *Stephen,* the firft King of that name, in honour of St. *Martin,* on a pleafant lofty hill of a confiderable length, who ftiled it the *Mons Sacer Pannoniæ.* The convent had the appearance of a caftle, and is furrounded by a large level heath, on which ftood formerly many villages and churches. The abbot holds immediately of the Pope. In 1594 the *Turks* made themfelves mafters of the convent, but in 1597 it was retaken by the *Imperialifts.*

4. The Gefpanfchaft of *Comorra, Comarum Varmegye, Comarienfis Comitatus,* is inhabited by *Hungarians, Germans, Bohemian-Sclavonians* and fome *Rafcians.*

Gutta, a town on the *Schutt,* at the eaft arm of the *Danube.* The inhabitants are fifhermen. In 1624 fome fmall fortifications were added to this place.

Comorra,

Comorra, Comorren, Comarom, Comarno, in 1745, made a royal free town, lies below the *Schutt,* betwixt the *Danube* and *Wag,* which join here. In this town is the court-houfe of the Gefpanfchaft with a Jefuits college and fchool. Nigh it is an impregnable fortification, furrounded to the weft with deep moats, and on the north and fouth by the *Danube* and *Wag,* which meet to the fouthward. It was built by *Ferdinand* the firft, and the *Turks* have never been able to take it, though they befieged it in 1594, and in 1663 renewed their attempt.

Tata, Dotis, a town ftanding amidft waters and fwamps, had formerly a fplendid royal caftle, which was confiderably embellifhed by *Matthias Corvinus,* together with a fine garden and abbey of Benedictines; but at prefent is greatly reduced, and its caftle rebuilt out of the ruins of the former. The adjoining hill, which contains quarries of white and red marble, is an inconvenience to the caftle, which it commands. In 1543, *Soliman,* Emperor of the *Turks,* facked the city, which was, in fome meafure, afterwards rebuilt. In 1558 the *Turks* made themfelves mafters of it a fecond time. In 1566 it was retaken by the *Imperialifts,* who loft it again in 1594. In 1597 they recovered it, but held it only fix months, the *Turks* again becoming mafters of it the following year. But the town has undergone ftill greater changes. It is now in the jurifdiction of Count *Efterhafy.*

Szony, or, according to others, *Sene,* a large town on the weft banks of the *Danube,* where, in a part by the inhabitants called *Pannonia,* are to be feen the remains of the old town of *Bregetium,* which afford many curious antiquities.

Alwos, a populous large village on the fame fide of the *Danube,* in the Diftrict of *Comarro* and famous for an aqueduct made there in the year 1747 by Mr. *Mikowiny,* principal architect to her Majefty.

Nefzmely, a fmall place where, in 1439, the Emperor *Albert* died. The inhabitants are almoft all of the Reformed religion.

5. The Gefpafchaft of *Gran, Efztergon Varmegye, Strigonienfis Comitātus,* lies on both fides the *Danube,* and is inhabited by *Hungarians, Germans* and *Bohemians.*

Batorkezy, a large, well peopled town in the jurifdiction of Count *Palfy.* The country about it abounds in corn and wine.

Purkau, or *Barakau,* formerly a fortrefs on the *Danube.* The *Imperialifts* befieged it in vain in 1594, but in 1684 gave the *Turks* a terrible defeat here, and afterwards took the town fword in hand. It lies directly oppofite to *Gran,* and is looked upon as a part of it.

Gran, Efztergom, Oftrihom, Strigonium, a royal free town on the *Danube,* where the river *Gran* joins it. The neighbourhood delightful. It was formerly the refidence of the head Archbifhop and Primate of the Kingdom, and had alfo a chapter; which laft, fince 1543, is removed to

Tirnau,

Tirnau. The Archbifhop of *Gran* too refides now at *Prefburg.* This city, notwithftanding, is ftill ftiled a metropolis, and has a Jefuits college and fchool. It confifts properly of the royal free town, the caftle fituate on a high rock, and in which is the cathedral, the *Wafferftadt* on the *Danube,* the *Rafcian* town, the *Thomasberg* and the *Jungeftadt.* All thefe parts are fo fortified as to be of mutual defence to each other. King *Stephen* was born in this town in the year 909 and buried in the cathedral built by him. In 1543 the city was for the firft time maftered by the *Turks,* and in 1595 again wrefted out of their hands. In 1604 it withftood the efforts of the *Turkifh* army, but in the following year was obliged to furrender. In 1683 it was yielded up to the *Imperialifts.* Here are natural warm baths.

Balna, a town formerly opulent and famous, but now poor and inconfiderable.

6. The Gefpanfchaft of *Stuhl-Weiffenburg,* or *Alba Regalis, Szeke Fejor Varmegye, Albenfis Comitatus.* Its inhabitants are *Hungarians* and a few *Germans.*

Cfoka, Ko, Monedulæ Petra, anciently a fortified caftle on a hill, but at prefent a heap of ruins.

Mohr, a large and well inhabited place, having the appearance of a town.

Cfik-var, an old caftle on the river *Carvitz.*

Stuhl-weiffenburg, or *Alba Regia, Szekes-Fiejer-var, Alba Regalis,* a royal free town fituate in a marfhy ground occafioned by the river *Sarwitz.* This town is not only ftrong by its fituation, but had alfo formerly confiderable works, which were difmantled in 1702. From the town run three very large caufeways or moles, among which are churches, houfes, gardens and meadows; infomuch that the inhabitants of thefe fuburbs, as they may be called, are more in number than thofe of the town itfelf. This was formerly the place for the coronation of the Kings, and generally alfo of their interment; but at prefent its beft houfes are gone to ruin and the town is extremely decayed: It has, however, ftill a Jefuits college and gymnafium. In 1490 it was taken and plundered by King *Maximilian.* In 1540 the *Imperialifts* made themfelves mafters of it, and three years afterwards the *Turks.* In 1583, 1598, and 1599 the *Imperialifts* attempted it in vain; but in 1601 carried it, though the *Turks* in the following year regained it. In 1688 it fell again into the hands of the *Imperialifts.*

Sar-kereftztur, a large thriving village in a fruitful country.

Ertfe and *Adon,* two large populous villages fituate in a rich country on the banks of the *Danube.*

7. The Gefpanfchaft of *Tolna, Tolma Varmegye, Tolnenfis Comitatus,* lies on the *Danube* and *Sarwitz,* and is inhabited by *Hungarians, Rafcians,* a few *Germans* and *Bohemians.*

Simon's

Simon Thurn, Simon Tornya, Simonis Turris, a well peopled town at the conflux of the *Sarwitz* and *Siw* ; has a ſtrong caſtle, and is the place of meeting for the Geſpanſchaft. In 1686 it was taken by the *Imperialiſts.*

Foldvar, Fodvar, a pretty good town ſituate on the *Danube,* with an ab_ bey dedicated to St. *Helena.*

Paks, a large place on the *Danube*: Tillage the chief employment of its inhabitants. In 1602 it was taken by the *Imperialiſts* and ſet on fire.

Tolna, a middling town on the *Danube,* which gives name to the Geſ_ panſchaft, was formerly more conſiderable than at preſent.

Bata, Bata Szek, a place of ſome figure, ſtands in a fruitful country, and adjoins to a ſplendid abbey of St. *Michael* the Archangel.

Szekzard, Sexard, a pretty town on the river *Sarwitz,* with a caſtle and abbey conſecrated to *Chriſt* the Redeemer. The red wine of this place is in great repute.

8. The Geſpanſchaft of *Baran, Barania Varmegye, Baranienſis Comitatus,.* lies betwixt the *Drave* and the *Danube.* The inhabitants are *Hungarians, Raſcians* and a few *Germans.*

Five Churches, Pets, Quinque Eccleſiæ, an epiſcopal city, deriving its name from five churches which it formerly contained. It is an epiſcopal ſee with a chapter and a Jeſuits college, and moſt delightfully ſituated in a good wine country. It begins to increaſe by degrees. It is not, however, ſo ſtrong, either by nature or art, but that the Emperor *Soliman,* in 1543, made an eaſy conqueſt of it. In 1664 the *Imperialiſts* recovered it by aſſault.

Pets Varadja, Petſvarda, an ancient and rich probſty.

Dombovar, a caſtle lying in ruins and ſituate on a ſmall iſland in the river *Sarwitz,* a ſtream celebrated in the *Turkiſh* wars.

Siklos, a caſtle, the priſon of the Emperor *Sigiſmund,* which ſtands on a mountain, with a town lying below it, belongs to Count *Bathyani.* In 1543 the *Turks* got poſſeſſion of this place by the cowardice of the com_ mandant, but the *Imperialiſts* recovered it again in 1686.

Mohatz, a ſmall town on the *Danube,* remarkabe for the unfortunate defeat of *Lewis* the ſecond King of *Hungary,* in 1526, when, with twenty thouſand men he engaged three hundred thouſand *Turks* headed by their Emperor *Soliman*; and, after the rout, was ſuffocated by the fall of his horſe in the muddy water of the little boggy brook *Caraſſus :* but better known for the ſignal victory gained there over the *Turks* in 1687.

Darda, Tarda, a town not far from the *Drave,* ſtrengthened formerly with fortifications for the defence of the bridges of *Eſſeck,* but taken by the *Imperialiſts* in 1686.

9. The Geſpanſchaft of *Simegy, Varmegye,* lies betwixt the *Drave* and the lake of *Platten.* Is inhabited by *Hungarians* and *Raſcians,* with a few *Ger_ mans* and *Bohemians.*

Platten Lake, *Balaton,* abounds in fiſh, and is ſurrounded, for the moſt part, by fine vineyards. Its length is twelve *Hungarian* miles, and from two to five wide. It abounds alſo in otters and beavers.

Kapos-var, formerly a very ſtrong caſtle on the river *Kapotz,* yet taken in 1555 by the *Turks,* and recovered by the *Imperialiſts* in 1556; was beſieged in vain by the latter in 1599, but taken in 1664 and again in 1686. It is little better at preſent than a village.

Samogy-var Simeginum, or *Simegia,* was likewiſe once a caſtle of note, ſituate on a high mountain not far from *Platten* Lake. The *Imperialiſts* made themſelves maſters of this place in 1605, but at preſent it is a heap of ruins. However it ſtill gives name to the county.

Szigeth, a ſtrong fortification, properly a triple town with a double caſtle, lies betwixt two waters; and, excluſive of its being ſurrounded on every ſide with a moraſs, has moats, walls and baſtions. In 1556 it was beſieged in vain hy the *Turks.* In 1566 *Soliman* made a ſecond attempt, and, though he died during the ſiege, the *Turks* made themſelves maſters of the town. The brave governor, Count *Nicholas Zriny,* or *Serini,* with the ſmall remains of his garriſon, made a ſally from the citadel and died ſword in hand. In 1664 the *Imperialiſts* were obliged to draw off their troops from before it: but in 1689 took it by ſurrender.

Babocſa, Babotſa, formerly a place of very great ſtrength, but taken by the *Turks* in 1555; and demoliſhed in the following year by the *Imperialiſts.* It was repaired three years after by order of King *Ferdinand,* and loſt by them in 1566; and, in 1595, by the *Turks,* who diſpoſſeſſed the *Imperialiſts,* however, of it in 1600. It has no garriſon at preſent.

Koppan, once a fortreſs, but now diſmantled.

10. The Geſpanſchaft of *Szala, Szala Varmegye, Szaladienſis Comitatus,* lies on the frontiers of *Styria,* and is inhabited by *Croats, Vandals* and *Hungarians.*

Lindva, called *Alſo-Lindva, Limbach,* a ſpacious and well built town in a fruitful corn and wine country, belonged formerly to the *Banſiers* of *Alſo-Lindva,* now extinct, and at preſent to Prince *Eſterhaſy.*

Katz-Caniſa and *Deſnia* are two towns ſurrounded with water, by which they are frequently overflown.

Cſaka-Tornya, Cſack-Thurn, is a conſiderable fortreſs ſtrengthened with a moat, wall and bulwark, and takes its name from the *Cſaki* family. It is ſituate in the midſt of a peninſula abounding in corn and wine.

Strigova, Stridova, a town ſeated in a pleaſant valley betwixt hills which are covered with vines, at no great diſtance from the river *Mur;* is by ſome ſuppoſed to be the ancient *Stridonium,* the place of St. *Jerom's* nativity.

Orinoſdin and *Palaſtraw* were formerly well fortified towns, ſome remains of which are to be ſeen at preſent on the aforementioned peninſula.

Legrad,

Legrad, Legradinum, a strong place near the conflux of the *Mur* and *Drave* upon the *Murakoz,* was erected for the defence of the peninsula, but its fortifications at present lie in ruins, insomuch that it is now but a small open town.

Neu Serinwar, Novi-Zrin, a strong place, or rather fort, on the river *Mur,* where it joins the *Drave,* was built in 1661 by Count *Zrini* or *Serini,* by which means the communication betwixt *Canischa* and other fortresses was entirely cut off; for which reason the *Turks* laboured hard to destroy it, and in 1663 vigorously attempted it, but were repulsed by the brave Count *Nicholas Serini,* who kept possession of it till 1664, when it was entirely demolished, insomuch that a mean village only is now to be seen here. It is called *Neu-Serinwar,* and lies over-against *Serinwar,* or *Old Serinwar,* a castle built of free-stone in a morass not far from *Canischa* on the river *Mur.*

Canischa was formerly a strong place, situate at a small distance from the *Mur* and *Drave* in a deep morass, which covers the frontiers of *Styria.* In 1566 it was besieged in vain by the *Turks,* but in 1600 was taken by them. In 1601 the *Imperialists* attempted in vain to take it. In 1603, 1604 and 1660 it was destroyed by fire. In 1702 it was dismantled by order of the Emperor *Leopold,* and is at present an inconsiderable little town.

Kesthely, a place of tolerable extent, commodiously situated at the foot of Mount *Szigligeth* in the territory of *Festetitz.*

Szala-var was formerly a strong castle giving name to the Gespanschaft, but at present is only a poor town on the river *Szala,* in which, however, some monuments of a *Roman* colony are met with.

Fenek-var, a castle situate in a low place, where are still to be seen the remains of a large *Roman* colony.

Sziligeth, a castle not far from the town of *Csobanz,* standing on a high rock on the *Platten* Lake.

Csobanz, a town poorly inhabited, with a castle standing on a high hill near the *Platten* Lake.

Tapoltza, a town formerly fortified with a double wall against the *Turks,* but at present an open place.

Sumeg, a tolerable town with a castle lying in ruins.

Egerszeg, by some called formerly *Szaladin,* a place of some strength, but at present a defenceless town in which the meeting of the Gespanschaft is held.

Sent Grot, Fanum S. Gotthardi, is to be distinguished from another place of the same name in the Gespanschaft of *Eisenburg.*

Kopornak an ancient abbey of *Benedictines.*

Sechi Sigeth, a town situate in a good corn and wine country.

11. The Gespanschaft of *Eisenburg*, *Vasvar Varmegye*, *Castriferrei Comi-tatus*, in the frontiers of *Styria*, is inhabited by *Germans*, *Vandals*, and *Hungarians*. It contains the following places of note.:

Rakitsan, a town and jurisdiction belonging to Count *Bottyan*, close by the river *Mur*, situate in a delightful plain on the frontiers of Lower *Styria*.

Mura-Szombath, a pretty good town, with a castle, seated in a level fruit-ful country, inhabited by those tribes of the *Sclavonians* commonly called *Vandals*.

Upper Lindau, *Felso-Lendva*, a populous town belonging to Count *Na-dasdy*, lying on the little river *Lindau*, whence it receives its name, and noted for its excellent wines.

Sanct Gotthard, *Szent Groth*, *Fanum S. Gotthardi*, a pretty good town on the river *Mura*, with an abbey founded by *Bela* the third. In 1664 the *Turks* were defeated by the *Imperialists* in a bloody battle fought here.

Dobra, a castle standing on a very high mountain, with a small town below it.

Rothen Thurn, *Voros-var*, a castle in an open plain on the banks of the *Pinka*, belonging to Count *Erdody*.

Monyorokerek, a castle and town seated on an eminence, and making one of the titles of the *Erdody* family.

Rormend, a town on the river *Raab*, in a plentiful neighbourhood, with a fine castle belonging to Count *Bottyani*. During the *Botschky* wars, in 1605, this place was reduced by famine, but recovered soon after and lost again. In 1621 it was retaken by Count *Bottiani*.

Eisenburg, *Vas-var*, *castrum ferreum*, once a very strong castle, and the principal in the whole Gespanschaft to which it gave name. But both its fortifications and buildings are demolished, and the county meeting now re-moved to *Stein Am Anger*.

Glussingen, *Nemeth-Ujavar*, a populous walled town with a castle stand-ing on a very high rock, which is wholly detached from all the other moun-tains. This place formerly belonged to *Laurence* Duke of *Sermia*, whose large possessions, upon his decease, devolving to the crown, King *Lewis*, about the year 1523, conferred it on *Francis Bottiani*, then Ban of *Dalma-tia* and *Illyria*, in whose family it still remains. It is most delightfully situated among woods, corn-fields and rising grounds, which are covered with vines.

Pinkafeld, a handsome town with a castle situate in a pleasant spot on the river *Pinka*.

Borostyan, *Borostyanko*, a castle on a high hill, with a town lying below it on the river *Pinka*.

Steinam Anger, *Szombath Hely*, *Sabaria*, a large populous town, in a de-lightful plain on the river *Gunz*; built out of the ruins of the ancient *Ro-*

man

man. Sabaria, but in a very different manner. In the adjacent country have been found ſeveral rudera and great numbers of old coins. This was the native place of St. *Martin*, biſhop of *Tours*, in *France*; has a chapter, and the provincial meetings are held here.

Rohonz, a populous town belonging to *Count Bottiani*. Its inhabitants follow huſbandry and handicrafts.

Vep, an old caſtle, having a ſtrong wall, belongs to Count *Erdody*.

Monoſtor, *Bors-Monoſtra*, an abbey near *Gunz* belonging to the *Præmonſtratenſes*, and founded by Count *Bors*.

Gunz, *Roſzeg*, *Ginſium*, a royal free town, in a pleaſant fruitful country, on a river of the ſame name, having a caſtle walled and moated after the *Hungarian* manner. It ſtands at the foot of a hill which is covered with vines. Here the high court for the circle below the *Danube* is held. It has alſo a Jeſuits ſchool and college. It continued a long time mortgaged to *Auſtria*. In 1532 it ſuſtained a ſiege againſt *Soliman* Emperor of the *Turks*; and, in 1621, Count *Bottiani* attacked it with as little ſucceſs.

Sar-var, a ſtrong fort lying betwixt the *Raab* and *Gunz*, which unite here. It belonged formerly to Count *Nadaſdy*, but at preſent is the property of Count *Dreſkovitz*.

Szent-Marton, *Fanum S. Martini*, a mean town in the *Bottyani* juriſdiction.

Domolk, a fine abbey of *Benedictines*, with a village at the foot of *Saghill*, a mountain noted for its delicious wines. Here is a celebrated image of the Virgin *Mary*.

12. The Geſpanſchaft of *Weſzprin*, *Weſprim Varmegye*, *Veſzprimienſis Comitatus*, inhabited by *Hungarians*, and a few *Germans*.

Czeſznek, *Tſcheſnek*, a caſtle on a high rock under the woody hills of *Bakoni* foreſt, belongs to Prince *Eſterhaſy*.

Papa, a pretty large and as well inhabited town on the river *Marzal*, with a ſeat on a plain lying near it. This place formerly had a double wall moated as well as its caſtle; which rendered it a good fortification. In 1594 it fell into the hands of the *Turks*, but was again taken from them in 1597 by Duke *Maximilian*. In 1600 the *Turks* laid fruitleſs ſiege to it. In 1702 it was entirely diſmantled.

Devetſer, a large village at the foot of *Somlyo-hill*, which produces fine wine.

Palota, a ſquare caſtle, with a town of the ſame name, in *Bakoni* foreſt, was formerly a pleaſure-houſe; and being fortified with a high wall and a broad moat, in 1565 and 1603, baffled all the efforts of the *Turks*; but, in 1593, was taken by ſurrender and voluntarily quitted again about five years after. The preſent proprietor of it is Count *Zichy*.

Veſzprim, *Veſprin*, a city ſtanding on an eminence, has a biſhop and chapter; but its fortifications having been razed in 1702, it is now become

an

an open place. On the death of *Matthias Corvinus*, in 1490, it was taken by the *Germans*, but, in 1551, besieged again by the *Turks*, who made themselves masters of it. In 1565 the Christians got possession of it. In 1593 it was recovered by the *Turks*, who, in 1598, were obliged once more to evacuate it to the Christians. In 1655 the *Turks* attempted it in vain; but, in 1663, they pillaged and set fire to it, though in their attack on the castle they were repulsed with great loss.

Vasony, Waschony, Nagy Vasonyko, a well peopled town, with an old castle, built by *Paul Kinss*, and now belonging to Count *Zichy* of *Vasonko*.

Tihany, a tolerable town, with a castle on a steep rock, stands on a peninsula in *Plattem* Lake, amidst a fine corn and wine country; and has an abbey of *Benedictines*.

UPPER HUNGARY;

HUNGARIA SUPERIOR,

WHICH makes the East part of the Kingdom, and borders on *Poland, Transylvania* and *Walachia*, consists of two circles, which are,

　I. The circle on this side the *Teisse Circulus, Cis. Tibiscanus*, containing eleven Gespanschafts. The inhabitants *Hungarians, Bohemians, Sclavonians, Germans* and *Russians*.

　1. The Gespanschaft of *Sips, Scepes Varmegye, Scepusiensis Comitatus*, is one hundred and six *Hungarian* miles in length, six broad, and about twenty-eight in circumference, being almost every where woody and mountainous, but interspersed, especially towards the middle, with delightful plains, fruitful fields, pastures, and rivers which abound in fish. The *Carpathian* mountains are here at their greatest height. Other remarkable mountains too here are the *Ochsenberg*, the *Konigsberg*, or *King's Mountain*, (so called from King *Matthias Corvinus*, who dined on the summit of it in 1474) the *Magura*, the *Ihla*, the *Reberg* or *Rauberberg*, and the *Bransko*. The forest of *Leubitz*, also is worthy notice. From these mountains issue the following rivers: The *Popper (Popperad)* which receives its source from the *Popper* Lake, in the western part of the *Carpathian* mountains; from whence

whence it runs into the *Dunavetz* in *Poland.* The *Dunavetz,* or *Dunojetz,* rifing in the northern fummits of the *Carpathian* mountains, and difcharging itfelf into the *Viftula.* The *Kundert, Hernat, Hernadus,* which fprings at the foot of *Konigsberg,* at the junction of the Gefpanfchafts of *Gomor* and *Lipto,* and runs into the *Teiffe.* The *Golnitz* has its fource in the *Ochfenberge,* and falls into the former. Other rivers of leffer note I omit. If the air be cold here it is yet very healthy ; and though no wine be produced in this country, it abounds notwithftanding in corn, as wheat, and more particularly fine barley, which is the beft in this part of *Hungary.* It produces alfo excellent peafe, and the inhabitants raife flax with great profit. It abounds likewife in good pafturage. The wild beafts here are bears, linxes, wild boars, ftags, wolves, foxes, marmottes (*mures alpini*) hares, and particularly chamois ; but the catching of the latter is very difficult : martins, beavers and red deer it has none. The inhabitants are but little addicted to mining, agriculture turning out more to their advantage, though fome copper and iron be dug here. It is peopled by *Germans* and *Bohemian-Sclavonians,* but the gentry are for the moft part *Hungarians.* We fhall firft take notice of

The towns within *Polifh* lordfhips. Thefe are fuch as, in the year 1412, were given in mortgage by King *Sigifmund* to *Uladiflaus Jagello,* King of *Poland.* The inhabitants are almoft all *German Lutherans,* and governed by a Vice-ftaroft, who refides at *Lublyo* ; but the magiftrates of thefe cities have the choice of a general prefident, who is dignified with the title of Count, but fubordinate to the Vice-ftaroft.

Lublyo, anciently called *Lublyan, Lubowna, Lublau,* is a fort on a mountain near the river *Popper,* and the refidence of the *Polifh* Staroft. In 1553 it was deftroyed by fire, has been feveral times befieged, and particularly in the *Huffite* war. Beneath it lies a large populous town of the fame name, the principal indeed of all thefe places, and particularly famous for its weekly markets and fairs. Pilgrimages are made here.

Pudlein, Podolin, Podolinetz, a pretty good town, with a caftle, on the river *Popper,* enjoys a good trade ; and has a fchool of the *P P. fcholarum piarum.* The adjacent country is none of the moft fertile. On a hill without the town ftands a chapel, dedicated to St. *Anne,* noted for pilgrimages, and a medicinal fpring near it.

Kniefen, Quafdo, a fmall town on the river *Popper,* with pretty good corn land.

Ober-Raufchenbach, Felfo-Rusbach, a town with a petrifying fpring.

Bela, a pretty large town, in a moft delightful plain; not far from the *Popper* ; but by frequent fires reduced very low.

Laibitz, Lebitz, Lubitza, a town on a river of the fame name ; formerly four times as large as at prefent, though the houfes ftill are about two hundred

dred in number. It has fuffered confiderably by fire in 1669, 1680 and 1708. The inhabitants follow agriculture and a wood trade.

Matzdorf, Mattharotz, Matejowaicze, Matthæi villa, a middling town on the river *Popper,* in 1718, and at other times, almoft totally doftroyed by fire.

Fuleck, Welka, a poor town, every day declining though feated in a large and fertile country.

Georgenberg, Szombathely, Spiſka Czobota, Mons S. Georgij, a well built town and delightfully fituated on the river *Popper.* It has been frequently deftroyed by fire, but is always handfomely rebuilt.

Deutfchendorf, Poprad, a pretty large town, in a moft delightful fituation; bounded on one fide by the river of the fame name, and on the other by a fpacious plain. Before the fire in 1718 it was much more wealthy than at prefent. The inhabitants apply themfelves chiefly to agriculture.

Michel'sdorf, Strafa, from a flourifhing condition funk gradually into ruin.

Menhardfdorf, Verbowo, a very mean place in every refpect.

Duranfdorf, or *Durlfdorf, Twarozna,* a town, but having more the appearance of a village, enjoys a large territory with plenty of wood.

Rifdorf, Ruſzkonotz, Ruſkinowce, a fmall town, in a barren neighbourhood, the chief employment of the inhabitants is wooden ware.

Neudorf Nowa Weſz, Iglo, the beft built and inhabited of all thefe towns, was formerly fortified with a wall, and lies on the river *Hernath.* Part of the inhabitants follow tillage, and the neighbouring copper and iron mines find employment for the reft.

Kirhdorf, Varallja, Szepes Varallja, Podhrad, a handfome town below the caftle of *Scepus,* has a confiderable fair on *Afcenfion-day.*

Vallendorf, Wlaſki, Olaſzy, a town fituate on the river *Herneth,* was originally a colony of *Italians* or *Latins,* as appears from the name, but its prefent inhabitants are all *Germans.* The country around it is pleafant and fertile,

2. The towns, caftles, feats and places of the *Lower* or *Greater Sitz,* or *refidence,* as it is called.

Dunavetz, an old caftle; fo named from a river which runs by it, ftands on a high and fteep rock. It is the moft northward of all the caftles, and is about a cannon-fhot diftant from the *Poliſh* caftle of *Schornftein,* which lies on the other fide of the river. It belonged formerly to Count *Zapolia,* but King *John* made a grant of it to *Jerom Laſky.* It has fince paffed through feveral proprietors to the noble houfe of *Johannelli.* In 1687 this caftle was taken by *Tekely.*

Altdorf O Faln, a little town, having a good trade, fituate not far from the borders of *Poland.*

Fridman,

Fridman, a caftle on the river *Dunavetz,* belonging to Count *Bargotzy.*

Roth Klofter, a convent of *Antonines,* fituate on the river *Dunawetz,* formerly belonging to the *Carthufians,* but at prefent to the *Camaldunenfes.* Near it is the village of *Lechnitz.*

Toportz, a village making a good appearance, with a feat of the *Gorgey* family : is an article-place where the Proteftants are allowed divine fervice.

Landok, a feat of the *Lufinfky* family near the *Carpathian* mountains, having marble and alabafter quarries near it, and formerly an abbey of the Præmonftratenfes.

Kafmark, Kefmark, Kefmarek, called alfo *Kaiſſerfmark,* a royal free town not far from the *Popper,* ftrengthened with a wall and towers, and one of the moft ancient towns in *Hungary.* In it are three churches, and a little way out of the town a proteftant oratory. The inhabitants are very induftrious in trade, tillage and handicrafts. *Matthias Corvinus* diftinguifhed this town with very fingular marks of his favour. In the *Huſſite* war, in 1433, it was plundered and laid in afhes, but in 1436 the *Polifh* and *Hungarian* nobility had a meeting here. It was frequently taken during the civil wars of the feventeenth century; and in 1576, 1720 and 1721, burnt down. The caftle near it was formerly furrounded with a double wall, five ftrong towers and a moat. In 1702 the inhabitants of *Kafmark,* having purchafed it of the *Tekely* family, laid it entirely in ruins.

Nebrer, or *Strafka,* a caftle on the river *Popper,* belongs to the family of *Horvath Stanfith* of *Gradetz.*

Batisfalva, a town with a feat in a fertile plain, at the foot of the *Carpathian* mountains ; the original of the ancient family of the *Mariafi,* and an article-place of the Proteftants, who exercife their religion here without moleftation.

Gronitz, Hranovitza, formerly a mine-town, but at prefent only a large village though called a town, belongs to Count *Nyari.*

Cfawnich, a feat and village on the river *Hernath.*

Letanfalva, Lettenfdorf, Letanowtze, a confiderable village with a fruitful territory, belonging to the chapter of *Scepus.* Here was formerly a convent of *John* the Baptift of *Letfanko,* and afterwards a convent of *Carthufians,* which being deftroyed in 1543, the Monks removed to *Leutfchau.*

Leutfchau, Lotfe, Lewotfe, a royal free town and the capital of the county, ftands on a hill. Its walls are remarkably thick, and have twelve towers on them. The church is alfo a fine building, and dedicated to St. *James.* Here is likewife a Jefuits college, a gymnafium and convictorium for noblemen. The inhabitants are moftly *Germans,* but it retains very little of its former profperity. It was founded in 1245, and called *Leutfchau,* there being formerly a watch-tower here for *fchauen dieleute, i.e,* to eye the people, or to keep a look-out againft the plundering *Tartars,* by whom it was deftroyed in 1285, but foon rebuilt. In the years 1332, 1342, 1432, 1441,

1550 and 1599 it was demolished by fire; and in 1600 miferably ftripped
of its inhabitants by peftilence.   In 1601 it was facked by the *Heyducks,* and
in 1602 by *Sigifmund Bathory.*   In 1605 *Botfkay's Heyducks* made themfelves mafters of it; in 1619 *Gabor Bethlen,* and afterwards *George Rakotzy.*
In 1682 Count *Tekely* took it, and in 1703 *Francis Rakotzy,* from whom
it was again recovered in 1710.   In 1494 a treaty was concluded here betwixt the Kings of *Hungary* and *Poland.*   One of the firft printing preffes
brought into *Hungary* was erected here by a family of the name of *Brever.*

Markusfalva, Markfdorf, Markuffowtze, a town and feat belonging to
the *Mariafi* family, who are likewife lords of the feat and large village of
Sepanfalva Sepenfdorf, Scepanowtze.

Zipfer-haus, Skepus, Szepes-var, Spifky Zamek, Scepufienfis arx, a caftle
which gives name to the county, ftands on a high rock, and is fortified both
by art and nature; but the walls threaten ruin.   Here *John Zapolya* was born.
In the *Botfky* wars in 1664 it held out a vigorous fiege, but in 1703 was
taken by *Rakotzy's* party, who loft it again in 1710, and at prefent it is the
property of Count *Cfaky.*   Below it lies the town of *Varalja,* of which mention has been made above; and on a hill near it the chapter of the cathedral
of St. *Martin* of *Scepus,* which is built in fuch a manner that it has the appearance of a fmall town.   In it the Jefuits have alfo a gymnafium, and at
a fmall diftance is a petrifying fpring.   In a mountain alfo near the caftle is
a cavern, in which, during winter, the water is fluid, but fo frozen in
fummer that quantities of ice are brought from thence for the cooling of
their liquors.

Wagendreffel, a mine-town, in a vale, yielding plenty of iron, lies on the
river *Golnitz,* and is the property of the *Mariafi* family.

Schwedler, a populous mine-town, having a very rich minage of copper,
belongs to the treafury.

Schmolnitz Czomolxock, a mine-town having fome fine copper works
which bring a great deal of gain into the treafury.   The whole metallick
Diftrict about the mountains fo abounds, both inwardly and outwardly, with
fulphur, that the vitriol, or copperas, not only gufhes out in the mines, but
breaks forth alfo from the furface of the earth.   Its contents are a much
richer and finer ore than that in *Herrengrund.*   During the *Botfkay* wars,
in 1604, this induftrious place was laid in afhes by the *Heyducks.*

Stoos, a tolerable mine-town, but deals only in iron.

Jekelfalva, the feat of a family of that name, lies on the river *Golnitz,*
and has iron mines.

Krompach, a mine-town on the *Hernath,* famous for the excellence of its
iron, and belonging to Count *Cfaky.*

Voikotz, a fmall village, famous for its medicinal fprings.

3. Count *Cfaky's* jurifdiction confifts of thirteen towns and twelve villages.
The towns are

Nagy-

Nagy-Szalak, Great-Schlagendorf, a small place on the river *Popper.*

Milenbach, a town pleasantly situated in a fertile territory.

Eisdorf, Zakotz, a small town.

Rabsdorf, or *Kaposztafalva,* on the river *Hernath. Sehmegen, Szmisany, Kirn, Kuriman, Sperndorf, Illyesfalva, Balensdorf, Harigotz, Densdorf, Danisotz,* all small towns.

Donnersmark, Csotortokhely, Stwartek, oppidum S. Ladislai, or *Quintoforum,* the original place of the Counts of *Henkel,* has a nunnery.

Velbach, or *Eylenbach,* a pretty good town.

Einsiedle, Remete, Mnisky, on the river *Golnitz,* has good iron mines.

Golnitz, a metal town lying on a river of the same name. Its iron mines of greater repute than the former, and in itself still more populous.

Baldotz, a town not far from *Zipserhaus,* having a warm bath and a mineral spring.

4. The places of the Upper *Sit* Residence, or, as it is called, the *Residence of the ten spearmen, Sedes decem lanceatorum,* an appellation derived from the ancient tenure by which the nobility in this country held their lands, who were obliged to attend their Kings, armed with lances, whenever the public welfare, or a war against any common enemy required. But they never fought under any other than the King himself, and, even then, always near his person, being, as it were, his body-guards. *Bela* the fourth added to their privileges, by which means they have their Vicecount, and under him a judge of the nobility; but one Præfect, or Count, presides over this as well as the Larger or Under-residence. The scattered villages of this residence are fourteen, the most remarkable of which are the following.

Abrahamsdorf, Abrahamfalva Abrahamowtze, a very populous large village in a fruitful soil.

Komarotz, a single house only, but having an excellent mineral spring.

Horka, having also a mineral spring.

Kiszots, noted for a spring of a petrifying quality.

Bethlehemfalva, anciently one of the titles of the *Torna* family.

2. The Gespanschaft of *Saros, Saros Varmegye, Sarosienfis Comitatus:* The inhabitants *Hungarians, Bohemians, Sclavonians, Germans* and *Russians.*

Makowitza, a castle on a high hill on the confines of *Poland.*

Tarza, Tarifszsza, a seat and large village on a river of the same name.

Zborow, a spacious town with three seats.

Siebenlinden, Hethars, Septemtiliæ, a pretty large town on the borders of *Scepus.*

Zeben Szebeny, Sabinow, Cibinium, a royal free town, of no great extent, on the river *Tártsa.* It surrendered to the forces of Count *Botskay* in 1604, and in 1663 to the *Turks.*

Kapi-var,

Kapi-var, a well peopled town having two feats.

Bartfa, Barthfeld, Bardiow, Bartpha, a royal free town on the fkirts of the *Carpathian* mountains, carries on a large wine trade with *Poland.*

Saros, a well inhabited town ftrengthened with a fortification and giving name to the county. In its neighbourhood are two lethiferous fprings.

Sebes, a pretty good town with a feat.

Eperies, Eperjes, a royal free town on the river *Tartfa,* having, befides its moats, a wall well fortified with towers. This is the court town for the circle on this fide the *Theis,* has a Jefuits college and fchool with a *Collegium illuftre* of *Lutherans,* founded by the proteftant *Lutheran* States, and inaugurated in 1667. It was under a prohibition for fome time, but in 1751 the Emperor gracioufly ordered it to be reftored to its ancient privileges. In 1604 this town fubmitted to *Botfkay.* In 1684 Count *Tekely* was driven out of the country, and the year following the place capitulated to the *Imperialifts.*

. *So-var, Saltzburg,* a pretty good town not far from *Eperies,* and fo called from its falt fprings, which are much celebrated.

3. The Gefpanfchaft of *Zemplin, Zemplin Varmegye, Zemplinienfis Comitatus,* inhabited by *Hungarians, Germans, Bohemian-Sclavonians* and *Ruffians.*

Sztropkow, a town, near which is a ftately palace belonging to the *Pethoi* family.

Kraina-Wyffi, Ukrainia fuperior, a hilly tract adjoining to the *Carpathian* mountains, and inhabited by *Ruffian* colonies.

Homonna, a town and caftle in a delightful fituation on the river *Labortza.*

Vranowo, a pretty good town with two feats.

Nagy-Mihalyt, a handfome well built town feated in a fertile country.

Kerefztur, a pretty large town in a fruitful country, noted for its excellent wines.

Terebes, a pretty good town belonging to the *Prinis* family, not far from which is a caftle lying in ruins.

Zemplin, which gives name to the Gefpanfchaft, once a fortified citadel, now only a town on the river *Bodrogh,* and the place where the *Hungarians* firft fettled.

Lelefz, a town having a convent and probfty of Præmonftratenfes, in which the records of the nobility are kept.

Szerdahely, a town with a feat.

Uihely, Sator Ujhely, a well built town, in a pleafant fituation, producing excellent wines.

Tarkany, a town on the *Theis* with a caftle.

Patak, Saros-Patak, a well built town on the river *Bodrog,* where formerly was a famous caftle belonging to Count *Rakotzy,* which is now difmantled.

mantled. The Proteftants have a gymnafium here, as well as the Jefuits, founded by *Sigifmund Rakotzy* on *Comenius*'s plan.

Talya, Mad, Targal, Toltfva, Benye and *Satoralja*, are fix towns on the rivers *Bodrog* and *Theis* noted for their excellent wines, which equal the *Tokay* in goodnefs.

Szerents, a town with a palace and caftle feated in a pleafant fertile country.

. *Tokaj*, a confiderable town pleafantly fituated near the conflux of the *Theis* and *Bodrog*, had formerly a ftrong caftle, which was demolifhed in the *Rakotzy* war. Here is a gymnafium belonging to the *P P. piarum fcholarum*, but it is more celebrated for the rich wine it produces, which, in flavour and ftrength, exceeds all other *Hungarian* wines; this charaĉter, however, is properly applicable only to the wine produced on the promontory of *Mezes-Male*. In 1527, and 1534, the Emperor *Ferdinand*'s troops took the town. In 1565 it was again maftered. In 1566 it was befieged in vain by *John Sigifmund*, Prince of *Tranfylvania*; and, in 1598, deftroyed by fire. In 1605 Prince *Botfkay* obliged it to furrender, after whofe death it fell again into the hands of the Emperor *Rodolphus*, who conferred it on Count *George Turzon*. In 1621 and 1624 it furrendered to *Bethlen Gabor*, but foon fell again into the Emperor's hands. In 1681 it was taken by Count *Tekely*, and in 1703 by *Rakotzy*.

4. The Gefpanfchaft of *Ungh, Ung Varmegye, Comitatus Unghenfis*, is inhabited by *Hungarians* and *Ruffians*.

Kraina Nyffi, Ukrajnia inferior, is the name of a hilly traĉt at the foot of the *Carpathian* mountains, inhabited by a *Ruffian* colony.

Orofzveg, a large village on the river *Labortza*, is the refidence of a *Greek* bifhop united with the *Roman* Catholics.

Vinna, a town having a caftle on a very fteep mountain and feveral feats.

Ungh-var, a ftrong fort on the river *Ungh*, with a town, in which the Jefuits have a gymnafium. From hence the county took its name, and, according to fome, the *Hungarians* alfo. In 1564 it made a ftout defence againft King *John*.

Szeregyna, a town having an old caftle belonging to the *Barkotz* family.

Palotz, a pretty good town with a caftle, the property by inheritance of the laft mentioned family.

Rapos, a well built town.

5. The Gefpanfchaft of *Abaujvar, Abaujvar Varmegye, Abaujvarienfis Comitatus*, is poffeffed by *Hungarians, Bohemian-Sclavonians* and *Ruffians*: The towns moftly inhabited by *Germans*.

Szepfi, a large town and privileged place formerly furrounded with a wall.

Cafchau, Kaffa, Koffice, a royal free town and fortrefs on the river *Hernath*, or *Hundert*, furrounded with a triple wall, a moat and bulwark. It

is

is a well built place, and in it is the royal chamber of *Zip*, a publick school, a feminary and noble convictorium of Jefuits. The wine, beer and air of this place are very bad; the laft fo even to a proverb. In 1400 it was befieged in vain by the *Poles*. In 1535 King *John* took it by ftratagem, and in 1556 it was deftroyed by fire. In 1604 its large and beautiful parifh-church was taken from the *Lutherans*, incenfed at which, and other oppreffive mea-fures, the *Cafchians* took part with *Stephen Botfkay*, who died here in 1606. In 1619 it furrendered to *Bethlen Gabor*; in 1644 to *George Rakotzy*; in 1681 to *Tokoly*, and in 1685 to the *Imperialifts*.

Uj-var, *Aba-Ujvar*, *Arx Nova*, a citadel erected by the Statholder *Aba*; but now a heap of ruins. From this place, however, the Gefpanfchaft takes its name.

Jafzo, a well built town environed with mountains and a wall. It has a caftle and a probfty of *Præmonftratenfes*.

Szeplak was formerly an abbey of *Benedictines*, but at prefent ferves for the *Tyrnau* feminary.

Upper and *Lower Metzenfeif*, *Felfo*, or *Metzenzofi*, are two large towns poffeffed by an ancient colony of *Saxons*, who employ themfelves in minage and agriculture.

Szalonz and *Fozer* are two caftles feated on a high hill, now lying in ruins.

Gonz, a large well built town, having formerly a ftately caftle, of which now only the remains are to be feen.

Nagy-Ida, a caftle on an eminence, with a town below it under the ju-rifdiction of Count *Cfaky*.

Boldog-ko, *Bodoko*, a caftle demolifhed in the *Rakotzy* war, and given to the college of Jefuits at *Leutfchau*. The wine of this place is excellent.

Szikfzo, a town famous for a good annual fair and two defeats here of the *Turks*.

Regetz, a ftrong citadel in a delightful fituation.

6. The Gefpanfchaft of *Torner*, *Torna Varmegye*, *Tornenfis Comitatus*, lies at the foot of the *Carpathian* mountains, being fmall and hilly, and (in common with the whole *Carpathian* chain) remarkable for natural curiofities: the inhabitants *Hungarians*.

Torna, once a fortified place, now an open little town, but having a fine feat belonging to Count *Keglevich*. It gives name to the Gefpanfchaft.

Szadvar, a caftle on a very high rock, but razed by order of the Em-peror *Leopold*, is under the jurifdiction of Prince *Efterhafy*.

Szélitze, little better than a village, but remarkable for a wonderful cavern in a mountain in its neighbourhood. The country too is hilly, and abounds too much in woods to produce a great deal of grain. The air alfo is fharp and cold. The aperture of this cavern, which fronts the fouth, is

eighteen

eighteen fathoms high and eight broad, and confequently wide enough to receive the fouth wind, which generally blows here with great violence. Its fubterraneous paffages confift entirely of folid rock, ftretching away farther fouth than has yet been difcovered. As far as it is practicable to go, the height is found to be fifty fathoms and the breadth twenty-fix. But the moft unaccountable fingularity in this cavern is, that in the heart of winter the air is warm on the infide; and when the heat of the fun without is fcarce fupportable, freezing cold within. When the fnows melt in fpring, the infide of the cave, where its furface is expofed to the fouth fun, emits a pellucid water, which congeals, immediately as it drops, by the extreme cold. The icicles are of the bignefs of a large cafk, and fpreading into ramifications form very odd figures. The very water that drops from the icicles on the ground, which is fandy, freezes in an inftant. It is obfervable alfo, that the greater the heat is without, the more intenfe is the cold within; and in the dog days all parts are covered with ice. The inhabitants make ufe of it for cooling the warm fprings. They thaw it alfo and drink the water. In autumn, when the nights grow cold, and the diurnal heats abate, the ice in the cave begins to diffolve, infomuch that by winter no more ice is to be feen: the cavern then becomes perfectly dry and a mild warmth. At this time it is furprifing to fee the fwarms of flies and gnats, as alfo of bats and owls, and even of hares and foxes, that make this place their winter retreat till in the beginning of fpring again it grows too cold for them. Above the cavern the hill rifes to a very great height, and on the fouthern afpect produces plenty of rich grafs. A fuller defcription, with a phyfical inveftigation of this cavern, is to be feen in the *Hamburg* Magazine, Vol. iv. P. 60; being a tranflation from the learned *Matthew Bel.*

7. The Gefpanfchaft of *Gomor, Gomor Varmegye, Gomorienfis Comitatus*: The inhabitants *Hungarians, Bohemian-Sclavonians* and *Germans.*

Murany, a fort on a very high and fteep rock, fortified with a wall, and having only one way to it, belongs to the *Kokar* family. In the rock are found magnets.

Tirgarten, a village at the foot of *Kralowe Hore,* out of which iffues the river *Gran.*

Katko, a town inhabited by tanners.

Jolfva, Jelfava, Alnovia, a town with a feat, peopled by tanners and fhoemakers; belongs to the *Kohar* family.

Tfetneck, a large opulent town, with a feat near it, in a plain. It thrives by its brafs and iron works, and is the conftant refidence of a *Lutheran* fuperintendent.

Dobfchau, Dobfcha, a mine-town, having numbers of *Germans* among its inhabitants. This place is noted for its iron, afbeftos, cinnabar and paper.

Alfo-Sajo,

Alfo-Sajo, Sajo inferior, on the banks of the *Sajo*, has the appearance of a town. A great deal of cinnabar is dug in its neighbourhood.

Krofzna-Horka, Grofna-Horka, a good old caſtle in a moſt delightful ſituation, belonging to the *Andraſi* family.

Rofenau, Rofnno-Banya, Rofnawa, a populous mine-town, ſtanding pleaſantly on a plain on the banks of the *Sajo*, among mountains: has a Jeſuits college and ſchool, and flouriſhes in works of copper, quickſilver and cinnabar, handicrafts and trade. It is under the juriſdiction of the Archbiſhop of *Gran*.

Berzetin, a town not far from the *Sajo*, and noted for being the reſidence of many noble families.

Pelſotz, a pretty good town on the *Sajo*. The county meeting is held here.

Gomor, once a fortification, now a mean town ſubſiſting by agriculture, ſtands on the river *Sajo*, in the juriſdiction of Count *Cſaſki*, and gives name to the Geſpanſchaft.

Keveny, a little town following handicrafts and tillage.

Gombafzeg, a pretty large town in a fertile country.

Rimafzets, a town having ſeveral conſiderable fairs.

Hainatſko, a caſtle on a high hill belonging to the *Vetſeji* family.

Putnok, a fort firſt erected againſt the *Turks*, but now without a garriſon.

8. The Geſpanſchaft of *Borfod, Borfod Varmegye, Borfodienſis Comitatus :* inhabited by *Hungarians, Bohemian-Sclavonians* and a few *Germans*.

Szendro, a pretty good town, but formerly of greater conſideration than at preſent : has a caſtle.

Dedeſſa, a ſmall town and poorly inhabited.

Borfod, once fortified, but the works being demoliſhed is now an open town, yet gives name to the Geſpanſchaft.

Dios Gyor, a caſtle ſtanding pleaſantly among woods on the ſide of a hill ; was built by *Mary* the firſt, but rather as a place of pleaſure than defence. It is now a heap of ruins. Below it, however, ſtands a town, of the ſame name.

Mifkoltz, a large populous town, noted for its excellent wine, and the many people of faſhion among its inhabitants ; as alſo for the probſty of the *Topoltzi* in its neighbourhood. It lies in a good corn country near the river *Sajo*.

Onod, a town and caſtle on the *Sajo* ; a place of note in the wars, and particularly for the meeting held there in 1707 by the chiefs of the *Rakotzy* party.

Cſerepes Foldes, formerly a fort run up in haſte, but now a pitiful village.

Kereſztur, a pretty good town in a level country. Its neighbourhood has been the ſcene of ſeveral battles.

Kovefd,

Koveſd, Mezo-Koveſd, a poor town; agriculture the occupation of its inhabitants.

Aſzalo, a ſmall town following the ſame occupation as the former.

Szent Peter, Fanum S. Petri, a town in a fertile country, in the cultivation of which the inhabitants are not negligent.

9. The Geſpanſchaft of *Heves, Heves Varmegye, Heveſſienſis Comitatus.* Its inhabitants *Hungarians, Bohemian-Sclavonians, Raſcians* and a few *Germans.*

Szarvaſko, an old epiſcopal caſtle.

Erlau, Eger, Agria, an epiſcopal city ſurrounded with old walls and bulwarks, and having a caſtle ſeated on a hill. The buildings were formerly handſome, but very much altered by the numberleſs calamities of war it has undergone. It emerges, however, gradually, being the reſidence of a biſhop who has a large income, and of a college and ſchool of Jeſuits, beſides the good wine and warm bath in its neighbourhood. It was firſt built by King *Stephen* the Pious; and, in 1552 made a gallant defence againſt the *Turks*; but, in 1596 was obliged to ſurrender to them by capitulation. In 1606 it was ſurpriſed by the *Imperialiſts,* who commited great outrages, but failed in their attempt on the caſtle.

Gyongyos, a populous town on a river of the ſame name, in a large plain, at the foot of the mountain of *Matra,* has a Jeſuits gymnaſium, and is famed for its yearly fair and wine.

Gyongyos Pata, a mean ſmall town at the foot of the mountain of *Matra,* and facing the afore-mentioned place.

Sirok-var, a town and caſtle near the ſource of the *Torna,* in the juriſdiction of *Nya.*

Paſzto is at preſent a mean town, agriculture the ſupport of its inhabitants.

Hatvan, formerly a conſiderable fortreſs on the river *Zagiwa*; but the works falling to ruin it is now become an open place. Its neighbourhood is fertile. In 1544 the owners of this fortreſs, out of fear of the *Turks,* ſet the caſtle and town on fire, but the latter rebuilt it. In 1596 the *Imperialiſts* having taken the town by aſſault demoliſhed it; but it has been rebuilt again.

Heves, formerly a fortreſs which gave name to the county, but now only a heap of ruins.

Oers and *Detko,* two ſeats in the modern taſte.

10. The Geſpanſchaft of the *Jaſi,* or *Philiſti,* is united to that of *Heves,* though under the ſame juriſdiction as the country of the *Cumani*; namely, under the Statholder, and not ſubject to the laws of the Geſpanſchaft. The inhabitants *Hungarian-Jaſi,* or deſcendants of the *Metanaſtaſ-Jaſi.*

Arock-Szallas, a good town in a level fertile country.

Fenſzaru, a town in the ſame fruitful territory near the river *Zagyva.*

Szent Gyorgy, Fanum S. Georgij, a good town in an open fertile plain.

Doſa and *Jukohalma,* two large villages in a wide plain.

Apati, a village, more like a town, ſeated in a good graſs and corn land.

Mihalytelek, a town ſeated like the former in a fertile territory.

Jaſz-Bereny, a large well built town, ſurpaſſing all the reſt in greatneſs and fertility of ſoil.

11. *Great Cumania,* or the *Kuner Land, Kunſag, Cumanorum Majorum Regio,* enjoys the ſame privileges with the *Jaſi;* and, though annexed to the Geſpanſchaft of *Heves,* is together with its towns under the Statholder. The *Hungarian Cumani* derive their original from the *Cuni.*

Madaraſz, a populous town in a fertile country, has good paſturage but wants wood.

Kolbaſz-Szeck, a populous town in a level country; yet wants not wood.

Kartzag, Ujſzalas, a large populous town, conveniently ſeated for tillage and graziery.

Kis-Ujſzalas, a ſmall town; its territory alſo ſmall, but fertile.

Kunhelyſeg, Cunorum Sedes, a ſpacious town, ſupported by tillage and grazing.

12. The Outward Geſpanſchaft of *Szolnok, Kulſo Szolnok Varmegye, Szolnok Exterior Comitatus,* peopled by *Hungarians* and *Raſcians.*

Poroſlo, a large populous town in a fruitful country. The employment of its inhabitants tillage and graziery.

Janos Hidgya, Johannis pons, a town with a probſty of *Præmonſtratenſes.*

Szolnock, a good fortification at the conflux of the *Jaſyva* and *Theis,* gives name to the county.

13. The Geſpanſchaft of *Cſongrad, Cſongrad Varmegye, Cſongradienſis Comitatus.* The inhabitants *Hungarians* and *Raſcians* with ſome *Germans.*

Szentes, a large village, having the appearance of a town, and ſtanding in a good graſs and corn country.

Cſongrad, an old fortification, with a large town, ſituate at the junction of the *Koros* and *Theis.* The county named from it.

Vaſarhely, large and populous, ſtands on the ſwamp of the *Hod,* which produces corn and affords good paſturage for cattle. The property of *C*ount *Karoly.*

Mind-ſzent, Fanum Omnium Sanctorum, a town in a fertile country.

Szeged, a royal free town, at the conflux of the *Maros* and *Theis,* is ſtrong, and enjoys a good trade, particularly in oxen, never wanting plenty of fiſh, and the neighbouring country fruitful. Here is a gymnaſium belonging to the *PP. piarum ſcholarum.* In 1503 all its defence was a moat and rampart; but falling ſome time after into the hands of the *Turks,* they erected a brick fort. In 1686 the *Imperialiſts* diſpoſſeſſed the *Turks* of it, defeating alſo the forces that were haſtening to its relief.

11. The

11. The circle on the farther fide of the *Theis, Circulus Trans-Tibifcanus,* confifts of fifteen Gefpanfchafts. The inhabitants *Hungarians,* and in fome parts *Ruffians, Walachians* and *Rafcians,* with a few *Germans* and *Bohemian-Sclavonians.*

1. The Gefpanfchaft of *Beregh, Beregh Varmegye, Beregienfis Comitatus.* Inhabited by *Hungarians* and *Rafcians.*

Helmez, a middling town lying among hills and belonging to the probfty of *Lelefzi.*

Munkats, a caftle almoft impregnable, feated on a high and fteep rock, in a fpacious plain, the natural ftrength of which is increafed by art and labour. It is the capital of a lordfhip, formerly bearing the title of a dutchy. Beneath it, on the river *Lætortza,* is a town which is the refidence of a 'Greek' bifhop united with the *Roman* church, and a convent of the order of St. *Bafil.* In 1688, after a blockade of three years, this famous caftle furrendered to the *Imperialifts ;* Count *Tekely's* lady, who had conducted this long defence, being carried to *Vienna ;* and, befides the *Tekely* family, vaft treafures in money were found there. In 1703 this was the place of rendezvous to the *Rakotzy* party, but by fuppreffion of that revolt this caftle efcheated to the crown.

Beregh Szafz, was at firft a military fortification giving name to the county, but afterwards became a confiderable town ; and, in allufion to a *Saxon* colony eftablifhed there, was called *Beregh Szafz;* but the prefent inhabitants are *Hungarians.*

Vary, a town of the beft figure.

Mufai, a town not far from the *Theis.*

2. The Gefpanfchaft of *Ugots, Ugotfa Varmegye, Ugotfienfis Comitatus,* is inhabited by *Hungarians, Ruffians,* and a few *Walachians.*

Nyalab, formerly a caftle, adjoining to the large village of *Kiralyhaza* on the *Theis,* at prefent lying in ruins.

Nagy-Szolos, a pleafant town, not far from the *Theis,* having a few feats, and belongs to the *Pereny* family.

Kanko-Var, a caftle on a high hill, but falling to ruin.

Ugots, Ugogh, the remains of an old fort which gave name to the county.

3. The Gefpanfchaft of *Maramarus, Maramaros Varmegye, Maramarufienfis Comitatus.* The inhabitants *Hungarians, Ruffians, Walachians* and a few *Germans.* Its chief commodity is falt.

Hofzfzu-mezo, Campilung Oluhe Pole, Campus Longus, lying on the *Theis,* and famous for an *Hungarian* colony eftablifhéd there, but expofed to frequent inundations.

Ur-mezo, Campus Dominorum, a large plain near the above-mentioned place.

Rhona,

Rhona, Rhona-fzek, a falt work producing a confiderable quantity of that commodity, lies betwixt two falt mountains; the product of which is carried from hence up the *Theis* through *Hungary.*

Szigeth, a pretty good town, in which the courts of the Gefpanfchaft and provincial meetings are held.

Maramarus, an old fort, now lying in ruins, from whence the county takes its name.

Vifk, a *German colony,* but the origin unknown.

Tetfo, a mean place on the *Theis.*

Hufzt, a caftle lying on a very high rock, and ftrong both by nature and art; having a town at the foot of it. At a little diftance from the *Carpathian* mountains, lies the fource of the *Theis.* In 1556 the *Tranfylvanians* made themfelves mafters of the caftle; and, in 1605 it was reduced by *Stephen Botfkay.*

4. The Diftrict of *Kovar, Kovar Videke, Kovarienfis Diftrictus,* lies on the frontiers of *Tranfylvania.* Its inhabitants *Hungarians,* but in the hilly country *Walachians.*

Kovar, a town with a mine-fort in ruins: once accounted one of the frontier fortreffes.

Verkefz, a middling town, not far from the foot of *Korvar* fort; having many perfons of rank among its inhabitants.

Kapnik, fmall, yet one of the mine-towns, and having rich mines of gold.

5. The Gefpanfchaft of *Middle Szolnok, Kozep, Szolnok Varmegye, Szolnok Mediocris Comitatus,* is peopled by *Hungarians,* and in the vallies by *Walachians.*

Hadad, the chief town, with a caftle, which, in 1564, was taken by *John Sigifmund,* Prince of *Tranfylvania.* It belongs to the *Vefeleny* family, which takes one of its titles from hence.

Szilagy, a town with a caftle on a high hill.

Tafchnad, populous and feated in a delightful country.

6. The Gefpanfchaft of *Krafzna, Krafzna Varmegye, Krafnenfis Comitatus.* Inhabited by *Hungarians,* and on the mountains by *Walachians.*

Krafzna, formerly a fortrefs, now an open place; deriving its name from the river on which it ftands, and giving it to the whole county.

Sanct Margaretha, one of the beft places in this Diftrict.

Somlyo, a town on the river *Krafzna:* the birth place of *Stephen Bathory,* King of *Poland.*

Zilah and *Mifke,* two mean places.

7. The Gefpanfchaft of *Szathmar, Szathmar Varmegye, Szathmarienfis Comitatus.* Inhabited by *Hungarians* and a few *Germans.*

Matolts, a pretty good town on the river *Samos.*

Gyarmath,

Gyarmath, a town on the fame river.

Hungarian Alftadt, Fefo Banya, a metal town lying betwixt hills abounding with rich mines.

Hungarian Neuftadt, Fefo Banya, Uj Banya, a metal town, having a gold mine.

Nay-Banya, Kapnik-Banya, Rivulus Dominarum, a metal town, now one of the royal free towns, formerly belonging to the Queens. The gold and filver mine-works are of great produce, and the money coined here is diftinguifhed by the mark N.B. The Jefuits have a college and gymnafium here.

Aranyos Medgyes, a confiderable town, formerly fortified with a caftle.

Erdod, formerly a ftrong caftle, but was razed to the ground in the *Tranfylvanian* wars. It is at prefent a town which feems to have given name to the illuftrious houfe of *Erdodi,* though others deduce it from *Erdodi* in the Gefpanfchaft of *Tolna.*

Szathmar Nemethi, a royal free town, being properly two towns; namely, *Szathmar,* on an ifland in the river *Samos;* and *Nemethi,* oppofite to it on an arm of the river: but, in 1715, both were erected into one town; the firft of which is fortified. In 1334, or 1335, it was taken, pillaged, and laid in afhes by the troops of the Emperor *Ferdinand* the firft. In 1562 it was befieged by the *Turks,* and fet on fire, but not taken. In 1564 King *John* took it; in 1605 *Stephen Botfkay;* in 1660 *Barfkay,* Prince of *Tranfylvania,* made himfelf mafter of it; and, in 1662, the *Imperialifts.* In 1681 the malecontents laid fiege to this place; and the Reformed held a national fynod here in 1646.

Cfenger, a fmall town, having a caftle.

Nagy Karoly, a fpacious town, with a grand caftle, in which the *Patres piarum fcholarum* have a college and gymnafium, the property of Count *Karoly.*

8. The Gefpanfchaft of *Szabolts, Szabolts Varmegye, Szaboltfenfis Comitatus.* The inhabitants all *Hungarians.*

Kis-varda, Varadinum Minus, a fortrefs near the *Theis,* feated in a morafs; its works are at prefent demolifhed, but it is the refidence of many people of rank.

Etfed, Echet, a fortrefs impregnable by reafon of its fituation among bogs and marfhes, but demolifhed in 1701, together with Little *Echet.*

Bathor, a town adjoining to the feat of *Nyir-Bathor;* whence the ancient family of *Bathor,* to whom it belongs, receive their name.

Kallo, a fort of confequence, but having no garrifon and running to ruin.

Nanas, Dorog, Hathaz, Vamos-Pertz, Bofzormeny, Szobofzlo, Polgard; feven towns belonging to the *Heyducks,* which *Matthias* the fecond, in confideration of their bravery, exempted from the county jurifdiction. They are

are greatly dwindled for want of inhabitants, notwithſtanding their ancient privileges were renewed in 1746.

Szabolts, an old fort in ruins, after which the county is called ; as the fort itſelf received its name from *Szabolts,* the celebrated *Hungarian* commander.

Cſak-var Cſage, an old ruined caſtle ; the original place of the *Cſaki* family, who derive their deſcent from the above-mentioned *Szabolts,* one of the ſeven *Hungarian* leaders who entered this country in the ninth century.

9. The Geſpanſchaft of *Bihar, Bihar Varmegye, Bihariienſis Comitatus.* Its inhabitants *Hungarians* and *Germans.*

Debretzen, a royal free town on a fine plain ; large and populous, but having neither walls nor gates. Both the Reformed and the *Patres piarum ſcholarum,* have a gymnaſium here. The country produces excellent paſturage, which is induſtriouſly followed. For twelve *Hungarian* miles, and more, not a wood nor mountain is to be ſeen, the whole being one continued heath ; whence, conſequently, they muſt be greatly diſtreſſed for fuel. In 1564, 1565, and 1640, it was burnt down.

Szent Job, Fanum S. Jobi, famous for an abbey of the bleſſed Virgin, and an old caſtle ſurrounded with a wall and palliſadoes. In 1604, when it belonged to *Stephen Botſkay,* this place was taken and plundered ; which excited him to take part againſt the Emperor. In 1660 the *Imperialiſts* left it ; and, in 1661, the *Tranſylvanian* peaſants who had revolted took it ; and, notwithſtanding the *Imperialiſts* ſoon after laid it in aſhes, it continued a conſiderable time in the hands of the *Turks.*

Telegd, a mean town on the river *Koros.*

Great Varadein, Varad, Nagy Varad, a metropolitan city, on the river *Koros,* ſurrounded with good fortifications. This place was formerly much celebrated for the relicks of the cannonized King *Ladiſlaus* ; is now the reſidence of a biſhop and chapter, and has alſo a college of Jeſuits. The town itſelf is not large, but has three ſuburbs of very conſiderable extent. The adjoining fortreſs is a regular pentagon, well fortified, beſides a deep and broad moat. In 1566, and 1613, it was taken by the *Tranſylvanians.* In 1598 it held out a ſiege againſt the *Turks,* but, in 1660, was taken by capitulation. They continued maſters of it alſo at the peace in 1664, but the *Imperialiſts,* in 1692, diſpoſſeſſed them. About a mile from the city is an excellent cold bath.

Szekely-hid, Pons Siculus, formerly a fortification commanding the bridge over the *Verettyo.* In 1660 it held out againſt the *Tranſylvanian* peaſants ; in 1664 it was ſurrendered by the *Imperial* garriſon to the *Tranſylvanian* Prince *Abaffi* ; but, on the enſuing peace, diſmantled.

Bihar, a very ancient town ; giving name to the Geſpanſchaft.

10. The Geſpanſchaft of *Zarand, Zarand Varmegye, Zarandienſis Comitatus,* inhabited wholly by *Hungarians.*

Zarand,

Zarand, an old fort, of which now only the name remains; but giving title to the Geſpanſchaft.

Deſzna, formerly a caſtle on a high hill, of which alſo the name only remains at preſent.

Halmagy, a ſmall caſtle on the river *Koros*, with a town of the ſame name.

Vilagos-var, *Vilagos-vara*, a caſtle lying in ruins.

Boros-ieno, once a fortification erected againſt the *Turks*, but now an open place, yet famed for its wines.

11. The Geſpanſchaft of *Bekes*, *Bekes Varmegye*, *Bekeſienſis Comitatus*; contains a wide heath, through which flows the river *Koros*; but having only a few towns. The inhabitants are *Hungarians* and *Bohemian-Sclavonians*, who adopt the cuſtoms of the former.

Bekes, anciently a fort, now a town on the river *Koros*, and giving name to the Geſpanſchaft.

Gyula, formerly a fort, and in 1566 taken by the *Turks*, but now evacuated. It lies near a town of the ſame name, which ſtands on an iſland in the river *Koros*.

Cſaba and *Szarvas*, two large villages built in the preſent century by a colony of *Bohemian-Sclavonians*.

12. The *Turuntal* Geſpanſchaft, *Torontal Varmegye*, *Torontalenſis Comitatus*. No account was taken of this province under *Charles* the ſixth; but, in 1747, her preſent Majeſty, *Maria Thereſa*, reſtored it to its ancient privileges. The inhabitants are *Hungarians*.

Tſur, *Thurn*, the principal town in this large Diſtrict, whence it takes its name.

Betſe, *Betſetek*, a town on the *Theis*, by ſome placed in the *Bannate* of *Temeſwar*.

13. The Geſpanſchaft of *Arad*, *Arad Varmegye*, *Aradienſis Comitatus*. The inhabitants *Hungarians* and *Raſcians*; and in the highlands *Walachians*.

Varadja, *Thot Varadja*, once a ſtrong caſtle, but now in ruins.

Radna, a pretty good town, ſupported by tillage.

Uj Arad, *Neu-Arad*, a ſtrong caſtle on the river *Maros*, and the reſidence of a *Greek* biſhop.

O-Arad, *Old Arad*, a town ſituate in a fertile country; but once in a better condition than at preſent.

Solymos, a frontier caſtle ſeated on a high rock; now deſerted and lying in ruins.

14. The Geſpanſchaft of *Cſanad*, *Cſanad Varmegye*, *Cſanadienſis Comitatus*. Its inhabitants *Hungarians* and *Raſcians*.

Cfanad, an episcopal see on the river *Maros*; formerly flourishing and well fortified, but afterwards difmantled; though at prefent, indeed, it begins to thrive again. In 1595 the *Turks* got poffeffion of this place.

Egres, a *Ciftercian* abbey founded by King *Andrew* the fecond.

15. The *Bannate* of *Temefvar, Temefvari Banat, Banatus Temefvarienfis*; bounded by the rivers *Maros, Theis* and *Danube,* and watered by the *Temes,* which is joined by the *Beg,* or *Beyhe.* In 1552 the *Turks* became mafters of it, and retained it at the peace of *Karlowitz,* in 1699; but loft it in 1716, after keeping poffeffion of it one hundred and fixty-four years; and, in 1718, formally ceded it to the Emperor at the peace of *Paffarowitz*; which ceffion, one fmall Diftrict excepted, granted to the *Turks,* was ratified, in 1739, at the treaty of *Belgrade.* Its government is divided into the civil and military jurifdiction.

Temes-var, Temefia, a fortrefs of importance on the river *Beg,* ftrengthened by art; but its chief defence confifts in a morafs formed around it by the above river. It is the capital of this *Banate*; the refidence of a governor and *Greek* bifhop, and has alfo a college and gymnafium of Jefuits. In 1551 the *Turks* befieged it in vain; but, in 1552, carried it. In 1596 and 1597 it held out two fieges againft the *Tranfylvanians.* In 1690 it was blockaded by the *Imperialifts,* but Prince *Eugene* took it in 1716 after a furious fiege.

Lipp, Lippa, a fmall ftrong fort on the river *Maros,* but now fuffered to run to decay. *George Margrave* of *Brandenburg* was the firft who walled and fortified this little place. In 1551 the *Turks* made themfelves mafters of it, but it was wrefted again from them the very fame year. In 1552 the *Imperial C*ommandant, dreading *Turkifh* cruelty, abandoned it, after having firft fet fire to the place; but the *Turks* extinguifhed the flames, took poffeffion of the fort, and kept it till 1595, when it was taken from them by the *Tranfylvanians*; who fiding afterwards with the *Turks,* the *Imperialifts* made themfelves mafters of it in 1603 : but, in 1614, it was loft again.

Fatfad, a fort on a high hill.

Lugos, Lugofch, Lugofum, a ftrong town feated among mountains; but taken, in 1552, by the *Turks.*

Caranfebes, once a fine flourifhing city, lying among mountains, and being the great magazine for all *Turkifh* goods, which were carried from hence by land to *Tranfylvania*; but now little better than a village.

Clafura, Claufura, a mountain on the borders of *Tranfylvania*; giving name to the neighbouring country.

Almas, a fmall open place, and gives name to the neighbouring country.

Uj-Palanka, Neu-Palanka, a frontier fortification, or fort, on the *Danube,* frequently taken, but left by the *Imperialifts* in 1716.

Warfchetz,

4

Werfchetz, Werfitz, a pretty good town, whence the adjacent Diſtrict takes its name.

Pantzowa, a place of conſiderable ſtrength lying oppoſite to *Belgrade,* and not far from the influx of the *Temes* into the *Danube,* was taken by the *Imperialiſts* in 1716.

Betſe Betſletek, a town on the *Theis.*

Denta, a fort run up in haſte on the river *Temes.*

Cſako-var, a ſmall town on the river *Temes.*

TRANSYLVANIA.

§. 1. TRANSYLVANIA is a part of the ancient *Dacia,* and that which, on account of its ſituation, was called the Northern ; and, in alluſion to the *Roman* government the conſular *Dacia.* This country probably receives its *Germam* name *(Siebenburgen)* from ſeven celebrated forts, or caſtles, in it. The *Latin* name *(Tranſilvania)* is derived from its lying beyond the foreſts with which the *Carpathian* mountains are environed; and the *Hungarians,* for the ſame reaſon, call it *Erdely,* which ſignifies a woody, mountainous country. To the north it borders on *Hungary, Poland* and *Moldavia* ; caſt of it lies the *Moldavia* ; to the ſouth *Walachia* and the Banat of *Temeſwar* ; and to the weſt *Hungary.*

§. 2. It is environed on all ſides with mountains, whence it enjoys a temperate air and wholeſome ſpring and river water ; and, though both mountainous and woody, yet has, notwithſtanding, fields ſo rich and fertile, that it is deſtitute of none of the neceſſaries of life. The mountains of *Tranſylvania* run from north to ſouth, branching out likewiſe eaſt and weſt. They terminate at the centre of the country in hills cloathed with vineyards and rich in mines. The former lie in the ſouthern parts of the country, but in ſome places the northern blaſts prevent the grapes from ripening thoroughly. In the hills of *Rudny, Siculi, Weiſſenburg, Abrugh, Hunyad, Torotſkoi,* and others, metals and minerals are found ; thoſe of *Deeſi Koloſi, Tordi, Vizakni, Homorod, Paraidi* and *Paullini* yielding rock-ſalt. Among the ſprings are, 1. The medicinal ; as thoſe of *Hunyadi,* which are warm ; thoſe of *Weiſſenburg,* which are cold ; of *Cſiki* and *Olach-falvi,* which are ſulphureous ; as alſo the ſprings of *Somlioi* and *Verebis.* 2. The ſubterraneous, warm and ſalt ſprings ; ſuch as thoſe of *Salzburg.* 3. Mineral ſprings ; namely, thoſe of *Homorodi* and ſome others. 4. Petrifying waters; as thofe of *Alma* in the *Udvarhely Sitz*. The chief rivers are, the *Szamos*, in *Latin Szamofius*; which taking its fource in the confines of *Biftritz* and *Maramarus*, after receiving the Little *Szamos* near *Dees*, purfues its courfe into *Hungary*. The *Maros, Marufius, Merif-cus*, having taken its fource in the northern mountains of *Sicily*, traverfes the middle of the country, and being joined by the, fmall rivers of *Aranyos* and *Kochel*, enters *Hungary* in the fouth-weft. *Aluta, Olta*, takes its rife at the foot of the *Carpathian* mountains, in the northern parts of *Sicily*, and paffes through the eaft and fouth regions into *Walachia*. *Tranfylvania* pro-duces grafs, medicinal herbs, grain, pulfe, vines and wood: minerals; namely, gold, filver, copper, iron, quickfilver, cinnabar, folar antimony, fulphur, vitriol, rock-falt, falt-petre, red ochre and chalk: animals; fuch as horfes, fheep, bees, and all kinds of wild beafts, birds and fifhes; par-ticularly buffaloes, wild-affes, elks, bears, wild boars, linxes, chamois, martens, ermines and beavers. Metals and falt are exported from hence to *Hungary*, and what the country does not afford, particularly the pro-ducts of art, are imported here out of *Germany* and *Turkey*.

§.3. With refpect to the different nations inhabiting *Tranfylvania*, they are, 1. *Hungarians*; who fuit perfectly with the natives of the country in their natural genius, underftanding, language and difpofition. 2. The *Sikli*, or *Sekli*, in *Latin Siculi*; a people defcended from the *Scythian Huns*, and confequently rather to be called *Scythuli*. They enjoyed formerly fome particular privi-leges on account of their antiquity, which have fince, however, been gra-dually abolifhed. They fpeak, indeed, the *Hungarian* language, but their pronunciation is broad, and they retain alfo ftill fome old *Hun* words. They differ likewife in their manners from the *Hungarians*. 3. *Saxons*; who are partly the remains of the *Gepidi*, partly the defcendants of thofe *Saxons* who, in 1154, were brought by King *Geyfa* the fecond out of *Germany* into *Tranfylvania*, and fome of whom were fettled in the royal demefnes; others reckoned the King's property. With refpect to their natural difpofition and ancient cuftoms, which they ftiffly adhere to, they are indeed widely different from the two beforementioned nations, but refemble the modern *Hungarians* much, and, like them, apply themfelves to learning, war, arts and trade. Their language agrees very well with that fpoken in Lower *Saxony*. The womens drefs differs from the *Hungarian*. Thefe are the principal nations in this country: the reft, who are reckoned foreigners (and who, in order to be made free of any place, muft unite themfelves with one of the abovementioned nations) are, 1. *Germans*; who fpeak the fame language with the other natives of *Germany*, but a different one from that of the *Saxons*. 2. *Walachians*; the reliques and defcendants of the ancient *Roman* colonies; and who, for this reafon, ftyle themfelves *Romunius* or *Romuin*; i. e. *Romans*. Their language has a great deal of the *Latin* in it, but vitiated with a mixture of the *Sclavonian*. They refemble alfo the ancient

cient *Romans* very much in their diet and drefs, conceive a liking for the *Italian* language, and the Woywodes have their phyficians and fecretaries from *Italy*. The few too who ftudy ufually go to the univerfity of *Padua*. They refide in the mountainous parts of the country, and wholly follow agriculture. From the firft time of their embracing the Chriftian religion, they have profeffed themfelves of the *Greek* community; but fince *Tranfylvania* has been fubject to the houfe of *Auftria*, the Jefuits have been perpetually endeavouring to bring them over to, and unite them with, the *Romifh* church, under the ambiguous title of the *Græci ritus Uniti.* The qualifications of the *Walachian* clergy extend no further than to be able to read and fing; and when any of them would fhine, he makes a tour to *Buchereft* in *Walachia*, where he learns a little politenefs and to fpeak in an ornamental ftyle; but, for any thing elfe, returns as illiterate as he went. The commonality of the *Walachians* are fo extremely ignorant, that hardly one in twenty of them can repeat the Lord's Prayer. See *Acta Hift. Ecclefiaft.* Vol. x. S. 110; and Vol. xi. S. 694; and Vol. xii. S. 60. 3. *Armenians*; who have a particular language, and apply themfelves chiefly to trade. 4. *Rafcians.* 5. *Bulgarians.* 6. *Greeks*; who are moftly traders. Here are alfo *Jews* and *Gypfies* who refemble the *Hungarians.*

§. 4. The Catholics confift of the *Hungarians*, the *Siculi*, and a few *Saxons*: they have fome rights and privileges in common with the church in *Hungary*, and are under a bifhop who refides at *Weiffenburg*, and is fubordinate to the Archbifhop of *Colotza.* The Reformed are wholly *Hungarians* and *Siculi.* Thefe have a fuperintendent over them, who, by the laws of the Kingdom, is the fecond in rank. The Proteftant *Lutheran* church confifts entirely of *Saxons* and fome few *Hungarians*; is divided into nine diocefes called chapters, and governed by a bifhop, who is always the upper preacher at *Berethalom.* This church, with refpect to the number of its members, has fo greatly the majority, that we reckon, in general, twenty-five Proteftants to one Catholick. The *Socinians*, or Unitarians, were formerly the prevailing fect in *Tranfylvania*, but are at prefent greatly dwindled. They have, notwithftanding, a fuperintendent. Thefe four churches are eftablifhed by the laws of the Kingdom. The *Greek* eaftern church, which the *Walachians* and *Greeks* profefs, is protected by particular privileges granted from the Prince of the country; as alfo that part of it which is united with the *Roman* Catholick church, and that diffenting from it. Over the former is a bifhop who has priefts under him; but the latter is fubject, with refpect to ecclefiaftical matters, to the bifhops of *Walachia*; and under thefe alfo, befides the *Greeks*, are the *Armenians*; fome of whom differ in point of publick worfhip from the *Greeks*, but others acknowledge the *Roman* Catholick church.

§. 5. The Catholicks have a college of Jefuits at *Claufenburg*, and feveral gymnafia. The Reformed, Proteftants and *Socinians*, have likewife gymnafia and fchools in which the elements of the fciences are taught; after

which,

which, such students as are of the Reformed religion go to *Switzerland* and
Holland; but the *Saxons* send theirs into *Germany*.

§. 6. That *Transylvania* was formerly a part of *Dacia*, has been already
observed above: but, exclusive of the *Dacians*, in this country dwelt also the
Getæ; which last were a people of *Thrace*. The Emperor *Trajan* made
war upon the *Dacian* King *Decebalus*, and having conquered and slain him
reduced his country into a *Roman* province; by which achievements he acquired the sirname of *Dacicus*. He established a *Roman* colony at *Sarmize-
thusa*, the capital of the country, and gave it the name of *Ulpia Trajana*.
Some monuments of it are still to be seen at *Varhely*. Of this ancient
Roman colony, not only the great numbers of inscriptions found on stone are
evident proofs; but likewise the *Walachians*, who, as has been shewn
above, are their descendants. In the third century, *Dacia* was again lost
under the reign of *Gallienus*; and, though reduced again by *Claudius*, yet
was neglected by *Aurelian*; upon which it shook off the *Roman* yoke and
recovered its former liberty. The country afterwards fell into the hands of the
Goths, and about the beginning of the fifth century was subdued by the *Huns*.
It was taken indeed from the *Goths* and *Gepidæ* by the assistance of the *Romans*, but, in the sixth century, the *Avari*, and in the ninth, the *Hungarians*, made themselves masters of it. In 1004 King *Stephen* the first made
Transylvania a province of the Kingdom of *Hungary*, in which state it continued, under the government of its Woywodes, till the death of King
Lewis the second in 1526. Upon this, one part of the *Hungarians* chusing
John Von Zapolya, Woywode of *Transylvania*, for their King; and the
other *Charles* the fifth, brother to *Ferdinand*; a bloody war commenced
betwixt both parties; which was terminated at last by the peace of *Waitz*,
in 1535, when it was agreed, that *John* should have *Transylvania* and some
parts of *Hungary* as a principality, which, upon his death without issue,
was to descend to King *Ferdinand*; and, in case he left any heirs, to continue to them as a principality. At his decease, his son, *John Sigismund*,
was, by the assistance of the *Turks*, established in the possession of this principality; and after him *Stephen Battory*, King of *Poland*, was elected Prince
of *Transylvania*; to whom succeeded, in the government of this province,
Sigismund Battory his kinsman, who abandoned the *Turks* to side with
Hungary. This Prince promised the Emperor, *Rudolph* the second, to
resign the principality to him in exchange for *Oppel* and *Ratibor*, and an
annual pension of fifty thousand rixdollars; but, instead of keeping his
word, he took possession again of *Transylvania*, and soon drove out his un-
cle, Cardinal *Andrew Battory*. Taking party afterwards with the *Turks*,
the Emperor *Rudolph*, out of resentment, assisted *Michael*, Woywode of
Walachia, to gain the principality of *Transylvania*; but he soon fell under
the suspicion of the *Imperial* court, and was forced to resign his principality to *Basta*, the Emperor's general. Upon this great disturbances arose.

The

The *Transylvanians*, averse both to *Michael* and the *Imperial* General, threw themselves into the hands of their old Prince *Sigismund Battory*, who found himself under a necessity of ceding the principality to the Emperor, for some lands to be given him in *Silesia*. Headed, however, by *Bethlen Gabor*, the *Transylvanians* persisted in their opposition to the Emperor; and violent measures being taken for obtruding the *Romish* religion upon them, they chose *Stephen Botskay*, a *Lutheran* lord, for their Prince; for whom all *Hungary* instantly declaring, the Emperor was obliged, in 1606, to declare him Prince of *Transylvania* and Governor of *Upper Hungary*. Next to him succeeded *Sigismund Rakotzy*, who soon resigned; upon which *Gabriel Battory* obtained the principality in 1608, who being murdered, *Gabriel Bethlen Gabor* succeeded to the government in 1613, and wrested the Kingdom of *Hungary* from *Ferdinand* the second; but he quitting it again in 1621, became a Prince of the *Roman* Empire; which title also he was obliged to renounce in 1624. After his decease, in 1629, the principality devolved to *George Rakotzy*; who making himself a tool to *Sweden*, invaded *Hungary*, and afterwards made war upon *Poland:* but by this last attempt gave so much umbrage to the *Turks*, that they compelled him to lay down the government; which he accordingly did in 1658. Upon this the States chose *Francis Redey*; and *Rakotzy* again attempting the government was opposed by the *Turks*, who likewise, instead of *Redey*, created *Achatius Barskay* Prince; but he being unable to gain ground against *Rakotzy*, resigned the principality to *John Kemeny*, once *Rakotzy*'s general. For this he was arrested by the *Turks*, who prosecuted the war against *Rakotzy*, and at last wounded him in a battle near *Clausenburg*, of which he died. *Barskay* was soon after deposed by the States, and *Kemeny* substituted in his stead; but the *Turks* opposed his election as invalid, and, in 1661, created *Michael Apaffi* Prince; on which *Kemeny* took part with the Emperor: but being defeated in 1662, he perished in the flight by a fall from his horse. *Apaffi* met with better fortune, and, by a peace made in 1664, preserved his dignity under protection of both Emperors. In 1687 the *Austrians* and *Hungarians* over-run the whole Principality, which, in 1689, formally submitted to the Emperor; but the succession was confirmed to Prince *Apaffi* and his line. On his death, in 1690, *Tekely* invaded *Transylvania*, but was soon driven out of it; and *Michael Apaffi* the second, succeeded his father. In 1699, *Transylvania*, by the peace of *Carlowitz*, continued subject to *Hungary*; and though, towards the beginning of this century, *Francis Rakotzy* laid claim to it, yet he was soon forced to more peaceable measures, and *Michael Apaffi* dying in 1713 without issue, *Transylvania* was absolutely annexed to *Hungary*.

§. 7. The government of *Transylvania* is wholly different from that of *Hungary*, and, by the joint consent of Prince and people, as appears from the *Apbrobata*, *concordata* and *diplomata*, formed into an aristocratical go-

vernment;

vernment; and, fince the year 1722, rendered hereditary to the Princes and Princeffes of the houfe of *Auftria*. Formerly the Prince of *Tranfylvania* fucceeded to the government by free election, but fince the year 1722 by inheritance; and though his power is connected with that of the King of *Hungary* and Archduke of *Auftria*; yet his government and privileges differ from both.

§. 8. The States of *Tranfylvania*, with refpect to the number of its nations, are divided into *Hungarians*, *Siculi* and *Saxons*; with refpect to its different religions, into Catholicks, Reformed and Proteftants; and formerly alfo into Unitaiians; but politically, after the example of *Hungary*, into prelates, nobility, gentry and royal towns. The prelates are the bifhop, the abbots, probfts, regular canons and two Jefuits. The magnates confift of the great officers of ftate, the counts and barons. The gentry are *Hungarians* and *Siculi*, but the royal towns *Saxons* only.

§. 9. The Principality of *Tranfylvania* is governed, in the name of the Prince and nobility, by the Diet, the office of ftate, the royal government; the exchequer, the affembly of *Hungarian* counts, the tribunals of juftice and the magiftrates of the *Siculi* and *Saxons*.

The Diets, *Comitia Provincialia*, meet by fummons from the Prince at *Hermanftadt*, and are divided into the upper and lower table. At the upper table fits the government with the prelates, counts and barons: at the lower, the King's council, with the deputies of the *Hungarian* counts, the tribunals of the *Siculi* and royal *Saxons* hold their conferences. At both fits a prefident reprefenting the Sovereign, who lays the King's intentions and propofals before the States.

The *Tranfylvania* office, which draws up and fends away the Prince's orders, is at *Vienna*; but, with regard to publick affairs, has no manner of connection either with thofe of *Hungary* or *Auftria*.

The high government, which refides at *Hermanftadt*, fuperintends the affairs of the Principality, whether temporal or fpiritual. At the head of it is a governor with the counfellors of the three nations and alfo of the three religions, the Catholicks, the Reformed and Proteftants.

The exchequer, to which appertains the care of the publick revenue and demefne, is divided into the *Tranfylvania* and mine office. The refidence of the former is at *Hermanftadt*, and that of the latter at *Abrug-Banya*. The *Hungarian* nation in *Tranfylvania* is formed after the model of the *Hungarian* Gefpanfchafts, being divided into Counties and Diftricts, and under the government of counts, vifcounts, judges of nobles and affeffors, who hold meetings on publick occafions. The *Siculi*, in their polity, are divided into feven high tribunals and as many judges, and are governed by a count. This dignity belonged formerly to the Woywode of *Tranfylvania*, but is now annexed to the Prince, who is reprefented by Counts elected for that purpofe. The *Saxon* nation is divided into feven fuperior and four inferior tribunals, and is governed under the King by a count of the country, judges and magiftrates.   §. 10. The

§. 10. The arms of *Tranſylvania* are divided into three parts by two in-
dentations diverging downwards. In the firſt are the ſeven *Hungarian* caſtles
in a field *or*; in the ſecond, in a field of gules, the eagle of the *Siculi*; and,
in the third, the *Saxon* ſun and moon in a field alſo of gules.

§. 11. The revenue ariſes from the contributions, cuſtoms, metals, minerals, rock-ſalt, royal demeſnes, eſcheats and confiſcations; and are levied
by the treaſury. *Tranſylvania* formerly could bring from eighty to ninety
thouſand men into the field; but at preſent the whole force of that Principality confiſts of ſix regiments, under a commander in chief, for the defence
of the country.

§. 12. With reſpect to the adminiſtration of juſtice in civil and eccleſiaſtical
affairs, civil cauſes are tried, in the Prince's name, in the ſuperior and inferior
courts, and each of the three nations has its particular court. In the royal free
towns belonging to the *Saxons*, the burghers cauſes are firſt heard before the
judge of the town, and afterwards diſcuſſed by the town council, from
whom there lies an appeal to the meetings of the towns, in which the national court preſides; and a further appeal lies from this court to the *Tabula
Regia*. In the *Hungarian* Geſpanſchafts the cauſes of the gentry are firſt
tried by their judge, and then brought before the whole nobility; from
whom alſo lies an appeal to the *Tabula Regia*. In the courts of the *Siculi*,
who have their peculiar cuſtoms and privileges, cauſes are firſt heard before
the King's judge, and in dubious caſes carried up to the
King's lieutenant; and from him to the *Tabula Regia*. The royal table,
or *Tabula Regia*, is the chief court of juſtice, and has a preſident with pro-
thonotaries and aſſeſſors; yet even from this board cauſes may be carried up
to the government, and from thence removed again to the Sovereign. In
church affairs there is but one court, which is at the reſidence of the biſhop
of *Tranſylvania*; from whom appeals lie to the metropolitan Archbiſhop,
from him to the Pope's nuncio, and from thence to the court of *Rome*.

Tranſylvania, according to the number of its natives, is divided into the
Hungarian Geſpanſchafts, the tribunals, the *Sedes* of the *Siculi*, and the
country.

1. The ſeven *Hungarian* Geſpanſchafts comprehend the weſtern parts of
Tranſylvania, and extend themſelves betwixt *Hungary*. The tribunals of
the *Siculi*, and the country of the *Saxons*, extend ſouthward; and to theſe
likewiſe belongs the Diſtrict of *Fogar*.

1. The Inner Geſpanſchaft of *Szolnock*, *Belſo Szolnok Varmegye*, *Szolnok
Interior Comitatus*, inhabited by *Hungarians*, *Walachians*, *Armenians* and a
few *Germans*; contains

Kozar-var, a caſtle on the little river *Szamos*, belonging to the *Keyekeſi*
family.

Kaplan, a caſtle on the Leſſer *Szamos*, belonging to Count *Haller*.

Dees,

Dees, Des, a good town, at the conflux of the Great and Little *Szamos;* famous both for its falt works, and for being the refidence of the Counts of *Bethlen.*

Szent Benedeck, Fanum S. Benedicti, a caftle on a rifing ground on the Leffer *Szamos,* belonging to the *Kornifi* family.

Bethlen, a caftle on the Greater *Szamos,* fortified with a wall and towers and giving title to the Count of *Bethlen.*

Ketteg, a well inhabited town belonging to the *Mikefi* family.

Szamos Uj-var, a good trading town, inhabited by *Hungarians* and *Armenians:* it lies near a fort erected by *George Martinufius* on the Leffer *Szamos.*

2. The Gefpanfchaft of *Doboka, Doboka Varmegye, Dobocenfis Comitatus;* its inhabitants *Hungarians* and *Walachians:* contains

Szent Mihaly, Szent Mihaly Teleke, a caftle in a village of the fame name; an hereditary eftate of the *Tormai* family.

Dobotza, Doboka, a fmall town on the Little *Szamos,* whence the Gefpanfchaft takes its name.

Apafalva, Apafifalva, a very large village from which the illuftrious houfe of *Apaffi* is fo called.

3. The Gefpanfchaft of *Kolos, Kolos Varmegye, Kolofienfis Comitatus;* inhabited by *Hungarians, Walachians* and a few *Saxons.*

Buza, a caftle betwixt *Dobok* and *Claufenburg,* the property of the *Cfaki* family.

Bontzida, a caftle ftanding pleafantly among eminences covered with vines.

Kolos, a town on a plain, noted for its falt mines and giving name to the Gefpanfchaft.

Szamosfalva, having two feats, and making one of the titles of the ancient family of *Mikoli.*

Claufenburg, Colofvar, Claudiopolis, Kolofvaria, feated in a plain on the Leffer *Szamos,* and capital of the *Hungarian* Diftrict; is a large populous town having a great many houfes of ftone, and ftrong walls fortified with towers. Over the *Portina* gate is ftill to be feen an infcription in honour of the Emperor *Trajan.* The Jefuits have a college here, and the Reformed a gymnafium, as alfo the *Socinians,* who are very numerous, and have a particular printing-houfe. Till 1603 they were in poffeffion of the cathedral, which was taken away from them and given to the Jefuits, whofe college and church they had pulled down. In 1601 *Sigifmund Bathory* befieged it in vain. In 1603 it was taken by the Prince who, but a little before had been dethroned. In the fame year *Bafta,* the *Imperial* general, difpoffeffed him of it. In 1659 a decifive battle was fought near this town betwixt *Rakotzi* and the *Turks,* in which the former was mortally wounded. In 1662 *Apaffi* blockaded

blockaded it by the aſſiſtance of the *Turks*; and, in 1664, a mutiny of the garriſon occaſioned the loſs of the place.

Monoſtor, Kolos Monoſtor, Monaſterium, a *Benedictine* abbey near the town, in which is a famous image of the Virgin *Mary.*

Gyalu, Feneys Gyalu, a metal-town with a caſtle; formerly belonging to the biſhop of *Tranſylvania,* now to Count *Banffi.*

Ezeres, an abbey, belonging to the *Mikeſi* family.

4. The Geſpanſchaft of *Torda, Torda Varmegye, Tordenſis Comitatus;* inhabited by *Hungarians* and *Walachians:* contains

Diſnajo, a ſtately caſtle ſeated on the *Maros.*

Vecs, Vets, a fine town with a ſeat belonging to the *Kemeny* family.

Gergeny, Gorgeny, once a fortification, but its works were demoliſhed in the *Rakotzian* wars. It belongs at preſent to the *Kaſznoni* family.

Regen, a fine town ſtanding high and pleaſant.

Szent-Ivany, Vaida-Szent-Ivany, a caſtle in a delightful ſituation, formerly the reſidence of the Woywodes of *Tranſylvania.*

Garnyeſzeg and *Szent Gyorgy,* two inconſiderable ſeats, the former of which belongs to the *Tekely* family.

- *Torda,* an open. but large and well built town, pleaſantly ſeated on the river *Koros;* is the capital of the Geſpanſchaft and remarkable for its mine-works of ſalt and the old *Roman* ſhaft of braſs. The *Hungarian* language is ſaid to be ſpoken here in great purity.

Kereſztes-Mezo, Prat de la Trajan, Pratum Trajani, a plain extending to the *Iron-gate,* where the Emperor *Trajan* gained a very ſignal victory over the *Dacians* under their King *Decebalus.*

Obſ. Betwixt this and the following Geſpanſchaft lies the *Arany* tribunal of the *Siculi;* a deſcription of which ſhall be given in the ſequel.

5. The Geſpanſchaft of *Kukollo, Kukollo Varmegye, Kukolienſis Comitatus,* inhabited by *Hungarians* intermixed with *Walachians;* contains

Rotnod, Ebeſſalva, two caſtles environed with walls and towers.

Bonyha, a fine caſtle belonging to the *Bethleni* family.

Kukollo-var, a ſtrong ſtately caſtle, on the river *Kukollo,* giving name to the Geſpanſchaft.

Szent Miklos, Fanum S. Nicolai, a fine town on the river *Kukollo* having two caſtles.

6. The Geſpanſchaft of *Weiſſenburg, Gyula-Fejer-var Varmegye, Albenſis* or *Alba Julienſis Comitatus.* Its inhabitants *Germans, Hungarians* and *Walachians;* contains

Torotzko, a metal-town, famous for its iron and ſilver mines, being capital of a barony of the ſame name; belonging to the *Tarotzki* family.

Enyed, Nagy Enyed, a creditable town where the Reformed have a gymnaſium. Great quantities of *Roman* coins have been found in the neighbouring grounds and mountains.

Hofzfzu-Afzu, a town and caftle belonging to the *Tomai* family.

Balaffalva, Blafendorf, a *Walachian* town, the refidence of the bifhop of *Walachia.*

Szent Kiraly, a fine feat belonging to the *Banffi* family.

Tovis, a town on the river *Maros,* having a convent of *Paulines.*

S. *Michael,* or *Totfalu, Caftrum S. Michaelis,* a caftle on a high rock runing to ruins.

Weiffenburg, Fejer-var, Gyula Fejer-var, Alba Julia, now *Karlfburg, Karoly-var, Alba Carolina,* a ftrong well built town on the river *Maros,* pleafantly feated amidft corn-fields and eminences covered with vines ; has a bifhop and college of Jefuits. It is called *Alba Julia* from *Julia Augufta,* mother to the Emperor *Marcus Aurelius Antoninus* ; and *Karlfburg,* in honour of *Charles* the fixth, by whom it was confiderably improved.

Zlatna, Zalakna, Little Schlatten, Auraria Parva, a metal-town, having rich gold and filver mines ; known fo long ago as the times of the *Romans* and *Dacians,* but of greater produce formerly than at prefent.

Borberek, a caftle on a high rock, fortified with towers ; with a town of the fame name on the river *Maros.*

Alvintz, Bints, a caftle formerly of figure, where Cardinal *George Martinufius* died ; but has fuffered much by fire.

Homorod, remarkable for its falt-works.

Brad, a town and capital of a Diftrict of the fame name.

Koros Banya, Chryfii Auraria, a metal-town at the fource of the *Koros.*

Great Schlatten, Abrug-Banya, Abrud-Banya, Awrud, Auraria, a confiderable place and chief of the metal-towns ; the refidence alfo of the mine-office, and feated amidft mines of gold and filver.

Ofenburg, Offen-Banya, a metal-town ; fo called from the fmelting *Oefen,* or furnaces.

7. The Gefpanfchaft of *Hunyad, Hunyad Varmegye, Hunyadenfis Comitatus,* inhabited by *Hungarians* and *Walachians* ; contains

Folt, a palace on the river *Maros,* formerly belonging to the Knights-templars, but at prefent to the *Folti* family.

Rapot, a palace, near the river *Maros,* famous for its mineral water, and belonging to the family of *Kafzoni.*

Arany-var, a ftrong fine caftle.

Branvitfka, a caftle belonging to the *Jofiki* family.

Ille, Illye, a caftle fortified by art and nature.

Dobra, a fort.

Nemeti, a feat with a large village.

Deva, Decidava, a creditable fpacious walled town, having a caftle ftanding on a high rock.

Hunyad, an old ruined fortrefs, whence *Johannes Huniades* was fo called, giving name to the county. Below it lies a town having good iron-works in its neighbourhood.　　　　　　　　　　　　　　　　　*Hatzeg,*

Hatzeg, a town possessed by *Walachians,* from whence the valley of this name stretches itself eight miles to the river *Syl,* the mountains and open pass of *Volkany.*

Nalatz, a seat of the *Nalatzi* family.

Varhely, i. e. the place of a town or a castle, being the name given to *Sarmitz,* or *Sarmizagefusa,* anciently the capital of *Dacia,* where *Trajan* planted a *Roman* colony, calling it *Ulpia Trajana.* The rudera of this colony lie in the above-mentioned valley of *Hatzeg,* and among them have been found several gold and silver coins with other antiquities.

Pofteni, a strong fortification, defending the narrow pass of the *Iron-gate,* in Latin *Porta ferrea ;* by the *Hungarians* called *Vas Kapu,* by the *Greeks Acontifma,* and by others *Orla Pafs.*

8. The District of *Fogaras* lies indeed in the *Saxon* territory, but belongs to this part, and contains

Fogaras, a good town, with a castle on the river *Aluta,* giving name to the District. In 1661 it opened its gates to Prince *Kemeny's* troops, but in the same year stood a siege against the *Turks.* On *Kemeny's* death it submitted to *Apaffi.*

Several *Walachian* large villages.

II. The territory of the *Siculi, terra Siculorum,* lies in the north-east part of *Tranfylvania,* and contains seven Tribunals.

1. The Tribunal of *Cfik, Cfik Szek, Cfikienfis Sedes,* together with the annexed Tribunals of *Gyergy* and *Kafzon ;* containing the following remarkable towns.

Cfik-Szereda, the capital, having a good trade and strong castle.

Miko-var, a strong castle ; giving name to the *Miko* family.

Somlyo, a large village with a gymnasium.

Szent Miklos, Fanum S. Nikolai, a town of good figure, and the principal in the whole Tribunal of *Gyergy,* near the source of the rivers *Maros* and *Aluta.*

Kafzony, a pretty good town, capital of the Tribunal of *Kafzony.*

2. The Tribunal of *Haram, Harom Szek, Haramienfis Sedes;* is subdivided into the Tribunals of *Orba, Kezdy* and *Sepfy ;* of each of which we shall give a particular account.

3. The Tribunal of *Kefdy, Kezdi Szek, Kefdienfis Sedes,* contains

Kefdo Vafarhely, on the river *Aluta,* and capital of the District.

Kefdy Szent-Kelek, Fanum S. Spiritûs Kezdienfe, a fine strong fortress feated on a high rock.

Beretzk, a pretty good town leading to the narrow pass of *Ojtos.*

4. The Tribunal of *Orbai, Orbai Szek, Orbacenfis Sedes ;* in which are

Zabola, a town, with a castle ; belonging to the Counts *Mikes* and *Kalnoki.* The mountains of this place produce rock-salt.

Kovafna, celebrated for its medicinal baths.

Papolz, Papulum, reckoned among the principal places of the District.
Zagon, a large village.

5. The Tribunal of *Sepſe, Sepſi Szek, Sepſienſis Sedes,* with Little *Miklos-vars;* contains

Szent Gyorgy, Fanum S. Georgii, a middling town on the *Aluta.*

Korofpatak, a town with a handſome caſtle belonging to the *Kalnocki* family.

Uzon, a town belonging to the family of the *Mikeſi.*

Bikfalva, a town, noted for the narrow paſs on the frontiers of *Moldavia.*

Hyefalva, a town which, together with the foregoing, belongs to the *Mikeſi* family.

Miklos-var, a town and caſtle, whence the Tribunal takes its name.

6. The Tribunal of *Udvarhely, Udvarhely Szek, Udvarhelyenſis Sedes,* with the two Tribunals annexed to it; namely, thoſe of *Kerefztur* and *Bar-dutz :* in which are

Almas, Homorod Almas, a Diſtrict having ſeveral ſubterraneous caverns with water in them of a petrifying quality.

Kereztur, Szekely-Kerefztur, a town, environed with high mountains, from which the Tribunal ſo called has its name.

Bardutz, Pardutz, a well inhabited town, producing plenty of ſalt, and from whence the Tribunal of that name is ſo called.

Udvarhely, the capital on the river *Kukollo,* and the reſidence of Jeſuits.

7. The Tribunal of *Maros, Maros Szek, Maroſienſis Sedes,* contains

Szent Pal and *Szent Demeter,* two caſtles; the former belonging to Count *Gyulaſi,* the latter to the *Redli* family,

Maros Vaſarhely, Agropolis, a double town ; one of which ſtanding upon a riſing ground, is ſurrounded with a wall; being populous and well built; the other is ſeated at no great diſtance from it, and lies in a valley, where the Reformed have a gymnaſium which was formerly at *Weiſſenburg.*

8. The Tribunal of *Aranys, Aranyas Szek, Aranyenſis Sedes,* is ſituate among the Geſpanſchafts of *Hungary,* and lies betwixt thoſe of *Torda* and *Kukoli* on the river *Aranyas;* containing

Bagyon, Bagyona, a good town near the *Maros.*

Kerefztes, Kerefztes Mezeje, a field receiving its name from the ſhocks of corn produced on it, and noted for a defeat of the *Turks.*

Szent Mihaly, Fanum S. Michaelis, a pretty large town on the river *Aranyas.*

Gerend, a fine ſtrong caſtle belonging to the *Kemeny* family.

Varfalva, a town and caſtle on the *Aranyas.*

Fel-Vingi, Foldvinz, a good town.

III. The royal country of the *Saxons, Fundus Regius Saxonicus,* is, with reſpect to its government, divided into ſeven Tribunals and two Diſtricts ; but, conſidered otherwiſe, conſiſts of five Diſtricts : namely,

Noſnerland,

1. *Nofnerland, Biftritzienfis Diftrictus,* is feparated from the others, and ftretches northward to the fartheft extremity of *Tranfylvania.*

Rudna, a metal-town well peopled.

S. Georg, Fanum S. Georgii, a well inhabited town.

Little Biftritz, a large village, in the road to the narrow paffes of *Tartar* and *Tirmenitz.*

Biftritz, Beflertze, Nofenftadt, a royal free town, founded in 1006; the capital of the Diftrict. It is fortified with walls, towers and moats; the *Patres Piarum Scholarum* and Reformed having a gymnafium there. It ftands on the little river *Biftritz,* in a very fpacious level plain, but neither boafts a good air nor water. The hills, which environ the valley, produce vines. In the year 1602 the *Imperialifts* made themfelves mafters of this town.

Metterfdorf and *Durbach,* together with *Treppen* and *Leichnitz;* all fmall free towns,

2. *Weinland* contains,·

Schefburg, Szeges-var, a royal free town, part of which ftands on a hill, the reft at the foot of it, and built in 1196. The boufes are handfome. It is the capital of the Diftrict, and is divided into the Upper and Lower town; the former of which is called the caftle, and is better fortified than the latter.

Berethalom, or *Birthelem,* a fpacious town, the refidence of the Proteftant bifhop. The church ftands on a high rock, and its neighbourhood produces good wine.

Medgyes, Medwis, Media, Megyefinum, a royal free town, and capital of a Tribunal of the fame name. The inhabitants follow tillage and the culture of vineyards, but the wine produced there is none of the beft.

3. The Diftrict on this fide the foreft, *Ante Sylvanus Diftrictus,* contains *Raifmark, Szerdahely,* a handfome town and capital of the province.

Müllenbach, Szafz-Sebes, a royal free town, pleafantly feated in a plain, was built in 1150, and fortified with a wall. From hence the tribunal fo called takes its name.

Bros, Szafz-Varos, a royal free town on the *Maros,* capital of the Tribunal of the fame name, and lying in a fertile country. The inhabitants diftinguifh themfelves by their improvements in agriculture. It was incorporated by the *Saxons* among the *German* towns, in preference to *Claufenburg,* where *Socinianifm* prevails.

Kenyer Mayo, Panis Campus, a plain where the *Chriftians* gained a victory over the *Turks,*

4. *Altland* contains

Keps, Kohalom, a pretty fpacious town with a caftle.

Great-Sing, Nagy-Sing, a tolerable town.

Los-Kirchen, Uj-Egyhaz, a free town.

Hermannſtadt; Szebeny, Cibinium, a royal free town; feated in a plain, being large, and well built, and fortified with a double wall and deep moat, which render it in general very ſtrong. It is alſo the principal place of this Diſtrict, of the *Saxon* colony, and of the whole country. It is governed by the royal chamber, the tribunal of appeals and the Diet; beſides which, the commanding general and royal governor of the *Saxon* nation have their reſidence here. The air, however, is unwholeſome. It borrows the name of *Hermannſtadt* from its founder *Hermann,* but receives its *Latin* and *Hungarian* appellation from the river *Cibin,* on the banks of which it lies, and which, in theſe parts joins the *Aluta.* In it is a gymnaſium of *Lutherans.*

Talmatz, a town, giving name to a Diſtrict.

Szeliſt, a town, whence a Diſtrict annexed to that of *Talmatſy* takes its name,

5. *Borzeland,* or *Wurzeland, Burciæ Diſtrictus,* receives its name from a root borne in its coat of arms, and lies eaſtward on the frontiers of *Moldavia,* containing

Marienburg, a pretty large town on the *Aluta.*

Kronſtadt, Corona, Braſſo, Braſſovia, the next in rank to *Hermannſtadt,* conſidered not only with reſpect to the number of its inhabitants, but alſo its appearance, is the principal place of this Diſtrict, and celebrated for its trade. It is fortified with walls, towers and moats; is feated among pleaſant mountains, and has a Jeſuits college and gymnaſium of *Lutherans.* In the town itſelf reſide none except *Germans,* but in its three large ſuburbs are *Bulgarians, Hungarians, Saxons* and *Siculi.* In the year 1421 the *Turkiſh* Emperor, *Amurath* the ſecond, laid this country waſte, and carried off the chief magiſtrate of *Kronſtadt.* In the years 1516, 1531, 1579 and 1588, violent earthquakes were felt here. In 1529 *Peter,* Woywode of *Moldavia,* beſieged the town and plundered the caſtle.

Zaiden, a free town, enjoying great privileges.

Roſenau, Roſnyo, a free town enjoying the ſame privileges, and leading to the narrow paſſes of the *Themis* and *Turzburg,* lying betwixt the mountains of *Walachia.*

HUNGARIAN ILLYRICUM,

O R

The Kingdoms of SCLAVONIA, CROATIA and DALMATIA.

§. 1. BY *Illyricum*, in its moſt confined ſenſe, was formerly meant a tract along the *Adriatick*; but the meaning has ſince been extended to the whole country betwixt the *Adriatick* and the *Danube*, even as far as the *Black-Sea*. To the former belong *Dalmatia* and *Liburnia*, with the ſmall province of *Japidia*. The more extenſive meaning of the word was firſt known in the fourth century, and the country comprehended under it, called *Great Illyricum*, being far the greateſt part of the *Roman* provinces in the eaſtern diviſion of *Europe*; and, according to its government, divided into Weſt and Eaſt *Illyricum*: the former including *Dalmatia* and the three *Pannonias*, with *Savia* and the *Noricum Mediterraneum* and *Ripenſe*: the latter conſiſted · of *Macedonia*, *Achaia*, *Theſſaly*, *Epirus* and *Crete*; *Dacia Ripenſis* and *Mediterranea*; *Moeſia prima*, *Dardania* and *Prævalitana*. In after times, numbers of the *Sclavi* having over-run the *Roman* provinces, *Great Illyricum* was extended to the whole country betwixt the *Adriatick* ſea and *Danube*, from the river *Save* to the *Scodras* and *Hemus*, in *Thrace*, and from thence to the *Euxine* ſea. Its provinces are *Pannonia*, *Savia*, *Dalmatia* and the two *Mœſias*.

§. 2. *Illyricum* ſtill contains the countries which belonged to it in the middle ages; and, according, to its ſeveral ſovereigns, is divided into the *Hungarian* and *Turkiſh Illyricum*. Of the latter we ſhall treat in *Turky*. The former reaches from the *Danube* to the *Adriatick*, betwixt the rivers *Drave*, *Save* and *Unna*; containing *Sclavonia*, *Croatia* and *Dalmatia*.

§. 3. The country betwixt the *Drave* and *Adriatick*, according to the temperature of the air and the nature of the ſoil, is very fruitful; but the maritime places are not accounted wholeſome. It produces grain, wine, oil, and other things neceſſary for cloathing and ſubſiſtence in plenty. Its chief rivers in the *Hungarian Illyricum* are, the *Drave*, *Dravus*, which, after watering the upper part of *Sclavonia*, falls into the *Danube* at *Peter Waradein*: The *Save*, *Savus*; which riſes in *Carniola* and, paſſing through the middle of *Sclavonia*, joins the *Danube*: The *Culpa*, *Colapis*, iſſuing from *Croatia* and falling into the *Save* near *Siſſek*: The *Unna*, which ſeparates *Croatia* and *Boſnia*, terminating the caſt part of the country and alſo joining the *Save*: The *Kerka*, *Naranna* and *Maracca*, which, together with the ſmall rivers of *Dalmatia*, diſcharge themſelves into the *Adriatick* ſea.

I

§. 4. The

§. 4. The inhabitants are of *Sclavonian* extraction, and, according to the different Provinces, divided into a different people; among which are *Hungarians, Germans, Venetians* and *Turks.* The principal nations are, the *Sclavonians,* who make as it were one people, with the *Servians* and *Rascians,* having a mixture of *Hungarians* and *Germans* among them, and inhabiting *Sclavonia :* The *Croats,* who dwelling in *Croatia* and *Sclavonia,* are augmented by colonies from *Germany* and *Walachia:* The *Dalmatians,* · of whom the *Uscocs, i.e.* deserters, a kind of Christian Refugees from *Bulgaria, Servia* and *Thrace* ; and the *Morlachians,* more properly *Mauro Walachians,* make a part. They have also *Venetian, Turkish* and *Albanian* colonies among them. The several nations of *Illyricum* use the *Sclavonic* language, which is divided into the *Dalmatian, Croatian* and *Rascian* dialects; but at present the *Croatians* and *Rascians* speak *German* and *Hungarian;* the *Dalmatians, Italian* and *Turkish* ; the *Walachians,* who have settled themselves in *Illyricum,* retaining their own speech. The *Croats* and *Rascians* dress after the *Hungarian* manner ; the *Dalmatians* like *Venetians* and *Turks :* but some still adhere to the ancient garb. The *Illyrians* in general apply themselves to trade, agriculture or war ; but to these arts the *Dalmatians* add navigation, in which they are very bold and expert.

§. 5. The only religion publickly tolerated here is the *Roman* Catholick, under the government of three Archbishops and twenty bishops. Most of them have *Hungarian* titles without revenues. Of the eastern *Greek* church in *Illyricum,* and some parts of *Hungary,* is one Archbishop and ten bishops. These nations concern themselves little about study ; *Zagrad,* however, has an accademical gymnasium ; and among the *Croats* and *Rascians* are · found many good geniuses who entertain a passion for literature, and endeavour to promote it.

§. 6. The government of *Sclavonia* and *Croatia* is connected with that of *Hungary* and *Stiria,* being hereditary in the Archducal house of *Austria. Croatian-Hungary* is under the jurisdiction of the Vice-roy, or Ban, of *Croatia, Sclavonia* and *Dalmatia,* who governs it by the laws of *Hungary* and the provincial acts of *Illyricum. Croatian-Stiria* has a governor in *Stiria,* in military *Croatia* and on the *Adriatick* coast.

The government of *Dalmatia* is threefold. That of *Venetian Dalmatia* is administred by certain proveditors in the name of the Republick of *Venice* ; *Turkish Dalmatia* is governed by a Basha deputed by the Grand *Signior,* and the government of *Ragusa-Dalmatia* is lodged in a rector and magistrate under the protection of *Hungary, Turky* and *Venice.*

§. 7. The prelates, nobility, gentry, and royal free castellans, enjoy the same privileges with the *Hungarians.* At the Diets, which consist of the four orders of the province, all deliberations run in the Sovereign's name. At the *Hungarian* Diets the States of *Illyria* appear by representatives.

. §. 8. The

§. 8. The adminiſtration of juſtice in *Sclavonia* and the Bannat of *Croatia* is the ſame as in *Hungary*; the free towns having inferior courts, from whence cauſes may be removed to the royal treaſurer. The like have alſo the other towns from whence there lies an appeal to the Bannat court, which is ſo called from the ban, or prorex who is preſident thereof: and this at certain times hears cauſes brought to them from the Geſpanſchafts, holding conſultations on other impoitant matters; and ſometimes, when the cauſes require a further inſpection, diſmiſſes the litigants to the *Tabula Regalis* at *Peſt*; from whence they may proceed farther to that of the ſeven men.

§. 9. The publick revenue ariſes from the contributions, cuſtoms, trade, tillage and graziery; and are divided betwixt the King of *Hungary*, the *Venetians*, the *Turks* and *Raguſians*, according to the extent of their ſeveral dominions. The military force of the *Hungarian Illyrians* never appeared to greater advantage than in the laſt war, when from *Croatia* only, no leſs, than fifty thouſand men were brought into the field, and twenty thouſand from *Sclavonia*. We ſhall now treat particularly of each Kingdom.

The Kingdom of SCLAVONIA.

Slavonia, Sot-Orſzag, Slowenſka Zeme, in *Latin Slavonia,* lies betwixt the *Drave* and the *Save,* terminating eaſtward on the *Danube,* and weſtward on *Carniola,* being about fifty *Hungarian* miles in length and twelve broad, and formerly making part of the *Savian Pannonia,* or *Pannonia* betwixt the rivers. In the middle ages it was called *Sclavonia* from the neighbouring *Slavini,* or *Slavi*; and was at firſt divided into Upper and Lower *Sclavonia*; and afterwards into the Bannat and Generalſhip of *Sclavonia*. *Upper Sclavonia* makes a part of *Croatia*: *Lower Sclavonia* extends from the *Danube* to *Croatia,* through the eaſtern parts of the country, and, in the year 1746 was re-annexed to the Kingdom of *Hungary* by the preſent Queen *Mary Tereſia*. The inhabitants are *Servians* or *Raſcians, Croets* and *Walachians,* with ſome *German* and *Hungarian* colonies.

1. *Bannat Slavonia, Bannalis Slavonia,* in 1746, was diſtinguiſhed into three ſeparate counties; and the county of *Walpo,* or *Walko* divided among the reſt.

1. The County of *Verowitz, Verotzei Varmegye, Verotzenſis, Verovitzenſis Comitatus*; to which was added the greateſt part of the county of *Walpo,* cohtains

Verotze, Verowitza, Verowititza, Verucia, on the *Danube*; formerly a very ſtrong and conſiderable place, giving name to the County, and the reſidence alſo of the Bannfier, but now an open place.

Vukin, Butſin, the capital of a lordſhip of the ſame name.

Orowitza, Drahowitza, the capital alſo of a lordſhip.

Walpo, a ſtrong fortreſs near the *Drave,* giving name to the County ; and, in 1547, heroically defended againſt the *Turks* by the lady of *Peter Perenius,* governor of *Sclavonia.*

Naſzitz, a lordſhip, appertaining to the *Piacſevi* family.

Eſſek, capital of the County, and ſeated near the *Danube* on the river *Drave.* The moſt remarkable thing here is the large wooden bridge erected over the *Danube,* and moraſſes, to the length of above an *Hungarian* mile. It was built in 1566 by *Soliman,* Emperor of the *Turks,* twenty thouſand ´men working inceſſantly at it. In the year 1529 this place was taken by the *Turks.* In 1537 the *Imperialiſts* were obliged to draw off from before it; but, in the year 1600, they met with better ſucceſs. In 1664 Count *Serini* burnt the bridge down to the ground, but it was ſoon rebuilt again by the *Turks.* In 1685 the *Hungarians* again ſet fire to part of it; and, in the fol-lowing year, they entirely deſtroyed it by the ſame means. In the year 1687 they attempted the town in vain; but, after the battle of *Mohatz,* the *Turks* voluntarily evacuated it.

Erdod, a caſtle and town which, together with the lordſhips of *Szarvas,* and *Kolovar,* lies on the *Drave;* and is remarkable for being a part of Count *Palfi's* title and coat of arms.

Dalya, Dail, the capital of a lordſhip, and belonging to the Archbiſhop of *Carlowitz.*

Noſtar, Nuſtar, a ſmall place giving name to a lordſhip.

2. The County of *Sirmi, Sirmia Varmegye, Sirmienſis Comitatus,* was formerly a dutchy. In it are

Valko-var, or *Buko-var,* on the *Drave;* formerly an important fortreſs in the juriſdiction of Count *Elſz,* but now a defenceleſs place.

Cſernigrad, or *Tarka-vara,* on the *Drave,* ſtill retaining ſome traces of a fortreſs.

Illok, Ujlack, capital of the County, with a caſtle on an eminence, fa-mous for being the burial-place of St. *John Capiſtranz,* a zealous diſciple of St. *Francis.* It belonged formerly to the duke of that name, but at preſent to Prince *Odeſchalchi.* In the year 1494 it was taken by King *Uladiſlaus,* and in 1526 by *Soliman* Emperor of the *Turks.*

Bakmonoſtra, Banoſtra, formerly noted for being the reſidence of the biſhop of *Sirmi,* but having now little or no appearance of its former grandeur.

Kamenitz, Kamanitz, Kamanich, a pretty good town, formerly in a bet-ter condition than at preſent.

Irik, a town at ſome diſtance from the *Drave.*

Szalankemen, Zalankemen, a walled town, having a caſtle, oppoſite to which the *Theis* falls into the *Danube.* In the year 1691 the Chriſtians, with great loſs, obtained a ſignal victory here over the *Turks.* In 1716 another action happened here between the ſame parties.

Semlin,

Semlin, Zemlin, a fortress at the conflux of the *Save* and *Danube;* the property of Count *Schonborn.*

3. The County of *Poſſeg, Poſſegai Varmegye, Poſſeganus, S. Poſſegienſis Comitatus;* in which are

Dioko-var, a fortification and lordſhip, belonging to the biſhop of *Boſnia.*

Kutivea, a rich abbey, belonging to the Jeſuits.

Platernitz, the capital of a lordſhip.

Poſſeg, a town of ſome trade and figure, ſeated in a fertile country, and the county town.

Cſerneck, Preſztowetz (Preſztorock) Welika, Pakratz, Siratſch, Podhorje and *Kuttina,* the chief places of lordſhips of the ſame names. Of theſe the ſecond, third and fourth belong to the *Trenek* family; the fifth and ſixth to the treaſury, and the ſeventh to Count *Erdod.*

2. The Generalſhip of *Sclavonia, Slavonia Militaris,* has for its governor the commander in chief of the *Illyrian* troops.

1. The Upper Prefecturate of the limits of the *Danube* and *Save, Præfectura Confiniorum Danubii & Savi Superior;* contains

Eſſek, a ſtrong fortreſs, and, on account of its garriſon and the *Sclavonian* regiment of horſe cantoned in its Diſtrict, reckoned among the military towns.

Peterwardein, Peter-vara, ſtrongly fortified, and ſtanding on the *Danube,* oppoſite *Belgrade.* In the year 1526 the *Turks* made themſelves maſters of it, but quitted it in 1687, when the *Hungarians* took poſſeſſion of the place. It is chiefly remarkable for the glorious victory obtained near it over the *Turks* in the year 1716 by Prince *Eugene.* This Diſtrict is the ſettled quarters of the *Peterwardein* regiment of foot.

Carlowitz, a military town, but the reſidence of the *Greek* biſhop of *Sclavonia;* famous for the peace concluded there with the *Turks* in 1699, and alſo for its red wine.

Mitrowitz, Demitrowitz, a military town on the *Save:* near it, for the defence of the frontiers, lies the *Sermi* regiment of horſe. Here ſtood the ancient *Sirmium,* once the celebrated capital of *Illyricum.*

2. The Lower Prefecturate of the confines of the *Save, Prefectura Confiniorum Savi Inferior.* Its remarkable places are

Ratſcha, Ratzka, a fort at the conflux of the *Save* and *Drina,* taken in the year 1716 by the *Imperialiſts.*

Brod, a ſmall fort on the *Save* for the protection of ſhipping. Its juriſdiction the quarters of the *Bradi* regiment of foot.

Gradiſk, a ſtrong town on the *Save,* making a good appearance, and divided into *Sclavonian* and *Croatian.* For the defence of the frontiers the *Gradiſki* regiment of foot is conſtantly quartered here.

　　　　　　　　　　　　The

The Kingdom of CROATIA.

Croatia, by the *Hungarians* called *Horwath Orfzag*, reaches from the *Drave* to the *Adriatick*, terminating eaftward on *Sclavonia* and *Bothnia*, weftward on *Stiria* and *Carniola*, and making a part of the ancient *Illyricum*. The *Croatians* derive their origin from the *Slavi*, and came into this country in the time of the Emperor *Heraclius*. Their ancient name was *Hruatæ*, or *Hrouatæ*, of which the *Greeks* made *Chrobatæ*. In the middle ages they had Kings of their own, who, for fome time, were fubject to the Emperors of the caft, and ftiled themfelves Kings of *Croatia* and *Dalmatia*. In the eleventh century *Croatia* and *Dalmatia* devolved to the King of *Hungary*, and the *Croats* have continued ever fince under the dominion of that Monarchy, though not without frequent attempts to recover their independency. The *Croats*, of all the *Illyrian* nations, have the greateft affinity in their language to the *Poles*. *Croatia*, with refpect to its fituation, is divided into two parts; namely, into that under, and that beyond, the *Save*. With regard to dominion, into *Hungarian* and *Turkifh*; and with refpect to its government, into the Bannat and Military.

1. *Croatia* on this fide the *Save*, *Croatia Cif-favana*, which is alfo called *Upper Sclavonia*, is inhabited by *Croats*, a few *Rafcians*, *Greeks* and *Walachians*; and contains

1. The County of *Warafdin*, *Varafdiai Varmegye*, *Varafdinenfis Comitatus*; of which the Counts *Ardodi* of *Monyorokerek* are perpetual chief Counts. In it are

Vinitza, *Vinea*, an old caftle with a town, in which is a feat, not far from the *Drave*, on the frontiérs of *Stiria*, belonging to the *Kegelwich* family.

Warafdin, or *Little Warafdin*, a royal free town on the river *Drave*, ftrengthened with a caftle and bulwark, ftands in a large plain, and was built by King *Andrew* the fecond, and his fon *Bela* the fourth, who erected it into a town. Betwixt it and a high mountain is a warm bath formerly called *Aquæ Jafæ*, but afterwards *Thermæ Conftantinianæ*.

Luidbring, *Lubrig*, a town on a hill where they pretend to fhew fome of our Saviour's blood.

Klanetz, a caftle feated on a mountain, with a fmall town, the burial-place of the Counts of *Erdod*.

Koprauitz, *Kopranitza*, a fortified town with a moat.

2. The County of *Krentz*, or *Crofs*, *Crifiai Varmegye*, *Crifienfis Comitatus*, is united with the following under one count. In it are

Kreutz, i. e. *Crofs*, *Koros-Vafarhely*, in *Latin*, *Crifium*, a royal free town, which is fortified, and has many privileges; being divided on account of its fituation, into the Upper and Lower.

Velo-var, ftill retaining fome veftiges of a caftle.

Welika,

Welika, or *Kralowa Welika*, *Regia Magna*, a small town on the frontiers of *Sclavonia*, famous under the government of *Mauritius* in *Pannonia*, and at that time stiled *Clara* and *Magnana*.

3. The County of *Zagrab*, *Zagrabiai Varmegye*, *Zagrabienfis Comitatus*; in which are

Iwanitz, *Ibanitz*, a strong fortress on the river *Korgs*, formerly the refidence of the Kings of *Sclavonia*.

Verbowetz, *Berbowetz*, or *Goritza*, a pretty good town, taking its name from the mountains.

Chafma, *Tfafma*, a small town where King *Colomann* lies buried.

Zagrab, *Agran*, a royal free city on the *Save*, having a bifhop and chapter, and is the capital of the whole Kingdom of *Croatia*.

Nagy, *Kis Tabor* and *Harvotfka*, castles belonging to the *Ratka* family.

4. The County of *Zagor*, *zagoriai Varmegye*, *Zagorienfis Comitatus*, extends to the *Adriatick*, containing

Krapina, a small town and castle on a river of the same name at the confines of *Stiria*, in which the records of the *Keglewich* family, are kept.

Lupoglava, or *Lepoglava*, a small town with a convent of *Paul*, the Hermite, and the burial-place of the regents *(proceres)* of *Sclavonia* and *Croatia*.

Zagoria, a tract of land environed with mountains.

Turopole, an open country, extending from the *Save* to the *Adriatick*: Among its inhabitants are feveral gentry who have particular privileges.

5. To the jurifdiction of the Generalfhip of *Warafdin*, *Prefectura Militaris Varafdienfis*, belong

The Fort and castle of *Warafdin*, the capital place of the Generalfhip.

The fort of the town of *Kreutz*.

Iwanitz, a fortress in the County of *Zagrab*.

St. George, *Fanum S. Georgij*, a fortified town on the *Save*.

Petrina, a fortress in Bannat *Croatia* mentioned below.

2. *Croatia* beyond the *Save*, *Croatia Trans-Savana*, or *Proper Croatia*, is divided into *Hungarian* and *Turkifh*.

a. Hungarian *Croatia*, on this fide the *Unna*, confifts, 1. of *Military Croatia*, *Militaris Croatia*; in which are

Karlftadt, *Carlowitz*, a confiderable fortrefs betwixt the rivers *Kulpa* and *Corona*, and the capital of this Generalfhip.

Krifanitz, or *Tunn*, a fortified town.

Barillowitz, a town having a garrifon.

Sichelburg, a castle feated on a lofty eminence near *Carniola*, the capital of a large lordfhip.

Sluin, *Sluni*, a fortified town, the nearest to *Dalmatia*, and giving name to the ancient Counts of *Sluni*.

Oguli, capital of a large and pleasant District.

2. *Banat Croatia, Banalis Croatia*; in which are

Petrina, a fortified town between the river *Kulp,* or *Kalapi* and *Petrina,* belonging to the jurisdiction of *Waradin,* but built by the *Turks* in the year 1592. In 1593 it held out a siege against the *Hungarians,* but in the following year was taken and demolished; but rebuilt again by the *Turks,* and, in the year 1599 again taken by the *Hungarians.* The year following the *Turks* besieged it twice in vain.

Siffek, a town at the influx of the *Kulp* into the *Save* fortified with a wall and moat. In the years 1591, 1592 and 1593, it withstood the *Turks;* but in the last year, fitting down a second time before it, they became masters of the place and burnt it the year following; but it was rebuilt by the *Hungarians.* On this spot the ancient city of *Silesia* is said to have stood.

Chrastowitz, Hrastowitza, a fortress not far from the *Save,* under the Jurisdiction of the bishop of *Szagrab.*

Dubitz, a frontier fortification, on the river *Unna* garrisoned by *Croats.*

Busin, a castle but one *Hungarian* mile from the *Unna,* giving title to Count *Keglevich.*

Koftanitz, or *Caftanowitz,* a strong castle environed by the *Unna.* In the year 1557 it fell into the hands of the *Turks.* In 1594 the *Hungarians* wrested it from them, but the *Turks* recovered it the very next year.

Zrin, a frontier fortification on the river *Unna;* so called from the Counts *Zrini,* if, as some suppose, they did not receive their name from the town. In the year 1540 it was besieged in vain by the *Turks,* but was taken in 1576. The *Hungarians,* in the year 1579, recovered it, but soon after lost it a second time.

Great and *Little Kladuffa* are two towns; the former lying on an eminence, the latter among morasses.

Grozdanfko, a castle on the river *Unna,* once made a grand figure, being the residence of the Counts of *Zrini,* and famous also for its silver mines.

Novi, a fortified town on the *Unna,* divided into *Caftel Novi (Neu-Caftel)* and *Tudor Novi, (New Tudorow.)*

Krupa, Kruppa, a castle on the other side of the *Unna,* in the year 1565 besieged by the *Turks,* and taken after a very stout resistance.

b. *Turkish Croatia* lies on the other side of the *Unna;* and to it belongs *Nowigrad,* a pretty good town.

Blagai, Blagifkithurn, a castle now lying in ruins, but formerly belonging to Count *Urfini.*

Vihits, Bihats, a fortified town surrounded by the *Unna.* In the year 1592 the *Turks* took it, but lost it again in 1594.

The Kingdom of DALMATIA.

Dalmatia Hung. Dalmatiai Orſzag.

Dalmatia, or, as it is written in old coins and inſcriptions, Delmatia, takes its name from its ancient capital Delmium, or Delminium, which the *Romans* took and deſtroyed in the 597th year from the building of the city. If the *Romans* brought it under the yoke, *Dalmatia* ſhook it off no leſs than five times, and for the ſpace of two hundred and twenty years, to *Auguſtus*'s reign, gave them a great deal of trouble. On the diviſion of the provinces between *Auguſtus* and the Senate, *Dalmatia* fell to the Senate as one of the proconſular provinces, but they voluntarily ceded it to the Emperor, who appointed a quæſtor over it. At the demiſe of *Conſtantine* the Great it was reckoned among the weſtern parts of *Illyricum*. It ſuffered extremely by the inroads of the northern Barbarians, and the *Goths* reduced it in their way to *Italy*. After this, *Juſtinian,* Emperor of the Eaſt, conquered *Italy* and alſo *Dalmatia*; but, in the year 1548, the *Slavi* entered it, and, about the end of *Heraclius*'s reign, eſtabliſhed themſelves in it. The country had then its particular Kings, of which *Zlodomir*, or *Zaromyr*, the laſt, dying without iſſue, left the Kingdom to his conſort, who bequeathed it to her brother St. *Ladiſlaus*, King of *Hungary*; ever ſince which it has been dependent upon that crown: but the *Venetians* are maſters of the maritime parts. In the wars which the Kings of *Hungary* had both with the *Venetians* and the turbulent *Dalmatians*, they were for a conſiderable time ſucceſsful; but, in the fifteenth century, the *Venetians* reduced the whole Kingdom of *Dalmatia*, though they have ſince been diſpoſſeſſed by the *Turks* of a conſiderable part. At preſent the *Hungarians*, *Venetians*, *Turks* and *Raguſans* ſhare it amongſt them. The *Dalmatians* uſe the *Sclavonian* language and cuſtoms, and profeſs the *Roman* Catholick religion.

The rivers in *Dalmatia* have indeed no long courſe, but are moſtly navigable. The country is as it were ſtrewed with mountains, but theſe not unfruitful; olives, vines, myrtles, and a great variety of palatable and wholeſome vegetables growing upon them, beſides treaſures of gold and ſilver ore within them. It has alſo many fertile plains; and, beſides a ſufficiency of horned cattle, feeds large numbers of ſheep. The air is temperate and pure.

1. *Hungarian Dalmatia* lies in the upper part of the *Adriatic* ſea, and contains part of the ancient *Liburnia*. Before we deſcribe it, we muſt previouſly ſpeak of the *Uſcocks* and *Morlachins*. The *Uſcocs*, galled by *Turkiſh* oppreſſion,

oppreſſion, made their eſcape out of *Dalmatia*, and hence obtained the name of *Uſcocks* from the word *Scoco*, a runaway or deſerter. They are alſo called ſpringers, or leapers, from the agility with which they leap rather than walk along this rugged and mountainous country. *Chiſſa* was their chief ſettlement; but the Turks, in the year 1537, taking that place, they retreated to *Zengh*, which was granted them by the Emperor *Ferdinand*; on account, however, of their robberies and other violences, they were ordered, in the year 1616, to remove and ſettle at the place appointed for them in a mountain of *Carniola*, four *German* miles in length and two broad, betwixt the rivers *Kulp* and *Brigana*. In the centre of this mountain ſtands *Sichelberg* caſtle; to the governor of which all the *Uſcocks* are ſubjects. Some of them live in ſcattered houſes, others in *Wenitz* and large villages. They are a rough ſavage people, large bodied, intrepid and given to rapine, though their only viſible employment be graziery. Their language is the *Walachian.* In their religion they come the neareſt to the *Greek* church, but ſome are *Roman* Catholicks. They have an Archbiſhop, biſhops, popes, or prieſts, and coluges or monks. The prieſts are not prohibited marriage, but their wife muſt be of a good family, and at her deceaſe they are not to marry again. Their children are not baptized till adults; and none among them go to confeſſion under thirty years of age. *Morlachia* is a part of *Liburnia* reaching from the juriſdiction of *Zengh* near *St. George* to the county of *Zara*; or, according to others, from *Vinodok* to *Novigrad*, being in length fifteen *German* miles, and in breadth five or ſix. The whole country is full of high mountains. The inhabitants are a branch of the *Walachians*, and deemed *Mauro-ulaha*, i. e. *Black Latins,* or a black ſkinned people; of which the *Italians* have made *Morlachian*. They are a large, ſtrong, robuſt people, and living amidſt barren mountains, inured to toil and hardſhip. Their chief employment is the breeding of cattle, and the greateſt part of them profeſs the *Greek* church. Some of them, at preſent, are under the protection of *Hungary*, others are deſcendants of the *Venetians*; and there is hardly a place of any ſtrength in *Dalmatia* which is not governed by *Morlachians*. The two caſtles of *Seliſſa*, or *Lopur*, in *Morlachia*, by the ancients called *Lopſica* and *Ortopola* (*Ortpla, Ortopula*) which is alſo known by the name of *Starigard*, are of very conſiderable antiquity.

Obroazzo, on the river *Zermagna*, has a ſtrong caſtle ſtanding on a high hill which, ſince the peace of *Carlowitz*, belongs to the houſe of *Auſtria*.

Hungarian Dalmatia conſiſts of five Diſtricts, moſt of which are under the Generalſhip of *Carlſtadt*.

1. The Diſtrict of *Zengh* lies on the confines of *Iſtria*, and contains

Bukari, Bokari, Bukariza, a ſmall, though pretty good town, with a harbour, from which the *Golfo di Bukaria*, contiguous to it, takes its name. From this place great numbers of cattle are ſhipped for *Italy*. It belonged

formerly

formerly, together with fome other places in thefe parts, to the Counts of *Zrini*; but *Peter Zrini* enering into a treafonable confpiracy, for which, in the year 1671, he loft his head, this town, with the other places, became forfeited to the crown.

Prundel, or *Brinye*, a frontier fortrefs, on a rocky eminence, in the middle of the country; belonging formerly to the Counts of *Frangipani*.

Zengh, Segnia, Senia, a royal free town, fortified both by art and nature, lies near the fea, in a bleak, mountainous, barren foil. The bifhop of this place is a fuffragan of the Archbifhop of *Spalatro*. Here are twelve churches and two convents. The governor refides in the old palace, called the royal caftle; and in the upper fort, *Forteza Nehai*, which ftands on a rifing ground fronting the town, lives the deputy-governor. This place belonged formerly to Count *Frangipani*, and near it dwell the *Ufcocks*.

2. The Diftrict of *Ottofchatz*; containing

Modrus, Modrufa, Merufium, an ancient, well built, epifcopal town, with a caftle on the *Lecko*, at the foot of the vaft mountain of *Capella*. It was formerly the capital of a County of the fame name.

Ottofchatz, a frontier fortification on, or rather in, the river *Gatzka*, which abounds with fifh. That part of the fortrefs where the governor and the greateft part of the garrifon dwell, is environed with a wall and fome towers; the reft of the buildings, which are but mean, ftand on piles in the water, infomuch that one neighbour cannot vifit another but in a boat.

Prozor, a fortified caftle, on a pretty high hill, near *Ottofchatz*, having a good well.

Fortetz, or *Fortctza* of *Ottofchatz*, is fituated upon a hill on the other fide of the *Gattzka*, but belongs to the jurifdiction of *Zengh*.

3. The County of *Lyka*, or *Licca*, lies between that of *Corbau* and *Mor-lachia*, being poffeffed principally by the *Turks*. The inhabitants are called *Lykani*. The caftle of *Oftrowitz*, *Oftrowitza*, is the principal, but has fuffered much in the *Turkifh* wars. In this County were formerly the towns of *Nova* and *Kibnik*.

4. The County of *Corbau*, *Corbavienfis Comitatus*, lies on the river *Unna*, and once belonged to the *Guffits* family, now refiding in *Carniola*. The *Hungarians* poffefs the weftern part, but the *Turks* the eaftern, and the latter keep a ftrong garrifon in the caftle of *Udbinya*. This place was in all probability the ancient epifcopal fee of *Corbau*.

5. The *Zwonigrod*, or *Serman* Diftrict, contains the town of *Zwonigrod*, capital of this County, though of no great figure, and *Carlopage, Campus Carolinus*, a military ftation.

II. *Venetian Dalmatia* abounds in caftles and fortified places, but few of modern ftructure. It comprehends the other part of *Liburnia*, called *Ba-nadego*, and confifts of

1. The Continent ; on which lies

Nona, in *Latin Ænona*, a very ancient town and strong fortification, hav_
ing a harbour, and being the see of a bishop suffragan to the Archbishop of
Spalatro. It is almost wholly surrounded by the sea; and, on the *Zara* side, is
a large morass and lake called *Lago di Nona*. This place was formerly the
capital of *Zupania*.

2. The County of *Zara*, the governor of which formerly stiled himself
a duke. In the year 1200 it was reduced by the *Venetians*, who losing it
again, in 1409 purchased it of *Ladislaus*, King of *Naples*; since which it
has continued in their hands. It contains

Zara, Jadera, the principal town in this County, and indeed the ca_
pital of *Venetian Dalmatia*; a town surrounded on all sides by the sea, saving
that it has a communication eastward with the Continent by means of a
draw-bridge commanded by a fort. It is reckoned one of the best forti_
fications in *Dalmatia*, and deemed almost impregnable. The citadel is di_
vided from the town by a very deep ditch hewn out of a rock. The har_
bour, which lies to the north, is capacious, safe and well guarded. The
rain is carefully preserved in cisterns to supply the want of fresh water. In the
castle resides the governor or proveditor of *Dalmatia*, whose office is only
triennial. It was formerly only a bishop's see, but, in the year 1145, an
Archbishop was established here. St. *Simeon* is patron of the city, and his
body lies in the cathedral, which is dedicated to him, in a coffin covered with
a crystal lid. Some *Roman* antiquities are found here.

Bibigne S. Caskiano and *Torrette* are small places.

Zara Vecchia, or *Old Zara*, now a village, but in the time of the *Ro_
mans* a place of considerable figure, received a new set of inhabitants by
a numerous colony of that people. In the middle ages it was called *Bel_
grad*, or *Alba Maris*. Here was also a bishop's see, which, on the demo_
lition of the town, was removed to *Scardona*; at present desolate, and its
inhabitants consisting only of a few peasants.

Aurana, Laurana, Brana, one of the most delightful places in *Dalmatia*;
on a lake of the same name; had formerly a rich convent of *Benedictines*,
the incomes of which, about the year 1217, were given away to the Knights
Templars by *Andrew* the second, King of *Hungary*, who instituted a com_
manderie here. About this time also the place was fortified. The suburbs
are large. It continued, for some time, in the hands of the *Turks*, but in
1684 they were dispossessed of it.

Zemonico, Zemunicho, Xemonico, a fortress razed in 1647.

Nadin, a strong castle in the centre of the County on the summit of a
mountain. The *Turks* were never able to make themselves masters of this
place till the year 1539. In 1682 they lost it; and, in the year 1684, it
fell into the hands of the *Venetians*.

Novigrad, a fmall town and caftle with a·lordfhip of the fame name.
The bay on which it ftands receives its name from it, and runs feveral
miles up into the country. In the year 1646 the *Venetians* were difpoffeffed
of this place, but recovered it again and razed it the following year.

Tenen, *Tininium*, or *Knin*, formerly called *Clim*, a fortified town on the
extremities of *Bofnia* and *Dalmatia*, feated on a hill in a good country, is
not large but of fome importance by reafon of its elevated fituation. It is
environed on both fides with two very broad and deep natural moats, formed
by the rivers *Kerka* and *Botifniza*, which precipitate themfelves not far from
thence from the hills; and is a bifhop's fee. In the thirteenth and fourteenth centuries it was the capital of a County. In the year 1522 the *Turks*
took it. In 1649 it was recovered by the *Venetians*, and for the moft part
demolifhed, but rebuilt again by the *Turks* in the year 1652, who loft it,
however, again to the *Venetians* in 1688.

Dernis, near the river *Ticola*, was formerly a confiderable fortrefs, but is
now an infignificant place. In the year 1638 the *Turks* quitted it and the
Venetians fet it on fire: the former, however, rebuilt it, but were obliged
to evacuate it in 1684.

Sebenico, one of the ftrongeft towns on the gulph of *Venice*, having four
citadels and a very large harbour, fince the year 1298 has been a bifhop's
fee. The cathedral of St. *John*, in the caftle, is a fine marble ftructure.
From the year 1412 it has been perpetually under the dominion of the *Venetians*, though no lefs than four times warmly befieged by the *Turks*.

Trau, *Tragurium*, a town on a peninfula, but feparated by a canal cut
through it from the continent. It is divided into the New and Old; the
latter of which has a double, the former only a fingle wall. Its three
towers are alfo a good defence to it. Nothing can be pleafanter than the
fituation of this place: the north fide is covered with beautiful gardens, and
on the ifland of *Bua* it has fine fuburbs, having a communication with it by
means of a ftone bridge, and being joined to the continent by three of wood.
It is the refidence of the *Venetian* proveditor, who bears the title of Count;
and alfo of a bifhop, fuffragan to the Archbifhop of *Spalatro*. The harbour, which is formed by a bay, has depth of water enough for the largeft
fhips, which ride there fheltered by two capes. It abounds alfo in fifh,
particularly fine fardines. It is an ancient *Roman* colony of the Emperor
Claudius; and fo early as the year 997 put itfelf under the protection of
Venice; on which it has been continually dependent ever fince the year 1420,
though not without many changes and difaftrous revolutions.

Cliffa, a ftrong town on an eminence, near which, betwixt two
lofty fteep rocks, is a narrow vale, through which lies the road from
Turkey into *Dalmatia*, and particularly to *Spalatro*, to which this place
ferves as a barrier. It has no water but what falls from the clouds, and one
little fpring fituate in a village at the foot of a fort. In the year 1646 the

Venetians

Venetians got poſſeſſion of it. The neighbouring country is covered with fine vine and olive yards.

Salona, Salonae, the reſidence of the firſt Kings of *Illyricum.* *Auguſtus* laid this place waſte ; but *Tiberius* rebuilt it, and ſending a *Roman* colony thither named it *Martia Julia,* making it the capital of *Illyricum,* as it was the place where the queſtor kept his court, and the rendezvous alſo of the *Roman* fleet. *Dioclefian* frequently made it the place of his retirement. The receiver, who collects the revenues ariſing from the mines, now reſides here. It was once famous for its purple, and likewiſe for making of helmets, coats of mail, &c. It had alſo a biſhop ; was ſituate in a moſt delightful plain, and in circumference containing ſome *Italian* miles, extended to a ſmall bay which forms its harbour. Through it ran the river *Salona,* ſo noted for trouts. At preſent it is a very ſmall town of little or no figure.

Spalatro, Spalatrum, a ſea-port on a peninſula, fortified with good baſtions of free-ſtone, but commanded by the neighbouring mountains, is the ſee of an Archbiſhop who is primate of *Dalmatia* and *Croatia.* It is alſo the ſtaple where all merchandize paſſing from *Turkey* to *Italy* muſt be tranſported. Near its harbour, which is very large and deep, is erected a large Lazaretto for performing quarantine. Among the *Roman* antiquities to be ſeen here, *Dioclefian's* palace is particularly worth viewing. This city, ever ſince the year 1420, has continued firm to the *Venetians.*

Sing, a caſtle of importance, on a mountain, built by the *Turks* in oppoſition to that of *Cliſſa :* the fortifications, indeed, are not much in the modern taſte, but on three ſides of it the rock is almoſt inacceſſible, and the fourth ſide is fortified with a thick wall and two ſtrong entrenchments, to which there is coming up by ladders. Here are two large ciſterns which are always kept filled with fine freſh water. The *Venetians* taking this place in the year 1686, in order to ſecure it to themſelves, greatly ſtrengthened its fortifications.

Almiſſa, Alminium; the ancient *Peguntium;* is an epiſcopal city, in the dutchy of *Chulm,* ſeated on a rock betwixt two high mountains at the mouth of the *Tettina,* and with its cannon commands any ſhips that are coming in. It was formerly ſo notorious a place for piracy, that the neighbouring towns of *Spalatro, Trau* and *Sobenico* joined with *Venice* to deſtroy it. Since this the town has not been able thoroughly to recover itſelf ; to which its long continuance in the hands of the *Turks* was another obſtacle.

Macarſca, or *Primorge, Primoria,* a province on the *Adriatic* about fifty-four *Italian* miles in length, is divided into Upper and Lower, and contains inacceſſible mountains. Its moſt remarkable places are *Kagoſnizza, Cucich; Breghli Dolgni, Brehgli Gorgnie, Baſt Vellccobardo, Macao, Macarſca,* (which, beſides being the reſidence of the governor, has a harbour) *Cottiſina, Tuceni, Podgara* and *Dracnizza;* the caſtle of *Igrani, Swagnſchia, Drivenich,*

Drivenich, a fortrefs, on the fummit of a mountain, *Zaztrogh*, *Prifl*, *Cot Lanzan*, or *Gradaz* and *Pozza*.

Duare, a town on the defile of *Rodobiglia*, on the eaft fhore of the *Tettina*, in the neighbourhood of *Almiffa*, on a very high mountain and fortified with old towers.

Narenta, or *Narenza*, anciently *Naro*, *Narona*, *Narbona*, was in former times a handfome town, the capital of *Dalmatia*, and one of its principal fortreffes. It was governed by a *Roman* proconful and council, and under its jurifdiction were comprehended many diftant places. In fucceeding times the *Sclavonians* feated themfelves here, and, under the title of *Narentani*, molefted the navigation of thefe parts till the year 987, when the *Venetians* made themfelves mafters of the town. They had their own chiefs for a long time, but, in the year 1479, became fubject to the *Turkifh* yoke. Of the ancient town not the leaft trace is faid to remain; but in this neighbourhood is another place, called *Narenza* from the river of the fame name, which the *Venetians* fortified but difmantled again in the year 1716.

Norin, a ftrong tower on the river *Narenta*.

Metrovich, a territory inhabited by numbers of *Greeks*, and having here and there a few houfes and towers.

Opus, a triangular ifland formed by the two branches of the *Narenta*: the air fomewhat unwholefome on account of the moraffes, but having an excellent fifhery. The fort which the *Venetians* have here commands the river on both fides, and opens a pafs into the dutchy of *Herzegowina*.

Citluch, or *Ciclut*, a ftrong fortification furrounded with walls in the ancient tafle, ftands on a rocky eminence on the right fide of the river *Narenta*. This place was built by the *Turks* in the year 1559, and at firft called *Sedaiftan*, that is, *an entrance for the Turks*; but afterwards ftiled by the neighbouring country *Citluch*; that is, *a walled place*. It confifts of the old and new town and the fuburbs, which are large. In the year 1694 it was maftered by the *Venetians*.

Gabella, that is, *the cuftom-houfe*, the falt duty having been formerly paid here, lies over-againft *Citluch*, to the fide of the river *Narenta*, and is a fmall fortrefs.

Herzegowina, i. e. *the Dutchy of St.* Saba, that faint having been buried here, is a territory formerly twelve days journey in length and four broad. Both the *Venetians* and *Turks* poffefs it at prefent; the former of which are mafters of the following remarkable places.

Caftel Nuovo, *Caftrum Novum*, capital of the Dutchy, and the beft fortification in all *Dalmatia*. It was anciently called *Neocaftro*, and was built by a King of *Bofnia*, in the year 1373, on a high rock adjoining to the fea. It is better fortified by nature than art, being an irregular quadrangle, fecured towards the fea by fhelves and inacceffible rocks. In the upper town is the caftle of *Sulimanega* and the fortified town of *Haftavich*; but its beft

3   fortification

fortification is the citadel of *Cornigrad*, which lies about fix hundred and fifty paces north of the town. In the year 1687 it was taken by the *Venetians*, and is now a place of confiderable trade.

Rhifano, Rhizana, Rhizinium, an epifcopal city having a very ftrong caftle, the rock on which it ftands being almoft inacceffible. It was taken notwithftanding by the *Venetians* in the years 1538, 1647 and 1684.

Peraflo, a good caftle lying near the fea.

Cattaro, Cataro, Catarum, according to fome, is the ancient *Afcrivium.* Befides its ftrong walls it has a caftle on an eminence; the many fhady mountains which on all fides furround it being alfo a good defence, but rendering it very dark. It is a bifhop's fee, and has been fubject to *Venice* ever fince the year 1418.

Budoa, Butua, a fmall city regularly fortified, having a bifhop, and a caftle near it called St. *Stephen.*

Paftrovichio is the capital of a province of the fame name.

2. Of iflands. The principal of thefe are

1. *Ofero,* on which ftands a city of the fame name. The bifhop of this place is fuffragan to the Archbifhop of *Zara.* The *Latin* writers term it *Abforus* and *Civitas Aufarienfis. Pliny* gives it the name of *Abfyrtium,* and *Ptolomy* that of *Apforrus.*

2. *Cherfo, Crepfa, Crexa,* which is called a County, has a communication with the former by means of a bridge. It is a woody tract, and breeds great numbers of cattle. The town of the fame name is pretty populous and enjoys a good air with a convenient harbour.

3. *Vegia, Veglia,* feparated from the continent by a narrow channel, was purchafed, in the year 1478, of Count *John Frangipani.* The town of *Veglia, Vegia, Curictum,* or *Curicum,* lies in the weftern part of the ifland; has a harbour and caftle, in which refides the *Venetian* count, or governor, and a bifhop, fuffragan to the Archbifhop of *Zara.*

4. *Arbe,* by *Ptolomy* called *Scardona,* is pleafantly feated, abounding in figs and the fmaller kinds of cattle, and producing alfo moft delicious wine. Its capital is an epifcopal city.

5. *Pago, Infula Paganorum,* which gives the title of count, is cold and barren, has little wood, but makes a good kind of falt. Befides the town of the fame name it has alfo a few caftles, and the remains of the ancient city of *Chiffa,* or *Keffa,* are ftill to be feen here.

6. *Selbo,* or *Selva,* and *Ulibo,* are fmall places.

7. *Great Ifole,* called alfo *Lantano* and St. *Michael,* contains feveral caftles and towns, fuch as *Oglier,* or *Uglian, Locara, Caglie, Codizza* and St. *Euphemia.* The power to which thefe and the adjacent iflands belong, is always mafter alfo of *Zara,* which faces it. The air is very clear and healthy.

8. *Pafma,* the northern parts of which are well cultivated.

Ifola

9. *Ifola Longa*, to the north of which lies *Bofava*, *Drague*, *Berbigno* and other places.

10. *Mortero*, or *Mortara*, celebrated for its wine, and abounding alfo in melons and olives. It has a deep fecure harbour lying betwixt two iflands. The town of the fame name is fituated in a valley betwixt two high hills of various culture, and was formerly called *Colento*.

11. *Brazza*, *Labraza*, or *Brac*, is fo called from the town of *Brazza*. The *Venetian* governor refides at St. *Peter*, which lies to the weft, near *Milna* harbour.

12. *Lefina* is a rocky and barren tract; but the town of the fame name, in the weftern part, though fmall, is well fortified, and ftands very pleafantly. For the fecurity of fhipping, a mole has been carried along the harbour. Bread and wine are very cheap here. It alfo produces figs; and from hence all *Italy* and *Greece* are fupplied with fardines. *Varbofca* and St. *George* are thinly peopled. The *Venetians* purchafed this ifland in the year 1424, and it is faid to be the celebrated ifle of *Pharos* or *Pharia*.

13. *Liffa*, or *Iffa*, a fmall ifland, once famous for the extenfive commerce of its inhabitants, and long before as an arfenal of the *Romans*.

14. *Corzola nigra*, *Corcyra*, is a principality which the *Venetians* made themfelves mafters of in the year 1386. It is very convenient for the building and repairing fhips, abounding in timber of all kinds. In the city of the fame name, and which is the only one on the ifland, refide the governor and a bifhop. It is fortified with ftrong walls and towers, has a fine harbour, and produces alfo plenty of good wine. The *Turks* attempting to make a defcent here were, in the year 1507, repulfed by the women, who behaved with heroick refolution when their daftardly hufbands had fled up into the country through fear.

III. *Ragufan Dalmatia.*

Ragufa is an ariftocratical ftate formed nearly after the model of that of *Venice*. The government is in the hands of the nobility, who are at prefent greatly diminifhed; and the chief of the republick, who is ftiled rector, is changed every month and elected by fcrutiny, or lot, in two different manners. During his adminiftration he lives in the palace, wears a ducal habit, namely, a long filk robe with white fleeves, and his falary is five ducats a month; but if he be one of the *Pregadi*, and affift at appeals, he receives a ducat *per diem*. Next to him is *il configlio de i dieci*, or the council of ten. In the *configlio grande*, or great council, all noblemen above twenty years of age are admitted; and in this council alfo are chofen the perfons who conftitute the board of the *Pregadi*. Thefe laft fuperintend all affairs civil and military, difpofe of all employments, and receive and fend envoys. They continue a year in office. *Il configlietto*, or the little council, confifts of thirty nobles, looks to the polity, trade and revenues of the ftate, and decides appeals of fmall value. Five proveditors confirm by a majority of

rotes

votes, the proceedings of the adminiftration: Civil caufes, and thofe more particularly relating to debts, are firft heard before fix fenators, or confuls, from whom there lies an appeal to the college of thirty; and from them again, in particular cafes, to the council. In criminal caufes fpecial judges are appointed. There are alfo three commiffioners appointed for the woollen trade; a board of health, confifting of five nobles, whofe care it is to preferve the city from all contagious diftempers; and four patrons of eminence manage the taxes, excife and mint. The revenues of the republick are faid to have amounted formerly to about a ton of gold; but being unable to defend itfelf, they have procured them feveral protectors, the chief of whom is the Grand Seignor. It is faid of them that they pay tribute to the *Turks* out of fear; to the *Venetians* out of hatred; to the Pope, Emperor, *Spain* and *Naples* out of refpect and political views. The tribute to the *Porte*, with the expences of the annual embaffy, is about twenty thoufand zequins*. The *Turks* are very ferviceable to them, bringing hither all kinds of neceffaries, efpecially fire-arms and military ftores. They keep fo watchful an eye over their freedom, that the gates of the city of *Ragufa* are allowed to be open only a few hours in the day. They wholly profefs the Catholick religion, but the *Greek*, *Armenian* and *Turkifh* perfuafions are tolerated for conveniency. The language chiefly in ufe among the *Ragufans* is the *Sclavonian*, but the greater part of them fpeak alfo the *Italian*. The citizens are almoft to a man all traders, and this place diftinguifhes itfelf by the finenefs of its manufactures. Silk is allowed to be worn here only by the rector, the nobiles and the doctores. Its territory is but fmall, and all the places of note in it are

Ragufa, the capital. The ancient *Ragufa* was built long before the birth of *Chrift*. It became afterwards a *Roman* colony, and was demolifhed in the third century by the *Scythians*; fo that at prefent it is but a fmall place. It was anciently called *Epidaurus*. The new city was built, on the demolition of the old, in the place where it now ftands, being enlarged from time to time. In former days it was called *Raufis*, or *Raufa*, but at prefent is ftiled *Pabrovika* by the *Turks*, and *Dobronich* by the *Sclavonians*. It is not very large in circumference, but is neverthelefs well built, being the feat of a republick and an Archbifhop's fee. It extends towards the fea, and both the city and harbour are defended by fort S. *Lorenzo*. Were the rock of *Chiroma*, which lies in the fea and belongs to the *Venetians*, fortified, it would be impregnable. The air is wholefome, but the foil fo barren that the inhabitants receive the greateft part of their neceffaries from the neighbouring *Turkifh* provinces. The circumjacent iflands are all fertile, pleafant, well inhabited, and embellifhed with fine towns, ftately palaces and beautiful gardens. The city is very much fubject to earthquakes, from which it has more than once fuffered incredible damage, efpecially in the years 1634

' A zequin, or chequin, is about nine fhillings and two pence. See *Par aire*'s table.

and

and 1667, in the latter of which ſix thouſand men were deſtroyed. A great fire too breaking out at the ſame time, the place was ſo demoliſhed that it did not thoroughly recover it for twenty years afterwards.

Gravoſa, ſituate on the peninſula of *Sabioncello,* having an excellent harbour, which is the beſt on all this coaſt. The entrance to it is very commodious, broad, deep, and well ſecured, being ornamented with raviſhing proſpects of the adjacent mountains, which are covered with fine vineyards, gardens and ſummer-houſes, where the *Raguſans* ſeek and find pleaſure.

S. Croix, Portus St. Crucis, lies alſo on the peninſula of *Sabioncello.*

Stagno, Tittuntum, ſmall but populous and well fortified, is a biſhop's ſee and has a commodious bay.

Meleda, Melita, or *Melitena,* an Iſland which, according to the opinion of ſome learned men, is that upon which St. *Paul* was ſhipwrecked ; but this is not probable. On it is a ſmall town of the ſame name, together with a few villages ſeated in a very fruitful ſoil and fine paſtures.

The three iſlands of *Calamota, Iſola Mezzo* and *Guipana,* which lie betwixt *Meleda* and *Raguſa,* are called *Elaphites.*

S. Andrea is a ſmall and pretty well inhabited iſland, having a town of the ſame name.

IV. *Turkiſh Dalmatia* extends through the country of *Herzegowina* from *Boſnia* to *Albania,* and contains the following places.

Scardona, in the *Sclavonian* language *Skardin,* an epiſcopal city on the river *Kerka,* is ſurrounded with walls and defended by two ſmall forts. It was anciently a place of conſiderable note, the *Romans* having erected a tribunal here for the ſeveral tribes of the *Japidæ* and the fourteen *Liburnian* towns ; and, in the year 1120, the biſhoprick was removed hither from *Jader.* In the year 1352 it fell into the hands of the *Venetians,* to whom it was ſold in 1411, and continued till 1522 under their dominion. In the laſt-mentioned year the *Turks* wreſted it from them ; after which they were ſeveral times driven out of it, but as often recovered it. The ruinous walls of the old caſtle and citadel are ſtill to be ſeen.

Clinowo, Kliuno, a well built town, on a riſing ground, which the *Turks,* in time of war, generally made their rendezvous, and alſo kept their magazines of proviſions and military ſtores in.

Maſter, an open town on the *Viſera,* over which is ſtill to be ſeen an old *Roman* bridge of ſtone.

Herzegowina, Arcegovina, a ſpacious and well fortified town, where reſides the *Turkiſh* beglerbeg.

Trebigne, Tribulium, Tribunium, or *Tribunia,* a mean city on a river of the ſame name, formerly the capital of a province, and ſtill a biſhop's ſee.

Popocco a fmall Diftrict, difficult of accefs by reafon of its fituation betwixt two long mountains, but remarkably fruitful in grain, wine and exquifite fruit, though generally overflown in autumn. This country put itfelf under the protection of *Venice* in the year 1694.

Clobuch, a caftle on a high rock having but one afcent to it, and that fo
narrow, that it feems impregnable by any other means than by famine.

Great and *Little Melanto*, two fea-ports, but of fmall importance.

Zenta, a County, once a part of the Kingdom of *Servia*, but afterwards
fell into the hands of the Counts of *Balza*. It is divided into the Upper
and Lower: Its chief towns being *Dugla, Drivafto, Scutari, Dagno* and
Podgoriza.

Medon, a town on the *Lago di Scutari*, built out of the ruins of *Djoclea*.

Aleffio, anciently *Liffus*, or *Liffum*, an epifcopal city on a mountain, famous
for being the burial place of the brave *Scanderberg*, who died and was interred there in the year 1467.

TURKEY

TURKEY

IN

EUROPE.

—————————————————

INTRODUCTION

TO

TURKEY IN EUROPE.

§.1. EXCLUSIVE of the feveral maps of *Hungary* and the *Danube*, in which *Turkey* in *Europe* is exhibited, there are alfo fome particular maps of it; but both the one and the other very faulty and defective. The lateft, and hitherto the beft, map of the whole *Turkifh* empire, is that of Mr. *Franz*, publifhed in the year 1737; who has built on what, from his own perfonal obfervations, he knew to be true and accurate in *De l'ifle* and *Hafe*. It is to be met with in *Homannifch-hafifchen Gefellfchafts-Atlas*, but its political divifion agrees better with the former than prefent ftate of *Turkey*. There are alfo maps of fingle provinces of the *Turkifh* empire; and of this fort fome have been publifhed at *Conftantinople* in the *Turkifh* language, at the new printing-houfe of the Effendi.

§. 2. The origin of the name of *Turkey* may be feen in §. 10. This Empire confifts of *European*, *Afiatick* and *African* poffeffions; but I fhall here defcribe only the *European*, though I fhall adapt the Introduction to the whole *Turkifh* dominions.

§. 3. *Turkey* in *Europe* is a part of the ancient Chriftian Empire of the Eaft; and at prefent borders towards the eaft on *Poland*, *Ruffia* and *Afia*; northward on *Croatia*, *Sclavonia* and *Tranfylvania*. It is bounded weftward by the *Adriatick* and *Dalmatia*, and fouthward by the *Mediterranean*. The extent is not to be afcertained with any precifion.

§. 4. The air of *Turkey* in *Europe* is healthy in itfelf, but the peftilence is often brought there from *Egypt*, and has more than once fwept away one fifth of the inhabitants of *Conftantinople*; yet, by the prevalence of cuftom, and of the *Turkifh* doctrine of fatality, they give themfelves no great concern about it. The provinces are univerfally fruitful, though with fome difference, infomuch that both agriculture and graziery turn to great profit there; vaft quantities of all kinds of excellent grain and fruits being exported every year by fhipping: but of this we fhall fpeak more at large in

the

the feparate defcription of each province. All the neceffaries of life are
equally good and cheap in *Turkey*. Its chief rivers are the *Save*, the *Danube*, the *Niefter*, the *Nieper* and *Don*, which have been fpoken of before in
Hungary and *Ruffia*; as likewife the *Theis*, in fpeaking of which fome of
the provinces in *Turkey* have been defcribed in the beginning of this
volume.

§. 5. The number of its inhabitants is greatly difproportionate to the extent and goodnefs of the country, which may be chiefly attributed to peftilence, polygamy and war. This accounts for fuch large tracts of fine foil
lying wafte; though with this the avarice of the governors is partly culpable.
The inhabitants confift of various nations; namely, of *Turks*, *Greeks*, *Armenians*, *Servians*, *Bofnians*, *Bulgarians*, *Walachians* and *Tartars*, with no fmall
number of *Jews*, efpecially at *Conftantinople* and *Sclavonia*. The *Turks* are
ftigmatized among the Chriftians as a flothful, ftupid and inhuman people;
but they are by no means fo wicked and dreadful a fet of creatures as Popifh
writers have induftrioufly reprefented them. - *Turkey* is not without men of
parts, probity and honour; nor benevolent, liberal, temperate, converfible
and ingenious people. In fhort, there is here, as in all other countries and
nations, a mixture of good and bad. *Dricefch* fays that, in compaffion and
love towards our neighbours, the *Turks* excel all the reft of mankind; and
this affertion is confirmed by feveral other travellers. One ftriking mark of
their charity are the haans, or publick inns, by the *Afiaticks* called *Cara-
vanfari*, which are to be met with almoft in every little village. In thefe
a traveller, of whatever religion or country he be, may continue three days
gratis; and in not a few he is found alfo in victuals. The *Turks* are very
fond of erecting thefe buildings, juftly accounting it a work of charity and
acceptable to God. To the flaves and fervants who are about them they
behave very commendably, and frequently better than the Chriftians do to
theirs. In the firft years of their fervitude thefe people fuffer moft, efpecially if young, the *Turks* endeavouring, partly by blandifhments and partly
by feverity, to bring them over to their fect; but thefe trials being happily
over, captivity is no where more tolerable than among the *Turks*; infomuch that, if a fervant underftand any art or trade, the only thing he can
want is his freedom, being fupplied with every thing elfe he can wifh
for.

With refpect to the external conftitution of the *Turks*, they are generally
robuft and well-fhaped, of a good mien, and patient of hardfhips, which
renders them fit for war, to which indeed they inure themfelves from their
youth. Perfons of rank feldom train up their children to any thing elfe, out
of a notion that no glory is comparable to that acquired in war. In their
drefs, manner of living and cuftoms, they are particular. They fhave
their heads, but wear long beards, of which they are extremely careful,
except thofe in the feraglio and military men who wear only whifkers.

The

The turban, or *Turkish* band, worn by the men is white, and confifts of long pieces of thin linen made up together in feveral folds. No one but a *Turk* muft prefume to wear a white tuiban. Their clothes are long and full. They fit, eat and fleep, agreeably to the cuftom of the Orientals, on the floor, on cufhions *(fophas)* matraffes and carpets. Rice is their moft general food, and coffee their ufual drink, wine being interdicted (§. 7.) They expend great fums on fountains, and no country indeed affords fo fine ; and thofe not only in the towns, but in the country, and other folitary places, for the refrefhment of travellers and labourers. Their moft ufual falutation is to bow the head a little, laying the right hand on their breafts; but to perfons of rank they ftoop fo low as to touch and kifs the border of their veft. In war time the left hand is the place of honour among military men, but in time of peace this diftinction ceafes among the officers of ftate and relations. The fair fex here are kept under a rigorous confinement. The *Arabick* word *Haram*, which fignifies a facred or prohibited thing, is, in its fulleft fenfe, ufed both of the women's habitation and of the women themfelves. The feraglios are improperly termed harams, the word feraglio fignifying no more than a palace. The nobility among the *Turks* are the chief military officers, judges and ecclefiafticks. The *Turkifh* commonality enjoy the greateft liberty ; and to the ticklifh orders of the Porte, which muft be anfwered with the head, thofe only are expofed who hold confiderable pofts ; or, as the phrafe is in *Turkey*, eat the Sultan's bread.

The *Greeks*, who are the ancient inhabitants of the country, live intermixed with the *Turks*, and in feveral places, particularly the iflands, out number them. In *Conftantinople* alone it is computed there are no lefs than four hundred thoufand. They are accuftomed to fervitude, and prefer living under the *Turkifh* exactions to the fpiritual tyranny of the Pope : but they muft be very cautious of not giving even the leaft colour for fufpicion of holding correfpondence with the enemies of the *Ottoman* Porte, or of meditating a fedition. In cafe alfo of a war with any of the Chriftian powers, it is ufual, for the greater fecurity, to difarm them. All *Greeks*, from the age of fourteen, pay annually, at the beginning of the *Turkifh* feaft of *Beiram*, a head-money *(charatfchi)* which is about a ducat, and receive a note of it. The ecclefiafticks are affeffed higher, a deacon paying two ducats and an archimandrite four ; but the bifhops, archbifhops and patriarchs pay large fums, generally as much as the arbitrary avarice of the grand vizir and bafhas fhall think proper to require. The duties on traders are eftimated to the value and price of the commodities they import. The *Turks* everywhere lay hold of all opportunities of extorting money from the *Greeks*, but efpecially from their clergy. In return for this tribute they enjoy the protection of the *Ottoman* Porte, and are maintained in the quiet poffeffion of their properties; infomuch that no *Turk* is to infult them, take any thing from them, or intrude themfelves into their houfes againft their
will ;

will; and, in cafe of any fuch injuries, they are certain of expeditious juftice againft the delinquents. The *Greek* women are exempt from all taxes, as are likewife great numbers of other *Greeks* who ferve in the navy or elfewhere. It fometimes happens indeed that a *Greek* girl of diftinguifhed beauty is taken away and carried to the feraglio; but it is a miftake in any to fay that Chriftian children are in general forced away from their parents to be brought up in *Mahometanifm*: when any thing of that kind is done, it muft be in the provinces which are at a diftance from *Conftantinople*. There are feveral particulars to be attended to by the *Greeks* and other Chriftians living among the *Turks*, but they would take up too much room here.

Foreign Chriftians, under the protection of an envoy, and who are included under the general title of *Franks*, pay no head-money. Of the other nations mentioned above, a proper account will be given in the defcription of the particular provinces.

§. 6. The principal language in *Turkey* in *Europe* is the *Turkifh*, but the *Greeks* fpeak alfo the modern *Greek*. The *Servians, Bofnians* and *Bulgarians* fpeak the *Sclavonian*; the *Walachians* and *Moldavians*, the *Walachian*; and the *Tartars* the *Tartarian* language, which comes very near the *Turkifh*. The Literati generally ufe the *Arabick*.

§. 7. The *Turks* are of the *Mahometan* religion, and appropriate to themfelves the name of Moflemim, which has been ftrangely corrupted into Mufelman, fignifying perfons profeffing the doctrine of *Muhammed*, which he calls *Iflam*. They term themfelves alfo Sonnites, *i. e.* obfervers of the oral traditions of *Muhammed* and his three fucceffors; and likewife true believers, in oppofition to *Ali's* adherents, whom they nickname *Schiiten*; that is, a wicked and abominable fect; and thefe confift of the *Perfians* and others. Their rule of faith and practice, like that of all *Muhammedans*, is the Koran, or, with the article *Al*, the Al-koran; the method and contents of which cannot be fpecified here. Some externals of their religion are the prefcribed ablutions, or purification both of their whole body *(Ghoft)* or particular parts *(Wodu)*; and thefe are always to be performed before their devotions; prayers, which are to be faid five times every twenty-four hours, and with the face turned towards *Mecca*; alms, both enjoined *(Zacat)* which are two and a half *per cent.* and voluntary ones *(Sadakat)*; fafts, either indifpenfible, as that of the whole month of *Ramadan*, or *Ramazan*, which is followed by the *Beiram*, a feafon of feftivity; or voluntary ones, particularly on the day called *Afhura*, which is the tenth of the month *Moharram*; the pilgrimage to the *Caaba*, or houfe of God, at *Mecca*, which every *Muhammedan* muft perform once at leaft in his life time either perfonally or by proxy. Among the binding traditions not mentioned in the Koran is circumcifion, the time for which is betwixt the fixth and feventeenth year, but generally the thirteenth. Drinking wine, indeed, is manifeftly prohibited in the Koran, yet the *Turks* make ufe of it occa-
fionally

fionally without any fcruple, though inftead of it they generally ufe fheibet, a liquor made of honey, fpices and the juice of fruits. Other interdicted things are games of chance, prophefying with arrows, and certain foods, as blood, pork, or any beaft dying of ficknefs, or killed by a wild beaft, or a fall, or a ftroke; likewife all worfhipping of idols, ufury, and fome fuperftitious and pagan practices. Polygamy is indeed permitted, but the Koran enjoins that no man fhall have more than four wives and concubines; and to exceed that number is the particular privilege of the prophet and his fucceffors. Divorces are alfo allowed of, but no one may take back again the wife he has once repudiated, till fhe has been married to another and afterwards divorced by him. Friday is the day fet apart for divine worfhip. The churches of the *Turks* are called mofches, and the fmaller ones mefcheds. The chief ecclefiaftick is the mufti, which fignifies an expounder of the law: and his office is of fuch dignity, that when he comes to court, the Emperor himfelf rifes from his feat and advances feven fteps to meet him. He alone has the honour of kiffing the Sultan's left fhoulder; whereas the grand vizir, with a more profound inclination too of the body, kiffes only the edge of the Emperor's veft, who advances only three fteps to meet him. By the law the mufti is to be confulted on all emergencies, in thofe more particularly relating to peace and war; but at prefent this peculiar regard fhewn him is little better than form and derifion: for were he either to give a difagreeable interpretation of the law, or in his opinion in council prefume to traverfe the Emperor's inclinations, he would be immediately depofed, and his place fupplied by one of a more compliable difpofition. On conviction too of treafon, or any other heinous offence, he is put into a mortar, kept for that purpofe in one of the feven towers at *Conftantinople*, and pounded to death. As the mufti of the *Turks* may be compared to the Pope, fo a cadalifker, who is alfo a fecular perfon, is not unlike a patriarch; a mola is an archbifhop; a cadi, who is alfo a layman, may be accounted a bifhop, and an iman a prieft. The chief employment of the latter is praying. The *Turks* have alfo their convents and monks, under the general name of dervifes; of which the principal are the Bektafhi, Mebelevi, Kadri and Segati; a principal part of whofe forms of worfhip confift in certain religious dances. Scheihk is a prelate or abbot. It remains to be obferved, that the *Turks* avoid all appearance of propagating their religion by violence, fire and fword; and indeed the Chriftians, and the feveral fects refiding amongft them, enjoy full liberty of confcience, living there in much greater tranquillity than among fome who ftile themfelves Chriftians.

The head of the *Greek* church in *Turkey* in *Europe* is the patriarch of *Conftantinople*, who is chofen by the neighbouring archbifhops and metropolitans, and confirmed by the Emperor or grand vizir. He is a perfon of great dignity, being the principal of all the *Greek* patriarchs, and head and director of the eaftern church. His income amounts to no lefs than one hundred and twenty thousand guilders, of which he pays the one half by way of annual tribute to the *Ottoman* Porte, adding six thousand guilders besides by way of present at the feast of *Beiram*. Subordinate to him are seventy archbishops and metropolitans, and a much greater number of bishops. An archimandrite is the director of one or more convents, which are called *Mandren*, and rank above an abbot, of which each convent has one. The monks, priests and students excepted, are obliged to follow some handicraft, and lead a very austere life. The most celebrated are those on mount *Athos*. The *Greeks* have but few nunneries at present. The secular clergy are subjected to no rules, as the regulars are who perform divine worship. The first is the lecturer, the second is the chanter, the third the under deacon, the fourth the deacon, the fifth the priest, and the sixth the archpriest. They are allowed to marry, but that only before ordination, and then only once, and that with a virgin. These secular ecclesiasticks never rise higher than an archpriest; the bishops, metropolitans, archbishops and patriarchs being chosen from among the monks.

The *Armenians*, who are of the same opinion with the Monophysites, with regard to one nature in Christ, but differ widely from them in many other points, and agree rather with the *Greeks*, have many churches in this country. The *Roman* Catholicks and *Jews* have also the free exercise of their religion, and the *Swedes* have been permitted to build a *Lutheran* church at *Constantinople*.

§. 8. The *Turks* are not without all kinds of learning, having some schools, colleges and academies, by them called *Medaris*; but these are not to be compared with those amongst us, and their management of them also is very different. In our days a *Turkish* printing-office has been set up at *Constantinople* by *Ibrahim Effendi*, who, after great opposition, obtained leave to print all kinds of books, except on matters of religion. Accordingly, among others, he has published some maps, and books of history and geography. He is also said to have a considerable knowledge of the *Latin* tongue. Literature, however, is not so rare among the *Greeks*, who, near their churches, not only have schools for instructing children in the principles of religion, reading, writing and learning by heart the psalms and passages of scripture, but have also universities, in which are taught grammar, *Latin* and the mathematicks, with the *Aristotelian* philosophy both natural and moral. These are said to be at *Demotica*, in the island of *Patmos*, *Jannina* and other places. Divinity is taught at the patriarch's palace at *Constantinople* by a chaplain of the patriarch's and some assistants, but particularly on mount *Athos*, which seems to be the pillar of the *Greek* church, and in other parts by the bishops, who are men of capacity, and take this trouble voluntarily upon them. Physick the *Greeks* learn either from the *Arabick*, *Jewish* or Christian physicians residing among them; or else go to the universities in *Germany*, *Holland* or *England*. The state of

learning.

learning, indeed, among the *Greeks* is at a low ebb in comparison of what it is with us; but be it alfo remembered, that they are deftitute of the fame means and opportunities of mental improvement.

§. 9. The *Turks* are not without manufactures, and thofe very curious and beautiful. The inland trade too, which the provinces, towns and inhabitants carry on with each other and with foreign nations, is very confiderable; but it is moft through the channel of *Jews* and *Armenians*. The *Turks* export indeed, both by land and water, the products of the country, and other goods, from one province to another, but not to foreign Chriftian countries: Great numbers of *Dutch*, *Englifh*, *French*, *Italian* and *Swedifh* fhips, as well as of other trading nations, repairing in great numbers to the harbours of *Turkey*, where they import their goods, and purchafe thofe of the country. They have alfo their envoys and refidents at *Conftantinople*, and their confuls in other ports. The exports from *Turkey* are filks, carpets, goats hair and wool, camels hair, cotton-yarn, dimety, burdets, waxed linen, fhagreen fkins, blue, red and yellow *Morocco* leather, coffee, rhubarb, turpentine, ftorax, gums, opium, galls, maftich, emery, *Lemnian* bole, pomegranate fhells, fponges, dates, almonds, wine, oil, figs, raifins, mother of pearl, box-wood, wax, faffron, &c. The traffick of the human fpecies, however fhocking, is another confiderable article in *Turkey*; for they not only fell there flaves of both fexes, but alfo beautiful young girls who are bought up particularly by the *Jews* in *Circaffia*, *Georgia*, *Greece* and elfewhere; their parents and relations readily parting with them in hopes of raifing their fortune.

The gold and large filver coin of all countries are current in *Turkey*, more efpecially the crofs-dollars of *Burgundy* and the *Dutch Lion* dollars, which they term *aflan*. The proper coins of the country are, firft, thofe of gold; namely, the altines, or ducats, which are about feven fhillings a-piece; the zechinos, worth about nine fhillings. Secondly, the filver; *viz.* the folota *(zelote)* worth about two fhillings and two pence farthing; the krip, about eleven pence; the grofh, or grofche, about three pence; the para, worth three afpers, an afper about one penny halfpenny: a purfe contains five hundred rixdollars, or one hundred and eight pounds, fix fhillings and eight pence.

§. 10. The *Turks* are of *Tartarian* or *Scythian* extraction, this appellation being firft given them in the middle ages as a proper name, it being a general title of honour to all the nations comprehended under the two principal branches of *Tartar* and *Mongul*, and therefore never ufed as a proper name of any *Scythian* or *Tartarian* people: nor do even the *Turks* appropriate it peculiarly to themfelves, both the *Monguls* and *Tartars*, properly fo called, reckoning it a mark of honour to them; the word *Tur*, as an adjective, fignifying fublime and preeminent, and as an appellative a governor. *Turci*

Q 2   therefore

therefore may import both the governor of a hord (*Ki* among the *Tartars* signifying a hord or company) as well as the hord itself. The *Scythian*, or *Tartarian* nation, to which, as I have before observed, the name of *Turks* has been peculiarly given, dwell betwixt the *Black* and *Caspian* seas, and became first known in the seventh century, when *Heraclius*, Emperor of the east, took them into his service, under whom they so distinguished themselves by their fidelity and bravery in the conquest of *Persia*, that the *Arabian* and *Saracen* Caliphs not only had particular bodies of them for guards, but their armies were likewise filled with them. Thus they gradually got the power into their hands, and set up and dethroned Caliphs at pleasure. Some governors also of this nation have wholly revolted from the dominion of the Caliphs. This happened about the ninth century. By this strict union of the *Turks* with the *Saracens*, or *Arabs*, the former were brought to embrace the *Mahometan* religion ; so that they are now become intermixed and have jointly enlarged their conquests. The *Turks*, however were superior and subdued the *Saracens*. Concerning the origin of the *Ottoman* empire, the substance of it, according to Prince *Cantemir*'s history, is as follows.

Genhizkan, *Zingis-chan*, at the head of the *Oguzianian* horde, issued out of Great *Tartary*, and made himself master of a vast tract of land near the *Caspian* sea, and even of all *Persia* and *Asia* the Lesser. Incited by his example and success, *Schach Solyman*, Prince of the town of *Nera*, on the *Caspian* sea, and lord of *Meruschahjan*, in the year 1214, passed mount *Caucasus* with fifty thousand men ; he made his way through *Azerbejan*, or *Media*, as far as the borders of *Syria* ; and, though stopt there in his career by the *Genghizkan Tartars*, yet, in the year 1219, he penetrated a second time into *Asia* the Lesser as far as the *Euphrates*. The report of his conquests reaching the court of *Persia*, *Solyman* and his people were spoken of as *Turks*, a name common to all the *Scythians* who came out of *Tartary* under the command of *Genghizkan*. *Othman*, his grandson, made himself master of several countries and places in Lesser *Asia* belonging to the *Grecian* empire ; and having, in the year 1300, at the city of *Carachifer*, taken upon him the title of Emperor of the *Othmans*, called his people after his own name. His residence he fixed at *Yenghifcheri*, and, exclusive of many other towns, in the year 1326, took *Prusa* in *Bithynia*, now called *Burfa*, which his son and successor *Orchan* made the seat of his empire. *Orchan* sent his two sons, *Solyman* and *Amurath*, on an expedition into *Europe* ; the former of whom reduced the city of *Callipolis*, and the latter took *Tyrilos*. *Amurath* succeeding his father in the government, in the year 1360 conquered *Ancyra*, *Adrianople* and *Philippopolis* . In the year 1362 he instituted the Janizaries, over-run *Servia*, and fell upon *Macedonia* and *Albania*. His son and successor *Bajazet* was very successful both in *Europe* and *Asia*, defeating the Christians near *Nicopolis* ; but, in the year 1401, he

himself

himſelf was routed and taken priſoner by *Tamerlane*. His ſons diſagreed, but *Mahomed* the firſt held the ſovereignty, and his ſon *Amurath* the ſecond, diſtinguiſhed himſelf by ſeveral important enterpriſes; and, particularly, in the year 1444, gained a ſignal victory over the *Hungarians* near *Varna*. *Mahomed* the ſecond, and the greateſt of all the Emperors, in the year 1453, made himſelf maſter of *Conſtantinople*, reducing the whole *Grecian* Empire under his dominion. He ſubdued twelve Kingdoms and two hundred towns. *Bajazet* the ſecond and *Selim* the firſt enlarged the *Turkiſh* Empire in *Europe*, *Aſia* and *Africa*. *Solyman* the firſt is not more famous for his victories over the *Hungarians* than his body of laws. The ſucceeding Emperors were not ſo fortunate. *Mahomed* the fourth, in the year 1669, ſubdued *Candia*; and, in 1683, laid ſiege to *Vienna*, but in *Hungary* met with ill ſucceſs. In the reigns of *Solyman* the ſecond, *Achmet* the ſecond, and *Muſtapha*, the *Hungarians* and *Venetians* were ſo ſucceſsful againſt the *Turks*, that *Muſtapha* the ſecond, in the year 1699, was glad to come to a peace at *Carlowitz*. *Achmet* the third, in 1718, concluded the treaty of *Paſſarowitz*, and *Mahomed* the fifth, by the peace of *Belgrade*, in the year 1739, re-annexed *Servia*, a part of *Walachia* and *Chozim* to the Empire.

§. 11. The proper and moſt uſual ſtyle of the *Turkiſh* Empire, is the *Ottoman* Kingdom or Empire, ſo called from the firſt founder; likewiſe the *Ottoman* Porte, with the following epithets; the ſublime Porte, the ſublime ſultanian Porte, the Porte of juſtice, majeſty and felicity. The appellation of Porte is ſaid to be derived from the large porte, or gate, built by *Mahomed* the ſecond at the entrance of the ſeraglio, or imperial palace, at *Conſtantinople*; though the Orientals in general call a royal palace the King's porte and gate. The Emperor alſo ſtyles himſelf *Chan*, or *Kan*, which ſignifies a Prince, or Sovereign, and is ſynonimous with the *Arabick* word *Sultan*, which is more frequently uſed either ſimply or with the article *El*, or with the addition of Great Sultan. The Princes of the *Crimea* are alſo ſtyled Sultans by the *Ottoman* Porte.

§. 12. The Emperor's title, according to the cuſtoms of the caſt, is very prolix and magnificent, as the following ſpecimen abundantly ſhews. " We, the ſervant and lord of the moſt honoured and bleſſed cities, the venerable houſes and ſacred places before whom all nations bow; of *Mecha*, which God delights to honour, of the reſplendent *Medina* and the holy city of *Jeruſalem*; of the imperial and deſirable cities of *Conſtantinople*, *Adrianople* and *Burſa*, Emperor; alſo of *Babylon*, *Damaſcus*, of the fragrant paradiſe and the incomparable *Egypt*; of all *Arabia*, *Aleppo*, *Antioch*, and many other highly celebrated and memorable places, cities and faithful vaſſals, Emperor; Emperor of Emperors, the moſt gracious and all-powerful Sultan, &c."

§. 13. The *Turkiſh* arms are a creſcent, by ſome deduced from Old *Byzantium*, many of the coins of that city being ſtamped with the moon; but others

evidently

evidently prove that it was ufed before the conqueſt of Conſtantinople, and was probably retained from the ancient Arabians.

§. 14. In the fucceſſion to the Empire no regard is paid to age or birth-right, the Turks thinking it fufficient if, in their elections, they keep to the Ottoman family. Women are excluded from the throne. The government is indeed purely monarchical, but, if the Emperor indulges not the humours of the people, and eſpecially of the mutinous Janizaries, he is not only in danger of being depoſed, but alſo of being put to death.

§. 15. The Emperor's council of ſtate is ſtiled Galibe Divan, or Divan Galibe, and meets twice a week in the Emperor's palace; namely, on Sundays and Thurſdays. The Grand Vizir fits as preſident, having the Kadinlaſkier of Romelia at his right hand, and that of Natolia at his left. The mufti alſo aſſiſts when expreſly fummoned. All the other lubbe-vizirs have likewiſe a ſeat here, and next to them ſtand on one ſide the teſterdar, or high-treaſurer, the reis-effendi, ſecretary of ſtate, and other commiſſioners of the Calem, or Exchequer; but the military officers, ſuch as the aga of the Janizaries, the aga of the Spahi's, the aga of the Siluds, &c. fit within the Divan. The Sultan hears what paſſes from an adjoining chamber which looks into the Divan. The ſeveral members, on council days, wear a particular habit. When the Sultan convenes a general council, to which are fummoned all the great perſons of the Empire, the clergy (ulema) the military, and other officers, and even the old and moſt experienced foldiers; ſuch a Divan is called Ajak Divani, the whole aſſembly ſtanding.

The higheſt officer, and next to the Emperor, is the Grand Vizir, whoſe income, without any breach of probity, may amount to ſix hundred thouſand dollars a year, excluſive of preſents and other perquiſites. On his coming towards the Emperor the Sultan advances three ſteps to meet him, whilſt he makes a profound reverence and kiſſes the edge of the Sul-tan's veſt: but if his dignity be great his danger is no leſs, it being the uſual policy of the Emperor's to ſhelter themſelves from the clamours of the people by throwing the whole blame of any male-adminiſtration on this officer, and giving him up to the publick reſentment. The Grand Vizir's com-miſſary is called the Kaimakan, and is choſen by the Sultan out of ſuch vizirs as are permitted to carry three horſetails. Whilſt the Emperor reſides at Con-ſtantinople or Adrianople he is without any power, but if only eight hours from the city, his authority is little leſs than that of the Grand Vizir. When the Emperor takes the field he nominates a Kaimakan, who, in caſe the Grand Vizir be at the diſtance of eight hours from the Emperor, has the full power and management of all affairs, excepting that he is not to act contrary to the Grand Vizir's inſtructions, nor caſhier or behead the old baſhas. This Kaimakan is not to be confounded with the governor of Conſtantinople or Adrianople, who bears the ſame title.

§. 16. The

§. 16. The high court of juftice is held in a large hall of the Grand Vizir's palace, called *Divan Chane*. The Grand Vizir being prefident, is obliged four times a week, namely, on *Friday, Saturday, Monday* and *Wednefday* to fit in the Divan and adminifter juftice to the people; unlefs, which feldom happens, he be hindered by affairs of a very important nature. In this cafe his place is fupplied by the Chiaoux Bafchi, *i. e.* the mafter of requefts. On *Fridays* the Grand Vizir has for affiftants the two kadiulaf-kiers of *Natolia* and *Romelia*; the former of whom fits at his left only as a hearer, but the other fits at his right and in quality of judge. On *Saturday* the Grand Vizir's affiftant is the galata or mollati, judge of the fuburbs of *Galata*, or the *Pera* judge. On *Mondays* the ejub mollafi (the judge of the fuburb of St. *Job* at *Conftantinople*) and the ifkindar mollafi. Laftly, on *Wednefday*, the iftambol effendi, or judge of the city of *Conftantinople*. The bills, or reprefentations of the parties *(Arzuhal)* are read, and the affiftants give their opinion of the matter. If their verdict be agreeable to the Grand Vizir, it is written on the arzuhal and the Grand Vizir fubfcribes it; if he difapprove it, he himfelf pronounces a decree and orders a copy of it to be given to the parties. Thus proceffes are foon brought to a proper iffue if the judge has rightly underftood the matter. *Kadi* is a word ufed for all judges of a province or any particular place.

§. 17. A Beiglerbeg is a Viceroy with feveral provinces under his command, the name itfelf fignifying a Prince of Princes. The three principal are the Beiglerbeg of *Rumili*, who refides at *Sophia*; the Beiglerbeg of *Natolia*, the feat of whofe government is at *Kutabia*; and the Beiglerbeg of *Damafcus*, or *Scham*, who keeps his court in the city of that name. Under thefe are the bafhas, or governors, whofe pofts indeed are very confiderable but precarious; and fubordinate again to thefe are the fangiacs, who may be termed deputy-governors. All thefe are likewife military officers.

§. 18. The national revenues are returnable to two treafuries; the difchazine, or publick treafury, is under the management of the tefterdar, or high treafurer, who has under him twelve offices, to which all the revenues of the empire, arifing from tributes, cuftoms, *&c.* are returnable; and out of thefe the army is paid. The treafurer is allowed the twentieth of all the money brought into the treafury, which muft bring him in at leaft two hundred thoufand dollars yearly, one fourth of which he pays to the kiet-chudabeg, or kiehaja, who is the Grand Vizir's commiffary and above the tefterdar. The money of this treafury is called *Deitulmali Muflimim*, *i. e.* the publick money of the muffelmans; and is not to be touched by the Emperor but in the greateft exigency, much lefs for his private occafions. The Sultan's private treafury *(Ifchazine)* which he difpofes of according to his own pleafure, is under the care of the hafnadar bafchi, who, next to the kyflar, ranks firft in the feraglio. Prince *Cantemir* fays that, in his time, twenty-feven thoufand purfes, amounting to thirteen million and a half of rix-dollars rixdollars, were annually returned to both treasuries. The confiscations of the estates and effects of the bashas, and other officers, together with the money arising from the escheats of *Turks* dying without male issue, make also a very considerable article.

§. 19. The Janizaries, properly *Jengitsheri*, a word compounded of *Jengi* new, and *Ischeri* a soldier, are the flower of the *Turkish* land-forces; they are all infantry, and were first formed out of captive Christians by the Emperor *Amurath* the first. Their number is generally forty thousand, divided into one hundred and sixty-two companies, or chambers, called *Odas*, in which they live together at *Constantinople* as in a convent. The recruits are termed *Yamagi*. The Janizaries are of a superior rank to all other soldiers, but they are also more arrogant and factious; and it is by them that the publick tranquillity is mostly disturbed. Every one receives three aspers a day with a certain quantity of mutton, rice and butter; which eatables, are not given to them, but dressed and set before them in their Odas. In time of peace they have no muskets. The Capis are also infantry; the Spahis light-horse, and the Timar Spahi, or the old and preferred Spahis, instead of pay have villages in several of the provinces, and are obliged, according to their income, to bring at least three slaves with them into the field. The foot, under the command of a basha, are still called by the ancient name of Segban, and the horse Seraje. The tributary Princes, as the Cham of the *Crim Tartars*, and the Princes of *Moldavia* and *Walachia*, are obliged also to send auxiliaries. The whole *Turkish* army makes above three hundred thousand men. The Janizaries and Spahi have their aga, or general. A serafkier, or commander in chief of the whole army, must be a basha of two or three horse-tails. These horse-tails are the insignia of grandeur in *Turkey*, and carried before the great men, It is only the principal bashas, and other persons of the highest distinction, who are allowed three. The Beiglerbegs and Sangiacs have also their several commands.

§. 20. The *Turkish* navy is laid up at *Constantinople* near the arsenal, and consists of about forty large ships; but in time of war, when the fleet is to put to sea, the *Turks* receive auxiliary ships from *Algiers*, *Tunis* and *Tripoli*: They buy up also or hire merchant-ships; and thus raise a fleet of one hundred and fifty sail, exclusive of galleys of two, three and four banks of oars. The admiral is styled *Capudan*, or *Capitan Bascha*.

We now proceed to give an account of

I. The countries in *Europe* subject to the *Ottoman* Porte; among which, besides *Dalmatia*, already described, is

TURKISH

TURKISH ILLYRICUM;

§. 1. WHICH ſtretches from *Sclavonia* to *Romania* and *Bulgaria*, betwixt *Croatia*, *Dalmatia* and the *Danube*. It has ſeveral mountains, and among theſe the *Monte Argentorato* is particularly remarkable: Its navigable rivers are the *Save*, the *Verbas*, the *Boſna*, the *Drino*, the *Morau* and the *Ibar*, not to mention the *Danube*, which conſtitutes the northern boundary of *Servia*.

§. 2. The country is fitted both for agriculture and graziery, producing grain, wine, and provender for all kinds of cattle : and the mountains, particularly thoſe of *Boſnia*, contain ſilver.

§. 3. The inhabitants are of *Sclavonian* extraction, and from the middle ages have been divided into *Servians*, *Boſnians* and *Ratzians*, though without any difference in their ſpeech and manners. Their language is the *Sclavonian* but called the *Illyrian*, and near of kin to the *Ruſſian*. They are properly communicants of the *Greek* church, but *Mahometaniſm* has gained confiderable ground among them, eſpecially as they are abſolute ſtrangers to learning. Their letters for writing are the *Ciruli*, which are alſo uſed by the *Ruſſians*. *Servia* makes great quantities of cottons.

§. 4. *Boſnia* and *Servia* were formerly united to *Hungary* by a perpetual compact, the former being then governed by a Ban and the latter by a Prince, or Deſpota ; but now both make a province of the *Turkiſh* Empire, which appoints Beiglerbegs and Sangiacs over them. *Turkiſh Illyricum* conſiſts of two Kingdoms ; namely, *Boſnia* and *Servia*.

I. *Boſnia*, called alſo *Rama*, derives both theſe names from the rivers *Boſna* and *Rama*; or the former perhaps from the nation of the *Boſſeni*. To the north it is ſeparated from *Sclavonia* by the river *Save*; eaſtward, by the *Drino* from *Servia*; to the ſouth a chain of mountains ſeparates it from *Dalmatia*, and weſtward it is divided from *Croatia* by the river *Verbas*. It is forty *Turkiſh** miles in length, fifteen in breadth, and conſiſts of three Sangiakſhips.

1. The Sangiakſhip of *Banialuck*, contains

Banialucka, a ſtrong fortreſs, the reſidence of a Beiglerbeg, and famous for a battle fought there betwixt the Chriſtians and *Turks* in the year 1737.

Jaitzo, formerly capital of the Kingdom, at preſent a frontier fortification towards *Croatia*.

Verboſania, *Varboſina*, a good town raiſing itſelf by trade and handicrafts.

* 66⅔ make one degree of the equator.

Dubitza, a town furrounded with a wall and pallifadoes.

2. The Sangiakſhip of *Orbach* contains

Strebernick, in *Latin Argentina,* a mean town, taking its name. from the filver mines worked there.

Orbach, the principal town in this Sangiakſhip.

Prifrendi, Prifereno, a pretty good town, having a biſhoprick.

3. The Sangiakſhip of *Sarali* contains

Sarajo, Seraglio,, a noted trading town on the river *Bofnia,* burnt by the *Hungarians* in the year 1697.

Swornick, or *Zwornick,* a town on the *Drin* fortified with a wall and caſtle.

II. *Servia,* ſo called from the *Serbij,* and ſometimes alſo termed *Rofcia* from the river *Rafca,* is fixty *Turkiſh* miles in breadth and thirty long. At the treaty of *Paſſarowitz,* in the year 1718, the greateſt part of it was ceded to the *Roman* Empire; but at the peace of *Belgrade,* in 1739, the *Imperialiſts* were obliged to give it up again to the *Ottoman* Porte. It was formerly divided into Proper *Servia* and *Rafcia*; and to the former, which makes the upper part, towards the *Danube,* belongs the Bannat of *Mafovia.* It conſiſts at preſent of four Sangiakſkips.

1. The Sangiakſhip of *Belgrade* lies betwixt the rivers *Drino, Save* and *Danube,* and contains

Belgrade, Greek Weiſſenburg, Nandor Fejervar, Alba Græcorum, a celebrated and important place and fortreſs, fituate at the conflux of the *Save* and the *Danube,* and confiſting of the citadel, the waſſerſtadt and the rafcianſtadt. It was formerly accounted the barrier and key of *Hungary,* to which it was firſt annexed by the Emperor *Sigiſmund.* In the years 1440, 1456 and 1494, it was befieged in vain by the *Turks,* but in 1521 taken. They kept poſſeſſion of it till the year 1688, when the *Hungarians* recovered it. In 1690 it fell again under the *Turkiſh* yoke, from whence it was unfuccefsfully attempted to be wreſted in the year 1693, but in 1717 accompliſhed. The *Hungarians* were again obliged to evacuate it in the year 1739, though not till they had firſt demoliſhed all its outworks, leaving nothing ſtanding but the old walls and ſome fortifications infeparable from them.

Szabats, a ſmall fortreſs on an iſland in the *Save.*

Viſnitza, a little town on the *Danube.*

Krotzka, Grofka, a ſmail town, near which, in the year 1739, the *Hungarian* army was defeated hy the *Turks.*

Rudnick, Rudnitza; a place of little importance.

Valjava and *Bedka,* two towns on the *Kolubra.*

2. The Sangiakſhip of *Cemender, Senderow Veg, Szendro,* an old faſhioned fortification on the *Danube.* In the year 1438 the *Turks* made themſelves maſters of it; and, in 1688, the *Hungarians:* theſe, in the year 1690, were again difpoſſeſſed by the *Turks,* and thefe again in 1718 by the *Hungarians.* It contains

Haſſan,

Haſſan-Baſſa-Palanka, a fort betwixt the rivers *Jeſſava* and *Morau*; ſo called from *Haſſnan* the *Boſnian*,' and *Palanka*, which ſignifies a fort or fortified place. Here are medicinal waters and baths.

Paſſarowitz, a town on the *Morau*, well known for the peace concluded there in the year 1718 betwixt *Charles* the ſixth and *Achmet* the third.

Ram, a town and caſtle oppoſite *Vj-Palanka*, in the juriſdiction of *Temeſwar*.

Columbatz, by the *Turks* called *Gugerzinlika*, a fortreſs of ſome conſideration on an eminence near the *Danube*.

Kirdap da Talia, a territory on the *Danube*, near which, betwixt the rocks on both ſides, is a vortex having very high waves. A little below it is

Tachtali, a dangerous part on the *Danube*, where the circling water, after falling from a rocky precipice, forms a whirlpool. This is occaſioned by the nature of the two ſhores, a high rock on the *Servian* ſide projecting a great way into the ſtream, which ſtriking againſt it with great impetuoſity recoils as it were againſt the oppoſite rocks on the *Walachian* ſhore; inſomuch that a veſſel miſſing the right channel is in great danger of being overſet. Beyond this nook, the *Danube* ſlackening its courſe, expands itſelf into a wide curve, in which curve, or elbow, lies the iſland of *Poretſch.* From hence we come to

Sip, or fort *Elizabeth*, oppoſite *Orſowa*. At a ſmall diſtance farther lies the *Demikarpi*, i. e. *the iron gate*, commonly called *Cataractæ Danubii*, which is alſo the name of the neighbouring country, where the *Danube* enters a ſtrait betwixt mountains, purſuing its courſe over a rocky bottom. The waves and agitations cauſed by the frequent obſtructions the ſtream, which is here very rapid, meets with, toſs a ſhip with ſuch violence that, unleſs the ſteerſman be very expert, and well acquainted with the place, it is highly dangerous; and ſtill more ſo in going upwards, which can only be done by the help of ſails. In the year 1737 the *Imperialiſts* were obliged to ſink their ſhips here for want of wind to waft them againſt the ſtream. In this narrow paſs the houſes ſtand within a palliſadoe, and are ſaid formerly to have been barricadoed with an iron chain, which has given riſe to the appellation of Iron gate.

Fetiſlan, in the *Raſcian* tongue *Kladowo*, a conſiderable town on the *Danube*, within a little of which terminates the chain of mountains which begin near *Vipalanka*, and from hence to *Windin* the *Danube* runs betwixt two plains. Here are every where opportunities of laying bridges over, though the river be of a great breadth. About one *Turkiſh* mile and a quarter from *Fetiſlan* are to be ſeen

The remains of *Trajan*'s bridge, which I ſhall ſpeak of under the article *Walachia*.

3. The Sangiakſhip of *Kratowo* contains

Niſſa, *Niſſus*, *Niſſena*, a pretty conſiderable place, conſiſting of the up-per and lower fortreſs. The *Niſſa*, from which it takes its name, runs through the middle of it. It is fortified with a wall and rampart. The houſes, as in all other *Turkiſh* towns, are very ſmall and built only of wood and mortar, ſo low too that one might reach up to moſt of the roofs. In 1737 it was taken by the *Hungarians*, who loſt it the year following.

Giuſtandil, *Juſtiniano*, a caſtle on the frontiers of *Greece*. It ſeems to derive its name and origin from the Emperor *Juſtinian*.

Procupia, *Procopia*, a pretty town, ſo named from the biſhop *Procopius*. The *Turks* call it *Urchup*.

Kratowo, capital of this Sangiakſhip, where ſeveral of the royal family of *Servia* lie buried.

Priſtin, an epiſcopal ſee, pretty large, and in a flouriſhing condition.

4. The Sangiakſhip of *Scupi*, containing

Uſcup, *Scupi*, *Scopia*, an open but large well built town, on the confines of *Albania* and *Boſnia*, being the reſidence of an archbiſhop. It enjoys a great trade, particularly in a very good kind of leather, and ſtands pleaſantly at the foot of mount *Orbilus*, on the river *Vardan* or *Apius*.

Novibazar, *Janibazar*, *Novobardum*, or *Novus Mercatus*, capital of the Sangiakſhip and formerly of ancient *Raſcia*.

Sitnitza, a little place.

Ibar, a ſmall town on a river of the ſame name.

Fochia, a place of ſome figure, on the river *Lim* near the frontiers of *Boſnia*.

Uſitza, a ſtrong fort, taken by the *Imperialiſts* in the year 1737.

The plain, or heath, of *Coſſova*, thought to be the *Campus Merulæ*, and by the *Turks* called *Rigo Mezo*, lies on the borders betwixt *Raſcia* and *Bul-garia*, being very large, and famous for a battle fought there in the year 1479, betwixt *Lazarus*, *Deſpota* of *Servia*, and *Amurath* the firſt, Emperor of the *Turks*, to the diſadvantage of the Chriſtians. Dr. *Brown* ſays that *Amurath*'s tomb is ſtill to be ſeen there. On this heath *Hunniades* alſo was defeated by *Muhammed*.

BULGARIA.

BULGARIA.

§. 1. THIS country terminates northwards on the *Danube*, eaftward on the *Black-fea*; and is bounded to the fouthward by Mount *Hæmus*, which feparates it from *Romania*; and weftward by *Servia*. It is fo named from the *Bulgarians*, a branch of the *Sarmatæ*, and was formerly called the *Lower Myfia*. The *Danube*, which runs through this country for the fpace of eighty miles, receives the *Ifter* at *Axiopolis*: The other river is the *Ifcha* or *Ifchar*, which rifes in Mount *Hæmus* and falls into the *Danube* near *Nicopolis*.

§. 2. At the foot of the mountain which divides *Bulgaria* from *Servia* is a warm bath where the water gufhes out in a ftream about the bignefs of a man's body; and, but fixty paces from it, in the fame valley, is a fpring as cold as ice; the fmell however manifefts that they both contain nitrous and fulphureous particles. On this mountain is a *Greek* convent for monks of the order of St. *Bafil*. In the frontiers of *Servia* betwixt the mountain of *Suha* and the river *Niffava*, are feveral warm baths whofe waters are of a fulphureous quality, and iffue from the mountain being deeply tinged with the red fand and ftones thereof. At the foot of mount *Witofcha* a few miles on this fide *Sophia*, towards the borders of *Romania* are alfo four warm baths of great repute in this country; and the mountain, exclufive of its iron mines, is covered with villages, corn-lands, meadows and vineyards.

§. 3. The country in general is very mountainous, but the levels and vallies are extremely rich and fruitful, producing wine and corn even to fuperfluity. The mountains too are alfo far from being barren, affording, in particular, excellent pafturage; as that of *Stara Plamina*, which reaches as far as *Widin*, being towards its fummit quite bare and defolate; but in the middle and lower part extremely fertile. Among the natural curiofities of this country are alfo to be reckoned the vaft number of large eagles in the neighbourhood of the town of *Babadagi*, where the Archers all over *Turkey* and *Tartary* fupply themfelves with feathers for their arrows, though thefe feathers are in number only twelve, and thofe only in the tail, fit for their ufes: the common price of thefe feathers is a *Lion-dollar*.

§. 4. The inhabitants formerly fo renowned for their martial achievments now give themfelves up to graziery, agriculture, and handicrafts. Their language is *Sclavonick*, differing only a little from the *Servian* in pronunciation. Some of the inhabitants are *Greeks*, others *Mahometans*. The *Greek* church here has a patriarch, though not acknowledged as fuch by the other patriarchs; and three archbifhops.

§. 5. The

5. The country being governed by four Sangiaks is divided of courfe into four Sangiakfhips.

1. The Sangiakfhip of *Bidin* or *Widin* contains

Widin, Widyna, Bidinum, Bodon, Vodenum, by the ancients called *Viminacium,* a ftrong fortification on the *Danube,* and a bifhop's fee. In 1739, the *Hungarians* made a fruitlefs attempt upon this place.

Drenowatz, Melkowatz, two little towns.

Gradifte, a pretty large town on the borders of *Servia.*

Chiprawaz, a pretty town and the refidence of a bifhop.

Kliffura, Zelezna, and *Copilowatz,* three pretty good towns, where not long fince refided a great number of *Albanian* merchants of the *Roman* catholic religion : but in 1700 they were ordered to remove.

Muftapha-Bafcka-Palanka, a fortrefs, having a rampart and quadruple wall built of freeftone with eight towers : this place is capable, however, of but little refiftance being commanded by the adjacent mountains.

Scharkioi or *Scherkui,* a town furrounded on all fides with a morafs; having a caftle of the fame name feated on a mountain, near which runs the river *Niffava,* increafed by two others, namely, the *Dufchtina* and *Sredorek* rivers.

Lefcoa, Lefcovita, a town on the *Lyperitz.*

Colombotz or *Golombotz,* a well fortified caftle feated on a mountain, at the foot of which is the ftrong pafs of *Urania.*

Catfhanitz, a fortrefs commanding the pafs into the mountains.

2. The Sangiakfhip of *Sardic* contains *Sophia,* by the *Bulgarians* called *Triaditza,* a pretty large and populous trading town, well built, but open, the ftreets alfo narrow, uneven and dirty, being paved only in the footways. Every houfe, however, has a garden well planted with trees and fruit-bufhes. The *Ifcha* or *Bojahe* in certain places wafhes the town, and in others runs quite through it. The greateft part of the traders here, as in other places, are *Greeks* or *Armenians.* It is the refidence of a Beiglerbeg and was built by the Emperor *Juftinian* out of the ruins of the ancient city of *Sardica.*

Samcova, a town in the mountains.

The Emperor *Trajan*'s gate ftands among hills, where the fteep rocks and dreadful precipices fcarce admit of any accefs. It was erected by that Emperor in commemoration of his marching his army through this country, having made himfelf a road through places before impervious. It confifts of two ftone pillars, with an arch over them, reprefenting a large open gate. This building is now very ruinous, and confifts of hewn ftone and bricks. The curious in antiquity have been too bufy in taking off the ftones, which has greatly defaced this ftately monument. In the mountains leading to this gate are feveral iron-works and a boiling fpring.

Ternowa,

Ternowa, Ternobum, formerly the capital of *Bulgaria,* a royal feat, and fortified, though at prefent but a mean place. It was alfo the refidence of a patriarch, and has ftill an archbifhop, who is ftiled archbifhop of *Ternowa* and all *Bulgaria,* and even patriarch.

3. The Sangiakfhip of *Nicopoli* contains

Nicopoli, Nigepoli, a large town on the *Danube,* defended by a caftle, noted in hiftory for the firft unfortunate battle fought there betwixt the *Chri-ftians* and *Turks,* in 1396.

Preflaw, the ancient *Marcianopolis,* built in honour of *Marciana* fifter to the Emperor *Trajan.* Its prefent name fignifies *an eminent City.*

4. The Sangiakfhip of *Siliftria,* a town on the *Danube,* large and forti-fied ; being alfo a bifhop's fee. It ftands but a little way from the remains of the wall erected by the *Grecian* Emperors againft the inroads of the Bar-barous nations. Very few of the inhabitants of this place are *Turks.* Its great antiquity is manifeft from the nature of its walls, which have all the appearance of *Roman* and not *Turkifh* architecture. It was alfo called *Do-roftolus,* Δίϛϱα, Δϱίϛϱα, Δϱίϛϱου. and Δϱήϛϱα, and contains

Dobrucia a town built without the limits of the abovementioned wall.

Axiopoli, formerly a town fituate at the place where the *Danube* changed its name into that of the *Ifter;* but now hardly the name of the place remains.

Kerfowa, a little town on the *Ifter,* where it inclines towards its fource.

Betwixt the feven branches or mouths through which the *Danube* or *Ifter* enters the *Black-fea* are a like number of iflands, the four fouthern-moft of which belong to *Bulgaria;* but the three northern ones to *Beffarabia.*

Chiuflenge, Profliwitza, Iftropolis, a pretty good town on the *Black-fea,* formerly a very powerful place.

Tomifvar, Tomis, once the capital of *Leffer-Scythia* and the place of *Ovid's* exile, being feated on a bay of the fame name.

Varna, a town on the *Black-fea,* noted for the defeat there of *Vladiflaus* King of *Hungary* in 1444, by *Amurath* the firft Emperor of the *Turks.*

Dionyfiopoli, a fmall town, formerly capital of *Lower Mæfia.*

Mefembria, lying at the foot of mount *Hæmus,* formerly a bifhop's fee.

Obf. The Diftrict of *Dobrudfche* extending from *Doreftero* to the mouth of the *Danube* is one entire plain interfperfed neither by rivers nor woods; though at the end of it, not far from *Doreftero,* is a wood by the *Turks* called *Dali Orman,* i. e. *Fools-wood.* The inhabitants, who derive their de-fcent from *Tartarian* emigrants, but are now called *Pfchkias,* are noted for their fingular hofpitality, which is fo great that when a traveller of any re-ligion or country, paffes through any of their villages, all the houfe-keepers of both fexes come out to falute him, intreating him in the civileft manner to take up his lodgings with them, and kindly accept

of

of what God has been pleafed to beftow. The perfon whole invitation the traveller accepts entertains him and his horfes, if they exceed not three, for the fpace of three days; and that too with a cordiality and cheerfulnefs, which can fcarce be paralleled. He fets honey and eggs before him, in both of which this country abounds; and bread baked under the embers, but of a very fine fort. They raife alfo a little houfe, for the reception of ftrangers in particular, with couches round the hearth for travellers to ufe as they think proper.

R O M A N I A.

§. 1. **T**HIS country, which is either called *Romania* from the *Romans*, or from *New Rome* (*Conftantinople*) the feat of the eastern part of the *Roman* empire, and is known among the *Turks* by the name of *Rumili*, is the ancient *Thracia* of which fuch frequent mention is made in the *Greek* and *Latin* hiftorians. To the North it terminates on mount *Hæmus*, to the Eaft on the *Black-fea*, the *Hellefpont* and *Propontis*, or the fea of *Marmora*; being bounded to the fouthward by the *Archipelago*, and caftward by *Macedonia* and the river *Strymon*.

§. 2. The country is for the moft part level though interfperfed with fome large and remarkable mountains, the moft confiderable of which is mount *Hæmus*, dividing the country to the North from *Bulgaria*. The next in Bignefs is *Rhodope* celebrated by the ancient poets for the cataftrophe of *Orpheus*. Mount *Pangæus* feparates this country from *Macedonia*, and *Orbelus* lies at no great diftance from the river *Neftus*. *Hæmus* and *Rhodope* are two long ridges of mountains, extending from the frontiers of *Macedonia* to the *Black-fea*. The rivers of note here are

The *Maritz*, by the ancients called *Hebrus*, which takes its rife in mount *Hæmus*, and traverfing *Romania* falls into the *Ægean* fea.

The *Carafu Meftro, Neffus*, or *Neftus*; receiving its fource in mount *Rhodope*, from whence it difcharges itfelf into the *Ægean* fea.

The *Strymon* which rifes in mount *Pangæus* and runs alfo into the *Ægean* fea.

§. 3. The territories among the mountains are cold and barren; but thofe near the fea pleafant and fertile; producing all kinds of grain with other neceffaries, particularly rice which grows here in great plenty, and is remarkably good.

§. 4. This country was anciently divided into feveral independent kingdoms; the inhabitants of which were the *Dolonci*, the *Denfeletæ*, the *Beffi*, the

the *Biflones*, the *Odomantes*, the *Cicones*, the *Edoni*, the *Brigi*, the *Thyni*, the *Pieres*, the *Odryfi*, the *Satrii*, the *Crobyziæ*, the *Midiæ*, the *Sapæi*, and *Celtæ*.

The *Thracian Cherfonefus* was alfo governed by its own Kings. The prefent inhabitants are *Greek*, defcendents of the ancient *Thracians*, with a mixture of *Turks*. The flourifhing ftate of. the fciences and beaux arts among the *Greeks* was chiefly owing to the *Thracians*; but at prefent there is fcarce a perfon of any eminence in literature in all *Romania*.

§. 5. The country is governed by three Sangiaks; and confequently divided into as many Sangiakfhips.

1. The Sangiakfhip of *Kirkeli*, lies to the North near mount *Hæmus*, and contains the following places.

Jetiman, a large town not far from *Trajan*'s gate.

Bafartfchick, a noted town among the *Turks* feated on the river *Maritz*, into which, at this place, falls another ftream which runs round the town. It is univerfally well built, the ftreets being broad and clean, carries on a great trade, and is delightfully fituated.

Philippopoli, a pretty large town ftanding on two points of land, though formerly only one mountain, befides which here are three more. Upon one of the points ftands a quadrangular tower, which was once a fortification, but at prefent ferves only as a watch tower. The *Maritz*, which begins here to be navigable, feparates the town from the lower fuburbs. This place is the refidence of a *Greek* archbifhop. The town was firft founded by *Philip* father of *Alexander the Great*, from whom it received alfo its name. In 1360, the *Turks* made themfelves mafters of it. The neighbouring country abounds remarkably in rue.

Muflapha-Bafcha-Kiupri, a town, by others called *Tzgupri Cupruffi*; takes its name from a very beautiful bridge erected here over the *Maritz*, by *Muflapha Bafcha*. This bridge confifts of twenty arches, all of freeftone, and runs out a confiderable way at each end. It is faid to have coft four hundred purfes, or two hundred thoufand rix-dollars. The foil in this neighbourhood is fertile.

Kirk-Ekklefie, both a town and a Diftrict, formerly called Τισσεράκοντα Εκκλησιαι, or forty churches, from the number of Chriftian churches it contained. This place is twelve leagues from *Adrianople*, has at prefent neither walls nor churches, and but very few Chriftian inhabitants, but great numbers of *Jews* who removed hither from *Podolia*, and fpeak a corrupt kind of *German*. Their chief employment is the making of butter and cheefe, to which they affix a particular mark whereby the *Jews* at *Conflantinople*, to whom they fend it, know that the whole is clean and made by *Jews*.

2. The Sangiakfhip of *Byzia* extends eaftward from the foot of mount *Hæmus* to the fea of *Marmora*; and contains the following places

Viza, *Byzia*, a mean town, but anciently the refidence of the Kings of *Thrace*, and ftill has a *Greek* bifhop.

Adrianopoli, or *Hadrianopoli*, by the *Turks* called *Edrene*, a large city fituate in a plain on the *Maritz*, and partly furrounded with hills, on fome of which the city alfo ftands. It takes its prefent name from the Emperor *Hadrian*, or *Adrian*, who founded or reftored it; for it was anciently called *Ufcudama*, and was the capital of the *Beffi*. In the year 1360, Sultan *Amurath* firft took it from the Chriftians, from which time it became the feat of the *Turkifh* Emperor till *Conftantinople* was reduced. It is of a circular form, furrounded with a wall and towers; has good houfes, but narrow and unequal ftreets. The Emperor fometimes vifits it, either for pleafure or when he is not quite fafe at *Conftantinople*. The feraglio ftands in a moft delightful fituation, having a beautiful country on the one fide, and being feparated on the other from the city by the river *Caradare*, or *Arde*, which here falls into the *Maritz*. The objects moft worthy of attention in it are fome mofques, the roofs of which are covered with copper, having alfo lofty fteeples, and colonades, with pedeftals and chapiters of caft brafs, beautiful marble gates of exquifite fculpture, delightful fountains, ftately porticoes with gilded balls on top, and curious tapeftry; all which make a very grand appearance. Its commerce, to which the river which waters the city is of great fervice, has drawn hither people from all various nations. It is alfo the feat of a *Greek* bifhop, and, in the year 1754, fuffered greatly by fire. The adjacent country is very fertile, whence the town enjoys plenty of all kind of neceffaries, and the wine, in particular, is reckoned the beft in *Turkey*.

Hapfa, *Hapfala*, a very large haan, or publick inn, where travellers are not only lodged but entertained *gratis*.

Burgas, *Bergafe*, a celebrated market town, having a haan, the fame as that of *Hapfa*.

Haznader Tfchiftick, an imperial palace, three quarters of a mile from *Conftantinople*. Near it lies

Taut (Dawud) Bafcha, being alfo an imperial palace, built by the haznader, or commiffioner of the Sultan's treafury, and the place where the Sultan always alights in his way to *Adrianople*. This is likewife the cuftomary rendezvous of the *Turkifh* army.

Conftantinople, by the *Turks* called *Iftambol*, and the refidence of the Ottoman Emperor. Its original name was *Byzantium*, but *Conftantine* the Great, the firft Chriftian Emperor, rebuilding it entirely, called it after his own name; and, in the year 330, it was confecrated as the feat of the *Roman* Empire. It continued the refidence of the Chriftian Emperors in the eaftern part of the *Roman* Empire till the year 1453, when it was taken by the *Turks* after a fiege of fifty-four days; ever fince which it has been the feat and capital of their dominion. It ftands, like antient *Rome*, on feven hills;

and,

and, by an expres order inscribed on a stone pillar, was therefore called *New Rome* ; but so little of these remain at present, that *Constantine* would now scarce know his own city again. Though ancient *Byzantium* was reckoned the most delightful, and, at the same time, the most convenient place for trade in the whole world ; yet of the present *Constantinople* it may be said, that nothing can exceed its situation and neighbourhood. It is of triangular figure, having the continent on one side, and on the other two the sea; namely, to the southward the sea of *Marmora* and the *Hellespont*, and eastward the issue of the *Black* sea. To the north lies its harbour, which is both convenient and of very large extent, being formed by an arm of the strait which runs north-west up into the country, and is joined by a river : the fortifications, however, are too antique and ruinous to make any tolerable resistance against an army. The city makes a grand . appearance from without, rising gradually from the shore in form of an amphitheatre, but is not equal within to the ideas which may be formed of it. It is of very uncommon extent, having twenty-two gates, six of which are on the land side, and sixteen towards the sea ; but the streets are narrow and slippery, running along a declivity, and most of the houses low, being built only of wood and mortar, but crowded with inhabitants. The best houses stand in places which are least subject to any great concourse of people, and where the city is most thinly inhabited, as the finest buildings are without the city near the harbour. The seraglio, which, together with its gardens, lies at the point of the triangle, near the canal and harbour, is a mile and a half round, being rather a collection of palaces and apartments joined together by the Emperors, as their several fancies led them, than one single building. The roof, like all the other palaces of the Grand Seignor, is covered with lead. The main entrance is of marble, and by the *Turks* called *Capi. i. e.* the porte, or rather gate ; and from hence the *Ottoman* Empire receives that name. Through this gate we enter the first court, in which are the mint, the infirmary and other buildings. The second court is called the Divan court, the great council chamber being there, with the kitchin, the treasury and stables. To the north, adjoining to the Divan, is the seraglio, properly so called, through which is a narrow passage leading to the audience-chamber, which is of amazing magnificence, particularly the throne. Thus far Ambassadors are permitted to come, and this is the *ne plus ultra* of all foreigners, though not a few travellers have taken upon them to say that they have penetrated even into the women's apartments. Betwixt the two mosques of Sultan *Solyman* and *Bajazet* is the old seraglio, in which are shut up the wives of the deceased Sultans. The palaces of persons of rank among the *Turks* make no great appearance without, but the inside never fails of being very splendid. Among the mosques, or *Turkish* churches, the most magnificent and celebrated is that of Saint *Sophia,*

phia, which fronts the great gate of the feraglio.   It was built by the Emperor *Juſtinian*, and the *Turks* hold it in the ſame veneration as did the Chriſtians, the Grand Seignor going there in perſon every *Friday*, and it is very ſeldom that a Chriſtian is admitted into it.   The pavement, walks and walls are all covered with marble, excluſive of a vaſt number of pillars of porphyry, marble and *Egyptian* granate.   The revenue belonging to it is ſaid to be ten thouſand guilders a day, and it contains one hundred thouſand perſons conveniently.   Round it are ſeveral chapels, being burial places of the imperial family.   The other moſques of Sultan *Achmet*, Sultan *Muhammed*, Sultan *Selim*, Sultan *Solyman*, Sultan *Bajazet*, and three more, are alſo very fine.   The *Greeks* have thirty churches, and the *Armenians* a great many. The *Roman* Catholics have alſo more than one, and the *Swediſh* nation has been permitted to build a *Lutheran* church here.   Among the ſeveral curioſities of *Conſtantinople* is the ancient *Hippodrome* by the *Turks* called the *Atmeïdan*, not far from the moſque of Sultan *Achmet*, and in which ſtands a pyramid of *Theban* marble inſcribed with hieroglyphics; the coloſſus alſo, or pillar conſiſting of ſeveral ſquare blocks of free-ſtone; a triangular pillar of caſt braſs, repreſenting three ſerpents folded in each other, but having the triple head fallen off; a column in honour of the Emperor *Arcadius*, ſtanding on the ſeventh hill, in the road leading from *Adrianople* to the *Hippodrome*; the ſlave market, and the building where they are kept, which is not far from the laſt mentioned pillar; and the formidable ſtate priſon of the ſeven towers, to which lately an eighth has been added.   It is built of fine free-ſtone at the ſouth end of the city, and environed by a wall with ſeveral ſmaller towers, ſome of which, in the year 1754, fell to ruin. Laſtly, the market places, which the *Turks* call Bezeſtens, being ſquares, or exchanges, built on piazzas, where the *Turks*, *Jews*, *Greeks* and *Armenians* meet for traffick.   The Janizaries have alſo their dwelling within the imperial palace, and live in one hundred and ſixty-two odas or chambers. The number of people in *Conſtantinople* is, by *Otter*, computed to be eight hundred thouſand.   *Athanaſius Doroſtanus*, as cited by *Elſner*, affirms that the *Greeks* alone make four hundred thouſand, but the *Armenians* not half ſo many.   The palace of the *Greek* patriarch here ſtands on a hill, about two hundred paces from the harbour, near the patriarchal church dedicated to St. *George*.   Such is the police of this large city, and ſo ſtrictly is good order maintained, that a foreigner, taking a Janizary with him, may walk about any where without the leaſt inſult or moleſtation.   The plague, which viſits it every year, frequently makes dreadful havock among them, though this is in a great meaſure owing to the diſorderly and filthy manner of living among the people.   The city has often ſuffered by fires, ſome of which have deſtroyed from fifty to ſeventy thouſand houſes.   In the year 1754 it ſuffered much by an earthquake, a fire breaking out at the ſame time.

time. In 1755 and 1756 two other terrible conflagrations happened here.

On the weft fide of the city are the fuburbs of *Ejup*, or St. *Job*. The country along the ftrait into the *Black* fea is covered with towns and villages, feats, gardens, meadows, vineyards and woods. The chief officers of the court generally refide there in fpring, fummer and the beginning of autumn, both for the benefit of the frefh air, and in order to be near the Emperor, who fpends the fummer at *Befiktatfch*, which we fhall prefently make mention of. Next to *Conftantinople*, on the oppofite fide of the harbour, lies

Galata, a fuburb furrounded with walls, towers and moats. The inhabitants of this place are chiefly *Greeks*, *Armenians*, *Franks* and *Jews*, who choofe it for their refidence, for the fake of living more at liberty than in the city. The *Roman* Catholics have a few churches here, and the *Greeks* fix. Here alfo are the warehoufes of the merchants; and near them, clofe to the harbour, is the *derfana*, *terfkahanne*, or dock. In this part alfo is the place called *Caffun Bafcha*; and fomewhat more to the eaft St. *Demetrio*. At no great diftance from it is

Pera, which is alfo a handfome fuburb, ftanding on an eminence, and the quarter where the Chriftian envoys refide; though inhabited principally by the better fort of *Greeks*. The air is healthy, and the profpect pleafant. Afcending from hence we come to

Tophana, fo called from the cannon foundery there, and which may be looked upon as one of the fuburbs of *Conftantinople*. It lies directly fronting the imperial feraglio.

Farther eaftward, on the ftrait where the *Black* fea communicates with the fea of *Marmora*, are the feats of *Funduklu*; *Befiktafch*, an inclofed imperial palace for women, and the cuftomary fummer refidence of the Emperor; *Ortakoy*, *Czanaklimana*, *Kurutfchefme*, *Arnantkoy*, and *Kajolar*. Next to thefe is a ftrong caftle commanding the entrance to *Conftantinople* from the *Black* fea, and oppofite to it is alfo another in *Natolia*. Farther on are the feats of *Baltilimani*, *Emirkoy*, *Ifteinia*, *Jenikoy*, *Therapia*, *Riffelkoy* and *Belgrad*; the laft of which is a *Greek* village, fituate in a wood, where the Grand Signior has *kiofkes*, or fmall pleafure houfes, which he fometimes vifits. The foreign minifters have alfo their places of retirement here. Omitting fome other places, at the iffue of the *Black* fea, both on the *European* and *Afiatick* fide, ftands a ftrong caftle; and, not far from it is not only feen a lighthoufe for the fafety of mariners, but, on an eminence about thirty paces from the fea, ftands the remains of *Pompey's* pillar, and near it *Ovid's* tower. Next is

Selivrea, *Selybria*, *Selymbria*, a celebrated port on the fea of *Marmora*, with an old ruinous caftle, ftanding on an eminence, formerly very ftrong, and with the houfes near it called the upper town. In the fuburbs is an imperial

granary,

granary, into which the corn of the province is brought. This place is the refidence of a *Greek* bifhop.

Heraclea, anciently *Perinthus,* on the fea of *Marmora,* formerly a large city, but now a mean place. Here are ftill to be feen the remains of an amphitheatre built in the time of the Emperor *Severus.* It is alfo the refidence of a *Greek* archbifhop.

Rodofto, a trading town lying on the fea of *Marmora.*

3. The Sangiakfhip of *Gallipoli* extends from mount *Rhodope* to the *Archipelago,* and is the fouth-weft part of the Diftrict. To it belongs

The *Thracian Cherfonefus,* a peninfula, environed on the fouth by the *Archipelago,* weftward by a bay into which falls the fmall river of *Melas,* and towards the caft by the ftrait which the ancients called the *Hellefpont,* and on the north it is joined to the continent by a tract of land, the breadth of which was reckoned by the ancients to be about thirty-feven ftadia. It contained formerly eleven towns, but at prefent the following are the principal places of note.

Gallipoli, formerly *Callipolis,* a well inhabited town, with a fpacious harbour, feated on the celebrated ftrait which divides *Europe* from *Afia,* by the ancients called the *Hellefpont.* This is the firft *European* town which the *Turks* made themfelves mafters of.

Sefto, Seftos, once a fortified caftle on the *Hellefpont,* oppofite *Abydos,* in *Natolia.* Farther on to the fouth and the *Ægean* fea lie

The celebrated *Dardanelles,* or two caftles which command the whole ftrait, and are the key to *Conftantinople.* One of them ftands on a peninfula in *Europe,* and the other oppofite to it in *Afia.* That in *Europe* confifts only of one round tower with fome outworks, and is likewife of fmaller importance; clofe by it ftands a village. Both thefe caftles were built by *Mahomet* the fecond in 1452. In the year 1656 the *Venetian* fleet forced their way through them and drove the *Turkifh* fleet afhore. All fhips coming out of the *Archipelago* are fearched here. On a rock in the middle of this ftrait ftands a tower, properly confifting of two, on which the *Turks* have fome fmall cannon. To mariners it ferves for a mark to fteer by, but the *Turks* ufe it as a watch-tower. In the middle of the rock is a frefh fpring.

Cardia, a fmall place on the weft fide of the peninfula; and, on the bay into which runs the river *Melas.* This place, according to *Pliny,* was fo called from its being built in form of a heart.

Trajanapoli, a fmall town on the *Maritz.*

Demotica, Didymotichus, a town on the *Maritz,* where *Charles* the twelfth, King of *Sweden,* fpent fome time in the year 1713. It is the refidence of a *Greek* Bifhop.

Polyftilo,

Polyſtilo, Aſperoſa, Aſtrizza, a mean place on the *Archipelago,* anciently called *Abdera,* and one of the moſt celebrated cities of *Thrace.* It was noted formerly for its gold and ſilver mines.

Obſ. The following parts of *Turkey* in *Europe* belong to *Greece.* I was extremely deſirous of deſcribing them according to the political diviſion founded on the ſyſtem of the *Ottoman* Porte, but the want of the necef-ſary helps and vouchers rendered that impracticable; ſo that, with other geographers, I muſt retain the diviſion and appellation of the ſeveral coun-tries made uſe of among the *Greeks :* though of their ancient ſtate I ſhall only ſpeak curſorily, this work being intended not for an ancient, but a modern, geography.

M A C E D O N I A.

TO the north it is bounded by the river *Neſſus* or *Neſtus,* eaſtward by the *Archipelago* ; ſouthward it joins *Theſſaly* and *Epirus,* and to the weſt *Albania.* The figure of it is very irregular, but the ſituation advan-tageous and the air clear, ſharp and wholeſome. The ſoil is for the moſt part fertile, and the maritime coaſts, in particular, abound in corn, wine, oil, and every thing that can be deſired for uſe or pleaſure. In the inland parts are ſeveral uninhabited waſtes. It had mines formerly of almoſt all kinds of metal, but particularly of gold. Among the many large moun-tains in this country is the chain of the *Scardi,* which traverſes the northern part of it. *Pangæus* was formerly noted for very rich ſilver and gold mines. The mountain of *Hæmus* joins the *Scardi* ſeparating this country from *Romania.* Mount *Athos* is one of the moſt celebrated mountains in the whole world, and ſhall be particularly deſcribed in the ſequel. Of woods, and all kinds of timber, here is a proper plenty ; and the many fine bays in this country are a great convenience to trade. The moſt remarkable of theſe are the *Golfo di Conteſſa (ſinus Strymonicus) Golfo di Monte Santo (ſinus ſingiticus)* and the *Golfo di Salonichi, (ſinus Thermæus.)* The principal rivers are, The *Platamone (Aliacmon)* which runs into the bay of *Salonichi* ; the *Viſtriza (Erigion)* which mingles with the follow-ing ; *viz.* the *Vardar (Axius)* the greateſt river in all *Macedonia,* taking its riſe in the *Scardian* mountains and falling into the bay of *Salonichi.*

The *Strymon* riſes in *Romania,* or *Thrace,* and diſcharges itſelf into the *Golfo di Conteſſa.*

Beſides

Befides the bays formed by the *Vardar* and *Strymon*, there are fome others of note, as thofe near *Achrida (Lychnidus, Prefpa)* with another between the bays of *Salonichi* and *di Conteffa.*

Macedonia having been formerly inhabited by feveral nations had a great number of towns. The moft remarkable places in it at prefent are the following

Heraclea, Heraclea Sintica, anciently *Sintia,* a fmall town on the river *Strymon.*

Philippi, a village having but few houfes, ftands near the ruins of the an- cient celebrated town of that name. Its inhabitants confift only of a few poor *Greeks,* yet is it the refidence of a *Grecian* bifhop, who ftyles himfelf Metropolitan of *Philippi* and *Drama,* and has feven bifhops under him. The city of *Philippi* ftood on a hill betwixt the rivers *Neffus* and *Strymon* on the borders of *Thrace,* to which, in its moft ancient times it belonged. It was at firft called *Crenides,* or fpring-town, from the many fprings iffuing out of the hill on which it ftood; afterwards *Dathos,* or *Thafus,* from the *Thafij* who built it, and laftly *Philippi,* from *Philip* of *Macedon,* who, after re- ducing it, rebuilt it with confiderable improvements; and from that time it belonged to *Macedonia.* Near this place *Caffius* and *Brutus* were de- feated by *Octavius* and *Mark Antony.* Under *Julius Cæfar* and *Auguftus* it was a *Roman* colony. At prefent it lies wafte, though ftill boafting fome curious remains of antiquity, particularly an amphitheatre. The Apoftle *Paul* has written an epiftle to the Chriftians of this place.

Conteffa, a fmall place giving name to the bay into which the river *Strymon* runs.

Emboli, Amphipolis, Chriftipolis, now a defolate place on the river *Stry- mon,* but formerly famous as an *Athenian* colony. The fecond name is the moft ancient; the third it received from the Chriftians, and the firft has been beftowed upon it by the *Turks.*

Mount *Athos,* commonly called *Monte Santo,* lies on a peninfula running out into the *Ægean* fea, and is, indeed, a chain of mountains extending the whole length of the peninfula, being feven *Turkish* miles long and three in breadth; but it is only one fingle mountain which is properly called *Athos.* Its uncommon height appears from the accounts of *Plutarch* and *Pliny,* who affirm that when the fun is at the fummer folftice, probably a little before its fetting, the mountain cafts its fhadow as far as the market-place of *Myrrhina* in the ifland of *Lemnos,* which, in the beft maps, is fifty-five *Italian* miles diftant; whence the height of mount *Athos* may be inferred to be about eleven ftadia. On it are twenty-two convents, befides a great number of cells and grottos, with the habitations of no lefs than fix thoufand monks and hermits; though the proper hermits, who live in grottos, are not above twenty: the other monks are anchorites or fuch as live in cells. It is evi- dent from *Ælian* that anciently the mountain in general, and particularly the

fummit,

fummit, was accounted very healthy and conducive to long life; whence
the inhabitants were called *Macrobii,* or long lived. We are further in-
formed by *Philoſtratus,* in the life of *Apollonius,* that numbers of philoſo-
phers uſed to retire to this mountain for the better contemplation of the
heavens and nature; and, after their example, it unqueſtionably was, that
the monks built their φρουλιϛήρια, or cells. The monks called ἁγιορῖται, or
ἁγιορεῖται, *i. e.* inhabitants of the holy mountain, are ſo far from being a ſet of
ſlothful people, that, beſides their daily offices of religion, they perform all
manner of work, cultivate the olive and vineyards, are carpenters, maſons,
ſtone-cutters, cloth-workers, taylors, *&c.* They live alſo a very auſtere
life, their uſual food, inſtead of fleſh, being vegetables, dried olives, figs,
onions, fruit, cheeſe, and, on certain days and *Lent* excepted, fiſh. Their
faſts are many and ſevere; which, with the healthfulneſs of the air, renders
longævity ſo common there, that many of them live above a hundred
years. In every convent are two or three ſtudying monks, who are exempted
from labour, but uſe exemplary diligence among the many writings to be
found in their libraries. Here it is the *Greeks* properly and chiefly learn
their divinity. The monks are in high eſteem for the orthodoxy of their
doctrine and the ſanctity of their lives. Theſe convents and churches have
bells, which are no where elſe allowed the *Greeks*; and are alſo environed
with high and ſtrong walls planted with cannon, againſt any ſurprize from
Corſairs. Beſides churches and convents, the mountain has alſo a town
called *Kareis,* inhabited alſo by monks, and the reſidence of the *Turkiſh*
Aga, who commands here in the name of the Boſtangi Baſcha, to defend
the place againſt the Corſairs. In this town a market is held every *Saturday*
among the monks and anchorites, which laſt bring hither knives and little
images, with what money they thus earn purchaſing themſelves bread; but
the monks carry them about every where and receive alms for them. The
mountain is under the protection of the Boſtangi Baſcha, to whom it annually pays twelve thouſand dollars, and almoſt a larger ſum is further to be
paid at *Salonichi* to the uſe of the Sultan himſelf. This heavy tribute is
diſcharged by alms and the liberal contributions of *Ruſſia* and the Princes of
Walachia and *Moldavia.* No fowls or cattle are kept upon this mountain,
though, on paying a conſideration, graziers are allowed to fatten their cattle here. On this chain of hills formerly ſtood five cities, *viz.*

Aiomama, a mean place, noted only for the bay to which it gives
name,

Salonichi, Theſſalonica, a celebrated trading city at the end of the bay of *Salonichi,* and, at preſent, the moſt conſiderable place in *Macedonia.* This city was
formerly called *Halia* and *Therma,* but *Caſſander* building it anew, gave it
the name of his ſpouſe *Theſſalonica,* who was ſiſter to *Alexander* the Great.
To its admirable ſituation for trade is probably owing all the regard which
the ſeveral conquerors of *Macedonia* have ſhewn it. The advantages de-

rived from it are such as are scarce to be met with elsewhere; and as it at-
tracted the encomiums of the ancients, so it has the admiration of the
moderns. Nor is it distinguished only by the greatness of its traffick, but is
also very remarkable for the stately remains of its ancient grandeur, such as
triumphal arches (of which there is one here almost entire, erected in honour of the Emperor *Antoninus*) churches of an extraordinary beauty and
statelinefs, now converted into *Turkish* mosques, particularly that of St.
Demetrius, which confifts of one church built over another, and having in
it above a thousand pillars of jasper, porphyry, &c. In this and other
churches are the monuments of several celebrated personages; and without
the city are great numbers of antique fragments with inscriptions. Numbers
of coins too are also frequently found here. It is the residence of a *Turkish*
Basha, and likewise of a Catholick and *Greek* archbishop, who has eight suffragans under him. In the year 1313 the city was sold to the *Venetians*,
who were dispoffeffed of it about eight years afterwards by *Amurath* the
second. The Christians here were formerly so confiderable that St. *Paul*
has addreffed them in two epiftles.

Jenizza, anciently *Bunonus*, *Bunomia*, *Pella*, at the mouth of the river
Vardar, or *Axius*, a town now in ruins, but noted for being the birth-
place of *Philip* and his son *Alexander*. In its neighbourhood formerly stood
the monument of *Euripides* the celebrated tragedian.

Chitro, *Sitron*, *Pydna*, on the bay of *Salonichi*, noted for being the place
where the mother, spouse and son of *Alexander* the Great were murdered by
Caffander, and for the victory obtained in its neighbourhood by *Paulus Æmi-
lius*, the *Roman* Conful, over *Perfeus* King of *Macedonia*.

Veria, *Beroea*, a place mentioned in the Acts of the Apoftles.

Aliffone, a town with a *Greek* convent.

Sarvitza, a town standing partly on a mountain and partly on a plain,
with a caftle on a high rock being a ftrong pafs.

Sarijiole, a place of no confideration.

Ædeffa, *Ægæa*, fituate on the river *Viftriza*, or *Erigonius*, anciently the
capital of the Kingdom of *Macedonia*; and, till the time of *Philip*, the
refidence and burial-place of its Monarchs.

Ochrida, *Achrida*, *Giuftendil*, anciently *Lychnidus* and *Lychnidia*, a large
trading town on the bay of the fame name, and the feat of a *Greek*
archbifhop.

Eccifo Werbeni, famed for its mineral waters.

Pirlipe, a place fituate among high mountains of the fame name, which
glitter like filver, and befides talc abound also in good metals and minerals.

Krupulik, i. e. *Bridge-town*, feated on the river *Pfinia*.

Kaplanik, or *Tiger-town*.

Comonava, a town having a *Greek* convent near it.

ALBANIA.

A L B A N I A.

ALBANIA, or *Arnaut*, comprehends the old *Grecian Illyricum* and *Epi-rus*. The former was added to *Macedonia* by *Philip*, the word *Epirus* fig-nifying the continent. It is to *Epirus* that *Italy* owed its first apricots, whence it accordingly called them *mala Epirotica*. The inhabitants make very good foldiers, but have fcarce any notion of learning among them; yet are they very fkilful in laying aqueducts, and without any mathematical inftru-ments meafure heights and diftances with all the exactnefs of a geometri-cian. Their method of treating hernia's is alfo remarkable, but very rough. The chief rivers in *Albania* are

The *Bojana.*

The *Black Drin*, called alfo the *Drino Nigro* and *Caradrina*, which falls near *Alefjo* into a bay of the *Adriatick.*

The *Argenta.*

The *Siomini*, in *Latin Paniafus.*

The *Chrevafta*, in *Latin Apfus.*

The *La Pollonia*, in *Latin Laous, Aeas, Aous.*

The *Delichi*, in *Latin Acheron*, of which fuch frequent mention is made in the old poets.

Its lakes are *Lago di Scutari*, containing fome iflands, and, among other rivers which difcharge themfelves into it, is, particularly, the *Moraca*, which abounds in fifh.

Lago di Plave, having a communication through the river *Zem* with that of *Scutari.*

Lago di Hotti, having alfo a communication with that of *Scutari.*

Lago Sfaccia, &c.

Here we fhall take notice of the following places, *viz.*

Scutari, Scudari, Scodra, a large and fortified town on a lake of the fame name, with a caftle, enjoys a great trade, is the refidence of a Beiglerbeg and archbifhop, and was formerly the feat of the Kings of *Illyricum.* In the years 1474 and 1478 it was befieged in vain by the *Turks*, but the *Ve-netians*, in 1479, furrendered it.

Drivafto, in *Latin Trivaftum*, a fmail town on the river *Chiri.*

Antivari, Antibarum, a fortification on the *Adriatick* taken by the *Ve-netians* from the *Turks* in the year 1573.

Dulcigno, Dolcigno, Ulcinium, Olchinium, a town with a good harbour and ftrong caftle. The inhabitants are noted for their maritime depredations. In the year 1571 this place fell into the hands of the *Turks.*

Aleffio,

Aleſſio, in *Latin Lyſſus,* a town not far from the *Drin,* where it diſcharges itſelf into the ſea. Here the valiant *Scanderbeg* died and was buried.

Croja, not far from the mouth of the *Drin,* noted for being the birth-place of *Scanderbeg.*

Durazzo, Epidamnus, Dyrrachium, a ſmall ſea-port, on a peninſula, having a pretty good harbour and caſtle. Its firſt name, *Epidamnus,* de-notes the corruptions of its inhabitants, who were ſo infamous for fraud, treachery and voluptuouſneſs, that the *Romans,* when they became maſters of the town, changed its name to that of *Dyrrachium,* whence its preſent name is derived.

Pollonia, Pallina, Pirgo, ſaid to be the ancient *Apollonia,* ſo famed for its delightful ſituation and excellent laws. In latter times it has been one of the ſeats of learning, but is now reduced to ſo deplorable a condition, that writers are not thoroughly agreed about its ancient name.

Vallona, Aulon, anciently fortified, ſtands on a bay, having a ſpacious but not very ſecure harbour. In the year 1464 it was taken by the *Turks,* from whom the *Venetians* recovered it in 1690, but were again obliged to reſign it the following year.

Canina, a town and ancient fortification.

Monti della Chimera, formerly the *Ceraunian,* or *Acroceraunian,* moun-tains, were reckoned the boundaries betwixt the *Ionian* and *Adriatick* ſeas; and ſo called becauſe often ſtruck with lightning.

. *Chimera,* once a ſtrong town, and noted for its warm baths. At preſent. it is but a mean place.

Butrinto, anciently *Buthrotum,* a ſmail place, on a lake of the ſame name, belonging to the *Venetians.*

Delfino, the beſt town in *Epirus,* ſeated near mount *Pindus,* and the re-ſidence of the *Turkiſh* governor.

Larta, a large and well peopled town on a bay of the ſame name, be-longing to the *Venetians.*

Voinizza, a ſmall fortification on the ſame bay, the property of the *Ve-netians.*

Preveſe, a town with a harbour belonging to the *Venetians.* Hereabouts muſt have ſtood the city of *Nicopolis,* founded by *Auguſtus* after the victory of *Actium.*

THESSALY.

THESSALY, or *JANNA.*

THESSALY, now called *Janna* by the *Turks,* derives its name from
King *Theffalus,* but was more anciently called *Æmonia,* from *Æmon,*
father of *Theffalus;* *Pelafgia,* from *Pelafgus,* grandfather to *Æmon*; and
Pyrrhæa, from *Pyrrha, Deucalion's* wife. It lies between *Macedonia,* the
Archipelago and *Greece,* properly fo called, or *Livadia* and *Albania.* It was
fometimes annexed to *Macedonia,* fometimes divided from it, and then again
united with it. The celebrated mount *Pindus,* now *Mezzovo,* or *Mezzo
Novo,* feparates it from *Epirus,* or a part of the prefent *Albania.* Among
its once celebrated twenty-four mountains the moft remarkable is *Olympus,*
now called *Lacha,* which, for its uncommon height, is celebrated by the ancient Poets, and made the refidence of the Gods. It is fuppofed by them
to reach up to the heavens, and yet is not much above an *Englifh* mile in
height: *Petras,* anciently called *Pelion,* being one thoufand two hundred and
fifty paces in height: *Offa,* which, together with *Nephele,* was, according
to the fabulous accounts of antiquity, inhabited by the *Centaurs,* whom *Hercules* flew, or drove out, *&c.* Here alfo are fituated the plains of *Pharfalia,* and between the mountains *Olympus, Pelion* and *Offa,* is the delightful valley of *Tempe,* which was fo adorned with the gifts of nature, and fo
delightfully watered by the gently winding ftreams of the tranfparent *Peneus,* now the *Salampria,* that it was reckoned the garden of the Mufes.
The country is certainly fertile to exuberance, and feems to exceed all
other parts of *Greece.* It produces oranges, citrons, lemons, pomegranates,
grapes of an uncommon fweetnefs, excellent figs and melons, almonds,
olives, cotton, *&c.* and cheftnuts take their *Latin* name from the town of
Caftanea in *Magnefia,* whence they were firft brought into the colder climates
of *Europe.* It was noted anciently for its breed of cattle and horfes, from
which, and the extraordinary fkill of the *Theffalians* in horfemanfhip, in all
probability the fable of the *Centaurs,* who are faid to have been half men
and half horfes took its rife. The modern *Theffalians* are a well made
fpirited people. The moft remarkable places in this country are

Lariffa, Larfo, by the *Turks* called *Genifahar,* or *Jeng-ifchahir,* the
capital, ftanding on the river *Peneus,* in a hilly and very delightful part of
the country. It is a good trading city and the fee of a *Greek* archbifhop.
Here the celebrated *Achilles* was born. In the year 1669 the *Turkifh* Emperor held his court here.

Tornovo,

Tornovo, a fpacious and pleafant city in which are eighteen *Greek* churches and three *Turkifh* mofques.   The prefent bifhop is under the archbifhop of *Lariffa*.

Janna, *Jannina*, a well inhabited town, from which the country receives its prefent name.

Zeiton, a town on a bay of the fame name.

Armira, a town on the *Golfo dell Armiro*, thought to be the *Eretria* of the ancients.

L   I   V   A   D   I   A.

UNDER this name at prefent is comprized ancient *Greece*, properly fo called, to which belonged the little Kingdoms of *Acarnania*, *Ætolia*, *Ozolæa*, *Locris*, *Phocis*, *Doris*, *Epiknemidia*, *Bæotia* (now *Stramulippa*) *Megara* and *Attica*.   It reaches from the *Ionian* fea to the *Archipelago*, containing anciently feveral places of note.   The principal rivers in this country, which is for the moft part mountainous, are, the *Sionapro*, anciently called the *Achelous*, feparating the *Arcananians* from the *Ætolians* ; the *Cephiffus*, falling into the lake of *Copai*, which it properly forms, and the *Ifmenus*, which probably ran into the *Afopus*, a river difcharging itfelf into the *Archipelago*.   Mount *Oeta*, in *Bæotia*, noted for the pafs of *Thermopyle*, which was not above twenty-five feet broad, and derived its name from the warm baths in its neighbourhood.   In *Phocis* are feveral celebrated mountains, as namely, *Parnaffus*, facred to *Apollo*, and of which all the poets are fo full, and *Helicon* and *Cythæron*, both confecrated to the Mufes, and confequently very famous among the poets.   The following places are the moft remarkable at prefent.

Lepanto, anciently *Naupaĉtus*, a town on a. mountain along the fhore of the gulf of *Lepanto*, formerly called the gulf of *Corinth*.   On the fummit of the mountain ftands a fmall caftie.   The town is furrounded by a fertile country çovered with olive and vineyards, corn-fields and plantations of oianges, citrons and lemons.

The *Dardanelles*, or two caftles defending the narrow entrance into the gulf of *Lepanto*.   Here the *Venetians*, in the year 1571, gained a moft fignal viĉtory over the *Turkifh* fleet.

Delphi,

Delphi, Delphos, now *Castri,* two *Turkish* miles north of the gulf of *Le-panto,* standing on a bare mountain. At present it is but a poor place, having about two hundred houses, but was anciently very noted for the temple and oracle of *Apollo.*

Livadia, a large populous town on the gulf of *Lepanto,* built round a mountain which terminates in a peak, having on it a castle. It carries on a pretty good trade.

Megara, a mean place, not far from the *Golfo de Engia,* once capital of a particular republick.

Athens, now *Setines,* the ancient capital of *Attica,* at first called *Cecropia,* from *Cecrops* its founder, but afterwards known by the name of *Athens,* which it derived from the goddess *Minerva.* Exclusive of its power, grandeur and opulence, it was highly celebrated for the incorruptible fidelity of its citizens; and for being the nursery of the most eminent philosophers, statesmen and orators; for its multitude too of great commanders no other city has ever equalled it. It was governed at first by Kings, then by Archons, but afterwards fell successively under the power of the *Persians, Macedonians* and *Romans.* In later times it came under the dominion of the *Turks,* from whom it was taken by the *Venetians.* In the year 1455 the *Turks* again took it. In 1687 the *Venetians* recovered it, but in the last wars betwixt those two powers the *Turks* got it into their hands. These vicissitudes have indeed greatly diminished its splendour; but, both within and without the city, in its present state, are many remains of its ancient grandeur, which proclaim the great perfection of *Athenian* sculpture and architecture. The inhabitants, at this time, are reckoned about ten thousand, three parts of whom are Christians, and have a great number of churches and oratories: The *Turks* too have five mosques here. It is the residence of a *Greek* bishop. Among the many great and small remains of ancient stately edifices, those more particularly worthy of notice are reckoned to be the temple of *Jupiter Olympius,* and, above all, the *Parthenion,* or the stately temple of *Minerva,* now converted into a *Turkish* mosque, and accounted the finest piece of antiquity in the whole world. In the last *Venetian* war this building suffered much by the cannon. *New Athens* is a part of *Athens,* and, as a compliment to the Emperor *Adrian,* was called by his name. The two rivers *Ilissus* and *Eridanus,* which water the plain in which *Athens* now stands, are very small; the former being diverted into several canals for watering the olive-yards, and the latter lost amidst the many branches into which it is conveyed over the country. *Athens* had anciently three harbours, of which *Phalareus* and *Munichia* lay to the eastward, and *Pyræus* to the west of a small cape. The latter being an inclosed spacious harbour, with a narrow entrance, is still much resorted to, and by the *Greeks* called *Porto Drago,* but by the *Italians Porto Leone,* from a pillar there in memory of a lion which was carried from hence to *Venice.*

Lepsina,

Lepſina, formerly the celebrated *Eleuſis,* ·now lying in ruins.

Stibes, Stives, the ancient *Thebes,* diſtinguiſhed for its many pompous temples, palaces, and other coſtly ſtructures; but at preſent without any remains of its ancient grandeur, except that it is ·the ſee of a *Greek* biſhop.

The *M O R E A.*

THE *Morea* is almoſt an iſland, but joins to the continent, or country of Proper *Greece,* by a narrow neck of land called the iſthmus of *Corinth,* ſo famed for the *Iſthmian* games celebrated there in honour of *Neptune.* It is but a peninſula, anciently called the *Peleponneſus,* and in more remote times *Aegialea* and *Apia.* It conſiſted of the following ſmall Kingdoms : *Sicyon, Argos* and *Mycene, Corinth,* Proper *Achaia, Arcadia* and *Laconia.* Its preſent name of the *Morea* is ſaid to be derived from *Morus,* a mulberry-tree, either from its reſembling the leaf of that tree in ſhape, or from the great number of mulberry-trees it produces. The chief rivers in it are the *Carbon,* anciently the *Alpheus* ; the *Pirnaza,* anciently the *Paniſus* ; the *Eurotas,* now called the *Baſilipotamo, i. e.* King's river, and falling into the *Golfo de Calochina.* Of the lakes, the moſt celebrated among the ancients was the *Stymphalis,* noted for the many ravenous birds which frequented it ; and the *Phineus,* for being the ſource of the river *Styx,* whoſe waters were of ſuch remarkable coldneſs as to freeze thoſe to death who drank of it. It alſo corrodes iron and copper, for which very extraordinary qualities the old poets have made it the river of hell. It has not only ſeveral mountains but alſo many fertile and delightful plains. At the treaty of *Carlowitz* the *Turks* ceded all the *Morea* to the *Venetians,* but, in the year 1715, again diſpoſſeſſed them of it. It is divided into four Diſtricts.

1. *Saccania* or *Romania Minor* contains the ancient cities of *Corinth, Sicyon,* and *Argos.* The moſt remaikable places of this Diſtrict are

Corinth, Corinto, by the *Turks* called *Gereme,* a celebrated town at the foot of *Acro-Corinth,* on which ſtands a caſtle, having a moſt beautiful proſpect on every ſide. The original name of this city was *Ephyra* with the addition of *Heliopolis,* or city of the ſun, and *Bimaris,* which ſignifies a place lying betwixt two ſeas. It was anciently one of the fineſt cities in all *Greece,* and abounded in ſtately buildings, ſuch as temples, palaces, amphitheatres, portico's, monuments, baths and other woiks; all ornamented

with

with pillars, cornices and pedeftals, whofe fingular decorations gave rife
to the appellation of the *Corinthian* Order, with numbers of ftatues by the
greateft mafters. But its edifices being demolifhed at prefent, and the
fpot on which it ftood filled up with little gardens and fields, it looks much
more like a village than a city: And what is ftill worfe, is frequently expofed to the defcents of the *Corfairs*. It is however the feat of a *Greek*
archbifhop. The Chriftians here from fome paffages in St. *Paul's* Epiftles,
appear to have been very numerous.

The remains of the ancient city of *Sicyon* on the river *Afopus*.

Nemea, a village, famed for the ancient *Nemæan* games.

Argos, a mean place on the river *Najo* or *Inachus*, formerly a fplendid
capital, is the fee of a bifhop, and defended by a citadel.

Mycene, a village, once the capital of a Kingdom.

Napoli di Romania, in *Latin Neapolis*, anciently *Nauplia*, is a city and
fortification on a peninfula, which ftretches into a bay, from the town
called *Golfo de Neapoli*. It is an archbifhop's fee, and has a good harbour.
In 1715, the *Turks* made themfelves mafters of this place.

2. *Braccio di Mania* or *Tzakonia*, comprehends the ancient *Arcadia* and
Laconia, and contains the following places, viz.

Leontari, formerly *Megalopolis*, and *Dorbo*, the ancient *Mantinea*, two
confiderable cities of old, but at prefent mean places, efpecially the
latter.

Mifitra, fituate on the river *Eurotas*, or *Bafilipotamo*, anciently called
Sparta and improperly *Lacedæmon*, the capital of *Laconia*. It confifts of
a caftle, the city, properly fo called, and below it lie two large fuburbs.
It is alfo a bifhop's fee.

Malvefia, by the modern *Greeks* called *Monembafia*, by the Turks *Me-
newtfche*, the ancient *Epidaurus*, lies on the *Golfo de Neapoli*, and is the
ftrongeft fortification in all the *Morea*. Its lufcious wine called Malmfey
was very famous in former times. It has a pretty good harbour, and is the
feat of a *Greek* bifhop.

The promontory of *Malio*, anciently *Malea*, lies on the fouth-eaft point
of land.

Colochina, is a fmall town, giving name to the *Golfo de Colochina*.

Maina, a place, with a Diftrict belonging to it, on the fouth part of the
land. Its inhabitants, and thofe in the neighbourhood called *Mainotti*, are
the defcendents of the ancient *Lacedemonians*, and to this very day the moft
valiant among all the *Greeks*; and though, by eftimate, their whole military
force does not exceed twelve thoufand men, they have never been conquered, nor even made tributary to the *Turks*. Their country is on all
fides furrounded with mountains, and their modern name is derived from
μάνια, *i.e.* madnefs, from their cuftom in battle of rufhing in upon the enemy
as if actuated by a phrenzy.

Cabo Matapan, anciently called the promontory of *Tanara*, ftretches fouthwards to a great diftance in the fea, and has two harbours, one called *Achilleus*, the other *Pfamatheus*.

3. *Belvedere* comprizes the ancient *Elis* and *Meffenia*, and in it are the following places of note, *viz.*

Coron, a confiderable fortified town, having a harbour on the bay of *Coron*: *Sagara* and *Colone*, two fmall-fea-ports.

Modon, anciently *Methone*, a place of confiderable trade and fortified, having alfo a harbour. It is the refidence of the governor of the *Morea* and a bifhop's fee.

Navarin, anciently *Pylus*, a fortified trading town on the fea. The harbour here is reckoned the beft and moft fpacious in all the *Morea*.

Arcadia, formerly *Cypariffa*, a mean place, from which a bay takes its name.

Langanico, *Olympia*, *Sconri*, feated on the river *Carbon*, anciently a city of great note, near which, on the *Olympian* plain, were celebrated the games fo called, which were firft inftituted by *Pelops* in honour of *Jupiter*, and afterwards revived by *Atreus* and *Hercules*. They were held every fifth year with great folemnity, amidft an infinite number of fpectators, and lafted for five whole days together. From thefe fpectacles the computation of time in *Greece* by Olympiads took its rife. In this city alfo was a very fine temple of *Jupiter Olympius* with a celebrated image of that god fifty ells high, which was reckoned one of the feven wonders of the world. Near it was alfo a famous grove dedicated to the fame god, but dwindled at prefent into an inconfiderable place.

Belvedere, by the *Greeks* called *Callofcopium*, is fituate on the fpot where the ancient capital of *Elis* ftood. The town received its prefent name from the delightful places round it.

Caftel Tornefe, is a fmall town delightfully fituated on an eminence near the fea.

4. *Chiarenza*, *Clarenza*, contains *Achaia*, properly fo called, together with the following places, *viz.*

Chiarenza, *Clarenza*, a pretty good town which fuffered much in the laft *Venetian* war.

Patras, *Patraffo*, in *Latin Patræ*, a town and caftle on a mountain in a bay; the refidence of a *Greek* archbifhop.

THE

The I s l a n d s in *G R E E C E.*

THE *Grecian* iflands may be divided according to the feas in which they lie; namely, in the *Archipelago,* the *Mediterranean,* the parts about *Candia* and the *Ionian* fea, as it was anciently called.

1. The iflands in the *Archipelago,* formerly called the *Ægæan* fea, but receiving its prefent name from *ἀρχ*@ chief, and *πέλαγ*@ a fea, it being the longeft and principal fea in thofe parts. It feparates *Europe* from *Afia,* to the north and weft, wafhing *Romania, Macedonia* and *Greece*; and to the caft *Natolia* or *Afia Minor.* It is interfperfed with many large and fmall iflands, comprized by ancient geographers under two general names. Thofe which form a kind of circle round *Delos* are called *Cyclades, i.e.* the circle iflands; but thofe which lie farther off from *Delos,* and are fcattered all over the *Archipelago,* they call *Sporades, i.e.* the fcattered iflands. Having no account into what kind of government thefe iflands are divided by the *Turks,* we are obliged to keep to the partition made of them by the ancients; but yet, for the fake of method, we fhall treat of thofe iflands which lie neareft to the coafts of *Europe* (to which all the *Cyclades* and fome of the *Sporades* belong) under the article of *Europe*; referving, likewife, in this place the others which lie on the coaft of *Afia,* and conftitute the greateft number of the *Sporades,* to the fourth part of the defcription of the earth, under the article *Natolia.*

1. *Samondrachi,* anciently *Samothrace,* and ftill, in older times, *Melites, Leucofia* and *Leucania* from its white colour; *Saocis,* from a very high mountain of the fame name; *Electria* and *Dardania,* from *Dardanus,* which laft is the moft ufual appellation. The ancient name of *Samothrace* is more properly tranflated *the country of the Thracians,* who formerly inhabited thefe territories; *Same* in the *Perfian,* old *Scythian, Thracian, Lithuanian, Finni,* and other cognate languages, fignifying *earth,* or *country.* It lies not far from the coaft of *Romania,* and was anciently famous for the worfhip paid there to certain deities called *Cabiri,* who were held in fuch high veneration that it was looked upon as prophanation even fo much as to mention their name. The town of *Samondrachi* lies on a high mountain commanding a profpect of its fpacious harbour.

2. *Embro, Lembro,* anciently *Imbros,* is a mountainous ifland overgrown with woods harbouring wild beafts and game. It has four villages, one bearing the fame name as the ifland, and defended by a caftle. It had formerly a town of the fame name, confecrated to the *Cabiri* (fee the preceeding article) and to *Mercury.*

U 2   3. *Thaffus,*

3. *Thaſſus, Thaſos,* in the *Golfo de Conteſſa,* was alſo anciently called *Aeria,* or *Aethria,* being famous, even to a proverb, for its rich gold mines, its fertility and excellent wine and marble. The place of the ſame name lies to the north of the iſland.

4. *Stalimene* (from εἰς τὰν λήμνον) formerly *Lemnos,* abounds with mountains and vallies, which in ſome places are well cultivated and produce all ſorts of fruit. The eaſtern part of the iſland is dry and barren; on the contrary, the weſt and ſouth coaſts are fertile, having a great number of ſprings. In it are two high mountains, by the ancients called *Meſchilæ,* which being volcanos occaſioned the name of *Aethalia* to be given anciently to this iſland. It was formerly ſacred to *Vulcan,* whom the inhabitants worſhipped as their patron; and has always been famed for a certain kind of earth, or bole, which is called from the place *terra lemnia,* and from the ſeals, or particular marks ſtampt upon it, bears likewiſe the appellation of *terra figillata.* It is looked upon as an excellent medicine againſt poiſon, the bite of a ſerpent, wounds and the dyſentery. It was dug up formerly with many religious ceremonies, in all probability, firſt introduced by the *Venetians,* and conſiſting in this, *viz.* that the principal *Turkiſh* and Chriſtian inhabitants of the iſland meet on the ſixth of *Auguſt,* and on that day only, at a chapel called *Sotira,* half way betwixt the village of *Cochino* and the mountain where the earth is found, and from thence proceed in proceſſion to the top of the mountain, where the *Greek* prieſts read the Liturgy; after which, certain perſons appointed for that purpoſe begin to dig, and as ſoon as they diſcover a vein of the deſired earth, give notice of it to the prieſts, who fill ſmall hair bags with it and deliver theſe to the *Turkiſh* governor and other officers preſent. When they have taken up as much as they think proper, they fill up the place again and return back in proceſſion as before. Some of the bags are ſent to the *Turkiſh* Emperor and the reſt marked with his ſeal, or with theſe two words, *Tin Imachton,* i. e. *the ſealed earth,* and ſold by the ſangiak, or his deputies, to the inhabitants and foreign merchants. The ſangiak muſt give an account to the Grand Sultan's treaſury of the money annually produced from it, and the inhabitants are capitally puniſhed if they keep this earth in their houſes, export, or any wiſe trade in it without his knowledge and permiſſion. In this iſland formerly was a famous labyrinth, conſiſting of a ſtately building ſupported by forty pillars of an uncommon height and thickneſs. The two principal places, and once towns, here are *Cochino,* formerly *Hepheſtias,* and *Lemno,* or *Stalimene,* the ancient *Myrina.* It is the reſidence of a *Greek* biſhop.

5. *Pelagniſi,* or *Pelagiſi,* anciently *Haloneſus,* is a very ſmall iſland.

6. *Sciatho, Siatta,* formerly *Sciathus,* is uninhabited by reaſon of the Corſairs, but had anciently two towns.

7. *Piperi,*

7. *Piperi*, formerly *Peparethus*, by others called *Opula, Lemene* and *Seraquino*, was famed of old for its excellent olives, but the wine of this place muſt be kept fix years before it becomes palatable.

8. *Icus*, a very ſmall iſland reckoned by ſome among the *Cyclades*, and ſituate near *Negreponte*, but others place it between *Sciatho* and *Sciro*. It had formerly two towns, whence it was alſo called *Dopolis*.

9. *Sciro*, anciently *Scyrus*, full of mountains and rocks, and conſequently ſtony and barren, as its name ſhows. The remarkable ſtone of *Sciro* which, when whole, floats on the ſurface of the water, but when broken ſinks to the bottom, is deſcribed by *Pliny*. Here are alſo marble quarries. The little town of *Sciro* is the reſidence of a biſhop, and the *Greek* families in the Iſland are computed at about three hundred.

10. *Negropont*, anciently *Eubœa*; afterwards, from its capital, called *Egrippos*, whence the word *Negropont* ſeems to be derived, as formed by the weſtern Chriſtians who firſt came to this iſland, probably from their not underſtanding theſe *Greek* words εἰς τὸν Ἐγρίπον to *Egripos*. In the moſt ancient times it was called *Chalcodotis*, or *Calcis*, *Macrá*, or *Macris* (i. e. *the long*) *Ellopia*, *Abantis* and *Oche*. *Strabo* makes its length ſeven hundred ſtadia, and its greateſt breadth one hundred and fifty. It is divided from the continent by a ſtrait formerly called *Euripus*, but now ſo narrow at the capital that a galley can hardly paſs through; and is joined to the continent by a bridge, being thought to have been once ſo by an iſthmus. The *Euripus* was anciently much celebrated for the ſtated irregularities of its motions. The learned Jeſuit *Babin* obſerves, that in the firſt eight days of the month, and from the fourteenth to the twentieth, incluſively, and alſo in the three laſt days, it is regular both in its ebb and flood; but, on the other days of the lunar month, very irregular; the ebb and flood returning ſometimes eleven, twelve, thirteen and fourteen times within twenty-four or twenty-five hours. This irregularity, which has baffled the reſearches both of ancients and moderns, gave riſe to a proverb among the *Greeks*. Such is the fertility of the level parts of this iſland, that it abounds in a very extraordinary manner in grain, wine, oil, and all kinds of excellent fruits. It has alſo ſeveral mountains which, for a conſiderable part of the year, are covered with ſnow. Of theſe the higheſt is *Oche*. Among the capes the moſt remarkable are *Capo D'oro*, called alſo *Capo Chimi*; *Capo Figara*, the ancient name of which was *Caphareus*; and *Capo Liter*, formerly *Cenæum*. In the firſt ages, when navigation was in its infancy, ſailing round the firſt of theſe capes was reckoned dangerous by reaſon of the many rocks and whirlpools along the coaſt. This iſland had formerly a great number of conſiderable cities, but the only places now worth notice are the following, *viz.*

Egripos,

Egripos, or *Negropont*, the capital of the island, which seems to be, so called from the *Euripus*, near which it lies, being probably seated on the ruins of *Chalcis*, the ancient capital of the island. The admiral of *Turkey*, who is also governor of the island and the adjacent parts of *Greece*, resides here; and the harbour is seldom without a fleet of gallies. It is also the seat of a *Greek* bishop.

Castel Rosso, anciently *Carystus* or *Caryste*, at the foot of mount *Oche*, is a bishop's see and populous. In its neighbourhood were formerly some quarries of marble with amianthus or asbestos.

Oreo, a small village, but remarkable for keeping up the remembrance of the ancient *Oreos*.

11. *Andros*, one of the most fruitful and pleasant islands in the *Archipelago*, having a prodigious plenty of wine, oil, barley and all kinds of delicious fruits, and being watered with innumerable springs; but its greatest riches are silk. The ancients called it also *Cauros*, *Lasia*, *Nonagria*, *Epagris*, *Antandros* and *Hydrusia*. On the island are between thirty and forty villages, and four or five thousand inhabitants, most of them *Greeks* with a colony of *Albanians*. The town of *Arna* has an harbour, and is the residence of a cadi and aga, as also of a *Romish* and *Greek* bishop. At some distance from it may be seen the remains of a large and strong wall with several pillars, cornices and pedestals of broken columns, on which stood statues and various inscriptions, some of which mention the Senate and people of *Andros*, and the priests of *Bacchus*; whence it is concluded that here stood the flourishing city of *Andros*.

12. *Macronisi*, *i. e.* the long island, was formerly called *Helena*, *Macris* and *Cranæ*, on account of its rocks and inequalities. It is both barren and uninhabited, being covered with a deep sand and has only one scanty spring, but was formerly peopled, and produces larger and finer plants or herbs than are to be found in any parts of the *Archipelago*.

13. *Coluri*, formerly *Salamis*, and likewise *Pitiussa*, *Seiras* and *Cychria*, lies in the *Golfo d' Engia*, being divided from the main land by the strait of *Perama*. Besides the small town of *Coluri*, which has a harbour, there are two villages on the island, one called *Ambelachi* lying in that part where the ancient city of *Salamis* stood, as the remains still shew. This island is famous for the important victory gained there by the *Greeks* over the *Persians*.

14. *Ægina*, called also anciently *Oeone* and *Myrmidonia*, lies likewise in the gulf of *Engia*, which takes its name from it, being, by the corruption of mariners, also called *Engia*. The ancient inhabitants, on account of their industry in cultivating the soil, were styled *Myrmidons*, *i. e.* Ants. Not far from the town of *Engia*, which is said to consist of about eight hundred houses, and has a castle, may be seen the remains of a magnificent structure, which was probably one of the celebrated temples which formerly graced this island.

5

15. *Pòrus*, anciently *Calabria*, lies next to the *Morea*, and is remarkable for being the place whither *Demoſthenes* was baniſhed.

16. *Zia, Cia, Cea,* formerly *Ceos*, called alſo *Hydruſſa*, was famous among the ancients for its fertility, paſtures and figs. The town of *Zia*, which gives name to the iſland, lies upon an eminence in the neighbour-hood of the ancient city of *Carthæa*, of the ruins of which, as well as of the ancient city of *Julis*, ſome remains are ſtill viſible. Thoſe of the laſt take up a whole mountain, and by the inhabitants was called *Polis, i. e.* the city. Near this place are to be ſeen the ruins of a ſtately temple. On the iſland reſides a *Greek* biſhop. The harbour lies to the north-weſt ſide, and is capable of receiving the largeſt ſhips.

17. *Joura,* formerly *Gyarus, Gyara,* or *Gyaræ*, is the moſt deſolate and unpleaſing place in the whole *Archipelago*. The *Romans* uſually baniſhed delinquents to this iſland.

18. *Tine,* formerly *Tenos, Hydruſſa, Ophiuſa,* very mountainous, but abounds in many parts with excellent fruit, and ſtill more in ſilk. The wine of this place was much admired by the ancients. Beſides the town, which is defended by a caſtle, this iſland contains betwixt thirty and forty populous villages. Here is alſo a *Greek* and *Latin* biſhop.

19. *Mycone* produces corn, wine, figs and ſome olives, but has little water or wood. The inhabitants are moſtly *Greek* Chriſtians, and have magiſtrates of their own religion; but a *Turkiſh* officer comes hither every year and col-lects the tribute paid to the Porte. A cadi ſometimes alſo viſits them and holds courts of juſtice here. In this iſland are upwards of fifty *Greek* churches, and ſeveral convents of monks and nuns, though not very well filled. Among theſe the convent of nuns of *Paleo Caſtro*, which lies in the middle of the iſland, is the principal. There may be in the whole country about five hundred ſea-faring people, many of whom live by pyracy. The *Myconians* are bald at twenty or twenty-five years of age. The city of *Mycone* lies on a large harbour, and is an open place containing about one thouſand inhabitants.

20. *Trogoniſi, i. e.* the iſland of goats, from the great numbers of thoſe animals formerly found there. It is a very ſmall place.

21. *Dèlos*, formerly a celebrated iſland, but at preſent a deſert rock, being a ſecure place of retreat for Corſairs. The *Greeks* call it *Dilli*, or *Dele*, in the plural number, comprizing under this title the iſland of *Rhenæa*; and giving to the Leſſer *Delis* of the ancients the appellation of *Delos*, but to the Greater *Delis* that of *Rhenæa*. *Dèlos*, as being the imaginary birth-place of *Apollo* and *Diana*, was held in the higheſt veneration by all nations, even by the *Perſians*. The oracle of *Apollo* in this iſland was the moſt celebrated in all the world. Here are ſtill to be ſeen the remains of the celebrated pillar of *Apollo*, and ſeveral ſtately fragments of the renowned temple and once ſo conſiderable city of *Delos*.

22. The

22. The Great *Sdili*, or *Deli*, formerly *Rhenæa*, *Rhenia*, or *Rheno*, lies near the laſt mentioned iſland, and has fine paſtures, but through fear of the Corſairs is uninhabited, being widely different from its former condition, of which its many ſtately ruins are a melancholy proof.

23. *Syra*, *Sira*, *Siro*, formerly *Syros*, is mountainous, but produces plenty of barley, wine, figs, cotton and olives, as alſo good wheat. The air is moiſt and cooler than in the neighbouring iſlands. The inhabitants, a few *Greek* families excepted, are *Roman* Catholics. The town of *Syra* is built round a little ſteep mountain. Between it and the harbour may be ſeen the remains of auguſt buildings which made a part of the ancient city of *Syros*. Near the caſt coaſt lie three ſmall iſlands called *Gadroniſi*.

24. *Thermia*, anciently *Cythnus*, *Ophiuſa* and *Dryopis*, receives its preſent name from the many hot ſprings met with here. It is not ſo mountainous as ſome of the other iſlands, and the ſoil, when well cultivated, produces very large quantities of barley, wine and figs. The iſland alſo affords plenty of honey, wax, partridges, a great quantity of ſilk, and as much cotton as the inhabitants require for their own uſe. The *Greek* Chriſtians in this iſland are computed at ſixteen thouſand. In the city of *Thermia* is a biſhop with about fifteen or ſixteen churches and ſeveral convents. On the iſland are ſtill viſible the ruins of two cities; one of which, on the ſouth coaſt, muſt have been of extraordinary ſplendour.

25. *Serpho*, *Serphanto*, formerly *Seriphus*, is rather a barren rock than an iſland, and accordingly the *Romans* made it a place of baniſhment for the vileſt malefactors. It abounds, indeed, in mines of iron and magnets, producing alſo great quantities of onions. The inhabitants are all *Greeks*.

26. *Siphanto*, *Siphno*, formerly *Siphnus*, *Meropia* and *Acis*, enjoys a very wholeſome air, good water and a fertile ſoil, producing great plenty of fine fruit with a ſufficiency of grain for the ſupport of its inhabitants: here is alſo no want of tame and wild fowl and other game. It had once rich gold and lead mines, but of the former the inhabitants now know nothing, and the latter are not worked. The number of people here is computed at five thouſand, who inhabit five or ſix villages, and are moſtly *Greeks*; but they have above five hundred chapels, four convents of monks and two of nuns. On a rock near the ſea ſtands a caſtle, and in the iſland are five ſecure harbours; namely, *Faro*, *Vati*, *Chitriani*, *Chironiſſe* and *Gulança*.

27. *Argentiere*, formerly *Cimolis*, is full of rocks, mountainous and barren. It is ſaid to contain ſeveral ſilver mines. The whole iſland is covered with a kind of clay, called *Cimoli* earth, uſed for waſhing and bleaching of linen. It contains but one village.

28. *Prepeſinthus*, a ſmall iſland betwixt *Siphanto* and *Melos*.

29. *Milo*, *Melos*, conſiſts almoſt entirely of one hollow, porous rock, which is macerated, as it were, by the ſea water. The heat of a continual ſubterraneous fire is felt here merely by putting the hand into the holes of the

rock;

rock; and on the island is a place ever burning in which the neighbouring earth fmokes like a chimney, Great quantities of alum and fulphur are found here. Alum grows here in certain natural caverns in the form of flat ftones betwixt nine and ten inches thick. It produces alfo plume-alum, alkaline fublimated and diffolved, and diftils in drops from the rock. Sulphur is found in one particular place perfectly pure and as it were fublimated; namely, in a cavern, the bottom of which is full of ful-phur which burns continually. The water in the lower grounds is not fit for common ufe: At the foot of a mountain betwixt the town and the har-bour, are baths and fome fprings fo hot as to burn one's finger. Here is alfo a purgative fpring. It abounds in iron mines, and though the furface of the ifland be, in general, mountainous and rocky, it is not without many de-lightful plains, and the foil is very fertile. Here is great plenty of grain of all kinds, exquifite fruits, efpecially grapes, melons and figs; with honey, flefh, fowl, game and fifh; but the air is none of the moft wholefome. The inhabitants are *Greeks*, given up to voluptuoufnefs, without any thoughts of the danger which threatens their habitation. It has alfo a *Greek* and *Latin* bifhop. The town of *Milo* may contain five thoufand fouls, and were it not for its extreme filthinefs, would make a tolerable appearance. About half a *Turkifh* mile from it is an excellent harbour.

30. *Antiparos*, anciently *Oliaros*, in fome parts fertile, but having only one village. It is remarkable, however, for a grotto, which is, indeed, a mafter-piece of nature, being about forty tofes high and fifty broad, and every where entertaining the eye with an infinite variety of figures of a white, tranfparent, cryftalline marble, reprefenting vegetables, marble pillars, and a fuperb marble pyramid; all which *Tournefort* looks upon to be natural.

31. *Paros*, formerly *Platea*, *Paëlia*, *Minoa*, or *Minois*, *Demetrias*, *Zacyn-thus*, *Hyria*, *Hileaffa* and *Cabarnis*, once opulent and powerful, now con-fifting only of one thoufand five hundred families, but having plenty of grain, wine, game, and, in particular, of good cattle. It is famous for its ex-traordinary white marble, and had alfo moft excellent artifts for working it. Thefe celebrated antiques called the chronicle of *Paros*, are marbles, hav-ing *Greek* infcriptions performed on this ifland, and bought, in the year 1627, by *Thomas Howard*, earl of *Arundel*, who, in 1667, made a prefent of them to the univerfity of *Oxford*, and they are now called the *Arundelian* or *Oxonian* marbles. The infcription is the moft authentic piece of ancient chronology, being made two hundred and fixty-four years before the Chri-ftian æra, and contains a fpace of above three hundred years. The town of *Parichia* feems to ftand on the ruins of the ancient *Paros*, the walks and houfes being decorated with feveral fine marble remains of that city, befides the monuments in the adjacent country. The *Panagia*, or *Madonia*, with-

out the city, is both the largeſt and moſt ſplendid church in the *Archipelago.* In this iſland, beſides a great number of churches and chapels, are ſeveral conſiderable villages. A large fleet may lie conveniently and ſecurely in the harbour of St. *Maria,* but the common anchoring place for the *Turkiſh* navy is *Drio,* on the weſt ſide of the iſland.

32. *Naxia,* the ancient *Naxos,* called alſo *Strongyle Dia, Dionyſias, Callipolis* and Little *Sicily,* the laſt of which names was given it on account of its reſembling *Sicily* in fertility; and it was called *Dia,* or *the Divine,* from the religious worſhip of *Jupiter* (Ζιύς, διòς) obſerved there. It is the moſt fertile iſland in all the *Archipelago,* and its wine ſtill maintains its former excellence. Beſides this, its plains are covered with orange, olive, lemon, cedar, citron, pomgranate, apple, mulberry and fig trees. It was alſo famed for a kind of marble which the *Greeks* called *Ophites,* being green ſpeckled with white like a ſnake. Some mountains on the weſtern coaſt afford emery of ſo much virtue that the adjoining cape is, by the *Italians,* called *Capo Smeriglio,* or *Cape Emery.* The people on the whole iſland do not exceed eight thouſand, and there is little harmony betwixt the *Greeks* and *Latins,* who have each an archbiſhop here. The inhabitants, like moſt of theſe iſlands, have the choice of their own magiſtrates, but ſometimes a cadi takes a circuit among them, and to him appeals lie. It contains betwixt forty and fifty villages, and but one town, which ſtands on the ſouth ſide of the iſland, and is defended by a caſtle. About a muſket-ſhot from it, on a rock near the ſea, ſtands a beautiful marble portal amidſt a heap of fragments of marble and granite, ſuppoſed to have been a temple of *Bacchus.*

33. *Amorgus,* or *Morgos,* enjoys a good ſoil, and produces fine wine. The town is built on the ſide of a rock with a caſtle near it. About three miles off, by the ſea ſide, ſtands a large *Greek* convent. Its beſt harbour is on the ſouth ſide.

34. *Caloyero, Cheiro, Skinoſa* and *Raclia,* deſart iſlands and rocks.

35. *Nio* produces ſcarce any thing but corn, has ſome convenient harbours, and its inhabitants are good ſeamen.

36. *Sikino.* Here grows the beſt wheat in the *Archipelago.* It has alſo plenty of figs. Its town, which is but ſmall, ſeems to hang over the ſea.

37. *Policandro,* ſtony, and affords but a ſcanty ſubſiſtance to its inhabitants, who endeavour to make it up by a cotton trade. The town may contain ſomewhat above a hundred *Greek* families.

The

The _Mediterranean_ I s l·a n d s lying about _Candia_, or in the _Candian_ Sea.

1. C _ANDIA_ itſelf, anciently _Crete_, called alſo _Aeria_, _Idæa_, _Curete_ and _Macaron_, or _Macaroneſus_, i. e. _the fortunate iſland_, in alluſion to its fertility and the purity of the air, is one of the largeſt iſlands in the _Mediterranean_, being in length about ſeventy _Turkiſh_ miles, and in breadth, in ſome parts, fifteen. Above half the iſland is covered with rocky, barren mountains, the moſt noted of which are _Pſiloriti_, formed from ὑψηλον ὀϱος, and the ancient _Ida_, being the higheſt in the lſland, though one continued barren rock, and for the greateſt part of the year covered with ſnow. The only thing it produces is the tragacantha, or buckthorn, famous for its gum. This mountain commands a view of both ſeas. _Sethia_, or _Laſthi_, formerly _Dicte_, is part of the _White_ mountains, as they were called, but now, from a neighbouring town, receive the name of the _Sfachia_ mountains. The vallies and level country are remarkably fertile in grain, fine red and white wine, oil, ſilk, wool, honey, wax, &c. The quince tree was firſt brought into _Italy_ from the town of _Cydonia_, and therefore, by the _Romans_, called _Malum Cydonicum_, or the _Cydonian_ apple. Here are tame creatures of all kinds, game and wild-fowl, but no ſtags or wild beaſts. A very conſiderable part of the country lies waſte. The inhabitants are _Greeks_, who have an archbiſhop, _Armenians_, _Turks_ and _Jews_. Ancient bards make frequent mention of the one hundred cities of this iſland, to which, on that account, was given the epithet of _Hecatompolis_, but it had many more, above one hundred and twenty occurring in old writings. Under the Emperor _Valentinian_ the firſt, above a hundred towns were deſtroyed and ſome overthrown by an earthquake. This iſland had originally its own Kings, but the republican form was afterwards introduced. Its firſt foreign !maſters were the _Romans_, under whom it continued beneath the eaſtern Emperors till the year 823, when the _Saracens_ ſubdued it ; from whom it was recovered again in the year 962. The _Genoeſe_ ceded it to _Boniface_, marquis of _Montferrat_, who, in the year 1204, ſold it to the _Venetians_. In 1644 the _Turks_ made a deſcent here, and, after a ſiege of twenty years, reduced the whole country, which, except a few forts, the _Venetians_ were obliged to cede to them at the peace in the year 1669; and, in 1715, the _Turks_ made themſelves maſters alſo of the _Venetian_ forts, ſince which time the iſland has been all their own. Under the _Venetian_ dominion the iſland was divided into four governments or Diſtricts.

1. The Diſtrict of _Canea_; _Territorio della Canea_, containing

Canea, a fortification, but ill preferved. The town makes a good ap‑
pearance, but the harbour is expofed to the north wind. This town pro‑
bably ſtands‑on the ancient ruins of *Cydonia,* once the moſt opulent city in
Crete.

Suda, a ſmall Iſland with a fort ſtanding on a bay, to which it gives
name.

Sfachia, a little town on the fouth fide, feems to be the old *Phaiſtos.*

Caſtel Selino, fituated alfo on the fouth fide.

The White Mountains mentioned above are covered, like the others, with
fnow, and are alfo called *Leuci.*

Garabufe, a ſmall iſland near a point of land on the north fide of the
iſland, where is a caſtle which, in the year 1690, was betrayed by *Aloyſius*
the commandant to the *Turks.*

2. The Diſtrict of *Rettino, Territorio di Rettimo* ; in it. are

Rettimo, a well peopled town, having a citadel and harbour, but choaked
up with fand. It was formerly called *Rethymna,* or *Rethymnia.*

Caſtel Milopotamo, a fea-port on the north fide of the iſland.

Arcadi, a convent on the fpot where formerly ſtood the ancient town of
Arcadia.

Caſtel Amari, near the centre of the iſland.

Pſiloriti, the ancient *Ida,* a celebrated mountain which we have before
taken notice of.

3. The Diſtrict of *Candia, Territorio di Candia,* containing

Candia, the preſent capital, being a fea-port on the north fide of the iſland,
in a plain, at the foot of a mountain. It ſtands on the ruins of *Heraclea,* which,
by fome, is not improbably held to have been the fame with *Matium.* From
this city it is that the iſland derives its preſent name, though others deduce
it from *Candidus* white, its foil being of that colour. It is faid alfo, that the
Saracens having built a town at the place where they had intrenched them‑
felves, called it *Chandar,* which, in their language, fignifies a fort, and
by time became gradually corrupted to *Candia.* This city was almoſt totally
deſtroyed in the tedious blockade and fiege carried on here by the *Turks* from
the year 1645 to 1669, infomuch that it is now only the melancholy
ſhadow of a wealthy place. The harbour too is fo choaked up as to be
practicable only to boats. It is, however, the fee of a *Greek* archbiſhop.

The Labyrinth, as it is called, runs under a ſmall mountain at the foot
of mount *Pſiloriti,* or *Ida,* in a thoufand intricate meanders, without any
appearance of regularity, and feems a work of nature, or only a vaſt fub‑
terraneous cavern. The entrance is a natural aperture feven or eight
paces broad, but fo low in fome places that there is no going in without
ſtooping. The ground is very rugged and uneven, but the top level, con‑
fiſting of a horizontal layer of ſtones. The principal paſſage, in which
there is not fo much danger of bewildering one's felf as in the others, is

4	about

about one thoufand two hundred paces in length, and reaches to the end of the maze where are two fpacious chambers. The moft dangerous part of the main walk is about thirty paces diftant from its mouth; if a perfon happen to take any other courfe he is immediately loft amidft the innumerable windings of this maze, and then it is no eafy matter for him to find his way out. For this reafon travellers always provide themfelves with guides and torches. There is little appearance of its having been a ftone quarry, and whether it was the famous ancient labyrinth of *Crete* is far from being fatisfactorily proved. It is quite dry.

At the beginning of the plain of *Mefforia*, the moft fertile in the whole ifland, on a little river which can be no other than the *Lethe*, are to be feen the ftately remains of *Gortyna*, or *Gortyn*, once a city of confiderable figure.

Caftel Nuovo, *Caftel Bonifacio*, *Timeni*, *Caftel Mirabello*, are all forts.

Spina longa is a fea-port with a citadel.

The Diftrict of *Settia*, *Territorio di Settia*, contains

Gierapietra, a village on the fpot where formerly ftood *Hierapytna*, a fortified town.

Settia, a fortified town on a bay called after its name.

Gotzo, formerly *Gaudos* and *Claudan*, lies fouth from *Candia*, and belongs to the order of *Malta*.

Gaiduronifia lies alfo to the fouthward of *Candia*.

Chriftina, the ancient *Letoa*, in a like fituation.

Standia, formerly *Dia*, lies north of *Candia*, being rather a rock than an ifland. Its prefent name is derived from εἰς τὴν Δίαν.

Scarpanto, the ancient *Carpathus*, is one hundred furlongs in length and two hundred in circumference.

Stampala, once called *Aftypalæa*, had formerly a town and a celebrated temple of *Apollo*.

Namphio, mountainous, and having fcarce either plants or herbs, but feveral fine fprings with plenty of wax and honey, and great numbers of partridges. The inhabitants are all *Greeks*.

Santorin, i.e. *Sant Erini*, St. *Irene* being its patronefs, anciently called *Califta* and afterwards *Thera*. It is fruitful, particularly in barley and wine, which has the colour of *Rhenifh*, is ftrong and brifk; and cotton, which grows in great plenty upon a fhrub fomething like a currant-bufh, and not, as in the other iflands, planted again a-new every year. It produces alfo fome wheat. The inhabitants, who are computed to be ten thoufand, are all *Greeks*; but one third of them of the *Latin* church, and fubject to a *Latin* bifhop. This ifland formerly had feven confiderable towns, but at prefent the five following are the only places worthy of notice, *viz.*

Apanormia,

Apanormia, having a fpacious harbour in fhape of a half moon, but its depth unfathomable, which renders it of little ufe to fhips as they cannot anchor in it.

Scaro, or *Caftro*, having a caftle on an inacceffible rock.

Pyrgos, feated on a mountain, the people whereof live in caves hewed out of rocks of pumice-ftone.

Emperio, or *Nebrio.*

Acroteri.

This ifland is much talked of in natural hiftory. It appears to be only a pumice-ftone incrufted with a furface of fertile earth, and, according to the accountsof the ancients, rofe in a violent earthquake out of the fea as a volcano. The fame origin had four other iflands near *Santorin*, and the fea here is of fuch depth as not to be fathomed by any founding-line. The firft, which lies in the harbour of *Apanormia*, and was formerly called *Hiera*, or *Automate*, but at prefent *Megali Cammeni*, *i.e.* the great burnt ifland, rofe out of the fea after a violent earthquake in the year 906, before the birth of *Chrift*; and, in the beginning of the eighth century, a new ifland fuddenly appeared with a terrible explofion which fhook all the neighbouring iflands, and adhered to the ifland of *Hiera*, which thereby became confiderably enlarged : at the fame time fhowers of pumice-ftones were caft up to a vaft diftance on all fides. The fecond ifland, which lies a little without the harbour, and was formerly called *Therafia*, but now, from the whitenefs of its earth, *Afpronifi*, fprung from the fea in the firft century after the Chriftian æra. In the year 1573 a fudden fire burft out of the fea and foon produced a new ifland, to which was given the name of *Micri Cammeni*, *i.e.* the fmall burnt ifland. In the years 1707 and 1708, betwixt this ifland and Great *Cammeni*, a fourth ifland rofe like a volcano, attended with an eruption of fire, concuffions, a crafh like the loudeft thunder, and a noifome fmoke, and has been gradually increafed by additional rocks.

Cerigo, formerly *Cythera*, the ifle betwixt *Candia* and the *Morea*, is mountainous and rocky, fo that it produces little corn, wine or oil. It has alfo no great number of inhabitants, but does not want for fheep and fowls both tame and wild. The principal town is on the fouth fide of the ifland, having a caftle on a very fteep declivity ; but the harbour below it lies open to the fouth winds. This ifland was in a particular manner confecrated to *Venus.*

The

The Islands *in the Sea, anciently called the* Ionian.

AMong thefe are,

1. *Le Sapienze*, formerly the *Sphagi* Iflands, which are three iflands lying oppofite *Modon* in the *Morea*. The largeft was formerly called *Sphacteria*, being famous in ancient hiftory for the Victory obtained there by the *Athenians* over the *Lacedemonians*. The circumjacent fea is called the fea of *Sapienza*.

2. *Strivali*, formerly *Plotæ*, *i. e.* the 'Floating Iflands;' afterwards named *Strophades* from their fuppofed circumvolutions; are two iflands at prefent inhabited by *Greek* Monks.

The following iflands belong to the *Venetians*.

3. *Zante*, formerly *Zacynthus*, a pretty large ifland, producing great plenty of ftrong wines, raifins, oil, figs, &c. but very fubject to earthquakes. On the fouth and weft fide is a large chain of mountains; with a mountain alfo towards the eaft, and another north; fo that it is almoft every where environed by mountains. The centre is a delicious plain. Near *Chieri* a fea-port on one fide of the chain, at about two hundred paces diftant from the fea, is a refinous fpring. It exports alfo falt. This ifland contains a great many towns and villages in it; and the *Greeks* who conftitute the majority of its inhabitants, have above forty churches, befides convents and a bifhop. Another part of the inhabitants are *Roman* Catholicks who have three convents and a bifhop. The remainder are *Jews*. The Republic of *Venice* has a proveditore and two counfellors. The chief place on the ifland is *Zante*, a large and populous town with a caftle ftanding on an eminence.

4. *Cephalonia* or *Cefalonia*, called alfo *Cephalenia*, in the firft Ages. *Samus* and *Epirus Melæna* or *Black Epirus*, had formerly four towns, the capital of which was *Samos*, fuppofed to have ftood at the place which the *Italians* call *Porto Guifcardo*. This fruitful ifland produces currants, oil, red wine, Mufcadel-grapes, citrons, oranges, pomegranates, and grain. Moft of the trees bear twice a year, namely, in *April* and *November*; but the fruits of the laft month are fmaller. Among its harbours that of *Argoftoli* is the beft. In 1214, or according to others in 1224, the *Venetians* firft got poffeffion of it. In 1479, it was taken from them by the *Turks*; but in 1499 again recovered. They have here a proveditore and two counfellors. The moft remarkable places are.

Cefalonia, a bifhop's fee.

Affo, a fortification built in the year 1595, on a high eminence environed with the fea and fteep rocks.

2   5. *Val*

5. *Val di Compare,* the prefent name of the ifland of *Ithaca,* noted for being the country of *Ulyffes.*

6. *Curzolari,* formerly the *Echinades,* were five fmall iflands on the coaft of the ancient *Acarnania,* which conftitutes a part of the modern *Livadia*; but many ages fince feveral of them are become united with the continent by means of the fand which the river *Achelous* carries into the fea. *Dulichium,* once a part of *Ulyffes'* dominions, is by *Strabo* reckoned among the *Echinades.*

7. *Santa Maura,* anciently called *Neritis,* and fince *Leucas,* in former ages a peninfula connected with the continent of *Acarnania.* But the *Carthaginians,* or according to others, the *Corinthians* fevered it, in fo much that at prefent, betwixt the ifland and the continent, there is a channel about fifty paces broad. It enjoys an uncommon plenty of grain, wine, oil, citrons, pomegranates, almonds, and other fruits, with fine paftures. The inhabitants are *Greeks* fubject to a bifhop. In this ifland were formerly three confiderable towns, with a moft magnificent temple of *Venus.* The beft harbours in it are *Englimeno, Demata,* and *Santa Maura.* The town of *Santa Maura,* which gives name to the ifland, contains about five or fix thoufand inhabitants; and being fituated in the water and defended by walls and towers is fo fortified, that there is no convenient approach to it either by land or water. Beyond its works, in a morafs, are two well inhabited iflands or fuburbs. The clufter of little iflands betwixt this and the continent communicate by bridges. In 1473 the *Turks* difpoffeffed the *Grecian* emperor of the town and ifland. In 1502 the *Venetians* made themfelves mafters of it; but afterwards furrendered it again to the *Turks,* and in 1684 recovered it from them. In 1715, the town and ifland fell a fecond time into the hands of the *Turks*; but the following year the *Venetians* drove them out, and have ever fince kept a proveditore here.

8. *Pachfu* and *Antipachfu,* anciently, Paxi or *Paxæ,* two fmall iflands betwixt St. *Maura* and *Corfu.*

9. *Corfu,* anciently *Drepane, Scheria, Phæacia* and afterwards *Corcyra,* takes its prefent name from the ancient caftle of *Corypho* ftanding on a mountain, and is fo called from the *Greek* word κορυφὴ, a hill. It is feventy *Italian* miles in length, and from *Capo Barbaro* to *Capo Balacrum* thirty broad; but in moft other places fcarce twelve: It was famous in former ages for the beautiful gardens of king *Alcinous.* Its fouthern parts are barren, mountainous and deftitute of water; but the northern coaft is very fertile in all kinds of delicious fruits, excellent wine, grain, olives, &c. This ifland was for a long time under the Kings of *Naples,* but in the thirteenth century the inhabitants fubmitted to the *Venetians,* who, from that time, have continued in poffeffion of it. And as it is a place of great importance to them, they conftantly keep a fleet of gallies and other fhips

in

in the harbour of *Corfu.* The Ifland is divided into four parts called Balias, or governments under a Balio, namely,

1. *Alefchimo,* eaftward; which contains twenty-eight villages and ten thoufand people. *Potami* has the appearance of a town. Towards the fouth-weft fome remains of the ancient town of *Gradichi* are ftill vifible.

2. *Mezzo,* the middle part, contains not lefs than twenty-five thoufand people, and befides thirty towns, *Corfu* the capital of the country, and the feat of the government, which confifts of a Balio, a proveditore and a capitano, two counfellors, the grande capitano and caftellano. Here is alfo an archbifhop. It is a good fortification; being compofed of the city, the large fuburbs of *Caftrati,* the old citadel, with the water-caftle and the new citadel; but this latter is commanded by mount *Abraham* lying near it. The harbour lies under the old citadel. Here is an academy of the liberal arts and fciences. In 1537, and in 1715, it was warmly befieged by the *Turks,* but ftood it out.

3. *Agiru,* the weftern part comprehends twenty villages and about eight thoufand inhabitants; but the only remarkable place in it is *Caftle St. Angelo,* which lies on the fouth cape called *Palacrum;* and beneath it ftands a ftately caftle called *Paleo Caftrizza.*

4. *Oros,* the north part contains twenty-five places, and about eight thoufand people. In it is

Caffopo, formerly *Caffiope,* once a confiderable town, but mean at prefent. It has a church for the devotions of thofe who happen to put into its harbour. The fort, which formerly ftood near it, is now in ruins.

Obf. It contains likewife feveral towns and Diftricts in the tributary countries, fuch as *Chomitz, Bender, Oczakow,* all belonging to the *Ottoman* porte.

COUNTRIES *in* EUROPE *under the protection of, and tributary to, the* Ottoman *Porte.*

Among thefe are the countries of Chriftian Princes, *viz.*

WALACHIA.

§. 1. WALACHIA in its moft extenfive import comprizes that part of ancient *Dacia* and *Cumania,* or the tract of land within *Bulgaria, Servia, Hungary, Tranfylvania, Red Ruffia,* and *Tartary;* and confequently alfo *Moldavia;* but in a more confined fenfe, in which we alfo at prefent take it, it contains only *Walachia,* as it is called, *beyond the mountains,* which is feparated from *Moldavia* by a vaft chain of hills and the river *Sereth,* being about forty-five *Hungarian* miles in length from weft to eaft; and from fouth to north in fome places forty, in others fcarcely fourteen broad.

§. 2. *Walachia* is fo called from the *Walachians*, who are its inhabitants, as fhall be fpoken of in §. 4. But they themfelves name it *Romulia*; and the *Hungarians Havafalfoldgye*.

§. 3. The air is temperate, the foil very fruitful, particularly in grain, wine and melons; graziery here too is very confiderable; but its principal reputation is for excellent horfes. The country is watered by a confiderable number of large and fmall rivers, the greateft part of which run from north to fouth, difcharging themfelves immediately into the *Danube*, or in conjunction with other rivers. The principal of thefe are the *Aluta*, which rifes in the mountains of *Tranfylvania*, and divides *Walachia* into two unequal parts, namely, the weft and caft; the *Jalonitz*, which has alfo its fource in the borders of *Tranfylvania* and the *Sereth* or *Strech*, the boundary on the fide of *Moldavia*.

§. 4. The *Walachians*, confidered as inhabitants of the country are defcended from the old *Roman* colony fettled here by the Emperor *Trajan*. To omit mentioning the teftimonies of ancient hiftorians, this appears evidently not only from their language, which is a barbarous *Latin* intermixed with feveral foreign words, but likewife from their cuftoms, and manner of diet, for inftance their thick pottages and onions of which they are very fond; their drefs, and great liking to the *Italian* language, the *Italians* themfelves and whatever belongs to that country. But they are intermixed with the *Sclavi* and *Pazinacitæ*. By way of diftinction, however, and to fhow their origin from the *Romans*, they call themfelves *Romunii*. Hiftorians are not agreed about the derivation of the Word *Walach*. Some taking it to be from one *Flaccus*, others from the *Greek* Βαλλειν, and others again deriving it, but without the leaft probability, from the *Tartar*. Others who come nearer the mark, imagine it to be derived from the word *Wallen*, i. e. ' Wander,' and accordingly tranflate it *Wanderers*. But the word *Wlach* in *Sclavonick* fignifies an *Italian*; and has a manifeft refemblance with the word *Welcher*, which fignifies alfo an *Italian*. The *Saxons* in *Tranfylvania*, ufually call the *Walachians Bloch*, which comes ftill nearer to *Wlach*, and is of the fame origin and import.

§. 5. The *Walachians* profefs the eaftern *Greek* religion; and as in writing they ufe the fame letters with the *Ruffians*, fo they agree with them in all their religious ceremonies. The commonality are moft wretchedly ignorant; and even the higheft attainments which the ecclefiaftics themfelves aim at, feldom go beyond reading and finging well. *Buchereft* is a kind of univerfity to them; whither they go to learn a polite deportment, the elegancies of the *Walachian* language and ceremonies of the church. The perfons of rank among the *Walachians* are fo fond of the *Italian* language that they apply themfelves to it more than their mother-tongue; and generally fend their fons to ftudy at the univerfity of *Padua*. Great numbers of *Mahometans* live alfo intermixed with the *Walachians*.

§. 6. With regard to the history of this people, it is well known that the *Romans* after their decisive victory over *Decebalus* King of *Dacia*, made themselves masters of his kingdom. *Trajan* sent thither several *Roman* colonies, who not only cultivated the land, but built them towns which they embellished with fine edifices. His successor, however, in the empire transplanted the greatest part of them into *Mæsia* and *Thracia*, where mingling with the *Bulgarians, Thracians, Servians* and *Ligurians*, they came to speak a new language or jargon. These kingdoms which lie on the *Danube* afterwards constituted part of the dominions of the emperors of the east. In process of time the *Walachians* moved farther north to the borders of *Podolia* and *Ruffia*, where they applied themselves to agriculture and breeding of cattle. The conversion of the *Bulgarians* and their neighbours to Christianity was followed in the nineteenth century by that of the *Walachians*, who embraced the *Grecian* doctrines. Towards the beginning of the twelfth century a numerous colony of *Walachians* under the conduct of one *Nigers* or *Negrovot*, for the sake of pasturage, religion, and other motives, quitting *Burzeland* and the rest of *Tranſylvania*, passed over those mountains which inclose *Burzeland* towards the south, and settled in the modern *Walachia*, founding the towns of *Tergoviſto, Buckereſt, Longenau,* and *Piteſto S. Georgi*. They choose their own princes whom they stile Waywodes or Despots. The Kings of *Hungary* becoming powerful made several attempts on the *Walachians*, and in the fourteenth century obliged them to pay tribute. But in the year 1391 and 1394 they were greatly haraſſed by the *Turks*, who in 1415, also laid the whole country waste with fire and sword, compelling *Dan* the Waywode to pay them an annual tribute. It was the year 1608 before the *Walachians* could rid themselves of this burden, when they put themselves under the protection of the Emperor of *Germany*. But the treaty of *Carlowitz* resigned them up again to the *Turkiſh* dominion. In the beginning of the present century, they suffered various calamities by the plague, war, and the many revolutions among its princes. At the treaty of *Paſſarowitz*, in 1718, the western part of *Walachia* as far as the river *Aluta* was ceded to the Emperor, but lost again in 1739.

§. 7. *Walachia* is governed by a Waywode or Prince stiled also the Hoſpodar, who is a vassal of the *Ottoman* porte, and whose yearly tribute generally amounts to fifty-eight or sixty thousand ducats. The arms of *Walachia* are a black eagle with both its feet on a mount, and in its beak is a crofs erect with the sun on one side of it, and on the other the moon, in a field argent.

§. 8. It has been observed above (§. 3.) that the river *Aluta* divides *Walachia* into two parts, which are

1. *Walachia* on this side the river *Aluta*, constituting the western part of the country, and called also the Bannat of *Severin*, to which belong

Meadia or *Mihalid*, a fort on the river *Czerna*. Near it, in 1738, an action happened betwixt the *Imperialists* and *Turkish* troops.

Baja, a town.

Old and *New Orsowa*; the former a pretty good town, but the latter only a fortification in the *Danube*.

Severin, Czærny, a kind of a little town on the *Danube*; but formerly fortified and the capital of a Bannat; being so called from the Emperor *Severus* its founder.

Zernigrad, or *Tschernigrad, Mauro Castro*, (μαυρον καςρον) i. e. ' *Black Castle*,' stood on an eminence near the *Danube*. Its ruins are covered with thick bushes. About a quarter of an hour farther downwards we have a view of the *Danube*. On it are to be seen

The remains of the stone pillars of a bridge, in all probability that which the Emperor *Trajan* laid over the *Danube*, in order to attack *Decebalus* King of the *Dacians* with greater advantage. According to Count *Marsigli* the river here is not one thousand paces wide; and the two first pillars of the bridge standing seventeen fathoms and a half asunder, he concludes that there must have been twenty-three in all; and that the whole length of the bridge was four hundred and forty-three fathoms. He further affirms that the masonry of the pillars was of common quarry stone lined with bricks. or tiles; and that, in all probability, the twenty-two arches with the whole upper-part of the bridge were made of oak. The Emperor *Adrian* not only caused the upper-part of the bridge to be removed, but likewise all the masonry above the water to be demolished. Ancient writers greatly exaggerate in their accounts of this bridge. Half an hour below it lies.

Tschernetz, at a little distance from the *Danube*.

Krajova, a small town.

Sidova, a fort near the *Aluta* and *Danube*, where some place *Trajan's* bridge.

Ognile Mari, in Latin *Salinæ magnæ*, salt-springs.

Renmik, Rebnik, Rebnitz, a pretty good bishop's see on the river *Aluta*.

Citatesva, a fort on the *Aluta* facing the red tower in *Transylvania*.

2. *Walachia*, on the other side of the river *Aluta*, is situate on the eastern part of the country, and contains

Langenau, Campolungo, in Latin *Campus Longus*, a genteel, well peopled. town, but suffered greatly in the *Turkish* wars of 1737 and 1738.

Bakow, Bratskow, the neighbouring country pleasant and fertile: it is also the residence of a *Roman* catholick Bishop.

Tergovisto, Tervis, the capital of *Walachia*, lying on the river *Jalonitz*. It is a town of good trade and has some fortifications. Here is also a palace belonging to the Waywode.

Buckereſt a fortified town on the *Dembrovitz*, the uſual reſidence of the Waywode; and an archiepiſcopal fee: Here is alſo an academy for the ſons of perſons of quality.

Jalonitza, a town on a river of the ſame name.

Braila, Braeli, Ibraeli, a little town ſeated on the *Danube*, with a fortified caſtle, having ſeven towers, taken in the year 1711 by the *Ruſſian* General *Ron*, but by order of his Prince immediately abandoned by him.

M O L D A V I A.

§. 1. **I**T has been already obſerved in *Walachia* (§. 1.) that *Moldavia* is ſometimes comprehended under the name of *Walachia*, conſidered in which light that part of it lying on this ſide the mountain is called *Walachia*. This country takes the name of *Moldavia* from the rivulet *Moldau*, which runs from the upper parts into the country, mingling itſelf with the river *Sereth*. It was formerly called *Bogdania*, the reaſon of which is explained in §. 3. Its length extends from weſt to eaſt, that is, from the river *Sereth* to the *Dnieſter*, for betwixt thirty and forty *Hungarian* miles, but its breadth from ſouth to north is about ſeventy.

§. 2. The country has ſome very fertile lands but a conſiderable part of the eaſtern diviſion lies uncultivated, conſiſting chiefly of deſarts; and the weſtern is very mountainous. Its principal rivers are the *Sereth*, already mentioned under *Walachia*. The *Pruth*, which has its ſource on the borders of *Tranſylvania* and *Poland*, running through *Moldavia* from north to ſouth.

The *Dnieſter*, which forms the boundary towards the eaſt and north. All theſe three rivers receive ſeveral ſmall ſtreams in their courſe, and the two firſt fall into the *Danube* or *Iſter*, which is the boundary to the ſouthward, but the laſt diſcharges itſelf into the *Black* ſea.

§. 3. The inhabitants are of *Walachian* extraction and profeſs the *Greek* church; but many of them are alſo *Mahometans, Ruſſians, Poles, Raſcians*, and *Armenians*. Towards the cloſe of the twelfth century a *Walachian* colony came from *Tranſylvania* and ſettled in this country. Their leader, named *Bogdan* eſtabliſhed their civil and eccleſiaſtical government, and for the ſupport of the latter obtained an archbiſhop and other eccleſiaſticks from the patriarch of *Conſtantinople*. He was the firſt prince of *Moldavia* and laid the foundation of the principal towns, for which reaſon the country was originally called from him *Bogdania*. The increaſe of the power of the King of *Hungary* was a misfortune to this country, the inhabitants, after ſevere ſtruggles, being made tributary to that monarch in the fourteenth century. In the year 1280, the *Turks* firſt made an attempt upon *Moldavia*.

§. 4. This

§. 4. This country has a Prince, or Woywode, of its own, who is alſo ſtyled Hoſpodar, and is a vaſſal of the *Ottoman* Porte, to which he is obliged to pay an annual tribute. The *Moldavian* coat of arms is an ox's head ſable, in a field or.

Moldavia is divided into the Upper and Lower.

1. *Upper Moldavia* reaches almoſt to *Jaſſy*, being bounded towards the eaſt by the river *Dnieſter*; to the north partly by that ſtream and partly by *Pokutia*, which is a part of the Kingdom of *Poland*; and weſtward by *Tranſylvania*. In it are the following places.

Chotzin, Coczin, a town on the *Dnieſter*, well fortified both by nature and art, and belonging to the *Turks*. In the year 1739 the *Pruſſians* made themſelves maſters of it, having firſt beaten the *Turks* out of their entrenchments, which they had thrown up near the fort. In the years 1621 and 1674, the *Turks* were defeated here by the *Poles*.

Soroka, a town on the *Dnieſter*.

Czudno, a town on the river *Pruth.*

Soczowa, or *Sotſchowa*, formerly capital of the country, at preſent only a middling town.

Herlow, a town on a rivulet which falls into the *Pruth.*

Stepanowze, Stepanowitz, Stepanowka, a town on the river *Pruth*.

Obſ. In the upper parts, towards the north, reſide the *Lip Tartars.*

2. *Lower Moldavia* borders, to the weſt, upon the mountains of *Tranſylvania*, which run along the road called *Tetras*, and belong to *Moldavia* and *Walachia*. Its ſouthern boundary is the *Danube*; to the ſouth-eaſt it is bounded by *Beſſarabia*, and to the caſt by the *Dnieſter*. It contains ·

Jaſſy, the capital and reſidence of the Woywode, ſeated on the river *Pruth*, being a ſpacious and fortified town lying in a country abounding in wine. In the year 1753 the whole town, with the palace of the Hoſpodar, ſome *Roman* Catholick convents, a rich walled church, and the Proteſtant *Lutheran* church, which had been newly built, were all deſtroyed by fire. In the years 1711 and 1739 it was taken by the *Ruſſians*.

Huſzi, Hus, Huz, a ſmall town on the *Pruth*, where *Czar Peter* I. in 1711, after an unſucceſsful battle, which laſted three days, made a peace with the *Turks*.

Faltſchii, a town on the *Pruth*. In its neighbourhood, amidſt very thick woods, are ſeen the ruined walls and foundations of houſes which run in ſtraight lines. Theſe, probably, are the remains of the ancient large city of *Taiphali*, mentioned by *Herodotus*, and on the ruins of which *Faltſchij* was built.

Galatſch, a trading town on the *Danube*.

Nemes, a very ancient town, ſtanding on a high mountain by a river of the ſame name.

Obſ. *Beſſarabia*, of which we treat in the following part, belonged formerly to Lower *Moldavia*.

Several Colonies of TARTARS *and their Diſtricts.*

MENTION has been made above, under the article *Bulgaria*, of the *Dobrutzi*, and likewiſe, under Upper *Moldavia*, of the *Lip Tartars.* Excluſive of theſe, from the northern branch, through which the *Danube* enters the *Black* ſea, to the river *Don*, in a winding tract of land, along the *Black* ſea and the ſea of *Aſoph*, to the extent of above one hundred miles, live ſeveral tribes of *Tartars*, generally comprehended by geographers under the title of *European Tartars*. But this appellation is by no means proper, there being many other *Tartars* in *Europe*, and therefore I choſe the above title to this ſection as being more accurate. Of this large country, which is a part of the ancient *European Scythia*, the *Tartars* made themſelves maſters in the firſt half of the thirteenth century, as ſhall be more fully ſhewn under the article of the *Crim Tartars*. Some of them wander about in hordes, or clans, whilſt others are ſettled in towns and villages. Some alſo are immediately dependent on the *Ottoman* Porte, and others ſubject to the Crim *Chan*, who is himſelf a vaſſal to the Grand Seignor. Having not been able to procure any creditable accounts of the reſpective limits of the *Turkiſh* ſovereignty, and the authority of the *Chan* over theſe *Tartars*, I chooſe to leave this point in ſuſpenſe rather than determine blindly. Geographers alſo vary greatly in their diviſion of theſe *Tartarian* Diſtricts; and, under a want of ſufficient authorities, it ſeems beſt, in deſcribing the whole country along the *Black* ſea and the ſea of *Aſoph*, from the mouth of the *Danube* to that of the *Don*, to begin from the weſt, or the *Danube*, proceeding according to the main rivers which interſect the country, and mentioning what is moſt remarkable in the ſeveral Diſtricts formed by theſe large ſtreams; which are the *Dnieſter*, the *Dnieper* and the *Don*; but theſe have been already treated of in *Muſcovy* and *Poland*. The *Dnieper* is joined by the *Bog*, or *Bug*, anciently called the *Hypanis*. Tracing it thus from the weſt, we meet with

1. *Beſſarabia*, lying on the *Black* ſea, betwixt the north branch of the *Danube* and the *Dnieſter*, and by the *Turks* called *Budriak*, or *Budſcack*. It belonged once to Lower *Moldavia*. The inhabitants, if not deſcendants of the ancient *Budini*, have at leaſt inherited their name. They are called *Budziaki*, *Bielgorod* and *Ackermann Tartars*, appellations derived from the town of *Budziak* and *Bielgorod*, which latter is alſo known by the name of *Ackermann*, The *Akermanni* are called *Axiacæ* by *Herodotus*, a word of which the etymology is *Ak-ſia*, i.e. *white water*; in alluſion to the *Dnieſter*, whoſe ſtreams are turbid and whitiſh. They are alſo ſtill called the *White Hordes*, and rove from place to place along the *Dnieſter*. Their uſual food is the fleſh of their oxen and horſes, cheeſe and milk, particularly that of mares. Of this Diſtrict the moſt remarkable places are

Kili,

Kili, by the *Moldavians* called *Kilia Nowa*, to diſtinguiſh it from a place of the ſame name, the ancient *Lykoſtomon*, but now no longer in being. It is ſeated on the northern branch of the *Danube* at its diſcharge into the *Black* ſea. This is the wideſt and moſt navigable branch of that river. The inhabitants apply themſelves to the making of ſalt.

Bielgorod, *Akerman*, or *Akkjirman*, by the *Moldavians* called *Tſchetate Alba*, are ſeveral names of one town and much of the ſame import: the firſt is in uſe among the *Turks* and *Ruſſians*, and ſignifies the *White Town*; the ſecond is compounded of *Ak* White, and *Kerman* a town or Caſtle; and the third is ſynonymous with *White-caſtle*. The old town lies at the fall of the *Dnieſter* into the *Black* ſea.

Budriak, a little place on the *Dnieſter* near the ruins of the ancient *Tyras*, or *Ophiuſa*.

Palanka and *Parkara*, ſmall towns on the *Dnieſter*.

Cauchan, near the *Dnieſter*, the head quarters of the *Akermann Tartars.*

Bender, a *Turkiſh* fortification on the *Dnieſter*, formerly called *Tizine*, but being made over to theſe *Turks* by one of the Princes of *Moldavia* they gave it the name of *Bender*, i. e. *a paſs*; and it has always a baſhaw for its governor.

Warnitz, near *Bender*, remarkable for being the place where, in the year 1709, *Charles* the twelfth of *Sweden* broke up his camp, and continued, till the year 1713, when the *Turks* were obliged to make uſe of force to get rid of him.

2. The country betwixt the *Dnieſter* and *Dnieper*, by ſome called *Oczakow Tartary*, is inhabited only along thoſe two rivers and near the ſea, the other parts being quite waſte and therefore called *Dzike Pole*, or the deſart plain. It affords, however, good paſturage but has not a ſingle tree upon it. In the year 1709, after the unfortunate battle of *Pultowa*, the above-mentioned warlike Prince travelled over it in his road to *Turkey*, ſuffering inexpreſſible hardſhips by the way: Near the *Dnieper* dwell the *Sietſch Koſacks*, or *Haidamacks*, of whom mention has been made under the article, Of the government of *Kiow* in *Ruſſia.* The moſt remarkable places here are

Oczakow, anciently *Ordeſſus*, by the *Turks* called *Kaahleh Ozi*, i. e. *the fortification of Ozi.* It is a very ſtrong place, ſituate at the influx of the *Dnieper* into the *Black* ſea. It receives its *Turkiſh* name from the *Dnieper*, by them called the *Ozi.* It lies on the declivity of a mountain with a caſtle above it. When Count *Munich* ſat down before it, in the year 1737, its fortifications were in excellent order, and its garriſon conſiſted of a large body of choſen *Turkiſh* forces; but the count, compelled by the want of fodder for the horſes and other cattle, riſqued an aſſault and carried it the third day after opening the trenches. The *Ruſſians* held the place till the following year, when they evacuated it; after having demoliſhed the works.

Kazikermen,

Kazikermén, or *Kizikermen,* i. e. *Caftle Kazi,* fituate on the *Nieper,* and built by *Muhammed* the eleventh; being taken, in the year 1695, by the *Ruffians,* they razed it, but it has fince been rebuilt.

3. The country betwixt the *Nieper* and efflux of the *Don* contains

I. A mainland along the *Black* fea and the fea of *Afoph,* inhabited by the Leffer *Nogay Tartars,* the remainder of whom belong to *Afia.* They wander from place to place, the bords keeping at thirty hours diftance, and frequently not fo far from each other. They feldom apply themfelves to agriculture. As among the *Crims,* horfe-flefh is their favourite food. They are fo hofpitable, that to entertain a traveller and his horfe is, with them, its own reward. If a fmall prefent of tobacco, or any thing elfe, be offered, they receive it with abundance of thanks, and never fail to make a return. The religion they profefs is *Mahometifm.* They are governed by Beys or Murfes of their own nation, or by fuch as the *Crim Chan,* their Sovereign, appoints from among them. With refpect to the limits of this Diftrict, from the *Nieper* to the *Don,* it was agreed at the treaty of *Belgradi,* in the year 1739, that a line fhould be drawn from the river of *Zalimy, Konfkich Wod,* which runs into the *Nieper* below the river *Samara,* betwixt *Kudac* and *Zaporozkaja Sietza,* or the *Zaporog Setze* (*Sietza* fignifies a pallifado, or wooden wall) to the river *Berda,* which falls into the fea of *Afoph.* The Diftrict within this line, or fouth near the *Crim,* continues under the Chan of *Tartary,* who enjoys a much larger Diftrict northwards, but dependent on *Ruffia.* In the firft Diftrict, and on the *Black* fea, are

Kinburn, a *Turkifh* fortification lying oppofite to *Oczakow,* and caft of the *Nieper,* where it difcharges itfelf into the *Black* fea. In the year 1736 the *Ruffians* took it and blew up the fortifications, but the *Turks* thought it worth rebuilding.

Abloe, Alfza, Cyganfkaja Dolina and *Kokzogar,* are little places betwixt the *Nieper* and *Black* fea.

In the fecond, or *Ruffian* Diftrict, and not far from the mouth of the *Don,* is the town of *Mius,* lying to the eaft of a fmall bay of the *Black* fea, into which runs the river *Mius.* The name of this place occurs in hiftory, having been formerly fortified. At the end of the above-mentioned fmall bay ftood

Fort St. *Paul, Taganroe, Lutic* and *Afoph,* which have been already mentioned, and the nature of the foil defcribed, in the account of *Bachmut,* under the article of *Ruffia* in *Europe.* The road along the fea of *Afoph* from *Perekop* to *Taganrae,* is three hundred and eighty-nine werftes* in length.

2. The *Crim* peninfula, *Cherfonefus Taurica,* in the *Tarkifh* maps *Kiram Athafi,* i. e. the *Crim* ifland, is of the fame figure with the *Morea,* and by the ancients reckoned to be nearly the fame in extent. It is environed by the *Black*

_____
* A hundred and four werftes make one geographical degree.

sea and the sea of *Asoph*, excepting at the very narrow neck of land, which joins it to the continent. The soil, in many parts, produces all kinds of grain, wine, &c. but the *Tartars* seem to despise agriculture, leaving it to their slaves and strangers. The favourite food of these *Tartars* is horse-flesh, milk and cheese, bread being little in use among them. Of all the *Mahummedan Tartars* these most resemble the *Calmucks*. The inland country was anciently possessed by the *Scythians*, who extended themselves northwards beyond *Perekop*, westward to the *Nieper*, and eastward as far as the river *Don*. The western and southern coasts of the *Crim* were inhabited by some *Greek* colonies, of which the town of *Chersonesus* was the most powerful. The east side of the *Crim*, as far as the *Don*, and the opposite country, or the tract from the *Don* along the sea of *Asoph* to the *Black* sea and mount *Caucasus*, was under the Kings of the *Bosphori-Greeks*, who were so called from the strait of *Bosphorus*. The *Scythians* proving troublesome neighbours to the *Greeks;* they sollicited the assistance of *Mithridates*, King of *Pontus*, who at last drove the *Scythians* out of the peninsula, forming the Kingdom of *Bosphorus*, which comprehended the whole peninsula, and the country facing it eastward to mount *Caucasus*. In the time of the Emperor *Dioclesian* the *Sarmatæ* were in sole possession of this Kingdom, excepting that the *Goths* had seated themselves on the west side of the peninsula and the whole tract of land lying northward from it along the *Don*. The peninsula afterwards came under the dominion of the Emperors of the east, though partly shared by the *Hunns*, who were succeeded by the *Chazars*, or *Cozars*, as these were by the *Pelowzers*. About the end of the twelfth century, the *Genoese* having made themselves masters of the *Black* sea and all its harbours, settled also in the *Crim*. In the thirteenth century the *Tartars* dispossessed the *Polawzers* of their country, particularly of the *Crim*, but the *Genoese* ports and castles baffled their undisciplined fury, especially the town of *Costa*, which held out till the year 1471, when it was taken by the *Turks* who reduced also the whole *Crim*, and appointed a Chan over it. In the year 1698 the *Russians* made an attempt on the *Crim* but gained only *Perekop;* in 1736 they penetrated a second time into the *Crim*, under the conduct of Count *Munich*, marched over the line which the *Perekop Tartars* had thrown up across the isthmus with a great number of towers garrisoned by one hundred thousand men for its defence; took *Perekop*, and opened a way into the *Crim*. In the years 1737, 1738 and 1739, the *Russians* renewed their enterprizes on the *Crim* with such success that, within these last four years, half of the *Crim* has been ravaged by them and likewise by the *Tartars*, whilst great numbers of the inhabitants have perished by famine and others abandoned the country.

The *Crim* has its own Chan, or Prince, who stiles himself, indeed, the Sovereign Chan of Lesser *Tartany*, &c. though in reality a vassal of the *Ottoman* Porte, and, on summons, must take the field with a considerable body as auxiliaries. The family of the Chan is fprung from *Mengli-Gæray-Chan*, fon to *Hadfey-Gæray-Chan*. The eldeft fon, or prefumptive heir to the Chan, has the title of Sultan *Galga*. The fecond Prince is called *Or Beg*, i. e. *lord of Or*, or *Perekop*; the third *Noràdin Beg*, &c.

The peninfula of *Crim* is thick fet with great and fmall villages, and contains, likewife, feveral large but very ill built towns.

1. To the Chan belongs

Perekop, a fortified place, on the ifthmus which joins the peninfula to the continent, and which has always been reckoned the key to the whole country. Its name is *Sclavonian*, and fignifies a cut made through a place, being derived from the ditch dug here, in very remote ages, acrofs the neck of land at the entrance of the *Crim*, for the fecurity of the peninfula, and which has been from time to time repaired, and of late fortified. On this account the *Greeks* called it *Taphros*, or *Taphræ*, and the *Turks* and *Tartars Or*, which have pretty much the fame fignification with *Perekop*. The houfes here are but mean. In the years 1638 and 1736 the *Ruffians* took it; at the laft of which times, the whole *Turkifh* garifon, confifting of two thoufand five hundred and fifty-four men, were made prifoners of war. Though the *Ruffians* demolifhed the place the *Tartars* took the pains to rebuild it. In the year 1738 the *Ruffians* again made themfelves mafters of it, but foon loft it.

Genitfchi, a fort, built, in the year 1736, by the *Ruffians*, on a very narrow point of land of the peninfula, near the ftrait into the *Dead-Sea*, or *Gniloe more*, as it is called; which is formed by fome bays of the fea of *Afoph*, and runs up into the country. They raifed alfo another facing it on the continent, but foon after demolifhed it.

Koflow, a town on a point of land in the fea of *Afoph*, on the weftern fide of the peninfula, having a fine harbour. It has alfo a ftone wall and towers, and its trade is very confiderable. The inhabitants are *Tartars*, *Turks*, *Greeks*, *Armenians* and *Jews*. The *Turks* import rice, coffee, dried figs, raifins, dates, cloth and filk-ftuffs, making their returns with flaves and grain. In 1736, both the garifon and *Turkifh* inhabitants abandoning it, the *Ruffians* took it without oppofition. Eleven werftes fouthwards is a ftagnant lake, in which, in fummer-time, falt is made.

Bakfchifaräi, an open town though the refidence of the Chans, lies on the weft fide of the peninfula, near the fea, and betwixt two hills which ferve the place inftead of walls. Its inhabitants are *Tartars*, *Armenians*, *Greeks* and *Jews*, and the houfes the beft built of any in the whole country. The Chan's palace is large and irregular, but not deftitute of contrivance. In the year 1736 the *Ruffians* made themfelves mafters of this place.

Achmetfchet, a town, and the refidence of Sultan *Galga*, or the Chan's eldeft fon, ftands on the largeft river in the country.

Karafbazar,

Karafbazar is large, and carries on a confiderable trade, yet has only low wooden houfes, but its four mofques are of ftone. The inhabitants are *Tartars, Armenians, Greeks* and *Jews*; and the horfe-fair held there is the moft confiderable in all the *Crim*. It is, by what I can find, called alfo *Cherfon, Kurfun, Corfun* and *Kurafon*. In the year 1737 the *Ruffians* laid it in afhes.

Crim Staroi, i. e. *Old Crim*, formerly a flourifhing city, now has the appearance only of a large village.

Sudak, a fea-port on the caft part of the country.

Arabat, a fmall port towards the eaft.

Kerfch, Gerfch, anciently *Panticapæum*, fince the *Bofporus*, is a confider-able town on a fteep mountain near the ftrait formerly called the *Bofporus*, but now, by foreign failors, the ftraits of *Caffa*, or St. *Jan's* mouth. It commands the entrance into the *Black* fea, is environed with a high wall and a caftle with feven towers to the fouth-eaft, and at the harbour a mole of ftone. Far the greateft part of the houfes are built alfo of ftone with flat roofs. It has twenty-two *Turkifh* mofques and as many *Greek* churches.

II. To the *Turks* belong

Genikola, a caftle on the ftrait near *Kerfch*, with a harbour, but not fit for fhips of burden.

Kaffa, Keffeh, the ancient *Theodofia*, a large trading fea-port town, fituate on the eaft of the peninfula, and fuppofed to have been built by the *Greeks* in the fifth century. In the year 1226, the *Genoefe* having got poffeffion of this place, enlarged and fortified it; but, in the year 1297, were driven thence by the *Venetians*, from whom however they quickly recovered it again. In the year 1474 the *Turks* took it from them. Whilft in the hands of the *Genoefe* its commerce was at fuch a height, that, if not fo large with refpect to the number of its inhabitants, it, in this particular, exceeded *Conftantinople* itfelf; but falling under the *Turkifh* yoke, its commerce declined greatly, and now confifts only in flaves brought hither for fale by the *Crim* and *Cuban Tartars*, the *Georgians* and *Mingrelians*. The inhabitants confift of *Greek*, Catholick and *Armenian*, Chriftians, *Turks, Jews*, &c. The Chriftians conftitute the majority, and enjoy a perfect freedom of religion. Here are ftill feveral defcendants of the old noble families of *Genoa*. The town is the largeft in all the *Crim*, contains about five or fix thoufand houfes, and is conftantly well garrifoned. North-weft of it are fome high mountains, at the foot of which it ftands in a very delightful and convenient fituation, but the harbour, befides a difficult bar, affords fmall fhelter in a fouth-eaft wind.

Balaklawa, a fort and harbour in a cape on the fouth-weft fide of the ifland.

THE

THE

KINGDOM

OF

PORTUGAL.

INTRODUCTION

TO THE

KINGDOM OF PORTUGAL.

§. 1. THE Kingdom of *Portugal* and *Algarve* is not only generally in-cluded in the maps of *Spain*, but there are alſo particular maps of each which are well executed. Not to mention the more ancient one by *Seccus* and *Tereyra*, which have been engraved, with ſome improvements, by *John Von Ram, Viſcher, Danckert, Schenck* and *Homann, Nicholas Viſcher* has publiſhed one more modern and very large, taken from a draught of *Placide*, a Monk of the order of St. *Auguſtine*; and *J. B. Nolin* has alſo put forth another, which was republiſhed in the year 1736 by *Homann*'s heirs. The firſt volume of *Luiz Caetano de Lima*'s *Geographia Hiſtorica* contains a ſmall general map of *Portugal*; and, in the ſecond, are ſix provincial maps, excluſive of the plans of particular towns.

§. 2. This country was formerly called *Luſitania*, but its boundaries were then different from what they are now. The name of *Portugal* is, by ſome, thought to ſignify *Portus Gallus*, or *Portus Gallorum*, on account of the multitudes of *French* which came to the city of *Porto*, on the river *Douro*, in order to aſſiſt the Chriſtians againſt the *Moors*. But others, with more probability, derive it from a town on the river *Douro*, by the ancients called *Cale*, but by the moderns changed to *Gaya*. Oppoſite to this place ſome of the inhabitants afterwards built a new town with a harbour, and gave it the name of *Portucale*, or *the harbour*, or *Port of Cale*; which, by an un-interrupted proſperity, proved the origin of the preſent flouriſhing city of *Porto*; and from hence the whole country has received the name of *Portugal*. The old name of *Luſitania* was aboliſhed, and the new one took place under *Ferdinand* the Great, King of *Caſtile* and *Leon*, who gave this country and *Gallicia* to his third ſon *Garcia*. The moſt ancient writing now extant, in which the name of *Portugal* is uſed for the whole Kingdom, is dated in the year 1069, and kept in a convent at *Arouca*.

This

§. 3. This Kingdom, which is the moſt weſtern part of *Europe* is bounded, on the ſouth and weſt by the *Atlantick* ocean : And on the north and eaſt by *Spain*. Its length from *Valença*, the moſt northern town in it, to *Sagres* the moſt ſouthern, near *Cape St. Vincent*, extends according to the inhabitants, above a hundred and twenty * *Portugueſe* miles ; and its greateſt breadth from *Peniche*, a ſea-port in *Eſtremadura*, to *Salvaterra*, on the frontiers of *Spain*, is about forty ſuch miles. But in the maps the former amounts only to ſeventy-five and the latter but to thirty *German* miles.

§. 4. The climate in *Portugal* is much more temperate than in *Spain*, tho' with ſome little difference in the ſeveral provinces. The northern parts feel a kind of a painful cold in winter, though this is chiefly owing to the rains of that ſeaſon ; and in the ſouthern the ſummer heats are very great. Both winter and ſummer, however, are very ſupportable, cooling ſea-breezes during the latter refreſhing the country, and the ſpring is extremely delightful. Though the ſoil be very fruitful, ſo much is agriculture neglected here, that above half the country lies waſte; and the inhabitants are ſupplied with a great part of their corn by importation. *Portugal*, however, abounds in excellent wine and oil, of the latter of which the greateſt part is made in the province of *Alentejo*. Olive-trees thrive better here near the ſea than up in the country. Here is alſo great plenty of fine honey, and conſequently of wax. The beſt honey, which is found in the fields, is almoſt of a white colour, and of a moſt agreeable flavour; and the wood honey here is more grateful to the taſte than that of other countries. *Portugal* alſo produces abundance of lemons, pomegranates, oranges, figs, raiſins, almonds, cheſtnuts and other fine fruits. It has likewiſe variety of ſea and river fiſh, and large quantities of ſea-ſalt.

§. 5. The country in many parts is mountainous : The chief of which in the *Entre Douro Minho* are *Bola*, *Gaviao*, *Geres*, St. *Catherina* and *Maram*. In the *Traz-Oz* chain are the mountains of *Momil*, *Montago*, &c. In *Beira* thoſe of *Alcoba* and *Eſtrela*; on the ſummit of which is a famous lake. In *Eſtremadura*, *As Cemas De Ourein*, and *Martinel*. In *Alentejo*, *Calderaon*, *Portel*, *Arra Bida* and *Tagro*; and in *Algarve*, *Monchique*, and *Caldeirac*. The mountains of *Portugal* contain all kinds of ores, as ſilver, copper, tin and iron. But the *Portugueſe* being ſupplied with metals from their poſſeſſions in other parts of the globe, and particularly with abundance of gold from *America*, no mines are worked in their own country. Gems of all kinds, as turquoiſes, and hyacinths are alſo found in the mountains; and particularly a beautiful variegated marble with many other curious foſſils of the lapideous kind, of which divers ſorts of works are made : Here are alſo very good mill-ſtones. On the hill of *Alcantara*, not far from *Liſbon*, is a remarkable mine of ſalt-petre, near which grows ſatyrion or ragwort.

* Of *Portugal* or *Spaniſh* miles, ſeventeen and a half make one degree of the equator.

From

From the mountains issue several streams and small rivers, which fertilize the vallies and fields; and either join the greater rivers in their course, or discharge themselves separately into the sea. The largest and principal of these are the following:

The *Minho* (in *Latin, Minius)* having its source in *Gallicia* a province of *Spain,* and emptying itself into the western or *Atlantick* ocean not far from the town of *Caminha.*

The *Lima, Limia, Belis,* by the ancients called also *Lethe,* or the *River of Forgetfulness,* the *Turduli* and *Celtæ* after losing their commander in chief, settling there, as if they had forgot their own country. This also takes its source in *Gallicia* falling into the sea below *Viana.*

The *Cavado* rises in the mountains of *Traz-os,* and empties itself into the sea below *Barcelos.*

The *Douro* having received its source in *Old Castile* in *Spain,* passes by the city of *Miranda* in *Portugal.* Its course is mostly among mountains: And after receiving the lesser rivers of *Coa, Sabor, Tua, Tavora, Paiva, Tamega* and others, it falls into the sea near the city of *Porto. San Joan de Pesquera* is the first town where this river becomes navigable. The waters of it are said formerly to have contained so much gold in it, that *John* III. caused a scepter to be made thereof.

The *Tejo* or *Tagus* has its source in the frontiers of *Arragon* and *New Castile:* In *Portugal* it is joined by the swift river of *Zezere* and others; widening greatly towards its mouth, so as to form several islands; and a considerable harbour at *Lisbon,* a little beyond which it discharges itself into the sea. This is the largest and principal river in the whole kingdom. In some places, as near *Santarem* for instance, it overflows and increases the fertility of the adjacent country. This river also affords gold.

6. The *Guadiana,* in *Latin* the *Anas,* rises in *New Castile,* and watering *Bagadoz* in *Portugal,* loses itself in the sea near *Ayamonte* and *Castro Maxim.* Its name signifies the river *Anas: Guadi* in the *Morish* tongue denoting a river.

All these several streams abound in fish; and the three chief, namely, the *Douro, Tejo,* and *Guadiana* divide the kingdom into three parts.

Some mineral springs have been found here, but are not used, except the baths of *Caldas* in *Estremadura,* which are famed for their virtues in venereal cases.

As *Portugal* is not without excellent pastures, and particularly in the country about *Montestrella,* and near *Ourique,* graziery in some places is very considerable: We see here an uncommon number of cattle and sheep; but in most places it is at so low an ebb, that the greatest part of their beeves come from *Spain.* The horses are not large, but very fleet. The *Portuguese,* however, breed more asses than horses, the latter being clandestinely imported from *Spain.*

The whole Kingdom contains nineteen cidades or cities; and five hundred and twenty-seven villas or smaller towns. The number of inhabitants may be pretty nearly computed. *Luiz Caetano de Lima* in his *Geografia Historica*, T. ii. p. 475 to 710 inserts a list of all the parishes in the cities, towns and villages of the whole kingdom; with the number of all the hearths and souls in every parish. This was communicated to him in the year 1732, by the Marquis de *Abrantes*, Censor and Director of the Royal Academy of History; who looked upon it to be very accurate. I carefully computed the parishes and the number of souls in them: And the several totals are as follow:

Entre Douro and *Minho*	963	parishes	432362 souls.
The mountains of *Traz-os*	549		135804
Beira — —	1094		551686
Estremadura —	315		293598
Altenejo — — —	356		268082
Algarve — —	67		60688

Consequently there are in the whole Kingdom 3344 parishes, and 1742230 souls.

But it must be observed that this list is thus far imperfect, as in all probability it does not include the ecclesiasticks, monks and nuns. These at most cannot be reckoned above 300,000; so that in all *Portugal* there are about two millions of people. The foreign voyages and colonies very much diminish the number of its inhabitants: And the multitude of convents there hinder the increase of them. It is customary among the *Portuguese* to betake themselves to sleep about noon, on account of the great heats; transacting most of their business in the morning and evening, or even at night. Their language is a compound of the *Spanish*, *Latin*, *Moorish* and *French*.

The nobility are very numerous: And many of them of the blood-royal, being descended from natural sons of the Royal-family. They were formerly more considerable than at present: Although so much of the ancient custom, by which the King conferred a maintenance on the Nobility, still remains, that he assigns them a pension from a certain fund to enable them to maintain their dignity, and this is a memorial of the ancient *Moradias*, as it was called, or attendance-salary. For such of the Nobility as become impoverished or disabled in the King's service, there is a publick foundation at *Belem*, not far from *Lisbon*, where they have every thing provided them in a very decent manner; being clothed at their entrance with the vestments of the Order of Christ.

The Nobility are divided into high and low.

The

The high or titled Nobility (*titulados*) confist of Dukes, Marquiffes, Counts, Vice-Counts, and Barons.

All thofe who are Grandees and ftiled Dons, like thofe of *Spain*, confift of three claffes, and receive from the royal treafury a penfion fufficient to fupport their refpective dignities. It is further to be obferved here, that the King frequently confers certain titles merely to give precedency, as the Marchionefs of *Alenquer*, of *Caftello-Melhor*, of *Atouguia*, *Santa-Cruz* and *Unhao*, with the Countefs of *Penalva*; a Duke's fons are alfo Grandees; and his daughters hold the rank of Marchioneffes. The Prior of *Crato* fits covered after the manner of a Count.

2. The inferior nobility or gentry are termed *Fidalgos*; and are incapable of bearing the title of Don, unlefs by permiffion from the King. The *Mocos Fidalgos*, or gentlemen-born, are accounted higher than citizens who acquire the title of *Cavallero Fidalgo*, without being ennobled thereby.

§. 9. That Chriftianity was propagated here by the Apoftles *James* and *John* fo early as the firft century cannot be proved. But in the fecond century it appears from *Tertullian* that it was received all over *Spain*, to which *Portugal* was then annexed: And in the third, new parifhes were formed. In fucceeding times the *Moors* and *Jews* increafed greatly, and lived intermingled with the Chriftians: And notwithftanding the fevere perfecutions under *John* II. whereby they were forced to an external acknowledgment of the *Romifh* church: And though all exercife of the *Jewifh* religion be prohibited by the fundamental laws of the Kingdom; yet there ftill remain great numbers of fecret *Jews* among the *Portuguefe*, and thofe too even among the nobility, bifhops, prebends, monks and nuns; and the very inquifitors themfelves. Many, unable to conceal themfelves, efcape to *Holland*; and there openly profefs Judaifm. The eftablifhed religion, however, of the country is the *Roman* catholick: And the inquifition, which was introduced by King *John* III. and has fince been fet up throughout all the *Portuguefe* dominions, *Brazil* excepted, is very active in detecting herefies and hereticks; and no lefs rigorous in punifhing them. Of the inquifition in the dominions of *Portugal* there are four efpecial high courts; namely at *Lisbon*, *Coimbra*, *Evora*, and *Goa* in the *Eaft Indies*, each of which is independent, though with fome fubordination to the fupreme council of inquifition. Impious and inhuman as this tribunal is, yet its feftivals, fort folemn burnings, called *Auto da fe*, or ' The act of faith,' afford the higheft delight to the bigotted *Portuguefe*, who whilft their fellow creatures, the fuppofed hereticks, are burning in the flames, fhout *Que grande clemencia! Bente abencoado fea o fanto officio*, i. e. *O what great goodnefs! praifed be the holy Office.* Even the principal Nobility reckon it an honour to lead the condemned criminals as mean fervants to the holy office. King *John* IV. however, in fome meafure curtailed the power of the inquifition, commanding that all its fentences fhould be laid before the parliament; and that the accused should be allowed council for making their defence. He likewise further enacted that only blasphemy, sodomy, poligamy, herefy, sorcery, Pagan customs, and the conversion of the *Jews* are to come under their cognizance. To be a *Christam Velho*, *i. e.* ' An ancient Christian,' or of an ancient Christian race, is accounted in *Portugal* a very great pre-eminence, and far above a *Christam novo*, or new converted Christian, or *a temparte de Christam novo*, a half-new converted Christian; by which last are to be understood those whose new converted ancestors married ancient Christians.

The number of convents in *Portugal* is said to be no less than nine hundred; and most of them very rich; but in multitude and opulence the Jesuits surpass all other orders. The most remarkable and wealthy abbey is that of the *Cistercians* at *Alcobaça*.

With respect to ecclesiastical persons, *John* V. at a very great expence procured a patriarch, obtaining in 1716 the Pope's consent to raise his court-chapel at *Lisbon* to a patriarchate. And in 1717, the new patriarch, who also retained the title of *Capellao Môr*, or *first Court Chaplain*, was installed with very great solemnity. In the year 1739, the Pope further granted that this patriarch should always be a cardinal, and of the Royal-family. The patriarchal church at *Lisbon* was erected into a palatinate; and the fourth part of all the ecclesiastical benefices in *Portugal* assigned for the support of the new canons; but in 1753 their revenues were greatly curtailed. Subordinate to the patriarch, as suffragans, are the bishops of *Leiria*, *Lamego*, *Angra* in the island of *Terceira*, and *Funchal* in *Madera*. Next are three archbishops who rank with Marquisses, and of them the first is the archbishop of *Braga*, who is primate of the kingdom and lord spiritual and temporal of his city and the neighbouring country, and stiles himself also primate of all *Spain*. His suffragans are the bishops of *Porto*, *Viseu*, *Coimbra*, and *Miranda*. The second is the archbishop of *Evora*, who has for his suffragans the bishops of *Elvas* and *Faro*. The third is the archbishop of *Lisbon*, whose suffragans are the bishops of *Portalegre*, *Guarda*, *Angola*, *Cape Verde*, and St. *Thomas*. The bishops hold the rank of Counts. Exclusive of those in *Europe*, the *Portuguese* have also archbishopricks and bishopricks in the other three parts of the world.

The King too besides the nomination of all bishops receives a fourth of their revenue; but this is generally given away in pensions. The Pope confirms the bishops; publishes his bulls in the kingdom without any previous consent of the King, and by his legate governs the clergy, who with respect to taxes and contributions depend on him. He has also the gift of many small prebends. The sums arising to them from these prerogatives, are by some thought to exceed the revenues of the crown, and the nuncio's have so good a time of it that they never fail of raising vast fortunes before they return to *Rome*.

§. 10. *Coimbra.*

§. 10. *Coimbra* and *Evora*, indeed, are univerfities, and *Lifbon* has a royal academy eftablifhed for *Portuguefe* hiftory, whofe motto, *Reftituet omnia* promifes mighty things, and fome good hiftorical pieces have been already publifhed by it. At *Santarene* is an academy of hiftory, antiquity and languages ; and, at St. *Thomas,* an academy of fciences on the footing of that at *Paris,* of which the King was pleafed to declare himfelf prefident : but, whilft the Papal power continues here at the enormous height it has hitherto been, fcience is like to make but a very indifferent figure. In 1746, an *Italian* capuchin publifhed a work, at *Valencia* in *Spain,* in the *Portuguefe* language, confifting of four volumes in quarto, entitled *Verbadeiro Methodo de eftudar,* &c. dedicated to the Kingdom of *Portugal,* fetting forth the wretched ftate of the fciences there and the contemptible manner in which they are taught. He even charges the *Portuguefe* with aiming at the perpetuity of ignorance and the triumph of barbarifm. He decries their fchools as the places of retreat for thofe errors which, by *Defcartes* and *Newton,* had been driven out of other parts of *Europe* ; and, at their univerfities, the mind, he fays, is obfcured with the abftrufe difquifitions of fcholaftic learning. According to him, *Galilæo, Defcartes, Gaffendi* and *Newton* are, in *Portugal,* the names of atheifts and heretics, not to be mentioned but with fome mark of execration ; and, laftly, he affirms, that the profeffor of anatomy, in a *Portuguefe* univerfity, is allowed only a fheep twice a year for demonftrating his lectures. If this be going too far, let us hear what a *Portuguefe* himfelf, *D'Oliveira,* fays. In the Preface to the firft volume of his Memoirs, he has thefe words : " In our country we live in " ignorance without knowing it ; but, on going out of *Portugal,* our eyes " feem fuddenly to open, and we immediately fee that ignorance in which " we were involved. The ignorance I fpeak of is evident : foreigners allow " us underftanding, docility, morals, difcernment, and a genius for com- " prehending what is commendable and good in the world ; but our con- " ceit, our gravity, our confined manner of living, which deprives us of all " freedom of thought, expofe us to juft cenfures, and give rife to thofe " hateful opinions other nations entertain of us. The main fource of our " ignorance, and a miferable caufe of offence to all nations, is the cuftom " in *Portugal* of prohibiting fuch numbers of books, *&c.*"

§. 11. The neglect of agriculture in *Portugal* has been taken notice of above, (§. 4.) and the like may be faid of all handicrafts, arts and manufactures ; for which laft of all kinds the country has the fineft materials : but the greateft part of thefe are difpofed of rough to foreigners, who are not wanting to fet a high price on their wares. The *Portuguefe,* indeed, make a little linen and varieties of ftraw work, and candy feveral kinds of fruits, particularly oranges. They have likewife fome coarfe filk and woollen manufactures ; but thefe are trifling articles, and fupport but a very fmall part of the nation. It alfo highly concerns all foreigners trading to *Portugal,*

and

and particularly the *English*, that the *Portuguese* apply not themselves to manufactures; and they are accordingly very induftrious in preventing it; of which the looking-glafs manufacture, fet up at *Lifbon*, was a memorable inftance. It is more agreeable, befides, to the political views of the court of *Lifbon* that the nation fhould be fupplied by foreigners rather than employ themfelves in providing for their feveral necefflities.

§. 12. The *Portuguefe*, indeed, carry on a very extenfive trade, but reap no great profit from it, being obliged to vend not only their own products but all the merchandize and riches brought home to them from their fettlements in other parts of the globe, and efpecially from *America*, to the *Europeans* trading with them, particularly the *Englifh*, in exchange for grain, and fmall and large quantities of manufactured goods of all kinds, with which they fupply both *Portugal* and its poffeffions abroad. The goods which they traffick in with foreigners are fea-falt, oil, wine, lemons, *Seville* and *China* oranges, figs, raifins, almonds, cheftnuts and other fruits, together with wool, filk and other materials for manufactures; but their chief commodities confift of imports from their own colonies, particularly from *Brafil*; fuch as fugars, tobacco, cocoa-nuts, ivory, ebony, *Brafil* wood, hides, fpices, drugs, gold, pearls, diamonds and valuable gems; and, excepting only thofe of *Brafil*, the Eaft and Weft *India* commodities may be had out of *Portugal* and at the firft hand. In 1755 a new trading company to *Grofpam* and *Maragnan* was eftablifhed here, and foreign merchants admitted to fhares.

The *Portuguefe* fhipping little frequent the other countries of *Europe* or the *Levant*, their voyages lying rather to the coafts of *Africa*, particularly the gold coaft, whence they carry Negroes to *Brafil*: they meet there alfo with fome gold and ivory. They trade likewife to their Eaft *India* colonies of *Goa*, *Din* and *Macao*; but this traffick, once fo very important, is now greatly declined: *Brafil*, however, continues a plentiful treafury to them. Foreigners are totally excluded from all commerce with *Brafil*; the *Portuguefe*, neverthelefs, carry on a confiderable clandeftine traffick with the *Spaniards*, which confifts chiefly in an exchange of gold and filver; and thus each King is defrauded of his fifth. From *Brafil* the *Portuguefe* bring not only fugar and tobacco, but alfo gold and diamonds. Of the quantity of gold annually brought in there and imported to *Lifbon*, fome conjecture may be made from the amount of the King's fifth, which the writer of *Anfon's* Voyage round the World makes for fome time paft to have been, *communibus annis*, one hundred and fifty arrobas, *Portugal* weights, or near three hundred thoufand pounds fterling. The gold brought every year to *Lifbon* cannot be lefs; and the whole annual produce of gold in *Brafil* may, without any exaggeration, be eftimated at near two millions of pounds fterling. The fleet, which fails every year to *Brafil*, goes and returns in feven or eight months; and, when homeward bound, is convoyed by fome

men

men of war which are fent to meet it. Ships from *Africa*, or the *Eaft In-dies*, come alfo in company.

§. 13. The long meafure of the *Portuguefe* is by barros and cavidos; a hundred barros make one hundred and fixty-four cavidos, and one hundred ninety-four and a half· *Hamburgh* or *Leipfic* * ells; but a hundred cavidos are equal only to fixty-one barros, or about one hundred and nineteen *Ham-burgh* ells. Of the *Portuguefe* weights I fhall mention only the arrobas and quintals. In *Portugal* an arroba is thirty-two pounds, at *Hamburgh* † twenty-nine or thirty. A quintal is four arrabas, and at *Hamburgh* makes one hundred and twenty-eight pounds. All fums of money are reckoned by millereis and crufadoes, which, however, are not real but only imaginary coins; twenty-four reis and an half make one grofche, confequently a mil-lereis; or one thoufand reis make a rixdollar and almoft feventeen grofche; one old cruzadoe is equal to four hundred, and a new one to four hundred and eighty reis. The filver coin of leaft value is a vintain, of twenty reis, a half tefton of fifty, and a whole tefton of one hundred reis; one real is equal to forty reis. Laftly, the gold coins are the moeda de oufo of four thoufand eight hundred reis, alfo half and quarter pieces; the dobra, of twenty-four thoufand reis, or a half of twelve thoufand eight hundred, of which alfo there are a half, a quarter, an eighth and a fixteenth piece.

§. 14. *Portugal*, or, as it was anciently called, *Lufitania*, paffed from the *Phenicians* and *Carthaginians* into the hands of the *Romans*, and, by the Emperor *Auguftus*, was made a *Roman* province. Towards the beginning of the fifth century, from the birth of *Chrift*, the *Alans*; about the year 440 the *Swabians*; and, about the year five hundred eighty-two, the *Vifi-goths*, made themfelves mafters of this country. In the eighth century *Portugal* was over-run by the *Moors* and *Saracens*, but gradually refcued from them by the *Chriftians*. *Henry*, born duke of *Burgundy*, performed fuch eminent fervices to *Alphonfo* VI. King of *Caftile*, againft the *Moors*, that he gave him his daughter *Therefa* in marriage, and, in the year 1093, created him earl of *Portugal*, and, in 1110, it became, by his father-in-law's will, his abfolute property. *Alphonfo Henriques*, his fon and fuc-ceffor, obtained a fignal victory, in the year 1139, over the *Moors* near *Ourique*. About the fame time too he affumed the title of King, and, in 1147, eftablifhed the order of *Avis*, Pope *Alexander* III. having, in the year 1179, confirmed his right to the throne. In the year 1181 he held

---

* The author feems here to make the *Hamburgh* and *Leipfic* ell the fame; whereas, accord-ing to the learned Sir *Jonas Moore's* table of foreign meafures, &c. compared with the *Englifh*, the *Hamburgh* ell is only one foot and 905 parts of 1000; or, one foot, ten inches and eight parts of an inch; whereas, that of *Leipfic* contains two feet, 260 parts of 1000; or, two feet, three inches and eight parts of an inch *Englifh*.

† The *Hamburgh* pound, according to Sir *Jonas Moore*, is 95 parts of 100 *Englifh*.

an affembly of the States at *Lamego*, in which the fucceffion to the crown was fettled. *Alphonfo* III. added *Algarve* to the crown of *Portugal.* Under King *Dennis* the order of *Chrift* was inftituted. The ligitimate male line of this family becoming extinct, in the year 1383, in the perfon of *Ferdinand, John* I. his father's natural fon, was admitted to the crown in 1385, in whofe reign the *Portuguefe* fettled in *Africa*, and difcovered the iflands of the *Azores*. In the year 1482, his great-grandfon *John* II. received the *Jews* who had been expelled *Spain*, and gave great encouragement to navigation and difcoveries. With an eye to future conquefts and difcoveries, in the year 1492, he entered into a firft convention with *Ferdinand* the Catholic King of *Spain*, and, in 1494, concluded a fecond, by virtue of which, all countries three hundred and feventy miles weftward of *Cape de Verde*, and the iflands of the *Azores* were ceded to *Ferdinand*, and the difcoveries eaftward agreed to belong to himfelf. Under King *Emanuel, Vafcus de Gama*, in the year 1497, difcovered the courfe to the *Eaft Indies*. In 1501 *Americ Vefpuce* took poffeffion of *Brafil*. In the year 1504 the firft Portuguefe fort in the Kingdom of *Cochin* was erected ; the war againft the *Moors* was vigoroufly profecuted in *Africa*, and under this King the glory and profperity of *Portugal* rofe to its higheft pitch. In the reign of *John* III. the caftle of *Diu* in *Afia* was built, and, in the year 1540, he admitted the newly founded order of the Jefuits, of which he was the firft member of all the *European* Princes. In *Henry* the cardinal the male line of this houfe alfo failed in 1580, and the fucceeding year the Kingdom was united to *Spain*; but under the Kings of this Monarchy the *Portuguefe* loft moft of their foreign acquifitions. The *Perfians*, in the year 1622, difpoffeffed them of the ifle of *Ormus*. The *Dutch* obtained great advantages over them in the *Eaft Indies*, took the *Molucca* iflands, over-run half *Brafil* in the year 1636, and, in 1637, made themfelves mafters of *St. Georgio del Mina*, in *Africa*. In the year 1639 their trade to *Japan* was ruined, and, in 1640, they loft *Malacca*, the chief town. At home too the *Portuguefe* were fo intolerably oppreffed that, in the year 1640, they fbook off the *Spanifh* yoke, electing *John*, duke of *Braganza*, for their King. This Prince, firnamed *John* IV. in 1654, drove the *Dutch* out of *Brafil*: but, in the year 1656, loft the rich ifland of *Ceylon*. *Alphonfus* VI. was dethroned by his brother *Peter*, who, in 1668, concluded a treaty with *Spain*, by which *Portugal* was declared an independent Kingdom and reftored to its ancient limits, *Spain* retaining only the town of *Ceuta* in *Africa*. Under King *John* the fifth the royal chapel of *Lifbon* was raifed to a patriarchate. *Jofeph*, the prefent King, has no male iffue yet.

§. 15. The King's eldeft fon, from the time of *John* the fourth, is ftyled Prince of *Brafil*, the other children and brothers being called Infants. King *John* the fifth declared his grandfon, the Prince of *Brafil*'s fon, Prince
of

of _Beira._ The King's title runs thus: _Joseph_, by the grace of God, King of _Portugal_ and the _Algarves_ on this and the other fide of the fea of _Africa_; lord of _Guinea_, of the conquefts, trade and navigation in _Ethiopia_, _Arabia_, _Perfia_, _India_, &c. In the year 1749, Pope _Benedict_ XIV. by a formal bull, conferred on the King the title of _Rex fideliffimus_, which was immediately inferted in all the public inftruments, the foreign powers making no difficulty of acknowledging it. But the proper meaning of it is not agreed on, fome rendering it the moft faithful, and others the moft orthodox or believing. It looks, however, as if his holinefs had only given his fanction to a title which the ancient _Portuguefe_ writers, _Francifcus de Albertinis_ for inftance, were frequently wont to confer upon their Kings.

§. 16. The arms of _Portugal_ are a fhield argent with five fmall fhields azure placed crofs-wife, on each of which are five filver pieces in the form of a St. _Andrew_'s crofs. On the border of the fhield are the arms of _Algarve_, being the feven ancient caftles of _Eflombar_, _Paderna_, _Aljafur_, _Albufeira_, _Cacella_, _Sagres_ and _Caftromarin_.

§. 17. The principal order of knighthood in this Kingdom is the order of _Chrift_, inftituted by King _Dennis_ foon after the abolition of the Knights Templars, and confirmed, in the year 1319 by Pope _John_ XXII. King _Emanuel_ added to the ftatutes, which were afterwards ratified, in 1505, by Pope _Julius_ the fecond, and are to this day a rule of conduct to the order. The badge of the order is a red crofs within a white one. The feat of the order is at the city of _Thomar_. It has four hundred and fifty-four commanderies. Concerning the origin of the order of St. _James_ writers differ, but it is faid to have been raifed, about the year 1030, from a fraternity to an order, and to have received the confirmation of Pope _Alexander_ III. It depended, at firft, on the _Caftilian_ grand mafter of the order, from whom it detached itfelf in the time of King _Dennis_, and Pope _Nicholas_ the fourth gave his fanction to this rupture, though it was not fully accomplifhed till the year 1290, when the _Portuguefe_ knights chofe themfelves another grand-mafter; but the _Caftilian_ grand-mafters, countenanced by fucceeding Popes, made a great pufh to bring them again under fubordination to them. The next in rank to the grand-mafter is the prior fuperior of _Palmilla_, who is invefted with an epifcopal jurifdiction and controls the city convent. To this order belong forty-feven fmall towns and places, and one hundred and fifty commanderies, befides the fplendid convent of _Santos o novo_, a little to the weft of _Lifbon_. The badge of the order is a red fword, in the fhape of a crofs, refembling the handles of ancient fwords.

The order of _Aviz_ is faid to have been inftituted fo early as the year 1147, by King _Alphonfo Henriques_, and to have been firft confirmed in 1162, and in 1201 received a fecond fanction. Its firft feat is likewife fuppofed to have been at _Coimbra_, whence it was removed to _Evora_ and afterwards to _Aviz_, which place _Alphonfo_ III. in the year 1211 conferred on it.

Whilst it continued at *Evora* it entered into a coalition with the order of *Calatrava*, and in this state they continued till *John* the second, under whom they separated. The next to the grand-master is the prior, superior of *Aviz*, who is both temporal and spiritual judge of the order, and, in quality of the former, judge also of the convent. Its commanderies are only forty-nine in number, and the badge belonging to it a green-cross in form of a lily.

These three orders are all religious, with liberty, however, of marriage to the knights, by virtue of a bull of Pope *Julius* III. in the year 1551. The Kings of *Portugal* are perpetual grand-masters. Under King *Alphonso Henriques*, somewhat before the year 1157, the knights of *Malta* came to *Portugal*. They have twenty-three commanderies here, among which is the priory of *Crato*; and three balias; namely, *Leça*, to which a revenue is annexed; *Acre*, merely honorary and alternately with *Castella*, having the office of the order; thirdly, the balia of *Negropont*. The government is lodged in an assembly consisting of a president and all *Cavalleiros professos* who have been three years at *Malta*.

§. 18. The government of *Portugal* is an unlimited monarchy, but, with respect to the imposition of new taxes, the settlement of the succession and other important concerns, the consent of the states, which consist of the clergy, the high nobility and the commons, is necessary. The clergy are represented by archbishops and bishops (§. 9.) the high nobility are the dukes, marquisses, counts, viscounts and barons (§. 8.) and the representatives of the commons consist of the procuratores, or agents, of the cities and towns. Among the commonalty, are likewise reckoned the lower nobility (§. 8.) and the masterships of the order, of knighthood. This assembly never meets but by proclamation from the King, in which they are styled *Cortes*. The last was held in the year 1697. Though the crown of *Portugal* be hereditary, yet the consent of the several states is also necessary to the succession of brother's children. The crown too devolves to the female line, but this right is forfeited if they marry out of the Kingdom. In the year 1641, it was confirmed by a manifesto of the states that, in cases relating to the succession, the *jus representationis* should be admitted. But this extends no farther than to brother and brother's children, after whom it devolves to the nearest of kin. The ordinance passed at *Lamego* (§. 14.) relating to the succession, is a fundamental law of the Kingdom; and, in the year 1641, a manifesto of the states was annexed to it.

§. 19. The highest state-assembly is the *Confelho de Estado*, or council of state, in which are transacted all the great affairs of the Kingdom, with the disposal, likewise, of all ecclesiastical and temporal posts and offices which are not immediately dependent on some other board or tribunal; but even these, at last, come before the council of state, particularly the nomination of all archbishops and bishops, viceroys, captains-general, governors
of

of the provinces, and of all other poſſeſſions belonging to the crown are determined here: peace and war, embaſſies, alliancies, &c. are all likewiſe canvaſſed in this council, which is ſaid to have been firſt inſtituted, in imitation of the *Spaniſh* court, by Queen *Catharine*, during the minority of King *Sebaſtian*. In the year 1732, it conſiſted of five eccleſiaſtics and an equal number of lay-officers. The ſecretary of ſtate alſo is a member of this council, and is aſſiſted by the official mayor and others.

The office of ſecretary of ſtate was reſtored to its ancient conſtitution by King *John*, on the twenty-ninth of *November*, in the year 1643, who divided the affairs appertaining to it into *merces* and *expediente*, on which account he is called *Secretaria das Merces e expediente*. Here the nominations to civil employments are divided, thoſe miniſters and ſecretaries excepted whoſe commiſſions are made out by the ſecretary of ſtate. It has alſo the diſpoſal of all military poſts, from captain to lieutenant-colonel included, with diſpenſations, grants of the commanderies of the ſeveral orders of knighthood, treaſury affairs and diſburſements, nomination of judges, ſentences of the grand-marſhal, &c. To this ſecretary of ſtate's office likewiſe belong all the King's bounties (called *merces*) together with orders of knighthood and penſions (called *tenças*) annual legacies belonging to churches (*capellas*) vacant or re-aſſumed lands and commandaries. The *Secretario das Merces e expediente* draws up alſo the paſſes for all foreign ſhips and *Portugueſe* merchantmen.

The *Secretario da Aſzinatura* lays before the King all patents (*alvaras*) *proviſoens*, *cartas* and *padroens*, which are delivered to him from any juridical perſon, for the King to ſign them; thoſe papers excepted which the ſecretary of ſtate and the ſecretary of bounties lay before his Majeſty.

Theſe three ſecretaryſhips are ſometimes diſcharged by the ſame perſon, of which we have an inſtance at preſent in that eminent ſtateſman *Diego de Mendoça Corte-Real*. The council of war (*conſelho de guerra*) was inſtituted on the eleventh of *December*, in the year 1640, by King *John* the fourth; and, in 1643, received its inſtructions, or plan, which conſiſt of twenty-nine articles. At this board are regulated all military affairs, or any thing elſe relating thereto. It aſſiſts alſo in the diſpoſal of all military offices from colonels to *governatores das armas* in the provinces, and the *capitaens generaes de exercitos*, to whom it ſends orders by the ſecretary at war. It recommends likewiſe to the employments of *juiz acceſſor*, *promotor fiſcal* of the council of war; the adminiſtrators and auditors general of all the provinces; fills up all the military poſts from a ſerjeant to a captain, excluſive of the latter, and has the ſuperintendency of the fortifications, magazines, quartering of ſoldiers, hoſpitals, artillery and other military matters. On great emergencies, the council of war is conſulted by the council of ſtate.

The council of the palace (*deſembargo do paço*) is the higheſt tribunal, to which cauſes may be brought from inferior courts by appeal. It nomi-

nates

nates to all offices belonging to the law, decides difputes of jurifdiction betwixt the lay and fpiritual courts; examines the briefs of Pope's nuncios,
and, befides a great variety of other bufinefs, draws up all laws, orders,
edicts, confirmations, privileges, grants, &c. This court confifts of a prefident and feveral counfellors *(called Defembargadores)* whofe number is not
limited ; five fecretaries *(efcrivaens da camera)* each of whom has his particular department, one being alfo ftyled *efcrivao do defpacho da mefa* ; a
thefoureiro-diftribuidor, and feveral other inferior officers. Under this tribunal is the *choncellaria mor da corte e reyno,* which confifts of a chancellor,
a *veador,* fome clerks, a *thefoureiro, porteiro* and other officers.

The *Cafa da fupplicaçao,* is the firft and higheft tribunal of juftice, being
without appeal in civil or criminal cafes. To its ordinary jurifdiction belong
the provinces of *Eftremadura, Alentejo* and *Algarve,* with the *Comarca de
Caftellobranco* in the province of *Beira.* To it alfo lie final appeals from the
cafa do civel do porto, in cafes which fhall be fpecified in the fequel. It confifts of forty-two officers in the following order ; a chancellor, ten *Defembargadores dos aggravos e appellaçoens,* or counfellors of appeal, two *corregedores,* belonging, in criminal affairs, to the court ; two *corregedores,*
belonging, in civil cafes, to the court ; two *juizes dos feitos da coroa e fazenda;*
two *ouvidores* of appeals in criminal affairs, a *procurador dos feitos da coroa,*
a *procurador dos feitos da fazenda,* a chancery judge, a jufticiary *promotor* and
eighteen *defembargadores extravagandes,* or fupernumeraries. A little alteration is fometimes made in this number.

The *Cafa do Civel e relaçao do Porto,* is the fecond *relaçao,* or fecond high
court of appeal, in the Kingdom, and has its feat at *Porto.* Under it are
the provinces of *Entre Douro e Minho, Traz os montes* and *Beira,* the *Comarca de Caftellobranco* excepted, which is fubject to the *cafa da fupplicaçao*
at *Lifbon.* In the year 1696, *Peter* the fecond ordered that all proceffes not
exceeding two hundred and fifty thoufand reis in immoveables, and three
hundred thoufand in moveables, fhould be finally determined by this court;.
But beyond that fum there lies an appeal from it to the *cafa da fupplicaçao.*
The *cafa do Civel* confifts of twenty-three officers ; *viz.* a chancellor, eight
defembargadores de Aggravos, two *corregedores* for criminal, and one for
civil affairs, one judge for matters pertaining to the crown and exchequer,
three *ouvidores do crime,* of whom one is alfo chancery-judge, one *promotor
da juftiça,* five *defembargadores extravagandes* and one crown-attorney.

The *Confelho da fazenda,* or treafury court, was eftablifhed by King *John*
the fourth on its prefent footing. The chief affairs belonging to it are divided into three claffes, with a *vedor da fazenda* for each. Of thefe one
fuperintends the finances of the Kingdom, the other thofe of *Africa,* the
Coutos and *Terças;* the department of the third are the *Indies,* the magazines and armaments. Befides thefe it has three *Vedores,* the tribunal
confifts of feveral counfellors, *miniftros de letras,* and *defembargadores,* with
others.

others *de capo Eefpada, i. e.* '.of the cloak and the fword :' But the number of, thefe is uncertain. It confifts likewife of a *procurador da fazenda*, and four ordinary clerks; to which are added fupernumeraries and feveral other officers. Subordinate to this court are divers others, as *O trebunal dos - coutos*, or, the chamber of accounts; *O tribunal da alfandega*, or, the cuf- tom-chamber ; *O tribunal da cafa da India e mina*, *O tribunal dos Almazens*, a *Tenencia* and a *Cafa da Moeda*; with the following *Cafas*, *O paçoda Ma- dèira*, *O confulado*, *Os portos fecos*, and a *Cafa dos cincos*; and laftly, in 1720, the *Junta do commercio* was annexed to this court.

§. 20. For the inferior adminiftration of juftice each of the fix provinces, of which the kingdom confifts, is divided into certain judicial Diftricts called *Comarcas :* Every Diftrict confifts of *cidades, villas* (§. 8.) inferior courts and jurifdictions, which are termed *Concelhos, coutos, Julgados*, and *Honras.* There is alfo another divifion of them. All jurifdictions are either *Correiçoens*, or *Ouvidorias*. The former belong to the crown, the latter to the *Donatarios*, who are partly ecclefiafticks and partly laymen. The judge whom the crown appoints for its Diftrict is ftiled *Corregedor ;* and the Do- natorian is ftiled *Ouvidor.* We fhall here make ufe of the laft divifion as being the moft intelligible, for a *Comarca* often extends to places of diffe- rerent *Correiçoens*, or *Ouvidorias.* The name is alfo ambiguous, fometimes denoting places belonging to the crown, at others comprehending a whole province. Sometimes again it fignifies the ecclefiaftical divifions of a Dio- cefe, as in the archbifhoprick of *Braga* for inftance, there are five *Comarca's*, which comprehend what the bifhop poffeffes in the *Traz-os Montes*, and in the Diocefe of *Porto.* The *Cabeça*, or capital of a *Correiçao* or *Ouvidoria*, is always the city or *villa* where the *Corregedor* or *Ouvidor* refides. In fuch towns too there is generally a *provedor* or upper-overfeer appointed for the *Comarca*, or *Correiçao*, or *Ouvidoria*, who takes cognizance of the execu- tion of wills, with a *Juiz de fora* or foreign judge : And frequently too a *Juiz dos orphaos* or a judge of the orphans. The cities have alfo their par- ticular magiftracy, which generally confift of one *Juiz de fora*, two or three *Vereadores*, one *Procurador do confelho*, with fome other inferior officers. Sometimes inftead of the *Juiz de fora* there is an *Efcrivao da ca- mera*, or *Thefoureiro da camera*, with a *Juiz do povo* or a judge of the people, not to mention other leffer variations. The *Roman* law with the gloffaries obtains in *Portugal*; befides which there are royal edicts, and the canon-law and Pope's mandates,

§. 21. The King's revenue arifes firft from the hereditary eftates of the houfe of *Braganza*, to which belong fifty *villas*. Secondly from the royal demefnes. In 1753, the King came to a refolution of re-annexing feveral demefnes and lands of confiderable extent to the crown, which, after the difcovery of the *Azores* and *Brafil* had been granted to private perfons, pro- pofing to give the prefent owners an equivalent in their ftead. Of this a
begin-

beginning has been already made in the estate of the late Viscount and present Count of *Affeca*. 3. From the customs of which that of *Lisbon* produces most. 4. From the taxes. 5. From the excise, which is very severe, and paid even by the clergy. 6. From the monopoly of *Brazil* snuff. In the year 1755, the monopoly of this snuff was farmed for 3,000,000 of crusadoes *per annum*. 7. From the coinage. 8. From the sale of indulgences, which the Pope renews to the King every three years by a special bull including three others, namely, a bull for the living, a bull for the dead, and a bull of composition; by virtue of which the paying of a certain part of an iniquitous gain renders the rest legal. 9. From the grandmasterships of the order of knighthood, which the King holds in his own hands. 10. From the ecclesiastical tithes in foreign countries. 11. From the duty of the fifth part of all the gold brought from *Brazil*, which every year amounts to 300,000 pounds sterling (§. 12.) and from the farm of the *Brazil* diamonds, and 12, from the confiscation of the estates and fortunes of persons condemned by the inquisition.

§. 22. The military forces of the *Portuguese* are but in an indifferent condition. In the year 1754, the King issued an establishment for his regiments in time of peace, by which the army, when complete, consists of

Twenty-one regiments of foot, Men	6300
Six of horse — — —	1500
Two other regiments of horse —	800
Four regiments of dragoons	1200
Two regiments of marines — —	2400
Two regiments of artillery —	1800
Total	14,000

The navy is not more formidable, as consisting only of twelve ships, and those but weakly manned, and gallies the King has none.

§. 23. *Portugal* contains six provinces and several islands in the *Atlantick* ocean. The *Portuguese* were the first of all *European* nations in discovering countries, and making settlements in them. But of all their ancient great conquests in other parts of the globe, they now only possess, 1. In the *Atlantick* ocean the *Cape de Verde* islands, *St. Thomé, do principe*, &c. 2. In *Africa, Magazan* fort on the coasts of *Morocco, Catcheo* or *Cacheo* on the Negro-coast; several forts in the kingdoms of *Loango, Congo, Angola* and *Monomotapa*, a fort in *Monoemugi*, the town of *Mosambique* in the kingdom of that name; and the town of *Sofola* on the eastern coast of the *Caffern*. 3. In *Asia* the towns of *Diu, Goa, Onor, Macao*, &c. 4. In *America, Brasil*, part of *Guiana* and *Paraguay*, but the description of these countries and places must be referred to the fourth volume of this work. The provinces of the Kingdom of *Portugal* are as follow, *viz.*

ENTRE

ENTRE DOURO E MINHO.

WHICH is so called from its being situate between the rivers *Douro* and *Minho*, by the first of which it is separated from the province of *Beira*, and by the second from *Gallicia* in *Spain*. Its extent from north to south is about eighteen *Portuguese* * miles; and from west to east twelve. It is not only very fruitful, but its rivers which empty themselves into the sea, and, its good harbours, particularly those of *Porto* and *Vianna*, are so convenient for trade, that in it are found two *Cidades*, namely, those of *Braga* and *Porto*, twenty-six *villas*, forty-six *Confelhos*, forty-four, or according to others forty-eight *Coutos*, and twelve *Behetrias*, *Honras* and *Julgados*; being in proportion to its extent the most populous province of the whole kingdom. With respect to its ecclesiastical state it contains two cathedrals, namely, *Braga* and *Porto*; five collegiate churches, *viz.* those of *Guimaraens*, *Barcellos*, *Cedofeita*, *Valença do Minho* and *Vianna*; a great number of abbies and convents for both sexes, most of them rich, with nine hundred and sixty-three parishes, in which, in the year 1732, were computed to be 432,372 souls, as appears from the list of the parishes given us in the *Geografia historica*. Among the commanderies of the orders of kighthood, the bailiwick of *Leça* is particularly deserving of notice, as it belongs to the order of *Malta*. The province is divided into six jurisdictions, three of which, containing the crown-lands, are called *Correiçoens*; and three, belonging to the *Donatarios*, *Quvidorias*.

The three *Correiçoens*, each of which is governed by a royal *Corregedor*, are

1. *Correiçao de Guimaraens*, containing four *villas* or small towns, twenty *Concelhos*, fourteen *Coutos*, four *Honras* and one *Julgado*. The remarkable places in it are

Guimaraens, a *villa*, owing its origin to a convent of *Benedictines* built there in the year 927, to which was immediately added a village, now increased to a town. The walls of this place are one thousand eight hundred and fifty paces in circuit. It is divided into the old and new town, contains five thousand inhabitants, four parishes; two of which are in the suburbs, one *Casa da Misericordia*, three hospitals and six convents; besides

---

* Seventeen *Spanish* or *Portuguese* miles and a half are equal to one degree of the equator.

I   another

another a little way out of the town.   It was originally the refidence of the
Kings of *Portugal*, and the birth-place of King *Alphonfo Henriques*.   The
Infant *Duarte*, fon to King *Emanuel*, had this town conferred upon him
together with the title of a dutchy in which he was fucceeded by his fon.
But at his death the title became extinct, and the place reverted to the
crown.   It is the *Correiçao* town and the refidence of the *Corregedor*, as alfo
of a *Provedor* and a *Juiz de'fora* for the *Comarca* of *Guimaraens*.   To the jurifdiction of this place belong ninety parifhes, and thirty thoufand and
twenty-eight fouls.

Amarante is a town on the river *Douro*, having only one parifh church;
its inhabitants about eleven hundred in number.   Here is a *Juiz de fora*,
who belongs alfo to the *Comarca*.

Canavezes, a *villa* on the river *Douro*, containing fix churches, and only
nineteen hundred inhabitants.

Povoa, a fmall town.

Twenty *Confelhos*, namely *Felgueiras*, with twenty parifhes.   *Unhao*, containing ten, and giving the title of earl.   *Santa Cruz de Riba Tamega*, with
twenty.   *Gouvea de Riba Tamega*, with eight.   *Geftaço*, with thirteen.   *Ceroilico
de Bafto*, with thirty-eight.   *Cabeceiras de Bafto*, with nineteen.   *Roffas*,
with two.   *Villaboa de Roda*, with one.   *Vieira*, with fix.   *Monte Longo*,
with fourteen.   *Ribeira de Soas*, with eleven.   *Povoa de Lanhofo*, with
twenty-one.   *S. Joao de Rey*, with three.   *Atey*, with one.   *Serva*, with
three.   *Ermello* with five.   *Ribeira de Pena*, with three: And *Villa-pouca
de Aguiar*, with thirteen.

Fourteen *Coutos*, which are thofe of *Abhadim*, *Fonte-Arcada*, *Mancellos*,
Moreira de Rey, *Parada de Bouro*, *Pedraido*, *Pombeiro*, *Poufadella*, *Refoyos
de Bafto*, *Taboado*, *Tibaens*, with four parifhes; *Travanca*, *Tuas*, and *Vimi-
airo*, with three.

Four *Honras*, and among thefe *Villacais*, with one parifh, and *Ovelha*
with two.

One *Julgado*, namely *Lagiofo*.

II. The CORREIÇAO DE VIANNA confifts of nine *villas*, which are

Vianna, lying not far from the mouth of the river *Lima*, being large,
well built and ftrong, and defended by the caftle of St. *Jago*, contains
feven thoufand inhabitants, two parifh churches, one *Cafa da Mifericordia*,
with an hofpital and feven convents : But its harbour is fit only for fmall
veffels.   It was founded by King *Alphonfo* III. and belonged formerly to
feveral proprietors under the title of an earldom ; but at prefent is the property of the crown.   It is the *Correiçao* town, and the refidence of a *Corregedor*, a *Provedor*, and a *Juiz de fora* ; as alfo the quarters of a *Meftre de
Campo General*.   To its jurifdiction belong twenty parifhes, and, according
to the late obfervations of father *Capaffi*, it ftands in latitude 41° 41".

Ponte

Ponte de Lima, a *villa* on the river *Lima,* on the spot where once stood the *Forum Limicorum* of the *Romans.* It has been twice destroyed, and was rebuilt in 1125 and 1360. It is a handsome city, has a collegiate church, a *Casa da Misericordia,* three hospitals and two convents, two thousand inhabitants and twenty-four parishes within its District. The *Juiz de fora* residing here belongs likewise to the *Comarca* of *Vianna.*

Ponte da Parca is a small town, containing only one parish and sixteen hundred inhabitants, but having twenty-five parishes within its district.

Souto de Ribeira de Homem, a small town.

Prado, a little place of five hundred and sixty inhabitants and one parish, with fourteen other parishes in its District.

Pica de Regalados, a town on a little river which falls into the *Cabado.*

Villa nova de Cerveira, a *villa* on the river *Minho.* It formerly stood in another place, and is said to have been built by King *Dennis.* It lies in a bottom environed with hills, is well fortified, and defended by fort *Azevedo* on the side of *Valença.* It contains six hundred inhabitants, one parish church, one *Casa da Misericordia,* one hospital, and on a mountain near it has a convent. King *Alphonso* I. erected it into a viscounty; but under *Philip* IV. it escheated again to the crown. It has since been raised to an earldom. In its District are thirteen parishes. The *Juiz de fora* here belongs likewise to the *Comarca* of *Vianna.*

Monçao, a *villa* on the river *Minho,* founded by *Alphonso* III. had a charter in 1261, is fortified, and contains six hundred inhabitants, one parish church, one *Casa da Misericordia,* one hospital and one convent. Its *Juiz de fora* belongs also to *Comarca,* and within its District are twenty-two parishes. Father *Capessi* places it in latitude 42° 5″.

Arcos de Valdevez, a small place containing about four hundred and forty inhabitants and one parish church, with forty-five parishes in its District. It confers the title of earl.

It contains likewise twelve *Concelhos,* which are *Lindoso* with one parish. *Pica de Regalados,* with seventeen. *Villa Garcia,* with four. *Entre Homem e Cavado,* with eighteen. *Bouro,* with twelve. *Suayo,* with three. *Santa Martha de Bouro,* with six. *Coura,* with twenty. *Albergaria de Penela,* with eleven. *Sotto de Rebordanes,* with two. *Santo Estevao da Faxa,* with two, and *Geraz de Lima,* with two.

Thirteen *Coutos,* namely, those of *Aboim de Nobrega, Azevedo Baldreu, Bouro, Cervaens,* or *Villar de Areas, Freiriz, Luzio, Manhente, Nogueira, Queijada,* which is united to *Boilhosa, Sabariz, Sa Osins* with four parishes, and *Souto.*

III. The CORREIÇAO DO PORTO contains one *Cidade* and three *Villas,* namely,

Porto, a *Cidade* on the river *Douro,* which a little below it empties itself into the sea. Its bar is somewhat dangerous on account of the sand-bank and rocks in it, except at high-water in winter. It is defended by the caftle of *S. Joao de Foz*, being fortified with an old wall and towers: In opulence, populoufnefs, beauty, and commerce it is the next city in the kingdom to *Lisbon*. It has four fuburbs and feven parifh churches, including the cathedral, above twenty thoufand feven hundred inhabitants, one *Cafa da Mifericordia*, fome hofpitals, twelve convents, among which is a Jefuits college, befides four other without the walls; and is the fee of a bifhop, under whom, exclufive of the city and its Diftrict, are four *Comarcas ecclefiafticas*, namely *Maya, Penafiel, Riba-Tamega*, and *Feira*, containing three hundred and forty parifhes: It is likewife the feat of a *Tribunal de Relaçao*, and a *Cafa do Civel*, which King *Philip* II. removed here from *Lisbon*. It has alfo a *Tribunal da Alfandega* and a mint. Is befides this the *Correiçao* town, and refidence of a *Corregedor*, a *Provedor*, a *Juiz de fora* and a *Juiz dos Orfaos*. This city owes its origin to *Cale* or *Gaya*, which lay oppofite to it, on a rugged mountain on the other fide of the river, being built by part of its inhabitants as a much more convenient fituation; and therefore called *Portucale* or *Porto de Cale*. It was afterwards, on account of its wealth and flourifhing commerce, erected into a bifhop's fee, who ftyles himfelf *Epifcopos Portucalenfes*, or *Portuenfes*. The whole Kingdom has received its name from hence. It was formerly fubject to particular Lords, but at prefent belongs to the crown. It ftands, according to father *Capeffi*, in 41° 10″ north latitude.

Villa nova do Porto, a fmail town feated on the fouth fide of the river *Douro*, oppofite to *Porto*, and not far from the old town of *Gaya*; with refpect to which it is called *Villa nova*. It was built in 1255, by King *Alphonfo*, contains about two thoufand nine hundred inhabitants, one parifh church, one *Cafa da Mifericordia*, one hofpital and one convent, with two others near the town.

Melres, and *Povoa de Varzim*, two fmail places; though the latter contains above eight hundred inhabitants.

Twelve *Concelhos*, which are thofe of *Gaya*, with twenty parifhes. *Gondomar*, with eight. *Aguiar de Soufa*, with feventeen. *Maya*, with fifty-four. *Refoyos de Riba de Ave*, with twenty-one. *Loufada*, with twelve. *Penafiel de Soufa*, with thirty-feven. *Porto Carreiro*, with three. *Penaguyao* with fourteen, and the title of earldom. *Bayao*, with eighteen. *Scalhaens*, with one. *Bem-viver*, with fixteen; and *Avintes* with one, and the title of earldom.

Seven *Coutos*, namely *Anfede, Entre ambos os rios, Ferreira, Meinedo, Paço de Soufa, Pendorada*, and *Villa-Boa de Quires*.

One *Julgado*, namely *Bouças*, with eight parifhes.

Five *Behetrias* and *Honras*, which are *Baltar, Barbofa, Frazao, Gallegos* and *Louredo*.

The

II. The three *Ouvidorias* belonging to the *Donatarios*, who nominate an *ouvidor* to each.

1. The *Ouvidoria de Barcellos*, confifts
Of feven *villas*; namely,

Barcellos, a *villa*, on the river *Cavado*, fortified with walls and towers: It has a collegiate and parifh church, a *cafa da mifericordia*, an hofpital and a convent; is the *ouvidoria* town, and refidence of an *ouvidor* and a *juiz de fora*, who are nominated by the royal family of *Bragança*. This place being the moft ancient earldom in *Portugal*, fince its firft erection into a Kingdom, King *John* raifed it to a dutchy; but the title is, at prefent, extinct.

Efpofende, a fmall town with a harbour, near the mouth of the river *Cavado*. The harbour, though practicable only for fmall fhips, has a fort for its defence. In this little place are one parifh church, one *cafa da mifericordia*, and one hofpital.

Caftro Laboreiro, *Famelicao* and *Rates*, are inconfiderable towns.

Villa de Conde is a pretty good fea-port with a fortified harbour.

Melgaço is a fmall town, near the river *Minho*, built in the year 1170, by King *Alphonfo Henriques*, and walled in by king *Dennis*; but its heft defence is a caftle which ftands on the north fide of it. It has a parifh church, a *cafa da mifericordia* and an hofpital. Its latitude is 42° 7".

Of three *Concelhos*; namely, *Larim*, *Portella das Cabras* and *Villa-Chaa*.

Of five *Coutos*; namely, *Cornelhaa*, *Fragofo*, *Gondutze*, *Palmeira*, or *Landim*, and *Villar de Frades*.

Of one *Julgado*; namely, *Vermoim*; and one *Honra*, viz. *Fralaens*.

2. The *Ouvidoria de Valença* confifts
Of three villas; namely,

Valença, a fmall but ftrong town on the river *Minho*, oppofite to the *Spanifh* fortrefs of *Tuy*, from which it is diftant a little above a cannon-fhot. It ftands upon an eminence, contains about eight hundred and fifty inhabitants; two parifh churches, of which one is a collegiate; one *cafa da mifericordia*; one hofpital, two convents, and is the *Ouvidoria* town, being the refidence of an *ouvidor* and a *juiz de fora*. This place was at firft called *Contrafta*, and was formerly erected into a marquifate by King *Alphonfo* the fifth, but foon after changed to an earldom. Under King *John* the fourth it reverted to the crown, and was afterwards annexed to the family of the Infantes. The title of *Marquis de Valença* was conferred by *John* the fifth on the count of *Vimiofo*. Its Diftrict includes ten parifhes.

Caminha, another fortified little town on the *Minho*, near its influx into the fea, where it forms a fmall ifland, on which is a fort and convent. It contains about one thoufand three hundred inhabitants, with one parifh church, one *Cafa da mifericordia*, two hofpitals, of which one is for difabled

foldiers,

foldiers, and two convents. It appears to have been built by King *Alphonfo* the third about the year 1265; was at firft a county and afterwards raifed to a dutchy; but, ever fince 1641, belongs to the Infantes. Its latitude, according to father *Capaffi*, is 41° 22'. To its Diftrict belong feventeen parifhes.

Valladares, a fmall place betwixt *Monçao* and *Melgaço*; contains only three hundred inhabitants, yet is an earldom and has a Diftrict of fixteen parifhes.

2. Of two *Coutos*, which are thofe of *Feaens* and *Paderne*.

3. The *Ouvidoria de Braga* confifts of

The *Cidade* of *Braga*, feated in a pleafant plain between the rivers *Cavado* and *Defte*. It derives its name from a certain kind of garment ufed by the ancient inhabitants. It is faid to have been built by the *Greeks*, but afterwards fell under the power of the *Carthaginians, Romans, Swabians, Goths* and *Moors*; and laftly of the Kings of *Leon*. The *Romans* gave it the title of *Augufta*, and the *Swabian* Kings honoured it with their refidence. The city, together with the civil and penal jurifdiction, belongs to its archbifhop, who is primate of the Kingdom; though with a right of appeal in criminal cafes from the archbifhop's *Ouvidor* to the royal court of *Relaçoens*. To this archbifhoprick belong five *comarcas ecclefiafticas*; namely, the *comarcas* of *Braga, Valença, Chaves, Villa-real* and *Torre de Moncorvo*. Its inhabitants amount to about twelve thoufand three hundred, with four parifh-churches, exclufive of its ancient large cathedral, and twenty-feven other churches in its Diftrict. It has likewife eight convents, one *cafa da mifericordia*, one hofpital and one feminary. Betwixt the church of *S. Pedro de Maximinos* and the hofpital, are fome ftately remains of antique buildings, particularly of an amphitheatre and an aqueduct. It is the *ouvidoria* town, and the refidence of an *ouvidor* and a *juiz de fora*. Father *Capaffi* places it in latitude 41° 33'.

Of thirteen *coutos*; which are, *Arentim*, with one parifh; *Cabaços*, with one; *Cambezes*, with one; *Capareiros* with one; *Dornellas, Ervededo, Feitofa*, with one; *Goivaens Moure*, with two; *Pedralva*, with two; *Provefende, Pulha*, with one, and *Ribatua*.

2. *T R A Z-O S   M O N T E S.*

THIS province is bounded to the north by *Gallicia*, to the caſt by *Leon*, and ſouthward alſo by *Leon* and the province of *Beira*, and weſtward partly by *Entre Douro e Minho*, and partly by *Beira*. It receives its name from its poſition with reſpeꝗ to the province of *Entre Douro e Minho*, lying on the other ſide of the mountain of *Marao*. In extent, from north to ſouth, it is thirty *Portugueſe* miles ; and from eaſt to weſt about twenty ; being, for the moſt part, mountainous. It is wild, barren and thinly inhabited, though bleſſed with fertile and delightful vallies, which produce rie, wheat, wine and fruits. Beſides the *Douro* it is watered by the ſmall rivers of *Tamega*, *Corgo*, *Tuela* and *Sobor*, which run all into the *Douro*. It contains two *cidades* and fifty-ſeven *villas*, many of which are alſo *coutos* or *julgados* ; and, according to *De Lima*'s liſt, five hundred and forty-nine pariſhes ; in which, in the year 1732, were computed one hundred and thirty-five thouſand, eight hundred and four ſouls. The greateſt part of this country conſiſts of the eſtates of *donatorios*. Here are alſo a great many abbeys, reꝗories and vicarages belonging to their ſeveral patrons, particularly to the royal houſe of *Braganza*, the marquis of *Villa-Real*, the marquis of *Tavora*, the archbiſhop of *Braga*, the *Bernardines*, the *Benediꝗines* and other orders. Laſtly, the orders of knighthood, and particularly that of *Chriſt*, have ſeveral commanderies here. It is divided into twenty-four juriſdiꝗions, of which two bear the title of *Coreiçaos*, and two are called *Ouvidorias*.

I. The *Correiçoens* are

1. The *Correiçao da Torre de Moncorvo*, the greateſt juriſdiꝗion in the whole province, and conſiſting of twenty-ſix *villas*, ten of which belong to the King and the others to ſeveral patrons ; as the family of *Braganza*, the family of *Villa-Real*, the marquis of *Tavora*, the *guedes* of *Miranda*, the lords of the *villa* of *Murça* and *Sampayos*, with the lords of *Villa-Flor*. The places in it are as follow, *viz.*

Torre de Moncorvo, a pretty good town, in a ſpacious plain, at the foot of mount *Roboredo*, lying betwixt the rivers *Douro* and *Sabor*. Beſides a caſtle, it is ſurrounded with a wall and ſome baſtions. Its inhabitants are about one thouſand three hundred in number, with a handſome pariſh church, one *caſa da miſericordia*, one hoſpital and one convent. It is the chief place of the *Correiçao*, and the reſidence of a *corregedor*, a *provedor* and a *juiz de fora*. The family of *Sampayos* are hereditary governors of the caſtle. Its Diſtriꝗ contains eleven pariſhes ; namely,

Freiro,

Freiro de Efpada na Cinta, a fmall town, containing about eight hundred and feventy inhabitants and one parifh church. In its Diftrict are three parifhes. The *juiz de fora* of this place belongs to the *Comarca.*

Monforte de Rio-Livre, a fmall place having only three or four hundred inhabitants, though its Diftrict contains twenty-five parifhes.

Anciaens, a place ftill more inconfiderable, containing only about one hundred and forty-five inhabitants. In its Diftrict, however, are fixteen parifhes.

Linhares, a fmail place; its inhabitants betwixt three and four hundred. This town is a *julgado.*

Villarinho da caftanheira, a little town, containing about feven hundred inhabitants, with a Diftrict of fix parifhes.

Cortiços, a fmall place having about two hundred inhabitants, and only two parifhes in its Diftrict.

Valdafnes, a fmall place; the inhabitants betwixt two and three hundred.

Sezulfe, an inconfiderable town, containing about two hundred and fifty-nine fouls.

Nuzellos, a very mean place belonging to the *Branganza* family; the inhabitants about fixty-two, and having only two parifhes in its Diftrict.

Pinho Velho, no better than the former. Its inhabitantss about fixty-three in number.

Lamas de Orelhao, a fmall place belonging to the family of *Villa-Real*; the inhabitants about two hundred and fifty, and having ten parifhes in its Diftrict.

Freixiel, little better than the former, its inhabitants not exceeding three hundred. It belongs to the fame family, and has only two parifhes in its Diftrict.

Abreiro, a little town confifting of betwixt two and three hundred inhabitants, and belonging alfo to the *Villa-Real* family. It has only one parifh in its Diftrict.

Mirandella, a fmall fortified town on the river *Tuela*, contains about one thoufand inhabitants, and only one parifh; but in its Diftrict are twenty-four.

Alfandega da Fé contains only one parifh, but its Diftrict confifts of fifteen.

Caftro Vicente, a fmall place, having one parifh and feven in its Diftrict.

Murça de Panoya, a fmall place, of about five hundred inhabitants, and Diftrict containing nine parifhes.

Torre de Dona-Chama, an inconfiderable place containing only three hundred inhabitants, but its Diftrict includes eleven parifhes.

Agua Revés, a fmall place, containing only about three hundred and fixty fouls.

Villa Flor, a fmall place of about eight hundred inhabitants, and having only ten parifhes in its Diftrict.

Chacim, ftill fmaller, having only five hundred inhabitants and one parifh in its Diftrict.

Villas Boas, an inconfiderable place, containing but three hundred and fifty inhabitants and one parifh in its Diftrict.

Frechas, a little town containing four hundred inhabitants.

Moz, a fmall place, having only two hundred inhabitants and one parifh in its Diftrict.

S. Payo, or *Sampayo*, the whole number of its inhabitants but one hundred and fixty-three, and only one parifh in its Diftrict.

2. The *Correiçao de Miranda* confifts

1. Of the *Cidade Miranda de Douro*, a fortified town on the frontiers of *Spain*, and feated in a barren mountainous country on the river *Douro*, at the conflux of the little river of *Frefno*. Befides its works it is alfo defended by a caftle and fort. But the only church in the city is the cathedral. It contains about feven hundred inhabitants, one *cafa da mifericordia*, one hofpi- tal and one feminary. To the bifhoprick of *Miranda* belong five *bigaira- rias*, or *arciprefados*; namely, *Aro, Braganza, Monforte, Mirandella* and *Lampa as*, which contain about three hundred and fourteen parifhes. It is the *correiçao* town, and the refidence of a *corregedor*, a *provedor da comarca*, and a *juiz de fora*. To the Diftrict of this city belong twenty-two parifhes, and its latitude, according to father *Capaffi*, is 41° 31'.

2. Of thirteen *villas*; viz.

Algozo, a fmall place on the river *Maças*, the inhabitants about four hundred, and containing twenty parifhes in its Diftrict. The *juiz de fora* of this place belongs alfo to the *comarca*.

Frieira S. Seritz and *Rebordainhos*, two inconfiderable places, containing only one fmall parifh each, and from one to three hundred fouls.

Vinhaes, a town confifting of about fix hundred and fixty inhabitants and two parifhes, with twenty other parifhes in its Diftrict.

Villar Secco da Lomba, a mean place containing fcarce three hundred in- habitants.· In its Diftrict are fix parifhes.

Val de Paço, a fmall place having only one parifh, containing about three hundred fouls. Its Diftrict confifts of three parifhes.

Failde and *Carocedo*, two fmall towns, having each a parifh, containing not above two hundred fouls.

Vimiofo, a fmall fortified town near the river *Maças*, having one parifh containing about feven hundred fouls. This place confers the title of count, and in its Diftrict are four parifhes.

Magadouro, a little place, containing one parifh, about four hundred and feventy inhabitants, and fourteen parifhes, in its Diftrict.

I

Panorroyas,

Panorroyas, or *Penas de Royas*, a mean town, confiſting of one ſmall pariſh of about one hundred and forty-five ſouls; yet its Diſtrict contains nine pariſhes.

Bempoſta, a little town, containing about four hundred inhabitants and one pariſh, with four belonging to its Diſtrict.

II. The *Ouvidorias* are

1. The *Ouvidoria da Braganſa*, confiſting of a *cidade* and ten *villas*, belonging to the houſe of *Braganza*; and lying in the *provedoria* of *Miranda* contains,

Braganſa, a *cidade* on a ſpacious plain near the river *Fervenſa*, which ſeparates it from *S. Bartholomew's* hill. It confiſts of one *cidade* and a *villa*. The former is fortified with towers, and within it has a good caſtle. The latter is alſo fortified, and Fort *S. Joao de Deos*, on *Caraſcal* hill covers them both; but is of little importance. *Braganſa* contains two pariſh churches and about two thouſand ſeven hundred ſouls, with one *caſa da meſericordia*, one hoſpital and four convents. It is alſo the *ouvidoria* town, being the reſidence of the *ouvidor* and *juiz de fora*. The former is the general magiſtrate for all the places of this province belonging to the houſe of *Braganza*. It has a variety of ſilk manufactures, and is one of the moſt ancient towns in the Kingdom. King *Alphonſo* the fifth, in the year 1542, erected it into a dutchy with fifty *villas* dependent on it; and *John* the ſecond, its eighth duke, was crowned King of *Portugal* by the name of *John* the fourth. The Diſtrict of this city includes no leſs than one hundred and twenty-three pariſhes; and its latitude, according to the above-mentioned ingenious father, is 41° 47'.

2. The following *villas*; viz.

Val de Nougeira, containing about one hundred and eighty-eight inhabitants. *Val de Prados*, three hundred and forty-five; *Villa Franca*, one hundred and fifty, and *Gueſly* four hundred and forty-ſeven. Theſe are four ſmall places, confiſting each of only one pariſh.

Rebordaos, a little place, containing one pariſh with three hundred and ſeventy-ſix ſouls, and having only one pariſh in its Diſtrict.

Outeiro, a ſmall place, containing about two hundred ſouls, with a ſtrong caſtle ſituate betwixt the rivers *Sabor* and *Maſas*. In its Diſtrict are ten pariſhes.

Chaves, a pretty good fortified town on the river *Tamega*, having two ſuburbs and two forts; one of which, called *Noſſa Senhora do Roſario*, has the appearance of a citadel, and contains a convent within it. The other is named *S. Noutel*. Between the town and the ſuburbs called *Magdalena*, is a *Roman* bridge of ſtone over the river *Tamega*, above ninety-two geometrical paces long, and ſomewhat above three in breadth. In the town are about two thouſand ſouls, one collegiate church, which is alſo the pariſh church, one *caſa da miſericordia*, two hoſpitals and two convents. With reſpect to its ſpiritual juriſdiction, it is ſubject to the archbiſhop of *Braga*.

This

This town was built by the *Romans*, and, as is imagined, by the Emperor *Vespasian*, who is supposed to have given it the ancient name of *Aquæ Flaviæ.* There are still visible in it not a few traces of its ancient grandeur and bigness. The latitude of this place is computed at 41° 46'; and in its District are fifty-three parishes.

Monte Alegre, is a small place, containing about four hundred inhabitants, with a fortified castle. In its District are forty-seven parishes.

Ruyvaes is a single parish of about eight hundred souls. In its District also is but one parish.

The *Ouvidoria de Villa-Real* consists of eight *villas* and one *honra.*

Villa-Real, the best and largest town in this province, is seated between two small rivers which empty themselves into the *Douro*. One of these is called the *Corgo.* The greater part of the houses stand without the walls, and the few within are called the *old town.* It contains two parish churches, one *casa da misericordia*, one hospital and three convents; is the residence of an *ouvidor*, a *juiz de fora*, and the *provedor* of *Lamega* visits it occasionally. With respect to its spiritual jurisdiction, it is under the vicar-general of the archbishop of *Braga*. It was built by King *Dennis*; made an earldom by *Alphonso* the fifth; raised to a marquisate by *John* the second; and to a dukedom by *Philip* the second. Under *John* the fourth it devolved again to the crown; and is an appenage of the Infantes.

Canellas, a small place near the *Douro.*

Abreiro, Freixiel and *Lamas de Orelhao* belong to the *provedoria de Moncorvo*, and as such have been mentioned before.

Almeida and *Ranhados* are situate in the province of *Beira*; in the *Comarca* of *Pinhel.*

Vimoioso has been mentioned before.

Sobrosa is likewise a *honra.*

B   E   I   R   A.

BE IRA is the largeſt province of the Kingdom, being bounded on the
north by the province of *Entre Douro e Minho* and *Traz-os Montes*; to the
eaſt by *Extremadura* and *Alentejo*; and to the weſt by the ſea.   Its extent
from caſt to weſt is generally computed at betwixt thirty-three and thirty-ſix
Portugueſe miles, and from north to ſouth about as many.   It is divided into
Upper and *Lower Beira*; the former being the northern part, and lying on
the ſea-coaſt; the latter lying towards *Spain* and *Extremadura*.   It produces
wheat, rie and millet; and, in ſeveral parts, excellent wine and oil in ſuch
plenty that conſiderable quantities of each are exported.   The mountain of
Eſtrella, the *Mons Herminius* of the *Romans*, in the *Correiçao da Guarda*,
is remarkable, and no leſs celebrated.   The aſcent from *Villa S. Romao*, at
the foot of it, to the ſummit takes up two hours and a half.   In ſeveral
places the mountain is ſeen to be hollow; and even the noiſe of a rapid ſtream
running through it diſtinctly heard.   It has alſo a fine quarry of alabaſter;
and on the top we are agreeably ſurpriſed with verdant paſtures and rivulets
of a very clear and pleaſant water; but the particular moſt worthy of attention in the whole mountain, is a lake environed with high rocks.   The
water of it iſſues out of the ground, is very clear and tepid, has a kind of
tremulous motion in the middle, and now and then ſmall veſicles are ſeen to
ariſe aloft from it.   From the ſtrong attraction in it to one certain place, it
is conjectured to have been an aperture through which it runs off again, and
which is the ſpring of another lake a little lower. And from theſe is formed
a river which takes its courſe to the foot of the mountain.   The ſnow of a
deep valley, in one part of this mountain, furniſhes *Liſbon* the whole
ſummer, though at the diſtance of above ſixty *Portugueſe* miles. The inhabitants have very ſtrange notions of this mountain and the lake.

The province of *Beira* contains four epiſcopal cities, two hundred and
thirty-four *villas*, fifty-five *concelhos* and ſome *coutos*.   It conſiſts of eight juriſdictions, ſix of which are *correiçaos* and two *ouvidorias*.   According to
De Lima's liſt, this province, in the year 1732, contained one thouſand
ninety-four pariſhes, and five hundred fifty-one thouſand, ſix hundred and
eighty-ſix ſouls.   *John* the fifth erected it into a principality, in honour of
his grandſon, the eldeſt ſon of the Prince of *Braſil*.   Its eight juriſdictions
are as follows:

1. The

The *cerreiçaco de Coimbra* confifts of the *cidade* of *Coimbra* and twenty-nine
villas. Many of the latter belong to particular proprietors, who appoint the
ouvidores; and as the bifhop of *Coimbra* is lord and count of *Arganil*; under his *ouvidor* are *Avoo, Coja, Santa Comba do Dao, Vacariça,* &c. The
dukes of *Cadaval* are lords and counts of *Tentugal,* and under their *ouvidor*
are *alvayazere, Buarcos, Pena Cova, Povoa de Santa Chriflina, Rabaçal,
Villa nova de Anços,* &c. Befides thefe donatarios, who are indeed, by
much, the principal, the dukes of *Lafoen* and the marquiffes of *Cafcaes* and
Marialva, with the counts of *Ericeira,* are lords of certain places in it, as
I fhall hereafter obferve.

Coimbra, is a *cidade* on the river *Mondego,* formerly called *Colimbria,* or
Conimbriga. It contains eleven thoufand nine hundred inhabitants and nine
parifh churches, the cathedral included, with one *cafa da mifericordia,* one
hofpital, eight convents and eighteen colleges. The bifhop of this place is
fuffragan to the archbifhop of *Braga*; but he has under him no lefs than
three hundred and forty-three parifhes, which are divided into the three
archdeaconries of *Vouga, Cea* and *Penella.* He is alfo count of *Arganil.*
Its univerfity was firft founded by King *Dennis* at *Lifbon,* in the year 1291;
but was foon removed hither; and has a rector, a *reformator,* or governor,
a chancellor, who is always *prior* of the convent of *Santa Cruz* in this city,
and other profeffors and officers. The number of ftudents is faid to be two
thoufand; but the author of *Lehrreichen Nachrichten* fays, that he was quite
amazed at the *Latin* ufed here in their difputations, it being little better than
a mixture of *Italian* and *Portuguefe.* The univerfity is a very magnificent
ftructure. The tribunal *do fanto officio,* which was erected here in the year
1541, is endowed with great privileges. This is the *correiçao* and *comarca*
town, and the refidence of a *provedor,* a *corregedor,* and a *juiz de fora.* It
has a bridge of ftone over the river *Mondego.* The place where old *Coimbra*
ftood is now called *Condexa a Velha.* It was formerly the refidence of the
Kings of *Portugal,* and feveral royal perfons are buried here. It has been
twice erected into a dutchy, contains within its Diftrict forty parifhes, and
is noted for its exquifite peaches. The latitude of *Coimbra* is computed at
40° 41'.

Efgueira is an old fmall town, containing about fifteen or fixteen hundred
inhabitants, with one parifh church, which is a vicarfhip and commandery
of the order of *Chrift,* one *cafa da mifericordia,* and one hofpital. It is the
provedoria town and the refidence of a *provedor* and a *juiz de fora,* but has no
corregedor, the *corregedor* of *Coimbra* coming hither. The places dependent
on this *provedoria* are *Aguieira, Anadia, Anjega, Affequius, Aveiro, Avelans de Caminho, Avelans de Cima, Bempofta, Brunhido, Cafal de Alvaro
Eixo, Eftarreja, Ferreiros, Ilhavo, S. Lourenço do Bairro, Ois da Ribeira,
Oliveira do Bairro, Paos, Preftimo, Recardaens, Sangalhos, Segadaens, Se-
rem, Soufa, Trofa, Villarinho do Bairro, Vagos, Vouga, Confelho de Fer-*

medo, and the *Couto de Efteve.* Of thefe fome belong not to *donatarios*, but are the property of the crown, and as fuch under the *correiçao* of *Coimbra*. The convent of *Lorvao* is in poffeffion of the civil jurifdiction of this city, but criminal affairs are managed by the King's officers. Its Diftrict contains but one parifh.

Arganil, a fmall town, containing one thoufand one hundred inhabitants, and one parifh ; is an earldom and belongs to the bifhop of *Coimbra*. In its Diftrict are four parifhes.

Goes is a fmall town, having betwixt fourteen and fifteen hundred inhabitants and only one parifh. Its Diftrict alfo contains but two.

Pombeiro is a fmall town of one parifh and one thoufand inhabitants. Its Diftrict alfo contains but one, yet is an earldom.

Botao is a fmall place containing about five hundred and fifty inhabitants.

Ançaa, an inconfiderable town with nine hundred inhabitants and one parifh, having five other parifhes in its Diftrict, and belongs to the marquis of *Cafcaes*.

Pereira, a fmall town of one parifh containing about one thoufand three hundred fouls.

Cernache, alfo fmall, containing only one thoufand inhabitants, and likewife one parifh.

Miranda do Corvo, a pretty good town on the river *Dueça*, is an earldom. It belongs to the dukes of *Lafoens*, and contains about two thoufand feven hundred inhabitants. In its Diftrict are two parifhes.

Pombalinho, a fmall town, having only one parifh.

Anciao, a little town, containing about one thoufand two hundred inhabitants and one parifh, belonging to the counts of *Ericeira*.

Mira, a fmall town, with a parifh containing about one thoufand fix hundred fouls.

Buarcos, a little town not far from the mouth of the river *Mondego*. The greateft part of this place was overthrown by an earthquake in the year 1752.

Villa novo de Anços, a fmall fea-port, containing nine hundred inhabitants and one parifh.

Villa Nova de Monçarros, containing one parifh and about fix hundred and fifty fouls.

Vacariça, a fmall town, having only one parifh but about one thoufand three hundred inhabitants. Its Diftrict includes two other parifhes.

Pena-Cova contains one parifh with nine hundred and forty inhabitants, and five other parifhes belong to its Diftrict.

Cantabede is a fmall town, but an earldom ; belongs to the marquis of *Marialva*, and confifts of one parifh with one thoufand two hundred fouls.

Celaviza,

Celaviza, *Carvalho* and *Fajao,* fmall places, containing from three to fix hundred inhabitants.

Coja, a fmall town having about eight hundred inhabitants and one parifh. Its Diftrict contains fix.

S. Combadao, or *Comba do Dao, Podentes* and *Avoo,* are fmall places, containing betwixt four and fix hundred inhabitants.

S. Sebafliao de Fradeira, containing about ninety-four fouls, and *Bovadella* three hundred, may be fuppofed to be but inconfiderable places.

Tentugal is a town belonging to the duke of *Cadaval;* and, though an earldom, contains, with *Povoa de Santa Chriftina,* but one parifh and about two thoufand fix hundred fouls.

Rabaçal, a fmall town, confifting of one parifh of about one hundred fouls. In its Diftrict are two parifhes.

Alvayazere, a fmall town, containing one parifh and near one thoufand inhabitants. Its Diftrict contains two parifhes.

S. Varao and *Fermozelhe* are *coutos* containing about one thoufand one hundred fouls.

Reguengo de Belide is a fmall parifh, having only one hundred and forty-feven fouls.

Gniayos, Alhadas, Outil, Tavarede, Cadima and *Zambujal, Mogofores* and *Cafal Combo* are all *coutos.*

Reguengo de Liceira is a fmall parifh of two hundred fouls.

The *coutos de Arazede do Bifpo* and *e Santa Cruz,* which is an earldom, make both but one parifh of one thoufand five hundred fouls.

Figueira is a town, and *Villa-Verde* a *couto* and earldom, incorporated into one parifh of one thoufand and fixty fouls.

II. The *ouvidoria de Montemor o Velho* contains the following *villas,* viz.

Montemor o Velho, a town on the river *Mondego,* with the addition of *Velho,* to diftinguifh it from *Montemor o Novo* in *Alentejo.* It contains about one thoufand nine hundred inhabitants, fix parifh churches, one *cafa da miferi-cordia,* four hofpitals, one convent; is the *ouvidoria* town and the refidence of an *ouvidor* and a *juiz de fora.* Here is alfo a *capitao* fuperior, who has the command of twenty-four companies of regular forces, which are quartered in this town and its Diftrict; to the latter of which belong eight parifhes.

Aveiro is a middling town, on a fmall bay, into which the river *Vouga* difcharges itfelf. It has a harbour fit for fhips of burthen. The bay is properly a canal from the fea which is increafed by the *Vouga,* and reaches from *Aveiro* to *Villa Ovar.* From the fea it is feparated by fand banks, and has feveral little iflands in which falt is made. This town confifts of five wards, the fourth of which is the moft ancient and principal, and is alfo walled in. The inhabitants amount to about four thoufand four hundred. It contains four parifh churches, all belonging to the order of *Aviz,* one

cafa

cafa da mifericordia, one hofpital and fix convents, with a *tribunal da Alfandega,* a judge, a fecretary, and other officers. Here is alfo a *juiz de fora,* and, by a particular indulgence, the *provedor* of *Efguiera* prefides here. *John* III. erected it into a dutchy, to which, on the death of the laft branch of the houfe of *Lancaftro, Maria de Guadalupe,* her fecond fon *Gabriel Ponce* of *Leao, Lancaftro* and *Cardenas,* Duke of *Banhos* in *Caftile,* was acknowledged as Duke, and in 1732 did homage for it to *John* V. Its Diftrict contains feven parifhes.

Penella, a middling cown containing about two thoufand fix hundred inhabitants and two parifhes; with three more belonging to its Diftrict.

The other *villas* are *Abiul, Brunhido, Cafal de Alvaro, Lourifal,* which is a marquifate; *Louzaa, Pereira Recardaens, Segadaens* and *Torres novas.*

III. The *Ouvidoria da Feira* contains

Feira, a fmall town feated in a pleafant fpacious valley, four miles from *Porto* and two from the fea. It contains about a thoufand inhabitants, one parifh church with a convent, one *cafa da mifericordia,* one hofpital, and is the *Ouvidoria* town and feat of the *Ouvidor,* who is appointed hy the donatarios. Within the Diftrict of this city are fifty-fix parifhes and two convents.

Ovar, having five thoufand eight hundred fouls and one parifh.

Pereira de Sufao, a fmall town with two thoufand three hundred inhabitants and one parifh.

Cambra, a fmall town containing one parifh of about eleven thoufand fouls. In its Diftrict are five more.

Caftanheira, a fmall place having about three hundred and thirty fouls in one parifh, its Diftrict alfo containing another.

IV. The CORREIÇAO DE VISEU confifts of the *Cidade* of this name, thirty-two *villas,* and thirty *Concelhos.*

1. The *Cidade* is *Vifeu,* delightfully fituated in a plain between the rivers of *Moudego* and *Vouga,* and fuppofed, to be the place where the old town of *Vacca* ftood. It contains two parifh churches befides its cathedral, one *cafa da mifericordia,* one hofpital, three convents, and is both the *correifao* and *comarca* town. It is alfo the feat of a *corregedor,* a *provedor,* and a *Juiz da fora.* The bifhop of this place is fuffragan to the archbifhop of *Braga.* In the church of *S. Miguel do Fetal,* which is without the walls, King *Roderigo* lies buried. Here are ftill remaining two ancient *Roman* towers. It was raifed to a dutchy by King *John* I. and its diftrict includes thirty-one parifhes.

2. The twenty two *villas* belonging to the *Donatarios* are

Alva, containing about two hundred and fixty fouls, *Banho* four hundred and forty, *Bobadella Candofa* four hundred and eighty, *Enfias* a hundred and twenty, *Lagares* three hundred and twenty, and *No-*
gueira

gueira five hundred and fixty, being all fmall places, and fo many diftinct parifhes.

Ferreira de Aves, a fmall town having fixteen hundred inhabitants and two parifhes belonging to its Diftrict.

Mortagoa, a fmall place with one parifh of five hundred fouls; but having eight in its Diftrict.

Oliveira de Conde, a fmall place not far from the river *Mondego*. Its parifh contains about one thoufand three hundred and forty fouls, and its Diftrict alfo has one parifh.

Oliveira dos Frades contains pretty near five hundred. *Oliveira do Ofpital* five hundred, *Penalva de Alva* fix hundred, *Perfelada* feven hundred, *Reris* fix hundred and fifty, *Sabugofa* feven hundred and fifty, being alfo an earldom; *Sandomil*, likewife an earldom, betwixt fix or feven hundred, *S. Pedro do Sul* nine hundred, *Taboa* near one thoufand, and *Trapa* five hundred and thirty inhabitants, all fmall places, and each making feparately a parifh.

Coja and *Santa Comba do Dao*, are two fmall places belonging to the bifhop of *Coimbra*, and therefore mentioned already.

13. The *Conçelhos* are, *Guardao*, of one parifh. *Befteiros*, of fifteen. *Rio de Moinhos*, of one. *S. Joao de Monte*, of one. *Mouras*, of one. *Freixedo*, of two. *Ovoa*, of one. *Pinheiro de Azere*, of one. *S. Joao de Areas* and *Sylvares*, of two. *Curellos*, of one. *Senborim* and *Folhadal*, of four. *Canas de Senborim*, of one. *Azurara*, of thirteen. *Tavares*, of five. *Moens*, of two. *Gafanhao*, of one. *Satao*, of two. *Gulfar*, of four. *Penalva do Cáftello*, of twelve. *Alafoens*, of thirty-feven. *Sever*, of five. *Sinde*, of one. *Azére*, of one. *Vide de Foz de Piodao*, of one. *Villa nova de Sobacco*, of one. *Sylvao*, of one. *Pavolide*, an earldom, and *Ranhados*.

Obf. *Alafoens* and one of thefe *Conçelhos* was erected into a dutchy in 1718, *Peter*, the fon of Don *Meguel*, legitimate fon to King *Peter* II. being created Duke of *Alafoens* by *John* V.

V. The CORREIÇAO DE LAMEGO contains one *Cidade*, thirty-three *villas*, and twenty-two *Conçelhos*.

1. The cidade *Lamego* is fituated in a low country environed with mountains, and lies not far from the river *Douro*. It is faid to have been originally founded by fome *Greeks* from *Laconia*; and was formerly, indeed, called *Laconi Murgi*, and afterwards *Urbs Lamacenorum* and *Lameca*, whence its prefent name of *Lamego* is derived. It confifts of three wards, contains about four thoufand four hundred inhabitants, two parifh churches, including the cathedral, one *cafa da mifericordia*, one hofpital, four convents; is the *Correifao* town, and the refidence alfo of a *corregedor*, a *provedor*, and a *Juiz de fora*. Its bifhop, who is fuffragan to the patriarch of *Lifbon*, prefides both over the old and new Diocefe. Thefe confift properly of the *Comarca de Riba de; Coa*, which King *Dennis* in 1296 annexed to it. The parifhes within them amount to nine hundred and ninety one, namely

feventy-

seventy-one abbies, seventy-five rectories or vicarages, and one hundred and forty-five parsonages, exclusive of some churches belonging to the *Bernar-dine* Monks. The city has lately been raised to an earldom; and is famous for the convention of estates held here by *Alphonso Henriques*.

-.2. The *villas* are.

Tarouca, a small place, having one parish of about eleven hundred souls, being an earldom, and containing four parishes in its District.

Bretiande, containing about three hundred and thirty; *Ocanhas*, or *Ucanha, Lazarim*, containing four hundred; *Lalim* four hundred, *Mondim* four hundred, *Passo* two hundred and twenty, *Lumiares*, which is also an earldom, containing a hundred and fifty; *S. Cosmado* four hundred and sixty, *Goujoim* two hundred, *Secca* two hundred and sixty, *Castello* four hundred, *Granja de Tedo* two hundred and thirty, *Arcos* one hundred and ninety, *Nagosa* one hundred and fifty, *Longa* three hundred and eighty, *Barcos* three hundred and eighty, *Taboasso* five hundred and thirty, *Cha-vaens* three hundred, *Moimenta da Beira* six hundred and fifty, *Leomi* seven hundred and forty, *Fragoas* a hundred and fifty, *Villa-Cova* five hundred and twenty, *Pendilhe* two hundred and eighty, *Varzea da Serra* a hundred and thirty, *Valdigem* five hundred, *Sande* four hundred and thirty, *Pa-rada do Bispo* one hundred and twenty, and *Fontilho* three hundred and thirty inhabitants, all small mean places; but each indeed a parish, though three of them only have Districts, and those too but one parish in them.

Castrodaire, a small town having betwixt seventeen and eighteen hundred inhabitants.

Armamar, a small town with two parishes, containing thirteen hundred souls.

Arouca, a small place with one parish of above fourteen hundred souls, and five other parishes belonging to its District.

3. The *Concelhos* are *Alvarenga*, of two parishes. *Aregos*, and *Barquieros*, of one. *Cabril, Caria*, of eight. *S. Christovao da Nogueira*, of one. *Ferreiras*, of three. *S. Martinho de Mouros*, of four. *Mossao, Paiva*, of nine. *Parada de Esther*, of one. *Pera* and *Peva*, of one. *Peso da Regoa*, of one. *Pineiros*, of three. *Resende*, of three, being also an earldom. *Ribellas*, *Sanfins*, of four. *Sinfaens*, of one. *Teixeira*, of one; and *Tendaens*, of one. To these belong also the *Couto da Ermida*, and the *Honra* of *Sobrado*, of one parish.

VI. The CORREIÇÃO DE PINHEL consists of fifty-five *Villas* and one *Concelho*. Among the *villas* are some belonging to Donatarios, which are exempt from the jurisdiction of the *Corregedor*.

Pinhel, is a town on a mountain, near which runs a small river of the same name, being fortified, and having six towers on its walls. The inhabitants are betwixt fifteen and sixteen hundred. It contains six parish churches, one *Casa da Misericordia*, one hospital, and one convent, is the

Correiçao

correiçao town, the refidence of a *corregedor* and a *juiz de fora:* And contains twenty-five parifhes in its Diftrict.

Almeida, is a regularly fortified town with a caftle on the river *Coa,* containing between twenty-one and twenty-two hundred inhabitants, one parifh church, one *cafa da mifericordia,* one hofpital, and one convent. It belongs to the Infantes, and in its Diftrict are two parifhes.

Francofo, is a fmall town with five parifh churches; but the inhabitants exceed not thirteen hundred. Its Diftrict contains thirty parifhes.

Figueiro du Granja has about three hundred and fifty inhabitants; *Matança* about four hundred and fifty; *Algodres* four hundred and fifty, with a Diftrict of eight parifhes; *Fornos* five hundred and eighty; *Penna Verde* fix hundred and forty, with a Diftrict of three parifhes; *Aguiar* five hundred, with a Diftrict of eight parifhes; *Sernanfelhe* fix hundred, with a Diftrict of five parifhes; *Guilheiro* two hundred and fifty; *Fonte Arcada* four hundred and forty, with a Diftrict of five parifhes; *Ponte* two hundred and thirty, being alfo an earldom; *Sindim* a thoufand; *Paredes* feven hundred, with a Diftrict of one parifh; *Vargeas* two hundred and fifty; *Trevoens* fix hundred and fixty, *Soutello* four hundred; *Tavora,* a marquifite, three hundred and thirty; *Paradella* two hundred, *Val Longo* two hundred, *Povoa* three hundred, *Penella* fix hundred and forty, *Sotto* nine hundred, *Sedavim* five hundred, *Horta* a hundred and twenty, *Nomao* three hundred and feventy, with a Diftrict of five parifhes; *Tofcoa* fifteen hundred, *Muxagata* five hundred and feventy, *Langroiva* four hundred and thirty, with a Diftrict of two parifhes; all mean places, and making fo many diftinct parifhes.

S. Joao da Pefqueira, is a fmall town on the river *Douro* containing four parifh churches, and about twelve hundred inhabitants. It is an earldom, and in its Diftrict are three parifhes.

Penedono, a fmall place on the little river of *Tavora,* contains two parifh churches and about fix hundred inhabitants. Its Diftrict alfo includes fix parifhes.

Marialva, a fmall place bearing the title of a marquifate, comprehends two parifh churches, containing only about three hundred fouls, and a Diftrict of eight parifhes.

Ranhados, is a fmall place belonging to the Infantes, having one parifh of about fix hundred and fifty fouls, and a Diftrict of two parifhes.

Moreira, a little place with two churches, and a Diftrict of fix parifhes.

Caftello Mendo, a fmall town containing about fix hundred and forty inhabitants, two churches and fifteen parifhes in its Diftrict.

Meda, containing about eight hundred inhabitants; *Cafteiçao* two hundred and eighty, with a Diftrict of one parifh; *Velofo* a hundred and feventy, *Lamegal* three hundred and fixty, *Alfayates* five hundred, with a Diftrict of two parifhes; *Villar Mayor,* which is alfo an earldom, four hundred and feventy, with a Diftrict of three parifhes; *Caftello Bom* three hundred and feventy, with a Diftrict of four parifhes; *Efcalhao* nine hundred and forty, *Caftello Rodrigo* two hundred and twenty, with a Diftrict of ten parifhes; *Almendra* feven hundred and ninety, *Caftello Melhor* two hundred and forty, being alfo an earldom; *Cinco Villas* two hundred and eighty; *Arreigada* three hundred and thirty, with a Diftrict of one parifh; *Azeite, Caftanheira, Ervedofa, Reygada, Valença do Douro, Val de Coelha, Touça,* all fmall places, but each a diftinct parifh.

Cacapito, a *Concelho.*

VII. The *Correiçao da Guarda* contains one *cidade,* thirty *villas,* and one *couto.*

1. The *cidade guarda* is fituate not far from the fource of the *Mondego,* on a part of mount *Eftrella;* and befides its caftle is fortified both by art and nature. It contains about three hundred inhabitants, has four churches and a magnificent cathedral, and one *cafa da mifericordia,* one hofpital, two convents, is the *correiçao* town, and the ufual refidence of a *corregedor* and a *juiz de fora.* The bifhop hereof who is alfo fuffragan to the archbifhop of *Lisbon,* prefides over two hundred and fixty parifhes, which are divided into fix Diftricts. King *Emanuel* raifed this city to a dutchy; but it has fince lapfed to the crown. In its Diftrict are forty parifhes.

2. The *villas* are

Jarmello, containing about feven hundred and feventy inhabitants, three churches, and one Diftrict of eight parifhes.

Manteigas, fmall, but having two churches and about thirteen hundred inhabitants.

Covilhaa, containing three thoufand five hundred inhabitants and thirteen churches. Divers manufactures of cloths, ferges and ftockings have been fet up here; but with little fuccefs. Its *juiz de fora* and *dos orfaos* belong to the *comarca,* and it has a Diftrict of forty-feven parifhes.

Celorico, a fmall town having three churches, about eleven hundred inhabitants, and a Diftrict of nineteen parifhes.

Gouvea, a marquifate, though a little place, containing nine hundred inhabitants, two parifh churches, and nine parifhes in its Diftrict.

Cea, a fmall place having about a thoufand inhabitants and one church; but its Diftrict contains ten. The *juiz de fora* of this place belongs alfo to the *comarca.*

Other inconfiderable places here, of which each makes a parifh, are *Valbelhas,* of about a hundred and feventy inhabitants, with a Diftrict of five parifhes. *Codeceiro,* of three hundred and thirty. *Vomo Telheiro,* of two hundred and twenty. *Baraçal,* of three hundred. *Açores,* of a hundred and fixty. *Linhares,* of five hundred, with a Diftrict of fix parifhes. *Mefquitella,* of eight hundred and eighty. *Mello,* of fix hundred and fifty. *Folgofinho,* of feven hundred and twenty. *Cabra,* of two hundred and forty. *Oliveirinha,*

rinha, of two hundred. *Santa Marinha*, which, together with *Caſtro Verde* makes but one pariſh, eight hundred and twenty, having alſo a Diſtrict of one pariſh. *S. Romao* eleven hundred. *Torrozello* three hundred and ſixty. *Villa Cova a Coelheira* two hundred and ſixty. *Vallazim* five hundred. *Loriga* four hundred. *Alvoco da Serra* two hundred and twenty. *Louroſa* ſix hundred. *Lagos* two hundred and fifty. *Midocns* ſix hundred and twenty, with a Diſtrict of one pariſh. *Seixo* ſix hundred and forty: And *Forno*.

3. The *Couto de Moſteiro* contains one pariſh of about nine hundred and forty ſouls.

VII. The CORREIÇAO DE CASTELLO-BRANCO conſiſts of twenty-two *villas*, namely,

Caſtello Branco, a town with a caſtle betwixt the two rivers *Ponſul* and *Vereza*, which run into the *Tagus*. This place belongs to the order of *Chriſt*, contains about three thouſand ſeven hundred inhabitants, two pariſh churches, one *caſa da miſericordia*, two hoſpitals and a ſtately palace, which in winter is the reſidence of the biſhop of *Guarda*. It is the *comarca* town, the ſeat of a *corregedor*, a *provedor*, and a *juiz de fora.* The *corregedor* is alſo *ouvidor* of the order of *Chriſt*. Belonging to its Diſtrict are nine pariſhes.

Alpedrinha, a ſmall place containing about nine hundred and fifty inhabitants and one church. The *juiz de fora* of this place and *Caſtellonovo* belong to the *comarca*.

Belmonte, is a little town containing about eleven hundred and forty inhabitants, two churches, and a Diſtrict of two pariſhes.

Sabugal, an inconſiderable place on the river *Coa*, having about ſeven hundred inhabitants, two churches, and fourteen pariſhes within its Diſtrict.

Penamacor, a ſtrong town with a caſtle ſituate in a barren mountainous country near the frontiers of *Spain*, has three churches, one *caſa da miſericordia*, one hoſpital, one convent and two thouſand three hundred inhabitants King *Alphonſo* erected it into an earldom, which is lapſed again to the crown. The *juiz de fora* of this place belongs alſo to the *comarca*; and its Diſtrict contains eight pariſhes.

Monſanto, an earldom, and ſmall town of about one thouſand inhabitants, two churches and a Diſtrict of three pariſhes.

Idanha a Velha containing about one hundred and twenty, and *Idanha a Noya* eight hundred inhabitants, are two little places on the river *Ponſul*; but the former was once conſiderable. It has alſo a Diſtrict of one pariſh, and the latter a Diſtrict of two.

Sarzedas, a ſmall town of about eighteen hundred inhabitants, one church and one pariſh in its Diſtrict, is an earldom.

E e 2　　　　　　　　　　　　　　　The

'The reft âre'but fmall places though diftinct parifhes, nâmely, *S. Vincente da Beira*, containing about feven hundred and feventy inhabitants, with a Diftrict of fix parifhes, being alfo an earldom; *Caftello Novo* containing four hundred and eighty, with a marquifite and a Diftrict of five parifhes. *Atalya*, containing two hundred and twenty. *Sortelha*, fix hundred and eighty, with a Diftrict of eight parifhes. *Touro* five hundred and fifty, with a Diftrict of three parifhes. *Proença a Velha* five hundred, with a Diftrict of two parifhes. *Bempofto* three hundred. *Pena Garcia* two hundred and fifty. *Salvaterra* three hundred and fifty, with a Diftrict of one parifh. *Segura* a hundred and forty. *Zibreira* three hundred and feventy. *Rofmaninhal* five hundred; and *Villa Velha de Rodao* about four hundred and fixty inhabitants, with a Diftrict of three parifhes.

ESTREMADURA or *EXTREMADURA.*

BORDERS to the north and eaft on *Beira*; to the fouth on *Alentejo*; and to the weft is bounded by the fea. Its extent from north to fouth is computed to be about thirty-nine *Portuguefe* miles, and from eaft to weft eighteen. Though others reduce the former to thirty-three and the latter to fixteen miles. It derives its name from a cuftom obferved by the Kings of *Leon*, during the dominion of the *Moors* in *Spain*, of nominating their conquefts of which the prefent kingdom of *Portugal* confifts, by their fituation confidered with refpect to the river *Douro*; all the countries beyond it being called *Eftrema Durii*, as *Alphonfo the Great* called the country, now the province of *Entre Douro e Minho, Extrema Minii*, being the moft remote with refpect to the river of that name. Through this province runs the large river of *Tagus*, which after forming a fpacious and fecure harbour at *Lisbon* lofes itfelf in the fea. The foil here is efteemed the moft fruitful in *Portugal*, as producing collectively what is found only feparately in the other provinces, particularly corn, wine, oil, millet, pulfe and fruits of all kinds. The country lying between *Lisbon* and *Abrantes* is a moft delicious plain, fo thickly fet with olives and other fruit trees that it makes a ravifl.ing appearance. This province alfo produces and exports great quantities of fea-falt. *Eftremadura* contains at prefent three *cidades*, one hundred and eleven *villas*, and, according to the accurate lift in *Lima*'s *Geografia Hiftorica*, three hundred and fifteen parifhes, which in 1732, were computed to contain obout 293,598 fouls, exclufive of the inhabitants of the parifhes of the caft part of *Lisbon* which are not included. It is divided into eight jurifdictions.

1. The

I. The Correiçao de Lisbon contains the capital of *Lisbon* with its Diftricts; and befides other magiftrates has five *corregedors*.

Lisbon or *Liffabon*, the metropolis of the kingdom extends from eaft to weft along the river *Tagus* near its influx into the fea; refembling an amphitheatre, and containing within its circuit feven mountains, the names of which are *S. Vincente de fora, S. André, Caftello, Santa Anna, S. Roque, Chagas,* and *Santa Catharina.* The length of the whole city is indeed little fhort of two miles; but its breadth is inconfiderable. The vallies of thofe mountains form ftreets of above a mile in length; but moft of thefe are very narrow, ill paved and dirty, and many of them alfo fteep and troublefome. This city is alfo continually increafing in beauty; and has fome houfes built in a very grand tafte. But whilft no care is taken to cleanfe the ftreets, and of lamps at night, it will always be very exceptionable. Since the erection of the royal chapel into a patriarchate, the city, with refpect to its ecclefiaftical jurifdiction, is divided into two Diocefes; namely, the weft and eaft. *Weft Lisbon* is under the patriarch; and contains twenty-one parifhes. *Eaft Lisbon* is fubject to the archbifhop, and contains fixteen parifhes. The Diftricts about the city are divided in the fame manner, twenty-five of their parifhes belonging to *Weft* and fixteen to *Eaft Lisbon.* This divifion of the ancient archbifhopric into two Diocefes was confirmed in 1716 by virtue of a bull from the Pope; but in 1741 was again annulled, and the whole city united into one Diocefe under the patriarch. The fuffragans to the patriarch and archbifhop have been fpecified above in §. 9. of the introduction. The churches of *Lisbon* are in general very fine; but that belonging to the patriarchate is of aftonifhing magnificence. The very ornaments and plate in it containing the treafure of feveral *Brafil* fleets. The pomp affumed by the patriarch on feftivals furpaffes even that of the Pope, if we except only the college of cardinals. The *Dominican* church here is very large; and on the day of an *Auto da fé* is the rendez-vous of the proceffion, and in it the fentences of the delinquents are publickly read. Not far from it is the holy-houfe or palace of the inquifition, the moft magnificent apartments of which are taken up by the inquifitor general of *Portugal,* who is prefident of that court. My purpofed brevity will not permit me to enter into a minute detail of all the rich churches of *Lisbon.*

The convents and colleges are fifty, *viz.* thirty-two for monks and eighteen for nuns. Of thefe the moft remarkable are *Santo Antao,* a college belonging to the Jefuits; *S. Bento,* belonging to the Monks of the order of St. *Benedict;* *S. Domingo Graça,* where the *Auguftine* hermits refide; and which, exclufive of a moft delicious profpect, has the appearance of a ftately palace. *S. Vicente* is alfo very magnificent, *&c.* Among other foundations I fhall only take notice of the *cafa da mifericordia* and the royal hofpital; the firft of which owes its original to King *Emanuel,* and is the moft opulent. The latter was founded by King *John*; but finifhed.

finiſhed by *Emánuel*, who, beſides many valuable privileges, endowed it with large revenues. The Diſtrict of the city contains twenty-three other convents, ſome of which ſhall hereafter be taken notice of.

The King's palace (in *Portugueſe* called the *Paço*) ſtands on the river; and is both ſpacious and convenient. It is called the *Paço da Ribeira* from its ſituation. On the eaſt ſide of it is a large ſquare for the bull-fights; and not far from it are the naval magazines. In the market-place, called *campo da Laa*, common malefactors are executed; and the unfortunate cɩeatures, ſentenced by the inquiſition, burnt. Near it is the granary, where all kind of grain is ſold; but the great ſlaughter-houſe ſtands ſtill nearer to the palace. *Corte Real*, a much ſmaller but more elegant palace than the above, lies on the weſtern ſbore of the *Tagus*. *Peter* II. formerly lived in it; but it has ſince been given to the Infant *Franciſco*. This I ſuppoſe to be the palace which was burnt in 1751. The *Paço da Bempoſta*, which is ſeated on the *campo de Santa Barbara*, belongs alſo to the Infant Don *Franciſco*. The *Paços da alcaçova* is a caſtle, and was the royal reſidence till the days of King *Emanuel*; but belongs at preſent to the Marquis of *Caſcaes*, as *alcaides mores* of *Liſbon*. The *paços dos Eſtaos* is now the reſidence of the inquiſitor-general. Beſides theſe royal edifices, there were ſeveral other very ſtately houſes belonging to the upper nobility, but the greateſt part of theſe buildings, as alſo of the whole city, has been deſtroyed by a moſt dreadful earthquake which happened on the firſt of *November* 1755.

The inhabitants of *Liſbon* do not, at moſt, exceed 150,000. According to the liſt given in the *Geografia Hiſtorica*, which I have ſo often cited, the pariſhes in the weſtern Dioceſe of this city amounted in the year 1732 to twenty-one, containing 83,319 ſouls. But thoſe of the ſixteen pariſhes of the eaſtern Dioceſe were not inſerted in it. The air of this place is tempe-rate and healthy. In 1721, a royal academy of *Portugueſe* biſtory was in-ſtituted in *Weſt Liſbon*. In this capital are alſo the following tribunals and offices, namely, *do conſelho de eſtado, do conſelho de guerra, do deſembargo do paço, da caſa da ſuppliçao, da meſa da conciencia, do conſelho da fazenda, da junta dos tres eſtados, do tribunal dos contos, do conſelho ultramarino, do conſelho da Rainha, do conſelho da caſa do Infantado, do conſelho da caſa de Bragança, do tribunal do ſanto officio, do tribunal da Alfandega, do tribunal* or *caſa da India, da Alfandega do tobaco, do tribunal da cruzado,* &c. The particular government of the city is lodged in a council, which con-ſiſts of a preſident, who is always a perſon of the firſt rank, ſix counſellors ſtyled *vereadores*, and ſeveral other inferior officers.

The trade of this place, and the navigation to and from it is ſo very confiderable, that the cuſtomhouſe which lies on the *Tagus* is the chief ſource of the King's *European* revenues. This is the grand magazine of all the goods which the *Portugueſe* fetch from their foreign colonies. The harbour is very large, deep, ſecure and convenient, and has two entrances; that

that on the north called the *corredor* lies betwixt the ſandbank, the rock of *Cachopos*, and fort St. *Julian*: The ſouthern entrance which is much broader and convenient, and called *careira da alcaçova*, is betwixt *Cachopos* and the fort of St. *Laurence.*

The city is walled round, having ſeventy-ſeven towers on the walls and thirty-ſix gates. It has ſo increaſed by degrees, particularly towards the weſt, that the old walls now divide the two Dioceſes. *John* IV. formed a deſign of ſurrounding the whole city with a high wall; but the work was diſcontinued after a prodigious expence. In the centre of this capital, on one of the mountains, ſtands a citadel which commands the whole place, and has caverns in it in which four regiments of foot are con-ſtantly quartered. Cloſe by the ſea, at the diſtance of about three *Portugueſe* miles from the city, both the entrances to the harbour are defended by two forts; that on the north ſtands on a rock in the ſea, and is called *S. Juliao*, but more frequently for brevity's ſake, *S. Giao:* The other, to the ſouth, is built on piles on a ſand-bank, and is named *S. Lourenço*, or *Cabeça ſeca*, but the fort is more commonly called *Bogio*. Two *Portugueſe* miles from *S. Julian* and one from *Lisbon* ſtands the fort of *Belem*, which commands the entrance into the city, where all ſhips coming up the *Tagus* muſt bring to and give an account of themſelves. Near it lies a little town, and di-rectly oppoſite to it, on the ſouth ſide, is the fort of *S. Sebaſtiao*, commonly called *Torre Velha*, or the *old fort.* It ſtands on the angle of a mountain, all along which, a little way on the other ſide of the city, the paſſage is de-fended from the beginning of the harbour by a chain of twelve forts. The proſpect at the entrance of the *Tagus* from the ſea cannot be exceeded.

Laſtly, to ſpeak of the name and hiſtory of the city, I can by no means acquieſce in the opinion of the *Portugueſe*, who derive it from *Eliſa*, a great-grandſon of *Noah*'s, who is ſaid to be its firſt founder, and that *Ulyſſes* was the reſtorer of it. It muſt, however, be admitted that the city was anci-ently called *Ulyſſæa*, or *Ulyſſipolis*, and hence was changed in time to *Oliſipo.* The *Romans*, as appears from ſome ancient inſcriptions, gave it the name of *Felicitas Julia*; but the *Goths*, upon their becoming maſters of it, called it *Oliſipona*, which the *Moors* altered to *Oliſibona*; whence, in proceſs of time, was formed *Lisbon.* In the year 1147 *Alphonſo* the firſt took it from the *Moors.* Under King *Ferdinand* the fourth, in the year 1373, and again in 1384, it was beſieged in vain by the *Caſtilians.* *John* the firſt made it his reſidence. The accurate father *Capaſſi* places its latitude in 38° 40', which is confirmed by *Couplet*, who adds only 25" to it, placing its longitude at 8° 2' 15" from the meridian of *Paris.* Of the pariſhes and convents in the Diſtrict of the city, the following are particularly worthy of notice, *viz.*

Belem, a pariſh, *Noſſa Senhora da Juda*, a monaſtery belonging to the monks of St. *Jerom*, and founded by King *Emanuel.* In its magnificent church, which ſunk of a ſudden, in the year 1756, ſeveral Kings and

Princes lay buried. Near it is a towń which, as well as the ſtrong fort here, has been already mentioned under the article of *Liſbon*.

At *Belem* is alſo a very noble foundation for noblemen impoveriſhed or diſabled in the King's ſervice.

Noſſa Senhora da Luz belongs to the order of *Chriſt*.

Santos o novo is a celebrated convent of *Jacobines*.

II. The CORRFIÇAO DE TORRES VEDRAS contains eighteen *villas*; viz.

Bellas, a little town containing about one thouſand two hundred and forty inhabitants.

Cafcaes, a fortified ſea-port, which from its ſituation on *Cape Da Roca*, in Latin *Promontorium Lunæ*, but now called *Cintra*, is one of the moſt delightful ſpots in the Kingdom, and contains about two thouſand inhabitants with two churches. It is too often frequented by merchant ſhips on account of the conveniency and advantage of ſmuggling here. From this town the marquis of *Cafcaes* receives his title, and in its Diſtrict are four pariſhes.

Collares, a ſmall town containing about one thouſand two hundred inhabitants.

Chileiros, an inconſiderable place having only two hundred and eighty inhabitants.

Mafra, a ſmall town, containing about one thouſand and forty inhabitants, near which, in a ſandy and barren place, King *John* the fifth erected a building of extraordinary magnificence. This was done in purſuance of a vow made in a dangerous fit of illneſs; and he propoſed alſo to found a convent, to be dedicated to the pooreſt friary in the Kingdom. Upon enquiry this pooreſt of convents was found at *Mafra*, where twelve *Francifcans* lived together in a hut. The King procured from *Rome* the draught of a building which ſhould far exceed the *Efcurial*, and the diſpoſition of it is as follows. In the center ſtands a ſuperb temple, conſiſting wholly of marble which was dug in the country of *Cintra*. Behind the choir is a houſe endowed with a large income for two hundred *Capuchins*, who officiate in this pompous church as chaplains. To the right of this building is a very ſpacious palace for the King, the royal family and the chief officers of the court. On the left is another palace, no way inferior to the former, for the patriarch and twenty-four canons, who have the privilege of wearing mitres. Twelve thouſand people were employed in raiſing this ſtructure; and, by certain computation, it coſt three fourths of the royal treaſury and of the gold of the *Brafil* fleets. At the diſtance of a quarter of a *Portugueſe* mile from the church ſtands an elegant houſe, with a ſmall wood, which has a beautiful effect in this ſandy waſte. The palace of *Mafra* looks towards the ſea and ſerves for a land-mark. Its Diſtrict contains two pariſhes.

Ericeira is a ſmall place containing about five hundred and ſixty inhabitants, and giving the title of count.

I

Cadaval,

Cadaval is a fmall place of three hundred and fifty inhabitants, with a Diftrict of eight parifhes. In the year 1649 it was raifed to a dutchy in favour of *Nuno Alvares Pereira de Mello,* marquis of *Ferreira,* whofe defcendants ftill enjoy it.

Villa Verde dos Francos contains only five hundred inhabitants.

Lourinha is a fmall place of about feven hundred inhabitants, with a Diftrict of two parifhes.

Alverca, a little place containing about one thoufand two hundred inhabitants, with a Diftrict of one parifh.

Alhandra, a town of about one thoufand three hundred and fifty inhabitants, with a Diftrict of two parifhes.

Villa Franca de Xira, a town containing about two thoufand nine hundred inhabitants, and a *juiz de fora,* who belongs alfo to the *comarca.*

Povos contains about fix hundred and thirty inhabitants; *Caftanheira* nine hundred; *Arruda* nine hundred and twenty, with a Diftrict of one parifh; *Sobral de Monte Agraça* one thoufand fix hundred inhabitants, with a Diftrict of one parifh; being all fmall places.

Torres vedras, i. e. *Turres veteres,* is feated in a low place among mountains, but its neighbourhood produces all kinds of fruits. It is one of the moft ancient towns of the Kingdom, no account being to be given of the æra of its building. To the north of the town runs the little river of *Sizandro.* It contains about two thoufand two hundred and fifty inhabitants, four churches, one *cafa da mifericordia,* one hofpital, three convents and a caftle; is an earldom and the *correiçao* town, and, as fuch, the refidence of the *corregedor,* a *provedor* and a *juiz da fora,* having nineteen parifhes belonging to its Diftrict.

III. The OUVIDORIA DE ALENQUER contains feveral eftates belonging to the Queen, and confifts of eight *villas;* namely,

Alenquer, a town on an eminence near a fmall river which falls into the *Tagus.* It is faid to have been built by the *Alans,* being anciently called *Alenker Rana,* i. e. *the temple of the Alans.* It contains about two thoufand five hundred inhabitants, five churches, one *cafa da mifericordia,* one hofpital and three convents; and is the chief town in the *ouvidoria* for the Queen's eftates, and the refidence of an *ouvidor,* who is alfo a *provedor,* and of a *juiz de fora.* For fome time paft it has been a marquifate, with thirteen parifhes belonging to its Diftrict.

Aldea-gallega da Merciana, is a fmall place oppofite *Lifbon,* contains about feven hundred and fixty inhabitants, with a Diftrict of one parifh.

Cintra, a town containing about one thoufand nine hundred inhabitants, four churches and an old caftle built in the *Moorifh* tafte at the foot of a mountain. In its Diftrict are fix parifhes. The beft air in all *Portugal* is thought to be in this place, which enjoys a moft pleafing coolnefs while *Lifbon* is fweltering with heat.

The mountain of *Cintra* confifts of large rocks of flint, fome of which are ten feet in diameter and lie on one another without any connection. It is very rich alfo in ore, and produces a great number of remarkable plants which the *Portuguefe* make little account of. It has alfo a vein of magnet. Near the fummit are feen the ruins of an old *Moorifh* town and fortrefs, with a refervoir under an arch which contains a very fine water to the depth of ten feet, which neither increafes nor diminifhes though it ftands very near as high as the fortrefs. Among the · waftes and rocks betwixt the mountain of *Cintra* to the top of *Cabo da Roca*, is a tract of three *Portuguefe* miles in length, which abounds in wolves and a fpecies of wild goats, by the *Portuguefe* called *cabreiros.*

Obidos is a town on a river which runs into the fea at a little diftance from it. This place is an earldom, contains two thoufand four hundred inhabitants, four churches and a Diftrict of fixteen parifhes.

Caldas, containing about eight hundred, and *Salir do Porto* one hundred and fifty inhabitants, are two little places near the fea. The firft is noted for its baths, which effectually cure all venereal diforders.

Chamufca is a fmall town of about one thoufand eight hundred and fifty inhabitants.

Ulme, a little place containing five hundred and eighty inhabitants, and one parifh in its Diftrict.

IV. The CORREIÇAO DE LEIRIA confifts of one *cidade* and twenty-one *villas.*

Leiria, a city in a delightful valley on the river *Liz,* at the conflux of the *Lena,* has a caftle on an eminence, with one church, befides its ftately cathedral, above three thoufand five hundred inhabitants, one *cafa da miferi-cordia,* one hofpital and four convents. In the year 1545 it was erected into a bifhop's fee, is the *correifao* town, the feat of *a corregedor,* a *provedor* and a *juiz de fora,* and was formerly the refidence of fome of the Kings. In its Diftrict are twenty-four parifhes.

Pombal is a town containing above three thoufand feven hundred inhabitants and two parifhes in its Diftrict. The *juiz de fora* of this place belongs alfo to the *comarca.*

Redinha is a little town of about two thoufand inhabitants.

Soure, a town of above three thoufand two hundred inhabitants, is alfo an earldom, with a *juiz de fora,* who belongs likewife to the *comarca.*

Ega, a fmall town and marquifate, containing about one thoufand one hundred inhabitants. To its Diftrict belongs one parifh.

Betalha, a town containing upwards of one thoufand eight hundred inhabitants.

Alcobaça, a fmall town feated between the little rivers *Alcoa* and *Baça,* contains about nine hundred and fifty inhabitants, with a convent of *Ciftercian* monks, which is the richeft abbey in the whole Kingdom. In its Diftrict is one parifh.

Cos,

Cos, containing about fix hundred and fifty, and *Mayorga* five hundred inhabitants, are both fmall towns.

Pederneira is a little town lying on the fea, having a harbour and above one thoufand three hundred inhabitants. In its Diftrict is one parifh.

Sella is a fmall town on the fea containing near one thoufand three hundred inhabitants.

Alfeizerao, which lies alfo on the fea, contains about feven hundred, and *S. Martinho* four hundred and eighty, inhabitants, being both fmall places.

Salir do Matto is a little place containing near five hundred fouls, with a harbour.

Alvorninha, or *Alburninha*, a little town, having about one thoufand five hundred inhabitants.

Santa Catharina contains only fix hundred, and has a Diftrict of three parifhes.

Turquel and *Evora* are two fmall places, the former of which contains little above fix hundred and fifty, and the latter about nine hundred inhabitants.

Algibarrota, or *Aljubarrota*, is a fmall town containing two parifhes and one thoufand fix hundred inhabitants. This place is remarkable for the victory obtained there by King *John* the firft over the *Caftilians*, in the year 1385.

Alpedriz is a fmall place, the inhabitants about fix hundred.

Peniche, a fortified town, with a harbour, containing three parifhes and about two thoufand eight hundred inhabitants, lies on a peninfula environed with rocks, being feparated from the main land by a canal which is always full at high water. It has a citadel and fort erected for its defence.

Atouguia, a fmall town on the fea, and an earldom, contains about one thoufand three hundred inhabitants, and is defended by a caftle.

V. The CORREIÇAO DE THOMAR contains twenty-one *villas*, exclufive of thofe belonging to donatarios. In it are

Thomar, a town in a pleafant plain not far from the ruins of the old city of *Nabancia*, from which it is feparated to the north by the river *Nabao*. The inhabitants are about three thoufand fix hundred, and, befides its two churches, one of which is a collegiate, it contains one *cafa da mifericordia*, one hofpital and four convents, among which that of the religious order of *Chrift*, fituate on a hill to the weft of the town, is the principal, and indeed the chief place of the order. The fuperior of the convent bears the title of prior and general of the order of *Chrift*. In the year 1752 the King founded an academy of fciences here on the footing of that at *Paris*, and was pleafed to declare himfelf prefident thereof, affigning a liberal income for its fupport. It is the *correiçao* town; but the ecclefiaftical jurifdiction in the town and Diftrict of the order belongs to the prelate of *Thomar*, who is invefted with epifcopal authority. This place belonged formerly to the

Knights

Knights Templars, and, on their fuppreffion, the order of *Chrift* fucceeded to their poffeffions, as alfo to the government of *Thomar.* In its Diftrict are eleven parifhes.

Pele, or *Pelle* and *Pias,* are two fmall places containing between five and fix hundred inhabitants. The latter has alfo a Diftrict of two parifhes.

Punhete is a fmall town of about one thoufand one hundred inhabitants, feated at the influx of the *Zezere* into the *Tagus.*

Maçao, a little town of about one thoufand four hundred inhabitants.

Antendoa, a fmall place containing about four hundred and fifty people.

Villa de Rey, betwixt fourteen and fifteen hundred, and having two parifhes in its Diftrict.

Sovereira Formofa is a little town confifting of above one thoufand fix hundred fouls, and having one parifh in its Diftrict.

Alvares, a town of above one thoufand one hundred inhabitants.

Pedrogaõ Grande, though but a fmall town, contains thirteen or fourteen hundred inhabitants. Its Diftrict too includes four parifhes.

Figueiro dos Vinhos is feated on the river *Aifo,* which runs into the *Zezere.* Its inhabitants between fourteen and fifteen hundred.

Dornas and *Aguas Bellas* are two fmall places containing betwixt five and fix hundred inhabitants; but the former has a Diftrict of two parifhes.

Ferreira and *Villa nova de Puffos,* are fmall places, having betwixt eight and nine hundred inhabitants.

Maçaas de Cominho is ftill fmaller, its inhabitants being only five hundred.

Arega is a fmall place of about feven hundred people.

Abiul, a little place containing above one thoufand five hundred fouls.

Ponte de Sor, an inconfiderable place of only fix hundred inhabitants, with a Diftrict of one parifh.

Alvaro is a town of betwixt eighteen and nineteen hundred inhabitants.

Obf. This Diftrict further contains the following *villas,* which belong to donatarios; namely,

Afzinceira, a very little place having about five hundred inhabitants.

Atalaya, a town containing betwixt thirteen and fourteen hundred inhabitants, being alfo an earldom.

Tanços is a marquifate, though containing only nine hundred inhabitants: Thefe three places belong to the count of *Atalaya.*

The two following conftitute a particular *ouvidoria,* and belong to the marquis of *Abrantes,* viz.

Abrantes, a town feated on an eminence on the *Tagus,* which is quite covered with gardens and olive-yards yielding a moft delicious profpect. The country likewife along the *Tagus,* between this town and *Lifbon,* is extremely

extremely fertile, and famous for the richneſs of its peaches. The town contains about three thouſand five hundred inhabitants, four pariſhes, one *caſa da miſericordia*, one hoſpital and four convents. The importance of its ſituation for the ſecurity of *Eſtremadura*, induced *Peter* the ſecond to have it fortified. *Alphonſo* the fifth raiſed it to a county, and, in the year 1718, *John* the fifth promoted it to a marquiſate in favour of *Rodrigo Annes de Sa Ameida* and *Menezes*, third marquis of *Fontes* and ſixth count of *Penaguiao*. It contains alſo fourteen pariſhes in its Diſtrict.

Sardoal is a town of about one thouſand eight hundred inhabitants and one pariſh in its Diſtrict.

VI. The OUVIDORIA DE OUREM contains ſeven *villas*, all belonging to the royal houſe of *Braganza*; namely,

Ourem, a town ſtanding on a hill which renders it on all ſides difficult to be approached. It formerly contained four pariſhes, but under *Alphonſo* the fifth they were all incorporated into one collegiate church. In it alſo is one *caſa da miſericordia*, one hoſpital and one convent. The houſe of *Braganza* has an *ouvidor* and a *juiz de fora* here. King *Peter* the firſt raiſed it to an earldom.

The other *villas* are *Agueda, Avellar, Chao de Couce, Maçaas de D. Maria, Porto de Moz,* and *Pouſa-ſlores.*

VII. The CORREIÇAO DE SANTAREM conſiſts of fifteen *villas*, all under the *cabeça* at *Santarem,* thoſe excepted which belong to the *donatarios*; namely,

Santarem, a town on the *Tagus* ſeated in a delightful plain, and environed with mountains which are beautifully interſected with vallies. In ſhape it reſembles a half moon, and is defended by a citadel erected in the modern taſte, commonly called *Alcaçova.* It contains thirteen churches, one of which is a collegiate church belonging to the order of *Aviz,* an academy of hiſtory, antiquities and languages, founded in the year 1747; one *caſa da miſericordia,* one royal hoſpital and two others, with eleven monaſteries and two nunneries; and is the chief town of the *correiçao,* the reſidence of a *corregedor,* a *provedor,* a *juiz de fora,* a *juiz dos orfaos* and a *juiz do tombo real,* who is always *deſembargador.* The name of *Santarem* is derived from *Santa Herena,* or Saint *Irene,* a martyr who lies buried here. It was alſo anciently called *Scalabis.* In the year 1146 it was taken from the *Moors.* Several Kings of *Portugal* have kept their court here; and its Diſtrict contains forty-five pariſhes.

Golcgaa, Aveiras debaixo, Almeirim and *Salvaterra de Magos,* are all ſmall places. At *Salvaterra* is a royal ſeat, where, in conformity to an ancient cuſtom, the Kings of *Portugal* reſide from the eighteenth of *January* to *Shrove-Tueſday.* The other *villas* belong to the *donatarios*; namely,

Torres novas, a marquiſate belonging to the eldeſt ſon of the houſe of *Abeiro.*

Aveiras,

Aveiras de Cima, bolonging to the counts of *Aveiras.*
Azambujeira, belonging to the counts of *Soure.*
Alcanede to the order of *Aviz.*
Alcoentre to the counts of *Vimieiro.*
Mugem to the dukes of *Cadaval.*
Lameroſa, or *Villa das Enguias,* to Emanuel *Telles de Menezes.*
Erra belongs to the counts of *Atalayda.*
Azambuja and
Montargil to the counts of *Val de Reys.*

V$_{III}$. The COMARCA DE SETUVAL contains three diſtinct juriſdictions, *viz.*

1. The *Correiçao de Almada,* to which belongs
Almada, a ſmall place ſeated on a bay of the *Tagus,* oppoſite to *Liſbon,* having a caſtle built on a rock.   This place is the reſidence of a *corregedor.*
Lavradio, a little place but a marquiſate.
Mouta, an inconſiderable place belonging to the counts of *Alvor.*

2. The *Ouvidoria de Setuval* belongs to the order of St. *Jago,* and contains
Setuval, by the *Dutch* called *S. Ubes,* a ſtrong town ſeated on a ſmall bay of the ſea, where the river *Sandao* diſcharges itſelf into it, having a harbour capable of receiving ſhips of any burden.   Beſides its old walls and towers it is ſtrengthened with eleven whole and two demi baſtions, with ſeveral other outworks.   It has likewiſe a ſtrong citadel called St. *Philip,* in which is a ſpring of excellent water; and the ſtrong fort of *Outao,* near the harbour, which alſo ſerves for a light-houſe.   Excluſive too of theſe, it has two ſmaller forts ; and in it are four churches, one *caſa da miſericordia,* one hoſpital, ten convents and an *academia problematica* founded by *John* the fifth.   It is the *comarea* town and the reſidence of an *ouvidor,* who is likewiſe *corregedor* of *Almada,* of a *provedor* and a *juiz de fora ;* as alſo of a tribunal *da Alfendega,* of another called *tabola real,* and of a ſait-office, which is here a very conſiderable branch of trade.   It is under the juriſdiction of the order of St. *Jago,* who nominate the perſons who compoſe the abovementioned *cabeça de Comarca,* and owes its foundation to the ruins of *Cetobriga,* a town formerly ſituated on the other ſide of the river, where *Troya* now ſtands, a very famous place in the time of the *Romans,* the name of which has been gradually altered to that of *Setobra* and *Setobala.* Being laid in ruins by the *Moors,* a company of fiſhermen ſome few years after, having built ſome houſes on the north ſide of the river called them *Setuval,* from the name of the old town.   Such is the origin of this place; which is ſo well known to the trading world.

The other *villas* are *Palmella, Coina, Barreiro, Alhos, Vedros, Aldea-Gallega, Alcochette* on the *Tagus ; Canha,* on a river of the ſame name which runs into the *Tagus ; Alcacere do Sal,* on the river *Sandao,* and *Grandola* on a river which joins the latter.

3. The

3. The *ouvidoria de Azeitao* belongs to the houfe of *Aveiro,* and contains the following *villas :* viz.

Azeitao, a little place.

Comora Correa, feated on the *Tagus.*

Sezimbra, lying on the fea near *Cabo de Efpichel.*

Torrao, on the river *Charrama,* which difcharges itfelf into the *Sandao,* contains about one thoufand two hundred inhabitants and a Diftrict of two parifhes.

Santiago de Cacem, fituated on a fmall bay, its inhabitants above one thoufand three hundred, and its Diftrict including eight parifhes.

A L E N T E J O.

THIS fifth province is one of the largeft in the whole kingdom. Its northern boundaries are *Eftremadura* and *Beira.* On the eaft it is joined to *Spain,* on the fouth to *Algarve,* and its weftern parts border on the fea. The length of this province from north to fouth is computed by fome at forty, and from eaft to weft at thirty *Portuguefe* miles ; but others reduce them both to thirty-four. It is called *Alentejo* from its fituation, becaufe, with refpect to *Eftremadura* and the other countries farther north, which were firft recovered, it lies *alem do rio Tejo,* that is, *beyond the river Tagus :* The two rivers in this province are the *Tagus* and *Guadiana.* It contains alfo fome mountains, but is for the moft part level and very proper for tillage. Wheat and barley are its principal products, though in many places it produces alfo plenty of wine, oil, fruits, game and fifh. Some parts of it afford gems and very curious utenfils; the white marble in particular of *Eftremoz* and *Vianna* ; the green ftone of *Borba* and *Villa-Viçofa* ; the white and red of *Setuval* and *Arrabida,* and the *Montemor* and *Eftremoz* utenfils are highly valued in *Spain.* The fertility of this province in the productions neceffary for an army fubjects it to the misfortune of being frequently the theatre of war; which it experienced among others in the two laft wars which were terminated by the peace of 1678 and 1715. And on this account it is that the King of *Portugal* has feveral good fortifications in it. The whole province contains four *cidades,* eighty-eight *villas,* three hundred and fifty-fix parifhes, and about two hundred and fixty-eight thoufand and eighty-two fouls belonging to them, being divided into eight jurifdictions.

I. The

I. The CORREIÇAO DE EVORA confifts of a *cidade* and eleven *villas*; namely,

Evora, a *cidade,* ftands, though not on a confiderable elevation, yet higher than the circumjacent country, which is almoft wholly environed with mountains. On the north and caft ftands mount *Offa*; to the fouth the mountains of *Portel* and *Vianna,* and to the weft *Montemaro.* The inhabitants amount to about twelve thoufand. It is divided into five parifh churches, of which the archiepifcopal cathedral makes one. It has alfo one *cafa da mifericordia,* one royal, with feveral other hofpitals, and contains, within its circuit and neighbourhood, twenty-one convents and colleges. It has been lately begun to be fortified with twelve whole and demi baftions, but is not yet finifhed. On the north fide it is defended by a quadrangular fort with four baftions and a like number of ravelines. It was anciently only a bifhop's fee; but, in the year 1540, was raifed to an archbifhoprick, under which are the bifhops of *Elvas* and *Faro.* It is the *correiçao* town, and the refidence of a *corregedor,* a *provedor,* a *juiz de fora* and *juiz dos orfaos,* and is alfo an univerfity. It was called in ancient times *Ebora*; and, on account of the great privileges conferred on it by *Julius Cæfar,* was diftinguifhed by the name of *Liberalitas Julia.* It was afterwards named *Elbora,* which was changed at length into *Evora.* In the year 1580 it furrendered to the *Spaniards.* Within its Diftrict are fifteen parifhes.

Eftremoz, one of the beft fortifications of the Kingdom, is fituate in a fertile pleafant country, having a caftle by no means inferior to a citadel. The town contains above fix thoufand five hundred inhabitants, three parifh churches, one *cafa da mifericordia,* one hofpital and fix convents. It is famed for its fine utenfils, and in its neighbourhood is found a beautiful kind of marble which, with the affiftance of a good polifh, refembles alabafter.

Vimieiro is a fmall town on an acclivity, containing about one thoufand fix hundred inhabitants. This place is an earldom, but has only one parifh in its Diftrict.

Canal is an inconfiderable place.

Pavia, a little town containing about feven hundred inhabitants.

Aguias, the inhabitants of which are between five and fix hundred.

Lavre, a fmall place containing between twelve and thirteen hundred inhabitants, ftands on a little river of the fame name.

Montemor o Novo is a town on the river *Canha,* having four parifhes and about four thoufand inhabitants. The *juiz de fora* of this place belongs alfo to the *comarca,* and within its Diftrict are twelve parifhes.

Montoito, a mean place.

Redondo, a town containing near two thoufand feven hundred inhabitants, with a Diftrict of four parifhes. Its *juiz de fora* belongs alfo to the *comarca.* This place is an earldom.

Vianna is a little town of between fourteen and fifteen hundred inhabitants. Its *juiz de fora* belongs to the *comarca*.

II. The OUVIDORIA DE BEJA contains a *cidade* and three *villas*.

Beja, a *cidade*, ftands high and is furrounded with fruitful fields. It was formerly called *Pax Julia* and alfo *Pax Augufta*. This city contains fix thoufand two hundred inhabitants, and is divided into four parifh churches, one *cafa da mifericordia*, one hofpital and feven convents. It is the refidence of the *comarca*, the feat alfo of an *ouvidor*, a *provedor* and a *juiz de fora*. *John* the fecond erected it into a dukedom, which belongs, at prefent, to the Infante Don *Francifco*. It was formerly a bifhop's fee, and now contains a Diftrict of twenty-one parifhes.

Moura is a fortified town, but its works fuffered confiderably in the laft war, and the caftle there was demolifhed by the *Spaniards*. It contains about four thoufand inhabitants, two parifh churches, one *cafa da mifericordia*, one hofpital and five convents; is the refidence of a *juiz de fora*, and has twelve parifhes in its Diftrict.

Serpa is a ftrong town, on an inacceffible eminence, containing two parifh churches, almoft four thoufand inhabitants, and a *juiz de fora*. The adjacent country is extremely fertile and beautifully variegated with plantations of fig and olive trees. In its Diftrict are feven parifhes.

Alcoutim is a fmall town on the *Guadiana*, and, though it ftands in *Algarve*, being fituate on the border of *Alentejo*, yet belongs to this Diftrict. It is defended by a caftle, and contains about one thoufand inhabitants and fix parifhes in its Diftrict. King *Emanuel* conferred on it the title of earldom, but it belongs at prefent to the Infantes.

Obf. *Beja*, as a *provedaria*, may be faid to contain alfo the following *villas*, moft of which are *donatarios*; viz:

Agua de Piexes, *Villa-Alva*, *Villa-Ruiva* and *Albergaria*, all fmall places belonging to the duke of *Cadaval*, and having their particular *ouvidor*.

Alvito, a little town of about two thoufand inhabitants, and a barony: *Villa nova de Alvito*, a mean place of about nine hundred; and *Aguiar*, ftill meaner, of about four hundred and fifty, belong all to the Count de Barao.

Vidigueira, a little town confifting of about two thoufand three hundred inhabitants, and an earldom, belongs, together with *Frades*, to the marquis of *Niza*.

Beringel, containing about one thoufand two hundred inhabitants, belongs to the marquis of *Minas*.

Faro is a very fmall place.

Ferreira, which is a marquifate, contains about one thoufand three hundred inhabitants with a Diftrict of one parifh.

Odemira, is a town of two churches and between nine hundred and one thoufand habitants. In its Diftrict are four parifhes, and its *juiz de fora* belongs to the *comarca*.

Oriola and *Ficalho* are two fmail places containing a few hundreds of inhabitants, but the former is an earldom.

III. The OUVIDORIA DO CAMPO DE OURIQUE, belongs to the order of *Sant Jago* and confifts of fourteen *villas*.

. *Ourique* contains about two thoufand inhabitants, one parifh-church, one *cafa da mifericordia*, one hofpital, and is the *comarca* town and the refidence of an *ouvidor*, a *provedor* and a *juiz de fora*. Near it, in the year 1149, was fought a memorable battle betwixt *Alphonfo Henriques* and the *Moors*.

. *Padroens* is a fmall place having about four hundred inhabitants and one parifh in its Diftrict.

. *Mertola*, a town feated on the *Guadiana*, contains about two thoufand four hundred inhabitants and a Diftrict of ten parifhes. The *juiz de fora* of this place belongs alfo to the *comarca*.

Almodovar contains about one thoufand eight hundred inhabitants, and in its Diftrict are five parifhes. The *juiz de fora* of this place belongs alfo to the *comarca.*

Villa nova de mil fontes is a little place of about one hundred inhabitants, having a Diftrict of one parifh.

Sines, or *S. Joao de Sines*, is a fmall fea-port containing about one thoufand four hundred inhabitants.

Collos is a little place of about fix hundred inhabitants.

Gravao is an inconfiderable town containing above four hundred inhabitants with a Diftrict of one parifh.

Caftroverde, a fmall town feated on the river *Corbos*, and having about two thoufand feven hundred inhabitants.

Entradas is a little place on the river *Corbos* containing betwixt fix and feven hundred inhabitants.

Pannoyas, alfo fmall, contains feven or eight hundred fouls.

Aljuftrel is a little town containing about one thoufand five hundred inhabitants, with a Diftrict of one parifh. The *juiz de fora* of this place belongs alfo to the *comarca*.

Alvalade, a little town of about one thoufand two hundred inhabitants, is feated between two rivers.

Meffejana, another little place, contains about one thoufand one hundred fouls.

IV. The OUVIDORIA DE VILLA VIÇOSA confifts of twelve *villas* and one *confelho*.

. *Villa Viçofa*, i. e. *the delightful town*, ftands in a moft fertile pleafant country, being handfome and particularly remarkable for a large and beautiful palace

in

in it. Befides its fortifications it has alfo a caftle. The inhabitants are about three thoufand feven hundred; and it contains two parifhes, one *cafa da mifericordia*, one hofpital and feven convents. It is the refidence of a *juiz de fora* appointed by the houfe of *Braganza*, and was formerly the feat of thofe dukes. In the year 1665 it held out a fiege againft the *Spaniards*, and within its Diftrict are three parifhes.

Evora Monte is a little town, feated on a rock, containing about eight hundred inhabitants. In its Diftrict are four parifhes.

Arrayolos contains two thoufand inhabitants, with four parifhes in its Diftrict.

Borba is a town confifting of about two thoufand feven hundred inhabitants and two parifh churches. It is feated on a hill, and its Diftrict contains two parifhes.

Monçarus is a fmall town on the *Guadiana* containing about one thoufand five hundred inhabitants. Its Diftrict confifts of four parifhes.

Villa Boim and *Villa Fernando* are two inconfiderable places.

Portel is a little town containing above one thoufand nine hundred fouls, with a Diftrict of feven parifhes.

Souzel is fmail, but contains about one thoufand three hundred inhabitants, with one parifh in its Diftrict.

Monforte is a fmall town.

Chancellaria and *Alter do Chao* are two little places.

Margem and *Lagomel* conftitute one *concelho*.

V. The CORREIÇAO DE ELVAS confifts of one *cidade* and fix *villas*; namely,

Elvas, a *cidade* and good fortification with a caftle called *Santa Luzia*. It ftands on an eminence, having the caftle above it, and contains three parifh-churches befides the cathedral, one *cafa da mifericordia*, one hofpital and feven convents. In its diocefe are fifty parifhes. It is the *comarca* city and the refidence of a *corregedor*, a *provedor* and a *juiz de fora*. One of its principal curiofities is a very large refervoir into which the water is conveyed through an aqueduct of a mile in length ; and, in the neighbourhood of the city, is of fuch height as to be fupported by a triple perpendicular arch. The inhabitants of this place and its Diftrict, which confifts of ten parifhes, are computed at about twelve thoufand four hundred and eight. In the year 1580 it was taken by the *Spaniards*, who, in 1659, fuffered a terrible defeat from the *Portuguefe* near this city. The country abounds in good wine and excellent oil.

Olivença, one of the beft fortifications in this province, lies in a charming plain. It has nine baftions, eight ravelins, with a caftle and fome other works ; as alfo two parifh churches, one *cafa da mifericordia*, one hofpital and one convent. It is an ancient earldom, and, including its Diftrict of four parifhes, contains about five thoufand three hundred fouls. The *juiz*

de

de fora of this place belongs alfo to the *comarca*. In the year 1577 the *Spaniards* made themfelves mafters of it..

Campo Mayor is a modern fortification, having four whole and five demi baftions, and two forts, both lying about a mufket-fhot from the town. It is feated in a wide plain, contains one parifh-church, one *cafa da mifericordia*, one hofpital, two convents, and not lefs than five thoufand three hundred inhabitants. The *juiz de fora* of this place belongs alfo to the *comarca*.

Ouguella is an inconfiderable place containing a few hundreds of inhabitants.

Barbacena, a little place of about feven hundred inhabitants, confers the title of vifcount.

Mourao, a little town on an acclivity near the *Guadiana*, contains a caftle, about one thoufand four hundred inhabitants, and a Diftrict of three parifhes. The *juiz de fora* of this place belongs to the *comarca*.

Terena is a little place of about nine hundred fouls, with a Diftrict of two parifhes.

VI. The CORREIÇAO DE PORTALEGRE contains one *cidade* and twelve *villas*, fome of which belong to the order of *Chrift*; namely,

Portalegre, formerly *Portus Alacer*, a city fortified after the antique tafte. with walls and towers. It contains about five thoufand fix hundred inhabitants, four parifh churches befides a cathedral, one *cafa da mifericordia*, one hofpital and five convents. It is alfo the *comarca* town, the refidence of a *corregedor*, a *provedor* and a *juiz de fora*, and, in the year 1550, was erected into a diocefe by Pope *Julius* the third.. The bifhop of this place has forty-one parifhes under him, and its Diftrict contains fix more;

Arronches, a fortified town, in a hilly country, ftands near the conflux of the two little rivers of *Alegrette* and *Caya*. The inhabitants are above one thoufand two hundred. It contains one parifh church, one *cafa da mifericordia*, one hofpital and one convent. In the year 1674 *Peter* the fecond raifed it to a marquifate. Its Diftrict contains fix parifhes, and the *juiz de fora* of this place belongs alfo to the *comarca*.

Alegrete, a marquifate, but yet a little place, contains not above nine hundred inhabitants.

Affumar, though an earldom, contains only fix hundred.

Niza is a little town of betwixt eighteen and nineteen hundred inhabitants, with two churches and a Diftrict of two parifhes. It is alfo a marquifate, and the *juiz de fora* of this place belongs likewife to the *comarca*.

Povca and *Meadas* are two fmall places appertaining to the counts of *Val de Reys.*

Apalhao, a little town of above one thoufand two hundred inhabitants.

Caftello

Caſtello de Vide, a town ſeated on a riſing ground, contains about five thouſand ſeven hundred inhabitants, with three pariſh churches. The *juiz de fora* of this place belongs alſo to the *comarco.*

Montalvao, is a ſmall place, but the inhabitants above nine hundred.

Aviz and *Villa-Flor,* an earldom ſeated on the *Tagus,* are two ſmall places, to be diſtinguiſhed from two others of the ſame name occurring in the ſequel.

II. The OUVIDORIA DE CRATO conſiſts of twelve *villas,* ſome of which lie in *Alentejo,* others in *Eſtremadura;* a few of them belong alſo to the *provedoria* of *Portalegre,* and others to that of *Thomar.*

Crato, is a ſmall town having one church, one *caſa da miſericordia,* one hoſpital, and one convent. It is the principal place of the priory of the order of *Malta,* and contains twenty-nine pariſhes under it. The prior determines in all cauſes both civil and criminal, and, by a *Nullius dioceſis Bull* from the Pope is exempted from the united juriſdiction of biſhops. This town is the ſeat of an *ouvidor* and a *juiz de fora,* and has ſix pariſhes in its Diſtrict.

S. Joao de Gafete, Toloſa, and *Amieira* are ſmall places.

Envendos, which is ſeated on the *Tagus, Carvoeiro, Certaa Cardigos,* and *Oleiros* are inconſiderable places.

Belver, ſeated on the *Tagus, Pedrogao Pequeno,* and *Provença a Nova* are all little places in *Eſtremadura.*

VIII. The OUVIDORIA DE AVIZ, contains ſeventeen *villas.*

Aviz, a mean town ſituate on a riſing ground near a river of the ſame name contains about fourteen or fifteen hundred inhabitants, one pariſh church, one *caſa da miſericordia,* one hoſpital, and one convent of the order of *Aviz,* the principal place of which it is, as likewiſe of the *ouvidoria.* It is alſo the reſidence of an *ouvidor* and a *juiz de fora.* The order of *Aviz* derives its name from this place, which firſt founded it in the reign of *Alphonſo,* and in the year 1211 obtained a grant of it. Without its walls is a large ſuburb, and its Diſtrict contains three pariſhes.

Fronteira, a marquiſate; *Cabeça de Vide, Veiros, Seda, Cano, Mora,* and *Galveas* an earldom, *Cabeçao,* and *Alandroal,* are all little places.

Figueira, Benavilla, Noudar, Alter-Pedrozo and *Jurumenha,* are mean places.

Benavente, is a town of above twelve hundred inhabitants, and two pariſhes in its Diſtrict.

Curuche, is a town of two thouſand four hundred inhabitants, and four pariſhes in its Diſtrict.

The KINGDOM *of* ALGARVE.

THE Kingdom of *Algarve* is bounded on the north by the province of *Alentejo*, from which it is feparated by the *Caldeirao* and *Monachique* mountains. On the eaft it borders on *Andalufia*, and to the fouth and weft is terminated by the fea. Its length from north to fouth is computed to be twenty-feven or twenty-eight miles; and from caft to weft only five or fix. The name is of *Moorifh* extraction, being unknown in *Spain* till the invafion of thefe parts; geographers, however, are not yet agreed, whether it fignifies a level and fruitful country, or a country lying weftward; or whether it fignifies a country lying very low, and at the utmoft extremity. It has three well known capes, viz. *Cabo de S. Vicente, Cabo de Carvoeiro,* and *Cabo de S. Maria.* Under the name of *Algarve* was formerly included a much larger tract of land than at prefent, ftretching not only along the whole coaft from cape St. *Vincent* to the city of *Almeria* in the Kingdom of *Granada*, but extending even to the oppofite part of *Africa*. It is reduced at prefent to the abovementioned boundaries; and though mention be made of the *Algarves* of *Algezira* in the titles of the King of *Spain*, yet the addition of *Algezira* limits the import of the name of *Algarves,* and is founded on the old fyftem of *Spanifh* geography after fhaking off the *Moorifh* yoke, or fignifies a ftrip of land along the fea from *Niebla* to *Almeria*; and that part of the oppofite coaft of *Africa* in which are fituated the towns of *Ceuta* and *Tangier*, and including even the Kingdom of *Fez*; whence confequently it bears no relation to the *Algarve* of *Portugal*. Befides a prodigious plenty of wine, oil, and corn, this province abounds alfo in all kinds of fruits, fuch as figs, grapes, and almonds. It contains four *cidades*, twelve *villas*, and fixty villages; fome of which are very populous. The parifhes amount to fixty-feven, and the inhabitants to 60,688.

This Kingdom belongs of right to the crown of *Portugal*, though claims have been made to it by the Kings of *Caftile* and *Leon*. Among the many proofs by which the *Portuguefe* fupport their right, the following are the principal: As early as the year 1188, King *Sancho* I. difpoffeffed the *Moors* of the town of *Sylves*, and in the fucceeding year, of many other Diftricts. From this time he took upon him the title of King of *Algarve*, as appears from feveral ancient inftruments, and particularly from a grant of his made to the convent of *Grijo* on the feventh of *July* 1190. The original of this grant is preferved at *Torro do Tombo*, and in it he ftyles himfelf *Sancius Dei gratiâ Portugaliæ & Algarvæ Rex*; having not only been acknowledged as

fuch

fuch by his vaſſals, but alſo by the Kings of *Leon* and *Arragon*, as appears unqueſtionably from ſeveral treaties cited by *Jeronimo Zurita*. *Alphonſo* II. proved that no limits were ſet to his conqueſts, not only by paſſing the river *Guadiana*, and taking the towns of *Serpa* and *Moura*, but by penetrating alſo into *Andaluſia*. In 1242, *Sancho* II. made himſelf maſter of the town of *Tavira*, and other places in *Algarve*, granting them to the order of *S. Jago* and its maſter : This grant was confirmed by a bull of Pope *Innocent* IV. which manifeſts him to have been Lord of the country. *Alphonſo* III. proſecuting theſe conqueſts, took *Faro* alſo from the *Moors* ; but in 1252, this Kingdom engaged him in a war with the King of *Caſtile*. *Alphonſo* X. claimed it, either as being ceded to him by *Sancho* II. who a few years before had made a tour to *Toledo*, where he met with an amicable and ſplendid reception, or as made over to him by *Aben Maffo*, or *Aben Afan*, the expelled *Mooriſh* King, in exchange for the county of *Niebla*. This war, however, terminated in 1253, in an agreement that the King of *Caſtile* ſhould hold the Kingdom of *Algarve* during his life, the property and ſovereignty of it ſtill remaining in the King of *Portugal*. At the ſame time alſo a marriage took place betwixt the King of *Portugal* and *Beatrice*, daughter to the King of *Caſtile*. Furthermore, in 1263, a new compact was made in which the King of *Caſtile* reſigned all poſſeſſion and claim to the Kingdom of *Algarve*, the King of *Portugal* engaging to aſſiſt him during his life with fifty ſpearmen. In this ſituation matters continued till the year 1266, when the Infant *Dennis* or *Dionyſius* voluntarily aſſiſted his grandfather againſt the *Moors* ; and came in perſon to *Seville*, in return for which the former exempted him from the ſuccour of fifty ſpearmen. Since that time there has been no farther diſpute about *Algarve*.

Algarve being of the above-mentioned extenſive ſignification, the Kings of *Portugal* ſtyle themſelves Kings of the *Algarves on this ſide and beyond ſea in Africa* ; though they poſſeſſed but a part of the *Algarve* on this ſide. Afterwards, however, they reduced the cities of *Ceuta*, *Tangier*, and many other parts of the *Algarves* beyond-ſea in *Africa*. The arms of *Algarve* have been mentioned already in §. 16. of the introduction.

This Kingdom is divided into three juriſdictions or *comarcas*, of which *Lagos* and *Tavira* are *correiçaoes*, as conſiſting of crown-lands. But *Faro* is no more than an *ouvidoria*; the Queen being donatary of its territories.

I. The CORREIÇAO DE LAGOS contains one *cidade* and ſeven *villas*, with fixteen villages, namely,

Lagos, a *cidade* on the ſouth coaſt in a bay navigable for the largeſt ſhips, into which runs a river. It has alſo a harbour, and is ſaid to ſtand on the ruins of the famous city of *Lacobriga*. Its ſituation will not admit of a regular fortification ; but the harbour is well ſecured by the two forts of *Bandeira* and *Pinhao*. The inhabitants are about two thouſand fix hun-

4   dred.

dred. It contains two parifh churches and four convents; is the refidence both of the governor and captain-general of *Algarve*, of a *corregedor*, a *co-marca*, and a *juiz de fora*; being alfo the *correiçao* town. To its Diftrict belong ten parifhes. The coaft betwixt this city and *Sagres* is defended by the forts of *Noffa Senhora da Guia, Santo Ignacio do Aftval, Vera Cruz da Figveira, S. Luiz de Almadena*, and *Noffa Senhora da Luz*.

Villa Nova de Portimao, is a fortified town, feated on a river which forms a fpacious and fecure harbour about half a mile in breadth and three fathom deep : It is defended by the forts of *Santa Catharina* and *S. Joao* : The fand-banks in it render the entrance very dangerous without the affiftance of a pilot. This town was built in 1463, contains above fixteen hundred inhabitants, one parifh church, one *cafa da mifericordia*, one hof-pital, and one convent; befides a college of Jefuits in one of the fuburbs. It is alfo the refidence of a *juiz de fora* belonging to the *comarca*. King *Emanuel* erected it into an earldom which belongs to the family of *Lancaftro*.

Sagres, is a fmall fortified place on a neck of land with a very fine har-bour. Its caftle is its coat of arms.

Villa do Bifpo, a little place of about five hundred fouls.

Algezur or *Aljezur*, is another fmall place; but having about eight hun-dred inhabitants, with a Diftrict of one parifh. Its late caftle is its arms.

Seixa, is a little place lying on the fea.

Paderne, a little place, having formerly a caftle, which now ftands for its arms.

Albufeira, is a little town of about nineteen hundred inhabitants, with two parifhes in its Diftrict. Its ancient caftle is its arms: And the *juiz de fora* of this place belongs alfo to the *comarca*.

II. The CORREIÇAO DE TAVIRA confifts of one *cidade*, three *villas*, and fifteen villages.

Tavira, a *cidade* feated on a bay having a harbour defended by two forts, which are pleafantly fituated, is divided by the river *Sequa* into the eaft and weft town. Befides its walls it is defended by a caftle, and contains above four thoufand feven hundred inhabitants, two parifh churches, one *cafa da mifericordia*, one hofpital, five convents, and is the *comarca* town; being the feat likewife of a *correiçao*, and the refidence of a *juiz de fora*. It is faid to be the *Balfa* of the ancients, and contains five parifhes in its Diftrict.

Loule, pleafantly fituated in a level country, is walled, and further de-fended by a caftle. The inhabitants of this place are above four thou-fand four hundred. It contains one parifh church, one *cafa da mifericor-dia*, a rich hofpital and three convents. It has alfo a *juiz de fora*, who belongs to the *comarco*, and is faid to have fprung out of the ruins of the old

old town of *Querteira*, which ſtood on the ſea near a river ſtill bearing its name. Its Diſtrict conſiſts of five pariſhes.

Carcella, a ſmall fortified ſea-port of ſix hundred inhabitants, has a caſtle in its arms.

Caſtromarim, a little town with fortifications, is ſeated at the mouth of the *Guadiana*, oppoſite *Ayamonte*. It contains about one thouſand inhabitants with a Diſtrict of two pariſhes; and has a caſtle in its arms.

Obſ. Mention has been made of the town of *Alcoutim* in *Alentejo* in the *Ouvidoria* of *Beja*, to the juriſdiction of which it belongs, though it ſtands in *Algarve*. With reſpect to eccleſiaſtical matters it is ſubject to the biſhop of *Faro*.

III. The OUVIDORIA or COMARCO of FARO contains two *cidades*, one *villa*, and thirty-four villages, namely,

Faro, a modern fortification with a caſtle, ſituate in a level country and on a bay, which furniſhes it with a harbour. It is ſeparated by a narrow arm of the ſea, commonly called *Barreta*, from *Cabo de Santa Maria*, which *Pliny* calls *Promontorium Cunium*. It contains four thouſand five hundred inhabitants, one church and a cathedral, with one *caſa da miſericordia*, one hoſpital, and four convents. The ſeat of this Dioceſe was firſt eſtabliſhed at *Oſſonoba*; from whence it was removed to *Sylves*, and afterwards hither. The juriſdiction of this city belongs to the Queens of *Portugal*, whoſe *ouvidor* reſides here, and by a particular privilege is alſo *provedor*. To its Diſtrict belong twenty-eight pariſhes.

Obſ. On the ruins of the preſent village of *Eſtoy*, which has a church, ſtood formely the epiſcopal city of *Oſſonoba*.

Sylves, a little town of about ſixteen hundred inhabitants, is ſeated on the river which joins the ſea below *Villa Nova de Portimao*, after receiving another at this place. It contains one church, one *caſa da miſericordia*, one hoſpital, and one convent; and belongs to the Queen. From the year 1188, when *Sancho* I. made himſelf maſter of it, to 1580, it was a biſhop's ſee. To its Diſtrict belong eleven pariſhes, ſome of which contain more inhabitants than many *villas*, namely, *S. Bartholomew de Miſſines*, which conſiſts of two thouſand; *Lagoa*, of betwixt twelve and thirteen hundred; and *Monchique*, of fifteen hundred. The church-village of *Eſtombar* had formerly a caſtle, which it ſtill preſerves in its arms.

Alvor, a ſmall place of four hundred and ſixty ſouls, and an earldom, lies betwixt *Villa Nova de Portimao* and *Lagos*.

Of the I S L A N D S,

IN the *Atlantic* ocean belonging to the crown of *Portugal*, thofe of *Porto Santo* and *Madeira* may be here moft conveniently treated of on account of their nearnefs; but the iflands of *Azores* are annexed to *Portugal*, as appears, among other things, from the defcription of *Angra* on the ifland of *Terceira*.

1. The ifland of *Porto Santo*, which lies about one hundred and forty miles from *Lifbon*, is five *Portuguefe* miles in length and two broad. It was difcovered firft, in the year 1419, by *Joao Gonçalves Zarco*. The principal town in it bears its name, and confifts of about fix hundred inhabitants, being pleafantly feated on a bay and land-locked on all fides, except towards the fouth and fouth-eaft. Among the fmall places here *Farabo* and *Feteira* are the moft worthy of notice.

2. *Madeira*, fituate about one hundred and fifty-two *Portuguefe* miles from *Lifbon*, and almoft the fame diftance from the ifland of *Terceira*, is eighteen miles long and fomething above four broad. It was difcovered on the fecond of *July*, in the year 1419, by *Joao Gonçalves*, who, from the vaft woods he found upon it, gave it the name of *Madeira*, i. e. *timber*. It is divided into two *capitanias*, or captainfhips; namely, *Machico* and *Funchal*.

1. The captainfhip of *Machico* belongs to the houfe of *Vimiofo*, and contains

Machico, a fmall town, lying on a bay of the fame name, and containing about two thoufand inhabitants.

Santa Cruz, a little town ftanding alfo on a bay of the fame name.

2. The captainfhip of *Funcal* belongs to the count of *Calblta*, and contains

Funhcal, a *cidade*, which is the principal town of this captainfhip and of the whole Ifland, being feated on a good bay. It is alfo the fee of a bifhop, fuffragan to the patriarch of Weft *Lisbon*. Towards the fea it is defended by five forts and one fortrefs, and towards the land by the caftle of *S. Joao de Pico*. It is the refidence alfo of a *juiz de fora*, who, with the title of *corregedor*, is the chief jufticiary on this ifland.

Ponta do Sol is a little town.

Calheta, alfo a little town but an earldom.

Next are the villages of *Camera de Lobos*, *Ribeira Brava*, and other like places.

Befides this one *cidade* and four *villas*, there are feveral villages on the ifland, which, put altogether, are divided into thirty-nine churches, and
have

have ten thoufand five hundred fire-places. They are further reckoned to contain one Jefuit's college, four convents of *Francifcans,* three of the order of *S. Clair,* one feminary, four *cafas da mifericordia,* feveral hofpitals, *&c.* The tythes of the Ifland, which belong to the King, as grand mafter of the order, amount, in the beginning of the year, to upwards of one hundred thoufand crufados. Its vines, the plants of which are brought from *Candia,* are greatly celebrated.

3. The *Azores,* or *Terceira* iflands, called alfo the *Flemifh* iflands, derive their firft name from the great number of hawks and falcons found there at their difcovery; the fecond from *Terceira,* the principal ifland among them, and the third from the *Flemings,* who difcovered it almoft at the fame time with the *Portuguefe.* They are nine in number, and, according to the chronology of their difcovery, ftand as follows:

1. *Santa Maria,* diftant about two hundred and fifty *Portuguefe* miles from Cape *St. Vincent,* in *Algarve,* was difcovered on the fifteenth of *Auguft,* in the year 1342, by *Gonçalo Velho Cabral,* being four miles long and three broad, with a harbour feated in a bay to the fouth-weft, defended by fome redoubts. The principal place on the ifland is

Porto, a fmall town with two convents, befides which there are a few parifhes and villages.

2. *S. Miguel* is the firft of the *Terceira* iflands in the paffage from *Lisbon,* and lies about two hundred and twelve *Portuguefe* miles from *Cabo de Efpichel.* It was difcovered by *Gonçalo Velho Cabral,* on the eighth of *May,* in the year 1444, which being the feftival of the appearance of St. *Michael,* occafioned its being called by this name. It is eighteen miles in length and two in breadth. Its two principal harbours lie on the fouth fide; namely, *Ponta Delgada* and *Villa Franca;* but are both defencelefs. It is the beft peopled of all the neighbouring iflands, the hearths being reckoned at ten thoufand nine hundred and eighteen, and the inhabitants at above fifty-one thoufand five hundred, exclufive of one thoufand three hundred and ninety-three ecclefiaftics, monks and nuns; and, though but lately cultivated entirely, is alfo the moft fruitful, particularly in corn and wine; infomuch that, one year with another, the wheat amounts to twelve thoufand bufhels, the millet to near as much again, and its wine to about five thoufand pipes. In the year 1474, *Ruy Gonçalves da Camera* purchafed the captainfhip of this ifland for thirty-two thoufand crufados, but a defcendant of his, namely the count of *Ribeira Grande,* made no lefs than thirty thoufand *per annum* of it. In the year 1753, however, the King reaffumed the captainfhip of *St. Miguel,* and, by way of equivalent, created the Count *das Ribeira,* marquis of *Villa Franca,* giving him fome lordfhips and commandaries of the orders of knighthood.

H h 2

With

With respect to its ecclesiastical state, it is divided into the three *ouvidorias* of *Ponta Delgata, Villa Franca* and *Rebeira Grande,* and contains one *cidade,* five *villas* and twenty-two villages.

Ponta Delgata, the capital, was only a village from the year 1445 to 1499, when King *Emanuel* erected it into a *villa;* and, in the year 1546, *John* the third raised it to a *cidade.* It is seated in a level country, on an open harbour defended by a fort, and contains about one thousand eight hundred and seventy-nine hearths, three parish-churches and seven convents, besides the palace of the *capitaens donatorios,* the town-house, the custom-house and poor-house. It is also the residence of a *juiz de fora,* from whom there lies an appeal to the *corregedor* at *Angra.*

Villa Franca, the most ancient town in the whole island, is so called from its being at first a free port. Before its harbour lies an island containing about a mile in circumference, and towards the sea the town is defended by a fort and some other works. It consists of one thousand eight hundred and thirteen hearths, has two parish churches and two convents, with nine places or villages belonging to its jurisdiction.

Rebeira Grande lies to the north in a large bay, being seated at the foot of a mountain. It was erected into a *villa* by King *Emanuel* in the year 1507. It contains one thousand four hundred and twenty-four hearths, two parish churches and two convents.

Nordeste, raised to a *villa* by King *Emanuel* in the year 1514, contains three hundred and thirty hearths and one church.

Agua de Pao was endowed with a charter by King *Emanuel* in 1515.

Alagoa, erected into a *villa,* in the year 1522, by *John* the third, contains six hundred and five hearths and two parish-churches.

Obf. The little island which appeared in the year 1720 betwixt *St. Miguel* and *Terceira* has since gradually vanished.

3. *Terceira* is so called as being the third which was discovered, though there is no certainty as to the particular year: this must have been, however, betwixt the years 1444 and 1450. It is full thirteen miles long and fix broad. The harbour of *Angra,* its capital, is two hundred and forty-five *Portuguese* miles distant from *Lisbon,* and about twenty-two from *Ponto* on *S. Miguel.* It consists of two captainships.

1. The captainship of *Angra* contains one *cidade,* one *villa,* and several lesser places, which are well inhabited.

Angra, a *cidade,* stands on the south coast of the island. Its harbour is formed by a bay situate betwixt two capes, the one lying west and the other east, about a quarter of a mile from each other, and both about twice that distance from the town. On the former stands fort *S. Sebastiao;* and, on a high mountain, called *Monte do Brasil,* fort *S. Joao Bautista,* with a bastion. The harbour is clear, spacious, and has good anchorage, being exposed only to the south-east. It is a populous city, having broad, straight,

I

clean,

clean, well paved ſtreets, a cathedral and five other churches, one poor-houſe, one hoſpital and eight convents. The biſhop of this place has all the nine iſlands of the *Azores* under him, and is himſelf a ſuffragan to the patriarch of Weſt *Liſbon.* The civil government is lodged in two ordinary judges, three *vereadores,* one *procurador,* one *eſcrivao da camera* and other officers. Here is alſo a *corregedor,* whoſe juriſdiction comprehends all the nine iſlands, but with right of appeal to the *relaçao* at *Porto.* Laſtly, it has alſo two *provedores,* one of whom preſides over the King's revenue, and the other ſuperintends the ſhipping. This city has the privilege of ſending a repreſentative to the *cortes,* or aſſembly of the ſtates. Beſides other fortifi‐ cations, the abovementioned caſtle of *S. Joao Bautiſta* is a good defence to it, having one hundred and ſixty pieces of cannon, moſt of them braſs. It was declared a city in the year 1533.

　　S. Sebaſtiao, a town, half a mile from the ſea, ſituate betwixt ſeveral mountains, is the oldeſt *villa* on the whole iſland. It is defended by ſix forts, and has four places within its Diſtrict.

　　2. The captainſhip of *Praya* conſiſts of one *villa* and ſeveral other places. The *villa* is *Praya,* which ſtands in a level country on a large and ſecure bay. It is walled round, and has four baſtions, with a church, three con‐ vents and one without the walls. It has likewiſe a poor-houſe under excellent regulations, two hoſpitals and a cuſtom-houſe.

　　3. St. *Jorge,* ſituate about eight *Portugueſe* miles from *Terceira,* is eleven miles long and one and a half broad, excluſive of the two capes. To the north it is little more than a high rock, but in the other parts the plains and eminences ſucceed each other. On the ſouth it has a harbour for ſmall veſſels, and is ſaid to have been diſcovered on St. *George's* day in 1450. The three *villas* on it are

　　Vela de Velas, the principal town, though ſmall, having only one church, one convent, and a harbour which has been already mentioned.

　　Villa do Topo is the moſt ancient place on the iſland, but of no conſidera‐ tion, and ſurrounded with a high rock.

　　Vila da Calheta is alſo a mean place.

　　The ſouth ſide of the iſland is well cultivated and inhabited, having, be‐ ſides the above-mentioned three *villas,* four other places; but the north ſide is ſo rocky as to be ſcarce ſuſceptible of tillage, whence, accordingly, there is only one place to be found in it.

　　5. *Gracioſa* lies directly eaſt and weſt, being three *Portugueſe* miles in length, and in its breadth two. The æra of its diſcovery is uncertain, but very probably was ſoon after that of St. *Jorgè.* On this iſland are two *villas,* viz.

　　Santa Cruz, the principal *villa* in the iſland, ſeated on a bay which forms a harbour called *Calheta,* being defended by a fort. It contains one church, one *caſa da miſericordia,* and one convent.

　　Praya lies on a bay which forms its harbour, being defended by a fort. This iſland is ſaid to have been ſo named from its remarkable fertility.

6. *Fayal*

6. *Fayal,* nine *Portuguefe* miles in length, in its greateft breadth three. It is faid, though not with any degree of certainty, to have been difcovered by fome mariners belonging to the ifland of *Terceira,* St *Jorge* or *Graciofa.* The chief place on this ifland is the *Villa de Horta,* fituated on its weftern coaft, and having an harbour landlocked on all fides except to the caft and north-eaft. It is defended alfo by a fort. Near it, to the fouth, lies another called *Portopin,* which is feparated from the main harbour by a narrow cape. The town contains one parifh church and five convents; and is defended by feveral forts, particularly *Santa Cruz.* It has been an earl-dom for fome time. The ifland contains ten other parifhes, and its parti-cular *ouvidor,* from whom the inhabitants may appeal to the *corregedor* of *Terceira.*

7. *Pico,* which is fixteen *Portuguefe* miles long and five broad, is vifible at a great diftance by reafon of its high mountain, to which it owes its name. It is faid to be three miles high. To the north this ifland is feparated from that of St. *Jorge* by a narrow water betwixt two and four *Portuguefe* miles over; and to the weft from *Fayal* by another, which is only a mile and-a-half in breadth. Its principal harbour is at-*Villa das Lagens;* the fecond, which is called *Magdalene,* and fronts *Villa de Horta,* is only for fmall veffels. The æra of the difcovery of this ifland is alfo uncertain. It car-ries on a great trade in its excellent wines: And another commodity be-longing to it is the wood with which it abounds, particularly cedar and *Teixos,* a firm red wood, which is highly prized. This ifland has its particular *ouvidor,* from whom alfo there lies an appeal to the *corregedor* of *Terceira.* Its principal place is the fmall port of *Villa das Lagens,* on the fouth; and the fecond *villa* is *S. Roque,* which is ftill lefs. On this ifland are alfo other well inhabited fubftantial places and parifhes; with a convent.

8. *Flores,* is ten *Portuguefe* miles long and three broad. The difcovery of this place is uncertain; but it has three roads and two *villas.*

Santa Cruz, the principal place is fmall, contains one church and one convent. It gives alfo the title of *C*ount to the eldeft fon of the Marquis of *Gouvea.*

Lagens, contains about three hundred hearths; and is confequently larger than the former.

Befides thefe this ifland contains fome other places.

9. *Corvo* lies northwards oppofite the ifland of *Flores,* from which it is divided by a ftrait of one *Portuguefe* mile in breadth. The whole circuit of this ifland is but three *Portuguefe* miles, having two fmall harbours, one church, and an infignificant place called *Noffa Senhora do Rofario,* which is fubordinate to the church of *Santa Cruz* on the ifland of *Flores.* The whole coaft is a chain of rocks.

THE

THE

KINGDOM

OF

SPAIN.

INTRODUCTION

TO THE

KINGDOM of SPAIN.

§. 1. OMITTING all mention of the maps of *Spain* publiſhed before the ſeventeenth century, the moſt uſeful one was that drawn by *Heſſel Gerard* from the obſervations of *Andreas d'Almada* profeſſor at *Coimbra*. In this map *Sanſo d'Abbeville* made ſeveral alterations, and theſe were the patterns followed by *David Funk*, *Nicholas Fiſcher*, and *Fr. de Witt*. *William de l'Iſle*'s map is little better than a copy of that executed by *Roderick Mendez de Sylva*; and *Matthew Seutter* has imitated him cloſely. *Homann*'s map of *Spain* and *Portugal*, as well that firſt as laſt publiſhed, is the completeſt, eſpecially the latter. There are alſo particular maps, of a very ſmall ſcale, of ſingle Provinces and Dioceſes of *Spain*; ſuch as the *Abbe Vayrac*, has inſerted in his *Etat preſent de l'Eſpagne*, and from his accounts, which are wholly borrowed from *Colmenar*'s *Delices de l'Eſpagne*, *Burgignon d'Anville*, has compiled his *Theatre de la guerre d'Eſpagne*: But his preſent majeſty intending a ſurvey of his Kingdom, and having for that end inſtituted a geographical academy at *Valladolid*, there is reaſon to hope, that we ſhall not long be without a more perfect map of *Spain*.

§. 2. *Spain* was formerly known by the name of *Iberia* and *Heſperia*, which, like that of *Spain*, is ſaid to be derived from three of its ancient Kings; but it has not yet been proved that theſe Kings were ever actually exiſting. The firſt people, perhaps, who came into this country, ſeeing nothing beyond the ocean which environs *Spain* on three ſides, imagined themſelves at the end of the world, and therefore aſſumed the name of *Iberians* from a *Phenician* word of that import; calling the country alſo afterwards *Iberia*, and even giving the name of *Iberus* to the largeſt river they met with. The *Greeks* named this country *Heſperia*, from its weſtern fituation, and in contradiftinction to *Italy*, which went under the fame appellation, called it *Hefperia the fartheft*. The moft ufual name it went by among the *Latins* was *Hifpania*, now altered by the inhabitants to *Efpana*, which *Bochart*, not without probability, derives from the *Phenician* word *Sepan* or *Sepana*, a *rabbit*; this country, according to *Varro, Galen, Ælian, Pliny* and *Strabo*, being fo over-run with thefe animals as to have fuffered greatly by them. There are yet fome coins remaining of the Emperor *Adrian*, on which a rabbit denotes *Spain*, intimating thereby, that from time immemorial this creature has bred peculiarly in that country.

§. 3. The limits of this Kingdom to the eaft are *France*, from which it is feparated by the *Pyrenean* mountains; to the north it is bounded by the *Atlantick* ocean, here called the fea of *Bifcay*; to the weft by the weftern ocean and *Portugal*; and to the fouth by the fame ocean and the *Mediterranean* fea. Its greateft extent from north to fouth is eight degrees, or about a hundred and twenty *Spanifh* miles *; but from caft to weft about a hundred and thirty-two; and in its lower part eighty-four.

§. 4. The air in moft of the provinces is pure and dry, but in *June, July* and *Auguft*, the days are unfufferably hot, efpecially in the middle of the country; in the night a traveller fhivers with cold. Towards the north, and in the mountainous parts the air, as ufual, is cooler than in the fouth, and near the fea contracts a moifture. It feldom rains; and the winter frofts are never fuch as to bind up the ground. Want of temperature in the heat and coolnefs of the air is the reafon that feed lies a long time in the ground before it fhoots up: Sometimes, indeed, a cool breeze, by the *Spaniards* called *Gallego*, iffues from the mountains of *Gallicia*; and this, without great precaution, occafions violent and frequently fatal colds. The country, in general, labours under a great fcarcity of corn, which is chiefly owing to their neglect of agriculture; for though the foil, be in many places extremely dry, and the growth of grain and other fruits very much obftructed in the day-time by the exceffive heats, and in the night by an intenfe cold, yet from hiftorians and geographers *Spain* appears to have formerly enjoyed a preat plenty of corn, fo that the prefent fcarcity of that commodity muft proceed from the neglect of tillage. On the other hand it abounds in the moft delicious fruits, fuch as pears, peaches, olives, figs, grapes, almonds, cheftnuts, lemons, oranges, pomegranates, &c. It produces alfo very good faffron. The *Spanifh* wines, particularly fack, are greedily bought up by foreign nations. The value of the wine and grapes annually exported out of the country about *Malaga* alone amounts to one million and a half of *piaftres*†. Several parts of the country alfo produce rue, and fugar-canes. *Spain* enjoys likewife great plenty of exquifite honey, and filk in

---

* Seventeen and a half to one degree of the equator.

† A *piaftre*, in all the different provinces of *Spain*, is an imaginary coin; in value about three fhillings and feven pence of our money.

abundance;

abundance; but little flax and hemp. Salt, especially towards the sea coast, is so plentifully procured, that considerable quantities thereof are exported: A great deal of sea-salt, in particular, is made in the maritime parts of *Andalusia, Catalonia,* and *Valencia;* likewise on the islands of *Majorca, Yviça,* and *Formentera;* where the sun serves instead of fire. But the principal salt-works are at *Mata* in *Valencia.* The kind of salt procured from the plant *kali,* which grows on the sea-shore, and is called *Soude de barille,* and *de bourdine,* it being used also in the making of soap and glass, is produced in such plenty in *Mercia* and part of *Granada,* that *Alicant* alone has exported in one year 4,111,664 pounds of *soude de barille,* and 770,960 pounds of the *soude de bourdine,* exclusive of a better kind of salt called *Agua azul;* besides no inconsiderable quantities exported from *Almeria, Vera, Quevas, Torre de las Aguilas, Almazarron, Cartagena, Tortosa,* and the island of *Alfacqs.* In this country are also to be seen innumerable flocks of fine sheep, part of which during winter feed in the levels, and in summer are driven up again into the mountains. From these *ovejas merinas* is produced the best wool. Others are kept always in one place; and the third kind make the fat sheep. *Ustariz* computes the number of shepherds in *Spain* at forty thousand. The best wool is that of *Old Castile;* though the *Spanish* wool in general is extremely fine and valuable. *Andalusia* and *Asturia* are particularly famous for their horses. There are also great numbers of mules; but few horned cattle. In *Andalusia* are caught the wild bulls for the bull-fights.

Among the many mountains in *Spain* the *Pyrenees* are the most remarkable. These separate *Spain* from *France,* and from the *Mediterranean* to the *Atlantick* ocean make a length of fifty *Spanish* miles; being in their greatest breadth not less than twenty-seven. They begin at *Vendres,* a sea-port in *Roussillon,* and extend to *Fuentarabia,* but under different appellations from the adjacent towns. Near *Roussillon,* they divide themselves into two branches, of which that dividing the country from *Languedoc* is called *Antipyrenée;* the other, between it and *Catalonia, Col de pertuis.* Exclusive of other mountains hereabouts, which are all branches of the *Pyrenees,* as *Mont Conigou, Sierra de Guara, Col de Prexa, Col de l'Argentière,* and *Port de Vieilla;* between *Gascony* and *Arragon,* lie the mountains of *Jaca* and St. *Christine,* and the famous *Pic de Midi,* which looks like a sugar-loaf standing on a table, and is of a prodigious height. In *Navarre,* betwixt *Pamplona* and St. *Jean de Pie de port,* are the mountains of *Adula* and *Roncevaux.* Over these *Pyrenean* mountains there are scarce five passages out of *Spain* into *France,* and those too narrow; one of them goes from St. *Sebastian* in *Guiposca* to St. *Jean de Luz;* the second from *Maya* in *Navarre* to *Annoa;* the third from *Taraffa* in *Navarre* to *Pie de Port;* the fourth through the county of *Comminges* in *Arragon;* and the fifth leads from *Catalonia* to *Languedoc,* over mount *Salses,* and through *Perpignan*. The very vallies between the mountains are covered with thick and lofty woods. *Sierra d' Occa*, or Mount *Idubeda*, is a chain of mountains extending from the *Pyrenees* to *Tortofa*. At the beginning it forms an arm, traverfing all *Spain*, from eaft to weft, as far as Cape *Finifterre*. To the fouth, below Mount *Cayo*, is another branch, rifing gradually, called *Mont Orofpeda*; and, at fome diftance from its firft acclivity, that is, near the fource of the *Tagus*, takes the name of *Sierra Molina*, which it changes farther fouth for that of *Sierra d' Alcaraz*. Here the chain turns off to the fouth-weft, feparating the Kingdom of *Granada*, and extends to the ftraits of *Gibraltar*, where the celebrated mountain, anciently called Mount *Calpe*, ftands oppofite to thofe of *Ceuta* in *Africa*.

These mountains yield great quantities of timber for fhipping, which is conveyed from *Arragon* and *Navarre* into the *Ebro*, by means of the little rivers *Cinca*, *Saburdan* and *Efca*, and forwarded down from thence into the *Mediterranean*. Pitch and tar are alfo made in *Arragon* and *Catalonia*.

The mountains of *Spain*, according to ancient writers, are very rich in gold and filver; but the *Spaniards* rather chufe that thefe metals fhould be imported from *America* than that they fhould have the trouble of fearching for them in their own country; and thus thefe treafures lie buried in the mountains, but the iron mines are worked here with great fkill and induftry. *Spain* wants not alfo other minerals, fuch as lead, tin, cinnabar, quick-filver, alum, vitriol, copperas, lapis calaminaris, &c. and likewife cryftal, diamonds, amethyfts and other gems. The mineral waters in it are *Lago*, *Orenze*, *Mondragon*, *Baiar*, *Alhama*, *Almagra*, and,

§. 5. The maritime parts, particularly thofe of *Gallicia* and *Andalufia*, abound in fifh, and among others in tunny, fturgeon, lampreys, falmon, haddock, &c. But for want of a proper improvement of thefe advantages, *Spain*, according to the computation of *Uftariz*, annually purchafes ftock and falt fifh from foreigners to the amount of above three millions of piaftres. The rivers in *Spain*, great and fmall, are reckoned at one hundred and fifty: fome of the largeft, as the *Minho*, *Douro*, *Tagus* and *Guadiana* have been already mentioned in *Portugal*, fo that I fhall only fpeak a few words of their feveral fources. The *Minho* rifes in *Gallicia*, near the town of *Caftro del Rey*; the *Douro* in *Old Caftile*, in a part of the mountains of *Idubeda*, called *Sierra de Cogollo*, near the town of *Aguillar del Campo*; the *Tagus* has its fource in *New Caftile*, on the borders of *Arragon*, in a mountain fome miles diftant from the town of *Albarafin*; the *Guadiana* iffues alfo from *New Caftile*, deriving its fource from fome lakes, or fwamps, there called *Las Lagunas de Guadiana*, and lying in a large level country known by the name of *El Campo de Montiel*, near *Canamoux*. At a fmall diftance from its fource it takes its courfe betwixt high mountains, and thus conceals itfelf for near three miles, till it fhows itfelf again in fome fens, termed *the eyes of the Guadiana*; but it is foon again hid amidft reeds and rocks, which

probably

probably gave occafion tò the miftake, of its lofing itfelf under ground ; whence, by the *Latins*, it was called *Anas*, or the *Duck*, though others derive this appellation from the *Arabic* word *Hanafa*, which fignifies to hide one's felf. The other great rivers which water *Spain* are

The *Guadalquivir*, a name derived from the *Moorifh* word *Vadalcabir* ; that is, *a large river*. It was formerly called *Perca*, and more anciently *Bætis* and *Tarteffus*. It receives its fource in *Andalufia*, where feveral little ftreams, iffuing from mount *Segura*, unite in a fmall lake, from whence this river flows. Its ftream is flow, and, from *Corduba* to *Seville*, is paffable only by fmall craft ; but, from the laft city to its mouth, navigable by fhips of burden, though dangerous on account of its many fand-banks. A few miles below *Seville* it forms a fmall lake, and dividing itfelf afterwards into two branches, which form an ifland at fome miles diftance from each other, thus empties itfelf into the bay of *Cadiz*. But the eaftern branch became choked up by degrees, fo that, at prefent, fome fmall veftiges of it only are to be feen.

The *Ebro*, in Latin *Iberus*, takes its fource in the mountains of *Santillana* in *Old Caftile*, on the confines of *Afturia*, from two fprings, the principal of which has given name to the town of *Fuentibro* ; that is, *the fpring of the Ebro*. It receives upwards of thirty brooks and rivulets in its courfe, and becomes navigable near *Tudela* ; but this continues no farther than *Tortofa*. The navigation of it befides is dangerous, on account of the many rocks in it. It difcharges itfelf however, at length, with great rapidity, into the *Mediterranean*, and at its mouth forms the little ifland of *Alfacqs*. The water of it is in fo great repute for drinking and wafhing, that it is fent in cafks to a confiderable diftance. It would anfwer the charge to make this river more navigable from *Navarre* down to its mouth.

§. 6. The Kingdom of *Spain*, according to *Uftariz*, contains about feven millions and a half of inhabitants, though it would fupport twice that number if properly cultivated, if its manufactures were encouraged, and its mines worked. In the times of the *Goths* and the *Moors* it contained between twenty and thirty millions of people, and might yet be very powerful if it had no poffeffions in *America*, but much more fo, were the Kingdom itfelf, and its *American* fettlements, under proper regulations ; whereas, now it is poor and bare of inhabitants. The ufual reafon affigned for this is the expulfion of the *Moors*. It is true, indeed, that when *Ferdinand* the Pious took *Seville* from them, in the year 1248, the feveral Diftricts of this Kingdom contained one hundred thoufand towns and villages, all very populous ; and, when *Ferdinand* the catholick reduced the Kingdom of *Granada*, it confifted of fifty fortified towns, befides an infinite number of leffer places, the greater part of which were afterwards demolifhed. This extirpation of the *Moors* was indifputably one principal caufe of the prefent thinnefs of its inhabitants. But this affected *Andalufia* and *Granada* only, and yet the

other

other provinces are not much more populous than these. That *America*
has ftripped the Kingdom of its inhabitants, *Uſtariz* will not allow, who
fays, that the greateſt part of thofe who went to *America* were of *Bifcay*,
Navarre, *Aſturia*, the mountainous parts of *Burgos* and *Galicia*, and yet
that thefe countries are ſtill the beſt peopled. On the contrary, the
Diſtricts of *Toledo*, *La Mancha*, *Guadalaxara*, *Cuença*, *Segovia*, *Valladolid*,
Salamanca, and other parts in both the *Caſtilles*, are the moſt thinly inhabited in all *Spain*. The caufes he affigns are, the total decay of handicrafts
and manufactures which formerly flouriſhed here, and the heavy taxes by
which they were exhauſted. Another grand caufe of the want of inhabitants, are the monaſteries, by which no lefs than two hundred thoufand
perfons are reſtrained from propagating their fpecies. If a family have
more than one or two fons, who at any rate are to be gentlemen, the relt
muſt be monks ; and though the great number of convents here be one of
the principal occafions of venereal debauchery, yet is the increafe of the fpecies much more hindered than propagated thereby; both fexes being, in their
early years, by thefe means deprived of their ſtrength and health. The way
too of living among the *Spaniards*, in other refpects, particularly in eating and
drinking, contributes likewife to their infecundity. In the ufe of fpices, particularly of pepper, they know no bounds. Their wines alfo are ſtrong and
inflammatory, and yet to thefe, after a meal, they add a very fiery fort of
brandy. Chocolate is a conſtant regale with them, morning and evening ;
and what can be the confequence of fuch a regimen but the drying up of
the nerves and juices ? On the other hand, they are no lefs immoderate
in the ufe of cooling foods and drinks, which are taken alternately with the
above-mentioned; and the conflict of fuch difcordant qualities muſt neceffarily bring on great diforders in the body. To thefe, laſtly, may be added
the copious perfpiration in the heat, the cold at nights, and the frequent
ſhifting of the hot and cold winds, which, exclufive of the irregularities
abovementioned, are fufficient to caufe ſterility. Leannefs here is fo general, that a fleſhy, corpulent man is hardly to be met with; and what is
worfe, there are few or no countries where lofs of fight is fo common.

Small too as the number of inhabitants of whom I have now fpoken is,
yet they are alfo very poor, though they live in a land capable not only of
fupporting them, but many millions more alfo in plenty; and which, befides its natural advantages, has prodigious fums poured into it from *America*.
Savala del Commercio de las Indas computes that, from the year 1492, when
America was difcovered, to 1731, above fix thoufand millions of *pezos* *,
or pieces of eight, in regiſtered gold and filver, have been imported into
Spain, exclufive of far greater fums unregiſtered, befides thofe received by
foreign merchants from the *Spaniſh* dominions in *America*, which are alfo

---

* An imaginary coin, in value about four ſhillings and two pence *Engliſh*.

thought

thought to exceed those imported into *Spain*. But reckoning only the first and registered sums, it appears that, one year with another, *Spain* receives from *America* above twenty-six millions of *pezos*, and yet continues poor and miserable. The money remaining in the country, besides its being of the worst kind, is also very disproportionate in quantity. *Ustriz* computes that all the coined and wrought gold and silver in *Spain*, that belonging to churches and even private persons included, amounts scarcely to one hundred millions of *piaftres*. The two main causes of this poverty are the want of industry in the *Spaniards* for agriculture, handicrafts and manufactures; which occasions the country to be annually drained of so many millions for corn and foreign goods. The second is, the unsatiable avarice of the clergy, who practise a thousand arts to bring the wealth of the country and its inhabitants into their hands. To them not only belong most of the towns and estates, and these, like their persons, are exempt from all publick taxes and imposts, but they also turn both living and dead to their profit; and yet their iniquitous and burdensome inventions, in such fordid pursuits, are implicitly complied with, and even supported by, the laity of all ranks. Their mendicant friars, who have divided all families among themselves, tax every one according to their condition; and, when they knock at the door, there is no refusing them, though they scarce vouchsafe to return thanks for what they receive. All wills are drawn up by them, and that commonly when the testator is near the last gasp; by which means they often impoverish both widows and children, assigning in the will their rightful inheritances to pious uses, as they most unjustly call them. The revenues they extort from this blinded people for the safety of their own souls and those of the deceased, as offerings to the saints or to the Virgin *Mary*, whom they have divided into a thousand virgins, under a thousand different names, as every where present; and, by several other strange kinds of devout pretences, exceeds that of all their immoveable possessions. Few marriages are made without their negociation; and, as by this means they come to be father confessors, they are the despotic lords of the whole family; the cash, their manner of living, their equipage, children and servants are all subject to their control. Having engrossed to themselves the education of youth, they prevail upon those of remarkable capacities, or large fortunes, to embrace the ecclesiastical state; and thus the possessions of the parents are annexed to the church. They make no small gain, besides, of their inns and taverns, as being the safest bawdy-houses in the country. Their commerce, which is free from all duties and payments, and carried on partly by privileges and partly clandestinely, is likewise a rich fund to them, especially if considered with respect to their astonishing failures without paying a shilling by way of composition, and by selling the sanction of their names for the merchandizes of others: an abuse this which the government has in vain strove to suppress.

In

In the *American* settlements, such is their influence over the people, that, should a King of *Spain* attempt to put any check upon their power or covetousness in those parts, or even in *Spain*, nothing less than a general revolt, with the loss of all *America*, would ensue. Lastly, the inquisition and crofs-bull put the finishing hand to the people's wretchednefs. More of this melancholy subject may be met with in a *German* book entitled *A brief hiftorical account of the true caufes of the ruin of the* Gothick *Kingdom in* Spain. S. 101—137.

The commons, or inferior nobility, ftyle themfelves *cavalleros* and *hidalgos*. The latter enjoy no precedence or privileges above the burghers, fome ancient families and knights of orders; but the higher nobility, including counts, marquiffes and dukes; who are ftyled the *titulos* or *titulados*; and particularly the *grandees*, who take precedence next to the King and the Princes of the blood, enjoy very great privileges. In the time of the *Goths*, the chief perfons of the Kingdom were called *magnates, optimates* and *proceres*: but *Aphonfo*, furnamed the Wife, ordering that all royal inftruments, inftead of the *Latin*, fhould be drawn up in the *Caftilian* language, the word *magnates* was tranflated *grandees*. They are divided into three claffes, but look upon themfelves all as on an equality. In point of ceremony, however, when the King nominates a grand, there is this difference, that, if he be of the firft clafs, he kiffes the King's hand and returns his Majefty thanks covered; if of the fecond, he does not cover himfelf till after the compliment; and, if of the third, he does not put on his hat till he has kiffed hands and is returned to his place among the *grandees* who are prefent. The nobility of thefe three claffes tranfmit the title of *Grandezza* to their defcendants; but in a nobleman nominated *grand* only *durante vitâ*, this title is no more than perfonal, and does not defcend to the children: and this, in fome meafure, makes a fourth clafs of *grandees*. The pre-eminence of a *grandee* does not, however, confift only in being covered in the King's prefence, this being common to others who are not *grandees*, as cardinals, nuncios, archbifhops, the ambaffadors of crowned heads, *&c.* but in many other privileges of honour. At an affembly of the ftates, or the nomination of a King and a Prince of the *Afturias*, they take place of all temporal ranks and pofts, the conftable and *amirant* of *Caftille* excepted. They receive the oath of fidelity, like the bifhops, before the King himfelf, and adminifter it to the *titulos de Caftilla*, i. e. *the counts and marquiffes*. Their eldeft fons alfo take the oath of fidelity. They enjoy all the privileges of a duke, though no dukes by title. On the Kings marrying a foreign Princefs, a *grandee* is always appointed for the proxy. At the deceafe of a Prince of the royal blood, he is laid on a bed of ftate and carried to the vault by a number of grandees. When the King takes the air on horfeback they ride next to his perfon, the mafter of horfe only excepted, who, by virtue of his office, rides by the King's fide. A foreign Prince,

on

on his coming to court, is received and conducted back by a *grandee*. They have free accels as far as the King's clofet, and, if they pleafe, even into it. Their wives alfo retain their privileges after the death of a *grandee*, and that too though they marry again. If they are *grandees* only by right of their wives, the title of *grandezza* is continued to them after their death. The King, in writing or fpeaking, ftyles them Prince, or coufin germain ; and, if they are viceroys, the word *illuftre* is added to it. At an audience of the Pope they are permitted to fit, and he gives them the title of *Sennoria*. No *grandee* is to be taken into cuftody for any offence, unlefs by expiefs order of the King ; and it muft be high treafon, or fome other heinous crime againft the ftate, to occafion the obtaining fuch an order. They and their eldeft fons are ftyled *excellency*. They account themfelves alfo equal to the Princes of the empire and *Italy*, and this has occafioned frequent difputes.

§. 7. The *Roman Catholic* worfhip is practifed in *Spain* with the greateft fcrupulofity and pomp. In no country is there more praying and ceremony, and lefs real Chriftianity. *God* and *Chrift* are much lefs known, and more feldom named, than the Virgin *Mary* and other faints. That the latter too is more refpected among the *Spaniards* than God himfelf, appears from their ufual compliment at parting, *Vay ufted con Dios*, i. e. *God be with you*; but it expreffes a much warmer cordiality, to fay *Vay ufted con lay Virgin*, that is, *the Virgin be with you*. The oath *boto a Dios*, i. e. *by God*, paffes for a trifle, but *boto a la Virgin*, i. e. *by the Virgin*, is looked upon as a moft impious expreffion : and thus it is in all countries where reafoning and the holy fcriptures are prohibited, and popery and prieftly power have an unlimited afcendancy. The *Spaniards* are mere flaves to the clergy, (§. 6.) but thefe fo artfully hoodwink them, that they perceive not the chains of their flavery, or if they perceive them bear-them willingly; and when galled by them, dare not fo much as vent a figh after freedom, much lefs ufe any overt endeavours to regain it. Under any difappointment, either of views of avarice or ambition, the clergy have the dreadful inquifition ready at hand, which feizes both on honour and life, fo that perfons of the moft unpolluted innocence account it a particular favour to come off only with the lofs of their fortunes. To be taken up for a *Jew* or a *Mahometan*, and confequently to be ftripped not only of all one has, but to be burnt alive, is fufficient not to have worked on *Friday* or *Saturday*, never to eat pork, and the like, though the informer be only fome menial fervant, moft of whom are fpies to the inquifition, and betrayers of the families in which they live. The court of inquifition was firft introduced, in the year 1478, by King *Ferdinand* the Catholic and his Queen *Ifabella*, at the fuggeftions of *John de Torquemada*, or *Torre cremata*, a Dominican, who was himfelf the firft inquifitor. At *Madrid* it is called the *Confejo de la fuprema e general inquificion*, and confifts of an inquifitor-general and fix counfellors, one of whom is always a *Dominican*, two judges, one fifcal, one alguacil mayor and feveral other officers

Vol. II.   K k   and

and affiftants. The poft of inquifitor-general, which is filled at prefent by the bifhop of *Ternel*, is of great power, dignity and importance. He is named by the King, but confirmed by the Pope ; is the only perfon whom the King' confults in filling up any vacancies of counfellors or inquifitors; and, with the approbation of the counfellors, he appoints the officers of the fubordinate courts of inquifition. The number of the *familiares*, or inferior officers and affiftants of the great council for executing its commands, and who are. difperfed all over *Spain* as fpies and informers, are computed at above twenty thoufand. Under this fupreme court are feveral other leffer courts of inqui-, fition, as thofe at *Seville, Toledo, Granada, Cordova, Cuença, Valladolid, Murcia, Lerida, Logrono, St. Jago, Saragoffa, Valencia, Barcelona* and *Majorca*; and alfo without the Kingdom, as in the *Canary* iflands, *Mexico, Carthagena* and *Lima*; each of which confift of three inquifitors, two fecretaries, one alguacil and other officers. However, before they can arreft a nobleman, a knight of an order, or an ecclefiaftick, the fupreme council is to be confulted ; to which alfo all the courts within the Kingdom fend every month a particular account of the condition of the forfeitures, and every year of all the caufes determined by them, and of the number of prifoners. An account of all the premifes once a year is fufficient from the courts without the Kingdom.

The archbifhops and bifhops in *Spain* are as follow :

1. The archbifhop of *Toledo*, who is primate of *Spain*, chancellor of *Caftile*, and, by virtue of his office, a counfellor of ftate. The prefent arch-bifhop is the Infante and Cardinal *Lewis Anthony James*, having for his co-adjutor in fpiritual affairs the titular archbifhop of *Pharfalia*. His fuffragans are the bifhops of *Cordova, Cuença, Siguenza, Jaen, Segovia, Cartagena, Ofma* and *Valadolid*, with the titular bifhop of the order of St. *James*.

2. The archbifhop of *Seville* who, at prefent, is alfo the abovementioned cardinal, his coadminiftrator for church affairs being the titular archbifhop of *Trajanopoli*. Within his province are the bifhops of *Malaga* and *Cadiz*, with *Ceuta* and the *Canary* iflands.

3. The archbifhop of *St. Jago*, under whom are the bifhops of *Sala-manca, Tui, Avila,. Coria, Placenfia, Aftorga, Zamora, Orenfe, Badajoz, Mondonnedo, Lugo* and the cividal of *Rodrigo*.

4. The archbifhop of *Granada*, who has under him the bifhops of *Gua-dix* and *Almoria*.

5. The archbifhop of *Burgos*, with the bifhops of *Pamplona, Carahorra* and *Palencia* for his fuffragans. ·

6. The archbifhop of *Tarragona*, fubordinate to whom are the bifhops of *Barcelona, Gerona, Lerida, Tortofa, Vique, Urgel* and *Solfona*.

7. The archbifhop of *Zaragofa*, having under him the bifhops of *Uefca, Barbaftro. Xaca, Tarazona, Albarrafin* and *Teruel*.

8. The archbifhop of *Valencia*, whofe fuffragans are the bifhops of *Se-gorve, Orihuela* and *Majorca*.

The bishops of *Leon* and *Oviedo* are immediately dependent on the Pope. The annual joint income of these several archbishoprics and bishoprics amounts to one million, three hundred and sixty-three thousand ducats; and that of the chapters is at least as much.

There are seven archbishops and thirty-one bishops also in *America.*

Amidst the great decrease of the inhabitants in *Spain* the body of the clergy has suffered no diminution, but has rather been gradually increasing; insomuch that *Uſtariz* computes the number of ecclesiastics and their servants at two hundred and fifty thousand. The Jesuits have great influence at court, and the King's own confessor is always one of that intriguing order.

The King nominates all archbishops and bishops, who are afterwards confirmed by the Pope. In the year 1753 an agreement was entered into between the Pope and King, wherein the former ceded to the latter the nomination to all small benefices; which has not only considerably strengthened the King's power over the clergy, but also keeps those vast sums of money in the country which used to be expended in journies to Rome for the soliciting of benefices. The King can also, as circumstances require, tax the ecclesiastical possessions according to his pleasure; which important article he owes to the address of the present Marquis *d' Enſenada.* The power of the Pope and his nuncio is notwithstanding very extensive here, though no bull can be published without a written permission from the King.

§. 8. The *Spaniards* are far from being wanting in disposition or capacity for the sciences; yet little progress is to be expected from them while they are debarred the use of their natural talents. The clergy being not very learned themselves, it is a point of policy with them to suppress all scientifical accomplishments among the laity; and, in order to keep them in ignorance and subjection, brand all literary researches with the name of heresy. This is so inveterate an evil that had it not been for the astronomy, philosophy, mathematicks, medicine, and other sciences introduced in the middle ages by the *Arabians* or *Moors* into the universities of *Cordova*, *Toledo*, and *Salamanca*, *Spain* and all *Europe* would have continued in the greatest ignorance. But the children of darkness among the ecclesiasticks, seeing these emanations of light, began to be apprehensive for their own power, and every where inveighed against the seats of the muses as the schools of hell, where the devil taught sorcery; tho' it is to these very *Arabian* doctors in *Spain* that the sciences which in *Europe* had been fettered by popery, owe the recovery of their life and vigour. No sooner, therefore, had the Christians gained the superiority in *Spain* than the clergy set about suppressing literature, in which they were attended with too great success. Even cardinal *Ximenes* himself, who would fain pass for a pattern of learning, at the taking of *Granada* most injuriously ordered five thousand *Arabick* manuscripts to be burnt, having, under pretence of securing them in a library, induced the learned among

the

the *Moors* and *Arabians* to collect them all over the country and commit them to his care. The modern clergy too are indefatigable in profecuting this deteftable fcheme of their predeceffors ; and though *Spain* contains no lefs than twenty-two univerfities, and feveral academies, among which is one at *Valadolid* for geography, yet are they fo conftituted and under fuch reftrictions that they can never arrive to any figure in literature. The bookfellers in *Spain* dare not keep a valuable book in their fhops, on any remarkable and intereſting fubject. The inquifition are too vigilant in fup-preffing them, and in concealing from the publick whatever may tend to open their eyes. They have pretences always ready for feizing foreign books, though they bear no manner of relation to religion. It is to be obferved alfo here, that moft of the books publifhed in the *Spanifh* lan-guage are printed out of *Spain*, few printing-houfes being to be feen there; and by far the greateft part of their paper is imported from *Genoa*.

§. 9. In this country there is a want even of the moft neceffary trades, and of the few to be met with here, the greateft part are in the hands of the *French*, who are very numerous in *Spain*, the natives themfelves, befides their averfion to work, difdaining to ftoop to handicrafts. They are not, however, wholly without manufactures, efpecially of filk and wool; but thefe fall far fhort of that flourifhing condition they might be brought to. Tradefmen and merchants are but little looked upon in *Spain*; for which reafon as foon as they have amaffed a competent fortune they leave off trade, procure a title, and fet up for perfons of quality. Some manufac-tures, after being fet up at a very great expence, have been oppofed and ruined by the inquifition and monks : Of this the fine manufactures of cloth, and other ftuffs at *Guadalaxara*, are a fhameful inftance. The great duties likewife on *Spanifh* ftuffs render them dearer than the foreign. Thus the *Spaniards* part not only with the products of their own country, but alfo with the treafures of *America* to foreigners; fupplying them with bread and other goods neceffary for conveniency and fplendor. According to *Uftariz*, the gold and filver exported one year with another amounts to fifteen millions of *piaftres*. The publick papers indeed inform us, that of late the *Spanifh* manufactures are in a very thriving way, and that they make very fine cloths, befides gold and filk ftuffs. But it is to be appre-hended that this profperity will not be lafting, or at leaft that it will be a long time before they are able to fupply the wants of their own country, fo as to exclude foreign manufactures.

Spain is extremely well fituated for trade and navigation. It might alfo be its own carrier in fupplying other *European* nations with its pro-ducts, the demand for which is fo general, and by thefe means enrich itfelf. But this vaft advantage they neglect, and leave it to other maritime nations, who turn it to a very good account. The *Spaniards* indeed deny them all accefs to their poffeffions in *America*; and are fo jealous of having
that

,that trade preferved wholly to themfelves, that no foreign fhips muft even approach their coafts: Yet of this commerce carried on in their own fhips they have the leaft profit, being little more than factors for the *French*, *Englifh*, *Dutch*, and *Italians*, who fend their goods to *America* by them, and have the greateft fhare in the returns of gold, filver, and other commodities; the *Spaniard*, under whofe name this trade is carried on, receiving a gratuity, or acting merely from friendfhip. Smuggling, which formerly was at a great height, and not a little encouraged by the clergy, has of late been in a great meafure fuppieffed by means of fome terrifying edicts. The clandeftine exportation of gold, to which no prohibition could put a ftop, is alfo greatly decreafed fince the year 1750, when the King permitted an exportation of filver on paying three *per cent*. The trade to *America* was formerly carried on by the flota and galleons. The flota, or plate-fleet confift of a certain number of fhips, fome belonging to the King and others to merchants: Thefe ufed to fet fail from *Cadiz* to *Mexico* about *Auguft*, unlading at *Vera Cruz*, and returning to *Spain* in eighteen or nineteen months. The *Avifos*, or advice boats, are light veffels employed to give notice of the condition and approach of the flota or galleons. The latter were two men of war called *Capitana* and *Almiranta*, which ferved as convoy for eight or twelve fhips bound to *Porto Bello*, and ufually putting to fea from *Cadiz* every *March* or *April*. Their firft port was *Carthagena*, from thence they failed to *Porto Bello*, and came again to *Carthagena*, from whence, by the way of the *Havannah*, they returned to *Spain*. But fince the years 1735 and 1737, the fleets and galleons have been difcontinued, and the trade to *America* carried on in regifter fhips, which any merchant may fend on permiffion obtained from the council of the *Indies*. Thefe fail from *Cadiz* directly to *Lima*, the harbour of which place is called *Callao*, *Buenos Ayres*, *Maracaibo*, *Cartagena*, *Honduras*, *Campeche* and *Vera Cruz*. But by thefe feveral exports thofe countries have been fo overftocked with commodities of all kinds, that fince the laft war the profits have been but fmall. It was refolved, therefore, in 1754, that in 1756 the navigation of the fleet and galleons fhould be reftored to its former footing. The affogue fhips are two veffels belonging to the King for bringing quick-filver on the King's account to *Vera Cruz*. In 1728, an exclufive charter was granted to a company for trading to the *Caraccas*, a permiffion to the inhabitants of the *Canary* iflands only excepted, who were therein to fend thither yearly one regifter-fhip, whofe cargo was to confift wholly of the product of thofe iflands. In the year 1756, another company was erected for trading to *Hifpaniola* and *Porto Rico*, and fending annually ten regifter-fhips to the bay of *Honduras*, and the ports of the province of *Guatimala*: Their ftock amounts to a million of *Spanifh patacons*, which are divided into fhares, and it has a houfe at *Madrid*, *Barcelona*, and *Cadiz*. The *Spaniards* carry on a very confiderable trade from *America* to

the

the *East-Indies*, especially to the coasts on the south-sea, betwixt the town of *Manila*, the island of *Lucon*, and the harbour of *Acapulco* on the coast of *Mexico*; but from this trade the Jesuits reap the greatest advantage. It is generally contained in one ship, which sails every year about *July* from *Manila* and reaches *Acapulco* in *December*, or about the beginning of the following year, and returns again from thence before the first of *April*. Its cargo, which belongs to the convents at *Manila*, and particularly to the Jesuits, consists of spices, *Chinese* silks, *Indian* stuffs, silk stockings, fine and coarse cottons, wrought gold and silver, and other smaller articles; but the return from *Acapulco* is for the most part made in money, and generally to the amount of two or three millions of rix-dollars. Sometimes, though seldom, two ships are sent.

. At *Cadiz*, *Seville*, St. *Lucar*, and port St. *Mary's*, books and accounts are kept in reals and *maravadis de plata*, *i. e.* of silver, which are at present no more than an imaginary coin; each real containing sixteen quartos or thirty-four maravadis, in *English* money about 5 *d.* ⅛. A pistole or doblon, which is an actual coin, contains forty of the above reals, and thirty-seven reals eleven quartos of actual *real de plata*. A true *real de plata* contains seventeen quartos or thirty-six maravadis, or also two *reals de vellon*, *i. e.* of copper, each real of copper containing eight and a half *quartos de vellon*. There are also *American reals de plata*, each of which contains twenty-one quartos one maravade, or two *reals de vellon* and a half; eight such reals make a piastre in actual coin, which is therefore called *peso d'otto reali*, or a piece of eight; or ten *reals de plata* of seventeen quartos ten reals. Of the above-mentioned imaginary piece of sixteen quartos and twenty *reals de vellon*. The ideal peso has eight *reals de plata* of sixteen quortos or fifteen reals, and two *maravadis de vellon*; whence it appears that the real peso's differ from the imaginary not less than 32 $\frac{13}{16}$ *per cent*. The imaginary pistole or doblon used in exchange makes thirty-two *reals de plata* of sixteen quartos. The imaginary *ducat de plata* is equal to three hundred and seventy-four silver maravadis, or eleven *reals de plata* of sixteen quartos. But in trade and exchange one silver maravade more. The King's accounts are transacted in *escudos de vellon*, *reali de vellon*, and *maravadis de vellon*; which are all imaginary coins. An *escudo de vellon* is ten *reali de vellon*, and thirty-four maravadis make one *real de vellon*, fifteen *reali de vellon* being equal to one piece of eight.

§. 10. The southern coasts of *Spain* were anciently frequented by the *Phenicians* out of commercial motives, which evidences that this land must at that time have been well peopled and cultivated. After them the *Carthaginians* came in a hostile manner and reduced the country; but these were in their turn dispossessed of it by the *Romans*. Towards the beginning of the fifth century it was over-run by the *Swabians*, *Alans* and *Vandals*; but these were soon subdued by the *Visigoths*, who entered *Spain* under the
command

command of their King *Atolphus*, or *Adolph*. *Witifa*, one of their Kings, dying in the year 711, the Kingdom was divided into factions, and the public revenues greatly diminifhed by the vaft wealth which the bifhops and clergy had accumulated. Such was the ftate of the nation when King *Roderic* afcended the throne; and to him is generally attributed the overthrow of the *Gothic* Kingdom in *Spain*, though his injury to the lady, or daughter of Count *Julian*, who is faid, out of refentment, to have inftigated the *Moors* to invade *Spain*, has not been duly proved. It is certain that *Spain* was delivered up to the *Moors* by the craft and treachery of Count *Julian* and *Oppa* archbifhop of *Seville*, who not only fpirited up the people clandeftinely to a revolt, but at the bloody battle of *Xeres* even deferted to the *Moors*; which turned the fcale againft the *Vifigoths*: and thus the *Moors*, who were alfo called *Arabians* and *Saracens*, became mafters of the Kingdom. This revolution happened in the year 714. *Pelayo*, a Prince of the *Vifigoths*, with a great number of the *Gothic* nobility, his followers, withdrew into *Gallicia*, *Bifcay*, and the mountains of *Afturia*. Other bodies of the fame nation difperfed themfelves in *Navarre*, *Arragon*, and the *Pyrenean* mountains; and this occafioned the large *Gothick* empire to confift of fuch a number of petty ftates: For *Pelayo* having in the year 716 obtained a victory over the *Moors*, the remaining *Goths* broke out into irreconcileable animofities, and feparately laid the foundations of the Kingdoms of *Leon*, *Navarre*, *Arragon* and *Sobrarbien*, and of the counties of *Caftile*, *Barcelona*, &c. Thefe fmall ftates and the *Moors* were continually at war, and agreed little better among themfelves, infomuch that their frequent alliances and feuds, render the hiftory of that people very intricate. The Kingdoms of *Caftile* and *Arragon* raifed themfelves above the reft. They were thrice united by marriage, but foon feparated again, till in 1472, a fourth and perpetual union took place by means of the marriage of *Ferdinand*, hereditary prince of *Arragon* with *Ifabella* heirefs of *Caftile*, who, in 1473, thus became King and Queen of *Caftile*; and on the demife of *Ferdinand*'s father in 1479, of *Arragon* alfo. To *Caftile*, at that time, belonged both the *Caftiles* and *Eftremadura*, together with *Andalufia*, *Murcia*, *Leon*, and the *Afturias*, *Navarre* and *Bifcay*, *Gipufcoa*, *Alava* and *Rioja*, as alfo *Galicia*. *Arragon* included *Arragon*, *Catalonia*, *Roufzillon*, *Valencia*, and the iflands of *Majorca*, *Minorca*, and *Yvica*. In the laft mentioned year *Ferdinand* erected the court of inquifition (§. 7.) and in 1491, by the conqueft of the city of *Granada*, put an end to the dominion of the *Moors* in *Spain*, which zeal was rewarded by the Pope with the title of Catholick King. His confort enabled the famous *Chriftopher Columbus* to go on his difcovery of the fourth and hitherto unknown divifion of the earth, which happened in 1492. The King, in 1494, annexed to the crown the three grand-mafterfhips of the orders of St. *James*, *Alcantara*, and *Calatrava*. In 1504, he acquired by ftratagem the whole Kingdom

of

of *Naples*; in 1509, conquered *Oran* on the coaft of *Africa*; and in 1512, made himfelf mafter of the Kingdom of *Navarre*, infomuch that at this time the feveral ftates of *Spain* formed but one body. In 1496, a marriage was concluded between *Philip* of *Auftria* and *Johanna* daughter to *Ferdinand*, the confequence of which was the uniting of the *Auftrian* dominions with the *Spanifh*. *Charles* I. or V. grandfon to *Ferdinand*, by this match in 1520, became both King of *Spain* and Emperor of *Germany*; but in 1556 refigned the empire in favour of his brother *Ferdinand*. His fon and fucceffor to the crown of *Spain* was *Philip* II. who alfo poffeffed *Milan*, with the feventeen provinces of the *Netherlands* and the county of *Burgundy*, and in 1581 fubdued *Portugal*. But the *Netherlands* afterwards revolted, and his prodigious fleet, or *invincible armada*, as it was ftyled, which he fent againft *Elizabeth* Queen of *England*, was in a great meafure deftroyed by ftorms and the *Englifh*. *Philip* III. weakened his country by expelling the *Moors*. Under *Philip* IV. *Portugal* fhook off the *Spanifh* yoke, and feveral other countries revolted. In the year 1648 he was obliged to acknowledge the united provinces of the *Netherlands* a free independent ftate : In 1659 he loft alfo the county of *Roufzillon*. *Charles* II. was difpoffeffed of a great part of what remained to him in other parts of the *Netherlands*, as likewife of *Franche Comté*; and by his death, in 1700, his family became extinct. *Philip* duke of *Anjou*, grandfon to *Louis* XIV. of *France*, afcended the throne of *Spain* by virtue of *Charles*'s will. But this occafioned thirteen years war betwixt *France* and the houfe of *Auftria*, till by the peace of *Utrecht* *Philip*'s poffeffion was confirmed by ceding *Gibraltar* and *Minorca* to *England*, and *Sicily* to the Duke of *Savoy*; the Emperor *Charles* VI. being able only to obtain the *Netherlands*, and certain dominions in *Italy*. In 1717, *Philip* took *Sicily* and *Sardinia* from the Emperor; but in 1720 acceded to the quadruple alliance, renouncing all right or claim to *France*, the *Spanifh Netherlands*, *Italy*, *Sicily* and *Sardinia*, which was confirmed in 1725, by the peace of *Vienna*, in which *Charles* VI. relinquifhed all pretenfions to the crown of *Spain*. In 1733, he again broke with the Emperor, depriving him of the two Kingdoms of *Naples* and *Sicily*, and procuring his fon Don *Carlos* to be invefted with them. In 1739, a war broke out betwixt *England* and *Spain*. King *Ferdinand* VI. at the treaty of *Aix la Chapelle* found means to procure for his half brother, Don *Philip*, the three Dutchies of *Parma*, *Placenza*, and *Guaftalla*.

§. 11. The King's title at large runs thus : *Ferdinand by the Grace of God, King of* Caftile, Leon, Arragon, *the two* Sicilies, Jerufalem, Navarre, Granada, Toledo, Valenzia, Galicia, Majorca, Seville, Cerdena, Cordova, Corfica, Murcia, Jaën, *the* Algafves *of* Algezira, Gibraltar, *the* Canary Iflands, *the* Eaft *and* Weft-Indies, *the iflands and continent of the Ocean, Arch-Duke of* Auftria, *Duke of* Burgundy, Brabant *and* Milan, *of*

2 Habfburg,

Habſburg, Flanders, Tirol *and* Barcelona, *Lord of* Biſcay *and* Molina, &c.
The ſhort title is *Rey Catholico de Eſpanna.* This title of Catholick King,
though before aſſumed by ſome Kings of *Spain* and other ſovereigns, was
in 1590, ſolemnly conferred on *Ferdinand* V. by Pope *Alexander* VI. Since
the year 1308, the hereditary Prince is ſtyled prince of the *Aſturias*, but becomes ſo always by creation. The other royal children are called *Infantes.*
The moſt uſual arms of *Spain* are a ſhield divided into four quarters, of
which the uppermoſt on the right-hand, and the loweſt on the left, contain a caſtle *Or*, with three towers for *Caſtile:* And in the uppermoſt on
the left, and the loweſt on the right, are three lions *Gules*, for *Leon;* with
three lilies in the centre for *Anjou.*

§. 12. The principal order of knighthood in *Spain* is that of the golden
fleece, which unqueſtionably received its origin from the *Burgundian* dominions, being inſtituted at *Bruges* in *Flanders* on the tenth of *January*
1430, by Duke *Philip the Good*, on the celebration of his marriage with
Iſabella of *Portugal.* The Duke at firſt alluded to the fable of *Jaſon's*
golden fleece; but afterwards, by advice of *Johannes Germanus*, chancellor
of the order, it was applied to the hiſtory of *Gideon.* At the ſecond feſtival
of the order held at *Liſle* on the 30th of *November* 1431, the ſtatutes
written in the *Burgundian* or old *French* were promulgated: In 1433, the
order was confirmed by Pope *Eugenius* IV. and in 1516, *Leo* X. ſhewed it
the ſame favour. By the marriage of the Arch-duke, afterwards Emperor
Maximilian I. with *Mary of Burgundy*, the hereditary dominions of *Burgundy*, and likewiſe this order, eſcheated to the houſe of *Auſtria.* And
though by a peace concluded in 1439, with *Charles* VIII. of *France*, he loſt
the dutchy of *Burgundy*, yet he retained the greateſt part of the *Burgundian*
hereditary eſtates, both for himſelf and afterwards for his ſon *Philip* King
of *Spain*, together with the grand-maſterſhip of this order, and a right to
this dutchy with the title belonging to it. And thus the Kings of *Spain*
have acquired this order and the ſovereignty from the houſe of *Auſtria* together with the territories of *Burgundy.* *Philip* II. a little before his deceaſe in 1598, reſigned the *Netherlands* to his daughter conſort to Duke
Albert of *Auſtria*, but with a clauſe excepting this order of knighthood, the
grand-maſterſhip of which was to continue to him and his ſucceſſors on
the *Spaniſh* throne; though this at the ſame time evidences that the ſovereignty of it is to be accounted an appendix to the territories of the *Netherlands* and *Burgundy*, whence accordingly when the conteſts and wars
concerning the ſucceſſion to the crown of *Spain* and its territories formerly
belonging to the *Netherlands* and *Burgundy* broke out in the beginning of
this century, both *Charles* III. (VI.) and *Philip* V. aſſumed the ſovereignty
of it. The Emperor, though he too at the peace of *Vienna*, in 1725, renounced the *Spaniſh* throne, yet he retained notwithſtanding the *Netherlands*,
conſtantly beſtowed this order, and his heireſs *Maria Thereſia* has conferred

it on her bufband *Francis* the firft ; whence it is now both in the hands of the Kings of *Spain* and the houfe of *Auftria*. The collar of the order confifts of an alternate range of golden flints and fteels, with fparks of fire on all fides, and a golden fleece pendent from it with this motto, *Autre n'aurai.* *Charles* the fifth permitted the knights, inftead of the collar, to wear only a fcarlet ribbon.

The three orders properly *Spanifh*, and which have a revenue annexed to them, are,

2. The order of *St. Jago di Compoftella*, inftituted in the year 1175, by *Ferdinand* the fecond, King of *Leon*, and has for its badge a red uniform crofs divided into twelve departments ; namely, thofe of *Ocanna*, *Merida*, *Villa nueva de los Infantes, Lerena, Xeres, Caravaca, Velez, Montanches, Segura de Leon, Hornachos, Segura de la Fierra* and *Caftilla vieja*. Thefe are fome-times governed by knights, fometimes by alcade mayors, who ufed to pay to the King an annual acknowledgment of three hundred and fixty-eight lances; but this is now altered to a fum of money. They include above one hundred and eighty parifhes, moft of which are fmall towns, and above eighty-four commanderies, whofe joint income amounts to two hundred and thirty thou-fand ducats *per annum*. This order has four convents who are brothers of the order, feven religious ones, one college, five hofpitals and fix hermitages.

3. The order of *Calatrava*, inftituted by *Sancho* the third of *Caftile*, has for its badge a red crofs divided into five departments ; namely, thofe of *Almagro* and *Campo de Calatravo*, *Martos*, *Almonaci de Zorita*, *Almodavar del Campo* and *Almaden*. The two firft are governed by knights, but the three others by alcade mayors. They contain above feventy-four parifhes, fifty-four commanderies, whofe annual income amounts to one hundred and ten thoufand ducats, and fix bailywicks. The annual acknowledgment of three hundred lances paid to the King by thefe commanderies and bailywicks, is here alfo changed to a fum of money. To this order belongs one convent of religious and one college.

4. The order of *Alcantara*, the badge of which is a lily placed crofswife, owes its inftitution to *Ferdinand* the fecond, King of *Leon*, and was at firft called *St. Julian de Pereyro*. This order is divided alfo into five de-partments ; namely, thofe of *Alcantara*, *Villa Nueva de la Serena*, *las Porofas*, *Valencia de Alcantara* and *Sierra de Gata*. This two firft are go-verned by knights, and the three others only by alcade mayors. They con-tain fifty parifhes and thirty-eight commanderies, which bring in two hun-dred thoufand ducats. Inftead of an acknowledgment of one hundred and thirty-eight lances, the commanderies of late pay money. To this order belong three convents of religious and one college.

The grand mafterfhip of thefe religious orders King *Ferdinand* the *C*atholic annexed, in the year 1494, to the crown, forming the *confejo real*

de

de las ordenes, a tribunal confifting of one prefident, fix counfellors, one fifcal, one fecretary, one contador-general, one alguacil mayor and one theforero.

5. The order of *Montefa* is of fmall confideration, being compofed only of nineteen commanderies, which none but natives of *Aragon* and *Valentia* can enjoy.

Spain, from the *Gothick* times to that of *Pelagius*, was an elective Kingdom; and, for two centuries after him, the throne was filled by the fuffrages of the ftates, who departed, however, in no inftance, from the royal family; but from *Ramir* I. to *Alphonfo* V. all the concern the ftates had in the creation of a new King, was to acknowledge him as a lawful and worthy fucceffor; and fince the latter, there appears not the leaft trace of an election, the crown always of courfe, without any form or ceremony, devolving to the neareft in blood. The Kings of *Spain* have fometimes limited the fucceffion to certain families, ranks and perfons; of which the firft inftance was *Philip* III. in the year 1619, and the fecond *Philip* V. in 1713. Females here are alfo capable of the crown, on failure only of the male line. In cafe of a total extinction of the royal family, it is an uncontroverted perfuafion that the right of electing a King would revert to the people. If the next heir be incapable of government, and efpecially when, on the deceafe of the former King, affairs are in confufion, the ftates are empowered to choofe five perfons to take the adminiftration upon them, among whom the Queen-mother is to hold the chief place.

On the inauguration of a new monarch, he is proclaimed in the church of the *Jeromites* at *Buen Retiro*, and receives homage from the ftates; but the unction and coronation has been difufed for fome centuries. The King's power is unlimited and the ftates of the Kingdom, confift of the clergy, the ancient nobility and the deputies of the towns. Since the beginning of the feventeenth century, or the time of *Philip* the third, the *Cortes*, or Dyets, have been difcontinued, no other affembly, fay the *Spanifh* writers, than conventions of the deputies, or agents of the towns, having been thought neceffary; in which are fettled the neceffary taxes and imports, without any infringement on the privileges of the ecclefiaftics and nobility, whom unqueftionably the Kings fummoned to thofe affemblies when the public good required it.

§. 14. The moft weighty affairs of ftate are difcuffed in the council of ftate, which confifts of a *decano*, or prefident, three other counfellors, a fecretary of ftate, befides the two *fecretarios de Eftado* and *del defpacho univerfal*, one of whom is *fecretario de eftadoy del defpacho de guerra, marina, Indies y hazienda, fuperintendente general del cobro y diftribucion de ella*; and the fecond is ftyled *fecretario de eftado y del defpacho de gracia y jufticia*.

The fupreme royal council *(confejo real y fuprema de fu mageftad)* or royal council of *Caftile*, is the higheft court of judicature, and divided into five inferior courts or chambers. The *fala primera de govierno* confifts of one prefident, feven counfellors, and two fifcals. The *fala fegunda de govierno* confifts of four counfellors. The *fala de mil y quinientas,*-i.e. *the chamber of the one thoufand five hundred*, confifts of five counfellors: the *fala de jufticia* of four ; and the *fala de provincia* alfo of four. This laft chamber has likewife its governor, two judges for the *compedencias*, two judges for the *commifziones*, and a judge for the *miniftros*. The provinces are diftributed among the feven counfellors of the *fala primera de govierno*.

To the chamber of *Caftile* belong a fifcal, a *fecretario del real patronato de Caftilla*, who tranfmits his report immediately to the King, and alfo receives orders immediately from him, a *fecretario de gracia, y jufticia, y eftado de Caftilla,* and a *fecretario de gracia y jufticia y real patronato* for *Aragon*.

The chamber of the *Alcaldes*, of the court and houfhold, which may alfo be claffed with the great councils, confifts of a *governador de la fala*, twelve judges and a fifcal.

The fupreme council of war *(confejo fupremo de guerra)* is compofed at prefent of four counfellors, one of whom is fecretary, one fifcal and three affeffors, who are members of the royal council of *Caftile*.

The fupreme royal council of the *Indies (confejo real y fupremo de Indias)* confifts of a *governador*, twenty-two counfellors, four fecretaries, two for *Peru* and two for *New Spain*, one accountant-general and other officers. This tribunal decides, without appeal, in matters belonging to the *Spanifh* feas and poffeffions in *America*.

The *confejo real de hazienda*, or council of the finances, is divided into four chambers. The *fala de govierno*, which fuperintends the revenues of the crown, confifts of one governor, twenty-two counfellors, one fifcal and one fecretary. The *fala de millones*, whofe branch is the excife, confifts of eleven counfellors, one fifcal and one fecretary. The *fala de jufticia* confifts of fix counfellors and one fifcal. The tribunal *de la contaduria mayor*, of fixteen members, of which two are fifcals. Befides thefe are alfo the *commiffaria* and *direccion de cruzada, real jonta de obras y bofques,* i. e. the *college of the royal buildings and forefts*; the *real junta general de comercio, moneda, de pendencias de eftrangeros y minas* ; the *real junta de facultades de vindedades*; the *real junta apoftolica* ; the *real junta del tabaco* ; the *real junta de abaftos*; the *real junta de la unica contribucion*, and the tribunal *del real pro-tho-medicato*.

§. 15. The principal tribunals in the provinces of the Kingdom are, the royal chancery of *Valladolid*; the royal chancery of *Granada* ; the royal council of *Navarre* ; the exchequer there ; the royal audience of *Corunna* in *Galicia*; thofe of *Seville*, with *Oviedo* and the *Indian* contracts *(audiencia real de la contratacion a las Indias)* of *Cadiz, Aragon, Valencia, Catalonia, Majorca* and the *Canary* iflands. Thefe courts have a prefident, affeffors, criminal judges *(alcaldes del crimen)* judges of gentlemen *(alcaldes de hijos-d' algo)* fiſcals, *&c.* The royal council at *Navarre* confiſts of regents, counſellors, fiſcals and alcaldes. The *audienzas* have a regent, alcaldes and other officers, whom I ſhall ſpecify in their proper place. Juſtice is adminiſtered in cities, and the police maintained there by corregedores, or regidores only and alcaldes. Towns and villages have alcaldes and bailies, who are ſubordinate to the judges of cities. A corregedor, or alcalde mayor, is never a native of the place of which he is magiſtrate. The title of *virrey*, or viceroy, appertains only to the governor of *Navarre*, the governor of a province being ſtyled *capitan-general*, and of a town *governador*. Theſe great officers take cognizance of all matters relating both to the police, war and juſtice. A viceroy and captain-general have the ſame power and authority. They appear with the ceremonial, and enjoy almoſt the ſame prerogatives as the King, the very governors of the towns being dependent on them. Beſides the royal edicts and laws enacted by the general aſſembly of the ſtates at *Toro*, as the fundamental and moſt important laws, juſticiary matters are decided in *Spain* by the old law books called *Fora* and *Fuero Juzgo*, *la partita* and the *Roman* law.

§. 16. The King's income ariſes principally from the *alcavala*, or tenth of every thing ſold; to which may be added the *cientos*, i. e. *the tythes and fourths of a hundred*; the exciſe on wine, oil, tallow, ſoap, paper, ſaltfiſh, *&c. (los millones)* the uſual aid of four hundred and forty-one thouſand, one hundred and ſeventy-ſix crowns, paid by all under nobility; the wine-gage money, which imports go under the name of provincial revenues; the ſtamp duties and the half annates. With reſpect to the penſions, their neat produce is inconſiderable: the duties on proviſions, which is fifteen *per cent.* thoſe on ſalt, tobacco, the poſt-office, the regulations of the crown of *Aragon*, and the croſs-bull, by virtue of which the clergy and laity are to pay a contribution towards carrying on a war with the infidels even when no ſuch thing is in agitation; indulgences; licences for eating butter, cheeſe, milk and eggs in *Lent*; the ſubſidies and tythes of church and abbey lands; the fourage of the orders of knighthood; the monies paid by thoſe orders in lieu of the lances and gallies which they were bound to furniſh; the grand-maſterſhips; the priorate of *St. Jago*; the remounting of the horſe belonging to the order; the taxes on downs, commons and other paſtures; the *Madrid* exciſe; the thirds, tenths and patrimonial rents of *Catalonia*, *Aragon*, *Valencia* and *Majorca*; the eccleſiaſtical payments for the military hoſpitals; the exciſe at *Navarre*; the quickſilver and other *American* revenues; the Weſt *India* trade; the coinage, *&c.* All theſe revenues have undergone great alterations. *Vayrac* gives an extract from the ſtate of the *Contaduria* mayor (§. 14.) which *Linſchot* and *Salazar* have tranſcribed; and, according to this eſtimate, they amounted, in the laſt century, to about thirty-two or thirty-three millions; but towards the cloſe clofe of it, they were fo reduced by bad œconomy, that, at the demife of *Charles* the fecond, they did not exceed feven or eight millions of livres. *Philip* the fifth made ufe of Prefident *Orry*, a *Frenchman*, for the improvement of the revenues, and by his dexterity they were again put on fo good a footing that the royal income, at prefent, amounts to forty-two millions of *efcudos de vellon*.

§. 17. The *Spanifh* land forces, at prefent, confift of ninety-fix thoufand five hundred and ninety-feven men. The Kingdom is alfo well defended on all fides: Towards *France* it has a fecure fence in the *Pyrenean* mountains; and the fea-coafts, befides forty-five towns, are lined with redoubts, forts and towers; and up the country the army of an enemy would be put to very great inconveniences and ftreights, and particularly the horfe, for want of forage.

§. 18. The ftrength of *Spain* confifts much more in a naval force than an army. For the protection of its *American* trade, clears the coafts of pirates, and ferves, on occafion, for the tranfportation of troops. It is alfo very well provided with almoft all kinds of naval ftores. *Aragon, Navarre, Catalonia* and the north coaft afford good timber; *Bifcay* and other parts abound in iron. At *Lierganes* and *Cerada*, not far from the fea towards *St. Andero*, are founderies for cannon and anchors; and for bombs, grenadoes and all kinds of bullets at *Fugui, Azura* and *Iturbiera*. Several places alfo make powder; and *Placencia*, in *Guipuzcoa* and *Valencia*, are famous for all forts of arms; as *Puerto Real*, which is not far from *Cadiz*, is for excellent cordage. *Cada* in *Galizia* makes both cordage and canvas, being fupplied with hemp from *Galizia, Granada, Murcia* and *Valencia*, but that not in a fufficiency for the demand; whence foreigners ftill have the advantage of fupplying them with moft of their canvas and cordage. Tar and pitch are made in feveral places of *Catalonia* and *Aragon*. The *Spanifh* fettlements in *America* have alfo very good fhip-timber, pitch and tar; for which reafon it would be more advantageous to them to build the greater part of their fhips at the *Havannah*. The *American* wood is befides more durable than that of *Europe*. The *Spanifh* navy received a fevere blow under *Philip* the fecond, and from that time has continued declining, till after the peace of *Utrecht*, when *Philip* the fifth was very intent upon reftoring it. It confifts, at prefent, of twenty-fix men of war, thirteen frigates, two packet-boats, eight chebecks and four bomb-ketches. The complement of all which amount to nineteen thoufand and fourteen men.

§. 19. The tranfmarine poffeffions of the crown of *Spain* are, firft,

1. In *Africa*, the towns of *Ceuta, Oran* and *Mafalquivir* on the coaft of *Barbary:* The expenfive keeping of which places ferves as a pretence for levying a confiderable part of the revenue of the crofs-bull *(§. 16.)* with the *Canary* iflands in the *Atlantick* ocean.

2. In

2. In *Afia* the iflands of *St. Lazarus*, the *Philippine* and *Ladrones*.

3. The greateft part of *South America, main-land*, as it is called, namely, *Tucumannia, Peru, Paraguya*, with the land of *Magellan* and *Chili*. In *North America, Mexico, New Mexico, California*, part of *Florida* and the ifland of *Cuba*, part of *Hifpaniola, Porto Ricco*, the *Caribbee* iflands, *Trinidad, Margaretba, Rocca, Orchilla, Blanca* and thofe of *Lucayan*.

§. 20. The Kingdom of *Spain* confifts of main-land and iflands. The main-land is divided into fourteen provinces, fome appertaining to the crown of *Caftile*, others to *Aragon*. The former are *Old* and *New Caftile, Bifcay, Leon, Afturia, Galicia, Eftremadura, Andalufia, Cranada, Murcia, Navarre*; and with thefe are alfo reckoned the *Spanifh* poffeffions in all the other parts of the world. The latter includes only *Aragon, Catalonia* and *Valencia*, with the iflands in the *Mediterranean*. Thefe provinces I fhall defcribe in order, beginning with the northern.

THE

KINGDOM

O F

GALICIA.

THIS province, which was formerly a Kingdom, is bounded on the south by *Portugal*; on the north and weſt by the ſea, being joined to the eaſt by *Aſturia* and *Leon*. It receives its name from the ancient *Gallazi*, the moſt powerful and numerous of the ſeveral nations who inhabited it. Its extent from ſouth to north is about thirty-four *Spaniſh* miles; and, in its greateſt length, from weſt to eaſt, it is about thirty. It is the moſt maritime of all the *Spaniſh* provinces, ⸳and enjoys, accordingly, the greateſt number of ſea-ports; among which *Corunna* and *Ferrol* are the moſt conſiderable. Cape *Finiſterre*, one of its promontories, the ·*promontorium Artabrum* and *Celticum* of the ancients, lies towards the weſt, and is well known to all navigators. The coaſts enjoy a temperate air, but in the inland parts it is ſomewhat colder and very damp. This country is ſo mountainous as to admit of few levels, and, the ſea-coaſt excepted, is but thinly inhabited in reſpect of the other provinces. It has no leſs than ſeventy rivers and ſmaller ſtreams; the principal of which are the *Minho*, already deſcribed in (§.4.) of the Introduction, and alſo in *Portugal*; the *Ulla*, the ſource of which lies almoſt near the centre of the country, in a Diſtrict called *terra de Ulla*, from whence it falls into the ſea below the little town of *Padron*; the *Tambra*, in Latin *Tamaro*, or *Tamaris*, which diſcharges itſelf into a bay not far from *Muros*; the *Mandeo* riſes not far from the *Ulla* and joins the ſea below *Betancos*. This province produces little corn, but makes up for that deficiency in wine, flax and lemons. It has alſo very fine paſtures. The ſea abounds in excellent fiſh, particularly ſardines and ſalmon, and a peculiar kind called *bezugos*. Its foreſts alſo afford good ſhip-timber. The univerſal poverty of the inhabitants obliges great numbers of them to ſeek a living in the neighbouring provinces, where they let themſelves out to the moſt

ſervile

fervile and laborious employments. For this humble induftry they are de-
fpifed by the other *Spaniards*. *Galicia* was raifed to a monarchy in the year
1066, by *Ferdinand*, King of *Caftile* and *Leon*; and *Ferdinand* the Catholic
quelled the mutinous temper of the *Galicians*, particularly of the nobility,
who for a long time had been wanting in due regard to the King's governor.
This province contains fixty-four cities and towns, but few of the latter are
any ways confiderable. The governador, or capitan-general, refides at
Corunna. The chief *civdades* and *villas* in it are as follow, *viz.*

Guardia, a fmall town in form of a half moon, as is alfo its little harbour at the mouth of the *Minho*. It is defended by a caftle ftanding on a
rock.

Bayona, a fmall town on a bay, which forms a convenient harbour. The
fea here abounds with fifh, and the Diftrict of land belonging to the city
with fine fruits. At the entrance of the bay lie certain iflands, formerly
called *the iflands of the Gods*.

Civdad Tui, a frontier place on the river *Minho*, feated on a moutain oppofite *Valença* in *Portugal*, is a bifhoprick of ten thoufand ducats *per annum*.
The air here is very temperate, and the country about it no lefs pleafant and
fruitful. In a war, betwixt *Spain* and *Portugal*, this is one of the three
places of rendezvous for the *Spanifh* troops.

Salvatierra, a little town on the river *Minho*.

Vigo, a town feated on a fmall bay, having a fort on an eminence, but
not capable of any long refiftance. It has alfo an old caftle, and ftands in
a very fruitful country: In the year 1702 the *Englifh* and *Dutch* fleets
forced their paffage in and made themfelves mafters of the *Spanifh* plate-
fleet when juft returned from *America*. In the year 1719 the *Englifh* again
got poffeffion of this place, but relinquifhed it after raifing contributions.

Redondela, or *Redondillo*, a little town on the abovementioned bay, having
a ftrong caftle.

Ponte Vedra, a town on a bay into which iffues the little river *Leriz*,
being poorly inhabited in proportion to its bignefs, but has a confiderable
fardine fifhery.

Padron, an ancient town at the mouth of the *Ulla*, which difcharges itfelf
here into a bay, is under the jurifdiction of the archbifhop of *S. Jago*.

Noya, a fmall town on a bay into which iffues the river *Tambra*, ftands in
a fruitful plain, and is noted for its fhip-building.

Muros, a town on the above-mentioned bay, has a collegiate church.

Finifterre, a little town near the celebrated cape of that name.

Mongia, a town on a bay, not far from which lies Cape *Coriana*.

Camarinas, a fmall place near Cape *Bellem*:

Malpico, an inconfiderable fea-port fronting the little ifland of *Syfarga*.

Civdad Corunna, anciently *Brigantium*, ftands on a peninfula in a fmall
bay, being divided into the upper and lower town. The former lies on a

hill, and is defended by fort St. *Diego.* The latter is called by the inhabi_
tants *Pexaria,* and lies at the foot of the mountain on a point of land which
is furrounded with water on three fides. It refembles in form a half-moon,
having a caftle at each end, one of which is called St. *Martin* and the other
St. *Clare;* but others give them the names of St. *Antony* and St. *Croix.* It
has a fpacious harbour, and is the feat of the *audiencia real* of *Galicia* ;
which, befides the governor and regente, confifts of feven alcaldes mayores,
one fifcal and other officers. Here is alfo a collegiate church, and in its
neighbourhood a valuable quarry of jafper.

Civdad Betanzos lies on the river *Mandeo,* which here forms a good
harbour.

Puentes is a little town feated on a bay.

Ferrol is a handfome town on a bay, which forms a very fine harbour.

Ortegal is a caftle, giving name to a cape lying near it.

Vivero is a fmall town feated on a mountain, wafhed by the little river
Landrova, which here forms a large harbour at its influx into the fea.

Ribadeo is a fmall town on a rock, at the mouth of a river of the fame
name, which makes it a good and fecure harbour. This place gives the
title of count to the duke of *Herzoge.*

Civdad Mondonnedo lies at the foot of a mountain at the entrance of a delightful plain. Its bifhop, who is alfo lord of the city, enjoys a yearly
revenue of four thoufand ducats, and is fuffragan to the archbifhop of
St. *Jago.*

Caftro de Rey is a fmall town near the fource of the *Minho.*

Lugo, anciently called *Lucus Augufti,* is an old epifcopal city, but was
formerly much larger than at prefent, being famous for its many warm
baths, fome of which are only tepid, but others boiling hot. The yearly
income of its bifhop, who is alfo fuffragan to St. *Jago,* is ten thoufand ducats.
In the year 564 a fynod was held here.

Sant Jago de Compoftella, the chief city of the whole province and an
archbifhop's fee, lies between the rivers *Tambro* and *Ulla,* in a moft fertile
plain, being furrounded with hills of a moderate height, which fhelter it
againft the nipping winds that blow from the mountains: it enjoys, however, a damp air. In this town are to be feen fine piazzas with feveral
monafteries of both fexes, and beautiful churches, among which the cathedral is particularly worth notice, as in it is kept the pretended body of the
apoftle *James* the younger, the titular faint and patron of all *Spain,* which,
towards the clofe of the ninth century, was difcovered by a divine revelation.
Subject to its archbifhop are twelve fuffragans, one thoufand eight hundred
and three churches, four collegiate churches, five archpriefts and one vicar-
age. His annual revenue is fixty thoufand ducats, and that of the cathedral
no lefs, but out of it he pays the King one thoufand eight hundred ducats
per annum. The bifhoprick erected here in the year 900, was raifed, in

1120, to an archbifhoprick. The order of *Sant Jago* takes its name from this city, which maintains alfo a certain number of knights of its own. The univerfity here was erected in the year 1532, and it has alfo a tribunal of inquifition.

Altamira, an earldom, belonging to the houfe of *Mofcofo*.

Porto-Marin, a little town feated on the *Minho*.

Sarria, a little town on the *Lugos*, where died *Alphonfo* XI. and the laft King of *Leon*.

Villa Franca, a little place on the confines of *Leon*.

Monforte de Lemos, the principal place of the earldom of *Lemos*, with a ftately feat of the count's, ftanding on a hill, at the foot of which runs the little river of *Cabe*. In the neighbourhood of this earldom is the high mountain of *Cebret*, upon which is an extraordinary fpring called *Lonzana*, the water whereof is fometimes very cold and at other times quite warm, ebbing alfo and flowing with the fea.

Orenfe, a city feated in a delightful country abounding in excellent wine and fine fruits, and lying on the *Minho*. One part of it ftands at the foot of a mountain, and in it we feel a fharp cold which is of long continu-ance, while the other, which lies on the fide of the plain, enjoys all the pleafures of fpring and the fruits of autumn, on account of the fprings here, which warm the air with their exhalations. Some of thefe fountains are fo moderately warm that a perfon may bathe in them; the water of others, on the contrary, is fo hot that eggs may be dreffed in them; but they are both falutary in feveral diftempers. Its bifhop, whofe yearly revenue amounts to ten thoufand ducats, is fuffragan to the archbifhop of *Sant Jago*.

Aravio is a little place having a caftle.

Cela Nova, a fmall town on the river *Lima*, near which grows plenty of cheftnuts.

Monte Rei, a little town, giving the title of count, with a good fort ftanding on a high hill, at the foot of which runs the *Tamega*. The neighbouring country produces abundance of flax and excellent wine.

Viana is the chief place of an earldom belonging to the houfe of *Pimentel*.

The PRINCIPALITY *of the* ASTURIAS.

THE principality of the *Aſturias* borders to the weſt on *Galicia*, to the
north on the ſea, to the eaſt on *Biſcay*, and to the ſouth on *Old Caſtile*
and *Leon*. Its extent from ſouth to north is about thirteen *Spaniſh*
miles, and from weſt to caſt twenty-ſeven. The air is tolerable, but the
country uneven and rugged. ·Towards the ſouth·it is ſeparated from *Leon*
and *Old Caſtile* by high mountains which˙are covered with woods, ſo that
the country is thinly peopled; yet the ſoil produces a pretty deal of grain,
plenty of fruit and excellent wine. Its horſes were of old very famous,
and much ſought after on account·of their ſpirit and goodneſs. The nobility
of this province value˙themſelves on their deſcent from the old *Goths*, and
on their blood not being adulterated by the mixture with˙*Jews* and *Moors*.
For after the unfortunate battle which the *Goths*, under King *Roderick*,
fought with the *Moors* near *Xeres*, the *Gothic* Prince, *Pelayo*, retreated with
a conſiderable· number of his nobility into the mountains of the *Aſturias*,
and there got a ſmall army together; but finding himſelf unable to face the
enemy in the·field,· he betook·himſelf,· with one thouſand brave *Goths*, to a
large cave in the mountain of *Auſena*, and upon the approach of the *Moors*
ſallied forth and totally routed them. On this ſpot a convent was afterwards
built, which was called *Santa Maria de Cobadonga*, from the name of the
cavern. · This defeat and bold ſtand made by the *Goths* is, to this˙day,· ſo
celebrated in *Spain*, that all the inhabitants of this mountain are·looked
upon as true *Goths*, and enjoy particular privileges; and, though they are
but peaſants, and go in great numbers from this mountain into the other
provinces of *Spain* to find work, yet they diſdain to be called by any other
title than that of *Godo*. Both great and ſmall give themſelves the appellation of *illuſtre Godo*, or˙*illuſtre montagnes*, and, amidſt their poverty, look
upon it as ignominious to intermarry with great and rich families of any other
race. They are likewiſe ſo much eſteemed, that other families frequently
give conſiderable ſums of money to marry among them; but the greateſt
honour of this Diſtrict ariſes from its belonging to the hereditary Prince of
Spain, who takes his title from it. It is divided into two unequal parts, and
hence ariſes the name of *the Aſturias*, as uſed in the plural number.

1. *Aſturia d' Oviedo* lies weſtward, and is the largeſt diviſion. It contains

Oviedo, anciently *Brigetum*, the capital of all *Aſturia*, and the only place
which bears the name of *Civdad*, being ſeated on a plain, in a kind of elevation, between the little rivers *Ove* and *Deva*; the firſt of which, has given
the name of *Ovetum* to the city, from whence comes *Oviedo*. Here the

                                                                    above-

above-mentioned *Gothic* Prince, *Pelayo,* held his refidence; for which reafon alfo the bifhoprick was removed hither from a neighbouring city called *Emerita,* and afterwards raifed to an archbifhoprick, but loft this honour. The bifhop of this place, who enjoys an annual revenue of twelve thoufand ducats, is immediately fubject to the Pope. The cathedral boafts a vaft number of reliques, which were brought hither from all parts of *Spain,* by way of fecuring them againft the *Moors.* In the year 1580 an univerfity was erected here, and, in 901, a council met in this place. The *audiencia real* contains one regente, four alcades mayores, one fifcal and one alguacil mayor.

Navia is a little town and harbour feated in a pleafant country.

Luarca, a little place having a fmall harbour.

Avila, a fmall place with a harbour.

Next is Cape *de las Pennas.*

Gyon is a little town fituate on a peninfula, being formerly the capital of *Afturia* and the refidence of the *Gothic* Prince *Pelayo.* His fucceffors ftyled themfelves Kings of *Gyon* till *Alphonfo* the Chafte took the title of King of *Oviedo.*

Villa Viciofa is a town feated on a bay forming a good harbour, into which the river *Afta* empties itfelf.

Riba de Sella is a little fea-port.

2. *Afturia de Santillana* lies eaftward, being the fmalleft divifion. In it are

Llanes and *S. Vincente de la Barquera,* two little fea-ports, the latter of which is defended by a caftle.

Santillana, in Latin *fanum Sanctæ Julianæ,* is the capital of this part of *Afturia,* and confers the title of marquis. It belongs to the duke of *Infantado.*

S. Martin, or *Sette Villas,* is a little town on a bay.

S. Ander, formerly *portus S. Emederi,* is a fmall, ancient and fortified fea-port, at the foot of a hill, feated in a pleafant country which abounds with fine fruits and wine, having a large, fecure and well fortified harbour; but, at its entrance, a dangerous rock called *Penna de Mogron.* The fuburbs are almoft wholly inhabited by fifhermen, whofe trade here turns to very good account, the neighbouring fea abounding in good fifh.

Liebana is a mountainous and rugged Diftrict belonging to the duke of *Infantado,* and confifts of five vallies; namely, *Cilorigo, Val de Prado, Vahebaro, Cereceda* and *Polanos.* The principal place, *viz.*

Potes, is of confiderable bignefs, being feated on the banks of the *Deva.*

Santogna and *Guarnifo* are two places, the chief occupation of whofe inhabitants is fhip-building. In thefe parts are alfo caft iron cannon.

SENNORIA

Sennorio de Biscaya, *or* Bizcaya.

In *Latin* Cantabria, Biscay.

THIS province is bounded to the weſt by the *Aſturias,* to the north‐ward by the ſea, which is here called the ſea of *Biſcay,* to the eaſt by the *French* lower *Navarre* and the *Spaniſh Navarre,* and to the ſouth by *Old Caſtile.* Its extent from ſouth to north is between five and twenty *Spaniſh* miles, and from weſt to eaſt about twenty‐nine. The air here is mild and temperate as in the other provinces, but the ſoil uneven and ſtony. In ſome parts nothing at all grows, but others produce wine and corn ſuf‐ficient for the uſe of the inhabitants, and have apples every where in ſuch plenty that they make excellent cyder of them, which, in ſome meaſure, ſupplies the want of wine. The ſea yields good fiſh of all ſorts; and the coaſt ſo abounds with oranges and lemons, that they ſell them very cheap. In the foreſts is excellent timber for ſhip‐building. It has alſo iron and lead‐mines. Its commodious ſituation on the ſea and in the neighbourhood of *France,* brings a very flouriſhing trade here, particularly in iron, all kinds of arms, and train‐oil. The inhabitants of theſe parts are reckoned the beſt ſoldiers and ſailors in *Spain.* They enjoy many privileges, of which they are alſo very jealous. Their dialect is particular, and has no affinity with the other languages of *Europe.* Till the year 859, the *Biſcayans* were governed by counts or governors appointed by the Kings of *Oviedo* and *Leon.* Under the latter they revolted, and choſe a chief for themſelves, which independency they retained till *Peter the Cruel* ſubdued them, and, under the title of a lordſhip, united *Biſcay* with *Caſtile.* This province con‐tains the following places, *viz.*

S. Antonio, a little ſea‐port near a mountain of the ſame name.

Laredo, a little town ſeated on a riſing ground, environed on all ſides with rocks. The harbour below it is very commodious. Great quantities of ſalt‐fiſh are exported from here.

Caſtro de Urdiales, a good caſtle with an armory.

Portogallete, a ſmall town, ſeated on a bay.

Bilbao, the capital of the province, lies in a plain environed by high mountains. The ſea‐tide, which flows here into the river *Ybaiçabal,* forms a ſecure harbour, which is very much frequented. The city is pleaſantly ſituated, enjoys a good air, ſtands in a fruitful country, is well built, and drives a great trade in iron, wool, ſaffron and cheſtnuts. Proviſions alſo are here cheap and plenty.

Vermejo

Vermėjo or *Bermeo,* a little place, with a good harbour.

Durango, a fmall but well inhabited town, lying in a deep valley be-tween high mountains. The inhabitants of this place make great quanti-ties of hard-ware and other iron works.

Hellorio, is a little place feated in a pleafant valley, being famous alfo for its manufactures of iron.

Ordunna, the only place in this province which bears the title of civdad, is fituate in a delightful valley environed with high mountains.

The Diftrict to the four towns extends itfelf along the province of *Old Caſtile.*

The PROVINCE *of* *G I U P U Z C O A*

Contains the following Places, *viz.*

MOTRICO, a fmall town on the fea-fhore:

Deva, a little town, feated on a river of the fame name, which here falls into the fea.

} Zumaia, a little place on the river *Viole,* which near it difcharges itfelf into the fea.

Guetaria, a little town, feated on a mountain near the fea, having alfo a caftle and a famous harbour.

Orio, is a little place, fituate at the mouth of a river of the fame name.

San Sebaftian, a town of confiderable bignefs, having a fecure harbour at the mouth of the little river *Gurumea,* by the ancients called *Menafcum.* It is feated at the foot of a mountain, which ferves as a defence to it againft the tempeftuous fea. The harbour lies within two redoubts, between which only one fhip can pafs at a time. Near its entrance alfo is a fort with a garrifon in it againft any furprife. The town is furrounded with walls, befides which it is defended with baftions and half-moons, and on the mountain, under which it lies, is a citadel. The ftreets are long, broad, ftraight and clean, the houfes neat and the churches fine. It en-joys a delightful profpect, having on one fide the fea and on the other a diftant view of the *Pyrenean* mountains. A confiderable trade is carried on in this place, particularly in iron, fteel, and wool. In the year 1728, a company was fet up here which trades in cacao to the *Caraccas.* In 1719, the *French* made themfelves mafters of this town.

2

Port *Paſſage* is a ſpacious excellent harbour with a narrow entrance, and ſheltered by mountains from all winds. In 1719, this place alſo was taken by the *French*.

Fuentarabia (in Latin *Fons rabidus*) is a little neat town on the ſea, for-tified both by nature and art. It has a fortreſs, ſtands like an amphitheatre on a hill, having alſo a pretty good harbour, and being towards the land environed by the *Pyrenees*, which are here called the *Sieras de Jaſquivel*. In 1638, it held out a ſiege againſt the *French*, for which the King ho-noured it with the title of a *Civdad*; but, in 1719, was taken by them. Near the city runs the river *Bidaſſoa* or *Vidaſſo*, which is here very broad, and is the boundary between *Spain* and *France*, being alſo by virtue of an agree-ment between *Ferdinand* the Catholic and *Lewis* XII. the property of both the crowns, ſo that the fare paid here by paſſengers, is divided among theſe two nations; the *Spaniards* taking of thoſe who croſs out of *France* to their ſide, and the *French* again from ſuch as paſs out of *Spain* to them. The mouth of this river is a little below the town.

The *Pheaſant Iſland* is worthy of notice, not only on account of the peace of the *Pyrenees*, which was concluded here in 1659, but likewiſe for the treaty of marriage tranſacted there between *Lewis* XIV. and *Maria Thereſia* the Infanta of *Spain*. In the year 1722, at this place were ex-changed the Infanta *Maria Anna Victoria*, and *Mademoiſelle de Montpenſier*, daughter to the Duke of *Orleans*, regent of *France*. It is called by the *French*, *l'Iſle de la conference* and *l'Iſle de paiz*.

Iron, called alſo *Iran*, is a little town ſeated among the *Pyrenees*, on the road leading over theſe mountains out of *Spain* into *France*; but is very narrow, and has precipices on each ſide. This is the firſt place belonging to *Spain*, after coming out of *France* by the abovementioned road.

Toloſa, *Toloſetta*, the capital of the province, is ſeated between two moun-tains in a delightful valley at the conflux of the *Araxas* and *Oria*, being not large, yet handſome.

Villa Franca and *Segura*, are two pretty little towns, ſeated on the *Oria*.

Plaſencia, a little town on the *Deva*, in the valley of *Marquina*, noted for the fire-arms made there.

Vergara, alſo a little town, but enjoying a conſiderable trade in iron and arms.

Mondragon, a little town on the *Deva*, ſituated on a hill. It has ſeveral medicinal ſprings, and makes good cyder, great quantities of excellent ap-ples growing in its neighbourhood.

Salinas, is a little town on an eminence lying on the *Deva*, near which are ſalt-ſprings.

Aſpeytia, a little town on the *Viola*, ſeated in a pleaſant valley. Near it are the Diſtricts of *Loyola* and *Onis*, which belonged to St. *Ignatius*, founder of the order of Jeſuits.

Heybar

Heybar and *Helgoybar* are two fmall places, in the latter of which cannon are caft.

Laftly, here is to be feen the mountain of *Adrian*, in *Spanifh* called the *Sierra de Adriane*, which took its name from the hermit *Adrian*. The road leading over it to *Alaba* and *Old Caftile* is very difficult to travellers. At the very beginning of it is a dark fpace between forty and fifty paces in length, cut through a rock, after which we afcend up a hill, which is reckoned the higheft among the *Pyrenees*. Thefe mountains are but little inhabited, a few fhepherds huts only being to be feen here.

The PROVINCE *of* A L A B A

I S pretty fruitful in rye, barley, and feveral kinds of fruits; and produces alfo tolerable wine. In it ˜are likewife very rich iron and fteel mines. This fmall province formerly had the title of a Kingdom, and at prefent contains the following places, *viz.*

Salvatierra, a little town.

Civdad Vitoria, the capital, feated on an eminence at the end of a pleafant valley, is pretty large, being environed with a double wall, and divided into the upper and lower town. The monafteries in it are magnificent, and the convent of St. *Francis* in particular is very large, whence that the general chapter of this place is ufually held in it. The city is much reforted to, and carries on a great trade in iron and fteel, and likewife a pretty confiderable traffick in wool, wine, and fword-blades.

Trevigno, a little town on a hill on the river *Ayuda*, having alfo a caftle, is the capital of the earldom of that name, and belongs to the Duke of *Najara*.

Pegna Cerrada, is fituate amidft very high mountains, and has a fortified caftle.

Onate, a fmall place, but has an univerfity.

Obf. The little territory of *Rioxa*, in fome maps is placed in *Bifcay*, and in others in *Old Caftile*, in which laft I have alfo defcribed it.

The KINGDOM *of* L E O N.

THIS Kingdom is bounded on the north by the *Afturias*, on the weft
by *Galicia* and *Portugal*, to the fouth by *Eftremadura*, and to the eaft
by *Old Caftile*. Its extent from north to fouth is about four-and-forty
Spanifh miles, and from weft to eaft between eighteen and twenty-five.
The foil produces all the neceffaries of life, particularly the Diftricts of
Vierzo and *Ledefma*. The wine too is pretty good: And in this province
are mines of turquois. The principal river in it is the *Douro*, which divides the province from weft to eaft, pretty nearly into two equal parts;
namely, the northern and fouthern, and after paffing through *Portugal*
difcharges itfelf into the fea. Here alfo are the little rivers of *Pifuerga*,
which takes its rife in *Old Caftile*; the *Carrion*, which rifes alfo in *Old
Caftile* and runs into the *Pifuerga*; the *Efla* and *Orbego*, both which have
their fource in the country about the city of *Leon*, unite below *Benavente*;
and afterwards join the *Douro*; the *Torto* and *Tera*, which run into the
Orbego and *Formes*, or *Rio de Salamanca*, which rifes in the *Sierra de Pico*,
falling on the frontiers of *Portugal* into the *Douro*. The remarkable towns
are as follow, *viz.*

Civdad Leon, the capital of the province, built by the *Romans* in the time
of the Emperor *Galba*, and called *Legio feptima Germanica*, from whence
the name *Leon* is derived. It lies between both the fprings of the river
Efla, at the end of a large plain which is bounded by the mountains of
the *Afturias*. The bifhop of this place, whofe yearly income is twelve
thoufand ducats, is immediately fubject to the Pope. The cathedral here
is famous not only for its beauty, but alfo for being the burial-place of feveral faints, thirty-feven Kings of *Spain*, and one Emperor. The city was
formerly larger, richer, and more populous than at prefent. It is the firft
city of any confideration which was retaken from the *Moors*, *Pelayo* making himfelf mafter of it in the year 722, when he fortified it; and from
that time till 1029 it had a royal feat.

Villa Franca, Cafabelos, and *Ponferrada;* anciently *Interammium Flavium*,
are three fmall towns feated in vallies between high mountains.

Civdad Aftorga, in Latin *Afturica Augufta*, ftands in a delightful level, on
the river *Aftura* or *Torto*, being an ancient city extremely well fortified both
by nature and art, but neither large nor populous. Its bifhop is fuffragan
to the archbifhop of *Santjago*, and has an annual revenue of ten thoufand
ducats. It was formerly the capital of the *Afturias*, but at prefent is only
the chief place of a marquifate erected here in 1465.

Sanda—

Sanabria, a small lake abounding in fish, belongs to the Monks of *S. Maria de Caslagneda*. The river *Ter* traverses it obliquely and with such rapidity as to cause a great agitation in the water. In the middle of it stands a small Island or rock, on which the Counts of *Benavente* have built a summer-house.

Benavente, a town on the *Esla*, giving the title of earl to the family of *Piementel*, who have a fine palace here with a well fortified castle, beautiful gardens, and a park. To this family also belongs

The earlom of *Mayorga*, the capital of which bearing the same name lies in a pleasant level:

Saldagna, a town seated in a delightful valley at the foot of the mountain of *Pegna de San Roman*, not far from the source of the *Carrion*, belongs to the Duke of *Infantado*.

Carrion de los Condes, a little town on the river *Carrion*, enjoys great privileges. Its District abounds in all the necessaries and pleasures of life.

Sahagun, a small town on the river *Sea*, has a good castle. The circumjacent country is very fruitful.

Torrequemada, or *Torquemada*, in Latin *Tnrris cremata*, a small town seated on the *Pisuerga*, belongs to the Duke of *Lerma* of the family of *Sandoval*.

Civdad Palencia, is a town on the *Carrion* in a fruitful country. The bishop of this place who has a revenue of twenty-four thousand ducats *per annum*, is suffragan to the archbishop of *Burgos*. The univerfity which either *Alphonso* VIII. or IX. erected here about the beginning of the 13th century, was in 1239 removed to *Salamanca*.

Civdad Medina de Rio secco, lies in a valley enjoying a good air, and is a flourishing town. In the year 1938 it had the title of a civdad, and is the principal place of a dutchy belonging to the amirante of *Castile*.

Villalpanda, is a little town in a pleasant and fruitful level, where the field-marshals of *Castile* have a fine palace and armory.

Dnegnos too is a little town seated on a hill, near the conflux of the *Pisuerga* and *Arlanzon*.

Cabeçon, is a little town standing on a mountain, with a fort erected on the river *Pisuerga*.

Simancas, *Septimanca*, a town near the *Douro*, being seated on an eminence at the end of a plain. The white wine of this place is very fine. It has a castle, wherein King *Philip* II. on account of its strength, ordered the archives of the Kingdom in 1566 to be kept.

Tordesillas, in Latin *Turris Syllæ*, is a small town, having a royal palace, in which *Johanna*, mother to *Charles* V. ended her days.

Toro, in Latin *Taurus*, is a town on a hill lying on the *Douro*, in a country abounding with grain, fruits and wine. Here the famous laws called *Leyes de Toro* were enacted in a diet of the Kingdom.

Civdad Zamora on the *Douro*, over which it has a fine bridge, lies in a fertile country, and is well fortified. The bifhop of this place, who has an annual income of twenty thoufand ducats, is fuffragan to the archbifhop of *Santjago.* In this city is kept the body of St. *Ildefonfo*, formerly bifhop of *Toledo.* It was called *Sentica*; but the *Moors* gave it the name of *Zamora*, or *Medinato Zamorati*, *i. e.* ' a town of turquoifes,' moft of the rocks in its neighbourhood containing gems of that kind.

Medina del Campo, in Latin *Methymna Campeſtris*, a very old town, has great privileges, particularly the following, that the inhabitants are free from all impofts, and fill up all the vacant employments both in church and ftate, &c. they enjoy alfo a confiderable trade.

Civdad Salamanca, on the river *Tormes*, lying partly on the plain and partly on the hills, is ancient, large, populous and rich. It contains fome beautiful churches, among which the cathedral is one of the fineft in all *Spain*, with fplendid convents, among which thofe of the *Dominicans* and *Francifcans* deferve particular notice, each of them containing two hundred Monks. The bifhop of *Salamanca*, who enjoys a yearly revenue of four-teen thoufand ducats, is a fuffragan to the archbifhop of *Santjago.* The uni-verfity, which in 1239 was removed hither from *Palencia*, is the moft noted in all *Spain.* The univerfity college is a very fpacious ftructure, and clofe to its ftately entrance is an infirmary for poor fick fcholars. In it are twenty-four colleges befides, in each of which live thirty ftudents, and among thefe the four moft confiderable are called the great colleges, and are appropriated to perfons of rank. The *Spaniards* term this city the mo-ther of virtues, fciences, and arts. Over the river *Tormes* is an old *Roman* bridge of ftone, and the ancient *Roman* way leading to *Merida* and *Sevilla* is alfo worthy of notice.

Alva des Tormes, a little town, but the principal place of the dutchy of the fame name, belongs to the family of *Alvarez.*

Pennaranda, the capital of the dutchy of that name, is feated among fruitful mountains.

Ledefina, a town on the river *Tormes*, fortified both by nature and art, is very old, and was formerly called *Bletifa.* Not far from this place is a warm bath on the bank of the river.

Civdad Rodrigo, an epifcopal city, ftands in a fertile country on the little river *Aguada* or *Agujar.* *Ferdinand* II. King of *Leon*, about the beginning of the 13th century founded this city in the place where *Mirobriga* anci-ently ftood. The bifhop hereof is fuffragan to the archbifhop of *Santjago*, and has an annual revenue of ten thoufand ducats. It is one of the three places of rendezvous for the *Caſtilians* when they are at war with *Portugal.*

Ginguello, is a little town.

Villa Franca, is a little town feated on the river *Tormes*, where they make good cloth.

5

The

The PROVINCE *of* ESTREMADURA.

THIS province is bounded to the north by *Leon*, to the weſt by *Portugal*, to the ſouth by *Andaluſia*, and to the eaſt by *New Caſtile*. Its extent from ſouth to north is pretty nearly about thirty *Spaniſh* miles, and from weſt to eaſt betwixt ſixteen and twenty-ſeven. The inhabitants are inured to its air and weather; but the ſummer heats are intolerable to foreigners travelling here. Thoſe who live at the foot of the mountains have good water; but in the plains we muſt put up with that which is taken out of pits dug in the ground. The ſoil abounds with grain, wine and fruits, and its paſtures are ſo good that great numbers of cattle are brought hither from other provinces to fatten. The rivers *Tagus* and *Guadiana* run quite through the country to *Portugal*, and in many places their ſtreams are joined by ſeveral ſmaller. *Eſtremadura* was a diſtinct province from all the reſt of the Kingdom, but in the preceding century it was united to *New Caſtile*: It is ſtill governed, however, by a particular captain-general, and in it are the following places, *viz.*

Belvis, a ſmall town with a caſtle, ſeated between mountains.

Bejar, a little town which in 1448 was raiſed to a dutchy, the title of which is in the family of *Sotomayor*. It ſtands in the midſt of a pleaſant valley between high mountains, whoſe tops are continually covered with ſnow. Among the fine ſprings in this place is one very cold and another very hot one, both which, however, are good for ſeveral diſtempers; the former by drinking the water, and the latter by bathing in it. In the neighbourhood of the town is alſo a remarkable lake which moſt certainly preſages bad weather by a very unuſual agitation.

Civdad Coria, in Latin *Cauria* and *Caurita*, lies on the little river *Alagon*, in a level abounding with all the neceſſaries of life. It gives the title of Marquis to the Duke of *Alba* of the family of *Alvares*. The biſhop of it is ſuffragan to the archbiſhop of *Santjago*, and enjoys a revenue of twenty thouſand ducats *per annum*.

The Diſtrict of *La vera de Plazencia*, (or Orchard of *Plazencia*) conſiſting of alternate vallies and mountains, is extremely delightful, and next to *Andaluſia*, the moſt fertile in all *Spain*; whence it contains no leſs than ſeventeen places, all well inhabited. The fineſt and moſt delicious fruits and vegetables, with wholſome and odoriferous plants grow here in the greateſt abundance. Here is alſo exquiſite wine, fine ſprings, and pleaſant brooks with excellent trout in them. Every thing in it, in ſhort, wears a ſmiling aſpect. In this Diſtrict not only ſtands the celebrated *Jeromite*
convent

convent of St. *Juftus*, which *Charles* V. in 1555, chofe for the place of his retirement, and where he alfo ended his days, but likewife

Civdad Plazencia, from which the Diftrict takes its name. -It is a beautiful well built city, feated on an eminence between mountains on the little river *Xerte*, being defended by a good caftle. *Alphonfo* IX. King of *Caftile*, built it about the year 1170. It was formerly in the poffeffion of particular lords under the title of a dukedom, but fince the year 1448 has been united to the crown. Its bifhop, whofe income amounts to fifty thoufand ducats *per annum*, is fubject to the archbifhop of *Santjago*. Under the jurifdiction of this city are the two following places, *viz.*

Pifaro, a little town, in the middle of a deep valley amidft high mountains. Its neighbourhood produces a great abundance of figs, lemons, and other choice fruits.

Xarabis, is a little town environed with whole woods of fruit-trees.

Almaraz, is a town feated in a fine plain on the *Tagus*.

Alcantara, a fortified town on the *Tagus* in a fertile country, takes its name (which fignifies a ftone bridge) from an old bridge, erected there in the days of the Emperor *Trajan*, over the *Tagus*, at the expence of feveral *Lufitanian* nations ; it was two hundred feet high, fix hundred and feventy in length, and twenty-eight in breadth, and on its account it was that the *Moors* built the city. In the year 1212, *Alphonfo* IX. King of *Caftile* took it, and gave it to the knights of the order of *Calatrava*, who afterwards received their name from it.

Valença d'Alcantara, is a fortified town on the confines of *Portugal*. In 1705 it was taken by the *Portuguefe*.

Caceres, is a little town lying on the *Sabor*, and well known for its fine wool.

Albuquerque, in Latin *Alba quercus*, is a town on an eminence defended by a very ftrong caftle, which ftands on a lofty mountain. It drives a good trade in wool and woollen cloth ; and was built about the middle of the 13th century. It gives likewife the title of duke, which in 1464 was conferred upon the counts of *Ledefma*, of the family of *Cueva*. In 1705 it alfo fell into the hands of the *Portuguefe*, who kept poffeffion of it till the peace of *Utrecht*.

Feria, a little town, and capital of the dukedom of this name, which in 1567, *Philip* II. erected for *Gomez Suarez de Figueroa*. It has a good fortrefs.

Albanga, a little town belonging to the order of *Santjago*, feated on an eminence, and defended by a very ftrong caftle ftanding on a rock.

Civdad Truxillo, or *Trugillo*, in Latin *Trogillium*, an ancient town ftanding on a hill, at the fummit of which is a good citadel. It acquired the title of a civdad in the year 1431.

Madrigalejo, a mean little village, but remarkable for being the place where *Ferdinand the Catholic* died.

Guadalupe, in Latin *Aquæ Lupiæ*, a little town, but well built, being ſeated in a valley abounding with wine, oranges, figs, and other fine fruits and trees. It is commodiouſly ſituated on a river of the ſame name, lying between mountains. The convent of *Jeromites* here has a celebrated image of the Virgin *Mary*.

Orelhana la vieja, a ſmall town on the *Guadiana*, with a caſtle, and giving the title of marquis.

Medellin, *Metellinum*, alſo a ſmall town on the *Guadiana*, and capital of an earldom belonging to the marquis of *Aytona*, of the family of *Moncada*. It lies at the foot of a mountain, on the top of which ſtands an old ruinous caſtle.

Civdad Merida, is ſeated on an eminence on the *Guadiana*, being a ſmall, ancient, fortified town, known in the times of the *Romans* and *Goths* by the name of *Auguſta Eremita*, and being much larger and more conſiderable than at preſent. In it are found ſeveral remnants of antiquity, particularly a triumphal arch, which the inhabitants call *Arco de Santjago*. Its environs are very pleaſant, and it abounds in wine, good fruits, and particularly in grain, with excellent paſtures.

Montijo, is an ancient caſtle, ſeated on an eminence. It gives the title of earl, and belongs to the family of *Portocarrero*.

Talavera de Badajoz, *Talabrica*, *Talaveruela*, is a town lying in a fertile country.

Civdad Badajoz, the capital of *Eſtremadura*, and a frontier towards *Portugal*, lies on an eminence on the river *Guadiana*, being divided into the upper and lower town. It is not large, but has good houſes, ſpacious ſtreets, fine churches, ſome monaſteries, and one college of Jeſuits. The biſhop of this place, whoſe annual revenue amounts to ſixteen thouſand ducats, is ſuffragan to the archbiſhop of *Santjago*. The fortifications are chiefly in the old taſte, but a few of the outworks are modern, more particularly a late modern caſtle called *S. Michael*, with the caſtle of St. *Chriſtopher* on the other ſide of the river, which principally covers the old *Roman* bridge of ſtone over that river, which is ſeven hundred paces long and fourteen broad, being quite ſtraight. In the days of the *Romans* this city was called *Colonia pacenſis* and *Pax Auguſta*, which laſt name the *Moors* corrupted to *Bax Augos*, and ſubſequent times to *Badajoz*. In the years 1658 and 1705 it was beſieged, but not taken. The country round it is very fruitful; and its ſheep produce a very fine ſort of wool.

Pegon, an iſland in the frontier river of *Caya*, which, about a ſhort *Spaniſh* mile from *Badajoz*, falls into the *Guadiana*. Here in the year 1729, the *Spaniſh* and *Portugueſe* Infantas were exchanged.

Valverde, is a little town, ſeated in a pleaſant valley.

Villa

Villa nueva de Barcarota is the capital of a marquifate with a very fightly caftle.

Xeres de Badajoz, or *Xeres de los Cavalleros,* belonging formerly to the knights of *Jerufalem,* had the title of a civdad given it by *Charles* V. Its neighbourhood is famed for graziery.

Azuaga, a fmall town, defended by a ftrong caftle, is a commandery of the order of St. *Jago.*

Zafra is a lordfhip belonging to the duke of *Feria.*

Medina de las Torres, a little town, but the capital of a dutchy of the fame name, belongs to the family of *Guzman.*

Civdad Llerena, or *Ellerena,* belongs to the knights of the order of St. *Jago,* by whom it was founded in the year 1241, and, in 1640, honoured by *Philip* IV. with the title of a civdad. The country around it is fruitful, and abounds, in particular, in fine paftures.

Villa de la Reyna is a commanderie belonging to the knights of *Santjago,* having a good caftle.

Salamea de la Serena, an old town, feated on a high mountain, having a ftrong caftle. It was anciently called *Ilipa,* as appears from the old monuments to be met with here. Its paftures around it feed great numbers of cattle, and the town belongs to the knights of the order of *Alcantara.*

A N D A L U Z I A.

THE name of *Andaluzia,* which it owes to the *Vandals,* is derived from *Vandalenhaus,* and formerly alfo extended itfelf over the Kingdom of *Granada,* then called *Upper Andaluzia;* whence, confequently, the *Andaluzia* we are now treating of is *Lower Andaluzia.* The ftrip of land lying on the fea from *Niebla* as far as *Almeria,* in the Kingdom of *Granada,* was anciently called *Algarve;* whence the King receives part of his title *(Rey de los Algarves de Algezira)* as is more particularly fhewn in the Defcription of the Kingdom of *Algarve* belonging to the King of *Portugal.* In former times it was called *Tarteffis* and *Bætica.* The *Andaluzia* I am now treating of is bounded on the north by *Eftremadura* and *New Caftile,* from which it is divided by a range of mountains called the *Sierra Morena;* to the weft by the *Portuguefe* Diftricts of *Alentejo* and *Algarve;* to the fouth, partly by the *Mediterranean* and partly by the ftraits of *Gibraltar;* and to the caft by *Granada* and *Murcia.* Its extent, from weft to eaft, is about fixty *Spanifh* miles, but its breadth very different, and, where largeft, not above thirty. The river *Guadalquiver,* by the ancients called *Bætis* and *Tarteffus,* traverfes the whole country, and the *Guadiana* feparates it to the weft from

the

the *Portuguese Algarve*. Of the other small rivers some run directly into
the sea; as the *Odier*, or *Odiel*; the *Tinto*, or *Azeche*, the water of which
cannot be drank, being noxious even to herbs and the roots of trees, and
having neither fish nor any living creature in it; and the *Guadalate*, or *river
of oblivion*. Others fall into the *Guadalquivir*; as, the *Guadiamar* and the
Xenil, which rises in *Granada*, &c. *Andaluzia* is reckoned the best province in all *Spain*, abounding in exquisite fruits of all kinds, honey, excellent wine, grain, silk, sugar, fine oil, numerous herds of cattle, particularly horses, metals, cinnabar, and a certain species of quicksilver. The
heat in summer is indeed very great, but the inhabitants generally sleep by
day, travelling and following their other employments in the night. In
other respects the air is good, being sometimes refreshed with cooling breezes.
This province is properly composed of three ancient Kingdoms, which, in
the King's titles, instead of the common names, are expressed by that of
Andalusia.

The KINGDOM *of* SEVILLE,

In which are the following places of note, *viz.*

PAYMOGO, a town, strong both by nature and art, situated on the
frontiers of *Portugal*.

Xeres de Guadiana, also a frontier place on the river *Chanca*.

S. Lucar de Guadiana, a town seated on a mountain near the river *Guadiana*, being defended on the side next the river by three towers, and on the
other side by two bastions. The tide, which runs up here, forms a small
harbour.

Ayamonte, a town situate at the mouth of the *Guadiana*, having a commodious harbour. It produces an excellent kind of wine, but not strong,
and gives the title of marquis to the families of *Zuniga* and *Guzman*.

Leye is a little town on a small bay into which falls the rivulet of
Saltes.

Guelvas, a little town seated betwixt the mouths of the *Odier* and *Tinto*,
and giving the title of earl.

Gibraleon is a neat little town lying on an eminence on the river *Odier*,
and the chief place of a marquisate belonging to the duke of *Bejar*.

Traigueros is a large handsome town, and the country around it abounds
in grain and wine.

Niebla is an ancient town and earldom on the river *Tinto*, belonging to the duke of *Medina-Sidonia*, whofe eldeft fon takes his title from this place.

S. Juan, a town feated on the *Trino*.

Palos, a little town fituate at the mouth of the *Trino*, in which the tide forms a tolerable harbour. From this place it was that *Chriftopher Columbus*, in the year 1492, fet fail on the difcovery of the new world.

Moguer, a little town lying on the *Trino*, which, in the year 1642, had the title of civdad conferred on it by *Philip* IV.

Lucena too is a fmall town.

Almonte is a neat town environed with a foreft of olives.

The civdad of *S. Lucar la mayor*, a little town feated on the *Guadiamar*, in a very fertile Diftrict called *Ajarafe*. In the year 1639 *Philip* IV. gave it the title of a civdad, and raifed it to a dukedom in favour of *Gafpar Guzman*, Count *Olivares*.

Gerenna, or *Jerenna*, a little place on an eminence hard by the river *Guadiamar*. In its neighbourhood is an uncommon number of ftones of furprizing dimenfions, which lie fcattered there in great diforder.

Cantillana is a little town which *Philip* III. erected into an earldom.

Almaden, a little place near which are mines of quickfilver.

Realejo is a large town.

Cazalla is a little town belonging to the duke of *Offuna*, and famous for its wine.

Gudalcanal, a mean place, having an old caftle, and being alfo a commanderie belonging to the order of *Sant Jago*. Near it are fome rich mines of quickfilver.

Villar Pedrofa and *Conftantina* are two little towns.

Pegnaflor, a little town, anciently called *Ilipula magna*.

Lora, formerly *Axalita*, or *Flavium Axalitanum*, is a little town, and a commanderie belonging to the knights of *Malta*.

Seville, in Latin *Hifpalis*, the capital of this country, bearing alfo the title of a royal city, lies on a large plain on the *Guadalquivir*. It is almoft round, and of confiderable extent, but not proportionate to the number of its inhabitants; and the ftreets are narrow, but the houfes fine, though in the *Moorifh* tafte, having alfo fplendid churches. It is the fee of an archbifhop, has a tribunal of inquifition and a mint, and carries on a large trade. Among its fuburbs, *Triana*, which lies oppofite to it on the other fide of the river, is the moft confiderable. In this city are reckoned twenty-nine churches, forty-four convents for monks, and thirty nunneries, all which are rich and well built, twenty-four hofpitals, and as many fquares or piazzas. The cathedral here is not only the largeft and fineft in all *Spain*, but *Labat* reckons it to be the largeft in the world, next to St. *Peter's* at *Rome*. It

1   has

has a great number of chapels, among which, that which serves for a vestry, and that where the chapter meet, are the finest. The steeple, or tower, of this church is said to be forty fathoms high. Of the convents, the most remarkable and magnificent are those of the *Francifcans, Nueftra Segnora de la Merred* and the *Dominicans.* Its univerfity was founded in the year 1504. *El collegio mayor* ftands near the King's palace. St. *Thomas's* college was built in the fifteenth century. The Jefuits college here is alfo worth feeing. The royal palace, called *Alcaçar,* ftands near the cathedral and is very fpacious. It was built by the *Moors,* fince which the Kings of *Spain* have alfo confiderably enlarged and decorated it; but the new works are inferior to the old. Some particular parts of this vaft ftrufture are very grand, but want the conveniencies of modern architefture. The exchange, which ftands behind the cathedral, and-is called the *Lonja,* is a large building. To the fuburbs of *Triana* we crofs by a bridge of boats. Near the entrance of it ftands the court of inquifition, which is an old ftrufture. Here alfo is the walk called *Alameda.* The archbifhop of *Seville* has a yearly revenue amounting to one hundred thoufand ducats. His fuffragans are the bifhops of *Maloga, Cadiz,* the *Canary* iflands and *Ceuta,* befides five civdades, containing about one hundred and forty-eight villas and towns, three collegiate churches, &c. In this city is an *audiencia real* confifting of one regent, or prefident, eight counfellors, and five *alcades de la Quadra de la Audiencia,* befides one fifcal. Since the year 1717, when the *audiencia real de lal contratacion a las Indias,* was removed from here to *Cadiz,* and, in 1726, eftablifhed there, *Seville,* and its once flourifhing manufaftures, have remarkably dwindled; infomuch that out of fixteen thoufand artificers in wool and filk, fcarce three or four hundred remain at prefent. This city was anciently called *Hifpalis, Spalis* and *Colonia Romulea.* Of *Spalis* the *Moors* made *Sbilia,* or *Ifbilia,* which, in procefs of time, was changed to *Sevilla.* The city and its Diftrift was more populous in the times of the *Moors* than at prefent. In the year 1729, a treaty was concluded here between *Spain, France, England* and *Holland.* The adjacent country abounds in wine, grain and feveral other fruits both for ufe and pleafure; and great quantities of oil, in particular, are made here. A long *Moorifh* aqueduft without the city merits particular notice.

The civdad of *Carmona* is a very ancient little city, feated on a hill, and formerly called *Carmo.* It had the title of civdad conferred on it by King *Philip* IV. The circumjacent country is extremely fertile, particularly in grain.

Alcala de Guadaira and *Utrera,* are two little towns.

Los Palacios, Palatium or *Pálantia,* is a little town, the inhabitants of which live by hufbandry. In thefe parts the roads are very bad; the tide, which runs thus far up the *Guadalquivir,* overflowing the grounds fo that the ftrip of land here which is expofed to inundations is not inhabited.

Alcantara,

Alcantara, or *Alcantarilla*, is a small town, on a gentle acclivity, where the *Romans* built a bridge over the morasses which is worthy of notice, having a tower and gates at each end.

Las Cabeças is a little town, situate near the beginning of a chain of mountains which extend to the south-east as far as *Malaga*. From hence to *Puerto S. Maria* are two roads, one of which lies eastward and goes through *Espera*, the other westward through *Lebrixa*.

Espera is a little ancient town on an eminence.

Lebrixa, an ancient pleasant town of confiderable extent. The country round it is pleasant and fertile, abounding in grain, wine and oil. This place was anciently called *Nabrissa*.

The civdad of *San Lucar de Barrameda*, in Latin *Lux dubia* and *Phosphorus facer*, or *Luciferi fanum*, a well built town, having a good harbour at the mouth of the *Guadalquivir*, lies on a hill. The harbour is difficult of access on account of a rock under water, but is defended by two batteries, and in the road a whole fleet may ride with safety. The town declines daily, and its principal trade is in salt.

Chipiona, in Latin *Cæpionis turris*, is a very ancient but poor place, lying on a rock near the sea.

Xeres de la Frontera is a large and pretty well built town, where several of the nobility reside. It has also a collegiate church. It was called *de la Frontera*, as being, when the *Moors* possessed *Cadiz*, the frontier town of the Christians. Not far from it stands a stately *Chartreux*, founded by a wealthy private person. The country round it is extremely fertile, producing, in particular, an excellent sort of wine, which is accordingly exported in great quantities. The horses bred here too are some of the best in all *Spain*. In the year 1713, near these parts, was fought the famous battle between the *Goths* and *Moors* which proved so fatal to the former.

Bornos, or *Bornes*, is a little town lying in a fruitful pleasant level, environed by high barren mountains.

Arcos, with the surname of *la Frontera*, formerly called *Arcobriga*, is an ancient fortified town standing on a high steep rock, at the foot of which runs the little river of *Guadalete*. This place gives the title of duke.

Rota is a little town with a harbour.

Puerto de Santa Maria, or *Port St. Mary's*, the capital of an earldom belonging to the duke of *Medina Celi*, enjoys a flourishing trade, and lies at the mouth of the river *Guadalete*, directly opposite *Cadiz*, whither we go from hence by shipping; and it exceeds that city in bigness. The streets also are broader and better paved, and the houses handsomer, but yet it contains scarce eight thousand inhabitants. It is only walled round, and the little castle which it has instead of a citadel, is but of small force. Here are great numbers of *French*, *English*, *Dutch*, *Genoese* and other merchants. In it are also made vast quantities of salt. Its harbour too is the rendezvous

of

of fome of the *Spanifh* gallies. In the year 1702, the *Englifh* and *Dutch* made themfelves mafters of this place without any oppofition. Beneath Port St. *Mary's*, on a point of land ftretching into *Cadiz* bay, ftands a tower and a battery called *S. Catalina.*

Matagorda, a fort on a neck of land, directly oppofite to *Puntal*, covers the harbour and bay of *Cadiz*. Near it alfo, to the caft, lies a fmall ifland, on which a fort is erected ; and higher up, on the main-land, ftands the fmall town of *Puertol Real*, on a bay in which are magazines.

Cadiz, a noted trading city, ftanding on an ifland on the north-weft end of a long mifhapen neck of land, extending itfelf from fouth-caft to north-weft, the weftern part of which is called *Cadiz*, but the fouth-caft the ifland of *Leon*. It is joined alfo to the continent (from which it is divided by a narrow canal, or arm of the fea) by means of the bridge of *Suaco*, both ends of which are defended with redoubts and fome other raifed works of earth. This ifland, from Fort *S. Catalina* to the ifland of *S. Pedro*, is about five *Spanifh* miles long; and from the fouth point, near the ifland of *S. Pedro*, to the northern one, near the bridge of *Suaco*, almoft two in breadth. It produces very little grain, but fome of the beft wine in *Spain* is made here. It has alfo fome paftures, and, on the fide next the harbour, a great deal of falt is made. The fifhery, likewife, is confiderable, particularly that of tunnies, which are commonly here between fix and eight, and frequently ten feet in length. The neck of land projecting from the ifland, which is in fome meafure quadrangular, is at its beginning very narrow, but widens with feveral inflexions, or angles, terminating at laft in two capes, the principal and weftern of which is called *Punta de S. Sebaftiano.*

The city of *Cadiz* is of pretty large circumference, yet is not the whole of this neck of land built upon, the delightful weft fide, called *Sancto Campo*, or *the church-yard*, being uninhabited, excepting that it has a fpacious hofpital and two chapels, not being fo commodious for fhipping as the caft fide. The greater part of the ftreets are narrow, crooked, ill paved and dirty; but a few of them are broad, ftraight and well paved. The houfes are moftly between three and four ftories high, being built with a quadrangular area, and many are really fine and ftately. Houfe-rent and provifions, in general, are dear here, and good water very fcarce. It is reckoned to contain thirteen convents, among which, the college of the Jefuits is the fineft in all *Anda-lufia*, but has only one parifh-church, which is the cathedral, though the fettled inhabitants here are computed at forty thoufand. The bifhop of this place is fuffragan to the archbifhop of *Seville*, with an annual revenue of twelve thoufand ducats. Here is the *audientia real de la contratacion a las In-dias* or the company trading to the *Indies*, which was removed hither, in 1717, from *Seville*. It was indeed, at the earneft requeft of the city of *Seville*, reftored to it again in 1725, but, in 1726, was the fecond time brought back to *Cadiz*. Both before and after the arrival of the *Spanifh American* flota, this city is crowded with ſtrangers to the number, it is ſaid, of fifty thouſand, who reſort hither on account of trade, which cauſes an extraordinary circulation of money with all its gay concomitants. *Cadiz* is the centre of the whole *American* trade, to which the *French, Engliſh, Dutch* and *Italian* merchants ſend their goods, which are ſhipped off here in *Spaniſh* bottoms to *America*, under the names of *Spaniſh* factors. Beſides theſe nations, all others who carry on any traffick by ſea have alſo their agents, correſpondents and factors here, and the conſuls of thoſe nations make a conſiderable figure. In this place none thrive better than merchants, who ſeldom riſque their own ſubſtance, but enrich themſelves at the charge of thoſe who remit them their effects; whence, whatever happens, they are no loſers. The *Spaniſh* goods exported from *Cadiz* to *America* are of no great value. The duty on foreign merchandize ſent hither would yield a vaſt revenue, and conſequently the profits of merchants and their agents would ſink without many fraudulent practices for eluding the duties. Both the harbour and bay of *Cadiz* are ſecure and ſpacious, the entrance being defended by Fort *Matagorda*, mentioned before (in §. 40) and by Fort *Puntal* ſtanding oppoſite to it on a point of that neck of land on which *Cadiz* is built. The *Spaniards* uſually call both forts *los puntales.* The entrance into the harbour betwixt theſe forts, and the points on which they ſtand, is reckoned by *Labat* to be five hundred fathoms wide. During the time of ebb, a good part of the harbour, which is ten *French* miles in circumference, is dry. The outer and furthermoſt bay, which begins between *Rota* (§. 38.) and *S. Sebaſtian*, and extends itſelf to *Puerto de S. Maria* (§. 39.) is divided into two parts by the rocks of *los Pueros* and *Diamante.* *Cadiz* is fortified with walls and irregular baſtions according as the land admitted them. On the ſouth ſide there is no approaching it on account of the high and ſteep ſhore; on the north ſide too the acceſs is dangerous by reaſon of many ſandbanks and rocks which lie under the water. The ſouth-weſt ſide, indeed, admits of landing, but is defended by Fort *S. Catalina.* On the ſouth ſouth-weſt point is a ridge of rocks, part of which, at full ſea, is covered with water. The outermoſt is a ſmall iſland, on which is a guard and light-houſe with two chapels, and here alſo ſtands Fort *S. Sebaſtian.* As the city, therefore, can be only attacked at the narroweſt part of the neck of land lying betwixt it and the ſouth-eaſt part of the iſland of *Leon*, it is alſo fortified on that ſide.

Cadiz was by the *Romans* called *Gades*, and by the *Phenicians Gadir*, or *Gaddir*, *i. e.* ' a hedge or fenced place,' and by ſome of the ancients corruptly called *Tarteſſus.* The *Tyrians* built the old *Gaddir*, and after them it fell into the poſſeſſion of the *Carthaginians*, from whom it came under the dominion of the *Romans.* It was recovered from the *Moors* in the year 1260. In 1596, *Cadiz* was plundered and burnt by the *Engliſh*, but rebuilt again by the *Spaniards.* In the year 1702, the *Engliſh* made
an other

another attempt upon it, but without fucccfs. The above-mentioned pillars of *Hercules*, which ftand near the beginning of the faid narrow neck of land, are, according to *Labat*, two ordinary round towers, which in all appearance were originally only windmills.

S. Petro is a little ifland lying betwixt that of *Leon* and the main-land, on which are to be feen an old tower, an hermitage, or chapel, dedicated to *S. Peter*, together with a few fifhermens huts. This is the old *Heracleum*, where once ftood a famous temple of *Hercules*. Its circumference is about four or five hundred paces.

Medina-Sidonia is a neat and pretty large town on a mountain. It is a very old place, and was anciently called *Affindum*, or *Affidonia*. At prefent it bears the title of a civdad, and is the capital of a dutchy belonging to the duke of *Medina-Celi*.

Conil is an ancient little town belonging to the dukes of *Medina Sidonia*. or *Celi*, who have here a pretty ftrong caftle. This place was formerly noted for its fifhery of tunnies, which once brought in eighty thoufand ducats, but is now reduced to eight thoufand.

Barbate is a little place on a fmall river of the fame name.

Vagel, or *Veger*, is a little town ftanding on a hill, the inhabitants of which fubfift chiefly by fifhing. It belongs to the marquis of *Medina-Celi*, and enjoys a fine profpect over the ftraits of *Gibraltar* to *Africa* and the *Atlantick* ocean.

Trafalgar is a promontory in the ftraits of *Gibraltar*, which feems to be the *Promontorium Junonis* of the ancients. The ftraits of *Gibraltar* are faid to be nine *Spanifh* miles long and above four in breadth, or about twenty-five and eleven *Englifh*.

Tariffa is a poor lonely town, on a fmall eminence in the Straits, fortified with old walls and towers. It has alfo a little caftle, in which the governor iefides, and is the chief place of a marquifate belonging to the duke of *Medina-Celi*. It was anciently called *Julia Traducta*, or *Julia Joza*; but its prefent name muft be derived from *Tarif* the *Moor*, whom I fhall have occafion to fpeak of again under the article *Gibraltar*.

Algeziras is a poor old town in the Straits, having a decayed harbour, and being reduced to a few fcattered houfes which lie in ruins. The word *Algezira*, in *Arabick*, fignifies an ifland, and the harbour being formed by two iflands it has been called, in the plural number, *Algeziras*. In thefe parts formerly ftood the town of *Kalpe*. Here the *Moors* firft landed and held the place almoft feven hundred years.

Betwixt the mountain and promontory, near *Algeziras* and the mountain at the foot of which lies *Gibraltar*, is a bay. The laft mentioned mountain is a high and fteep rock, joined to the main-land by means of a low neck of land about two hundred fathoms broad, and bounded to the weft by the above bay, to the eaft by the *Mediterranean*, where this rock

is

is of an uncommon height and almoſt of a perpendicular ſteepneſs ; yet towards the bay, or on the weſt ſide, the aſcent is not ſo difficult. It divides
itſelf into ſeveral parts, between which the ſea flows, and its capes, or points,
are defended with walls, bulwarks and towers in the old taſte. This rock
abounds with very wholeſome herbs, among which is the ranunculus. The
mountain, formerly known by the name of *Kalpe*, lies directly oppoſite
to *Ceuta* in *Africa*, which is alſo called the *Sierra Ximiera*, and *Sierra de las
monas*, i. e. *Apes hill*, but was formerly named *Abyla*. Theſe two moun-
mountains, as obſerved in the Introduction to the firſt volume, are not improbably thought to be the celebrated pillars of *Hercules*. Weſtward, at the
foot of the firſt mountain, towards the bay, lies

 Gibraltar, a famous town, called by the *Moors Gebel Tarif*, or *Taric*,
i. e. *mount Tarifs*, or *Taricks*, from the name of a *Mooriſh* general, who, at
the beginning of the eighth century, bringing over the auxiliaries of three
Mooriſh Princes to *Spain*, landed at the foot of this mountain, to which he
gave his name, and the town built afterwards on it was alſo called ſo from
him, *Gibraltar* being evidently an abbreviation of *Gebel*, or *Gibel Tarif*, or
Tarik. It is not ſo conſiderable for extent or beauty as for its ſtrength
and ſituation, which renders it one of the keys of *Spain*. It is accordingly provided with all the artillery, ſtores and forces neceſſary for its
defence. Excluſive of *Europeans* of moſt nations, here are *Jews*, *Turks* and
Moors, who are all permitted to enjoy a free trade. The harbour is formed
by a mole which is well fortified and defended. In the year 1704, the
united fleets of *England* and *Holland*, after a cannonading of only a few
hours, took it by ſurrender ; and, though the *Spaniards* endeavoured to
recover it the following year by a formal ſiege; and afterwards by a blockade,
they miſcarried in both, and at the treaty of *Utrecht* the *Engliſh* were confirmed in the poſſeſſion of it. In the year 1727, the *Spaniards* again attempted *Gibraltar*, but with no better ſucceſs than before. The inhabitants
carry on a very conſiderable clandeſtine trade with *Spain*. The author of
the *Inſtructive Accounts* ridicules the endeavours of the *Spaniards* in laying
a formal ſiege to *Gibraltar* and attempting it by force, this place being, ſays
he, impregnable, except by money or famine.

 S. Roch, near it is a new and not deep gold mine, which ſome years paſt
had been worked, but is at preſent diſcontinued on account of the vicinity
of *Gibraltar*. It muſt in all appearance have been very rich, the very mar-
caſite of it affording gold, a thing ſeldom ſeen in other mines.

 Eſtepona is a little town ſtanding high on the *Mediterranean*.

 Alcala de los Gazules, a very ancient town ſeated on a mountain.

 Zahara lies on a hill, having a very old caſtle on the top of it. It belongs
to the duke of *Arcos*, and gives the title of earl to his eldeſt ſon.

 Hardales, ſmall, and lying at the foot of a very high rock, on which is a
caſtle ſurrounded with fine fields and paſtures.

Moron

Moron is also small, and was anciently called *Arucci.*

Offuna, anciently *Urfao*, *Urfon*; and *Orfonna*, is pretty large and populous, and strongly situated, but has only one well, for the supply of the inhabitants. The neighbourhood too, for some miles, is wholly destitute of water. This place is a dutchy, and, in the year 1549, an university was erected here. It has also a collegiate church.

Marchena, anciently *Colonia Marcia*, is an old town, seated on an eminence in the middle of a plain. In the suburbs, which are larger than the town itself, is the only well in the town and neighbourhood. This place also gives the title of duke to the duke of *Arcos.*

Civdad Eeija is a little town on the river *Xenil*, and was formerly more considerable than at present. Its ancient name was *Aflygis*, or *Aftyr*; but a *Roman* colony being transported hither they called it *Augufta Firma.* It stands in a very fruitful country; and the inhabitants make great profit in particular of their wool, having large flocks of sheep. Between this place and *Offuna* are several fens and moraffes, by the *Spaniards* called *Lagunas.*

The KINGDOM *of* CORDOVA

IS much smaller than the former, but equal to it in fertility. It contains

Cordova, anciently *Corduba*, and *Colonia patricia*, a large and beautiful city on the river *Guadalquivir.* It enjoys a great trade, and the title of a *Civdad Real.* It lies at the foot of a ridge of mountains, which are a branch of the *Sierra Morena*; but the south side of the river is a wide plain. Within its circuit it contains several vineyards and gardens; but is not very populous in proportion to its extent. Its fine suburbs have the appearance of so many towns. The bishop hereof, who is suffragan to the archbishop of *Toledo*, has an income of forty thousand ducats *per annum.* The cathedral is antique, large, and magnificent; its roof being supported by three hundred and sixty-five stately pillars of alabafter, jafper, and black marble; and being built in the time of the *Moors* for a mofque, is even to this day called *Mezquita.* The episcopal palace is a large structure. The inquisition stands by the river side, and the King's palace at the west end of the city is very spacious. In the year 1589 it suffered very much by an earthquake. The adjacent mountains are covered with delightful gardens, vineyards, and plantations of lemons, oranges, figs, and particularly olives. They are also interfperfed with several pleafant vallies, having springs of good water. When the above-mentioned trees are in blossom, they diffuse a fragrancy all over the country. Befides the excellency of the wine, and the plenty of fruits and vegetables, this country breeds the fineft *Spanifh* horfes.

Lucena is a town dependent on the city of *Cordova.*

The civdad of *Andujar*, is a town of fome bignefs on the river *Guadalquivir*, having a caftle. Its-principal ·commodity is filk. The neighbourhood abounds in corn, wine, oil, honey, and all forts of fruits. It is alfo a fine fporting country. Not far from hence ftood· *Illurgis*, or *Illiturgis*, by the *Romans* called *Forum Julium* ; And the place is ftill named *Anjudar el Viejo.*

Porcunna, anciently *Obulco*, *Obulcula*, and *Municipium Pontificenfe*, is an old town, and a commandery of the order of *Calatrava.*

Martos is a little town, with a fortrefs, feated on a rock; being alfo a commandery of the order of *Calatrava.*

Alcaudete, is a town lying among mountains, being an earldom, and defended by a caftle.

Vaena, *Valna*, is a town rather large than confiderable; ftanding on a high mountain, and belonging to the Duke of *Sexi.* Near it is a moft delightful grove of lemons, oranges, dates and olives intermixed.

Palona is an earldom, belonging to the houfe of *Porto Carrero.*

Caftro Rio, very fmall, and ftands on an eminence by the river *Guamos.*

The civdad of *Luceria* lies in a country abounding with corn, wine, and oil.

The civdad of *Alcala Real* lies in a hilly country; but produces fine fruits and good wine.

Archidona, is a neat town, feated in a plain at the foot of a mountain.

The KINGDOM *of* JAEN

IS the fmalleft of the three: To it belongs

Tolofa, a little town.

Linares, fomewhat larger.

Caftona, anciently *Caftulo* or *Caftalo*, a mean place, formerly large, rich, and celebrated. Here are ftill to be feen the remains of a ftately aqueduct.

The civdad of *Baeza*, formerly called *Vatia*, is a good town feated on a hill, and raifed to an univerfity in the year 1533.

The civdad of *Ubeda* ftands in a country abounding in corn, oil, and fruits, particularly figs. In the 13th century the inhabitants were exempted from all imports and duties throughout all *Spain*, the Kingdoms of *Toledo*, *Seville*, and *Murcia* only excepted.

Caçorla

Caçorla, is a little town belonging to the archbifhop of *Toledo*, who has a governor here.

Jaen, a town at the foot of a mountain, and defended by a caftle, is populous, and has fome fine churches and monafteries: But the greateft devotion is paid here to St. *Veronica*. Its bifhop, who is under the archbifhop of *Toledo*, has a yearly income of twenty thoufand ducats. It was once the capital of a *Moorifh* Kingdom ; and the country, befides an uncommon plenty of corn, wine, oil, and fruits, abounds particularly in filk.

Mufuela, is a little town.

The KINGDOM *of* GRANADA.

THIS Kingdom, fometimes called *Upper Andalufia*, borders to the weft and north on *Andalufia* ; to the north-eaft on *Murcia*, being bounded to the eaft and fouth by the *Mediterranean*. Its extent from weft to caft is pretty nearly about forty-eight *Spanifh* miles; but in its breadth it varies greatly; being in fome places fcarce five miles wide; but from *Cabo de Gates* to *Cuefcar*, at leaft twenty. The principal rivers here are the *Xenil*, which rifes not far from *Granada* the capital, and paffing through *Andalufia* falls into the *Guadalquivir*. The *Guadalantin*, the fource of which is in the neighbourhood of *Guadix*, whence traverfing *Murcia*, it runs into the *Mediterranean*. Of leffer ftreams there is here a great number. The country is very mountainous, particularly towards the fea, though interfperfed with delightful vallies. The *Las Alpuxarras*, are mountains of a prodigious height, inhabited by a *Moorifh* race, who have embraced the *Roman* catholic religion, but ftill retain their national cuftoms and manner of living, and their language alfo is a mixture of *Arabic* and *Spanifh*. They are divided into eleven Diftrifts, by them called *Taus*; but by the *Spaniards Cabeça de partido* : The principal of thefe are *Toa de Orgiva* and *Taa de Pitros*, fituate betwixt the two fmall towns of *Pitros* and *Portugos*. The air and weather in this country is temperate and healthy. *Alhama* and *Alicun* are noted for their baths, and a fpring near *Antequera* is faid to be an excellent lithontriptic. The country in general is very fruitful, particularly betwixt *Xenil* and *Darro*, producing an exuberance of corn, wine, oil, fugar, flax, hemp, and exquifite fruits, fuch as pomegranates, citrons, lemons, oranges, olives, capers, figs, almonds and raifins, of which there are two kinds, namely, *pafferillas del Sol*, being thofe dried by the fun on the branches, and *pafferillas de Lexia*, which are dipt in a lye made of the afhes of the burnt branches, and afterwards dried in the fun. Here is alfo a great plenty of honey and wax. Several places in the mountains near

Antequera make falt, which the fun prepares from water conveyed into pans. The cultivation of filk here is alfo very confiderable : And of the galls, which are found in this country in vaft quantities, à dye is made for leather. The country alfo produces plenty of dates, of which great advantage is made, and even of the very acorns; which, to fay the truth, far exceed the fineft nuts. Befides good ftone for building, the earth affords garnets, hyacinths, and other gems. Great quantities too of fumach for preparing goats and fhamois fkins are exported from hence. Of the foude, made here, I have fpoken in the introduction, §. 4. In the time of the *Moors Granada* was the moft populous and heft cultivated province in *Spain*; but through the contempt in which the *Spaniards* hold agriculture, things are unhappily altered; yet may this country ftill vie in products with any other; and *Spain* has not a tract of land fo well inhabited as the mountains of *Las Alpuxarras*, which, exclufive of the many towns and villages interfperfed among them, are wonderfully improved by the induftrious inhabitants, who turn their vineyards and plantations to a very good account. The fea-coaft, for its protection againft the *African* Corfairs, is lined with high towers which command an extenfive profpect over the fea from the Straits of *Gibraltar* to the *Rio freyo, i. e.* ' the cold river.'

This province firft became a diftinct Kingdom in the 13th century, when the *Moorifh* King *Abenhud*, who refided at *Cordova*, having in the year 1236 loft his life and crown in a battle againft the Chriftians, his fubjects and followers betook themfelves to *Granada*, and chofe a new King, who made the city of *Granada* his capital and place of refidence. This Kingdom, which was the laft of the *Moorifh*, then contained thirty-two large towns, and ninety-feven fmaller, and continued from the year 1236 to 1492, when *Ferdinand the Catholic* reduced it and annexed it to the crown of *Caftile*. The places of note in it now are as follow, *viz.*

Granada, one of the largeft cities in *Spain*, faid to be near twelve hundred paces in circumference, ftanding partly on hills and on a level. It is fortified with walls and towers, and lies on the river *Xenil*, into which the river *Darro* falls, after watering a part of the city. This city is the capital of the Kingdom of *Granada*, the fee of an archbifhop, with forty thoufand ducats *per annum*, to whom the bifhops of *Guadix* and *Almeria* are fuffragans. Here is alfo a royal tribunal, to which, befides a prefident and fifteen counfellors, belong four criminal judges, four judges of nobles and two fifcals. The univerfity here was founded in the year 1531. It has alfo a court of inquifition; and, which is much better, carries on a very great trade in filk. It is divided into four wards, the principal of which called *Granada*, lies in the plain and in the vallies betwixt two mountains. Here the nobility, clergy, merchants, and moft wealthy of the citizens refide. Its publick and private buildings are very handfome, and the ftreets arched for the convenieney of fupplying the houfes with water by means of conduits.

In

In the cathedral lie buried feveral Kings, as namely, *Ferdinand* the Catholic and *Philip* I. with their Queens, &c. The King's tribunal is held in a large and ftately edifice, and oppofite to it is the *Alcaxeria*, or vaft building where merchants have their warehoufes. In the *Plaça mayor* are held the bull-fights. The fecond ward ftands on the mountain, and by the *Moors* of *Granada* is called *Alhambra, i. e.* ' the red;' but by the *Spaniards, La Sierra del Sol,* from its expofure to the rifing fun. The inhabitants confift of the defcendants of the ancient *Moors,* and of old *Spanifh* Chriftians, as they are termed. Here are two palaces, one built by the *Moors,* the other by *Charles* V. and *Philip* II. and both in their feveral kinds of becoming grandeur. The firft, which is very large, is environed with walls, towers, and baftions; and both ftanding high command a moft delightful profpect. A little above the old *Moorifh* palace is the magnificent and beautiful houfe of *Xeneralife,* which was alfo built by a *Moorifh* prince; and on the top of the mountain ftands a church dedicated to St. *Helena.* The third quarter, called *Albaycin* and *Alveycin,* was formerly accounted only a fuburb, and ftands on two hills. The fourth is *Antiqueruela,* which is fo called from its being inhabited by people who came thither from *Antequera,* and who work almoft univerfally in filk. When *Ferdinand* the Catholic made himfelf mafter of this city in 1462, the fhort method taken by cardinal *Ximenes* for the converfion of the *Moors,* was, that they fhould either be baptized or put to death. A little without the city are fome hofpitals and convents. The country is very fruitful, the air healthy, and the weather agreeable. The large plain extending itfelf on the fouth and weft fide of the city, called *La vega de Granada, i. e.* ' the orchard of *Granada'*, is full of pleafant towns and villages.

Santa-Fé, lies in this plain, and bears the title of a civdad, being, though but a little place, built by King *Ferdinand* the Catholic when he befieged *Granada.*

Loxa or *Loja,* is a town lying at the foot of a mountain near the river *Xenil,* and abounding in vegetables and fruits. The neighbouring mountains too afford excellent pafture for very numerous flocks of fheep.

Alhama, is a pretty town, feated in a narrow valley betwixt high and very fteep mountains. Its warm baths are celebrated, particularly as ftrengtheners, and the waters of them is alfo drank.

Antequera, a pretty large and lightly town, ftands partly on a level and on little rifings at the foot of the mountains. The ftreets are long and broad, and the houfes handfome. At one end of the city on a hill ftands a caftle. The fnow, rain, and fpring-water in the mountains run promifcuoufly into feveral pans, where the heat of the fun prepares it for falt.

Fuente de la piedra, or *Fuente de Antequera,* is a medicinal fpring of great virtue againft the ftone.

Monda

Mouda or *Munda*, is an ancient little town, feated on a hill.

Settenil, *Septenilium*, is a fmall town on a rock, out of which moft of the houfes are hewn : The neighbourhood is all pafture ground.

The civdad of *Ronda*, anciently *Arunda*, lies on a high fteep rock, environed by the *Rio Verde*, or Green river; to which there is a defcent of one hundred fteps hewn out of the rock.

Marbella, is a little fea-port on the *Mediterranean*.

Fuengirola, a little town, anciently called *Cilniana*.

Molina, formerly *Suel*, alfo a fmall town.

Cartama, formerly *Cartima*, an ancient town at the foot of a very high mountain, the north fide of which is quite barren; but all the other parts well cultivated and fruitful.

Malaga, is an ancient fortified fea-port at the foot of a fteep mountain, well built, populous, and having an harbour fuitable to its extenfive commerce. The chief exports of this place are wool, olives, oil, raifins, fack and other wines. It is defended by two caftles, *viz*. *Giblalfarro*, which is feated on a hill; and *Alcazzava*, lying below it. It is alfo a bifhopric in the province of *Seville*, and worth twenty thoufand ducats a year. Clofe by the city runs the little river of *Guadalquivirejo*. The *Phenicians* were the firft who built a town in thefe parts calling it *Malacha* or *Malaca*, from their great fale of fifh here. Not far from this city in the year 1704, an obftinate engagement was fought betwixt the *Englifh* and *Spanifh* fleets, to the difadvantage of the latter.

Velez Malaga, a town lying not far from the fea, in a pleafant plain environed with hills; has a caftle on an eminence. Its chief commodity is raifins. Near the harbour is a large building called *Torre de Velez*, which ferves at prefent for a cuftom-houfe.

The *Las Alpuxarras*, is a mountainous tract, of which I have fpoken in the general account of this Kingdom.

Puerto de Torres, is a little port at the mouth of the *Rio Frio*.

The civdad of *Almugnecar*, is a little town, having a good harbour defended by a ftrong citadel.

Salobregna, is a fmall town, on a high rock by the fea, with a ftrong caftle, where formerly the *Moorifh* Kings kept their treafure : It carries on a good trade in fifh and fugar.

Motril, is a good fea-port, having a confiderable fifhery, and abounding alfo in fine wine, fugar, and honey.

Veria, or *Beria*, formerly *Baria*, is an inconfiderable place.

Almeria, is a city lying on a bay, with a little river running into it. The country about it fertile, particularly in fruits and oil. Its bifhop is fuffragan to the archbifhop of *Granada*, with a yearly income of four thoufand ducats. Not far from hence the land projects fouth-eaftward into the fea, and

and forms a cape by the ancients called *Charideme*, by the moderns *Cabo de Gates*.

Muxacra, is a little town feated on a hill, and having an harbour.

Vera, formerly *Virgi*, is a little fea-port.

Alboloduy, alfo a fmall town.

Guadix, formerly *Acci*, and *Colonia Accitana*, is a pretty large town, feated on an eminence in the midft of a fpacious plain, terminating in mountains. Its bifhop is in the province of *Granada*, and his income eight thoufand ducats.

Baza, is a pretty good old town in the valley called *Hoya de Baca*, which produces wine, corn, flax and hemp.

The civdad of *Guefcar*, or *Huefca*, anciently called *Calicula*, a fmall town at the foot of the *Sagar*, lying betwixt the little rivers of *Guadadar* and *Dravate*, is a márquifate belonging to the Duke of *Alba*, and carries on a great trade in wool.

Velez el Rubio, is a country town.

The K I N G D O M *of* M U R C I A

BORDERS to the fouth-weft on *Granada*, to the weft on *Andaluzia* and *New Caftile*, to the north alfo on *New Caftile*, to the north-eaft on *Valencia*, and to the fouth on the *Mediterranean* fea. Among the feveral Kingdoms of which the *Spanifh* monarchy is compofed, this is the fmalleft, if *Cordova* and *Jaen* be not taken for diftinct Kingdoms. Its greateft breadth from fouth to north is pretty nearly about twenty-two *Spanifh* miles, and its greateft length, from weft to eaft, twenty. The principal river in this province is the *Segura*, anciently called the *Terebus*, *Straberum*, and *Sorabis*. It receives its fource in *New Caftile*, and taking its courfe from weft to caft traverfes this country and the Kingdom of *Valencia*, falling afterwards into the *Mediterranean* fea. The *Guadalantin* has its fource in *Granada*, and empties itfelf near *Almacaron* into the fame fea. The air is healthy and pure, and the country produces wine and corn, which are both good; but being very mountainous, not in any great quantity, fo that its chief commodities are fine fruits, fuch as oranges, lemons, &c. fugar, honey, filk, and grain, fuch as rice, peafe, &c. Of the foudo, in which it alfo carries on a great trade, I have fpoken in (§. 4.) of the introduction. In *Murcia* are the following places, *viz.*

The civdad of *Lorca* ftanding on a hill, which is watered by the river *Guadalantin*, being pretty large, but decayed, though fituated in a fruitful country.

couñtry. The inhabitants 'are chiefly new Chriſtians, 'as they are called, or baptized *Moors.*

The civdad of *Almacdrón* is 'a little ſea-port at the mouth of the river *Guadalantin,* having ſome mines of alum, which bring in a large revenue to the Duke *de Eſcalona* and the Marquis *de Vela.*

The civdad of *Cartagena* .is. a pretty good city, ſtanding on a bay, and having a caſtle ; but is greatly declined from its former grandeur, though its harbour is óne of the beſt in all *Spain.* The bay abounds ſo plentifully in mackerel, that the little lſland at the entrance of the harbour is thence called *Scombraria.* The biſhop of this place, who enjoys a revenue of twenty-four thouſand ducats, is ſuffragan to the archbiſhop of *Toledo.* The adjacent country produces à great deal of ſedge (in Latin *ſpartum*) from whence the city itſelf received the name of *Spartarium,* and the country that of *Spartarius campus.* Diamonds, rubies, amethiſts, and other gems are likewiſe found here ; but particularly many mines of alum. *Hannibal* or *Hamilcar* are ſaid to have been the founders of this city.

Palos, is a cape running eaſtward into the ſea.

The civdad of *Murcia,* capital of the province, lies in a pleaſant plain on the river *Segura,* being large and populous, and having ſtraight handſome ſtreets, with fine churches, convents, and other publick buildings. It is the ſeat of a court of inquiſition, and is defended by the caſtle of *Monte-Agudo,* which ſtands on an eminence without the city. The country abounds in fine fruits, and particularly in oil, ſugar-canes and ſilk.

Cantarilla, is a little place on the river *Segura.*

Mula, which is ſomewhat bigger, ſtands in a fruitful plain on the ſame river.

Totana, is a ſmall town belonging to the order of *Santjago.*

S. Crux de Caravaca, a little place ſeated on a ſmall river which runs into the *Segura ;* is famed for a miraculous crucifix.

Lorqui, Cieça, and *Calaſpara* are all inconſiderable places.

Tavara, or *Tobarra,* is a little town.

The KINGDOM of *VALENCIA*

I S bounded to the weſt by *Murcia* and *New Caſtile,* to the north by *Aragon* and *Catalonia,* and to the eaſt and ſouth by the *Mediterranean* ſea. Its extent from ſouth to north is pretty nearly about fifty *Spaniſh* miles; but from weſt to caſt, and that only in ſome places, is only twenty ; being much leſs in moſt. It is one of the beſt watered provinces in *Spain.*

All its rivers, large and small, run cast or south-cast into the *Mediterra-nean*. The *Segura*, already mentioned under the article *Murcia*, joins the sea not far from *Guàrdamar*. The *Xucar* comes from *New Castile* and, having received the rivers *Cabriel* and *Oliara* in its course, joins the sea near the town and cape of *Cullera*. The *Guadalaviar*, by the *Romans* called *Durias*, *Turias*, and *Dorias* is every where delightfully bordered with woods and flowers. It receives its source in the confines of *Aragon* and *New Castile*, losing itself in the sea below *Valencia*. The *Morviedro* runs into the sea below the place of the same name, as *Millares*, *Millas*, or *Miglias* does below *Villa Real*. The air is temperate and agreeable. The country, besides its extraordinary fertility in wine and fruits, produces also rice, flax, hemp, silk, honey and sugar; and, if the surface of many of its mountains want that fertility, this is well compensated by the minerals within. This province was much more populous formerly, though even at present, it is one of the best peopled. In the year 788, it was erected into a Monarchy. The most remarkable places in it are the following; *viz.*

The civdad of *Orihuela*, in Latin *Orselis*, standing in the midst of a de-licious plain on the *Segura*, and being environed with mountains. It is defended by an old castle, is an university and likewise a bishop's see subject to the archbishop of *Valencia*, and worth ten thousand ducats *per annum*. It is also the capital of a particular District.

Guardamar is a little sea-port, at the mouth of the *Segura*, famous for its exportations of salt.

Torra de las Salinas, *Dita Mata*, or *La Mata*, by the *Dutch* called *Alematte*, or *Alimatte*, carries on a great trade in salt, with which it is furnished by a small lake formed by saline springs; and, though the water be apparently weak-ened by rain, yet is salt made of it by the exhalations of the sun. The usual quantity thus produced, one year with another, is about nine hundred thousand fanegas, each fanega being one hundred and fifty pounds weight; but, when heavy rains are succeeded by excessive heat, it comes little short of one million and five hundred thousand. This is the most considerable salt-work in all *Spain*.

Elche is a handsome town, situate in a most delightful neighbourhood abounding in wine, dates, exquisite fruits, olives and very large palm-trees, and gives the title of marquis.

Aspe is an inconsiderable place.

Elda is a little town and the capital of a county.

Alicant, a good sea-port, fortified with a few bastions. Its castle stands very high, and this town is famous for its red and white wine, particularly the red. Meal, and every other necessary of life, is here very good. In the year 1706, it was taken by the *English*, but, in 1708, recovered again by the *Spaniards*, who, in the year 1709, retook also the strong castle. It exports foap, wine and annife feed; and along the coaft are feveral towers where a watch is continually kept againft the Corfairs.

Xicona is a little town, feated among hills, being defended by a caftle, and famous for its wine.

Villa Loyfa is a little fea-port near *Cape Finiftrat.*

Alcoy is a pretty little town on a river of the fame name. Near it, in the year 1504, were difcovered fome iron mines.

Biar is a little town noted for fine honey.

Contentayna is a town and an earldom feated on the mountain of *Mariola,* on which grows an uncommon quantity both of curious and efculent herbs.

Onteniente is a little town where, in the year 1752, was erected an academy of beaux arts.

Vellada is a fmall place.

Ayora the like, but its inhabitants are faid to fpeak *Caftilian* in its purity.

Milleres is alfo a little town.

Montefa, a fortrefs.

Xativa was once one of the moft beautiful towns in all *Spain,* but fiding with *Charles* III. in the year 1707, *Philip* V. ordered it to be demolifhed, and, inftead of it, a new town to be built called St. *Philip.*

Altea is a little fea-port trading in wine, flax, filk and honey.

Cape Artemus is the ancient *Artemifium,* but is now called *Cabo de S. Martin,* and alfo *Punta del Emperador.*

The civdad of *Denia,* in Latin *Dianeum* and *Artemifium,* from *Diana,* in honour of whom it was built; as alfo *Hemerefcopeum,* on account of an high tower there, lies at the foot of the mountain *Mongon,* on a declivity, ftretching to the fea. Here is a tower of extraordinary height from whence fhips are perceived at a vaft diftance, as alfo a caftle, very ftrong both by art and nature, with a convenient double harbour. This place gives the title of marquis.

Oliva is a little town, giving the title of earl to the duke of *Gandia.*

Civdad Gandia, lying not far from the fea on the river *Alcoy,* gives the title of duke to the houfe of *Borgia,* and has an univerfity founded in the year 1549.

Alzyra is a little town feated on the *Xucar,* but having a very extenfive filken trade.

Algemezin is alfo a fmall place.

The civdad of *Valencia,* which is the capital of the province, lies on the river *Guadalaviar,* in a very pleafant and fertile bottom, being large, beautiful, defended with fome fortifications and populous, and inhabited both by perfons of rank and merchants, the latter of whom drive a confiderable trade here. It is the fee of an archbifhop, whofe yearly revenue amounts to forty thoufand ducats, having the bifhops of *Segorve, Orihuela* and *Majorca* for

his

·his fuffragans. In the year 1470, an univerfity'was erected here, and it has a tribunal of inquifition with an *audiencia real*, in which, befides the gover-ɲnador and captain-general of the province, a regent prefides. Next to thefe are eight counfellors, four jufticiary or criminal officers and two fifcals. Several monuments of antiquity have been found here. The city is very ancient, and was firft built, in the fix hundred and fixteenth year from the foundation of *Rome*, by the veteran foldiers who had ferved under *Viriatus*; for which reafon the inhabitants ftyle themfelves *Veteres*, or *Veterani*. It was demolifhed by *Pompey*, but afterwards rebuilt. In the year 524, a general council was held here, and the *Moors* were in poffeffion of it till 1238. The chief export from this place is almonds.

Porto el Grajo is a little walled town fortified towards the fea.

Morviedro, in Latin *Muri Veteres*, is an ancient town, feated on a high rock near a river which takes its name from the town. It is the remains of the celebrated city of *Saguntum*, which was fet on fire by its own inhabitants when they chofe to do that rather than furrender to the *Carthaginians*, who lay before it. Its name is derived from the walls ftill exifting, which fhow the circuit of the ancient *Saguntum*. Here are alfo to be feen the remains of an amphitheatre and palace.

Civdad Segorve, in Latin *Segobriga*, is an ancient pleafant city feated near the *Morviedro* on a declivity betwixt mountains. It is the fee of a bifhop fuffragan to the archbifhop of *Valencia*, with an annual revenue of ten thoufand ducats. It gives the title of duke to the family of *Cordona*.

Xeriça is a fmall place, but yet the capital of a county, which, fince the year 1565, has been re-annexed to the crown.

Almenara, a little town and an earldom, fituate not far from the fea.

Carpefa and *Moneada* are both little places.

Burriana too a fmall fea-port fituate in a very fertile neighbourhood.

Villa Real is feated on the river *Millas*, but, in the year 1706, was almoft deftroyed for its attachment to King *Charles* III.

Honda is a little town on the river *Millas*, ftanding among mountains which produce feveral kinds of wholefome vegetables.

Villa Hermofa is alfo a fmall town on the river *Millas* giving the title of duke.

Viflabélla too a little place.

Adzeneta, fmall and feated on the mountain *Pegna Golofa*, on which grow multitudes of efculent plants.

Caftello della Plana is a very confiderable fea-port.

Oropefa, a little town, feated at the foot of a mountain near the fea.

Penifcola, or *Penofcola*, a fmall place on a high promontory called *Cabo Forbat*, is environed on three fides by the fea, and the difficulty of the accefs on the land fide renders it a place of great ftrength.

Las Cuevas, Salſadella and St. *Matheo* are three little towns; the laſt is
ſtrongly ſituated.

Morella, a little place, on the borders of *Aragon,* environed- by rocks
and high mountains, was pillaged in the year 1705 by the troops of *Philip*
V. the effects of which diſaſter are ſtill too viſible in it.

Talets and *Traiguera,* ſituate on the river *Servol,* are two ſmall towns.
Hoſtalet is a frontier village lying betwixt *Valencia* and *Catalonia.*

The KINGDOM *of* NEW CASTILE.

CASTILE is the principal and moſt opulent Kingdom in all *Spain.*
It is generally divided into the *Old* and *New;* the former part having
been recovered from the *Moors* ſome time before the latter. *New Caſtile,*
of which, in regard to the order of ſituation, we ſhall firſt ſpeak, is by ſome
called alſo the Kingdom of *Toledo,* being the centre of the Monarchy and
the King's reſidence. It is divided, on the north, by a chain of mountains
from *Old Caſtile;* and a like chain divides it alſo eaſtward from *Aragon* and
Valencia. On this ſide it borders upon *Murcia.* To the ſouth it is likewiſe ſeparated by a chain of hills from *Andaluzia;* and weſtward is bounded
by *Eſtremadura.* Its greateſt extent, from caſt to weſt, is about forty-
ſix *Spaniſh* miles; and its greateſt breadth, from north to ſouth, about
fifty. It enjoys a good air and is very fertile. The *Tagus, Guadiana* and *Guadalquivir* have already been ſaid to receive their ſources in
this province. The two firſt run from north-caſt to ſouth-weſt quite through
it, and greatly improve its natural fertility. The other rivers of conſideration
in this province are the *Xucar,* which makes its way through *Valencia* into
the *Mediterranean* ; the *Xarama,* which takes its riſe in a mountain of *Atiença,* and, after receiving the *Henares* and *Tajuna,* falls into the *Tagus* and
the *Guadarama* ; which riſing in the mountains of *Toledo* runs from north to
ſouth through the province, loſing itſelf below *Toledo* in the *Tagus.*

New *Caſtile* was formerly governed by earls, the laſt of whom, by name
Garcias, dying without iſſue, in the year one thouſand twenty-nine, bequeathed his dominions to his ſiſter *Nugna,* ſpouſe to *Sancho* the Great,
King of *Navarre,* who raiſed this province to a Kingdom. *Ferdinand,*
King of *Leon,* ſucceeded to it on the demiſe of *Henry* of *Caſtile,* and, in
the year 1217, both Kingdoms became united in his perſon ; but the laſt
and permanent union of the Kingdoms of *Caſtile* and *Aragon* was, by the
marriage of *Ferdinand,* Prince of *Aragon,* with *Iſabella,* heireſs of *Caſtile.*

2   I have

I have already obſerved, under the article *Eſtremadura,* that this country is incorporated with *New Caſtile,* which of itſelf conſiſts of three cantons; namely,

1. *La Mancha,* which is the ſouthern part of the country, and contains the following places; *viz.*

Almodavar del Campo, a town ſituate in a pleaſant valley at the foot of mount *Morena,* and defended by a caſtle.

Elviſo, a little town at the foot of the ſame mountain.

Miguel Turra, ſmall, but ſeated in a plain which abounds in corn, wine and oil.

El Convento de Calatrava, a place in the ſame delightful ſituation as the former, and capital of the order of *Calatrava.*

Almodavar del Campo, a little town belonging to the order of *Calatrava.*

Almagro, a little town belonging to the ſame order, and capital of the Diſtrict of *Campo de Calatrava,* having medicinal ſprings. Its environs are level and fertile.

Civdad Real is a pretty and well inhabited town, noted for its excellent wine; but its neighbourhood ſuffers ſometimes by the inundations of the *Guadiana.*

Calatrava, which gave name to the order ſo called, is ſeated on the *Guadiana.*

Villa Rubia De los Ojos is a ſmall place, and ſo called from the *ojos,* or eyes, of the *Guadiana* in its neighbourhood; for which ſee Introduction, (§. 5.)

Confuegra, a commanderie belonging to the order of *Malta,* ſtands at the foot of a mountain, and is defended by two caſtles.

Mora is ſmall but the capital of an earldom.

Orgaz is alſo ſmall and an earldom.

2. *La Sierra,* which lies in the caſt part of *New Caſtile,* and is ſo named from its numerous mountains, contains

Velez el Rubio, a little town, but formerly fortified, being a commanderie belonging to the order of *Santjago.*

Segura de la Sierra, alſo ſmall, but the richeſt commanderie of the order of *Santjago.*

Civdad Alcaraz, which ſtands in a very fertile country, and is defended by a caſtle ſeated on an eminence.

Montiel and *Villa Nueva de los Infantos,* both inconſiderable places.

Villena, a fortified town, and

Almanza, where, in the year 1707, the army of *Philip* V. defeated that of *Charles* III.

Chinchilla, Albazete, S. Clemente and *Mieſta,* all places of little note.

Alarcon, a mean town, on the river *Xucar,* built in the year 1178, and ſoon after razed by the *Moors,* but retaken again by *Alphonſo* IX.

Moya,

Moya, a town on an eminence on the river *Algarra,* being an earldom
and defended by a caftle.

Valeria, an ancient place, ftanding on a hill, and formerly of fome confideration.

Civdad Cuenca, anciently *Conca,* ftands high among mountains. Its
bifhop, who is fuffragan to the archbifhop of *Toledo,* enjoys a yearly revenue
of fifty thoufand ducats. This place is alfo the feat of a court of inquifition.

Caracofa, or *Caracena,* is the capital of a marquifate.

Molina is a fmall town feated amidft excellent paftures, which are full of
cattle, and particularly of fheep.

3. *Algarria,* which is the north part of *New Caftile,* contains the following places, *viz.*

Madrid, the metropolis of all *Spain,* and the refidence of the King. It
ftands in the centre of a large plain, which is terminated on all fides by
mountains. To the weft of the city, where it has neither walls nor moats,
it is watered by the *Manzanares,* which fwells in winter by the melting of
the fnows; but for the greateft part of the year is naturally fhallow, particularly in fummer. King *Philip,* however, was at the expence of a very
ftately ftone bridge over it, the whole length of which is eleven hundred
paces, and to the extent of feven hundred, it is twenty-two broad. This
bridge is called *La Puente Segoviana.* Over this little river, on the fouth
fide of the city, *Philip* V. built a much finer bridge, which is diftinguifhed
by the name of the *Bridge of Toledo.* The city is large, and adorned with
beautiful fquares; the ftreets too are wide and ftraight, having feveral fine
fountains in them with ftately convents and hofpitals. The churches too are
moft magnificently decorated, particularly the chapel of *S. Ifidorus;* fome of
the noblemens houfes are alfo very fplendid. The inhabitants are computed
at one hundred and fifty thoufand. Provifions of all kinds are both excellent
and reafonable in this city, and the continual refidence of the court here
caufes a biifk trade and circulation of money. One great nufance in it is
that of throwing, in the night time, all manner of filth into the ftreets,
which, in hot weather, muft occafion a very offenfive fmell *, and in
rains turn them to an infupportable mire. Hence it is that the city may
often be very fenfibly fmelt at the diftance of five or fix *Englifh* miles.
On each fide, however, of the ftreets are two narrow paffes where a perfon
may walk dry-footed. The air is beft and pureft in the court part of the
town. The large maiket wants nothing but cleanlinefs to make it a very
fine place. It is geneially occupied by flefh and green ftalls and the like;
and, being likewife the fcene for the bull-fights, all the houfes on each
fide have balconies for the conveniency of viewing thefe fpectacles. The
other fquares in *Madrid* are the Sun-maiket, the market *del la Sabada,* and

* General *Stanhope,* who was many years ambaffador in *Spain,* affures us to the contrary.
He tells us, and not improbably, that the intenfe heat of the fun fo parches up the filth, that
be never obferved any difagieeable fmell to arife from it.

that

that of *S. Domingo.* The King's palace ftands on an eminence in the weft fide of the city, commanding a delightful profpect of the river *Manzanares* and the country beyond it. In the year 1734, it was burnt down, but fince that has been rebuilt with greater magnificence. In this city are the colleges of which mention has been made in §. 14. of the Introduction, and three royal academies ; *viz.* the *real academia efpannola,* inftituted for the improvement of eloquence and the *Spanifh* tongue ; the *real academia de la hiftoria,* and the *real academia medica.*

Buen Retiro is a royal feat, eaftward of the city, built by *Philip* IV. and confifts of four main parts which form a complete fquare, with a pavillion at each angle. In the area ftands an equeftrian ftatue of *Philip* IV. It has likewife a large grove beautifully variegated with canals, fummer-houfes, parterres and fountains. Near this palace are alfo two moft delightful hermitages, called by the names of *S. Antony* and *S. Paul,* but they may more properly be accounted pleafure-houfes. In this charming recefs the King ufually fpends the whole fpring and a great part of fummer.

Cafa del Campo is a royal feat, ftanding on the other fide of the river to the weft of the city, directly facing the King's town palace, and is a very delightful place having a fine park. At the entrance of the garden ftands an equeftrian ftatue of *Philip* III.

Florida is another royal pleafure-houfe, lying near the former in an enchanting fituation.

Le Pardo is a royal feat, having a garden and park, being a large fquare building with a tower at each corner. Here it was that, in the year 1739, the marquis of *Villarias* and the *Englifh* minifter, Mr. *Keene,* figned the famous convention for accommodating all mifunderftandings betwixt thofe two powers. On a mountain not far from hence ftands a convent of *C*apuchins, which is vifited with great devotion on account of a fuppofed crucifix there.

Sarfuela is a pleafant royal feat. .

S. Ildephonfe, is a palace, five *Spanifh* miles from *Madrid,* in a delightful lonely country : part of it is alfo a convent where the prefent King's mother-in-law fpends her days. Here too King *Philip* V. retired on his refignation of the crown in the year 1724, and caufed the reliques of fome faints, which were highly revered, to be brought from the efcurial into this chapel.

The efcurial, which is feven *Spanifh* miles from *Madrid,* is a moft fuperb ftructure, but ftands in a wild barren country, being enclofed almoft on every fide with mountains. It was built by *Philip* II. purfuant to a vow, for the victory obtained at St. *Quintin* by his army over that of *France.* It was begun in the year 1557, and took up twenty-two years and immenfe fums of money before it was finifhed. It is indeed an amazing ftructure, and has the appearance of a town. The windows in it are computed at eleven thoufand. It contains twenty-two areas and feventeen

cloifters,

cloyfters, but a profound ftillnefs reigns throughout the whole. It forms an oblong fquare, and towards the mountains, or the weft, is the fartheft and principal fide, which has three ftately towers upon it; but that in the middle is the largeft and moft ornamented, having a grand portal fupported by eight pillars of the *Doric* order ftanding on each fide on a bafe of one hundred and thirty feet in length and an *Englifh* ell in height. This work, which is full fifty-fix feet high, terminates in the *Corinthian* order, fupporting four *Ionic* pillars of moft exquifite workmanfhip with two pyramids on each fide of them. Betwixt thefe four pillars are two rows of niches, and in the firft, which is over the gate, are to be feen the arms of *Spain*, and over them, in the fecond row, the image of *S. Laurence*, both mafter-pieces of fculpture. This gate leads to the church, the convent, the King's apartment and the college. It is twelve feet broad and twenty-four in height; whereas the two doors on the fides are only ten feet broad to a height of twenty. Thefe three doors, with the two elegant pieces of building at the end, form a very grand view. In the firft fide there are two hundred and twenty-five windows. The oppofite fide towards the eaft is of the fame length, but does not join in the middle. It has five doors and three hundred and fixty-fix windows. Round it is a terras fupported by arcades and inclofed with iron rails about three feet in height, beyond which is a fpacious area. On the fouth fide, which is not fo long as the former, are three hundred and fix windows placed in five rows; but the north, on account of the cold winds, has only one hundred and fixty, though, in all other particulars, anfwerable to that of the fouth. The infide of this palace is divided into three parts; the middle of which contains the church with a fpacious lobby; the two others, which form the ends, are divided into two main buildings, one eaftward and the other weftward; which are fubdivided again into four fpacious cloifters correfpondent to each other in dimenfions and architecture, with a fine marble fountain in the midft of each. On the right fide, facing the fouth, is the convent, which confifts of five cloifters, and is fo large as to take up the whole eaft fide. And the north too has the like divifions. Such is the extent of the eaftern part of this north fide, that it includes all the apartments of the royal family. The four fmall cloifters in the weft are affigned to the officers of the houfhold and ftudents. The convent belongs to the *Jeromites*, who are two hundred in number, and are fplendidly provided for. They hold lectures in all fcholaftic languages. At pafling through the great door to the firft mentioned weft fide, we come into a magnificent portico, which leads into a large court fronting the north part of the church. This portico, which is twenty feet in breadth and eighty in length, being adorned alfo with a very fine roof, feparates the convent and college; and over it is the library containing a large and choice collection of books. The two fides of the court are taken up by two large buildings four ftories high; that on

the

the right being the convent; and the other, on the left, containing the college halls and the King's apartment, which, exclufive of its admirable paintings, has little in it fuitable to the magnificence and grandeur of the other parts of the building; the liberality of the founder being chiefly taken up in enriching the church, which, as well as the whole building, is dedicated to *S. Laurence.* It contains forty chapels and as many altars. In the principal chapel is kept the church treafure; and in it likewife is the great altar, which, from the pavement to the roof, is of the fineft jafper. This chapel alone is faid to be worth five millions of piaftres. The veftry on one fide of the choir is a very large faloon, decorated with paintings by *Titian* and other celebrated mafters. In this are kept the veftments and furniture for the altar, with the plate, and, exclufive of many other valuable pieces, a gold crucifix fet with very large pearls, rubies, turquoifes, emeralds and diamonds.

The pantheon, or burial-place, of the Kings and Queens of *Spain*, lies under the high altar, and is, beyond difpute, the moft ftriking part of the efcurial. It was begun by *Charles* V. and religioufly carried on by *Philip* II. and III. till *Philip* IV. put the finifhing hand to it. Among many other remarkable pieces here, what moft attracts admiration, is the arms of *Spain* emblazoned in its feveral colours by all kinds of gems, which are enchafed with amazing fkill and beauty. The riches and fplendor of this vault, with the metals and gems glittering on all fides, render it indeed the pride of *Spain*; but, at the fame time, it is an affecting monument of the vanity and tranfitorinefs of all fublunary grandeur. The roof is fupported by fixteen pillars of jafper of different colours, behind which, in perfpective, are placed others of marble, the chapiters of all which are of gilt brafs. The chapel of the pantheon ftands at the end directly oppofite the door, and, exclufive of a multitude of other refulgent decorations, contains a gold crucifix richly fet with diamonds. The fpace on the fides of this chapel is difpofed in four rows of magnificent niches of equal bignefs, which are filled with twenty-fix coffins of black marble, containing the deceafed Kings and Queens, and being curioufly decorated with works of gilt brafs. The bodies of thofe Princes and Princeffes, and even of the Queens, who left no male infants, are not depofited here, but in two vaults on the fide of the pantheon. The efcurial, befides its uninterrupted profpect towards *Madrid*, has a fine walk extending to the village of *Efcurial*, being a length of about two *Englifh* miles and a half.

Cadahalfo is a fmall pretty town furrounded with delightful gardens and woods.

Efcalona ftands high, near the little river of *Alberce*, being defended by a large caftle. It is alfo a dutchy and has a collegiate church.

Maqueda is the capital of a dutchy. The neighbourhood of this town is covered with olive and vine-yards.

Talavera la Reyna, formerly *Libura*, or *Ebura*, is a handfome town feated in a valley on the *Tagus*, and fortified in the old manner with walls, towers and breaft-works. It was formerly appropriated to the Queen's revenue, and, at prefent, is famous for its earthen-ware. It boafts many inhabitants of quality, and contains alfo a collegiate church. Provifions of all kinds are very cheap here.

Talavera la Vieja is a town on the *Tagus*.

Cebola is a large town.

Toledo, an ancient fortified city, ftands high on the river *Tagus*. The ftreets are narrow and its fite very unequal; yet, exclufive of a great number of handfome private houfes, here are alfo feveral ftately edifices, with feventeen fquares and thirty-eight convents, of which, that of the *Francifcans*, built by *Ferdinand* and *Ifabella*, is the principal. Next to this is the convent of the Predicant Monks. It contains alfo many churches and fome hofpitals. The cathedral here is one of the moft confiderable and the richeft in all *Spain*; and, adjoining to it, is the archiepifcopal palace. The archbifhop, of whom mention has been already made in §. 7. of the Introduction, has a revenue, one year with another, of three hundred thoufand ducats: that of the cathedral is alfo one hundred and fifty thoufand, but with the defalcation of fixty-fix thoufand payable to the King. Beneath the jurifdiction of this archbifhopric are five cities, one hundred and nine towns, five hundred and fixteen villages, four collegiate churches, twenty-five archpriefts and thirty-fix convents, exclufive of two hundred and fixty-four other free convents in the towns and villages which belong to this archbifhopric. It has alfo an univerfity founded in the year 1475, and is the feat of a court of inquifition. The King's palace, by the inhabitants called *Alcaçar*, is a large building feated on a rock on one fide of the city, which it commands, as it alfo does the *Tagus*, which wafhes the bottom of this rock and the neighbouring country. *Toledo* is populous, and, exclufive of its fine fword-blades, carries on a great trade in wool, filk and ftuffs, but with the great inconvenience of going down to the *Tagus* for water. It has three bridges over that river. The adjacent country is quite dry and barren, except that part which borders on the *Tagus*. No lefs than eighteen councils have been held in this city.

The *Spanifh* hiftorians relate, that at *Toledo* formerly ftood a tower called *Hercules's* Chamber, concerning which a ftrange notion prevails among the people, that the opening it would prove the ruin of *Spain: Roderick*, however, in hopes of finding a treafure, ordered it to be opened, when all he found in it was a cheft containing a board, on which were painted fome men of a fierce appearance, with this *Latin* infcription, *By fuch a people will* Spain *foon be overthrown*. This was univerfally interpreted of the *Moors*; but the tower door was no fooner fhut than it funk into the earth and difappeared.

difappeared. Abfurd as this ftory is, yet feveral hiftorians and geographers mention it with abundance of gravity.

Yepes is a fmall place, noted for its wine and oil.

Ocanna, a little town, which, befides its wine and oil, makes fine earthen-ware.

Villa Rubia ftands in a fertile and well cultivated foil.

Arenjuez is a royal feat, chiefly to be admired for its delightful fituation. It lies about feven *Spanifh* miles from *Madrid* and fix from *Toledo,* being feated in a charming plain on an ifland formed by the rivers *Tagus* and *Xarama,* which are united above it by a canal. The fine gardens, par-terres, fountains, cafcades, grottos and walks on all fides of it, render it a paradife of natural beauties.

Anover is a fine town on the *Tagus.*

Illefcas, a little place on a fpacious pleafant plain.

Leganes is the capital of a marquifate.

Fuente Duegna, a little place lying on the *Tagus.*

Zurita is another little place on the *Tagus,* having a caftle, and being a commanderie belonging to the order of *Calatrava.*

Paftrana, the principal place of the dutchy of the fame name has a col-legiate church.

Arganta is a little town with a caftle.

Alcala de Henares, anciently *Complutum,* is a town feated on the river *Henares.* The houfes in it are well built and the ftreets handfome. It has a collegiate church and a celebrated univerfity, which was reftored in the year 1494 by Cardinal *Ximenes;* and to him is owing the inftitu-tion that the profeffors, at a certain age, are made prebends. The uni-verfity is a ftately building and contains a church, in which the above-mentioned celebrated cardinal lies buried. The town belongs to the arch-bifhop of *Toledo.*

Civdad Guadalaxara, anciently *Arriaca,* or *Carraca,* lies on a rugged eminence near the river *Henares,* but the adjacent country is fertile. Here the duke *de Ripperda* fet up feveral manufactures of cloths and other ftuffs, which feemed to promife well, but, on his difgrace, in the year 1726, they fell to decay.

Briuega, or *Brioca,* is a little town, feated on the *Tajuna,* and having a caftle. It trades in wool and ftuffs. At this place General *Stanhope,* commander of the *Englifh* forces, was, in the year 1710, obliged to fur-render himfelf and his men prifoners of war.

Tortofa is a little town.

Hita, anciently *Ceffata,* an old little town, ftanding on an eminence and defended by a fort.

Caducra, a fmall but pretty place.

Buitraga

Buitraga too is fmall, but ftands on à rock which is fortified both by art and nature, and belongs to the duke *de Infantado*, who has a very fine palace here.

Uzeda is the capital of a dutchy, and has a caftle.

Alcovendas is a little town in a barren neighbourhood.

The KINGDOM *of* OLD CASTILE.

THE name of *Old Caftile* was given to this province as being recovered from the *Moors* fooner than *New Caftile*. To the fouth it joins *New Caftile*, to the weft *Leon*; on the north it is bounded by *Afturia* and *Bifcay*, and on the eaft by *Navarre* and *Aragon*. Its figure is very irregular; and confequently its dimenfions very different. Its greateft length, namely, from *Valladolid* to *Taraçona* is pretty nearly about twenty-eight, and its greateft breadth about forty-five *Spanifh* miles. The principal rivers in it are the *Duoro* and *Ebro*, the firft of which is joined by the *Atayada*, *Andoja*, *Arebarillo* and *Pifuerga*, which alfo receives the *Arlança* and *Arlançon*. The fources of all thefe rivers are in this province, which is mountainous, and not fo fertile as *New Caftile*. The moft fruitful part in it is a tract called *la Tierro de Campos*, lying north near *Medina de Rio Seco* and *Palencia*. The wine produced here is excellent; and its plains are covered, with cattle, particularly with fheep, the wool of which is the beft in all Spain. It was formerly only a county, fubject to the Kings of *Leon*; but in 1016 was erected into a Kingdom. The moft remarkable places in this province are as follow, *viz.*

Pedraçu de la Sierra, a town feated on the river *Duraton*, having a caftle, in which *Francis* Dauphin of *France*, and *Henry* his brother, children to *Francis* I. were confined for four years. This was the birth-place of the excellent Emperor *Trajan.*

Sepulveda, is a little town on an eminence, feated amidft fteep rocks on the river *Duraton*: It was anciently called *Segobriga* and afterwards *Sepulvega*, which is eafily altered to its prefent name.

The civdad of *Avila* is an ancient middling city, feated in a fine plain, being environed with mountains, and producing excellent grapes, and other fine fruits. Its bifhop has a revenue of twenty thoufand ducats *per annum*, and is fuffragan to the archbifhop of *Santjago*. The univerfity here, which was founded in 1445, was confirmed in 1538, by Pope *Gregory* XIII. and afterwards enlarged. The convent of *Auguftines* here has a celebrated image of the Virgin *Mary*. In the town too are alfo fome manufactures.

Mengra—

Mengravilla, a village not far from *Avila*, contains ſalt mines.

Hontiveros or *Fuentiberos*, is ſituated in a delightful plain, abounding in corn, wine, fruits and ſaffron.

Arevalo, is a town at the conflux of the rivers *Andaja* and *Arebalillo*. It is ſurnamed the *Noble*, from the number of noble and diſtinguiſhed families which have received their birth here.

The civdad of *Segovia* is a city of great antiquity, and ſtill conſiderable. It ſtands high betwixt two hills, being large, populous, and handſome, having excellent manufactures of cloth and paper; and exporting alſo a very fine ſort of wool. The biſhop of this place is ſuffragan to the archbiſhop of *Toledo*, and has a yearly revenue of twenty-four thouſand ducats. The King's palace ſtands on a rock in the higheſt part of the city; and, excluſive of cannon, has alſo a garriſon. The duke *de Ripperda*, at his diſgrace in the year 1726, was brought as a priſoner to this place, where he continued till 1728, when he found means to make his eſcape. At the foot of it runs the river of *Atyada*. The *Roman* aqueduct here, called *Puente Segoviana*, is a moſt amazing fabric, being carried from one mountain to another to the length of three thouſand paces, and ſupported by two rows of ſeventy-ſix lofty arches: It paſſes through the ſuburbs, and conveys water over the whole city.

S. Maria La Real de la Nieva, is a little town ſtanding on a rock near the river *Atyada*, and noted for an image of the Virgin *Mary*.

Olmedo is another little town, in a very fertile and delightful plain.

Madrigal too is a ſmall town in a plain, abounding with corn and excellent wine.

Pennafiel is the capital of a marquiſate, giving title to the eldeſt branch of the houſe of *Oſſuna*. It lies at the foot of a high mountain, and is defended by a caſtle. The fertility of the ſoil is very extraordinary; and its cheeſe, in particular, is accounted ſome of the beſt in *Spain*.

Coca, a little town, ſeated on an eminence among mountains, belongs to the count of *Alcala*, and is called the Prince's priſon, *Philip William* Prince of *Orange* having been impriſoned in its caſtle.

Cuellar, anciently *Colenda*, is a little old town ſtanding high, and giving the title of marquis to the eldeſt branch of the ducal houſe of *Albuquerque*.

Atiença is a little town on a mountain of the ſame name, near which is ſome excellent corn-land, paſtures; and ſalt-ſprings.

The civdad of *Siguenza*, anciently called *Seguntia*; is an old city lying at the foot of Mount *Atiença*. It ſtands high, with the river *Henares* running at its bottom; and, beſides good walls, has alſo a caſtle. The yearly revenue of its biſhop, who is ſuffragan to *Toledo*, is forty thouſand ducats. The univerſity here was erected in 1471.

Fuencaliente, is a little town near the ſource of the *Xalon*.

Medina

Medina Celi, in Latin *Methymna cœleftis*, is the cápital of a dutchy, to which belong eighty other villages, and on the death of Duke *Lewis Francis*, devolved to his eldeft fifter the dutchefs dowager of *Feria*. This place was formerly very flourifhing.

Arcos is a town.

Monteaguda is a town, giving the title of earl to the marquis of *Almazan*.

Almazan, is a pretty village on a marquifate, where a fuppofed head of St. *Stephen* the protomartyr is worfhipped.

The civdad of *Ofma*, in Latin *Uxama*, is a little old town on the river *Duoro*. Its bifhop, who is fuffragan to the archbifhop of *Toledo*, has a yearly revenue of fixteen thoufand ducats. Here is alfo an univerfity. In the time of the *Romans* it was much larger than at prefent, but was deftroyed by *Pompey*; and the ruins of it are ftill vifible.

Burgo de Ofma is a little town on the other fide of the river fronting the city.

San Eflevan de Gormaz, is the capital of a fmall earldom belonging to the duke of *Efcalona*, and ftands on an eminence near the *Duoro*.

Aranda de Duoro, is a large handfome town on the river of that name.

Crugna is a little town, but an earldom, and defended by a caftle.

Roa is a little town, feated on a large plain near the *Duoro*, having a caftle belonging to the count of *Siruela*.

The civdad of *Valladolid*, in Latin *Valiboletum*, and anciently called *Pindia*, is a large, beautiful, populous city, fituate in a fpacious delightful plain, which is watered by the *Pifuerga*. It is one of the moft ftately cities in *Spain*, having long, broad, fine ftreets, with large and lofty houfes, fplendid palaces, fpacious and elegant fquares, piazzas and fountains, feventy convents, among which the *Dominican* convent of St. *Paul*, for its magnificent church, the college of St. *Gregory* and that of the Jefuits, are more particularly worthy notice. Here is alfo a fine royal palace near the *Dominican* convent, an univerfity founded in 1346, a geographical academy lately erected, a court of inquifition and a tribunal of juftice compofed of one prefident, fifteen counfellors, four criminal judges, one fupreme judge for *Bifcay*, four judges of nobility, two fifcals and an alguacil mayor. Yet the revenue of its bifhop, who is fuffragan to the archbifhop of *Toledo*, does not exceed twelve thoufand ducats *per annum*. It has very pleafant environs, with a large ftone bridge over the little river *Efcueva*, which runs through the city.

Lerma is a fmall town on the river *Arlanza*, with a palace belonging to the duke of *Paftrana* and *Infantado*, who is likewife duke of *Lerma*.

Civdad Burgos is the capital of *Old Caftile*, and ftands on a hill; but extends itfelf along the plain to the little rapid river of *Arlançon*, over which it has a good bridge. *Burgos* is large, but irregular, and moft of its ftreets

are

are narrow and crooked. It has, notwithſtanding, many fine ſquares, pub-lic buildings and noblemens houſes. The revenue of its archbiſhop is not ſhort of forty thouſand ducats a year; and his ſuffragans are the biſhops of *Pamplona*, *Calahorra*, and *Palencia*. The cathedral is large and ſplendid. In the convent of the *Auguſtines* is a magnificent chapel, in which extraor-dinary devotion is paid to a crucifix. Above the city ſtands the citadel. *Bega*, one of its ſuburbs, exceeds it in convents and hoſpitals; and, among others, has a very large one for pilgrims. This city was built in the ninth or tenth century, on the ruins of the ancient *Auca*.

Las Huelgas is a celebrated abbey called the *Noble*, from its one hundred and fifty nuns, who are all of good families. Under the abbeſs are ſeven-teen other convents, fourteen towns and fifty villages; ſome very large. She has alſo the diſpoſal of twelve commanderies. This abbey owes its foundation to *Alphonſo* IX. King of *Caſtile*.

Caſtro Xeres, capital of an earldom belonging to the houſe of *Mendoza*, ſtands high, and has a citadel.

Vivar is a ſmall place, famed for the birth of the hero *Roderigo*, ſur-named the *Cid*.

Melgiar de Ramenſa and *Aquilar del Campo*, are two little towns on the river *Piſuerga*.

Val de Porras, is a valley among the mountains of *Burgos*, which, exclu-five of its fine paſtures, produces fruits and corn; and is a merindade or bailiwic.

Eſpinoſa de los Monteros, is a little place among the above-mentioned mountains, being ſeated in the midſt of a pleaſant and fruitful valley, on the little river of *Trueva*.

Amaya is a ſmall town at the foot of a very high rock, where *Roderic* I. earl of *Caſtile*, held his court.

Miranda de Ebro is a little town, having a caſtle on a mountain, which produces excellent wine.

Brivieſca is a large town belonging to the houſe of *Velaſco*. In it is a convent of *Jacobines*, with a college.

Manaſterio de las Rodillas, is a village where the beſt *Caſtilian* cheeſe is made.

Betwixt this place and the city of *Burgos*, at the foot of an eminence in a delightful plain near the highway-ſide, lies a very fine and rich char-tereux.

The little Diſtrict of *Rioxa*, which ſome maps place in the province of *Biſcay*, lies properly in *Old Caſtile*, and takes its name from the *Rio*, or river of *Oxa*, which waters it. The air of this place is very pure and whole-ſome, and the ſoil produces grain, wine, and honey. In it are the follow-ing places, *viz.*

The

The civdad of *Santo Domingo de la Calzada*, which lies at the foot of a mountain near the little river *Laglera*, in a pleafant fertile plain.

Najara, a little town, belonging to the duke of that name.

Navarrete, *Guardia* and *Baſtiada*, three fmall towns.

Logronno, *Lucronium*, and *Juliobriga*, a fortified town on the river *Ebro*, feated in a large and delightful plain. In it is a court of inquifition. The neighbouring country abounds in vineyards, corn-fields, gardens, and woods of olive, fig, mulberry and other fruit-trees. Here are alfo good paftures.

Lara is a little town, having a good caftle.

S. Pedro d'Arlanza is a little town, having an old convent, much reforted to on account of a fuppofed miraculous image in it.

Berlanga, *Verlanga*, is the capital of a marquifate.

Soria is a town lying near the fite of the famous *Numantia*, of which fome monuments and remains are ftill to be feen at *Garay*.

Agreda ftands at the foot of Mount *Cayo*, in Latin *Cannus*, in the place where the ancient *Gracchuris* ftood.

Aguilar del Campo is a little town, having an old caftle, belonging to the family of *Mauriquez*, on whom it confers the title of marquis.

Civdad Calahorra, in Latin *Calaguris*, ftands high on the river *Ebro*. Its bifhop, who is fuffragan to the archbifhop, of *Burgos*, has a revenue of twenty thoufand ducats *per annum*.

The KINGDOM *of* NAVARRE.

THIS Kingdom, which by way of diftinction from *Lower Navarre*, belonging to the *French*, is called *Upper Navarre*, borders to the weft on *Old Caſtile* and *Bifcay*; and to the fouth on *Aragon*; to the eaft it is bounded by *Aragon* and the *Pyrenees*, which divide it from *French Navarre*, and alfo by the fame mountains towards the north. Its extent from fouth to north is about twenty *Spaniſh* miles, and from weft to eaft about eighteen. The name of this province is modern, being entirely unknown to the ancients. It is watered by three rivers all falling into the *Ebro*, namely, the *Aragon*, which iffues from the Kingdom of that time, the *Arga* and *Ega*. Here are two main roads leading over the *Pyrenean* mountains into *France*; namely, one from *Pamplona* along the valley of *Batan*, by the way of *Maya* and *Annoa* to *Bayonne*; the other, *viz.* that of *Roncevaux*, which is the beft, leading from *Pamplona* by the way of *Taraſſa* to *S. Jean piè de Port*. The greateft part of this province lying among the *Pyrenean* mountians, its chief wealth confifts in graziery. The mountains abound in

game

game of all kinds, as boars, ſtags, roe-bucks, wild-fowls, &c. The moſt remarkable of its vallies are *Roncal*, or *Roncevaux*, where *Charles the Great* loſt a battle againſt the *Spaniards*; *Batan*, and *Vara*. The King of *Spain* receives no revenue from this country, all the impoſts and duties being by compaͨt to be employed in the public ſervices. *Navarre*, from the year 718 to 1512, had its peculiar Kings of different families; but in the laſt mentioned year was reduced by King *Ferdinand* the Catholic, under the frivolous pretence, that *John d'Albert* its King, as ally of *Lewis* XII. of *France*, with whom Pope *Julius* II. was at variance, had been declared an enemy to the church, and excommunicated by that Pope. This province is divided into five merindades or jurifdiͨtions.

The civdad of *Pamplona*, anciently *Pompeiopolis*, or *Pompelo*, becauſe built by *Pompey the Great*, is the capital, and lies on a plain near the *Pyre-nees*. It is a place of tolerable extent, and has two caſtles, one within the city, the other without on a rock. Its biſhop is ſubjeͨt to the archbiſhop of *Burgos*, and has a yearly revenue of twenty-five thouſand ducats. It is alſo the reſidence of the Vice-roy of *Navarre*, and has an univerſity founded in the year 1608.

The civdad of *Tafalla* is a pretty and tolerably large city, ſeated on the riͤer *Cadaço*, and having a caſtle and an univerſity; its neighbourhood pro-duces good wine.

Marzilla is a pretty little place, not far from the *Aragon*.

II. Merindade de Olita contains

Civdad Olita, a pretty town ſeated in a fruitful country on the river *Ci-daço*, and formerly the reſidence of the Kings of *Navarre*. Beſides this it contains nineteen market-towns and twenty-ſix villages.

III. Merindade de Tudela contains

The civdad of *Tudela*, which is ſeated on the river *Ebro*, being on ac-count of its pleaſantneſs the reſidence of many perſons of quality.

The civdad of *Cafcante*, which is a ſmall place, but ſeated in a delight-ful plain.

Twenty-two market-towns and villages.

Obſ. The woody country on the other ſide of the *Ebro* is called *Bar-dena Real*.

IV. Merindade de Estella contains one civdad, twenty-four towns, and one hundred and ſix villages.

The civdad of *Eſtella*, in Latin *Stella*, is a handſome town ſeated in a pleaſant plain on the *Ega*, and having a caſtle.

Artajona is a little place on an eminence in a wine country.

Miranda and *Falces* are two ſmall places.

Peralta is a little place on a peninſula, formed by the river *Arga*, and produces an excellent ſort of wine.

Milagro is a little place feated on an eminence, and having a caftle, at the conflux of the rivers *Aragon* and *Arga.*

V. The Principality of Viana, formerly the title of the eldeft fon of the King of *Navarre,* derives its name from *Civdad Viana,* a tolerable place on the river *Ebro,* but feated in a very fruitful foil.

VI. Merindade de Sanguesa contains one civdad, twelve towns, and one hundred and fixty-eight villages.

Civdad Sanguefa, anciently *Iturijfa,* is a town feated on the river *Aragon.*

Xavier is a town noted for being the birth-place of the celebrated faint and miffionary of that name.

S. Salvador de Leyre is a large and fplendid abbey.

The Kingdom *of* ARAGON

IS bounded on the weft by *Navarre* and *Caftile,* to the fouth by *Valencia,* to the eaft by *Catalonia,* and to the north joins the *Pyrenean* mountains. Its extent from fouth to north is nearly about forty *Spanifh* miles; but its greateft length from weft to eaft not above twenty-fix. The river *Ebro* croffes the country from north-weft to fouth-eaft, fo as to divide it into two almoft equal parts. In its courfe it receives the following leffer ftreams; namely, from the north the *Cinca* or *Cinga,* which rifes among the mountains of *Bielfa,* and has a very fwift current; the *Callego,* anciently called the *Gallicus,* which iffues from Mount *Gavas,* near the county of *Bigorre;* the *Ifuela,* whofe fource is a little below *Huefca;* and feveral other fmaller ones as the *Aragon,* the *Riguelo,* the *Guerva,* the *Rio de Aguas,* the *Rio Martin,* and the *Guadalope.* From the fouth it receives the *Xalon* (in Latin *Salo*) which runs from *New Caftile,* the *Xiloca* and other fmaller ftreams, as the *Guadalaviar* and *Alhambra.* With all thefe rivers, *Aragon,* in general, is dry and barren, fome parts being even uninhabited. The foil is for the moft part fandy, mountainous and ftony, fo that where the rivers do not come, or where water is not conveyed by art, it produces nothing. In thofe parts, however, which are watered, we fee corn, wine, oil, flax, fruits and in fome places alfo faffron, which make up the whole riches of the country. *Aragon* formerly confifted of two parts; namely, the county of *Aragon* and the *Lande Sobrarbe.* At the beginning of the eleventh century they were both erected into a Kingdom, when *Sancho* the great King of *Navarre* divided his Kingdom among his children, giving to *Gonfalve,* *Sobrarbe,* and to *Ramir* the Kingdom of *Aragon.* *Ramir,* by the death of

Gonfalve,

Gonfalve, fucceeding foon to the Kingdom of *Sobrarbe* united it to his crown, whence the name became extinct. *Aragon* formerly enjoyed its particular conftitution and privileges; but the inhabitants having in the year 1705, efpoufed the party of the Arch-duke *Charles* III. *Philip* V. an_nulled all their privileges, and rendered them fubject to the laws of *Caftile*, by which the crown revenues from this Kingdom were confiderably aug-mented.

Civdad Saragoza, or *Saragoffa*, the capital of this country, ftands in a very fertile plain on the *Ebro*, by whofe windings this country is rendered fo fertile that it produces all manner of fruits in great abundance. The city is large and handfome, the ftreets long and broad; but very foul, and ill paved. The *Calle Santa* or *Calle de Coffo*, is the ufual airing-place of the quality, being the handfomeft and broadeft. This city contains feventeen large churches, with fourteen fplendid convents, exclufive of the leffer. Its cathedral is a large irregular building. The collegiate church of *our Lady of the Pillar* is the moft remarkable edifice here, both for its fuppofed mira-culous image, and alfo for its valuable treafures. The image is very fmall, but as it were doubly covered with coftly ornaments; being exalted very high on a pillar of the fineft jafper. Whoever attentively views the image, finds his eye dazzled as it were by the fun, with the multitude of filver lamps and wax-lights continually burning in the chapel, mingled on all fides with the reflexion of the gildings, jewels, and golden chande-liers, which are ftill a greater hinderance to the fight, fo that it is not at all times the image can be viewed. , Among the convents the moft remarkable is that of St. *Francis*, on account of its magnificent church. The yearly re-venue of the archbifhop of this place amounts to fifty thoufand ducats; and his fuffragans are the bifhops of *Huefca*, *Barbaftro*, *Xaca*, *Tarazona*, *Albar-racin*, and *Teruel*. The univerfity here was founded in the year 1474, and confirmed in 1478. *Philip* V. inclofed the inquifition-office within a cita-del. Here is held the *Audiencia Real* of *Aragon*, whofe prefidents, befides a judge, are the governor and captain-general. It is further compofed of eight counfellors, four criminal officers, two fifcals, and one alguacil mayor. A great many perfons of quality refide in this city. It enjoys alfo a very confiderable trade. The *Phenicians*, who were the founders of this city, called it *Salduba*. But a *Roman* colony being fent thither by *Auguftus* gave it the name of *Cæfarea* or *Cæfar Augufta*, of which its prefent name is a manifeft corruption. Near this place, in the year 1701, the forces of *Philip* V. were totally routed by thofe of *Charles* III.

Fuentes is feated in a fruitful plain, and was erected into a marquifate by *Ferdinand* the Catholic.

Quinto is fmall, but defended by a caftle.

Belchite is a little town, feated in a fruitful foil.

Saftago is fmall, but an earldom.

Hijar

Hijar or *Ixar*, a little town on the river *Martin*, lies at the foot of a hill, on which ftands a caftle, and confers the title of duke; the neighbouring country abounds in corn, wine, oil, filk and faffron.

Cafpe, an ancient town, having a caftle, is fituated at the junction of the *Guadalope* with the *Ebro*. *Alphonfo* II. having taken it from the *Moors* in the year 1068, conferred it on the knights of St. *John*. The country around it produces wine, grain and oil, with fome faffron and filk.

Alcanitz is a genteel town on the river *Guadalope*, having a collegiate church and a fortrefs. It belongs to the order of *Calatrava*.

Noncfpe is a handfome fortified town.

Frefneda, a little place.

Mont Roy, a town deftroyed in the year 1705, by the forces of *Philip* V.

Montalvan, a fortified town on the *Martin*, lies between two rocks, having a good citadel. It is the moft valuable commandery in this Kingdom belonging to the order of *S. Jago*.

Civdad Teruel, at the conflux of the *Alhambria* and *Turias*, or the *Guadalaviar*, is fituated in a pleafant level, being defended by a citadel. Its bifhop is fuffragan to the archbifhop of *Saragoffa*, and his income is twelve thoufand ducats *per annum*.

The civdad of *Albarracin*, anciently *Lobetum* and *Turia*, ftands on an eminence adjoining to the *Guadalaviar*. The yearly revenue of its bifhop is fix thoufand ducats; he is a fuffragan to the archbifhop of *Saragoffa*.

Mont Real, a little town having a caftle, ftands on the *Xiloca*.

Civdad Daroca, ftands on the fame river, being difficult of accefs. It has, notwithftanding, a collegiate church, and the adjacent country is very fruitful.

Arifa, a fmall fortified town, having a caftle on the river *Xalon*, gives the title of marquis. The country around it abounds in fruits, grain and wine; and affords alfo fome faffron.

Alhama is a village famed for its medicinal baths.

Ateca or *Texa*, is a little place.

The civdad of *Calatayud* is a large city, feated at the conflux of the *Xalon* and *Xiloca*, at the end of a charming valley which abounds in grain, wine, oil and fruits.

Almugna is a large handfome village, where the *Grio* falls into the *Xalon*.

Ricla is a little place, but an earldom.

Carignena or *Sarignena*, is a fmall town.

Epila is another little town feated on the *Xalon*; but the neighbouring country is very fruitful.

Muela is ftill fmaller, and ftands in a walte.

Alagon is a little place, ftanding on a peninfula formed by the rivers *Ebro* and *Xalon*.

Mallen

Mallen is a little place belonging to the order of *Malta*.

Civdad Borja, a pleasant gay place, having a castle, stands at the foot of an eminence near *Cayo*; and the country around it produces plenty of grain, wine, oil, hemp, flax, and most kinds of esculent plants.

Civdad Tarazona, anciently *Turiazo* or *Tyriasso*, stands not far from *Mon Cayo*, on the little river of *Queiles*, in a very fertile country. It is divided into the upper and lower town, the former of which is seated on a rock, and the latter in a level. It is a bishopric comprehended in the province of *Saragossa*, and worth twenty thousand ducats a year. In it are eight convents.

Taufte, a beautiful little town, seated on the *Riguel*.

Exea de los Cavalleros, a little town lying betwixt two rivers, one of which is called the *Ores*. Its furname was added to it in commemoration of the great services performed there by some *French* and *Gafcogne* cavalry, when *Alphonso* I. King of *Aragon*, took it from the *Moors*.

Luna, though an earldom, is but a small place, being seated amidst mountains.

Uncaftillo is a little place on an eminence, having a castle.

Sos is a large handsome village, having a palace, where *Ferdinand* the *C*atholic was born.

Tiermas, in Latin *Thermæ*, is a village, receiving its name from its warm baths.

Salvatierra is a little town.

Obf. The ancient county of *Aragon* was only a tract of land running among the mountains, and confisting of some fine vallies, as those of *Canfranc*, *Aifa*, *Aragues*, &c. interfperfed with towns and villages. The valley of *Tena* is one of the largest and best in the mountains of *Aragon*, being in summer exceeding delightful, and containing eleven villages, the principal of which are *Sallent*, *Panticofa*, *Pueyo*, and *Lanuça*. The first stands very high, near the river *Callego*, into which the river of *Agua Lempeda* difcharges itself from an eminence. From this place are two roads leading into *Bearn* in *France*; one through the vale of *Afpe*, the other through that of *Offeau*. But the former is both the shortest and best.

The civdad of *Xaca* or *Jaca*, is an ancient city at the foot of the *Pyrenees*, on the river *Aragon*, being seated in a delightful plain abounding with grain, fruit, and wine. The revenue of the bishop of this place does not exceed three thousand ducats a year, and he is suffragan to the archbishop of *Saragossa*.

San Juan de la Pegna is a stately and opulent convent.

Xavier is a little town on the river *Callego*.

Ancanego is still smaller, and stands on the same river.

Loarre is a large village at the foot of the *Pyrenees*, having a strong castle, in which Count *Julian*, who excited the *Moors* to invade *Spain*, died a prisoner.   *Ayerve*,

Ayerve, anciently *Ebellium*, is a large handfome village.

Civdad Huefca, anciently called *Ofca*, is a beautiful place feated on the river *Ifuela*, in a fertile level delightfully terminating on all fides in well cultivated eminences. It lies in the province of *Saragoffa*, and the revenue of its bifhop is thirteen thoufand ducats *per annum*. It has alfo an univerfity founded in the year 1354.

Tuera, or *Zuera*, is a little place on the river *Callego*.

Almudevar is a fmall place, but the neighbouring country abounds with grain, wine and faffron.

Villa Mayor is larger than the former, but feated in a very barren territory.

La Puebla de Alfinden is a handfome town, lying not far from the river *Ebro*, having a caftle on an eminence. The neighbouring country abounds in fine improvements.

S. Maria de Magalon, which is feated among the mountains of *Lefine*, is a church, ftanding on an eminence, and much celebrated for a miraculous image there of the Virgin *Mary*. Clofe by it is an inn*.

Offera is a little place lying on the *Ebro*.

Pina and *Vililla* are two country places. The laft is noted for its great bell, of which fo many fabulous ftories are told. The country betwixt thefe places and *Fraga* is called *the wafte of Aragon*, and very juftly, it being a barren heath feveral miles in extent every way, infomuch that from *Burjalajos* to *Fraga* there is not even a tree or any water to be feen.

Mequinença, formerly *Octogefa* and *Iclofa*, is an ancient town, ftrongly fituated, and having a caftle. It ftands in a fertile country betwixt the rivers *Segre* and *Ebro*.

Fraga, anciently called *Flavia Gallica*, is an old town ftanding high, betwixt mountains, on the river *Cinca*. It is ftrongly fituated, and, for its greater fecurity, and alfo for that of travellers, has always a good garrifon.

Monzon, in Latin *Montio*, is a ftrong town, on an eminence, feated on the river *Cinca*, and having a good caftle.

Civdad Barbaftro, in Latin *Barbaftrum*, is a town on the river *Vero*, which not far from hence joins the *Cinca*. The bifhop of this place is a fuffragan to the archbifhop of *Saragoffa*, with an annual revenue of eight thoufand ducats.

---

* I know not why our author, who is, beyond all difpute, a writer of the greateft judgment, as well as learning, fhould mention this as a circumftance worthy notice; for, whatever it may be in *Spain*, it is certain nothing is more common in *England* than to fee places of this kind erected as near as poffible to a church for the conveniency of trade; which gave occafion, I fuppofe, to thofe quaint lines of the ingenious conducter of the *Craftfman* in his Poem called *Oxford*:

Where-ever God erects a houfe of prayer,
The Devil always builds a chapel there.

Graus,

Graus, a little town on the *Eſſera*, was burnt down, in the year 1706, by the forces of *Philip* V.

Medianos is a little place on the *Cinca*.

The principality of *Sobrarbe*, which formerly bore the title of a Monarchy, conſiſts of ſeveral vallies, as, namely, thoſe of *Terrantona, Giſtain, Puertolas*, &c. The chief place in theſe is *Ainſa*, which is a little town in a plain on the river *Ara*.

San Quiles is a genteel place, ſituate at the foot of the *Pyrenees*.

Campo, a little town, ſeated on the river *Eſſera*, lies among mountains.

Seira, or *Sera*, a very handſome market-town, lies alſo among mountains.

The earldom of *Ribagorza* is ſeated among the *Pyrenees*, and includes ſeveral vallies, as, namely, thoſe of *Benabarri*, with a place of the ſame name ; *Venaſque*, in which is the little town ſo called, having alſo a good caſtle and garriſon ; *Tamarit* and *S. Eſtevan de Litera*, and other vallies. From *Venaſque*, by the way of the inn called *Hoſpitalet*, and farther on, through *Puerto*, lies a very narrow road among rocks into *France*.

The PRINCIPALITY *of* CATALONIA.

CATALONIA, to the weſt, is bounded by *Aragon* and *Valencia* ; to the ſouth and eaſt by the *Mediterranean*, being ſeparated to the north from *France* by the *Pyrenean* mountains. Its greateſt extent, from weſt to eaſt, is about twenty-eight *Spaniſh* miles ; and, from north to ſouth, thirty-ſeven. It was formerly larger than at preſent, but *France* has at different times curtailed it of the counties of *Rouſxillon* and *Conflans*, with a good part of *Cerdagne*, and long ſince of *Foix*. This fine country is watered by ſeveral rivers, ſome of which intermix, whilſt others diſcharge themſelves ſeparately into the ſea. Of theſe the firſt is the *Segre*, in Latin *Sicoris*, being the largeſt of all the *Catalonian* rivers. It takes its riſe in *Cerdagne*, near *Camaraſa*, receiving in its paſſage the *Noguera Pallareſa*, and, not far from *Lerida*, the *Noguera Ribagorzana* (in Latin *Nocharia Ripacurtiana*) and the *Cervera*. It unites afterwards with the *Cinca*, loſing itſelf, at laſt, near *Mequinencia*, in the river *Ebro*. The little river of *Corp* runs into the above-mentioned *Cervera*, and the *Noya* into the *Llobrégat* near *Martorel*. Of the ſecond kind, beſides the large river *Ebro*, paſſing only through a ſmall part of *Catalonia*, is the *Francoli*, which diſcharges itſelf into the ſea at *Tarragona* ; the *Llobrégat*, anciently called the *Rubriçatus*, which receives its ſource in the

the mountain of *Pendis* and mingles with the fea below *Barcelona*; the *Befos*, or *Betulus*, which alfo joins the fea not far from *Barcelona*; the *Ter*, *Thicis*, or *Thiceris*, which iffues betwixt Mount *Canigo* and *Col de Nuria*, and falls into the fea below *Toroella*; the river *Fluvia*, in Latin *Fluvianus* and *Cluvianus*, emptying itfelf into the fea near *Empurias*, and another *Llobrégat*, the entrance of which lies near *Rofes.*

Catalonia, though it does not produce fugar-canes, like the reft of the provinces of *Spain*, yet enjoys a very good foil with a pure wholefom air. The winters, in the northern parts contiguous to the *Pyrenees*, are attended with fome frofts and fnow; but in the fouthern, particularly along the fea-coaft, that feafon is very mild. Some few places only excepted, which extend themfelves into moft delicious plains, this country is almoft entirely mountainous; but thefe mountains are fo far from being barren, that they are covered with very valuable woods and fruit-trees. The country enjoys a conftant plenty of wine, grain, oil, pulfe and fruits. The feveral kinds of fiefb here are excellent. It produces alfo great quantity of flax and hemp. Marble, cryftal, alabafter, jafper, amethyft, &c. are likewife found here, together with gold, filver, tin, lead, iron, alum, vitriol and falt, though but very little copper; and the coaft has its coral fifheries. In a word, this province is one of the moft populous in all *Spain*, and contains one arch-bifhopric, feven bifhoprics, twenty-eight large abbies, one principality, two dutchies, five marquifates, feventeen earldoms, fourteen vifcounties and a great number of baronies. When the *Moors* had over-run the greateft part of *Spain* and began their attacks on this province, the *Catalonians* made an effort to fecure their freedom, and applied to *Charles Martel* of *France* for affiftance; that Prince accordingly, and his fon *Pepin*, vigoroufly affifted them in their wars againft the *Moors*. *Charles* the Great, being both King and Emperor, made himfelf fo redoubted by the *Moors*, that *Zaro*, governor of *Barcelona*, agreed to pay him tribute. On the death of this *Zaro*, *Bernard*, grandfon to *Charles*, was made earl and governor of *Catalonia*, with the title of a marquis and duke of *Spain*; but *Provence* and *Languedoc* being afterwards annexed to his government, he had a collegue of the name of *Geofroy*, whofe fon *Godfrey*, on *Bernard's* deceafe, was created governor of *Barcelona*, and, for his good fervices to King *Charles* the Fat, in his bloody wars with the *Normans*, created by him, in the year 884, hereditary count of *Barcelona*, which was to continue to him and his heirs for ever, though with this reftriction, that they fhould remain vaffals to the King of *France*. In the year 1137, Count *Raymond Baranger*, marrying *Petronilla*, heirefs of *Aragon*, united *Catalonia* to the crown of *Aragon*, and, in 1182, fhook off all dependency on *France*. *Catalonia* continued united to *Aragon* till the year 1640, when it fubmitted to *France*. In 1652, the King of *Spain* recovered *Barcelona* and fome other places; and, laftly, by the treaty of the *Pyrenees*, in the year 1659, faw himfelf again mafter of

5   all

all *Catalonia*. In 1705, the whole principality submitted to the Archduke of *Auſtria*, and ſtood firm to his cauſe, inſomuch that, though in the year 1713, he was obliged to evacuate *Catalonia*, *Majorca* and *Yvica*, for the ſafety of his conſort, and the perſons of rank whom he had left behind, yet did the inhabitants of *Barcelona* make all poſſible preparations for a further oppoſition, being determined either to maintain their privileges or die ; but, in the year 1714, *Barcelona* was obliged to ſurrender at diſcretion, and the whole country reduced to the ſubjection of *Philip* V. who aboliſhed all thoſe valuable privileges which they had ſo often aſſerted with a ſucceſsful intrepidity. Some divide this principality into *Old* and *New Catalonia*, including in the former the country betwixt the *Pyrenees*, which runs along the river *Llobregat* eaſtward to the ſea; and weſtward, the tract from *Llobregat* to the borders of *Valencia* and *Aragon*. The uſual diviſion in the county itſelf, is into fifteen *vigueres*, or juriſdictions, ſix of which lie along the coaſt; namely, thoſe of *Tortoſa*, *Monblane*, *Tarragona*, *Villa Franca de Panades*, *Barcelona* and *Gerona*, the latter containing alſo that of *Ampurdan*. Along the *Pyrenean* mountains are the two juriſdictions of *Campredon* and *Puicerda*, with the earldom of *Cerdagne*. The two *vigueres* of *Belaguer* and *Lerida* join to the frontiers of *Aragon* ; and, within the country, are thoſe of *Agramont*, *Tarrega*, *Cevera*, *Manreſa* and *Vique* : but not knowing, with any certainty, what places belong to each juriſdiction, I ſhall here inſert only the moſt remarkable.

Tortoſa is an ancient, fortified and large city, on the river *Ebro*, over which it has a bridge of boats. The avenue to it is defended by two baſtions and other outworks. It lies partly on a level and partly on a hill, being divided into the old and new town, of which the former is the largeſt. The ancient ſtrong caſtle, with which, beſides other fortifications, it is defended, ſtands on an eminence betwixt it and the old town, being built in form of a citadel. Here are many churches and convents, and the biſhop, who is ſuffragan to the archbiſhop of *Saragoſſa*, has a revenue of fifteen thouſand ducats a year. Its univerſity is inconſiderable, but the adjacent country abounds in grain and fruits. It produces alſo ſilk and oil, and makes a pretty kind of earthen ware. Near it are ſome mines and fine quarries of ſtone. The *Ebro*, being navigable, is a great conveniency to the trade of this city. It gives the title of marquis.

Alfacqs is the name of an iſland lying at the mouth of the *Ebro*.

Gineſtar, a country town.

· *Mora*, a little place.

Flixs too is ſmall, but well fortified both by art and nature, and ſtands on a peninſula, formed by a large circuit of the *Ebro*, which environs this place on three ſides ; and the fourth is defended by a mountain, fortifications and a ſtrong caſtle ſeated on an eminence. In this place is a cataract on the *Ebro*.

Tivica is a little town.

Cambrilla, a fea-port, the country about it fertile and pleafant.

Tarragona, which is a principality, anciently called *Tarcon* and *Tárraco,*. is an old fortified town, fituate on an eminence near the fea, near the mouth of the little river *Francoli.* It was formerly much larger, more opulent and more populous than at prefent; but is ftill the fee of an archbifhop who enjoys a revenue of twenty thoufand ducats a year, and has for his fuffragans the bifhops of *Barcelona, Gerona, Lerida, Tortofa, Vique, Urgel* and *Solfona.* The univerfity here was founded in the year 1532. It has a good trade and an harbour, but the latter, on account of its many rocks,. is impracticable for fhips of burden. The neighbouring country produces corn, oil, flax and very good wine; and both within and without the city are to be feen many antiquities.

Tamarit is an old caftle lying on the fea.

Villa Franca de Panades, a fine town.

Martorel, a little town at the' conflux of the *Noya* and *Llobregat,* belonging to the counts of *Benevento.*

Ignalada is a pretty little town feated on the *Noya.*

Sarreal, a fmall place, but having fome fine quarries of alabafter in its neighbourhood.

Monblanq, fituate on the river *Francoli,* is a pretty good town and the capital of a dutchy.

Sivrana is a fortrefs feated amidft rocks and mountains and of difficult: accefs.

Pradas is a little town and the capital of an earldom.

Pobledo, in Latin *Populetum,* is a little town. Its neighbourhood affords alum and vitriol.

Aitona, Hitona, is fmall, but the capital of a marquifate.

Lerida, anciently *Ilerda,* an old dirty city, ftanding on an eminence on' the river *Segre,* is fortified and well built, and the fee of a bifhop who' has a revenue of twelve thoufand ducats *per annum,* and is fuffragan to the archbifhop of *Tarragona.* In this town is a court of inquifition, an univerfity founded towards the beginning of the fourteenth century, and a: citadel. In the year 528, a council was held here.

Balaguer is a town fituate at the foot of a high mountain on the river *Segre,* and in a very fertile country.

Ager is a little place giving the title of vifcount.

Camarafa, another little place giving the title of marquis.

Belpuch is fmall, but famous for its convent of *Francifcans.*

Tarrega is a little town.

Cervera, a fortified little town, feated on a hill joining to a river of the fame name, has a caftle.

Hoftalets is a little place.

5 *Agramont,*

Agramont, a little town but capital of a jurifdiction.

Solfona, anciently called *Calea*, ftands high on the river *Cardonero*. The bifhop of this place, who is fuffragan to the archbifhop of *Tarragona*, enjoys a revenue only of four thoufand ducats *per annum*.

Berga, anciently *Berginium*, is a little town on the river *Llobregat*.

Cardona is a handfome town on an acclivity near the river *Cardonero*, being the capital of a dutchy and well fortified, with fome very profitable falt-works in its neighbourhood.

Manrefa, or *Minorifa*, anciently *Rubricata*, is old, and ftands at the junction of the *Cardonero* and *Llobregat*, being an earldom.

Mont Serrat is a famous convent of *Benedictines*, feated on a high rocky mountain of that name. It is much vifited by pilgrims on account of a miraculous image there of the Virgin *Mary*, which brings in a good revenue to the church-treafury: the pilgrims, however, are entertained for three whole days at the convent. On this mountain are alfo feveral hermitages.

Efparaguerra is a little town principally inhabited by cloth-workers.

Barcelona, or *Barcino*, the capital of *Catalonia* and the refidence of a governor, is a large, well fortified fea-port, divided into the old and new town. Moft of the ftreets are broad, well paved and clean. It has many fine and ftately houfes, a large and magnificent cathedral, with fome other beautiful churches and convents, and handfome fquares. The bifhop hereof is fuffragan to the archbifhop of *Tarragona* and has a revenue of ten thoufand ducats *per annum*. In it is alfo an univerfity, an academy of arts and fciences founded in the year 1752, a court of inquifition and an *audiencia real de Catalunna*, in which, next to the governor and captain-general, prefides the judge-regent. The other members are ten counfellors, five criminal officers and two fifcals. Its trade being confiderable, it has alfo a fpacious, deep and fafe harbour, one fide of which is defended by a large mole, and at the end of it is a light-houfe and a fmall fort with a garrifon. The other fide of the harbour is covered by the caftle of *Monjouy*, betwixt which and the city runs a line of communication, and on it is a ftrong fort commanding the entrance into the harbour. In the year 1753, the King ordered five other forts to be erected on this mountain. The city, till 1162, had counts of its own, but that year was united to *Aragon*. In 1640, it revolted from *Spain* to the *French*; but, in the year 1652, was again recovered by the *Spaniards*. In 1691 and 1697, it was befieged by the *French*, and in the laft year taken, but reftored again at the peace of *Ryfwick*. In the year 1705, King *Charles* III. made himfelf mafter of it. In 1714, it furrendered to *Philip* V. after a very obftinate defence.

Badelona is a little fea-port having a fortrefs.

Mataro,

Mataro, another, where, in the year 1708, King *Charles* III; confummated his marriage with the Princefs of *Brunfwick*.

Moncada, a little town on the river *Befos*.

Rocca, *Linas* and *S. Saloni*, are three little places. About a *Spanifh* mile from the latter is a harbour of the fame name, but frequented only by fmugglers, who barter for wine.

Pineda is a little fea-port feated at the mouth of the river *Tordera*.

Oftalric is a little town on the *Tordera*.

Centellas, a fmall town lying in a valley.

Vique, anciently *Aufonia*, ftands in a very fertile plain. The bifhop hereof is fuffragan to the archbifhop of *Tarragona*, with a yearly revenue of fix thoufand ducats. This place is alfo an earldom.

Mon-Seni is a high mountain, on which, befides wholefom plants and herbs, feveral forts of gems are found.

Blanes, anciently *Blanda*, is a little fea-port.

Toffa is ftill fmaller, and fo called from a promontory formerly known by the name of *Promontorium Lunarium*.

Palamos is a little fortified town on a bay which forms a good harbour. It ftands partly on a level and partly on an eminence projecting into the fea. Befides its fortifications, it has alfo a citadel, is an earldom, and; at a fmall diftance from it, lies Cape *Palafugel*, fo named from a fmall neighbouring town.

Gerona, anciently *Gerunda*, is an old fortified city of tolerable bignefs, ftanding on an acclivity adjoining to the *Onhar*, which, at a little diftance from hence, falls into the *Ter*. It is an earldom and a place of confiderable trade, yet the revenue of its bifhop, who is fuffragan to the archbifhop of *Tarragona*, is but three thoufand ducats *per annum*. Its univerfity alfo makes no great figure. In the year 517, a council was held here. The large jurifdiction, of which this is the capital, is reckoned the moft fertile tract in all *Catalonia*. This city, in the year 1694, was taken by the *French*; in 1697, by the *Spaniards*; in 1705, by the forces of *Charles* III. and, in the year 1711, again by the *French*.

Bagnolas, anciently *Aquæ Votonis*, is a fmall place:

Ampurias, a fea-port at the mouth of the *Fluvia*, once confiderable; but now mean. It is an earldom, and the Diftrict belonging to it is called *Ampurdan*.

Bafcara is a little town feated on the *Fluvia*.

Figuera alfo a little town.

Caftello d' Ampurios too a little town feated on a bay.

Rofes likewife ftands on a bay, is ftrong, and has a good harbour guarded by a fort. This is the only place which continued firm to *Philip* V. It owes its original to the ruins of the town of *Rhoda*, which ftood a little way off near Cape *Cruz*.   *Jonquieres*,

Jonquieres, in Latin *Juncaria*, is a mean place, fituate at the foot of the *Pyrenees*, being only five miles from *Pertus*, the firft place in *Rouffillon*, but the road from hence to it extremely bad.

Compredon is a town of fome ftrength feated on an eminence near the river *Ter*, with a citadel in the middle. In the year 1698, it was befieged and taken by the *French*.

Aulot is a little mean place feated on the *Fluvia*.

Ripol, in Latin *Rivipullum*, lies at the conflux of the *Frefero* and *Ter*, being fmall but having a fine convent of *Benedictines*.

Baga, anciently called *Bergufia*, is a little place feated among high mountains on the river *Llobregat*.

The earldom of *Cerdagne*, in Latin *Ceretania*, the capital of which,. called *Piucerda*, *Puteus*, or *Podius Ceretanus*, is large and fortified in the modern tafte. It ftands in a fine plain, at the foot of the mountains betwixt the rivers *Carol* and *Segre*.

Llivia, anciently *Julia Lybyea*, is a ftrong little town, having a citadel.

Urgel is an ancient city and earldom, feated on the *Segre*, in a plain environed with high mountains which are covered with vineyards. The bifhop of this place, who is fuffragan to the archbifhop of *Tarragona*, enjoys a yearly revenue of nine thoufand ducats.

Caftelbo is a little place giving the title of vifcount..

Noguera Pallarefa, a little town on a river of the fame name, yet the capital of a marquifate.

Tremp too is a fmall town, but has a great many gentry among its inhabitants.

The BALEARIC and PITHYUSÆ ISLANDS in the MEDITERRANEAN.

THE four iflands of *Majorca*, *Minorca*, *Yvica*, and *Formentera*, of which we have a good map executed by *Ottens*, formerly compofed the Kingdom of *Majorca*. Its original inhabitants are not determinable with any certainty. *Strabo* fpeaks of a colony of *Greeks* from the ifle of *Rhodes*, who fettled there. Afterwards the *Carthaginians* became mafters of them ; and from thefe they fell under the dominion of the *Romans* In the fifth century the *Vandals* poffeffed themfelves of thefe iflands : And towards the end of the eighth and the beginning of the ninth century thefe were diflodged

lodged by the *Moors*, who after a much shorter possession, were expelled by the Emperor *Charles the Great*, whose grandson *Bernard* held the sovereignty of them. But the *Moors* were not long in recovering their settlement here, and had their particular King over them. Numerous wars were carried on betwixt them and the *Catalonians*, with various success, till towards the end of the year 1229, *James* I. King of *Aragon* dispossessed them of the island of *Majorca*. In the year 1232 he also reduced *Minorca*; and in 1234 *Yvica*; so that the whole Kingdom of *Majorca* became annexed to the crown of *Aragon*. These four islands, with the other smaller ones lying near them, are by the ancients divided into the *Balearic* and *Pithyusæ*.

1. The *Balearic* islands are *Majorca* and *Minorca*, with some smaller. The *Latins* term them *Baleares* and the Greeks *Balearides*, which appellations some derive from the *Greek* and others from the *Phenician* language. Both etymologies agree in the same import, namely, *the islands of the Slingers*; the inhabitants excelling in a dexterous use of the fling, as the *Minorcans* do to this day. The *Greeks* also called these islands *Gymnesia*, or *Gymnasia*, from the inhabitants going naked in summer. We shall give a separate account of each.

1. *Mallorca*, or as foreigners pronounce, it *Majorca*, is the largest of these islands, being about fifteen *Spanish* miles in length and ten broad. Its four chief capes, which lie also towards the four cardinal points of the world, are *Pedra*, *Grosser*, *Salinas* and *Formentor*. This island is divided into two parts; that towards the north and west is mountainous, but not barren: The other lying south and east is level; and laid out in corn-land, pastures, vineyards and orchards. This island naturally abounds in corn, wine, oil, honey, saffron, large and small cattle, wool, cheese, fish, rabbits, partridges, deer, wild-fowl and horses, without any ravenous wild beasts. The whole island is encompassed with strong towers, from which an enemy may be descried at a distance. It has several good harbours and anchoring places. The air is temperate and wholesome; but the excessive heat and drought here frequently occasion a scarcity; though the island, in general, is well supplied with water. The inhabitants, in their manners and customs, resemble the *Spaniards*; but particularly the *Catalonians*. Persons of fashion here speak *Spanish*; but the language of the commonality is a medley of *Limosin*, *Greek*, *Latin*, *Spanish* and *Arabic*. The island maintains twenty companies of foot, five troops of horse, and two companies of matrosses for the defence of the capital; besides four regiments cantoned in other parts of the island.

Majorca, anciently *Palma*, which is the capital, lies on a bay betwixt two capes. It is fortified in the modern taste; is large, and has broad streets, spacious squares, stately stone houses, and twenty-two churches, besides chapels and oratories. The largest square here is that of *Born*, which

which is environed with grand buildings on every fide; from whence the principal inhabitants view the bull-fights and other fhows. The cathedral too is large and magnificent. In the town are fix hofpitals, and three other foundations for women. The bifhop hereof is fuffragan to the archbifhop of *Valencia*. It has an *Audiencia Real*, in which the *Comandante General* prefides; a court of inquifition, and an univerfity. The inhabitants are computed at ten thoufand.

Lluch Mayor is a town of about five hundred houfes; but labours under a fcarcity of water.

Randa is a mountain remarkable for a handfome church and college built on its fummit. It is alfo the birth-place of the famous *Raymund Lully*.

Sion is a mountain remarkable alfo for a church and a college.

Campos is a fmall town.

In the little bay of *Gavina*, near Cape *Las Covetas*, is fhipped the falt exported from this ifland. About five miles from it is Cape *Salinas*, which is fo named from the fens where the fea-falt is made.

Calafuguera is a little town.

Porto Pedro, a fpacious fecure harbour, defended by a fort.

Manacor, a little town.

Cape *Pedra* is alfo protected by a fort.

The lake of *Albufera*, *i. e.* ' the Little-Sea,' is a body of water of about twelve thoufand paces in circumference: And near it the fea forms another lake called *Grac-Mayor*, whofe waters mix with that of this lake.

Civdad Alcudia, contains about one thoufand houfes, and lies betwixt the two large harbours of Port *Mayor* and Port *Minor*, being defended by two forts.

Pollença; or *Pollentia*, is an old town of about feven hundred houfes, having a good harbour; and its wine, called *Montona*, is excellent.

S. Vincent is a little bay, commanded by a fort.

Calobra is the moft confiderable harbour in the whole ifland, both for its fecure entrance, and the fine country with which it is furrounded, and likewife for the fprings of frefh-water near it.

Llampayes has a pretty large harbour defended by two towers. In the year 1561 the *Turkifh* fleet attempted a defcent here but were repulfed by the inhabitants.

Soller and *Bunola* are two little towns.

Santelma is a port covered by two good fortreffes.

Andraig has a large harbour but expofed to the weft wind. It is defended by a fort erected at a little place called *Mola*.

Paguera is one of the beft harbours on the whole ifland.

Safellas and *Sineu* are two little towns.

Obf.

Obf. The fmall iflands lying about *Majorca* are

1. *Cabrera*, facing Cape *Salinas*, being fo called from the multitude of goats found there. It is all over mountainous and uninhabited, except its fpacious and fecure harbour, the entrance into which fronts *Majorca*, and is defended by a caftle, in which a fmall garrifon is always kept. This ifland is a place for exiles.

2. *Las Bledes*, which lies not far from the harbour of *Olla*, is of fome confideration, and was formerly very populous. In it is a good quarry of marble.

3. *Formentor* lies near Cape *Albacux*.

4. *Colomer* is not far from *Calafiguera*.

5. *Foradada*, where the fon of King *James*, who conquered *Majorca*, built a college on a hill for the inftruction of the monks of the order of St. *Francis* in the *Arabic* language, with a view to the converfion of the *Moors*.

6. *Pantaleu* is the place where the above-mentioned *James*, King of *Aragon*, landed, when he invaded the *Balearic* iflands.

7. *Dragonera* is about a thoufand paces in length, nine hundred broad, and one thoufand two hundred from *Majorca*. It is uninhabited, and all its product is an edible bird of prey called a *Spaniard*. On Mount *Popia* ftands a fortrefs. The name of it is probably derived from a kind of fnake called, in the *Catalonian* language, *Sargantana*.

8. *Mijana* is the principal of feveral fmall iflands lying about *Dragonera*.

9. *Moraffa* is the name of a clufter of ifles.

2. *Minorca*. The fouth fhore of this ifland is fmooth, but towards the north very rugged, which is occafioned by the violent blafts from that quarter. The fea has fo withdrawn from this ifland, that near the harbour are feveral new flats which are turned into garden-grounds. The whole furface of the ifland is about two hundred and thirty-fix fquare *Englifh* miles, which are nearly equal to one hundred and thirty thoufand one hundred and forty *Englifh* acres. The air is moift; the heat, in a *Farenheit* thermometer, placed in the fun, rifes only to one hundred and two, and, confequently, is not very intenfe. The quickfilver too is feldom known to fink under forty-one, which is fhort of the freezing point. The foil cannot be extolled for its fertility. The water alfo is hard, whence nephritic diforders are common here. The hedge-hog is reckoned in this ifland among venemous animals, being fuppofed in rutting time, in the fpring, to corrupt the water. Here is great plenty of fifh, particularly of wilks, which are of great fervice to the commonalty in *Lent*. One of the moft profitable commodities of the country is falt, which the fun prepares in cavities betwixt the rocks. Here are lead-mines but no flints; likewife a vaft variety of petrified fhells and fine marble. Some of its wine is excellent; and the inhabitants, when they had the *Englifh* for cuftomers, were faid to make

twenty-

twenty-feven thoufand pounds fterling a year of it. They have alfo a certain kind of cheefe of which they are extremely fond. Rabbits to this day
are to be met with in it in great plenty. Here is alfo wool, honey, wax, and
capers growing along the walls ; but, as well as of olives and cotton, there
are alfo plantations of this fruit. The palm-trees here bear no fruit. The
opuntia is very common and plentifully eaten, as are alfo the acorns. The
myrtles are of great advantage to tanners. Their chief neceffaries, however,
fuch as corn, beef, brandy, tobacco, linen, ftuffs, books, reliques, *Agnus Dei*'s,
&c. they are fupplied with from abroad ; and thefe collectively ftand them
yearly in not lefs than feventy-one thoufand two hundred pounds fterling.
They live moftly on vegetables, love dancing, and have fuch a turn for
poetry, that the very peafants challenge each other to trials of genius that
way. They are alfo very dextrous with their flings, and command their
cattle with them ; however they want induftry, whence they neglect many
profitable occupations in farming and trade. The houfes on the ifland are
computed at three thoufand eighty-nine, and the inhabitants at twenty-
feven thoufand. The *Englifh* took it from the *Spaniards* in the year 1708,
and kept it till 1756, when, after a very brave refiftance under General
Blakeney, they furrendered it to the *French,* who had invefted Fort St.
Philip with an army of fifteen thoufand men well provided with all military
ftores. The ifland is divided into four quarters.

1. The quarter of *Civdadella,* fo called from
Civdadella, or *Citadella,* the capital of the ifland and refidence of the
governor, is fortified and contains about fix hundred houfes.

2. The quarter of *Mahon* is fo called from
Mahon, an excellent harbour. Its entrance, indeed, is fomewhat
difficult on account of the feveral rocks in it ; but within it is land-locked
on all fides. It is defended by Fort St. *Philip,* and near it lies the little
trading town of *Mahon,* from whence it takes its name. The *Englifh* always
kept this place in a good ftate of defence.

3. The quarter of *Alajor* is fo called from a fmall place near it.

4. The united quarters of *Mercadal* and *Ferrarias* are alfo fo called
from too fmall places bearing the fame names.

II. The *Pythyufæ* iflands are fo called by the *Greeks* from the multitude
of pines growing in them. Of thefe the moft remarkable are,

1. *Ivica,* anciently *Ebufus,* being about five miles in length and four
broad. The foil is not unfruitful, but little cultivated, moft of the inhabitants being taken up with the falt trade, as being more profitable. It is
very mountainous, yet, befides pines, produces alfo fruit trees of various
kinds. The principal places in it are
Ivica, the capital, and a modern fortification, but much dwindled from
what it was in the times of the *Carthaginians* and *Romans.* It is the refi-

dence of the governor, from whom there lies an appeal to the *audiencia real* at *Majorca.*

S. Hilario is a little place on a bay.

Porto Magno, a mean place having a harbour. About this lies feveral other iflands.

2. *Formentera,* anciently *Ophiufa* and *Collubraria,* i. e. ' *the Adder Ifland,*' was formerly well inhabited, but at prefent is forfaken and defert, the *African* Corfairs continually fwarming about it. All that is to be feen here is a kind of wild afs, though the ifland contains in it fome harbours and good anchoring places.

THE

THE

KINGDOM

OF

FRANCE.

INTRODUCTION

TO THE

KINGDOM OF FRANCE.

§. 1. OF no Kingdom or nation have there been more maps made than of *France*. To omit the more ancient ones of *Poftel*, *Thevet*, *Plantius*, *Jolivet*, *Orontius Finæus*, &c. and, fince their days, thofe of *Vifcher fenior*, *Jollain* and *Taffin*, *Sanfon* has, in a particular manner, contributed to the improvement of the maps of that Kingdom ; but, fince the inftitution of the royal academy of fciences, which included geography in its feveral difquifitions, and particularly fince the meridional line at *Paris* has, with incredible labour and application, been drawn to the extremity both of the northern and fouthern borders, by thofe celebrated mathematicians *Caffini* and *la Hire*, the maps of *France* have arrived at a much greater perfection. Exclufive of many others, that publifhed in the year 1703, by *William de l' Ifle*, and again, in 1741, by *Homann*'s heirs, deferves particular commendations. Subfequent to him, *M. d' Anville* has executed fome maps of *France* with great accuracy, as may be feen in thofe of the *Abbé de Longuerve*'s *Defcription hiftorique et geographique de la* France *ancienne et moderne*, and in the laft edition of *Introd. à la defcript. de la* France, by *M. la Torce*. The new maps of *France*, by *Caffini*, are formed upon the neweft obfervations and the noble work of meafuring the larger and fmaller parts of the Kingdom, is ftill continued under the direction of *M. Caffini*, fubject to the infpection of *M. Trudaine*, who has not thought fuch an employment beneath the dignity of his character as minifter of ftate. Of the provincial maps of this Kingdom there are an infinite number.

§. 2. *France* derives its prefent name from the *Franks*, who, in the fifth century, came out of *Germany* into *Gaul* and made themfelves mafters of the whole country from the *Rhine* to the mouth of the *Loire* ; but its more ancient appellation of *Gaul* is derived from the *Galli*, or *Galati*, who

were

INTRODUCTION TO

were defcendents of the *Kalatai*, or *Celtæ*. The *Romans*, according to *Cæfar*, firft gave the name of *Gauls* to the *Celtæ*; but the various divifions and names of *Gaul* pertain to ancient geography.

§. 3. The late conquefts included, this Kingdom extends itfelf from the thirteenth to the twenty-fixth degree of longitude, computing the meridian from *Ferro*; and from the forty-third to the fifty-firft of latitude, inclufive of the firft and laft degrees in both calculations; whence its greateft length, from caft to weft, appears to be one hundred and twenty-eight, and its greateft breadth, from fouth to north, one hundred and thirty-five common *German* miles. To the fouth it is bounded by the *Mediterranean* fea and *Spain*, from which it is feparated by the *Pyrenean* mountains; to the weft, by the weftern ocean; to the north by the *C*hannel*, as it is called, and the Netherlands; and, eaftward, by *Germany*, *Switzerland* and *Italy*.

§. 4. The air is mild and wholefom, particularly the interior parts of the Kingdom. The winters in the northern provinces are cold and laft four or five months, though generally with great temperature; and thofe regions which border on the *Mediterranean* are not reckoned to enjoy fo wholefom an air as the other provinces. The country, in general, is fruitful, yet not without many barren tracts and mountains. *Guettard*, in his differtation on the nature and fituation of the foil in *France* and *England*, publifhed in the *Memoires de l' acad. roy. des fciences*, for the year 1746, and printed in the year 1751, is of opinion, from feveral books of travels, *Hellot's* Obfervations, publifhed with defigns, and Mr. *Buaches'* papers, that in *France* there are three fpecies of earth: that, about *Paris* and *Orleans*, in one part of *Normandy*, and as far as *London*, is all fand and affords throughout no other metal than iron; the fecond circle, within which the firft is comprehended, and which contains *Picardy*, *Champagne*, *Tourraine*, *Berry*, *Perche* and a part of *Normandy*, is of marle, without any other ftone than indurated marle, and iron which is its only ore; but the third circle, which contains the frontiers and mountainous parts, and extends over the greateft part of *England* and *Germany*, affords flate, ftones and all forts of metals. Thefe difcoveries deferve attention, and a further afcertainment by experiments. The chief mountains of *France* are the *Alps*, towards *Italy*; the *Pyrenees*, which border on *Spain*; and thofe of the *Cevennes* and *Auvergne*. The rivers here of greateft note are

The *Seine*, anciently *Sequana*, the courfe of which is in the Diftrict of *Montagne* in *Burgundy*. It becomes navigable at *Troyes*, and, being joined by the *Yonne*, the *Loing*, the *Marne*, the *Oife*, the *Eure* and other fmaller rivers, difcharges itfelf betwixt *Havre de Grace* and *Honfleur* into the *Britifh C*hannel.

---

* *i e.* St. *George's*, or the *Britifh* Channel.

The

The *Loire*, the ancient *Ligaris*, takes its rife in Mount *Gerbier le Jour*, on the borders of *Vivarais* and *Velay*, and at *Rouanne* becomes navigable, but, at its entrance into *le Forez*, is too fhallow for any veffel, not to men.tion the rocks in feveral parts of it, particularly above *Rouanne*, at a place called *Le Saut de Piney*. It receives the rivers *Allier*, *Cher*, *Indra*, *Creufe*, *Vienne* and *Mainne* in its courfe, communicates with the *Seine* by means of the canals of *Briare* and *Orleans*, and below *Nantes* falls into the weftern ocean.

The *Garonne*, anciently called the *Garumna*, receives its fource in the mountains of *Aure*, in the valley of *Aran* and county of *Comenges*. At *Muret* it becomes navigable, and being joined by the *Auriege*, the *Sare*, the *Gimone*, the *Tarn*, the *Rize*, the *Gier* and the *Lot* after its influx into the *Dordogne* at the *bec d' Ambez*, changes its name into the *Gironde*; and at *La tour de Gordouan* difcharges itfelf into the weftern ocean by two outlets of *Le pas des Anes* and *Le pas de Grave*. The celebrated canal by which this river, and confequently the weftern ocean is joined to the *Mediterra-nean*, fhall be fpoken of in *Languedoc*.

The *Rhone*, anciently the *Rhodanus*, has its fource in the mountain of *La Fourche*, in *Uri*, one of the cantons of *Switzerland*, and rapidly tra-verfes the whole country of *Valois*. Its water is whitifh, and near *Bouveret* it rufhes into the lake of *Geneva*, where it runs for above half a *French* mile, preferving its ftream pure, the grayifh tincture of which is eafily diftinguifhed from that of the lake, which is of a bluifh colour; but, at the diftance of one or two *French* miles, it is diftinguifhable neither in colour or motion from the lake, every part being of the fame colour and equally calm; whence the feveral writers, both ancient and modern, who have treated of this river, have erred in affirming that the *Rhone* traverfes the whole lake of *Geneva* without mingling with its waters, and iffues out again preferving its firft colour and rapidity entire. See *Journal Helvetique* for *April*, 1741; or the *Hamburg* Magazine, vol. x. p. 76. About a *French* mile or two above *Geneva*, where the bottom of the lake becomes more floping, the *Rhone* begins to renew its courfe. In the city of *Ge-neva* it recovers its ufual rapidity and name, being joined there likewife by the river *Arve*. Four miles below the lake of *Geneva* it is faid to precipitate itfelf into the clift of a rock a quarter of a *French* mile long, but, in its nar-roweft part, not above two or three toifes broad, with a depth from twenty to about twenty-five toifes. Inftead of the water of this river, it is faid, that in this fall we fee nothing but a thick mift, though I am inclined to think that, upon farther fearch, the river would appear not to lofe itfelf in a fub-terraneous cavern, but that it only runs through a narrow, fteep and rocky aperture in the mountains. It becomes, however, wider afterwards, and at *Seiffel* is navigable, receiving there the *Fier*, the *Ain*, the *Saône*, the *Ifere*, the *Sorgue* and the *Durance* into its channel; and, through the three out-lets

lets of *Le Gras de Sauze*, St. *Anne* and *Le grand gras* falls into the *Mediter-ranean* fea. This river is remarkable for the following particulars, *viz.* that contrary to the *Seine* and the other rivers of *France*, it increafes with the warmth of the weather, its water being higheft when the days are longeft, which proceeds, perhaps, from the melting of the fnow on the *Alps*; that after the junction of the river *Arve* fome fine gold is found amongft its fand; and that it has a kind of ebb and flood, though not certain and regular. See *Journal Helv.* for *May* 1741, and the *Hamburg Magazine*, vol. x. p. 256.

§. 5. *France*, in fome parts of the Kingdom, produces an exuberance, and in all a fufficiency of the neceffaries and enjoyments of life. In plentiful years it yields more corn than is neceffary for the fubfiftence of its inhabi-bitants; but from a neglect of laying up againft an unfavourable feafon, and the ill contrivance of the granaries for preferving the little which is laid up, a bad harveft is generally known to be fucceeded by a fcarcity; and in war time there has often been a great dearth of grain. Of wine this Kingdom enjoys great plenty, producing it in all its provinces. The vine was in all probability firft introduced here by the *Greeks*, who fettled at *Marfeilles*, and in other places on the *Mediterranean*. From *Marfeilles* it fpread into *Gallia Narbonenfis*, and the *Romans* tranfplanted it into *Gallia Celtica*. Among the feveral *French* wines that of *Champagne* is reckoned the beft, being a good ftomachic, racy, and in tafte and flavour exquifite, with an agreeable tartnefs. That of *Burgundy*, the beft of which grows about *Beaume*, has a very fine colour and a pleafant tafte. The wines of *Angers* and *Orleans* are alfo delicate, the only thing amifs in them being, that they are a little heady. In *Poictou* grows a white wine not unlike *Rhenifh*. The neighbourhood of *Bourdeaux*, and the lower parts of *Gafcogne* produce excellent wines, among which the *Vine de grave* is the moft com-mon. Its tafte is fomething harfh and unpleafant, does not foon intoxicate, nor offend the ftomach. *Pontack* grows in *Guienne*. *Mufcadel* and *Fron-tiniac* are the delicious products of *Languedoc*. Betwixt *Valence* and St. *Valliere*, along the banks of the *Rhone*, grows a very agreeable but roughifh red wine, having a tafte not very unlike that of bilberries; the name of it is *Vin d'Eremitage*, and it is in great repute for wholefomnefs. Laftly, not to mention others, fome parts about *Metz* yield alfo good wines, which are often put off for *Champagne*. This Kingdom alfo pro-duces plenty of fea and fpring-falt; the former is made in the fouth, but chiefly along the northern coafts, being both gray and white: The latter abounds in *Burgundy* and *Lorain*. The territories for oil of olive are *Pro-vence* and *Languedoc*. *Normandy, Languedoc, Provence* and *Orange* produce alfo faffron. There is likewife in every part of the Kingdom a fufficiency of grain, vegetables and fruit. The north parts in particular have large orchards for making of cider: And *Bourdeaux* exports whole fhip-loads of prunes.

prunes. Capers grow principally in the country about *Toulon*. Flax and hemp thrive in the *French Netherlands, Picardy, Bretagne, Maine, Dauphiné* and *Alface*; but all the lin-feed produced in this Kingdom comes from the north. Moft of the provinces, *Rouffillon* in particular, together with *Languedoc, Berry, Normandy, Bourgogne* and *Mefzin* abound in wool. Silk too is cultivated here with great induftry and advantage, efpecially in *Languedoc, Provence, Lionois* and *Dauphiné*: And during all the time of the attendance on the infects which produce this commodity, public prayers are offered up for the profperity of thefe ufeful animals : For a more particular account of this article I refer to §. 9.

The two chief provinces for horned cattle are *Normandy* and *Auvergne*. In thefe too great numbers of mules are produced; but for ftrong horfes, *Burgundy, Normandy*, and *Alface* breed the beft; though *Bretagne* and *Limofin* the moft. Game and wild-fowl are found here alfo in fufficient plenty. The people on the fea-coafts, thofe efpecially of *Bretagne* and *Piccardy*, employ themfelves to great advantage in fifhing, exclufive of what is caught in the rivers. The *Pyrenean* mountains furnifh them with what timber is neceffary for fhip-building; great quantities being alfo purchafed out of *Alface, Burgundy*, and *Lorrain*; but in the other provinces the fcarcity of timber and wood for fuel begins to be more and more felt.

Of minerals; *Languedoc* is faid to have veins of gold and filver. In *Alface* thefe metals are found in the fand of the *Rhine*; and the mountain of *Wafgau*, in that province, yields a filver ore of different kinds, which is worked; and further difcoveries are making after it. *France* boafts alfo of mines of copper, as at *Amiens, Abeville, Rheims, Troyes, Beauvais* and *Lorrain*. Of iron there is found a plenty at various places; and *Alface* has a mine of fteel, with others of lead. Pit-coal is principally found in *Hennegau*. All parts of the Kingdom make falt-petre. Here is alfo no want of marble; the quarries fince Monf. *Colbert*'s time, particularly in *Languedoc, Provence*, and *Bourbonnois*, where the principal works of this kind are, being kept continually open. *France*, indeed, does not produce many gems; but yet is not wholly without them, and *Languedoc* has a mine of excellent turquoifes. Bath and mineral waters of approved efficacy are found in feveral places, as at *Bagueres*, and in the valley of *Offau*, fome miles from *Pau, Alface, &c.* befides other remarkable fprings, fome of which, during the greateft heats in fummer remain quite cold, even when the fun-beams play upon them : There are others whofe waters caufe the teeth to drop out. Other remarkable particulars relating to thefe fprings fhall be mentioned and defcribed in their proper places.

§. 6. The qualities, manner of living, and language of the *French* are known to every one. I fhall only fay with refpect to the latter, that under the *Merovignian* Kings, the *Flemifh* was the language at court; and under the

Carlovinian the *German*. About the end of the ninth century the *German* language became extinct in *France :* And the *Roman*, which was the universal language of the Kingdom, became also that of the court. ‾ In the *Journal Helvetique* for *May* 1741, and the *Hamburg Magazine*, vol. x. p. 422, &c. is an essay well worth perusing concerning the question, *At what time the court of France ceased to be German*. The *French* language is formed out of that of the *Gauls*, *Romans*, and *Franks*; with the additions and refinements made in it from time to time by persons of genius, till it arrived to its present supposed perfection. Though the number of persons in *France* be computed at twenty millions, yet are not all the provinces sufficiently inhabited. The many and long wars of this Kingdom having not a little lessened the number of its inhabitants; and another disadvantageous circumstance attending it was the compulsive departure of so many *Hugenots*, who are computed at no less than eight hundred thousand. The peasants and burghers afford nothing remarkable. The nobility are divided into four classes. The first includes only the Princes of the blood, consisting of the house of *Orleans*, and the two branches of the *Bourbon* family, namely, those of *Condé* and *Conti*. Next to them immediately follow the natural children of the King who have been legitimated; and these precede all the great men of the Kingdom. To the second class belong the high nobility, such as dukes and peers of the realm, of which there were formerly only six spiritual and as many temporal; but at present there are fifty-five in all, including the six Princes of the blood, the King too creating as many more as he pleases. The Princes of the blood are peers by birth. The exaltation of a district to a dutchy or viscounty, is performed by letters patent. The principal duties and privileges of the peers are, that they assist at the unction of the King, attend when he holds a *Lit de justice*, and enjoy a seat in the parliament of *Paris*, which is from hence called *La cour des pairs*. 2. The other dukes, counts, and marquisses. 3. The principal officers of the court and other departments: The knights of the Holy Ghost, the governors of provinces, lieutenant-generals, bailiffs, *Seneçhaux d'epee*, &c. Certain families of distinction rank also among the nobility. The third class is composed of the common ancient nobility, who, in some provinces, as *Languedoc* in particular, are stiled nobles, but in most *ecuyers*. They are divided into the *Noblesse de race*, and the *Noblesse de naissance*. The first are those nobles whose ancestors have for a long time past been constantly held so; or who for a hundred years, at least, have been in possession of employments which confer nobility. The second are those nobles whose ancestors were created such, and from whose patent it appears that they were commoners. In the fourth class are reckoned the new nobility, whom the King has either raised by a patent, or conferred some post upon by which they are ennobled : The officers therefore of the crown, with the King's secretaries, the counsellers of the

par-

parliament of *Paris*, and the counfellors of other high tribunals in that city, are claffed under this rank. The *echevins*, or chief magiftrates of the feveral cities, are alfo ennobled by the King, and conftitute what is ludicroufly called the *belle nobility*, the affemblies in which the *echevins* are elected, being convened by the ringing a bell : The military nobility are alfo of this clafs. The nobles enjoy diftinct privileges from the commonality, being exempt from the poll-tax, the quartering of foldiers, and the duties of franc-fief, together with fome other privileges of lefs importance.

§. 7. The eftablifhed religion in *France*, fince the repeal of the edict of *Nantes*, or the charter of the reformed in the year 1685, is the *Roman* catholic; yet in *Dauphiné*, *Languedoc*, and other provinces there are fuch numbers of fecret Proteftants, that they are computed in the whole at no lefs than three millions; but thefe, on account of their religious affemblies, at which twenty, thirty, nay, even forty thoufand perfons were fometimes prefent, have of late been cruelly harraffed and perfecuted. In the year 1745 fome Proteftant preachers were hanged: Laymen were fent to the gallies, and women to the convents, and even children taken away from their parents; befides many other violences and oppreffions. From what I fhall hereafter fay concerning the freedom of all the fubjects of the Kingdom in religious matters, it will appear that though all proteftant worfhip in *France* were totally abolifhed and fuppreffed; yet might multitudes of Proteftants there ftill continue in the Kingdom, provided they would but conform to the civil reftrictions in matters of confequence. The clergy, indeed, are very active in perfecuting them; but outward compulfion is lodged in the hands of the temporal magiftrate only. In *Alface* the Proteftants are openly tolerated by virtue of a treaty; and in many places even the *Jews.*

The celebrated privileges of the *Gallican* church, or as in order to keep fair with the court of *Rome*, it is at prefent called the church of *France*, confifts in certain rights and privileges grounded upon the original conftitution of the church, while *Gaul* was yet fubject to the *Romans*, and the primary ecclefiaftical laws of Chriftianity, which have been ftrenuoufly maintained againft the pernicious innovations attempted by a more modern and incompetent power. It is not poffible to fpecify them all together, or exactly to determine the number of them. The heads of them are included in four propofitions, fet forth in the declaration of the affembly of the clergy of *France*, figned on the 18th of *March* 1682. 1. That the fovereign power in all temporal matters, is fubject neither immediately nor mediately to the mere fpiritual power of the church and the Pope, which extends only to things relative to falvation. Alfo no temporal power can be depofed by the church, nor the fubjects on any pretence whatfoever abfolved from their duties of fidelity and fubmiffion to it, and fupported and authorized in their refufal of fuch duties. 2. That the Pope is fubject to the general

neral councils of the church, which doctrine is not only confirmed by the fourth and fifth council of *Conftance*, but alfo fully ratified by the church; and therefore not only to be obferved in time of fchifm, but is alfo of perpetual and invariable force, and at all times to be conformed to. 3. That the power of the Popes is not unlimited, but connected with the canons, and circumfcribed within the boundaries fet by them : That they cannot permit, ordain, or command any thing contrary to them, or to the cuftoms and privileges of particular churches, grounded on them. 4. That in the decifion of all controverted points of faith, regard is to be paid to the fentence of the Pope, but to its validity and perpetual certainty, or infallibility, the affent of the catholic church is previoufly neceffary : In thefe propofitions, however, a change of the higheft importance has been made, not only the *French* court but even the bifhops and faculties of divinity in *France*, maintaining the Pope to be infallible in refpect of doctrine; the freedom of appealing likewife from the Pope to a general council being abrogated; but the privileges which the church of *France* ftill retains in confequence of thefe maxims, for ritual cuftoms, ufages, and rites demonftrable from hiftory, may be juftly comprifed under three heads.

The firft relate to the privileges of the temporal magiftracy and fovereign, to which belong eight claufes; but moft of them attended with a long train of confequences. 1. In temporal matters, and fuch as relate to government, neither the Sovereign, nor his officers, nor magiftrates are fubject to any church difcipline; not only as inflicted by the bifhops, but even by the Pope himfelf. And for this reafon principally it was that the bull *in cænâ Domini* was rejected in *France*. 2. The Pope has no other proper jurifdiction in *France* than fuch as the King is pleafed to grant him; and the nuncio's and *legates à latere* are only envoys from the court of *Rome* to that of *France*, and cannot by virtue of any papal mandate whatfoever interfere in any juridical matter, unlefs fuch mandate previoufly receive the King's fanction, and be formally regiftered in parliament. No procefs can be removed to *Rome*, thofe only excepted, which the *concordat* and royal ordinances have left to the decifion of that court; but this decifion is to be regulated by fuch laws and papal ordinances only, as have been exprefly and formally acknowledged in *France*. Further, no fubject can be fummoned to *Rome*; but in all cafes where appeals lie from the fpiritual courts to the papal fee, the Pope muft appoint delegates and judges *in partibus*. No papal Count Palatine, or apoftolical notary, can act without a royal licence; and no papal legitimation of illegitimate children is of any further effect than to enable to take holy orders. Laftly, no papal decree can be promulgated in *France*, much lefs is to be accounted legal, valid, and obligatory, till it has been examined and enforced by the folemn acceptance of the civil power. 3. The calling and holding of all ecclefiaftical affemblies, as likewife the confirmation of their decrees,

depends

depends on the civil power, abstractedly from any knowledge or assent of the Pope. 4. Though the King be empowered to enact laws which merely concern the ecclesiastical constitution, and to limit the exercise of their power, without calling a council or acquainting the Pope of it, yet without the King's permission and ratification, no church law, either by bishops or the Pope, subjects any one to any outward penalties or punishments, or even to church discipline. 5. The Pope can levy no money on any one, nor under any pretext whatsoever enjoin the payment of any, the fees and imposts allowed him by the *concordat* only excepted ; whereas the King can tax the clergy without standing in need of a papal mandate, whilst other princes must purchase this privilege by paying to the see of *Rome* a quota of such tax.

It is but very lately since the King enjoined the clergy to pay the twentieth penny; and in pursuance of that to deliver in an inventory of their estates and incomes. This, indeed, occasioned a great ferment, till in 1753 the clergy obliged themselves to pay the King an annual sum of twelve millions of livres under the name of a free gift, and this without being exempted from the usual free gift which they bring to the throne every five years. 6. No ecclesiastical foundations, much less any new orders of monks and nuns are to be introduced or erected without warrant from the King, and the rules of such order are subject also to the cognizance and amendment of the civil power. The King nominates all archbishops, bishops, abbots and priors, unless that with regard to the latter, for the maintenance of the severity of the discipline of the cloister, the convent itself be indulged in the choice of its superior. Of these offices, which are not looked upon as properly connected with the formal care of souls, one person may hold several, exclusive, however, of bishoprics. The Kings of *France*, at present, possess the right of nomination to all prelacies by virtue of the *concordat* made in 1515, betwixt *Francis* I. and Pope *Leo* X. 8. To the King belongs, throughout his whole Kingdom, the *Regale*, as it is called, taken in its strictest sense, which consists in this, that, during the vacancy of a bishopric, he himself has the management of it till it be filled up, together with the nomination of all the ecclesiastical officers, in the bishop's stead ; and the disposal of all the vacant benefices, parish-churches excepted. All ecclesiastical jurisdiction too is so far subordinate to the civil power, that appeals from any sentences of ecclesiastical courts are received in the parliament on any appearance of the least abuse of church power, or an infringement in any single article of these privileges, or of any contravention to the royal edicts. These *appels comme d'abus*, as they are called, are immediately heard in the first instance. 10. All execution of church discipline and exercise of ecclesiastical power is subject to the inspection of the temporal magistrate, insomuch that the parliament, without any formal appeal to it, or any other complaint, can examine and de-

2   bate

bate upon all writings and procedures of the clergy, without exception, when ſtriking at theſe privileges.

The ſecond head of the rights of the *Gallican* church relates to the privileges of the biſhops and clergy when aſſembled. Theſe may be reduced to the four following. 1. The biſhops are the proper judges of all matters of faith in their ſeveral Dioceſes, and conſequently may not only examine every other deciſion of them, thoſe of the Pope himſelf not excepted, but are alſo to watch and ſee that the doctrine of the church be maintained in its real purity, and not to give their conſent to new tenets, or deciſions of any controverſies relating thereto, till after the matureſt deliberation. 2. The power of the biſhops, with reſpect to divine worſhip and church diſcipline, as eſtabliſhed by the canons and granted by the ſovereign, cannot be reverſed by the Pope, either by general mandates or private injunctions. Thus the compoſing of breviaries, and other books of liturgy, as likewiſe the inſtitution of new feſtivals, with the alteration of ceremonies, belong to the biſhops. No papal diſpenſation or indulgence can be pleaded againſt their juriſdiction : No appeal from their ſentences is regular, unleſs to the archbiſhop and primate of the Kingdom : and from them to the Pope, whoſe examination of the affair muſt be tranſacted by domeſtic judges of his own appointment. This holds good likewiſe with reſpect to the prelates of orders, peculiar chapters, and independent churches. 3. All eccleſiaſtics are exempt from foreign pecuniary impoſitions and juriſdiction, their benefices are not ſuject to annates or contributions, nor can the fees of the chancery at *Rome* be raiſed to their grievance. Neither can their perſons be ſummoned to appear out of the land. 4. All eccleſiaſtics may apply for the protection of civil power againſt the abuſe of their eccleſiaſtical ſuperiors; and in this they are not to be obſtructed or puniſhed by any kind of church diſcipline.

The third principal head of their immunities relates to the privileges both of the members of the *Roman* catholic church in *France,* and the collective body of the inhabitants and ſubjects of the Kingdom : And theſe conſiſt in three points. 1. No perſon can be denied the ſacraments, be ſolemnly excommunicated, or in any other ſhape be moleſted by church diſcipline, but in a manner agreeable to the canons, as confirmed by the civil power. Thus all perſons may appeal againſt any abuſe of eccleſiaſtical power, and attempts made upon the general liberties. 2. Every one is at liberty to read the holy ſcriptures in their mother tongue ; which the clergy are not to prohibit or forbid. 3. No perſon is to be compelled by the clergy to divine worſhip, or be moleſted on account of his faith, provided he forbear propagating his opinions ; arraining the worſhip and doctrine eſtabliſhed by the ſovereign ; and treſpaſſing againſt the ordinances of the civil power, as conſidered with reſpect to prohibited religious aſſemblies ; and the marrying of perſons who require a formal certificate.

The

The *Janfenifts*, together with the followers of *Quefnel* and the oppofers of the famous conftitution *Unigenitus*, conftitute a particular religious party in *France*, but account themfelves within the pale of the *Romifh* church, and in this Kingdom muft be confidered as fuch by the Pope, although declared heretics by the conftitution *Unigenitus*, and excluded the commu-nion of the church; for were they in their own country debarred the com-munity of divine worfhip, it is to be apprehended, that they would one and all depart, as not a few have done already. In the mean time the civil power is endeavouring, though by gentle means, to put a period to this party.

In the whole Kingdom are eighteen archbifhops, one hundred and thirteen bifhops, twenty-two thoufand two hundred and ninety-one priefts, feven hundred and feventy abbeys for men, three hundred and feventeen abbeys and priories for women, exclufive of a great number of other con-vents, together with two hundred and fifty commanderies of the order of *Malta*; namely, two hundred for knights and fifty for the chaplains and *fervans d' armée*. Among thefe commanderies there are fix grand priories and four provincial commanderies. According to the lifts in the *almanach royal*, the yearly revenue of the archbifhoprics and bifhoprics is four millions three hundred and thirty-feven thoufand livres; the particular income of each, with its taxation at the court of *Rome*, fhall be fpecified in the ac-counts of the cities. The number of ecclefiaftics in *France*, is computed, in the whole, at above one hundred and ninety thoufand, with an income greatly exceeding one hundred millions of livres. The ecclefiaftical ftate is compofed of eighteen provinces, each of which confifts of an archbifhopric and the bifhoprics fubordinate to it, and receives its name from the arch-bifhopric. Thefe are as follow, *viz.*

1. The province of *Paris*, to which, befides the archbifhopric of *Paris*, belong the diocefes of *Chartres, Meaux, Orleans* and *Blois.*

2. The province of *Lion*, to which, exclufive of the bifhopric of *Lion*, belong the diocefes of *S. Claude, Autun, Langres, Macon, Chalon* on the *Saone* and *Dijon.*

3. The province of *Rouen*, containing, befides the archbifhopric of *Rouen*, the diocefes of *Bayeux, Avranches, Evreux, Séez, Lifieux* and *Coutances.*

4. The province of *Sens*, which contains, exclufive of the archbifhopric of *Sens*, the diocefes of *Trojes, Auxerre, Nevers* and *Bethlehem*, which was removed to *Clamecy.*

5. The province of *Rheims*, exclufive of the archbifhopric of *Rheims*, comprehends the diocefes of *Soiffons, Chalons* on the *Marne, Laon, Senlis, Beauvais, Amiens, Noyon* and *Boulogne.*

6. The province of *Tours* comprehends, befides the archbifhopric of *Tours*, the diocefes of *Mans, Angers, Rennes, Nantes, Quimper-Corentin, Vannes, S. Pol de Leon, Treguier, S. Brieux, S. Malo* and *Dol.*

7. When

7. The province of *Bourges*, to which, befides the archbifhopric of *Bourges*, belong the diocefes of *Clermont*, *Limoges*, *Tulles*, *Pu* and *S. Flour*.

8. The province of *Alby* contains, befides the archbifhopric of *Alby*, the epifcopal churches of *Rhodez*, *Caftrez*, *Cahors*, *Vabres* and *Mende*.

9. The province of *Bourdeaux*, to which, befides the archbifhopric of *Bourdeaux*, pertain the epifcopal fees of *Agen*, *Angoulefme*, *Saintes*, *Poitiers*, *Perigueux*, *Condom*, *Sarlat*, *Rochelle* and *Luçon*.

10. The province of *Auch*, exclufive of the archbifhopric of *Auch*, contains the diocefes of *Acqs*, *Lectoure*, *Comenges*, *Conferans*, *Aire*, *Bazas*, *Tarbes*, *Oleran*, *Lefcar* and *Bayonne*.

11. The province of *Narbonne* comprehends, befides the archiepifcopal fee of *Narbonne*, the diocefes of *Befiers*, *Agde*, *Carcaffone*, *Nimes*, *Montpellier*, *Lodeve*, *Uzes*, *S. Ponts de Tuieres*, *Aleth*, *Alais* and *Perginan*.

12. The province of *Touloufe* comprehends, befides the archbifhopric of *Touloufe*, the epifcopal fees of *Montauban*, *Mirepoix*, *Lavaur*, *Rieux*, *Lombez*, *S. Papoul* and *Pamiers*.

13. The province of *Arles*, exclufive of the archbifhopric of *Arles*, contains the feveral diocefes of *Marfeilles*, *St. Paul trois chafteaux*, *Toulon* and *Orange*.

14. The province of *Aix*, exclufive of the archiepifcopal fee of *Aix*, contains the bifhoprics of *Apt*, *Riez*, *Frejus*, *Gap* and *Sifteron*.

15. The province of *Vienne*, to which, befides the archbifhopric of *Vienne*, the diocefes of *Grenoble*, *Viviers*, *Valence* and *Dié* belong. The bifhops of *Geneve* and St. *Jeane de Maurienne*, are fuffragans of the archbifhop of *Vienne*.

16. The province of *Ambrun* comprehends, exclufive of the archiepifcopal fee of *Ambrun*, the bifhoprics of *Digne*, *Graffe*, *Vence*, *Glandeve* and *Senez*.

17. The provinces of *Befançon*, befides the archbifhopric of *Befançon*, contains the diocefe of *Belley* in *Bugey*. Subordinate alfo to this archbifhop are the bifhops of *Bafel* and *Laufanne*.

18. The province of *Cambray* contains, befides the archiepifcopal fee of *Cambray*, the feveral diocefes of *Arras*, *S. Omer* and *Tournay*. The bifhop of *Strafburg* is alfo a fuffragan to the archbifhop of *Maynz*.

Of thefe archbifhops the firft fixteen only, with their refpective bifhops, come to the affemblies of the clergy, the two laft, and their bifhops, as alfo the bifhop of *Strafburg*, with thofe of *Mets*, *Toul*, *Verdun*, all fuffragans of *Triers*; and *Orange*, fuffragan to *Arles*, being excluded. The general affemblies of the clergy are two, *viz.* ordinary and extraordinary: the latter have no ftated times of meeting, but depend on emergencies: the former, called alfo *contrats*, are held every ten years, each ecclefiaftical province fending four deputies; namely, two prelates and two abbés.

In

F R A N C E.

345

In *France* are nine *bureaux genereaux*, or *chambres ecclefiaftiques* and *fuperieurs*, which decide finally in all cafes or proceſſes coming before them by appeals from their refpective diocefes. Thefe fit at *Paris, Lyon, Rouen, Tours, Bourdeaux, Bourges, Touloufe, Aix en Provence* and *Pau*. To thefe nine *bureaux* appeal all diocefes, or biſhoprics, and on this account it is that they are called *bureaux diocefains*. Befides thefe, there is alfo a *chambre fouveraine du clergé de* France, to which, as the fupreme ecclefiaftical court, appeals are made concerning any impofitions of taxes, or from the fentences of the *bureaux diocefains*. The judges in it are three counfellors of parliament, with fome others who are deputed to officiate there. With refpect to religion, nothing further remains worthy obfervation, but that the decrees of the council of *Trent*, as far as they relate to church difcipline, are not admitted, though, in point of doctrine, received as the very ftandard of orthodoxy.

§. 8. The *French* have always diftinguiſhed themfelves in the arts and fciences. Painting was firft brought to great perfection among them under *Francis* I. when *Roux* and *Francifco Bolognefe* introduced all the beauty of that art into *France*. Since that time this Kingdom has produced feveral eminent mafters. In the year 1648, was inftituted at *Paris* the *academie royale de peinture et de fculpture*, which has proved a nurfery of admirable artifts that way; yet the anonymous author of a fmall treatife publiſhed in the year 1746, laments that the ingenious art of painting is on the decline in *France*, and fupported by a frivolous tafte which tends neither to the fupport or improvement of thefe liberal arts. Engraving, which *Italy* owes originally to *France*, has here been brought to a perfection unequalled in any other part of *Europe*; their works exhibiting whatever can be expected from the moft accurate ſkill and judgment. *France* has alfo made great advances in ftatuary; though in this point the palm is due to *Italy*. It has always been famous too for architects; and the great *Colbert*, in the year 1671, founded an *academie royale d' architecture*, who hold their meetings in the *Louvre*. If the *French* learned the art of ſhip-building from the *Engliſh*, they have fince been copied in feveral particulars by their mafters. Their reputation in military architecture is well known. In gunnery and fireworks they alfo excel. No nation has hitherto equalled the *French* in what is called the *beaux arts*, but in other fciences their application, and reputation fubfequent to it, begin vifibly to decline. The zeal and liberality of perfons of wealth and diftinction, for thefe laft hundred years, for the improvement of literature, is far more worthy of imitation than the ficklenefs of the *French* in modes and faſhions. At *Paris*, befides the ancient univerfity and *college royale*, are four academies; *viz. l' academie* Francoife, *l' academie des infcriptions et belles lettres, l' academie de chirurgerie* and *l' academie des fciences*, with three other academies for the education of young noblemen.

VOL. II. Y y . Exclufive

Exclufive of thefe, there are eighteen other univerfities in the whole Kingdom, with feveral academies befides for the fciences and *beaux arts.* -

§. 9. The *French* manufactures and productions are excellent, and fupported and improved with the greateft diligence and encouragement. It was
by the advice of the abovementioned *Colbert*, that wife encourager of manufactures and the liberal arts, that *Louis* XIV. erected the manufacture of the
Gobelins at *Paris*, which is fo called from two brothers *Giles* and *John Gobelin*,
who, under *Francis* I. found out the method of dying a moft beautiful fcarlet,
which afterwards came to be called after their names. The houfe of the
Goblins at *Paris* is full of the works of the moft excellent mafters in tapeftry,
filigree and fculpture; and in the article of tapeftry alone, no lefs than two
hundred perfons are employed from year to year. *Europe* affords nothing
equal to this manufactory, the quantity of the works, the inimitable delicacy of them, and the number of artifts, being great beyond credibility.
Indeed, the prefent flourifhing ftate of arts and handicrafts in *France*, is
principally to be afcribed to the inftitution of this manufacture. Such is
the tapeftry wove here, that it may be faid to equal that formerly made by
the *Englifh* and *Dutch*, who firft learned the art in the *Levant*, where it
had its original. Wool and filk too, are fo ingenioufly worked here as to
approach very near to life and nature; and the intertexture of the filken and
woollen threads are not inferior to the fineft ftrokes of the moft mafterly
pencil. There are two kinds of *French* tapeftry, *viz.* the *haute* and *baffe
lice*, though the difference betwixt them confifts principally in the manner
of weaving and not in the workmanfhip itfelf, the loom for the *haute lice*
being perpendicular, and that for the *baffe* horizontal. There are alfo confiderable manufactures of tapeftry at *Felletin, Beauvair, Arras, Auvergne,
Aubuffon*, &c. The *French* tapeftry and carpets being known in moft parts
of the world, and no lefs in requeft for their beauty, it may eafily be
imagined, that they bring great fums of money into the country; yet, in
many refpects, are they furpaffed by the filk manufactures, though the latter are not at prefent fo flourifhing as formerly. The *French* hiftorians inform us that filk manufactures were firft fet up in *France* in the reign of
Louis XI. yet, in the time of *Henry* II. filks were fo fcarce, even at court,
that this King was the firft who had ever been feen with a pair of filk
ftockings on, which were a part of his drefs at the marriage of his fifter.
Henry III. made fome good regulations with regard to the culture of filk,
but of no remarkable effect. *Henry* IV. caufed great numbers of white
mulberry-trees to be planted, particularly in *Tourayne*; but after his demife
this order was neglected. It was not till the time of *Lewis* XIV. that the
culture of filk arrived at any confiderable degree of profperity, and under
his government, the quantity of raw filk produced in the provinces of
Dauphiné, Languedoc and *Provence* only amounted yearly to one million,

eight hundred thoufand pounds weight. The city of *Tours* alone had eight thoufand looms and eight hundred mills conftantly employed, which afforded a comfortable maintenance to four thoufand people. In the city of *Lyon* were formerly eighteen thoufand looms, but, in the year 1698, thefe were decreafed to four thoufand: this city, however, is ftill diftin-guifhed for its manufactures of filk, and particularly its *armozins*, which are fo remarkable for their beautiful glofs, and were accidentally found out by one *Octavius May*. Black *armozins* are gloffed by a decoction of beer and *Seville* oranges. In other colours diftilled pumpkin-water is ufed. Very fine filk ftuffs of all kinds are wove alfo in the province of *Auvergne*, at *Caftel Naudary*, *Nifmes*, &c. but thofe of *Lyon* furpafs all others. Within thefe laft hundred years the *French* have alfo confiderably improved their woollen manufactures, the happy effect of inftructions from foreigners who fettled among them, affifted likewife by the clandeftine tranfportation of wool from *England*. The cloths and other woollen ftuffs made at *Abéville* being little inferior, either in finenefs or goodnefs, to the *Englifh* or *Dutch*. This place is famous alfo for its manufactures of linen and fail-cloth; its cannon founderies and foap. In *Upper Normandy* cloths are exquifitely died; *Rouen*, in particular, carries on a flourifhing trade in that commodity. *Bretagne* produces large quantities of hemp and flax, and abounds in manufactures of linen and fail-cloth. The province of *Berry* is noted for its linen and cloth, and *Auvergne* for fine thread lace, and all kinds of ftuffs; its paper too is accounted the beft in *Europe*. The town of *S. Flour* is diftinguifhed for tapeftry, cloth and knives; as *Nifmes* and other places are for their manu-factures of ferge and ftuffs. At *Cambray*, in *French Hennegau*, is made that fine linen called *Cambray* linen or *Cambrick*, which exceeds the whole world for beauty, as *S. Quentin* excels in its lawns. The *French* alfo make gloves and ftockings of fpider-filk, an invention owing to Monf. *Bon*. According to *M. Reaumur's* computation, twenty-feven thoufand fix hundred and forty-eight fpiders would be required to produce one pound of filk, whereas two thoufand three hundred and four filk-worms yield that quantity. The glafs manufacture alfo is of no fmall advantage to *France*. In the year 1688, one *Abraham Thevert* contrived a method for cafting better and larger plates than had been before known. By this method, which bears fome refemblance to that of fbeet-lead, looking-glaffes twice as large as thofe made at *Venice*, by blowing, are caft; but it anfwers, likewife, in all kinds of figures and decorations in that commodity. The principal ma-nufactures of glafs are at *S. Gobin* in *Picardy*, where they are caft and polifhed afterwards at *Paris*. *Languedoc* alfo makes glafs, but neither fo fine or white as the above. *Vincennes* has alfo a manufactory of very pretty porcelain; and, exclufive of thefe, *France* has many other in all kinds of fabricks and manufactures. After the repeal of the edict of *Nantes*, many thoufands of the Reformed, on account of the perfecution in

France, left the country and fettled in the *United Provinces, Germany* and *England*. By thefe means the *Dutch*, who formerly ufed to take goods of the *French*, to the annual amount of thirty-fix millions, exclufive of corn and falt, at prefent take fcarce four or five.

§. 10. The trade of *France* is very large. In §. 5 and 9 may be feen the variety and multitude of natural and artificial commodities produced in that country, which are certain to meet with a good market among foreigners. To repeat only the principal, as gold and filver brocades, gold and filver embroideries, fewing filk, filken ftuffs, fattins, crapes, cambrick, lawn, other fine linens, laces, toys and millinery ware, tapeftries of va- rious kinds and prices, fine cloths, plufh, woollen ftuffs of feveral kinds ; houfhold furniture, as matraffes, carpets, &c. fine hats, fail-cloth, looking- glaffes, drinking-glaffes, paper, parchment, foap, knives and other hard ware, wines of many forts, vinegar, brandy, corn, hemp, flax, walnut- tree wood, potafh, pitch, turpentine, oil of turpentine, linfeed and oil of olive, almonds, raifins, cheftnuts, figs, prunes, capers, nuts, honey, faffron, falt, mules, &c. For the inland trade the navigable rivers are a great conveniency to them, and thefe again have at a vaft expence been joined by canals, of which the *canale royale* in *Languedoc* is the moft noted. The coaft trade, from one port to another, muft alfo come into account, as be- ing the means of intercourfe betwixt the maritime provinces of the King- dom, whereby they reciprocally fupply each other's neceffities. The foreign trade, indeed, is an article of another confideration and extends to every part of the known world. The land trade is carried on with *Swit- zerland* and *Italy*, by way of *Lyon*; to *Germany* through *Metz* and *Sraf- burg*; to *Holland* by way of *Lifle*, and to *Spain* by way of *Perpignan* and *Bayonne*. The open and clandeftine commerce carried on with *Spain* is of vaft benefit to this Kingdom. The *French* ports on the Channel and the Weftern Ocean are frequented by all the trading nations of *Europe*. The trade with *England*, the *United Provinces* and *Italy* is of great advantage to *France*. Few *French* fhips are feen in the ports of the northern nations; which is fo far a detriment to *France*, as they might thereby be their own carriers and make confiderable profits by the returns, whereas now they have them from fecond or third hand. Their trade on the *Mediterranean* with *Afia* and *Africa* has, for fome years, been very confiderable, and the far greateft part of it is carried on by the city of *Marfeilles* only. They carry on alfo a confiderable traffick to *Guinea*, from whence, befides gold, ivory, &c. they import Negroes for the fupply of their *American* colonies, which now make a very great figure, and are the fource of a prodigious branch of commerce to them, not to mention the clandeftine traffick carried on with the *Spanifh* continent in *America*, in which the *Spaniards* themfelves are the chief agents. The *Eaft-India* trade is in the hands of a company whofe particular port is that of *Orient* in *Bretagne*.

The

The *French*, for fome time paft, have carried on a trade, befides that with *Europe*, with the other three parts of the world ; but, inftead of an equal courfe of profperity, it has met with feveral fufpenfions and interruptions, efpecially when in the hands of particular companies, who, not underftanding their real advantages, were profufe where they ought to have fhut their hands, and parfimonious when they might have been gainers by liberality. *Des Landes*, in his effay *fur la marine* and *fur le commerce*, p. 169, charges thefe companies with three effential faults : 1. That they had no fooner fowed than they were for reaping, without waiting the proper feafon, or confidering, that thofe fruits are beft that ripen lateft. Their eagernefs after gain would not let them poftpone an immediate trivial advantage to what, in time, would have proved of very confiderable benefit. This alfo put them foon out of conceit with any fingle thing. 2. Through a wild purfuit of exceffive gain, they imported a greater quantity of foreign goods than could be difpofed of in the Kingdom, not to mention that this was the ready way to lower their price and value. 3. In thefe companies private views always had the afcendant over public good. If the prefent *Eaft-India* company are clear of thefe faults, yet it is to be apprehended, that extending their trade as they do, totally exhaufting the national claffes (§. 19.) which are otherwife at a very low ebb, and employing all people indifcriminately and without diftinction, they may be involved in the fame deftiny with the former company. The *abbé de S. Pierre* in his *Mémoires Politiques* obferves, and *Des Landes* from him, that the balance of the *French* trade with foreigners amounts *communibus annis*, at leaft to one hundred and fifty millions of livres. In the year 1724, the King crected a board of trade, which confifts of counfellors of ftate and other commiffaries of his council, with twelve deputies elected out of the merchants of the principal trading cities in the Kingdom, and four *offices de confeiller*, who are admitted into the King's council as intendants of the commerce of *France*.

The following cities have the privilege of mintage, *viz.*

The cities,	their marks.	The cities,	their marks.
Paris,	A.	*Perpignan,*	Q.
Rouen,	B.	*Orleans,*	R.
Caen,	C.	*Reims,*	S.
Lyon,	D.	*Nantes,*	T.
Tours,	E.	*Troyes,*	V.
Angers,	F.	*Amiens,*	X.
Poitiers,	G.	*Bourges,*	Y.
La Rochelle,	H.	*Grenoble,*	Z.
Limoges,	I.	*Aix,*	&.
Bourdeaux,	K.	*Rennes,*	9.
Bayonne,	L.	*Metz,*	AA.

The

INTRODUCTION TO

The cities,	their marks.	The cities,	their marks.
Touloufe,	M.	*Strafbourg,*	BB:
Montpellier,	N.	*Befançon,*	C.C.
Riom,	O.	*Lolle,*	VV.
Dijon,	P.	*Pau,*	V or Q.

The feveral mints of thefe cities are under the infpection of the three *cours des monnoies* or mint-courts, at *Paris, Lyon* and *Pau.* In *France* books and accounts are generally kept in livres (pounds) fous and deniers. One livre * contains twenty fous, and one fol or fou twelve deniers. The laft coin is the fmalleft, and confifts of copper. Three deniers make a liard, which is alfo of copper. One *ecu,* or crown, is equal to three livres, fixty fous, or feven hundred and twenty deniers. One *Louis blanc,* or coined louis, in filver is double the value of an *ecu.* A coined louis in gold, or as it is called a *louis d'or,* is at prefent worth twenty-four livres. There are alfo double and half louis. A piftole is ten livres, the new *louis d'ors* weigh feven penny-weights fifteen grains, and are of twenty-two carats; on this footing the value may be about fix rix-dollars and a half.

§. 11. *France* was formerly inhabited by the *Celtæ,* on whom the *Romans* firft conferred the name of *Gauls. Julius Cæfar* reduced *Gaul* into a *Roman* province. In the fifth century the *Burgundians, Vifigoths,* and *Britains* came into *Gaul* and fettled in feveral parts of it. Thefe were followed by the *Franks* from *Germany,* who under their leaders *Merovick* and *Childerick,* poffeffed themfelves of a part of *Gaul.* But under *Clovis,* i. e. *Lewis,* extended their dominion from the *Rhine* to the mouth of the *Loire.* King *Clovis* was baptized in the year 496, and quelled the *Burgundians, Vifigoths* and *Britains,* which laft fubmitted to him. The Kingdom of the *Franks* was divided into two principal parts, namely *Auftracia,* or the eaft part, and *Neuftria,* or the weft. On the deceafe of *Clovis,* his four fons divided their father's Kingdom among them. *Theudebert,* fon of *Theodorick* King of *Auftrafia,* with the affiftance of his brothers in 534, put an end to the Kingdom of *Burgundy,* and *Clothar* the fourth fon to *Clovis,* in the year 558 again reunited the whole monarchy of the *Franks*; but it was afterwards feveral times difmembered and again reunited. Towards the middle of the feventh century the power of the *Maires du palais,* or *Majores domûs,* was grown to fuch an exorbitant height as to border on independency and. abfolutenefs. At the death of *Dagoberts* II. *Pepin* was chofen duke of *Auftrafia* and *Maire* of the palace of *Neuftria.* He dying in 714, his fon *Charles Martel* fucceeded him in thofe dignities. On the demife of this va-

---

* In order to furnifh the reader with a competent idea of thefe feveral coins at one view, the value of each of them, computed in *Englifh* money, is as follows, viz. one livre makes 10 d. one fol ⅔, a denier 2 ⅓, a liard ⅞, an ccu 2 s. 6 d. a louis blanc 5 s. a louis d'or 1 l. and a piftole 8 s. 4 d.

liant duke his two fons *Charlemagne* and *Pepin* divided the country betwixt them; it being agreed that the former fhould be duke of *Auftrafia*, and the latter of *Neuftria*. They created, indeed, *Chilperick* King; but *Charlemagne* retired into a convent, and *Pepin* after he had by the confent of the Pope and the States fent *Chilperick* and his fon *Theodorick*, the laft of the *Mero-vingian* race into a convent, was proclaimed King at *Soiffons* in the year 752. Thus a new family afcended the throne. *Pepin* deprived *Aiftulphus* King of *Lombardy* of the whole exarchate and beftowed it on the fee of *Rome*; which grant was confirmed and increafed by his fon *Charles the Fat*, who over-run the Kingdom of *Lombardy*, annexed that country to *France*, obtained feveral advantages over the *Saxons*; and on *Chriftmas-day* in the year 800 was at *Rome*, together with his fon *Pepin*, anointed Emperor of the *Romans*. His imprudent fon *Lewis*, undefervedly furnamed the Pious, committed a very great error in making the governments of his provinces hereditary; as he gave thereby a fatal blow to his own dignity, and that of his fucceffors. The ftates thus grew powerful, and the Kings of the *Carlovingian* houfe, by their indolence, weakened themfelves. The laft of this race was *Lewis* V. The court of *France* was now no longer *German*. *Charles* duke of *Lorain*, to whom of right the crown of *France* belonged, was excluded from it; and in the year 987 the fovereignty of this realm was conferred on *Hugh Capet*, a nobleman of great diftinction; the Kings of his family hurt them-felves by their many croiffades. *Philip* IV. or *the handfome*, very unjuftly put an end to the Knights Templars, and his third and laft fon *Charles* IV. died without male heirs. Upon this *Philip* VI. of *Valois* inherited in the year 1328 the throne, to which alfo *Edward* III. of *England* made preten-fions; and this oppofition involved thofe princes and their fucceffors in long and bloody wars. In the year 1361, *John the good* inherited *Burgundy*, which he beftowed on his youngeft fon *Philip*. *Charles* VII. difpoffeffed the *Englifh* of *Normandy* and *Guyenne*. *Lewis* XI. ruled arbitrarily, and on the death of Duke *Charles the bold*, took poffeffion of the province of *Bur-gundy*; *Provence*, with *Tculoufe* and *Champaigne*, defcending to him by in-heritance. In the year 1498 died *Charles* VIII. the laft of the firft line of the houfe of *Valois*, on which the crown defcended to *Lewis* XII. duke of *Orleans*, whofe coufin and fon-in-law *Francis* I. Count of *Angoulefme*, who had married *Clodia*, daughter of *Anne* of *Bretagne*, fucceeded him, and con-cluded the famous *concordat* (§. 7.) with *Leo* X. In the year 1549 *Henry* II. took *Boulogne* from the *Englifh*; and in 1552 *Metz*, *Tull*, and *Verdun* from the Emeror *Charles* V. Three of his fons fucceeded him in order: Under the firft, namely *Francis* II. fome ftrictures of the melancholy religious dif-turbances began to appear in *France*; and thefe, under *Charles* IX. broke out into two civil wars; after the conclufion of which on the 18th of *Auguft* 1572, was perpetrated the maffacre of *Paris*, which leaves an indelible ftain on the hiftory of *France* and Chriftianity. Under *Henry* III. the fa-mous,

mous *Ligue* formed by the violent Catholics threw the Kingdom into more terrible convulsions, the consequence of which was the violent death of the King in 1589. His legal successor was *Henry* IV. King of *Navarre* and duke of *Bourbon*, who had been bred a Protestant and turned *Roman* catholic, but secured his former brethren by the celebrated edict of *Nantes*. He too was murdered. The reign of *Lewis* XIII. was disturbed by several religious wars. In the year 1620 he annexed the Kingdom of *Navarre* to *France*: His prime minister Cardinal *Richelieu*, in 1633, founded the *French* academy, and weakened both the *Hugenots* and the power of the States. *Lewis* XIV. in his long reign, carried the reputation and power of *France* to the highest pitch. To his Kingdom he added the provinces of *Alsace* and *Roussillon*, with a considerable part of the *Netherlands*, *Franche Comté* or the earldom of *Burgundy*, and the principality of *Orange*. He enlarged also his dominions in *America* and *Asia*, and placed his grandson *Philip* of *Anjou* on the throne of *Spain*, whilst navigation and manufactures were improved under the auspices of his chief intendant of the finances Monf. *Colbert*. But the intestine disturbances of his Kingdom occasioned partly by the opposition to Cardinal *Mazarine*, by the repeal of the edict of *Nantes*, the persecution of the *Hugenots*, and likewise by the well known bull *Unigenitus Dei filius* (§. 7.) had very bad effects; divine Providence being pleased also by other means to humble this mighty and inflated monarch.

§. 12. The King's title is *Lewis* XV. *by the Grace of God King of* France *and* Navarre. The title of *Sire* (or Lord) is given to him by his subjects as a mark of his unlimited power. Foreigners stile him the *Most Christian King*, or his *Most Christian Majesty*. It is generally said that this title was conferred by Pope *Paul* II. on *Lewis* XI. about the year 1469. But *French* historians affirm that it is derived even from the time of *Childebert*, though under the *Merovindian* family it had lain dormant. The King's subjects are not to make use of it. The Popes also term the Kings of *France* the first born sons of the church, because at the time of the baptism of *Clovis*, he was the only Christian Prince then in being. Since the year 1349, when, *Humbert*, Count *Dauphin* of *Viennois*, by a double gift united his country to the crown of *France* and the royal family, the King's eldest son and presumptive heir to the crown is stiled *Dauphin*. He used to be called *Dauphin* of *Viennois*, but at present is stiled *Dauphin* of *France*, which title was first given to a son of *Lewis* XIV. who died in 1711. If the *Dauphin* dies before the King his father, the son whom he leaves is *Dauphin*. *Lewis* XIV. lived to see his great grandson, the present King, *Dauphin*. The eldest son of the present *Dauphin* is stiled duke of *Burgundy*; the second duke of *Acquitain*, which is the ancient name of *Guyenne*; the third of *Berry*, and the fourth of *Provence*. The rest of the royal children, as well sons as daughters bear the surname of *de France*, and the sons are again distinguished by particular titles, the second being stiled duke of *Orleans*; the third of

Anjou,

Anjou, and the fourth of *Berry:* The others are not yet fettled. The princeffes are called *Mefdames de France*.

The King bears *party per pale* two fhields united; the dexter fhield containing three lilies *Or*, in a field *azure* for *France*; the finifter, in a field of *gules*, a link-chain *Or*, difpofed partly crofs-wife and partly in a fmall fquare within a larger.

Among the appurtenances the moft remarkable is the banner over the *French* crown, with which the mantle is furmounted, the *French* military cry, *Mon joie S. Denis,* together with the *Auriflammeum Or,* which is the ftandard of the Kingdom.

§. 13. The higheft order of knighthood in *France* is that of the Holy Ghoft, founded in the year 1578 by King *Henry* III. The enfigns of it are a golden crofs with a white dove enamelled on one fide in the middle, and on the other the image of St. *Michael,* appendant to a blue ribbon, paffing from the right to the left fide. The knights alfo wear on the left breaft of their coats a filver crofs, with a dove embroidered *argent.* The temporal knights are alfo knights of the order of St. *Michael,* and the greateft part of them likewife of that of *S. Louis.* The three feftivals of the order are the *Circumcifion, Candlemas,* and *Whitfuntide.*

The order of St. *Louis* was inftituted by *Lewis* XIV. in *April* 1693. Its enfign is a gold crofs enamelled *argent,* and decorated with golden lilies. On one fide of it is a coat of mail with this infcription, *Lud. M. inftit.* 1693. On the other is a drawn fword with a wreath of laurel at its point; the motto, *Bell. virtutis præm.* The *grands croix* or great croffes wear it on a broad flame-coloured ribbon over the fhoulder, having alfo a gold embroidered crofs on their coat. The commandeurs wear it in the fame manner, but without the embroidered crofs. The other knights wear the crofs appendant to a narrow flame-coloured ribbon faftened to a button-hole. Thofe knights of the Holy Ghoft, who are alfo knights of the order of St. *Louis,* wear the crofs of the latter with a narrow red ribbon faftened to the blue ribbon near the crofs of the Holy Ghoft. Of the ten *grands croix,* each has a penfion of fix thoufand livres *per annum.* Of the ten commandeurs, each four thoufand. Of nineteen others, each three thoufand. Of thirty knights, each has two thoufand. Of thirty-two others, each has fifteen hundred. Of fixty-five others, each has one thoufand. And of fifty-four others, each eight hundred.

The order of St. *Michael* was inftituted by *Lewis* XI. in the year 1469, and revived in 1665 by *Lewis* XIV. The knights wear a gold chain of double fcallop-fhells with a medal reprefenting a rock, on which St. *Michael* is figured as encountering the dragon.

The religious order of St. *Lazarus* derives its origin from the *Holy Land,* after the reduction of which, by the *Saracens,* the knights returned to *France,* where *Lewis* VII. in the year 1137, beftowed fettlements on them at *Boigni*, *Orleans* and *St. Lazare* at *Paris*; which grant was confirmed by *St. Lewis* in 1265. This order is divided into two grand-maſterſhips, one of which reſides in *France*, and the other which is for *Italy* in *Savoy*. The knights wear a temporal habit, and may marry. *Henry* IV. in the year 1607; inſtituted the order of our Lady of Mount *Carmel*, and united it with the order of *S. Lazarus*; which union was confirmed in 1664 and 1672, by *Lewis* XIV. The Sovereign himſelf is grand-maſter of the three firſt orders, but of the fourth he names one.

§. 14. *France* is not without its fundamental laws, of which the principal is the *Salic*, which is an ordinance of *Charles* V. in the year 1374, relating to the term when the heir to the crown is of age, and another of *Charles* VI. in 1404, relative to the coronation ; but theſe depend on the King's pleaſure, whoſe power is unlimited. The crown is hereditary, and it is a maxim among the *French* politicians that the King never dies; the ſame inſtant that cloſes the eyes of the laſt King placing his ſucceſſor on the throne. The poſition that death takes the living by the hand, *le mort ſaiſit la vif*, holds good even in the ſucceſſion to the crown, independent of the conſent of the ſubjects; the unction, or coronation, though the latter is an immemorial cuſtom, and uſually performed in the cathedral of *Rheims* by the archbiſhop of that place, or if he be incapable, by his ſuffragan the biſhop of *Soiſſons*; and at this ſolemnity is uſed the oil-cruet, ſaid to have been brought from Heaven by an angel at the baptiſm of *Clovis* ; and beſides the crown and uſual ſceptre, another called *la main de juſtice*, is put into the King's left hand. Females are excluded from the crown by the *Salic* law ; and by an edict of 1717, the legitimated natural ſons are under the ſame incapacity. It enacts alſo, that on the total failure of the lawful line of *Bourbon*, the ſtates of *France* ſhall be at liberty to chooſe a King according to their own pleaſure. The Kingdom is indiviſible. During the King's minority, the adminiſtration is lodged in a regent, nominated by the former King, or in caſe of no ſuch nomination, conſtituted by the parliament. *Lewis* XIV. indeed, at his deceaſe in 1715, left a form of regency during the minority of his great grandſon and ſucceſſor, but the firſt Prince of the blood, *Philip* duke of *Orleans*, took the government into his own hands, the parliament having voted it agreeable to the laws of the nation. The King is of age at thirteen years and a day.

The ſtates of the Kingdom, or the clergy, nobility, and citizens had their general conventions till the year 1614, ſince which they have been diſcontinued. Out of a ſelect number of them, *Philip the Handſome* inſtituted at *Paris*, a perpetual aſſembly or parliament, whoſe power was ſo great as frequently to control the royal prerogative; but it has been from time to time ſo curtailed that the preſent parliaments are wholly ſubject to the crown. The ſtates of *Britany*, *Burgundy*, *Dauphiné*, *Provence*, *Languedoc*, and *French Flanders* ſtill enjoy the privilege of holding conventions where the King's
                                                                    demands

demands, especially when any new tax is in agitation, are discussed; an assessment made among themselves for the sums required, and the method of levying them determined.

§. 15. For the management of state affairs, both at home and abroad, the following chambers and officers are assigned, *viz.*

The supreme council of state, besides the King and prime minister, when there is one, is composed of the three secretaries of state and the comptroller-general of the finances.

In the *council des depeches*, or secretary of state's office, exclusive of the King, *Dauphin*, and prime minister, fit the chancellor, the president of the council of finances, the four secretaries of state, and the comptroller-general of the finances. Here all the affairs of the provinces are transacted: Commissions, letters and orders sent to the governors, commandants, and other officers of the several provinces and cities. The secretaries of state make the motion and every one expedites the resolutions taken on the articles of his department; the foreign and domestic affairs of state, as also those relative to the provinces and generalities, being divided among the members.

The council of finances, besides the King and chancellor, consists of a president, one of the six intendants of the finances, a member of the council of state, and the comptroller-general of the finances.

The council of trade, the members of which, besides the King and prime minister, are the chancellor, two secretaries of state, and the comptroller-general of the finances.

The privy council, which is also called the *confeil des parties*, is convened by the chancellor on such days as he pleases. In the King's absence there is always a chair set for him, and the arrets run *Le Roi en son conseil*; but when he is actually present, *Sa Majestie y etant* is added. This council consists at present of the chancellor or keeper of the great-seal, twenty-one ordinary counsellors of state, the secretaries of state, the comptroller-general of the finances, the intendants of the finances, who are all ordinary members, and twelve state counsellors who officiate only half yearly. The ordinary state counsellors have each a salary of 5500 livres, the half yearly 3300. In this council also assist twenty-two *maitres des requêtes*, who belong also to the parliament, and lay before it any affairs committed to them. These *maitres des requêtes* are at present eighty-eight.

The domestic affairs of the Kingdom are transacted in the *grand conseil*, erected in the year 1492 by *Charles* VIII. At first it took cognizance only of the finances and war, but in 1517 *Francis* I. assigned to it the decision of all causes relating to archbishoprics, bishoprics, and abbies. It has now also the power of determining all disputes of the other sovereign tribunals, in matters of jurisdiction and other contests. The president of it is the chancellor of *France*, assisted by several presidents, counsellors, and other officers.

La grande chancellerie de France, or the high court of chancery, confifts of the keeper of the great-feal, who is frequently the fame with the chancellor of *France*; of four grand audienciers, who perufe papers fent from the fecretary of ftate's office to be fealed, and make a report to the chancellor of it; of four comptrollers-general of the audience, who give the papers allowed to be fealed to an officer called *chauff-cire*, or chaf-wax, and receive them back from him; and four *gardes des rôles des offices de France*, who keep the regifters of all the offices in *France* which require the great-feal ; with feveral clerks and other officers.

§. 16. Juftice is adminiftered in this Kingdom by inferior, middle, and fuperior courts : To the inferior courts belong the *prevotés, maires, judicatures, chatellenies,* and other jurifdictions dependent on the crown or particular lordfhips. From thefe appeals lie to the bailliages (bailiwicks) or *Senechauffées* (precinct courts) and from thefe again to the *Prefidiaux,* or provincial courts, as to the middle tribunals. The laft pronounces definitively in certain fmall cafes both civil and criminal; but the more important cafes are cognizable, only by the fupreme parliament, as it is called, where all caufes are difcuffed in *dernier reffort.* The word parliament, at firft, denoted the King's tribunal, which confifted of the great men of the Kingdom, and where the King perfonally affifted ; but the great inconveniency arifing to the fubjects travelling from the remoteft provinces of the Kingdom to procure themfelves juftice, rendered fuperior tribunals neceffary to be erected in other parts ; and from the affinity of thefe tribunals to that of the Kings courts they were alfo called parliaments. The parliaments at prefent are twelve, namely, thofe of *Paris, Touloufe, Rouen, Grenoble, Bordeaux, Dijon, Aix, Rennes, Pau, Metz, Befançon* and *Douay.* Not unlike thefe are the fupreme courts at *Colmar* for *Alface,* and at *Perpignan* for *Rouffillon.* Thefe fourteen *cours fouveraines,* as they are called, ftill retain this veftige of their former grandeur, that the King's orders become not of general force till regiftered by them. The parliament of *Paris,* which formerly refembled a diet, and had very great power, (§. 14.) ufed to follow the court, till *Philip the Handfome,* in the year 1302, ordered that it fhould conftantly hold its feffions in *Paris.* The Princes of the blood, the dukes, counts, and peers of *France,* together with the archbifhop of *Paris,* fit and vote in this auguft affembly. It confifts of nine chambers, namely, the great chamber, the criminal chamber, or *la tournelle,* three chambers *des enquêtes,* and the two chambers *des requêtes du palais.* It regifters all edicts and explanatory ordinances of the King, and every thing relating to him, together with marriages, treaties of peace, &c. as alfo letters patents for erecting certain Diftricts into peerages, dutchies, marquifates, counties, &c. The principal officers of ftate, dukes, peers, Princes of the blood, and ecclefiaftics are tried by it. The King nominates the firft prefident of the great chamber, and the *procureur,* or attorney-general. The other feats or
offices

offices, in thefe nine chambers are ufually purchafed. For a fuller account of thefe parliaments we fhall refer to the defcription of *Paris* in the fequel. In, the prefent century it has received two confiderable mortifications ; namely, in the year 1720 and 1753, when, on account of the great power they claimed in their contefts with the bifhops, and of their refufal to regifter the King's edicts, they were banifhed to *Pontoife*; and, during its banifhment from *Paris* the laft time, the King erected a chamber of vacations in the convent of 'the *Auguflines*, which was compofed of eight *maitres des requetes* and twenty other perfons of the long robe, and invefted it with the fame power as the parliament, both with refpect to civil and criminal matters. Whether the late procedures of the parliament only had in view the maintenance of the liberties of the *Gallick* church (§. 7.) or was of abfolute neceffity for preventing any further dangerous encroachments, if not a total abolition of them, it is certain, the court is no lefs interefted to oppofe the parliament in the complete re-eftablifhment of the whole circle of thefe privileges, whatever appearance it may have of increafing the regal prerogative thereby, than of putting a check to thefe innovations of the bifhops.

With refpect to the laws by which juftice is adminiftred in *France*, in the provinces of *Guyenne, Languedoc, Provence, Dauphiné, Lyonnois, Forets, Beaujolois, Upper Auvergne,* and others, the *Roman* law is in ufe, and they are therefore called *les pais du droit ecrit* : whereas, the other provinces, which have their own cuftomary laws, are called *les pais coutumieres*. Thofe of principal note among the latter are about fixty ; but, including the juridical cuftoms of fingle places, they make about two hundréd and eightyfive. Befides thefe there are ftatutes of univerfal force, ftyled *ordonnances, edicts* and *declarations.* In the year 1666, under *Lewis* XIV. the civil and criminal proceffes were amended and reduced to a general uniformity. In *France* a ftrict diftinction is made betwixt the canonical and papal ecclefiaftical law ; and by the obligatory canons, or church ordinances, are underftood only thofe canons of the firft ages of Chriftianity and œcumenical councils, which have been confirmed by the formal acceptance and confent of the churches bound by them, which could not be tranfacted without the approbation of the civil power. Thus the interpolated decretals of the fee of *Rome* are abfolutely excluded. The King's ordonnances iffued for the protection of the liberties of the *Gallican* church (§. 7.) conftitute the moft confiderable part of the ecclefiaftical law in *France.*

§. 17. The King's revenue is both ordinary and extraordinary. The ordinary revenue comprehends, 1. the demefnes, which confift in lands, lordfhips and forefts. 2. The duty on wine *(les aides)* ; and thefe are the twentieth part of the wine fold by wholefale, and the eighth or tenth of that retailed. 3. The *gabelle*, or falt-duty ; with refpect to which, *France*, by virtue of an edict, is divided into three provinces. The province of the

great

great falt duty, which contains the departments of *Alençon, Amiens, Angers, Bourges; Caen, Chalons, Dijon, Langres, Laval, Mans, Moulins, Orleans, Paris, Rouen,* St. *Quentin, Soiſſons* and *Tours,* in all which falt is fold at a high rate. The province of the little falt-duty includes the departments of *Lyonnois, Dauphiné, Provence, Languedoc, Rouſſillon, Rouergue* and *Auvergne,* in which falt is at a much lower price than in any of the preceding. The provinces where the falt-duty does not take place, are *Poitou, Limouſin, Guyenne, Gaſcogne* and *Britany.* In the three biſhoprics of *Metz, Toul* and *Verdun,* and *Franche Comté,* the price of falt is different. 4. The tax *(taille);* which, in the generalities of *Montauban* and *Grenoble,* and in the three elections of the *Generalité* of *Bourdeaux,* namely, thoſe of *Lanes, Agen* and *Codom,* is paid according to the eſtates, without any regard to the quality of the poſſeſſors; but in all the other countries is perſonal, the nobility, clergy and certain officers being alone exempted from it. 5. The *capitation,* or poll-tax, the cuſtoms of all kinds, the duty on ſtamp paper and ſeveral other duties. 6. The gift of the clergy, of which mention has been made above in §. 7.

The extraordinary impoſitions, which are of what number and kind the King pleaſes. Among theſe is the augmentation of the tax for the ſupport of the army, the *taillon,* the tenth or twentieth part of the revenue of the whole Kingdom, in lands, houſes, offices, &c. and the erection and ſale of new offices. It was under *Lewis* XII. that the public ſale of offices began, inſomuch that, at preſent, the greateſt part of the employments in this Kingdom are the properties of particular families, and may be diſpoſed of again by them, the actual poſſeſſors paying only annually a certain ſum to ſecure to their families the hereditary enjoyment of theſe offices; but, as this does not riſe to any very conſiderable amount, the King frequently erects new employments, the ſale of which brings in very large ſums. The King's revenue, at preſent, is computed to be about three hundred millions of livres.

For the collection of theſe impoſts the Kingdom is divided into certain diſtricts, or juriſdictions called *generalités* and *intendances;* namely, thoſe of *Paris, Soiſſons, Amiens, Chalons* and *Sedan, Orleans, Tours, Bourges, Moulins, Lyon, Grenoble, Riom, Poitiers, Rochelle, Limoges, Bordeaux, Montauban, Rouen, Caen, Alençon, Auch, Bretagne,* or *Rennes, Touloufe, Burgundy,* or *Dijon,* and *Franche Comté,* or *Beſançon;* together with the *intendances,* which are *Rouſſillon* and *Cerdaigne, Metz, Alſace, Lorrain, Hainault* and *Maubeuge en Flanders.* The *generalités* are again ſubdivided into elections as they are called; but thoſe of *Bretagne, Touloufe* and *Montpelier* into *Recettes,* or contingencies. That of *Aix* into *Vigueries,* and thoſe of *Burgundy* and *Franche Comté* into *Bailliages.* All theſe ſmall diſtricts conſiſt again of pariſhes and every pariſh of hearths. The number of pariſhes excluſive of *Lorrain,* is computed at thirty-nine thouſand and forty-five, and the hearths at three millions feven hundred and thirteen thoufand five hundred and fixty. three. Over every *genƐralité* is a treafurer, who has his particular office, with an intendant who takes cognizance of all matters relative to the law, polity and finances of the province, and two receivers general befides other offices. Among the above-mentioned *generalités* are fix, which are called *les pais des etats*, where the ftates raife the fums required by affeffments made among themfelves (§. 14.) The others are ftyled *les pais d' eleƈions*, of the inftitution of which I have fpoken above.

The other branches of the King's revenue, as, namely, the cuftoms, wine-licence, falt and ftamp duty, tobacco, pofts, *&c.* are farmed to the company of farmers general, who have their under-farmers and receivers. The great officer for levying the taxes is the comptroller-general, who keeps a duplicate of all receipts and difcharges relative to the royal revenue.

The principal management and jurifdiƈion of the crown revenue and every thing relating to it, is committed to the following tribunals; namely, 1. For the unfarmed revenues are the eleven chambers of accounts at *Paris, Blois, Montpelier, Grenoble, Dijon, Rouen, Aix, Nantes, Pau, Dole* and *Lifle*; but that of *Paris* is the chief, and to it all who hold immediately of the King, take the oath of fidelity. Laftly, the revenue court, which judges and determines all difputes betwixt the fubjeƈs and the officers of the revenue, relative to the duties and affeffments. Thefe are held at *Paris, Clérmont, Montpellier, Montauban, Grenoble, Bordeaux, Rouen, Pau,* and in fome other places the parliaments aƈ in their ftead. Laftly, here are alfo nineteen *cours des eaux et forêts*, from all which there lies an appeal to the *tribunal des eaux et forêts* at *Paris.*

The ambition and extenfive views of *France,* with its various wars, have neceffarily burdened the fubjeƈs with enormous taxes and impofitions; and thefe being aggravated by the rapacity of the financiers and farmers, the people have often been reduced to the laft extremity of wretchednefs; from which, without the many refources nature and induftry has put into their hands, they could never have recovered: yet is the King's treafury, amidft all its vaft incomes, not only empty, but has more than once been involved in prodigious debts; infomuch that, in refpeƈ of them, *Lewis* XIV. at his death, might have been termed the Great, as he had been before for his aƈions and perfonal qualities.

§. 18. The land forces of *France* amount, at prefent, to above two hundred thoufand men, but with great numbers of *Germans, Switzers* and' other foreigners among them; and, on occafion, this large body can be augmented to above double that number. The *French* are alfo in no. inconfiderable repute for military atchievements; they have had heroic. commanders, and their engineers-and gunnérs are fcarce to be equalled. The number of fortreffes belonging to the Kingdom is very great; and

exaƈ

exact plans of them, as alfo of the foreign fortreffes, to the number of above one hundred and eighty, may be feen in the gallery of the *Louvre*, which we fhall not omit in the defcription of *Paris.* In this palace is alfo a royal academy for training up five hundred young gentlemen in the feveral branches of the art of war. Befides the magnificent building of the *Invalides* in the capital, where alfo difabled feamen are admitted, there are in the Kingdom above feventy other military hofpitals, which muft neceffarily be a great encouragement to enter into the fervice and behave well in it.

§. 19. The navy of *France* is alfo very confiderable, and for this it has many advantages. Its fituation is extremely commodious (§. 3.) the coafts are fafe and the maritime provinces fruitful; befides that it produces within its felf moft of the materials for fhip-building, though timber begins to grow fcarce. It boafts alfo fkilful artifts, who, know how to make the beft ufe of thefe advantages. Laftly, their marine is under good regulations, efpeci-ally with regard to the men, with refpect to whom, in the year 1681, the *claffes*, as they are called, were inftituted throughout all the maritime provinces, where the fea-officers and failors, and others belonging to the navy refide. In *Guyenne, Britany, Normandy, Picardy,* and the conquered and recovered countries, are four *claffes*; three in *Poictou, Laintonge, le pais d' Aunis,* the ifles of *Re* and *Oleron*; *Languedoc* and *Provence.* Each clafs ferves three or four years alternately, and thofe who are not in actual fervice may enter on board merchant-fhips. The number of feamen at firft regiftered was fixty thoufand, but *Deflandes* affures us that, at prefent, they are near feventy thoufand. This ingenious gentleman, however, finds fault with the *French* feamen, as deferting and going abroad on the leaft difcontent; but adds, that it is the national foible of his countrymen to be fond of novelties and, on every idle caprice, to quit the Kingdom without the leaft reafonable caufe. The art of fhip-building the *French* borrowed from the *Englifh,* and *Henry* IV. was the firft who ferioufly applied himfelf to the encreafe of the marine. *Lewis* XIII. by Cardinal *Richlieu*'s care, had a tolerable fleet, but, under *Lewis* XIV. the naval power of the *French* be-came formidable and was very advantageoufly employed, but before his death declined greatly, and in the laft war was fo weakened, particularly by two fignal victories of the *Englifh,* that it will require a great many years and vaft expence to reftore it to its former ftate. For fome years paft, indeed, fuch great diligence has been ufed in this article, that *France* has already one hundred and eleven men of war. Thefe are diftributed at *Toulon, Breft, Port Louis, Rochefort* and *Havre de grace,* exclufive of fifteen gallies which lie at *Marfeilles.* The navy is under the management and jurifdiction of the admiralty at *Paris.* Exclufive of the three companies of *gardes de la marine* at *Toulon, Breft* and *Rochefort,* which confift wholly of gentlemen verfed in navigation and the art of war, another company of gentlemen has been erected under the title of *gardes du pavilion amiral,*
who

who always attend the admirals and ferve only on board flag fhips. Thefe are chofen out of the preceding *gardes.* The King alfo maintains one hundred independent companies for the fea, each confifting of forty-five men, under the command of a lieutenant of a man of war. Thefe ferve as marines on board the fhips. The fea-ports are all well fortified.

§. 20. To the crown of *France,* exclufive of what it has in *Europe,* belong certain places in *Afia* on the coaft of *Coromandel,* among which *Pondichery* is, the principal. In *Africa* is the fort called *Baftion de France,* which lies in the Kingdom of *Algiers;* the fortrefs of *Arguin,* near the *white promontory;* the ifland of *Gorée,* near the *green promontory;* Fort *Joal, Vintan,* or *Vintain, Louis, Portendic, S. Jofeph, Albreda* and *Biffos;* the iflands of *Bourbon* and the *Ifle de France* in the *Indian* fea; and, in *America,* part of *Canada* and *Louifiana,* with a part of *Florida* and the iflands on the river of *S. Laurence,* and particularly the ifland of *Cape Breton.*

§. 21. The Kingdom of *France,* with the old and new territories belonging to it, is varioufly divided; namely, in refpect of its civil conftitution, into parliaments (§. 16.); of its finances, into *generalités* (§. 17.); of its ecclefiaftical conftitution, into archbifhoprics and bifhoprics, or ecclefiaftical provinces, as they are called (§. 7.); and, laftly, with regard to its military conftitution, it confifts of thirty-feven governments. The laft being the principal divifion, and that moft in ufe at prefent, I fhall make it the bafis of the defcription in view, and therein follow the order of the moft approved *French* geographers. Thefe governments are under the jurifdiction of particular governors, and, in their abfence, of general-lieutenants, fubordinate again to whom are the deputy-governors, or *lieutenants de roy.* The governors are obliged, by the duties of their pofts, to retain the feveral provinces and cities in a quiet obedience to the King, and to preferve the public tranquility; they have the command of the troops cantoned in their feveral governments, are to keep the fortreffes in a defenfible condition, and, when neceffary, affift the execution of juftice. The governors of the cities and fortreffes are not only independent of thofe of the provinces, but frequently have the command of a fmall diftrict lying near.

There being alfo a provincial divifion of *France,* it will be neceffary, for the greater perfpicuity in our defcription, previoufly to fpecify them. Thefe provinces are, 1. *L'Ifle de France.* 2. *Picardy,* which is divided into the *Upper* and *Lower.* To the *Upper Picardy* belong the countries of *Amienois, Sauterre, Vermandois, Noyonois, Thierache, Laonois, Soiffonois, Valois* and *Beauvairis;* the lower divifion contains the *pais reconquis,* or recovered countries, *le Boulonois, le Marquenterre, le Ponthieu* and *le Vimeu.* 3. *La Brie,* which is divided into *Brie Champénoife* and *Brie Françoife.* 4. *La Champagne,* containing *Upper* and *Lower Champaigne, Rhetelois, Argonne, Remois, Chalonois, Pêrtrois, Vallage, Baffigni* and *Senonois.* 5. The dutchy of *Burgundy,* which is divided into various large territories and diftricts;

namely, *Dijonois, Auxerrois, Auxois, Pays de Montagne, Chalonois, Maconois,* *Autunois* and *Charolois.* 6. *La Bresse,* which is divided into *Upper* and *Lower Bresse.* 7. *Le Bugey,* consisting of *Bugey* properly so called, *Valromey* and *le pays de Gex.* 8. *Le Dauphiné,* which is divided into *Upper* and *Lower.* *Upper Dauphiné* consists of *Graisivaudan, Royanez, Briançonnois, Ambrunois, Gapençois* and *le pays de Baronies.* *Lower Dauphiné* is divided into *Viennois, Valentinois, Tricastinois,* and *Diois.* 9. *La Provence* consists of three parts; namely, *Upper Provence, La Campagne* and *Lower Provence.* 10. *Languedoc,* consisting principally of the *Upper* and *Lower Languedoc* and the *Pays de Sevennes,* being subdivided into particular jurisdictions and districts. 11. *Foix,* containing *La comté de Foix* and the territory of *Donnesan.* 12. *La comté de Comenges,* to which also belongs the *comté de Conserans.* 13. *L'Armagnac,* which is divided into *Upper* and *Lower.* *Upper Armagnac* includes the four vallies of *Magnoac, Nestes, Barousse* and *Aure.* *Lower Armagnac* consists of the territories of *Armagnac, Causan, Fezenzac, Brullois, Lomagne, Gaure* and *riviere Verdun.* 14. The *comté de Bigorre,* to which belong the plains and mountain of *Bigorre,* with the territory of *Rustan.* 15. The principality of *Bearn,* to which belong the *Vicomtez* of *Bearn* and *Oleron.* 16. *Le pays de Basques,* consisting of the plains of *de Labour, Lower Navarre* and the *vicomté de Soule.* 17. *Gascogne,* which consists of five districts; namely, *Les Landes, la Chalosse, le Tursan, le Marsan* and *le pays d'Albret.* 18. *Guienne,* comprehending the districts of *Bourdelois, le pays de Medoc, les Captalats de Certes* and *de Buch.* 19. *La Saintonge.* 20. The territory of *Aunis.* 21. *Poitou,* which is divided into the *Upper* and *Lower.* 22. *Bretagne,* divided also into the *Upper* and *Lower.* 23. *Normandy,* likewise divided into the *Upper* and *Lower,* each of which includes seven jurisdictions; namely, *Caux, Bray, Vexin,* consisting of *Vexin Normand* and *Vexin François, Campagne, Ouches, Lieuvin, Roumois,* containing eight districts; namely, *Auge, Campagne de Caen, Bessin, Coutantin, Avranchin, Bocage, Les Marches, Seez, Argentan* and *Houlme.* 24. *Le Perche,* to which belong *Grand-perche, le perche Gouet* and *le pays de Timerais.* 25. *Beauce,* in which *Chartrain, Mantois, Hurepoix, Dunois* and *Vendomois* are generally included. 26. *L'Orleanois.* 27. *Le Gatinois,* to which belong *Gatinois François, Gatinois Orleannois* and *pays de Puisaye.* 28. *Le Nivernois,* to which appertain seven Districts. 29. *Le Bourbonois.* 30. *Le Forez.* 31. *Le Beaujollois.* 32. *Le Lionnois.* 33. *Le Rouergue.* 34. *Le Quercy.* 35. *L'Agenois.* 36. *Le Condomois.* 37. *Le Basadois.* 38. *Le Perigord.* 39. *L'Angoumois.* 40. *Le Limosin.* 41. *La Marche.* 42. *La Touraine.* 43. *L'Anjou.* 44. *Le Maine.* 45. *Le Blaisois.* 46. *Le Berry.* 47. *L'Auvergne.* 48. The *French Netherlands,* comprehending the *comtéz* of *Artois* and *Cambray,* a part of *Flanders, Hainault* and *Namur,* with a portion likewise of the dutchy of *Luxembourg.* 49. *Lorrain.* 50. *Alsatia.* 51. *Franche Comté.* 52. *Roussillon.*

THE

THE

GOVERNMENT

O F

P A R I S.

THE boundaries of this government are not eafily fettled, the *French* geographers themfelves being not agreed about them. In it however, befides the city, is included a part of the neighbouring country; and, exclufive of the governor, here are two general-lieutenants, one of which is appointed for the city, the other for the *prevoté* and *vicomté*, through its utmoft extent. Within the boundaries of this jurifdiction are contained four others; namely, the jurifdiction of the *Louvre* and *Tuilleries*, as alfo thofe of the *Baftile* and the royal hofpital of the invalids, the four governors of which receive their orders immediately from the King.

Paris, in Latin *Parifii*, *Lutetia* and *Lutetia Parifiorum*, the capital of the Kingdom, lies in the middle of the ifle of *France*, in a large plain on the river *Seine*, being a populous ftately city and of prodigious extent. It is of a circular form, being about two large *French** miles in diameter, and, including the fuburbs, in circumference fix; or, if we go according to its topographical projection, each quarter is three leagues in circumference, and the whole circuit of the city five. Its ftreets have been computed to be 912, containing upwards of 20000 houfes from four to feven ftories high; and this exclufive of churches, convents, chapels, colleges, communities, warehoufes and fhops. The number of its inhabitants cannot greatly exceed 400000, for, from the year 1728 to the year 1736, the annual bills of mortality were, at a medium, 17800; and it is calculated, upon very probable computations, that of twenty-five perfons in this

---

* Of the larger *French* miles, or leagues, twenty make a degree of the equator; of the fmaller twenty-five.

city

city one dies yearly; whence the above number of deaths make the number of living inhabitants 445000. The ſtreets are of a proper breadth, and were originally well paved with flint-by *Philip Auguſtus*, in the year 1223. The houſes are ſightly and towards the ſtreet uniform. *Paris* affords no good drinking water, the inhabitants being obliged to uſe either that of the *Seine*, which is fetid and occaſions dyſenteries, or another ſort of water which is ſtill worſe, being productive of the gravel and ſtone. The new regulations formed by M. *Turgot*, *prevôt des marchands*, for keeping the city and the ſtreets clean; conſiſt in a canal, lined with free-ſtone, ſix feet broad and about the ſame depth, which receives all the ſoil from thoſe parts of the city through which it runs. This canal is kept clean by means of a large reſervoir formed in a particular building for that purpoſe, which is ſupplied with water from all the ſprings in the neighbourhood. In it are ſix pumps, worked by a machine which is kept going by four horſes, and theſe diſcharge the water into a large baſon lined likewiſe with free-ſtone, whence it runs under ground through two ranges of pipes of a conſiderable bore, diſcharging itſelf with great violence into the above-mentioned canal, and thus carries off all the filth into the river *Seine*. At night the ſtreets are lighted with lamps for ſix months in the year. Here are alſo good regulations in caſe of fire, for the firſt preſident of the parliament, the *prevôt des marchands* and the *lieutenant de police*, the horſe and foot *patrouille*, and ſome companies of the *French* and *Swiſs* foot-guards are obliged to repair to the fire with all the expedition the diſtance of the place will admit; but the actual extinguiſhing of it is the care of the monks of the four mendicant orders, who, on ſuch occaſions, are to expend, for the benefit of the public, part of the ſtock acquired by their indolent beggary.

For a general idea of this city and its curioſities, it muſt be obſerved, that *Paris* is the capital of the Kingdom, the reſidence of an archbiſhop, of the principal parliament, of a chamber of accounts, of a court of taxes, of a court of mintage, of a ſupreme chamber of the tythes, of the court of the King's palace called *les requêtes de l' hôtel*, of two chambers of the *requêtes du palais*, of the office of the treaſurer of *France*, of a treaſury and demeſnes chamber, of the marble table, to which belong the court of the conſtable and marſhal of *France*, of the admiralty, of the court *des eaux* and *des forets*, of a court-leet of the palace, of an election, of a particular foreſt-court, of the *bazoche*, or juriſdiction of the clergy, of the parliament, of the juriſdiction of the clergy of the chamber of accounts, &c. all which ſeveral courts fit in the *palais* on the Iſland to which it gives name. Beſides theſe here alſo reſide the great council, which meets in the *hôtel d' Aligre* of the *chatelet*, which includes the court of the *prevôté* and *vicomté* of *Paris* with the *le Siege preſidial*, a court for civil cauſes, a court of police and criminal cauſes, and other inferior courts.

In it alfo are forty-feven parifh-churches, befides twenty others, three abbies, and twelve priories for men, feven abbies and fix priories for women, feventeen collegiate churches, of which thirteen have chapters; fifty convents and fraternities of ecclefiaftics and laics, forty-three nunneries, fourteen female communities, eleven feminaries, twenty-fix hofpitals, and forty chapels, an univerfity with four faculties and forty-three colleges, exclufive of fome others; fix academies, befides three others where young gentlemen are taught bodily exercifes; five public libraries, exclufive of feveral private ones; four royal palaces, four caftles, above one hundred hotels, fome of them very ftately; feventy-three market-places, fixty fountains, twelve bridges over the *Seine*, ten of which are of ftone, and eleven great gates.

To give a more particular defcription of this city, it confifts of three parts, *viz. La ville* lying on the north of the *Seine*; *La cité*, which is quite environed by that river, and *l'univerfité* fouth of the fame, with twelve fuburbs. In the year 1702, it was divided into twenty *quartiers*; concerning the moft remarkable particulars of which I fhall endeavour briefly to give a tolerable account.

1. *La cité* is the moft cleanly and wealthy *quartier* of all, confifting of three iflands formed by the *Seine*, namely, the *ifle du Palais*, the *ifle de Nôtre Dame*, and the *ifle Louviers*: The laft is fmall, contains only ftorehoufes for wood, fronts the arfenal and communicates with the *quartier* of St. *Paul* by means of a wooden bridge. The *ifle de Nôtre Dame* is fo called from the cathedral *de la Cité*, to which it belongs; and by means of the ftone bridge of St. *Mary* communicates with the *quartier* of *S. Paul*; by that of *la Tournelle*, with the *quartier de la place Maubert*; and by a wooden one with the *ifle du Palais*. In this ifle are fome fine hotels, with the parifh-church of St. *Louis*. The *ifle du Palais* confifts properly of *Old Paris*, and is fo called from the palace where the parliament fits, of which I fhall fpeak farther. Befides the above-mentioned wooden bridge which joins this ifland with the *ifle de Nôtre Dame*, it communicates likewife with other parts of the city by means of feven bridges of ftone. Of thefe the *pont neuf*, or new bridge, is the principal, and, indeed, of all the bridges in *Paris*, being a very ftately and beautiful fabrick. It lies on the northweft parts of the ifland, and is carried over both the branches of the *Seine*, infomuch that it joins this quarter of the city both with the univerfity and *Saint Germain*, and the quarters of *S. Opportune* and the *Louvre*. It was begun in the year 1578, and in 1604 finifhed. It confifts of twelve arches, is one hundred and feventy toifes in length, twelve broad, and has on it three divifions; that in the middle is five toifes broad, and is appointed for coaches and other carriages, on both fides of which is a foot-paffage raifed two feet high: Over the piles, on each fide, are alfo femi-circular lodgments, occupied by one hundred and feventy-eight fmall fhops belonging to the

King's

King's footment; but thefe obftruct a moft beautiful profpect. In the centre of the bridge is a fine equeftrian ftatue of *Henry* IV. in brafs, bigger than the life, and ftanding on a marble pedeftal. On this bridge is alfo the building called *La Samaritaine,* in which is a pump, which through feveral pipes fupplies the *quartier de Louvre,* and fome others of the city with water. It is fo called from the ftory of Chrift and the *Samaritan* woman, which is repiefented upon it. The *pont au change,* on the north fide of which ftands a ftatue of *Lewis* XIV. in brafs, and the *pont Nôtre Dame,* the firft ftone bridge in *Paris,* and on which is alfo a water-work, leads to the *quartiers de S. Jaques de la Boucherie* and St. *Opportune,* each having two rows of houfes upon it; thofe of the firft being four, and thofe of the others two ftories high. *Le pont S. Michael,* which leads to the *quartier de S. André,* and fronts the *pont au change* and the *petit pont,* which is oppofite to the *pont de Nôtre Dame,* are both likewife built on. Laftly, the *pont* St. *Charles* and *hotel Dieu* lead from the *hotel Dieu* to the univerfity quarter.

On this ifland ftands the cathedral of *Notre Dame,* which is indeed of *Gothick* ftructure, but is neverthelefs large and ftately, being fixty-five toifes long, twenty-four broad, and feventeen in height : On the infide are four rows of pillars, thirty in a row, with forty-five chapels built betwixt the outermoft rows and the wall. The pillars in the nave of the church are decorated with large, magnificent, beautiful pictures. The choir was fplendidly repaired and beautified at the expence of fome millions of livres by *Lewis* XIV. in purfuance of a vow of his father. Among the fepulchral chapels in this church, thofe of the Cardinals *de Retz* and *Noailles* are the moft remarkable. The two fquare towers belonging to it are thirty-four toifes high, and flat at the top, with a baluftrade of free-ftone, whence we have a moft grand profpect of the whole city and the neighbouring country. In that on the left hand are two large bells, one of which weighs 31,000 and the other 40,000 pounds. In this church are performed the obfequies of perfons of the royal family, and likewife the principal religious rites. Near it ftands the palace of the archbifhop, in which, in one of the halls, is the library of the advocates, founded there for the public ufe by Mr. *Riparfoud,* a counfellor of the parliament, and opened in the year 1708. Subject to the archbifhop's jurifdiction are four chapters, namely, the church of St. *Marcellus,* St. *Germain l' Auxerrois,* the collegiate church of *S. Honoré,* and the collegiate and parifh-church of *S. Opportune.* His yearly revenue amounts to 180,000 livres, and his taxation at the *Romifh* court is 4283 guilders. This archbifhopic was firft erected in the year 1622, till which time it was only a bifhopric.

On this ifland are alfo the following churches, *viz.* the collegiate church of *S. Denis du Pas,* of the convent of the Bleffed Virgin, the collegiate church of *S. Jean le rond,* which is the parifh-church of the laft mentioned convent, the church of St. *Marine,* which is the parifh-church of the arch-

archbifhop's houfhold; the collegiate church of *S. Agnan*, and the parifh-churches of *S. Chriflophle*, *S. Genivieve des Ardens*, *S. Landry*, *S. Denis de la Chartre*, which belongs to a priory of *Benedictines*; *S. Mary Magdalen*, having alfo the title of an archdeaconry, and being frequented by a famous devout fraternity; *Sainte Croix de la Cité*, *S. Pierre des Arcis*, *S. Pierre aux boeufs*, *S. Germain le vieux*, *S. Martial*, the church of the *Barnabites*, the priory of *S. Eloy*, and the priory and church of *S. Bartholomew*, the moft beautiful parifh-church in all this quarter, and ftanding near the *palais*.

This *palais* was formerly the refidence of the Kings, but *Lewis* XII: wholly refigned it to the courts of juftice. In the general fketch of this city I have already mentioned the courts which meet in this large building. The moft eminent among them is the parliament, which was fettled here in the year 1302, by *Philip the Handfome*, and confifts of the *grande chambre*, to which belong the firft prefident, nine upper prefidents, or *prefidents à mortier*, fo called from the fhape of their caps which refemble a moitar; twenty-one temporal and twelve ecclefiaftical councellors, three advocates-general, and one attorney-general. The next to thefe are the five *chambres des enquêtes*, which judge the appeals of written procefles, and to each of thefe belong three prefidents, with thirty or thirty-two counfellors; the *Tournelle criminelle*, or criminal court, in which the counfellors of the *grande chambre* and the *chambres des enquêtes* fit alternately; two *chambres de requêtes du palais*, in which certain privileged perfons only, fuch as the officers of the King's houfhold, and thofe of the Princes and Princefles of the blood, together with all others who have the right of a *committimus* are tried, and each of thefe confifts of three prefidents and thirteen counfellors. Befides thefe there is alfo a chamber belonging both to the parliament and the privy-council, called the *chambre des requêtes de l'hôtel*, compofed of *maitres de requêtes*, of whom there are at prefent eighty-eight, who are divided into four quarters, each ferving three months, namely, three in the *requêtes de l'hôtel*, and three in the King's privy council, or *confeil des parties* (§. 15. of the introduction.) Laftly, among thefe are alfo the *chambre de la Maree*, *i. e.* ' the fifh-chamber,' which fettles the price of fea and river-fifh, with the *balliage du pays*. The parliament is opened the day after *Martinmas*, with a folemn mafs celebrated by a bifhop, and continues fitting till the eighth of *September*, when a vacation-chamber is appointed till the next feffion, for criminal caufes and others which requires dif-patch.

Under the jurifdiction of this parliament are the *ifle de France*, *Beauffe*, *Sologne*, *Berry*, *Auvergne*, *Lyonnois*, *Forets*, *Beaujolois*, *Nivernois*, *Bour-bonnois*, *Maconnois*, *Poitou*, *Pays d'Aunis*, *Anjou*, *Augoumois*, *Picardie*, *Champagne*, *Maine*, *Perche*, *Brie* and *Touraine*. The chapel belonging to the palace built by *Lewis* IX. in the year 1247, though *Gothick*, is

very

very beautiful. Leaving the palace we come to the place Dauphiné, which
is of a triangular form. Lastly, on this island is the noble hospital of the
hotel Dieu, where 8000 infirm poor of both sexes are taken care of, and
attended by nuns of the order of St. Augustine.

2. The quartier S. Jaques dela Boucherie is a part of the city, and contains the parish-churches of S. Jaques de la Boucherie S. Leu and S. Gilles;
Les filles penitentes, who have a church here; the collegiate church of
S. Sepulchre, the hospital of S. Catharine, where poor women and maidens
are entertained three days by nuns of the order of St. Augustine, who also
are to look to the interment of any bodies accidentally found dead; and
the grande chatelet, which was formerly a castle, but at present the place
where the prevoté and vicomté of Paris, with their several jurisdictions, hold
their sessions for civil and criminal causes.

3. The quartier S. Opportune makes also a part of the city, and contains
the collegiate and parish-church of S. Opportune, the mint and the prison
called fort l'Eveque. In this part also is the street of la Ferronniere, where
on the 14th of May 1610, Henry IV. was stabbed by the execrable
Ravaillac.

4. The quartier de Louvre, or S. Germain l'Auxerrois, is likewise a part
of the city, and contains 1. The church and chapter of S. Germain l'Auxerrois, which being the parish-church of the King's palaces, the Louvre
and Thuillieres, is called also the royal parish-church.

2. The Louvre, in Latin Lupara, is commonly divided into the old and
new though properly it is but one building. The old part of this royal palace was begun to be built of stone by Francis I. in the year 1528, and was
finished by Henry II. in 1548. Succeeding Kings improved and enlarged
it till the time of Lewis XIV. who ordered it to be rebuilt on a new plan,
which, if completed, would have rendered it a most magnificent structure,
as appears from that part already built, though unfinished. The plan of
the whole building forms an equilateral quadrangle containing a court in
the centre sixty-three toises square. The principal of the four main wings
was built by Lewis XIV. as also the greatest part of the two others which
form the sides, together with a new front in that part lying next the Seine.
The four inner fronts it is said, according to the plan, were to have con-
sisted of eight pavillions and eight corps de Logis, supported by three rows of
pillars; but there is no likelihood of its ever being finished. In it, notwithstanding, are several stately chambers. In the large gallery, in particular, which is two hundred and twenty-seven toises or six hundred and
eighty-one ells long, and joins this palace to that of the Tuilleries, are to
be seen upwards of one hundred and eighty models of fortresses in France
and other countries, all performed with the utmost accuracy, and so natural as to represent the several cities therein described, with their streets,
houses, squares and churches, and likewise all their works, moats, bridges

and rivers, together with the very country adjacent, such as confisting of plains, mountains, corn-lands, meadows, gardens and woods. Some of thefe models may even be taken to pieces that the curious may be the better enabled to perceive the admirable conftruction of them. *Lewis* XIV. was the firft who ordered thefe models to be conftructed, and when any alterations are made in the fortreffes the alterations are alfo obferved in thefe models, or new ones conftructed with the moft fcrupulous conformity to the prefent ftate of the place. In this palace is alfo the King's printing-houfe, and it contains likewife the moft valuable collection of paintings: In it too we meet with feveral academies, *viz.* 1. The *Academie Françoife*, founded in the year 1633 by Cardinal *Richlieu*, and confirmed in 1637 by the parliament. The object of this foundation is the improvement of the *French* language. It confifts of forty members, who meet on *Mondays*, *Thurfdays* and *Saturdays*, in a hall of the *Old Louvre*. 2. The *Academie royale des infcriptions & des belles lettres*, inftituted for the advancement of polite literature. In this academy ancient monuments are explained, and the tranfactions of the Kingdom perpetuated by medals, infcriptions, &c. In the years 1699 and 1716, fome alterations were made in its conftitution. It confifts of four forts of members, *viz.* twelve honorary, and twenty who are penfioners, together with twenty-fix *affociés* and twelve ftudents. Of the twenty-fix *affociés* eight are foreigners, fix profefs no other kind of fcience, and form the clafs of the *affociés*. The twelve others, together with the twenty penfioners and the twelve ftudents, muft refide in *Paris*, Their meeting days are *Wednefdays* and *Saturdays*. 3. The *Academie royale des Sciences*, inftituted in 1666, has alfo its honorary members, its penfioners, affociates, and ftudents. 4. The *Academie royale de peintures & de fculpture*, founded in 1648. The mafter-pieces of the painters and fculptors admitted into it are difpofed in three halls, and marked with the names of the feveral artifts. Here are alfo the pictures of great numbers of celebrated painters; and among the ftatues feveral of *Gyps* taken from antiques. 5. The *Academie royale d'architecture* founded in 1671. We now quit the *Louvre* to take a curfory view of the other curiofities of this quarter: Of thefe there are three, *viz. Le petit Bourbon*, or wardrobe, where formerly refided feveral princes of the houfe of *Bourbon*, before that family afcended the throne. Here are kept the beautiful and rich hangings, and the fplendid houfhold furniture of the crown. Among the former the moft worthy of admiration are thofe done from the paintings of *Julio Romano* reprefenting the battles of *Scipio Africanus*, and purchafed by *Francis* I. for 20,000 rix-dollars, a very large fum in thofe days. The whole fet make above 20,000 ells, and receive their ineftimable value more from the largenefs and beauty of the work than from the richnefs of the materials, which yet are of filk and gold. The houfhold furniture confifts of tables, looking-glaffes, branches, clothes, &c. In it alfo are kept all kinds of ancient arms. 4. The great council meets in

the palace *d' Alegre* in the *rue S. Honoré*, which affords nothing elfe remarkable. 5. A convent and church of the predicant Monks in the *rue
S. Honoré*, and this is the principal houfe of that order, and the refidence of
its general. It has alfo a library containing no lefs than 20,000 volumes of
printed books and manufcripts. 6. The collegiate church of *S. Honoré*.

5. The *quartier du palais royal*, belonging alfo to the city, and containing
1. *Le palais royal*, built by Cardinal *Richlieu* betwixt the years 1629 and
1636, and therefore originally called the *hôtel de Richlieu*, afterwards *le palais Cardinal*. In the year 1639, he made a prefent of it to King *Lewis* XIII.
after whofe deceafe his dowager, with her fon *Lewis* XIV. in 1643, refided in it, on which account it has ever fince been called *le palais royal*.
Lewis XIV. firft conferred it on his brother *Philip*, and after his death on
Philip duke of *Orleans*. It contains pictures to the value of four millions of
livres purchafed by the regent duke of *Orleans*, who placed them in the chief
apartments, moft of which too are hung with red damafk. The foundation
of this collection of paintings were the cabinets of *Chriftina*, Queen of
Sweden, which after paffing through feveral hands, but always remaining
entire and complete, were bought by the regent for 400,000 livres. Here
are alfo great numbers of pieces by the moft celebrated ancient mafters, as,
namely, *Raphael, Rubens, Titian, Vandyke, &c.* In this palace too we meet
with the *Academie de mufique*, founded in the year 1669; the collegiate
church of *S. Thomas de Louvre*, and the church *de S. Nicholas de Louvre.*
3. The *hopital de Quinze Vingt*, with a church, and over the entrance a
ftatue of *S. Louis*. 4. The palace *des Tuilleries*, fo called from a tile-kiln formerly here, communicates by means of a gallery with the *Louvre*, but is
a feparate ftructure. It was begun in the year 1564 by Queen *Catherine de
Medicis*, and completed in 1600 by *Henry* IV. *Lewis* XIV. in 1664, caufed
the infide to be embellifhed. It confifts of five pavillions and four *corps de
logis*, ftanding all in a direct line, above one hundred and fixty-eight toifes
in length, and making a very grand appearance. In the pavilion adjoining to the river is a fpacious and magnificent theatre : Its garden is alfo
very beautiful, and in fair weather reforted to by aftonifhing crowds of
people. Behind the garden along the *Seine* is the *Cours la Reine*, confifting of three walks, near which on the right hand are the *Elyfian* fields,
which are alfo pleafantly planted with ftately rows of trees In fair
weather we fee here a furprifing number of coaches. In the riding-houfe
of the *Tuilleries* is the *Academie de la Gueriniere*, where young gentlemen are
taught all the academical exercifes. Behind the drawbridge of the *Tuilleries* is an equeftrian ftatue of *Lewis* XV. 5. The parifh-church of
S. Roche, where great numbers of perfons of note have chofe to be interred,
and among thefe, in particular, is the celebrated poet *Pierre Corneille*.
6. The convents of the *Jacobines* and *Feuillans*. 7. *La place de Louis le
Grand*, or *des Conquêtes*, is of an octangular form and furrounded with houfes
                                                                                  built

built in an uniform manner on each fide, moft of which make a fine appearance. In the centre ftands an equeftrian ftatue of brafs of *Louis le Grand*, which is very juftly accounted a mafter-piece in its kind, being all of one caft, though twenty feet two inches in height. No place affords its parallel, not even thofe of *Florence* and *Rome* put together. One *Balthazar Keller* was the cafter of this ftatue, which he executed from a model of *Jirardons*. The pedeftal is of white marble thirty feet high, twenty-four in length, and thirteen broad.   8. A convent of *Capuchins* and two nunneries.   9. In the fuburbs of *S. Honoré* is the parifh-church of *La Magdaleine*, with a priory of nuns belonging to the order of *Benedictines*.

6. The *quartier Montmartre* is a part of the city, and contains befides feveral fine hotels, four convents, among which is that of the *Auguftins dechauffés*, with a library and mufeum of antiques : The *place des victoires*, which is round, and in the centre ftands a ftatue of *Louis* XIV. of gilt brafs thirteen feet high, erected by the duke *de la Feuillade*, with this infcription, *Viro immortali:* The royal library in the *rue Vivien*, which in the year 1741 contained above 94,000 printed books and 30,000 manufcripts, with a cabinet of medals and a collection of copper-plates, the large folio volumes of which alone form a confiderable library; thefe accordingly are under their peculiar librarian, and among them is an incredible number of the moft valuable pieces.   In the churchyard of *S. Jofeph* lies *Moliere* the celebrated comedian.   In the fuburbs of *Montmartre*, on an eminence, ftands a rich nunnery of *Benedictines*; on this fpot too *S. Denis*, the firft bifhop of *Paris*, is faid to have fuffered martyrdom. *Monceaux* and *Porcherous* were formerly confiderable caftles.

7. The *quartier* of *S. Euftace* is a part of the city, and receives its name from the parifh-church of *S. Euftace*, in which is to be feen the tomb of that great minifter of ftate Monf. *Colbert*. The church *La Juftienne* is properly called *S. Marie Egyptienne*. The *hôtel de Ferme du Roi* is a handfome large houfe where the farmers-general meet, have their warehoufe, and where the duties are paid. The *hôtel de Touloufe* is one of the beft and fineft in all *Paris*; whereas the *hôtel de Soiffons* is remarkable only for its largnefs.

8. *Le quartier des Halles* makes a part of the city, and is fo called from the *Halles* where the beft eatables are fold, but is otherwife the dirtieft fpot in all *Paris*, remarkable only for them, the parifh-church *des S. Innocens*, and the fine fountain of that name.

9. The *quartier S. Denis* is a part of the city, and contains the hofpital *de la Sainte Trinité*, inftituted for one hundred boys and thirty-fix girls, who are all to be orphans and natives of *Paris*; but muft be healthy and learn trades; the parifh-churches of *S. Sauveur* and *Nôtre Dame de bonne nouvelle*, a nunnery, a community, and the gate of *S. Denis*, which was erected as a triumphal arch in honour of *Lewis* XIV.   In the fuburbs of

S. *Denis* is alfo the feminary or priory of S. *Lazare*, which in the year 1633 was united with the congregation *de la Miffion*, and the houfe *de Sæur de la Mifericorde*.

10. The *quartier* S. *Martin* forms a part of the city, and contains the collegiate and parifh-church of S. *Mary*, properly *Mederick*, a houfe *des Peres de la doEtrine Chretienne*, called S. *Julien de Menetries*; the parifh. church of S. *Nicholas de Champs*, in which are interred thofe eminent proficients in literature *William Bude*, *Peter Gaffendi*, and *Adrian de Valois*; the priory *de S. Martin de Champs*, whence the quarter takes its name; the convent of the *Madelonettes*, and the gate of S. *Martin*, erected in form of a triumphal arch in honour of *Lewis* XIV. In the fuburbs of S. *Martin* is the parifh-church of S. *Laurent*, the market-place of the fame name, a houfe of the bare-footed *Auguftines*, with a library, and the great hofpital of S. *Louis*.

11. *Le quartier de la Greve* is part of the city, and lies on the *Seine*. It contains, 1. The open place called *La Greve*, where all public rejoicings are celebrated, and malefactors likewife executed. 2. The *hôtel de Ville*, the tribunal of which confifts of the *Prevôt des Merchands*, four *echevins*, a King's attorney and King's counfellor, one fubftitute and a clerk of the court: Over its gate is an equeftrian ftatue of *Henry* IV. and in the court under an arcade one of brafs of *Lewis* XIV. in an erect pofture; in the court alfo is a kind of public chronicle, being an account in *Latin* and *French*, of the moft remarkable actions of *Henry* IV. infcribed in letters of gold cut on tables of black marble. 3. The hofpital *du S. Efprit*, the parifh-church of S. *Jean en Greve*, the church of S. *Gervais*, remarkable for its fine door, and for being the burial-place of feveral perfons of diftinction; the church-yard of S. *Jean*, the greateft green and fruit-market, and the hofpital of S. *Gervais* or S. *Anaftafius*.

12. The *quartier* of S. *Paul* or *de la Mortellerie*, forms a part of the city, and contains the parifh-church of S. *Paul*, in which many celebrated perfons lie interred; the nunnery of *Ave-Maria*; fome *hôtels*, particularly the *hôtel d'Aumone*; the convent of the *Celeftines*, the church of which is noted for fome fine monuments, in which are depofited the hearts of certain perfons of diftinction; the arfenal confifting of many fpacious buildings, with alfo a foundery, and a houfe for making falt-petre; but containing no great number of arms, nor even thofe it contains good. The beft part of it is an affortment of all kinds of ingenious fire-arms, and a very long and large mufquetoon of two barrels, which, is fo difpofed betwixt a pair of wheels that it can not only be removed to any place, but when put to the check, be pointed in any direction, and fired off like an ordinary mufket. It is loaded from behind, will pierce a thick board at the diftance of two hours, and for difcerning an object at that diftance has a telefcope fixed to the barrel. This piece was the invention of an engineer under *Lewis* XIV.

in

in order to take off the enemy's generals at a diftance, but was never ufed: The *Baftile* is an old caftle with eight towers and a redoubt, but too fmall to defend the city, and too low to command it, in fo much that now it ferves only for a ftate prifon. On any public rejoicings the cannon of this place are always difcharged.

13. The *quartier Sainte Avoye* forms a part of the city, and contains the church of the *Auguftines*, with the priory of *Sainte Croix de la Bretonnerie*; two other convents for men, a nunnery, and, among other hotels, that of *Soubife*.

14. The *quartier du Temple*, or *du Marais*, is likewife a part of the city, and fo called from *le Tempel*, an ancient building formerly belonging to the Knights Templars, but, at prefent, a commandery of the knights of *Malta*, and the refidence of the grand prior of the *French* nation. This being a privileged place, many artificers have fet up here, as being a place in which, though not freemen, they can carry on their bufinefs without moleftation. In this quarter are alfo two convents for men·and three nunneries, with the *hôpital des enfans rouges*, or ' red children,' fo called from their cloathing, and two hotels.

15. The *quartier S. Antoine* forms a part of the city, and here the *maifon profeffe des Jefuites* are poffeffed of the libraries of the learned *Menage, Guet* and *Guyet*. In its magnificent church are the hearts of *Lewis* XIII. and XIV. each preferved in a cafket of gold fupported by two angels of maffy filver and as big as life, hovering with expanded wings. In the *palais royal* is an equeftrian ftatue of brafs of *Lewis* XIV. within thirty-five pavilions, under which one is fheltered both from the rain and fun, and the adjacent hotels are all inhabited by perfons of the firft rank. The church *des filles de la vifitation* is fmall but fightly. In this quarter are alfo fome hotels with three convents and an hofpital. The gate of *S. Antony*, which is feen here, was erected as a triumphal arch to *Henry* II. In the fuburb *S. Antoine* is a foundling hofpital, an abbey *de S. Antoine des champs*, the fine looking-glafsmanufacture, where above five hundred perfons are employed in polifhing the plates which are caft at *S. Gobin*, as has been obferved in the Introduction, §. 9. The houfe of *Rambouillet*, from whence the ambaffadors of the Proteftant powers make their public entrance into *Paris*. The caftle of *Bercy*, the *hôtel des Moufquetaries*, the parifh-church of *S. Marguerite*, four nunneries and one convent of *Francifcans*. the monks of which are called *pique-puces*, or ' prick-fleas.' From the latter the minifters of the Catholic powers make their public entrance into *Paris*.

16. The *quartier de la place Maubert* makes a part of the univerfity, or of the third divifion of *Paris*, and is fo called from the univerfity and its· colleges, of which, confequently a general account can come in no where more properly than here. This univerfity was founded by *Charles* the Fat, and confifts of four faculties, *viz.* divinity, the·civil and canon law, phyfic

fic and the fciences. Its head is the rector, who is always chofen from the
faculty of the fciences. The election is renewed every three months, yet
many are known to enjoy that honour for feveral years. In public folem-
nities the rector takes place next to the Princes of the blood, and at the
King's funeral walks with the archbifhop of *Paris.* He prefides in the convocation of the univerfity, and his counfellors are the deans of the faculties
of divinity, law and phyfic, with the *procuratores* of the four nations, of
whom is compofed the faculty of the fciences. The univerfity court is conftantly held on the firft *Saturday* of every month, and likewife on occafion
of any mifunderftanding betwixt the members of the univerfity, but the
verdict is reported to the parliament. The patrons of the papal privileges
of the univerfity are the bifhops of *Beauvais, Sentis* and *Meaux,* one of
whom it chufes, who takes upon himfelf the full power in the Pope's name.
The chancellor of the churches of *nôtre Dame* and *S. Genevieve,* by apoftolic
prerogative, have the privilege of creating a licentiate of the faculties, and
of granting diplomas to teach in *Paris* and elfewhere; but favours of this
kind the chancellor of *S. Genevieve* ufually grants only in the faculty of philofophy. The firft of thefe faculties is the faculty of theology, which confifts of a great number of doctors, both fecular and regular, who are difperfed all over the Kingdom and even in foreign countries. The fenior of
the fecular doctors, refident at *Paris,* is the dean of the faculty. It has
alfo fix houfes and focieties, from which the doctors are called; the *Ubi-*
quifts excepted, who, inftead of adopting the name of any particular houfe,
ftyle themfelves only doctors of divinity, or doctors of the faculty of theology at *Paris.* The principal of thefe houfes, or focieties, are the *Sor-*
bonne and the college of *Navarre,* of which mention fhall be made in their
proper places. The faculty of the civil and canon law has two deacons,
one of whom bears only the title, but the other difcharges the duties of his
office. Ever fince the year 1679, the doctors of this faculty read lectures
in the hall of the college of *Cambray,* or of the three bifhops, where alfo
difputations are held for degrees of bachelor, licentiate or doctor. The
eldeft of the fix poffeffors which conftitute the *fexvirate* of the college is
ftyled *primicerius,* and every one above twenty years ftanding obtains the
title of *comes.* The faculty of phyfic has generally 100 *docteurs regens,* out
of whom, every two years, a prefident is elected who is ftyled *Doyen en*
charge, by way of diftinction from the other, who, on account of his age,
is ftyled *decanus.* The faculty of the fciences is the moft ancient; and, at
prefent, confifts of four nations; namely, thofe of *France, Picardy, Nor-*
mandy and *Germany.* Thefe nations are again divided into feveral provinces;
the firft into five; the fecond into five; the third is for *Rouen,* and the diocefes fuffragan to the archbifhopric thereof; and the fourth is divided into
two, namely, the continent and iflands; the former of which comprehends
Germany, Lorrain, Alface, Bohemia, Hungary, &c. and the latter *England,*
Scotland

Scotland and *Ireland.* Every nation has its particular chief here, who is ftyled *procurator*, and prefides in the convocations. They have alfo their cenfor, who attends to the obfervance of the ftatutes of every nation. To this faculty belong thirty-feven colleges, but fome of them are fhut up. The whole univerfity comprehends forty-three colleges, but in eleven of thefe only are lectures read, namely, in the colleges *de Navarre, de la Marche, du Cardinal Moine, de Beauvais, de Montaigu, des Graffins, des quatre nations,* or *Mazarine, de Harcourt, du Pleffis, de Lifieux,* and *de Cambray.* The annual income of the univerfity amounts to about 50000 livres.

The *quartier* next to be treated receives its name from the open place or fquare of *Maubert.* The *Carmelite* convent here contains in it nothing remarkable. The *college de la Marche* confifts of fix ftipendiates; that of *Navarre* belongs to the faculty of divinity, and was founded in the year 1304, by *Joanna of Navarre,* confort to *Philip* the Handfome. It contains four different foundations, *viz.* one for grammar, the fecond for the liberal arts, the third for the chaplains, and the fourth, which is very numerous, confifts of the bachelors of divinity. *Lewis* XIII. in the year 1638, added another fellowfhip of divinity. In thefe colleges, exclufive of the profeffors of humanity and philofophy, are four for divinity, two of whom read lectures in the forenoon and two in the afternoon. The houfe of *S. Charles,* or the *Peres de la doctrine chrétienne,* have a public library here, which was founded by *M. Miron* and opened in the year 1718. The hofpital *de la mifericorde* is inftituted for the maintenance of 100 female orphans of fix and feven years of age. In the parifh churches of *S. Medard, S. Martin, S. Hypolite,* the collegiate church of *S. Marcel,* and a convent of *Francifcans,* is nothing remarkable. The houfe of the *Gobelins* ftands in the fuburbs of *S. Marcel,* on a little river formerly called the *Bievre,* but now from that houfe, or palace, bearing the name of *Gobelins.* In § 7. of the Introduction, I have given a particular account of the much admired manufactures of this place. In the fuburbs of *S. Victor* is the *hofpital general,* which alfo goes by the name of *la Salpetreria,* that commodity having been formerly made here. The buildings belonging to this hofpital are of large extent. In it is a moft noble foundation for the female fex, near 7000 of whom are here provided for and live under the infpection of fixty fifters, fubordinate to whom are eighty governeffes and a confiderable number of maid fervants. Into this place are received, 1. The foundlings, who take up one ward. 2. Girls and young women, who few and knit. 3. A great number of bad women, who are compelled here to fpin wool. 4. Some hundreds of idiots of that fex, who live here in little houfes built in fafhion of a ftreet. 5. Many other poor women, fome of whom are kept here *gratis* and others pay a fmall matter. 6. Delinquents, who are

5   confined

confined but not put to work; and this, it feems, is an error common to all the houfes of correction in *France.* To this incomparable foundation alfo belongs the caftle of *Biceftre,* which is feated on an eminence at a fmall diftance from the houfes aforementioned, being well defended on all fides by a wall, which is of very confiderable circuit, and contains within it many large buildings and feveral open places. In this caftle are near four thoufand perfons of the other fex; *viz.* 1. Poor men, who are entertained here at free-coft. 2. Some who pay for their reception into the houfe. 3. Perfons difordered in their fenfes, whofe ward looks like a village, being built with regular ftreets. 4. Men and women afflicted with the venereal diftemper, to the number of twenty-five of each fex at a time. Common prifoners are kept here in a particular houfe walled in, and every one in a cell by himfelf. The chapel for this clafs is of a triangular form and extends in height through all the five ftories of the building. None of them are under confinement, the houfe being ftrictly guarded by foldiers who ftand centry, the caftle of *Biceftre* having always a garrifon of fixty men; but the moft remarkable thing here is the large well, and more particularly for its being a modern work. It is of a circular form, and, from top to bottom, lined with free-ftone. To the furface of the water it is 128 feet in depth, and twenty more to the bottom. Round the well, at the water's edge, is a gallery in which, on occafion, a perfon may be let down in order the more conveniently to take a furvey of the lower parts. The water of this well, which is drawn up by means of an engine worked by four horfes who are relieved every three hours, is received all day long into huge buckets, or rather cafks, each of which contains 1200 pounds weight of water, and out of thefe difcharged into a leaden bafon, whence it runs into a refervoir, where it is ufually fix feet in depth, and is farther conveyed from thence, by means of pipes, to all the parts of this vaft building, where it is generally wanted. The *hôpital de la piétie,* in the *rue St. Victor,* where poor children are brought up, conftitutes alfo a part of the *hofpital general,* and is the place where the governors ufually hold their meetings. Thefe three foundations, together with the *hôtel Dieu* have one common fund, amounting to full 2000000 of livres *per annum.* Its principal governors are the archbifhop of *Paris,* the *premier prefident* of the parliament, with the attorney-general, its fpiritual affairs being committed to a rector and twenty-two fecular priefts. Oppofite to the laft-mentioned hofpital lies that valuable fpot of ground called the King's phyfic garden, where we cannot but admire the infinite variety of the feveral plants and trees in it, for the keeping of which in proper order and its further improvement, his Majefty allows every year a fum of 13000 livres. Public lectures on botany, anatomy and chemiftry are read here *gratis;* and, among other things, is to be feen here the *mufæum* of natural curiofities

collected

collected by the celebrated traveller *Tournefort.* In the abbey of St. *Victor*
is a public library, which does not indeed contain any very confiderable
number of books, but has many amongft them which are ancient and fcarce,
together with fome curious manufcripts. In the little library, as it is called,
though it takes up feveral rooms, is a prodigious collection of maps and
copper-plates which fill a fpacious chamber, and additions are continually making to them. In this quarter are alfo the college and church of
the *Bernardines,* the parifh church of St. *Nicolas du Chardonnet* with a feminary, the *college des cardinals Moine,* together with the ufeful foundation of the *feminarie des bons enfans.*

The *quartier S. Benoit* forms a part of the univerfity. In it is the college of phyficians containing five profeffors. The *little chatelet* is a kind of
antique fortrefs and at prefent ferves for a prifon. The *ruë St. Jacques* is
principally inhabited by bookfellers, and here alfo is the church of St. *Yves*
with a priory. In the *ruë Beauvais* is the oldeft college for civil law with
four profeffors at the head of it. *S. John de Lateran* is a commanderie belonging to the knights of *Malta.* The royal college here was founded in
the year 1531, by *Francis* I. but, in 1610, was laid the foundation of the
prefent building, in which are twelve profeffors who teach the Oriental,
Greek and *Latin* languages, together with rhetoric, philofophy, mathematics, phyfic and law. It is independent of the univerfity as well as the college of *Lewis* the Great, formerly called the college of *Clermont,* which
belongs to the Jefuits, and in which, exclufive of the fathers, are upwards
of 600 penfioners. The valuable library here owes the greateft number of
its books to M. *Fouquet,* the minifter of ftate; and the well-known prefident *Harley* alfo left his library to them, which is kept in a feparate apartment. In the college of *Cambray,* or of the three bifhops, are two profeffors for law. The parifh church of *S. Benoit* is a collegiate church. The
college du Pleffis belongs to the *Sorbonne.* Exclufive of thefe are alfo, in
this *quartier,* the *colleges de Lufieux, de Montaigu* and *des Graffins,* together
with five others; but in thefe laft no fciences are taught; alfo four convents,
one feminary, fix churches and three abbeys; *viz. Sainte Genevieve* and
Val de Grace, the latter of which is very beautiful, and *Port Royal.* The
abbey of *S. Genevieve* was founded by King *Clovis,* whofe marble monument is ftill to be feen in the church; but it receives its greateft honours
from the remains of *S. Genevieve,* which are depofited here in a filver coffin
gilt with gold, and placed fo over the high altar that it can be taken down
on occafion of any folemn proceffion. In the library belonging to this abbey are about fixty thoufand printed books, and among them a few which
are old and fcarce. Here is alfo a pretty valuable cabinet of antiquities and
natural curiofities: and, laftly, in this quarter is likewife the royal obfervatory, which ftands in the higheft part of the city.

18. The *quartier S. André* belongs alfo to the univerfity. The archiprefbyterial parifh church of *S. Severin* is a very ancient building .- The convent of *Mathurins* is the place where the election of the rector of the univerfity is always held. Near it too is the hall where the bookfellers and printers hold their meetings, and where all books deftined for the prefs are examined. In the *rue de la harpe* is the moft remarkable piece of *Roman* antiquity in all *Paris*; namely, the remains of the palace which the Emperor *Julian* built for himfelf here, and which afterwards became the refidence of *Childebert* and fome other Kings of the firft race, as appears from certain ancient letters of this King ftill preferved, which are dated from this palace under the title of *Palatium Thermarum*, or *Le palais des Thermes*, or *des bains*. Its old walls make a part of fome private houfes, infomuch that nothing now remains eafily diftinguifhable from them, and of a piece with the reft of the building, excepting one fpacious lofty fquare vault above ground, which continues, even yet, firm and entire, though unfupported by any pillars, and encumbered with the load of a garden covered with fruit-trees, and a depth of foil equal to three mens height. Into this garden we may enter from the fourth ftory of the college of *Cluny*, to which it lies contiguous. According to the common opinion this vaft receptacle was a granary. In the *college d'Harcourt* the natives of *Normandy* (fee num. 16.) hold their meetings. The *forbonne* takes its name from *Robert de Sorbonne*, its founder, who erected it, in the year 1252; but it was Cardinal *Richlieu* who fettled its form of government and built its prefent handfome houfe and fine church, the latter of which is a mafterpiece of architecture. In it may be feen the admirable marble monument of this cardinal, on which the celebrated *Girardon* worked twenty years. The refort of ftudents to this college is fhort of what it ufed to be formerly, but its thirty-fix chambers belong to the oldeft doctors of the houfe and the fraternity of the *Sorbonne*. In the large hall were held the meetings of the faculty of theology. Its library is one of the largeft in all *Paris*, and contains feveral ancient manufcripts. In the parifh church of *S. Cofme* are interred feveral eminent perfonages. Near this is the building in which the royal academy of furgery, inftituted in the year 1731 and confirmed in 1748, hold their meetings on every *Tuefday*. On that next after *Trinity-funday*, in a general meeting of the faculty, the *Peyronian* prize of a golden medal worth 500 livres, is adjudged to the beft differtation on a particular given fubject. The King's firft phyfician is prefident of this academy. The convent of *Francifcans* here is the richeft in all *France*. The ftatue of St. *Louis*, which ftands over the principal entrance of the church, is highly valued. In this neighbourhood are alfo the *colleges de Premontré* and *de Grammont*, with the parifh church of *S. André des Arts*, the large convent of *Auguftines*, in the church of which, befides it being the convocation houfe of the *French* clergy, are held, when

the

the King is at *Paris*, the inftallations of the order of the Holy Ghoft; and, laftly, the play-houfe.

19. The *quartier* of *Luxembourg* is alfo a part of the univerfity. The number of convents and communities in this quarter is too great to be here enumerated, wherefore I fhall only mention the *noviciate* of the *Jefuits* on account of its fine church, and the convent of *Carthufians* on account of its beautiful paintings and monuments. The palace of *Orleans*, or *Luxembourg*, is a magnificent ftructure, and was rebuilt, in the fpace of fix years, by *Mary de Medicis*, fecond wife to *Henry* IV. during her widowhood, though fhe ended not her days here, but at *Collogne*, and that too in very great poverty. In its famous gallery are twenty exquifite paintings by *Rubens*, of which, one piece, which fymbolically reprefents the life of this Queen, is nine feet in breadth and ten in length, yet this celebrated painter completely finifhed them all in two years. The murder of *Henry* IV. is entirely omitted, and all we fee relating to his death is *Time* carrying him up to Heaven, where *Jupiter* and *Hercules* receive him into their arms. The garden belonging to this palace is large. In the *hotel des ambaffadeurs*, the ambaffadors extraordinary are entertained for three days; but thofe of very remote countries refide here all the time they ftay at *Paris*. In the fquare where the fair of St. *Germain* is held, are covered walks erected for the booths. The *hotel de comté* has nothing remarkable in it. The church of St. *Sulpice* is one of the largeft in *Paris*, and near it is a feminary. In the *rüe des Canettes* is the *academie de Jouan*, where young gentlemen are inftructed in bodily exercifes. In the *hôpital des petites maifons* are feveral forts of unhappy people, as, namely, 400 old and poor perfons, who are lunatics, and patients for the venereal difeafe and the itch. It is under the care of the *grand bureau des pauvres*, who is formally empowered to collect an annual tax, by way of alms, from Princes, perfons of diftinction, burghers, handicraftfmen and others, poor people excepted, and has its own court and officers. In this quarter is alfo an hofpital for incurables.

20. The *quartier de* St. *Germain des Prez* is the laft part belonging to the univerfity, being joined to the quarter of the King's palace by means of the royal bridge, as it is called, which is built of ftone over the *Seine*, and is about feventy-two toifes in length and eight toifes four feet broad. The abbey of St. *Germain des Prez*, founded by *Childebert* I. at the requeft of Bifhop *Germain*, is poffeffed by *Benedictine* monks of the congregation of St. *Maur*, and contains a very large library, which, next to the King's, is far the moft valuable. It takes up two large halls, and the ancient manufcripts in it, amounting to 8000 volumes, among which are feveral which are very curious, ftand in a particular hall. It is alfo daily increafing, and, though not intended for common ufe, yet every perfon of learning has free accefs to it. It has alfo a cabinet of antiquities. The college of the four

nations, or *Mazarin*, fo called from the illuftrious cardinal of that name
its founder, ftands on the *Seine*, and is under the direction of the *Sorbonne.*
The lectures read here are all *gratis*, and the library has been public ever
fince the year 1688. In *Univerfity-ftreet* is the *Academie de Dugart*, for the
inftruction of young people in bodily exercifes. In the hofpital *de la cha-
rité* are three large wards containing 150 beds, and likewife another for thofe
who are to be cut for the ftone. Out of thefe, patients who are on the mend-
ing hand are carried into the hofpital *des Convalefcens*, where they continue
till they are quite recovered. In this convent are fine *hôtels*, five convents,
befides one abbey and two communities ; but the moft remarkable building
in it is the *hôtel royal des invalides*, erected by *Louis* XIV. for the relief of
difabled officers and foldiers. This incomparable foundation derives its æra
from 1670, and the following year the ground was laid out for this fpacious and ftately ftructure, and in eight years the whole was finifhed. It lies on
the borders of the country, not far from the *Seine*, and forms a regular
quadrangle of feventeen acres. In it are five courts furrounded with lodgings
three ftories high, including the ground-floor. The middle area is about
four times as large as the others, and the buildings round it make a fine appearance. Here are alfo two rows of arched walks, over each other, forming
narrow galleries, by means of which one may go quite round the whole
building. The ftructures on the outfide are allo decorated with trophies and
all kinds of military embellifhments. At the end of the court, directly oppofite to the main entrance, is the inner door of the church. The infide of
thefe fpacious buildings contain nothing remarkable, except a great number
of chambers. Of the common foldiers, of which there are 3000, a certain
number always live and fleep in one ward ; but the officers, who amount to
about 500, live only three or four together. In a large hall here is held,
every *Thurfday*, a meeting relating to the affairs of the hofpital, in which
the comptroller-general prefides. The infirmary *(les infirmaries)* is feparated
by a court, and under excellent regulations. The church confifts of two
parts ; the inner one is for the ufe of the houfe ; and the outer one, which
is newly built, is, both infide and outfide, fo very magnificent, as fcarce to
have its equal in the Kingdom. With refpect to ecclefiaftical matters, this
foundation is under the direction of the preachers of the miffion of St. *Lazarus.* The penfioners here are employed fuitably to their refpective circumftances ; and fuch as are able are frequently exercifed in the ufe of
their arms, and ftand fentry at the doors of the houfe. This hofpital has
its feparate government.

A little below this houfe, on the banks of the *Seine*, is, at prefent, a royal
military academy, which it is faid is to be larger than the infirmary. In
this academy 500 young gentlemen are to be trained up in the art of war,
thofe having the preference in being admitted who lofe their fathers in battle.
In the year 1751, was publifhed a royal edict for this commendable purpofe.

Paris

Paris is the very centre of all tafte, wit and humour, and likewife the place where vanity and pleafure are at an uncommon height.

I fhall laftly conclude with a fuccinct biftory of this large city. In the time of the *Romans*, before the birth of *Chriſt*, on the Ifland of the Palace, as it is now called, ftood a city which went by the name of *Lutetia*, being inhabited by the *Pariſii*, a people of *Celtic Gaul*, who, on the progrefs of the *Roman* arms in *Gaul*, themfelves fet fire to it. The victors rebuilt the place, but it was only fmall, and ftill continued fo till the reign of the Emperor *Julian the Apoſtate*, who built himfelf a palace here, of which notice has been taken above in treating of the *quartier* of St. *André*. Under the *Frank* Kings it was inlarged, and *Clovis* made it his refidence, but the Kings of the *Merovingian* race dwelt in the neighbourhood, and thofe of the *Carlovinian* line let the city go out of their hands, fo that it came under the poffeffion of the predeceffors of *Hugh Capet*, and, on account of the increafe of its buildings, was divided into four *quartiers* about the year 954, when *Hugh Capet* becoming King, fixed his refidence in this city, as did all the fucceeding Kings after him. From thence forward it was not only enlarged from time to time, but *Philip the Auguſt* built a new wall round it, paved the ftreets and divided the city into three parts ; namely, *La cité*, *La ville* and *L'univerſité*. Its further enlargements caufed it, in the year 1416, to be ranged into fixteen *quartiers*, but *Louis* XIV. divided it into twenty, which new divifion was refolved upon in *December* 1701, but was not completed till the year following. On the twenty-fourth of *Auguſt*, 1572, it was in a moft deteftable manner ftained with the blood of 6000 *Hugenots*, the fignal for this moft horrible maffacre being given by the tolling of the great bell, which is ftill to be feen in the tower of the palace, where the parliament fits, being at the end of the *Pont au Change*, and defcribed above, N. 1. In the years 1589 and 1590, it was befieged in vain by *Henry* III. and IV. but, in 1594, readily opened its gates to the latter after his coronation.

In the neighbourhood of *Paris* are the following country feats and places, *viz.*

Madrid, a royal palace, lying one *French* league from *Paris*, in the wood of *Boulogne*, which *Francis* I. built in the year 1529, after the model of the palace of *Madrid* in *Spain*, where he was kept prifoner. This place is now the refidence of the King's forefter.

Haubervilliers, or *Nôtre Dame des Vertus*, fo called from its parifh church, is a borough fituate one *French* league from *Paris*.

Roiſſy, *Raincy*, *Groſbois*, *Freſnes* and *Charantonneau* are feats.

St. *Maur des Foſſez*, anciently *Bagaudarum Caſtrum* and *Monaſterium Foſſatenſe*, is a borough on the *Maine*, where the duke of *Bourbon* has a noble feat. The abbey of *Benedictines*, which formerly ftood here, was, in the

the year 1535, altered to a chapter under the title of a deanery, and is, at present, a board * estate, belonging to the archbishop of *Paris.*

Conflans, a fine seat near the conflux of the *Seine* and the *Marne,* belongs to the archbishop of *Paris.*

Ivry, Choify, Petit-bourg, La maison rouge and *Villeroi,* are all splendid palaces.

Arcueil is a village where, in the year 1624, was built an aqueduct about 200 toises in length, and, in the lowest part, twelve high. This supplies *Paris* with the fine † water of *Rougis.*

Iffy, a village containing some beautiful country-seats, but particularly that of the Prince of *Conty.* In this village is a parish church, an abbey of *Benedictines* and a seminary.

Seaux, Chaftenay, Chilly, Vaugien and *Dampierre* all pleasant seats.

Rambouillet, a little town having a noble seat belonging to the count of *Touloufe.* In the year 1711, this place was raised to a dutchy and peerage, to which also belong the caftle and foreft of *S. Leger,* with the ancient caftle of *Poiguy.*

Maisons, a fine seat moft delightfully fituated.

Claguy, a complete regular building, *Manferd's* chief mafterpiece, and belonging, at prefent, to the duke of *Maine.*

Ruel, a pleafant borough with a fine feat; the garden belonging to it greatly admired.

Mont Valerien, a rugged hill planted with vines, and faid to refemble Mount *Golgatha* in the *Holy Land.* On it ftands a church to which is a great refort of devout people, and near it is a fraternity called *prêtres du Calvaire.*

---

* By a board eftate our author means here an eftate peculiarly appropriated to fupply the expences of the table.

† How our author comes to talk here of *fine water,* after telling us above, page 364, that *Paris* affords none but what is very bad, is a little unaccountable, unlefs by the word *fine* here be meant *pellucid,* in which fenfe it may very juftly be applied to this water in oppofition to that of the *Seine,* which is turbid.

The G O V E R N M E N T *of the*

I S L E   D E   F R A N C E.

THIS government contains a country abounding in grain, fruits and
wine, but is of larger extent than the ſmall province of the *Iſle de France*,
as comprehending alſo a part of *Perche, Picardy, Brie, Gatinois, Beauce,*
and all *le Vexin François.* Excluſive of the governor, here is a *lieutenant
du Roi* and four ſub-governors; and, beſides theſe, ſome particular governors, of whom notice ſhall be taken in their proper places. The inferior
governments, and ſmall diſtricts of this juriſdiction, are as follow, *viz.*

I. The ſub-government of the *Iſle de France* proper. This little province is almoſt on all ſides environed by the *Seine,* the *Marne,* the *Oiſe* and
the *Aine.* Its moſt remarkable places are

Saint Denis, bearing the ſurname of *en France,* in Latin *Fanum S. Dio-
nyſii,* and anciently *Catolacum,* or *Catulliacum,* a town ſeated in a fruitful level
country, owing its origin to the celebrated abbey of *Benedictines,* founded
there by King *Klotar,* in honour of St. *Dennis,* ſo early as the year 600,
but greatly improved by *Dagobert* his ſon, and altered more than once in
ſucceeding reigns. The abbey is now rebuilt from the ground, and conſiſts of the fineſt free-ſtone, having alſo a garden laid out in a moſt elegant
taſte. The church, though *Gothic,* is nevertheleſs handſome, and contains
not only a very rich treaſury, in which are alſo kept the crown-jewels,
but is likewiſe the place of interment for the Kings of *France* and their families. The firſt King buried here was *Dagobert,* many of whoſe ſucceſſors
alſo lie here; but it was not till the *Capet* line aſcended the throne, that this
church became the common burial place of the Kings, and of this line only
three are wanting here; namely, *Philip* I. *Lewis the Young* and *Lewis* XI.
Among the monuments moſt remarkable in it, are thoſe of *Francis.*I. and
his family, *Lewis* XII. and his Queen, and *Henry* II. with his conſort and
children. No tombs have been erected here for *Henry* IV. or any of the
ſucceeding Kings. In it alſo are interred the following illuſtrious perſon-
ages, *viz. Bertrund du Gueſclin,* conſtable of *France* and Marſhal *Turennc:*
This convent has been without an abbot ſince the year 1692, the-abbots
board-lands amounting to 100000 livres, having, on the death of Cardinal
Retz, the laſt abbot, been tranſpoſed to the houſe of St. *Cyr,* and the cc-
cleſiaſtical juriſdiction of the town aſſigned to the archbiſhop of *Paris;* but
the convent, with its whole precincts, is immediately ſubject to the Pope;
and its prior is perpetual vicar-general to the archbiſhop. The preſent income of the abbey is 60000 livres, together with the lordſhip of the town,

and

and appeals from its court lie only to the parliament of *Paris*. Exclufive of the churches already defcribed, in this town are thirteen others, among which is the collegiate church of St. *Paul* and five convents.

Chelles, in Latin *Cala*, a little town near the wood of *Bondy*, anciently called *Laochonia Silva*. In it is a famous abbey of *Benedictines* founded in the year 660. Near this town too King *Childerick* was murdered.

Vincennes, in Latin *ad Vicenas*, this place lying twenty ftadia from *Paris*, which make near one 'French league. It is an old palace furrounded with towers, begun in the year 1337 by *Philip de Valois*, and finifhed by *Charles* V. *Lewis* XIII. pulled down part of it to make room for a new building which *Lewis* XIV. completed. Its ancient towers ftill ferve as a prifon for ftate-criminals. In it *Charles* V. founded a holy chapel after the model of that at *Paris*, with a chapter of fifteen canons. Here alfo is a priory of *Minims*, a menagerie, and a manufacture of porcelaine.

Montmorency, in Latin *Mons Maurenciacus*, an old little town, raifed to a dutchy and peerage in the year 1551. After the beheading of that excellent man Duke *Henry de Montmorency*, in the year 1632, under the title of the dukedom of *Enguien*, it was given to the houfe of *Condé*. In it is a chapter and convent, and hard by a very beautiful houfe built by the celebrated painter *Le Brun*, which, after him, came into the poffeffion of *Crozat*. To the dutchy of *Enguien* belongs alfo the feat of *Efcouen*, or *Ecouen*, where the duke of *Bourbon* refides.

Charenton, a large village on the *Murne*, where the Reformed had formerly their principal church, which was a fine building; but, on the revocation of the edict of *Nantes*, was foon demolifhed, and two convents built in its ftead.

Louvres, a borough.

Luzarches, a little town famous for its thread laces.

Dammartin, a little town.

II. The fecond fub-government includes

1. *La Brie Françoife*, containing

Lagny, in Latin *Latiniacum*, a little town feated on the *Marne*, in which are three parifh churches and an abbey of *Benedictines* belonging to the congregation of *S. Maur*. In the year 1142, a council was held here.

Brie (properly *Braye*) *Comte Robert*, in Latin *Braia Comitis Roberti*, a little town, lying on the river *Yerre*, receives its name from *Robert of France*, Count *Dreux*, and, after paffing through feveral hands, was annexed to the crown under *Francis* I. In it is a governor, a royal court of judicature, a caftelany and a diftrict, fubject to the *prevoté* and *vicomté* of *Paris*, with a falt-office, and a convent of *Minims*.

Corbeil, in Latin *Corbolium* and *Jofedum*, a town, lying at the conflux of the *Juine*, or *Effone*, with the *Seine*, by the latter of which it is divided into the old and new town. The former lies within the precinct of *Brie Françoife*,

and

and the latter, which is the largeft within the jurifdiction of *Hurepoix*. It has two bridges of ftone, one of which is over the *Seine*, and the other over the *Juine*; two fuburbs, one collegiate ·church, exclufive of three other parochial ones; two priories, two convents, one hofpital, and is the feat of a *prevoté* and *caftelany*. Its principal trade confifts in fkins. It had formerly counts of its own.

Rozoy, a little town, having one parifh-church and a convent.

Ville neuve S. George, a little town lying on the *Seine*. On the other fide of the river lies the village of *ville neuve le Roi*, in which is a fine feat.

Gevres, formerly *Trefme*, a little place, but a dukedom and peerage.

2. *Le Valois*, in Latin *Ducatus Vadenfis*, a dutchy, formerly called the *Comté* of *Crefpy*, and poffeffed by the counts of *Vexin* and *Amiens*. It was annexed to the crown by *Philip the Auguft*; in the year 1284, conferred by *Philip the Bold* on his youngeft fon *Charles*, and raifed to a dutchy in 1402 by *Charles* VI. It belongs at prefent to the duke of *Orleans*, and was by *Lewis* XIV. raifed to a peerage. In it are the following places of note, *viz.*

Crefpy, the capital of the Diftrict, lying betwixt two rivers, and carrying on a confiderable trade in grain and wood. It formerly made a much better figure than it does at prefent. In it is a collegiate church, exclufive of another church, and one convent; and it has alfo a governor, with a Diftrict and a country court of juftice, the officers of which are nominated by the duke of *Orleans*, and one election.

Senlis, in Latin *Sylvanectes* and *Auguftomagus*, a city feated on an eminence on the little river *Nonete*, is the principal election town, the refidence of a governor, a Diftrict-court, a *prevoté*, a country-court, a foreft-court, a falt-office, a *marechauffee*, and a venery. Its bifhop is fubject to the archbifhop of *Reims*, who has under his care one hundred and feventy-feven other parifhes, forty-four chapels of eafe, three abbies, nine priories, and nineteen infirmaries, with a yearly revenue of 18,000 livres : Its tax to the court of *Rome* is 1254 florins. In the city and its three fuburbs are fix parifh-churches, among which is the cathedral, a collegiate church and a royal chapel. The city has its own laws, and is defended with a wall, a dry moat, and baftions. In this place the wool for the manufactures of *Beauvais* is cleanfed, and this conftitutes its principal traffic.

Mont l'Eveque, an elegant feat belonging to the bifhop.

Chantilly, a pretty little town, which fince the year 1661 belonged to the houfe of *Condé*, has a fine feat, fronting the entrance of which ftands a noble equeftrian ftatue of brafs of the laft duke and conftable *Montmorency*.

Verneuil, a feat belonging to the houfe of *Bourbon Condé*.

La Verfine, a hunting feat.

Creil, a little town on the river *Oife*.

Pont S. Maxence, a little trading town on the *Oife,* in which is a royal *prevoté,* fubject to the court of *Senlis.*

Verberie, a little town on the *Oife,* in which feveral councils have been held. Near it is a mineral fpring, the water of which is cold and infipid, yet contains a kind of falt refembling the common fort.

Behfy and *Condun* are two little towns.

Compeigne, in Latin *Compendium,* an ancient town lying on the *Oife,* and the principal place of an election, having alfo a feat. In it are two parifh-churches, one chapter, one abbey of *S. Cornelius,* one college of *Jefuits,* and one alms-houfe. It has alfo a governor, who has his refidence here, and the jurifdiction is divided betwixt the King and the abbey. Its principal trade confifts in corn, wood, and wool. Some ecclefiaftical councils were held here in the years 757, 833, 1185, 1201, 1277 and 1329. Here too it was where, in 1430, the *Englifh* took the maid of *Orleans* prifoner; and at this place a treaty was alfo concluded in 1624 with the *Netherlands.* The Elector of *Bavaria,* in 1709, retired here.

Mouchi and *Blerancourt,* elegant feats.

La Ferté Milon, in Latin *Firmitas Milonis,* a little town lying on the river *Ourques,* which divides it into the upper and lower town. It receives its name from Count *Milon* its founder. In it are two churches, a manor and a *caftelany,* fubordinate to the court of *Crefpy.*

Villers-Cotterets, in Latin *Villaris ad collum Retiæ,* a little town at the beginning of the foreft of *Rets,* from which it receives its name. In it is a fine palace belonging to the duke of *Orleans,* one parifh-church and one abbey, a *prevoté* fubject to the court of *Crefpy,* and one governor.

Nanteuil, a large borough, having a regular caftle.

3. *Le Soiffonnois* conftitutes a part of *Picardie,* and contains

Soiffons, in Latin *Noviodunum,* and *Augufta Sueffionum,* the capital of the whole government, feated in a pleafant fruitful valley on the river *Aifne.* It is a pretty large and well built town, gives the title of count, is the refidence of a general-governor, a *generalité* and an *intendance,* a board of the finances, an election, a Diftrict, a court of juftice, a falt-office, a foreft-court and a *marechauffée.* Its bifhop, is fubject to the archbifhop of *Reims,* in whofe abfence he performs the ceremony of the King's coronation, and has three hundred and ninety-feven (others fay four hundred and fifty) parifhes and twenty-three abbies belonging to his Diocefe, with a revenue of 18,000 livres *per annum,* out of which he pays the court of *Rome's* taxation of 2400 florins. In this town, exclufive of the cathedral, are three collegiate churches, one college of the fathers of the oratory, fix abbies, feveral convents, a *French* academy, and an ancient caftle. Several councils were held here in the years 743 or 744, 853, 866, 941, 1078, 1092, 1120 or 1137, 1155, 1202 or 1210, and 1456.

Brenne

Brenne or *Braine*, a fmall town, lying near the little river of *Vefle*, gives the title of count, which is annexed to the dutchy of *Valois*. In it is a fmall abbey.

Vefly, in Latin *Veliacum*, a little town on the river *Aine*.

Coevres, a little town, which in the year 1645 was raifed to a dutchy and peerage, under the title of the dutchy of *Etrées*.

Humieres or *Mouchi le pierreux*, a little town and a dutchy.

4. *Le Noyonnois*, forms a part of *Picardie*, and comprehends

Noyon, in Latin *Noviomagus Veromanduorum*, *Noviomum*, and *Noviodunum*, a very ancient, pretty large and well built city, feated on the little river *Vorfe*, which at a quarter of an hour's diftance from this place falls into the *Oife*. It is the capital of an election, the refidence of a governor, a royal manor fubject to the juridical court of *Laon*, has a falt-office, a foreft-court, a *marechauffée*, and an old royal *prevoté*. Its bifhop is fuffragan to the archbifhop of *Reims*, is a count and peer of *France*, and has a Diocefe confifting of four hundred and fifty parifhes and feventeen abbies, with a revenue of 25,000 livres. Its taxation to the court of *Rome* is 3000 florins. In this city, befides the cathedral and a royal chapel, are ten parifh-churches, two abbies, two convents, one community, one feminary, and two hofpitals. Its principal trade confifts in grain. In it the famous *Calvin* was born. This city has been feveral times deftroyed by fire, and it likewife fuffered very much in the time of the *League* in *France*. In the year 1516 a treaty was concluded in this place betwixt *Francis* I. and *Charles* of *Auftria*, afterwards emperor.

Chauny, in Latin *Calniacum*, a town feated on the river *Oife*, which here begins to be navigable. It has a particular governor, a foreft-court, a royal *caftelany*, a court of juftice of its own, two parifh-churches, with three convents, and belongs to the marquifate of *Giufcard*.

5. *Le Laonnois* makes alfo a part of *Picardie*, comprehending

Laon, in Latin *Lugdunum clavatum*, and *Laodunum*, or *Lodunum*, a city ftanding on a fteep eminence in the midft of a large plain. It is well built, with beautiful ftreets, has likewife an old caftle, enjoys a wholfome air, has a governor and manor, which is the principal in all *France*; and the feat of an election, a court of juftice, a falt-office, a foreft-manor, a royal *prevoté*, and a *marechauffée*; befides which it has a cathedral, three collegiate churches, five abbies, two convents, three houfes of orders, one college, which is maintained at the expence of the city, one general hofpital, and an almshoufe. Its bifhop is fubject to the archbifhop of *Reims*, is the fecond duke and peer of *France*, has three hundred parifhes and twenty-four abbies in his Diocefe, with a revenue of 30,000 livres, out of which he pays the court of *Rome* a taxation of 4000 florins. The neighbouring country produces excellent wine.

　　　　　　　　　Corbigny,

Corbigny, or *S. Marcoul*, a little place, containing a celebrated church dedicated to *S. Marculſ*, with a priory of *Benedictines*. Here the Kings of *France*, after their unction, uſed formerly to perform *une neuvaine*, or prayers for nine days, when it is ſaid, they received a power of healing the King's evil. At preſent they perform this devotion by proxy, ſending thither one of their chaplains in their ſtead.

Lieſſe, or *Nôtre Dame de Lieſſe*, a little place containing in its church a pretended miraculous image of the Virgin *Mary*, is revered with particular devotion.

Coucy, in Latin *Codïeiacum*, a little town, divided into the *upper* and *lower*, the former of which ſtands on a hill, and is called *Coucy le Chatel*, the latter *Coucy la ville*. It belongs to the duke of *Orleans*, and is an ancient barony, which was raiſed to a peerage in 1400 and 1505. In it is a governor, a royal manor, a ſalt-houſe, a foreſt-court, and a priory of *Benedictines*.

Premontré, a large abbey, and the principal of the order of *præmonſtratenſes*.

III. The third ſub-government contains,

1. *Le Beauvaiſis*, which forms a part of *Picardie*, and comprehends

Beauvais, in Latin *Bellovacum* and *Ceſaromagus*, a city lying on the river *Terain*, being the ſeat of a *prevoté*, and a manor-court, which together with the foreſt-court, are dependant on the biſhop of this place; a province-court, a ſalt-houſe, and a *marechauſſée*. In it are a cathedral, ſix collegiate churches, thirteen pariſh-churches, three abbies, a general hoſpital, and an alms-houſe. The biſhop of this place is ſuffragan to the archbiſhop of *Reims*, is a count and peer of *France*, and has a Dioceſe conſiſting of twelve chapters, fourteen abbies, forty-eight priories, four hundred and forty-two pariſhes and three hundred chapels, with an annual revenue amounting to 55,000 livres, out of which he pays a taxation of 4600 florins to the court of *Rome*. In this town, ever ſince the year 1664, has been a conſiderable manufacture of tapeſtry, and in it are alſo made great quantities of ſerge and woollen cloth. What, however, weakens this city is that it is almoſt on every ſide ſurrounded by hills, yet in the years 1443 and 1472 it happily held out a ſiege, in the latter of which the very women diſtinguiſhed themſelves, and for this reaſon have the honour of preceding the men in the annual ſolemn proceſſion held on the tenth of *July*.

Clermont en Beauvaiſis, a town ſtanding high on the river *Breche*, and giving the title of count, is the principal place of an election, and an hereditary eſtate belonging to the houſe of *Bourbon*. It has a particular governor of its own.

Gerberoi, in Latin *Gerboredum* and *Gerberacum*, a mean little place ſeated on a mountain, at the foot of which runs the river *Terain*. The biſhop of *Beau-*

vais is lord of this place, and ftiles himfelf *Vidame*, *i. e.* (vice-*dominus*) of it, as did its former lords, though as feoffees to him. In it is a chapter.

Fitziems, or *Warti*, a little place, raifed to a dutchy and peerage in the year 1710.

Cagny or *Boufets*, a little place, in the year 1695 raifed to a dutchy, and in 1708, to a peerage under the title of *Boufets*. In it is a feat and a ftatue of brafs of *Lewis* XIV.

Bulles, a very little town, having a royal *prevoté*, fubjeɛt to the manor of *Clermont*. In this town is made a very fine fort of linen.

S. Leu, a little town, ftanding on a hill near the river *Oife*. In it is a priory of *Benediɛines*. Its neighbourhood produces good wine, but is better known for its excellent quarries of ftone.

Liancourt, a fine feat belonging to the duke of *Rochefoucault*.

2. *Le Vexin François* conftitutes a part of *Normandy*, and contains the following places, *viz.*

Pontoife, formerly *Briv-Ifara*, *i. e.* ' the bridge on the *Oife*,' a town feated on a hill on the river *Oife*, over which it has a bridge of ftone, whence the place receives its name. It is the capital of an eleɛtion, the feat of a *vicomté*, a *prevoté*, a royal *mairie*, a *caftelany* a falt-office, and a *marechauffée*. In it are two parifh-churches, one collegiate church, one abbey and one convent. In the years 1720 and 1753, the parliament of *Paris* was removed hither. This town is defended by a caftle.

Magny, a little town, in which is a royal manor, one parifh-church, one priory, three convents, and one hofpital. This place belongs to the houfé of *Neuville-Villeroi*.

Chaumont, in Latin *Calvus mons*, a little town, which takes its name from a bare mountain, upon which in the 12th century a fortrefs was ereɛted, is the capital of an eleɛtion, a manor and a foreft-demefne, containing one parifh-church and two convents.

La Roche Guyon is a little town and dutchy, having a caftle.

IV. The fourth fub-government comprehends

1. *Le Mantois*, or *Mantoan*, conftituting a part of the territory of *Beauce*, and containing

Mante, in Latin *Medunta*, a town lying on the *Seine*, over which at this place is ereɛted a fine broad bridge of ftone. It is the principal place of an eleɛtion and the refidence of a governor, the feat of a *prevoté*, a manor, a court of juftice, a falt-office and a *marechauffée*. In this town is one chapter and two convents, and not far from it ftand alfo another. Fronting the town lies the delightful ifland of *Champion* in the *Seine*.

Meulan, a fort and little town feated on the *Seine*, over which it has a bridge of ftone. The fort lies upon an ifland in the *Seine*, and contains a governor, one parifh-church and one convent: the town ftands on the main land having in it two parifh-churches and one convent.

Dreux,

Dreux, in Latin *Durocaſſes* or *Durcaſſes*, one of the moſt ancient towns in the Kingdom, lies at the foot of a mountain near the little river of *Blaiſe*. It is the principal place of an election and the ſeat of a royal office, a lieutenancy *de robe courte*, a foreſt-court, a ſalt-office and a *marechauſſée*. It has a governor who reſides in the caſtle, in which is a fine collegiate church, beſides two pariſh-churches and two convents in-the town. This place gives the title of count, and makes great quantities of woollen cloth. Near it, in the year 1562, happened the memorable battle betwixt the Papiſts and Proteſtants.

Montfort l'Amauri, a little town ſtanding on a hill, and erected into a dutchy in the year 1692.

Anet, a fine level near the conflux of the rivers *Eure* and *Aure*, belonging to the duke of *Vendome*, under the title of a principality.

S. Cloud, formerly *Nogent*, in Latin *Novigentum* or *Novientum*, a borough ſeated on an eminence near the *Seine*, and belonging to the archbiſhop of *Paris*. This place is a dutchy and peerage, to which it was raiſed in the year 1678 ; for which reaſon its biſhop bears the title of duke and peer of *S. Cloud*. In it is a collegiate church, one convent and a fine royal palace belonging to the duke of *Orleans*. Some of the paintings in it are admirable, and it has alſo a beautiful garden adorned with curious caſcades and fountains. This place is much reſorted to by the inhabitants of *Paris*, on account of its extraordinary pleaſantneſs. It has a manufactory of porcelain, and a fine bridge of ſtone over the *Seine*. In this place *Henry* III. was murdered in the year 1589.

Verſailles, formerly only one pariſh and priory, ſubordinate to that of *S. Magloire* at *Paris*, but having lords of its own, who were vaſſals to the archbiſhop of *Paris*, *Lewis* XIII. bought this place with a view of building himſelf a hunting-ſeat in it ; but *Lewis* XIV. erected here a moſt ſtately and ſpacious palace, of a regular conſtruction, the inſide of which is ornamented with paintings by the greateſt maſters, and other rich furniture. It has alſo a fine chapel. The grove is delightful beyond imagination, being embelliſhed with ſtatues, fountains, caſcades, and bowers. The large canal in it, through which the water of the river *Eure* is conveyed into this place, is thirty-two toiſes in breadth, and eight hundred long. In the middle, it is interſected by another canal about five hundred toiſes in length, of theſe two branches one goes to the little palace of *Trianon*, the other ſupplies the *menagerie*. Its fineſt water-works are defended with rails, and played off on extraordinary occaſions only, ſuch, for inſtance, as when ambaſſadors are there ; the others are open, and played all the ſummer continually. The *menagerie*, indeed, is principally deſigned as a receptacle for the rarer kinds of beaſts ; but it has alſo a very commodious and beautiful manſion-houſe divided into ſmall ſummer and winter rooms, extremely well contrived, which are ſeparated from each other by an octangular hall. Round

it

it are feven courts in which wild beafts are kept. Laftly, to this place belongs the fine palace of *Trianon*, which, though but one ftory high, makes
a very magnificent and pleafing appearance, its outfide confifting wholly of
variegated marble of exquifite workmanfhip. The garden is large, and
abounds in ftatues and water-works. The extenfive foreft belonging to it
includes feveral villages, caftles, and country-feats. *Verfailles* being the
ufual refidence of the court, it has given occafion to the building of a
town, the ftreets of which are ftraight, and confift of a great number of
hôtels. The large avenue to the caftle, which confifting of three *allées*, the
middlemoft of which is twenty-five toifes broad, and each of the two others
ten) divides it into the *Old* and *New Verfailles* ; the latter of which is more
beautifully built than the former. In the old town is a convent, and in
the new a parifh-church, with a large houfe for the miffionaries of *S. Lazarus*, to whom the church belongs. *Verfailles* has a governor of its own,
who is accountable to the King only, with a royal manor ; but this is fubject to the *prevoté*, and *vicomté* of *Paris*.

Marly, a village lying on the *Seine*, one *French* mile from *Verfailles*,
where *Lewis* XIV. built him a fine palace in a wood, confifting of one
large infulated pavilion, with two oppofite rows, each containing fix fmaller
pavilions, joined to each other by covered green walks, all exactly of the
fame dimenfions, and ftanding equidiftant. The garden here is not to be pa-
rallelled for hedge-work, covered greenwalks, and in particular for the richnefs and variety of its verdure, and the ftatues and water-works here rival
even thofe of *Verfailles*, excepting that the white marble-work is much
fpotted and damaged ; the regent, during the King's minority, having let
every thing here run to ruin : Yet is there ftill remaining in it a multitude
of fine objects fufficient to weary out the beft legs and the moft curious
eye. The water-machine, which lies on a branch of the *Seine* betwixt
Marly and *Lachauffée*, is compofed of fourteen wheels, and is without its
equal. It forces the water of the *Seine* up a tower or fquare edifice, flat on
the top, and ftanding on a hill about fix hundred toifes from the river, on
which is a refervoir. From this tower the water iffues into the aqueduct,
which is carried along over thirty arches of furprifing height, being three
hundred and thirty toifes in length ; from thence it runs through two iron
pipes three hundred and fifty toifes farther into the great refervoir at *Marly*,
the fuperficies of which is 18,700 cubic toifes, and the depth fifteen feet :
Laftly, from this place it is conveyed to *Verfailles*. This machine goes day
and night perpetually, and the yearly expence of keeping it in repair is no
lefs than 150,000 livres. Above fifty perfons are appointed to take care
of it, each having their particular employment affigned them ; and five
watch every night in order to be ready againft any fudden accident ; for
which purpofe it is no fmall conveniency that every pump, with its appur-
tinances can be ftopped in an inftant without the leaft obftruction to the
others. ·   *S. Ger-*

S. Germain en Laye, a well inhabited town, ſeated on an eminence along
the *Seine*, having very good air and water.   It owes its origin to a convent
founded there by King *Robert* in the wood of *Laye*, called, in Latin, *Le-
dia*, which in proceſs of time has been altered to *Leyia*, or *Layia*.   He
alſo built there a palace which the *Engliſh* deſtroyed in the year 1346.   A
new building was raiſed here by *Francis* I. which afterwards received the
name of the *old palace*, by way of diſtinction from one built there by
Henry IV. which was called the new.   *Lewis* XIII. added ſome conſiderable embelliſhments to the *old*, and *Lewis* XIV. enlarged it by the addition
of five ſtately pavilions, infomuch that, though no regular building, it
makes a grand appearance.   The garden offers nothing particular, but the
foreſt belonging to it is a moſt delightful place.   The *new palace*, as it is
called, fronts the *old*, and ſtands on the brow of a hill, with a garden of
ſix beds of earth ſupported by arches floping into the valley.   This building
is a true image of the tranſitorineſs of ſublunary things, the galleries in it
being turned into granaries, and the paintings ſcarce diſtinguiſhable for
filth.   The very alcove in which the redoubted *Lewis* XIV. was born is
a duſt-hole.   The grand ſtair-caſe leading into the garden is wholly in
ruins, and the arches for the beds in the garden, in many places, ſunk
into the ground.   The *old palace* was the reſidence of *James* II. after his
flight from *England*, and here too it was that he died.   The ſtreets of the
town are well paved ; the houſes lofty and handſome, with ſome large
ſquares and *hôtels*, a pariſh-church, a hoſpital and three convents.   It is
the ſeat of a royal *prevoté* ſubordinate to the *prevoté* and *viçomté* of *Paris*,
a *caſtelany*, a court of *venerie* and a foreſt-court.

S. Cyr, a convent of nuns of the order of St. *Auguſtine*, founded, in the
year 1686 by *Madame de Maintenon*, miſtreſs, if not the ſecret conſort, of
Lewis XIV. of which ſhe herſelf was abbeſs till the fifteenth of *April*,
1719, when ſhe died.   It contains fifty ladies of quality, thirty-ſix lay-
ſiſters, or *ſœurs converſes*, and 250 pupils, to whoſe admittance it is required that they be betwixt the age of ſeven and twelve, can prove their
nobility by the father's ſide for four generations, and have no defect in
body or mind.   Here they continue till twenty years of age, when ſuch of
them as are diſpoſed to be nuns are diſtributed among the royal abbeys,
where they are admitted *gratis*, and others married to gentlemen with a
portion of 400 piſtoles, beſides a certainty of preferment to the bridegroom ;
or are ſent back again to their parents.   On the death of any one of the
fifty dames, her place is filled up by election from among the young ladies.
The thirty-ſix lay-ſiſters inſtruct them in every branch of education becoming their ſex.   This foundation has a yearly income belonging to it of
180000 livres, 100000 of which ariſe from the extinction of the abbey of
S. Denys, excluſive of its lands and immoveables.   The convent is a very
ſpacious and ſplendid ſtructure.

Peiſſy,

Poiſſy, in Latin *Pinciacum,* a little town lying on the *Seine,* over which it has a large and beautiful bridge of ſtone. The diſtrict to which it belongs is called *Le Pinſerais.* The Kings formerly reſided here, and it is noted for being the birth-place of *S. Louis.* In it are a collegiate and pariſh-church, with a priory, in the church of which lie buried ſeveral Kings, Princes and Princeſſes. Excluſive of theſe, it has two convents, one hoſpital, a royal *prevoté* ſubject to the *prevoté* and *vicomté* of *Paris,* and a ſalt-office. In the year 1561, a fruitleſs conference was held in this place betwixt the Papiſts and Proteſtants. In this town, on *Thurſdays,* is a famous market for cattle.

Houdan, a little town on the *Vegre* having a particular governor of its own.

2. *Le Hurepoix* forms a part of *Beauce. French* geographers are not agreed about the places properly belonging to it, ſome among them reckoning *Melun, Corbeil, La Ferté, Alais* and *Fontainebleau* belonging to it, but others the following, *viz.*

Meudon, in Latin *Moldunum,* a market-town with a convent of *C*apuchins, and remarkable for the royal palace erected there. This place was the favourite reſidence of the only ſon of *Lewis* XIV. Its ſuperiority to all other royal ſeats, conſiſts in the beauty of its foreſt, in its large and lofty terraces, which are vaulted underneath, and the unequalled proſpect from thoſe terraces, where the eye, at one glance, commands a view of all the country round *Paris,* together with the courſe of the *Seine.* In this place alſo is an old and new palace. The largeſt of theſe two is the old, which contains a fine gallery decorated on both ſides with paintings repreſenting the military exploits of *Lewis* XIV. together with many ancient marble and braſs ſtatues. At the end of it is a ſaloon, the ſtatues in which are all modern, but executed by the greateſt maſters from *Rome.* The chapel in it was firſt founded by the *Dauphin,* who cauſed the pavement and the two altars to be made of the fineſt marble. He alſo projected the ſummer apartment. The new palace, which he cauſed to be built from the ground by the celebrated *Manſard,* conſiſts of one *corps de logis* only, and ſtands in the middle betwixt the large parterre and the wood above it, inſomuch that out of the ſecond ſtory a perſon may go directly into the wood, which is ſo interſected with hedges and allies, that, in many places, eight viſtas or more preſent themſelves at one view. Near the palace ſtands a fine convent of.Capuchins.

Dourdan, in Latin *Dordinga,* a town on the river *Orge,* in Latin the *Urbia,* has a particular governor of its own, with a *prevoté,* a foreſt-court, two pariſh-churches, a community, one hoſpital and a priory, and is famous for its ſilk and woollen ſtockings.

Montlehery, in Latin *Mons Letherici,* which latter was in the twelfth century changed to *Mons Leberici,* or *Leberii,* a little town ſeated on an eminence, and bearing the title of an earldom, has a royal jurifdiction, a *prevoté*, a *caftelany* and a priory. Near it, in the year 1465, a battle was fought betwixt *Lewis* XI. and *Charles* duke of *Berry*.

Arpaion, formerly *Chatres*, a little town, in the years 1720 and 1723, raifed to a dutchy, and the feat of a county court and *prevoté*; both jurifdictions belong to the duke of *Arpajon*.

Chevreufe, a little town belonging to the convent of *S. Cyr*, and formerly bearing the title of a dutchy.

La Ferté Alais, in Latin *Firmitas Alepia*, and *Firmitas Adelheidis*, a little place on the *Juine*.

Longjumeau, a little place having a priory.

Vaux le Villars, a moft beautiful feat and dutchy.

3. *Le Gatinois François*, fo called by way of diftinction from *Le Gatinois Orleannois*. The country of *Gatinois*, in Latin is called *Pagus Vaftinenfis*; and this part of it comprehends

Melun, among the *Romans* known by the name of *Melodunum*, or *Metiofedum*, an ancient town lying on the *Seine*, which divides it into three parts. The old town ftands upon an ifland, and is joined to the new ones by two bridges of ftone. That part of the town which lies in *La Brie* to the right of the *Seine*, is the principal. This town is the chief place of an election, the feat of a governor, a *vicomté*, a *prevoté*, a manor, and a provincial court of juftice, a falt magazine and a *marechauffée*, and gives title likewife to an archdeaconry belonging to the archbifhop of *Sens*. It contains one collegiate church, three parifh churches, two convents, one abbey, with another a little way out of the town. It carries on a trade with *Paris* in corn, meal, wine and cheefe. In the year 1419, it was befieged and taken by the *Englifh*, but ten years after the inhabitants drove them out and fubmitted to *Charles* VII.

Fontainebleau, in Latin *Fons Bliaudi*, owes this appellation to a dog, called *Bliaut*, found here drinking at a fpring. It is but a mean town, though it enjoys a particular governor of its own, has a royal *prevoté*, a court of venery and of forefts. This place firft became known under *Philip Auguftus*, but its prefent fame is owing to the royal palace there, which is very fpacious, being, indeed, rather an affemblage of four palaces together, and having five courts in it of different architecture. *Francis* I. *Henry* IV. *Lewis* XIV. and XV. having raifed them buildings here. The ftag-gallery along the orangery is particularly worthy notice. It is decorated with paintings of all the royal palaces and feats, between which are ftag's heads fet off with branches of a very uncommon fize. Under each ftag's head is an infcription fhewing in what wood and by what King the ftag was killed. Thus many a ftag is introduced fpeaking, and very politely fays, King *Charles*, *Lewis*, or *Philip*, did me the honour of taking me. In a corner at one end of the gallery is the place in which *Chriftina*, Queen of *Sweden*,

in

in the year 1654, had her maſter of the horſe taken off. The caſtle has four gardens, and about it lies a large foreſt conſiſting of eminences and levels. The eminences in it are of ſo particular a kind of rock, as, at a diſtance to look like a confuſed pile of large ſtones.

Moret, in Latin *Muritum*, an old little town on the banks of the *Loing*, conferring the title of count.

Montereau Faut Yonne, in Latin *Monaſteriolum Senonum*, a town ſeated at the conflux of the *Seine* and *Yonne*, and having a collegiate church.

Chateau Landon, the *Velaunodum* of the ancients, a town ſeated on the river *Loing*, and having a *prevoté* ſubject to the manor of *Nemours*, two churches, one abbey, one convent and an hoſpital.

Milly, a town ſeated on the little river *Ecolle* and having a collegiate church.

Nemours, a little town, ſeated on the *Loing*, owes its origin to an ancient caſtle by the *Romans* called *Nemus*, or *Nemoſium*, from its ſituation in a wood. It is the reſidence of a governor and the principal town of an election ; was raiſed by *Charles* VI. to a dutchy, and by *Lewis* XII. to a peerage. *Lewis* XIV. conferred this place on the duke of *Orleans*. In it is an ancient palace with a pariſh church, a priory, three convents, and, in the ſuburbs, another pariſh church and an abbey. The inhabitants trade in corn, wine and cheeſe.

Courtenay, a ſmall town ſeated on the little river *Clairy*, and belonging, at preſent, to the marquis of *Fontenilles*.

Eſtampes, in Latin *Stampæ*, a town ſeated on the river *Juine*, and bearing the title of a dutchy. It is the capital of an election and the reſidence of a royal bailiwick, a *prevoté*, a ſalt-office and a *marechauſſee*. It has its own law, and contains five pariſh churches, ſix convents and two chapters. In this town have been held three provincial ſynods and one national. Some include it in *Hurepoix*, others place it in *Beauce*. This diſagreement ariſes from a conteſt about it betwixt the *Iſle de France* and *Orleannois*, both of which exerciſe certain privileges in it.

4. *Le pays de Thimerais* makes a part of the province of *Perche*, and contains the following places, *viz.*

Chateau neuf en Thimerais, the capital of this country, a little town, and the reſidence of a governor and bailiwick ſubordinate to the juriſdiction of *Chartres*.

Senonches, a market-town bearing the title of a principality, and the reſidence of a court ſubject to that of *Chartres*.

Breſſoles, a market-town.

Bazoche, a little place, having a royal juriſdiction and *viſcomté*, and being alſo the reſidence of a barony ſubordinate to the court of *Chartres*.

Champron, a ſmall place.

The

The GOVERNMENT *of*
P I C A R D Y and *A R T O I S.*

THE name of *Picardy* is not to be met with in any monument or re-
cord till towards the end of the thirteenth century, yet the name of
Picard is of more ancient date. Inftead of dwelling on the uncertain and
contrary meanings affigned to it, we fhall only obferve that *French* geographers think it received its original at *Paris*, and was applied to the people
of this country on account of their vivacity and warmth of temper, *picard*
fignifying a paffionate wrangler. This province, fouthward, terminates on
the *Ifle de France*; weftward, on *Normandy* and the Channel; northward,
on the *pas de Calais*, *Artois* and *Hainault*; and, eaftward, on *Champagne*.
The country is level, produces wine, fruits of all kinds, plenty of grain and
great quantities of hay, particularly along the *Oife*. Wood being fcarce
here, moft of the inhabitants burn turf. The *Boulnois*, indeed, affords
pit-coal, but not fo good as that of *England*. Its principal rivers are
   The *Somme*, in Latin *Samara* and *Sumina*, which takes its rife in *Vermandois*. The whole extent of its courfe is confined to *Picardy*, and in it
to the *generalité* of *Amiens* only. At *Bray* it becomes navigable, and, after
receiving the leffer rivers of *Ancre*, *Elce* and *Auregne* in its courfe, divides
itfelf into twelve fmall channels, which, in the city of *Amiens* difperfe, and
unite again at another place which receives the veffels coming from *Abbeville* and *S. Vallery* laden with *Dutch* and *Englifh* goods. This river is very
deep and has no ferry, except betwixt *Abbeville* and *S. Vallery*, in a part
called *Blanquetaque*. It falls into the Channel. The *Oife*, in Latin *Ifara*,
has its fource in *Picardy* on the frontiers of *Hennegau*, and traverfing the
generalités of *Soiffons* and *Paris*, is joined by the *Verre*, the *Delette*, the
Aine, the *Terain*, the *Aronde*, &c. At *Fere* it becomes navigable, and at
· *Conflans* lofes itfelf in the *Seine*. The *La Canche*, in Latin *Cancius* and
Quentia, receives its fource in *Artois*, and at *Montreuil* becomes navigable,
emptying itfelf into the fea a little below *Staples*. The *Lauthie*, in Latin
the *Aetilia*, receives its fource on the borders of *Picardy* and *Artois*, falling
into the fea betwixt the mouths of the *Somme* and *Canche*. The *La Lis*,
in Latin *Legia*, rifes near the village of *Lyfbourg* in *Artois*, becomes navigable at *Vindres*, not far from *Aire*, and empties itfelf, near *Chent* into the
Scheld. The *Aa*, the fource of which is above *Rumilly le Comté*, not far
from *Therouanne*, is made navigable at *S. Omers* by means of fluices, and
difcharges itfelf into the Channel. The *Scarpe*, which, from its fource
near *Aubigny*, begins at *Arras* to bear veffels, and, not far from *Mortagne*,
in *Flanders*, mingles with the *Scheld*. The *Deule*, formerly little larger
than

than a brook, but now, by means of canals and fluices, made to procure a communication betwixt *Lens, Lifle, Douay* and the *Deule,* is become a confiderable river. The *Deule,* which bears a part in this communication, is the *Upper,* the *Lower Deule* taking its courfe below *Lifle* as far as *Lys.*

Near *Boulogne* is a mineral fpring, of which I fhall give an account in that place. The fituation of this province on the fea, its many navigable rivers and canals, together with the induftry of the inhabitants, render it the feat of a flourifhing trade. In it are made beautiful ftuffs of filk and wool, ferrets, coarfe linen, and that fineft kind of linen called *batift,* or lawn, and alfo foap. It carries on a large trade likewife in corn and pit-coal. In the governments of *Calais* and *Boulogne* are annually bought up 5 or 6000 colts, which being afterwards turned loofe in the paftures of *Normandy,* are fold for *Norman* horfes. In the woods here are feveral valuable glafs manufactures, particularly that of *S. Gobin* is much celebrated. The fifheries on this coaft are alfo very advantageous.

This province fell early under the dominion of the *French,* and *Amiens* was the refidence of the firft Kings who ruled in *Gaul.* In the year 823, *Lewis,* furnamed *the Pious,* appointed counts over it, who foon affected fo great power that they became almoft abfolute and independent. *Philip* of *Alface,* earl of *Flanders,* had the province of *Amiens* given him as a portion with his wife *Elizabeth,* countefs of *Vermandois,* and fhe dying without iffue, he kept poffeffion of it, on which King *Philip Auguftus* harraffed him with a war. It was agreed, however, betwixt them, that *Philip* fhould hold the country during his life, but that on his death it fhould defcend to *Eleanor* of *Vermandois,* countefs of *S. Quentin* and fifter to the count's late fpoufe, after whom it was to devolve to the King. *Charles* VI. mortgaged all the towns on the *Somme* to the duke of *Burgundy* for 400000 dollars, but *Lewis* XI. having redeemed them in the year 1463, *Picardy* has from that time never been alineated.

Of the county of *Artois* I fhall fpeak particularly in the fequel.

In *Picardy* are four bifhoprics, two provincial courts, two governments, fix manors, twenty *prevotés,* five courts of admiralty, four foreft-manors and four lordfhips. In civil matters it is under the parliament of *Paris.* The little diftricts of *Beauvaifis, Noyonnois, Laonois, Soiffonnois* and *Valois,* which belong all to *Picardy,* have been annexed to the government of the *Ifle de France.* With refpect to its military government, there are in *Picardy* and *Artois,* exclufive of the general governor, three general lieutenants; namely, one for *Upper Picardy,* another for *Lower Picardy,* and one for *Artois;* fix deputy-governors; *viz.* four for *Picardy* and two for *Artois,* with a multitude of particular ones. We fhall divide this province according to its deputy-governments, previoufly obferving that *Picardy* is divided into the *Upper* and *Lower.*

To

To the *Upper Picardy*, in which alfo may be reckoned the *Middle*, as it is called, belong the little diftrids of *Vermandois*, *Thierache*, *Santerre* and the county of *Amiens*; to the *Lower Ponthieu* and *Vimeu*, with the *Boulonnois* and the *Recovered country*. The fub-governments here are, at prefent, as follow, *viz.*

1. The fub-government of *Thierache* and *Vermandois*.

Thierache, in Latin *Theorafcia*, was fo called in the time of *Charles the Great*, from *Thierry*, lord of *Avennes*. The moft remarkable places in it are

Guife, in Latin *Gufia* and *Gufgia*, the capital of this fmall diftrid, which ftands on the *Oife*, having a caftle noted for holding out feveral fieges, of which that by the archduke *Leopold*, in the year 1650, was the laft. In this town is a deputy-governor, a particular governor, an election, a faltoffice, a fmall collegiate church and a convent. In the year 1527, *Francis* I. raifed it from an earldom to a dutchy and peerage, which appertain, at prefent, to the houfe of *Bourbon Condé*. The dutchy belonging to it is of large extent, reaching through *Picardy* as far as *Champagne*.

La Fere, a little town feated in a marfhy country near the conflux of the *Saar* into the *Oife*. In it is a governor, a royal jurifdiction, a manor and foreft-court united with that of *Marle* and *S. Quentin*, and a falt-office. It contains alfo two collegiate churches, one of which ftands in the caftle, an abbey, a convent and a college, with a powder-mill, a cannon foundery, an arfenal and barracks. It was formerly fortified, but its whole defence, at prefent, confifts in the fluices on the *Oife*, by means of which the adjacent country may be laid under water for fome miles. In the large wood, which receives its name from this town, are feveral glafs-houfes ;

S. Gobin, in particular, which has a feat in the middle of it, is highly famed for its excellent manufacture of looking-glafs, in which plates are fometimes caft 105 inches in height and fixty in breadth. They are carried from hence to the *Gobelins* at *Paris*, and there polifhed.

La Capelle, a little town, formerly fortified, and taken, in the year 1636, by the *Spaniards*.

Vervains, a little town feated on an eminence on the *Serre*, is a *caftelany* and marquifate, having a great corn trade, and being known in hiftory for the peace concluded there, in the year 1598, betwixt *France* and *Spain*.

Marle, a little town and earldom, ftands on an eminence, at the bottom of which runs the *Serre*, having an ancient caftle, a particular governor, a manor, an inferior foreft-court, and a falt-office.

Ribemont, or *Riblemont*, a little town ftanding high on the *Oife*, has a particular governor, is ruled by laws of its own, and is a royal *prevoté :* At the foot of the eminence ftands an abbey.

Moncornet, a town feated on a hill near the *Serre*, makes a coarfe kind of ferge.

Aubenton, a little town, having a falt-office.

2. Vermandois,

2. *Vermandois*, fo called from the *Veromandui*, its ancient inhabitants. In it are

Saint Quentin, anciently *Augufta Veromanduorum*, fo called from the body of *S. Quentin* which lies here. It is fortified, ftands on a rifing ground near the *Somme*, is the capital of an election and the refidence of a governor, a *prevoté*, a falt-office, a foreft-court and a *marechauffee*. It has alfo its particular laws. The collegiate church of *Saint Quentin* is one of the fineft in all *France*, and near it is an abbey of *Benedictines* belonging to the congregation of St. *Maur*, befides which it has alfo a collegiate church, and an abbey. In this town, and its neighbourhood, is made fo beautiful a lawn, that its trade therein is computed to amount annually to two millions of livres. In the year 1557 it was taken by the *Spaniards*, in confequence of the fignal victory which they obtained near it over the *French*.

Ham, a little town, having a caftle feated on the *Somme*, in a marfhy country, has a particular governor, a Diftrict, and a royal jurifdiction, a *vicomté*, a *caftelany* under St. *Quentin*, and a mayoralty, together with three parifh-churches, and one abbey.

Vermand, a market-town on the *Oumignan*, has a fine abbey of *Premonftratenfes*.

Saint Simon, in the year 1635 raifed to a dutchy and peerage.

Catelet, a little town, the fortifications of which in the year 1674, were razed to the ground. It lies in a Diftrict containing the borough of *Beaurevoir*, near which is the fource of the *Scheld*, with the abbies of *Mont Saint Martin* and *Honnecour*. This place was formerly annexed to the empire; but for above thefe three hundred years paft has belonged to the Diocefe of *Cambray*.

II. The deputy-government of *Santerre* contains

Peronne, a little but a very ftrong town, ftanding among marfhes on the river *Somme*. This place has been feveral times befieged, but was never taken. It is the refidence of a governor, a bailiff, an election, a manor-court, and a falt-office; has its particular rights, and contains five parifh-churches, among which one is a collegiate church; has alfo three convents, a college and an abbey. It is a place of great antiquity, and the Kings of the *Merovinian* race had a palace here.

Mondidier, in Latin *Mons Defiderii*, a little town feated on a mountain, being the refidence of a governor, an election, a *prevoté*, a bailiwick, a falt-office, and a *marechauffée*. It is governed by particular laws of its own, and contains a priory, together with three convents and a college; but is now much declined.

Roye, in Latin *Rauga*, a town near one of the fources of the *Moreuil*, in which is a governor, a *prevoté*, a bailiwick and a falt-office, with a collegiate church, three parifh-churches, two hofpitals, and one community.

Nefle,

Nesle, a little town, but the first marquisate in *France*, having above eighty fiefs belonging to it; is now in the possession of the house of *Mailly*.

Lihons, a little town, with a priory.

Albert, Anere or *Enere*, a little town.

Bray, a place of no consideration.

III. The deputy-government of *Amienois, Ponthieu,* and *Vimeu,* contains

1. The earldom of *Amienois*, comprehending a great part of the country formerly inhabited by the *Ambiani*; and being properly called *Picardy*. The bishops of *Amiens* formerly procured the lordship of this country to be conferred on them by the Kings, under the title of an earldom; but *Philip Augustus*, in the year 1185, re-united it to the crown. *Charles* VII. granted it to *Philip the Good*, duke of *Burgundy*; and *Lewis* XI. in 1477, united it again to the crown. The most remarkable places in it are

Amiens, in Latin *Ambianum*, and *Samarobriva*, the capital of *Picardy* and of the county of *Amienois*, bearing also the title of a *Vidamy*. It lies on the *Somme*; is the residence of the governor-general, a prefect, a deputy-governor, a *generalité*, an intendency, an election, a provincial-court, a court of mintage, a *prevoté* for *Amienois* and one for *Beauvaisis*, a tobacco-tax and salt-office, a forest-court, and a *marechaussée*. The streets here are straight and broad with handsome houses on each side. The cathedral is particularly magnificent, but principally values itself for containing, among other reliques, the head of *John* the Baptist. In this city is also a collegiate church, a chapter of St. *Nicolas*, fourteen parish churches, a seminary, a college of Jesuits, a general hospital, another hospital, several abbies and convents, a particular law and an academy of the arts and sciences, which, till 1750, had been only a private literary society. In it are also made great quantities of ferrets, half-silk stuffs, and soap. The bishop hereof is suffragan to the archbishop of *Reims*, and within its Diocese, exclusive of the cathedral, are twelve collegiate churches, twenty-six abbies, fifty-five priories, seven hundred and fifty rectories, one hundred and three chapels of case, forty-eight communities, six colleges, two general hospitals, and six alms-houses. The yearly revenue of its bishop is 30,000 livres; and his taxation at the court of *Rome* is 4900 florins. This city is defended by a very good citadel.

Conty, a borough lying on the little river *Seile*, and giving the title of Prince to the second line of the house of *Bourbon*.

Poix, a little town, dutchy, and peerage belonging to the duke of *Bouillon*. This place has no less than twelve rectories and nine fiefs dependent on it.

Doulens or *Dourlens*, in Latin *Donincum*, and *Doningium*, a town seated on the river *Authie*, is the capital of an election, the seat of a royal *prevoté*, and a salt-office. It has a strong citadel, together with three parish-churches, an abbey, two hospitals, and a community.

5   *Corbi*

Corbie, a little town on the *Somme*, containing five parishes, a fine abbey, an hospital and salt-office. The fortifications of this place were razed by *Lewis* XIII.

Pequigny, a little town on the *Somme*, having a collegiate church.

Rubempre, a little place, and lordship belonging to the count *de Maille*.

Ponthieu, in Latin *pagus Pontivus*, reaches from the *Somme* to the *Canche*, abounding in grain, fruits, and pasturages, being formerly an earldom, and still governed by laws of its own.

Abbeville, the capital of this District, lies in a pleasant and fruitful valley, where the river *Somme* separates into several branches, the town being divided by it into two parts; besides being farther watered by the lesser rivers of *Scardon*, *Sottins*, and *Corneille*, or *Taniere* : The sea-flood rises too here in the *Somme* to the height of about six feet. This town is large, and very conveniently situated for a fortification ; being also the capital of an election, the seat of a provincial court, a District, and a *prevoté*, with a bailiage, a *marechauffee*, a forest-court, a court of commerce, a court of admiralty, a salt-office, &c. It contains also one collegiate church, fourteen parish-churches, fifteen convents, a commanderie of the order of *Malta*, and a college. As it is very conveniently situated for commerce, it carries on a great trade in grain, oil, hemp, flax, cordage, soap, &c. The woollen manufactures set up here in the year 1665, by Mr. *Roberts*, a *Dutchman*, has succeeded so well, that at present its cloths are little inferior in fineness and goodness to those of *England* and *Holland*. In it are also made very beautiful baragons, together with mockettos, a kind of carpets, dimity, plush, coarse linen, spun wool, and fire-arms. Lastly, this place gave birth to the celebrated *French* geographers *Nicholas* and *William Sanson*, *Peter Du Val*, and *Philip Briet*.

Saint Riquier, formerly *Sentule*, a small town, seated on the little river *Scardon*, which receives its source near this place,. has a royal *prevoté*, and a celebrated abbey of *Benedictines*, to which it owes its name and being: In it are also two parish-churches, one chapel, and an hospital.

Drugy, a small place in which the *Abbé* of *S. Riquier* has a seat. The castle of *La Ferté*, to which is annexed a fine *castelany*, are dependents on this town.

Crecy or *Creffy*, a little place on the river *Authie*, but an old *castelany*, and the seat of a bailiwick and a *prevoté*, carrying on likewise a small trade in cattle, wool and hemp, is famous for the first battle betwixt the *English* and *French*, which was fought there in the year 1346, with great resolution, till the latter were totally routed with the loss of above 30,000 men. It gives name also to a large wood. Near this place lie *Etrees*, *Caumartin*, and a seat called *Thomas*.

Montreuil, a strong town on an eminence; three *French* leagues from the sea, is the residence of a bailiwick, dependent on that of *Amiens*, and the capital of a particular county, containing eight parifh-churches, one collegiate church, two abbies, and feveral convents.

St. *Paul*, the capital of an earldom of the fame name, belonging to the houfe of *Vendome*.

Rue, a fmall town, feated among marfhes on the little river of *Maye*, is the refidence of a falt-office, a royal bailiwick fubject to the court of *Abbeville*, has a governor, four parifh-churches and two convents, with a brifk trade for fifh, fheep, wool, horfes, and other cattle.

Pont de Remy, a little place on the *Somme*, over which it has a bridge leading to a fmall Ifland, on which ftands a fort. Belonging to it is a governor, a *caftelany*, and a priory.

Crotoy, a little town, one *French* league from the mouth of the river *Somme*, directly oppofite to *S. Valery*. Fifhing is the principal fupport of its inhabitants.

3. *Vimeu* or *Vimeux*, is properly a Diftrict of *Ponthieu*, and contains the following places, *viz.*

S. Valery, in Latin *Leuconaus*, a town feated at the mouth of the *Somme*, owes its original to a convent, founded there about the year 613, by *S. Valery*, which is at prefent a confiderable abbey of *Benedictines*, belonging to the congregation of St. *Maur*. In this town is a court of admiralty and a governor. It carries on alfo a great trade both for exports or imports, though without any proper harbour.

Sancourt, a village, remarkable only for a victory gained there in the year 881, by the *Franks* over the *Normans*.

Augft, in Latin *Auguftus*, a little town lying on the fea.

Le Bourg d'Ault, contains a court of admiralty and a falt-office, and belongs to the duke of *Orleans*. Its neighbourhood affords the beft frefh fifh of any in the whole channel.

Gamaches, a town and marquifate, having a caftle and a fmall chapter.

Oifemont, in Latin *Avimons*, a market-town, having a royal *prevoté* fubordinate to *Amiens*, and being alfo a commanderie of the order of *Malta*.

Azincourt, a borough lying near the *Breffe*, where the *French*, in the year 1415, though greatly fuperior in number, were defeated by the *Englifh*, who took more prifoners than their own army amounted to.

IV. *Le Boulonois*, and the *Recovered Country*, form alfo a deputy-government.

1. *Le Boulonois* is a particular government, wholly independent of the governor-general of *Picardy*, and extends from the *Canche* to the borders of *Flanders*, making one part of the ancient earldom of *Flanders*; fince which it has had earls of its own, who were vaffals to the counts of *Flanders*, and afterwards to the counts of *Artois*. *Lewis* XI. annexed this county to the crown, giving in lieu of it to its poffeffor *Bertrand de la Tour*, count of *Auvergne*, the county of *Lauragnais*, in *Languedoc*, and the jurifdiction he

tranftransferred to the church of the Virgin *Mary* at *Boulogne*, but received homage for it of the county, and prefented it with a golden heart of 6000 livres value ; and this the Kings⠀obferve even to the prefent time. In it are the following places, *viz.*

· *Boulogne*, anciently *Gefòriacum* or *Gifòriacum*, and fince *Bononia*, a feaport feated at the mouth of the little river *Liane*, with a harbour defended by, a fort, but its entrance,very difficult. Men of war come here no farther than St. *John's* road, and it is at time of flood only that merchant-fhips can get into the harbour. In it is a bailiage, a *prevoté*, a court of admiralty, a *marechauffée*, a foreft-court, and a governor. It is divided into the upper and lower town, which are at about one hundred paces diftance from each other, the latter being both the largeft and handfomeft, and inhabited principally by tradefmen. It contains, indeed, but one church, whereas the upper town, though fmall, contains the cathedral and town-houfe. In it is a feminary, a college, a convent, an hofpital, and fome religious houfes. The bifhop of this place is fuffragan to the archbifhop of *Reims* ; his revenue, 12,000 livres, and his taxation at the court of *Rome* 1500 florins. This Diocefe contains two hundred and feventy-feven parifhes, with one hundred and forty-feven chapels of eafe. At fome hundred paces diftant, in the road to *Calais*, is a mineral fpring of a ferrugineous tafte; and therefore called *La fontaine de fer*, *i. e.* ' the iron fpring.'

Etaples, in Latin *Stapulæ*, a little town feated near the *Canche*, and having a harbour, but fit only for fmaller veffels. In it is a *prevoté*, and its trade confifts in herrings and mackarel. According to the *Abbé Longuerve*, this is the port which *Cæfar* called *Tecius*, and which was afterwards altered to *Vicius*.

Monthulin, once a fmall fortrefs, but demolifhed by order of *Lewis* XIV.

Ambleteufe, a little town on the channel, having a governor, and is a free port. A harbour capable of containing thirty and forty-gun frigates, has been fet on foot here, but is not yet finifhed, though it feems a matter of little difficulty. This is the place where King *James* II. landed at his departure out of *England* in 1688.

Marquife, a borough, having a quarry of gray marble.

Bournonville, a little place yet a dukedom.

.. II. The *Recovered Country (le pais reconquis)* is fo called becaufe retaken, in the year 1558, from the *Englifh*, in whofe hands it had continued above 200 years. It confifts of two counties, is feven *French* leagues in length, two and a half broad, and fourteen in circumference, containing,

1. The county of *Guines*, which is four *French* leagues in length, and nearly the fame in breadth. The firft count hereof was *Sigfried the Dane*, who did homage for it to the count of *Flanders*. On the extinction of his male line, this country devolved to the *caftelany* of *Ghent*, from which Count *Arnulphus*, in the year 1282, fold it to *Philip* III. for 3000 livres,

and _Philip the Handfome_ fold it again to the count of _Eu._ King _John_ reaffumed it, but, in the year 1360, ceded it to the _Englifh,_ from whom,
in 1413, _Charles_ VI. retook it. _Lewis_ X. conferred it, as a fief, on _Charles
the Bold,_ duke of _Burgundy,_ after whofe death it was again re-annexed to
the crown. To it belong twelve fief-baronies and twelve fief-peerages. The
principal places in it are as follow, _viz._

Guines, a town, once only a village, feated among marfhes, and dependant on the abbey of St. _Bertin,_ till enlarged and fortified by _Sigfried the
Dane_; but its works have been razed by exprefs order.

Ardres, a fortified town, in a marfhy fituation, and bearing the title of
a principality, is the refidence of a royal _prevoté,_ which ferves alfo for the
county of _Guines,_ but fubordinate to that of _Montreuil,_ being alfo a royal
mayoralty. Near this town, in the year 1520, _Francis_ I. and _Henry_ VIII.
had an interview, accompanied with a tournament, where the appearance
was fo fplendid that the fpot on which it was held has ever fince been called
Le camp de trap d' or, _i. e._ ' the field of cloth of gold.' The government
of _Ardres_ is fubject to that of _Picardy,_ and contains nineteen parifhes, which
pay no capitation.

Liques, an ancient abbey of _Premonftratenfes_ not far from _Ardres._

Courtebonne, a marquifate.

2. The county of _Oye_ has fhared the fame fate with that of _Guines,_ and
contains

Oye, a borough.

Calais, a ftrong town and fea-port fituate in the ftrait called _le pas de
Calais,_ the breadth of which is about fix or feven _French_ leagues. It is the
capital of the _Recovered Country,_ and the feat of a bailiwick fubject to the
parliament of _Paris._ Its figure is an oblong fquare with one of the longeft
fides towards the fea and the other towards the country. Exclufive of its
regular fortifications, on its weft fide ftands a citadel, and the entrance to
its harbour is alfo defended by a fort. In the town is a fine parifh church,
and another in the fuburbs of St. _Pierre._ In it are alfo four convents, two
communities for the inftruction of youth, and two alms-houfes. The ftreets
are ftraight and well paved, being adorned with feveral houfes in the modern
tafte. The fine arfenal here was built by Cardinal _Richlieu,_ a ftatue of
whom in brafs ftands in the court. The harbour, among many other inconveniences, is of very difficult accefs. In times of peace, two packet-
boats pafs weekly betwixt this place and _Dover_ in _England._ It carries on
a good trade in wine, brandy, falt, flax, horfes and butter, and reaps great
advantage from the canal dug there in the year 1681, which opens a cheap
and convenient intercourfe betwixt it and St. _Omer's, Graveline, Dunkirk,
Bergues_ and _Ypres._ To the government of _Calais_ belong twenty-four parifhes.
This town is exempt from all taxes, but the affeffments for the repairs of the
fortifications and canals run high. It has a particular governor, a deputy-
governor,

governor, mayor, &c. of its own. *Edward* III. King of *England*, in the year 1346, took this town by a formal fiege, and the *Englifh* held it till 1588, when they were again difpoffeffed of it by the *French*. In the year 1594, it was taken by the *Spaniards*, who gave it up again, in 1598, at the peace of *Vervins*. In 1694, 1695 and 1696, it was bombarded by the *Englifh*.

The twenty-four parifhes here which have been faid to belong to the government of *Calais*, are, the fuburbs of St. *Pierre*, the villages of *Andre*, *Balinghem*, *Boningue*, *Boucre*, *Campagne*, *Coquelle*, *Coulogne*, *Efcales*, *Frethun*, *Guemp*, the town of *Guines*, the village of *Hames*, where was formerly a ftrong caftle, which, in the year 1558, was difmantled ;' *Hervelinghem* and *Mareq*, in the latter of which alfo was formerly a caftle and an abbey; *Nielle*, *Nouvelle Eglife*, *Ofquerque*, *Oye*, alfo called a fmall town, *Peuplinque*, *Pinen*, *Sangatte*, where lie the baronies of *Calimote*, St. *Trias* and *Vieille Eglife*.

Fort Nieulé, which lies a quarter of an hour's diftance from *Calais*, on the fide towards *France*, being built in the year 1680, ftands on piles, and is a regular oblong fquare, defended by four baftions and two half-moons, being joined to the citadel of *Calais* by means of a mole. The fluices here are of great fervice either for draining the country or laying all the neighbourhood of *Calais* under water in order to obftruct a fiege.

The County *of* A R T O I S.

THIS county, which forms a part of the *Netherlands*, borders, to the fouth and weft, on *Picardy*, being bounded to the north by *Flanders*, and eaftward by *Hennegau*. It is twenty-fix *French* leagues in length, and about half as much in breadth. It is one of the beft and fineft provinces in the whole Kingdom, and, exclufive of its great fertility in grain, carries on a confiderable trade in flax, hops, wool, cabbage and oil of turnep-feed, having alfo divers manufactures of linen. The principal rivers in it are the *Scarp*, the *Aa* and the *Canche*, all which have been already mentioned in *Picardy*. The name of this country is derived from the ancient *Atrebates* who dwelt in *Gallia Belgica*, and made no fmall figure in *Cæfar*'s time. It was long a part of *Weft Flanders*. In the year 1180, *Philip Auguftus* received it, as a portion, with *Ifabella* of *Hennegau*, a relation of *Philip* of *Alface*, count of *Flanders*. *Lewis* VIII. in the year 1226, erected it into a county in behalf of his brother *Robert*. *Margaret* of *Fland rs* brought it her fpoufe *Philip the Bold* of *Burgundy*, whofe male iffue poffeffed it till the days of Duke *Charles*, after whofe death *Lewis* XI. took poffeffion o it in violation of the jufter claim of *Mary* daughter to *Charles*; but, on her marriage

riage with *Philip*, archduke of *Auſtria*, *Charles* VIII. at the treaty of *Senlis*, in the year 1493, obliged himſelf to reſign the counties of *Burgundy* and *Artois* to him, as a fief to *France*, which was accordingly done.   It continued in the poſſeſſion of the houſe of *Auſtria*, and afterwards of *Spain*, till *Lewis* XIII. and XIV. re-united it to *France*; which annexment was confirmed by the treaties of *Nimeguen*, *Ryſwick* and *Utrecht*.   This country, together with *Picardy*, is now ſubjeﬅ to one governor-general; excluſive of whom, here are a general-lieutenant and two deputy-governors, one for *Arras* and *Bapaume*, and the other for *Aire* and St. *Omer*.   Beſides theſe, here are alſo ſeven particular governors.   In the year 1530, *Charles* V. inſtituted a provincial council here which, in civil matters, is ſubordinate to the parliament of *Paris*, and contains twelve courts of juſtice under it.   The raiſing of the royal revenues here is adminiſtered by conſent of the ſtates, who are ſummoned both by public and particular *lettres de cachet*, which each perſon muſt produce at his admiſſion.   Theſe ſtates conſiſt of the clergy, and among them the biſhops of *Arras* and St. *Omer*, with a great number of abbots, and the deputies from every chapter form a part; the nobility, to the number of about ſeventy; the commoners, who conſiſt of the council of *Arras*, and the deputies of the magiſtracy of the eight principal towns of the country.   The free gift required of them is, in ſome meaſure, ſettled at 400000 livres; but the charges of forage are more or leſs, according to the number of cavalry in the country.   No cuſtoms are paid here.

· I have already mentioned the twelve juriſdiﬅions, or diſtricts, into which the country is divided, and as ſuch under the provincial tribunal.   Theſe are,

    1. The government or diſtrict of *Arras*, to which belongs

Arras, the *Origiacum* of *Ptolemy* and the *Atrebatæ* of *Cæſar*, the capital of the country, ſituate on the *Scarpe*, being divided by walls, moats and a ſmall valley through which runs the little ſtream of *Crinchon* into two parts, the moſt ancient of which is called *La Cité* and the other *La Ville*. The town is large and regularly fortified, having alſo a ſtrong citadel.   Its biſhop, who is ſuffragan to the archbiſhop of *Cambray*, has a diocefe of 400 pariſhes with a revenue of 22000 livres a year, being taxed at the court of *Rome* in 4000 florins.   He is alſo lord of the city and preſident in the aſſembly of the ſtates.   The abbey of St. *Vaſt* here has a moſt magnificent chuꝛch belonging to it, and to it likewiſe appertain the ſmall, but very fertile diſtrict of *Le pays de Salve*, lying betwixt *Artois* and *Flanders*.   In it are alſo eleven pariſh churches, one ſeminary, one college of Jeſuits and ſeveral convents.   The large market-place here is ſurrounded with fine buildings, among which is the governor's houſe.   In this city is held the aſſembly of the ſtates.   It is alſo the reſidence of the governor and a foreſt-court.   The tapeſtry hangings made here, and ſo called as being invented in this place, want not for beauty, but muſt not be ſet in competition with thoſe of *Paris*, *Bruſſels* or *Antwerp*.   In the year 1477, it was taken by
Lewis XI.

Lewis XI. in 1493, by the Emperor *Maximilian*; in 1640, it was taken again by the *French*; but, in the year 1654, held out a liege againſt the *Spaniards.*

Buquoy, an earldom.

Henin-Lyetard, a little town and an earldom, in which is an abbey.

2. The juriſdiction of *Bapaume* receives its name from

Bapaume, in Latin *Bapalma*, a fortified town, ſeated in a barren country without rivers or ſprings, and having an old palace which gave riſe to the town, with a particular governor of its own, a royal and foreſt-court. In the year 1641, the *French* took it from the *Spaniards.*

Bourſy-Louverval, a place of no conſideration.

3. The diſtrict of *Aveſnes*, ſo called from the little town of *Aveſnes*, which lies on the borders of *Picardy*, two *French* leagues diſtant from *Dourlens.*

4. The diſtrict of *Heſdin* takes its name from

Heſdin, a ſtrong town ſeated on the river *Canche*, the reſidence of the court of the diſtrict, of a foreſt-court, and of a particular governor, being a regular hexagon and almoſt ſurrounded with marſhes. *Philibert Emanuel*, duke of *Savoy*, general to the Emperor *Charles* V. built this place in the year 1554, after deſtroying old *Heſdin*. In 1639, it was taken by *Lewis* XIII. who kept it after the peace of the *Pyrennees.*

Old Heſdin, now a little place, having only two churches, lies one *French* league from the above-mentioned fortreſs. It was a place of ſome ſtrength till deſtroyed by *Charles* the fifth's general.

Humieres, a little place giving the title of duke.

5. The county of St. *Paul* is large, was formerly a fief of the *Boulonnois*, and, after paſſing through ſeveral lords, fell under *Lewis* XIII. *France* being confirmed in the poſſeſſion of it by the treaty of the *Pyrennees*. In it are

St. *Paul*, a little town but the capital of this county.

Croix and *Guincourt*, two little places, the firſt of which gives the title of count.

6. The diſtrict of *Pas*, in a little place of the ſame name, contains

Crequi, a ſmall place giving name to a very illuſtrious ducal family now extinct.

Beauqueſnes, a town and royal *prevoté.*

·. *Huuchin*, a little place but a marquiſate.

Riſquebourg, ſtill of leſs conſideration but alſo a marquiſate.

7. The diſtrict of *Aubigny*. The town of *Aubigny* is divided into two parts; namely, *Aubigny le comté* and *Aubigny la Marche.*

8. The diſtrict of *Lens*, containing *Lens*, formerly *Eænae*, a mean little town ſeated on the river *Souchet*, being anciently fortified and ſeveral times

befieged. It has ftill a collegiate church, end near it, in the year 1648, the *Spaniards* were totally defeated by the *French.*

9. The advocacy of *Bethune* holds its feffions in

Bethune, a ftrong town fituate on the little river *Bietre.* It is the third town in the county and has a ftrong citadel ; the houfes in it, however, are but mean and the ftreets ill paved, but its market-place is both large and handfome. It contains one collegiate church, two parifh churches, two priories, one college of Jefuits, fix convents and one hofpital. In the year 1710, it was taken by the Allies, but again reftored at the peace of *Utrecht:* *Annezin,* a caftle.

Richebourg l' Advouë, and *Richebourg S. Vaft,* two little places.

10. The bailiwick of *Lillers* is the fmalleft of all, and holds its feffions at *Lillers,* a little town, formerly having its particular lords and being fortified.

Pernes, a fmall town lying on the *Clarence.*

11. The bailiwick of *Aire,* to which belong

Aire, in Latin *Aeria* and *Aria,* a confiderable fortrefs feated on the river *Lis* which divides it into two unequal parts. In it is a collegiate church, a college of Jefuits, feveral convents and two hofpitals, one of which is inftituted for foldiers. In the year 1641, this place was taken by the *French,* but foon recoverèd again by the *Spaniards.* In the year 1676, it was again taken by the *French,* and confirmed to them at the peace of *Nimeguen.* In the year 1710, it was taken by the Allies, but yielded up again to the *French* at the peace of *Utrecht.*

S. François, a fort lying a cannon-fhot diftance from *Aire,* and having a communication with it by means of a canal.

S. Venant, a little town, feated on the river *Lis,* was difmantled, but its fortifications have been fince repaired.

Fauquemborg, alfo a fmall place.

12. The royalty of *Terouenne,* or *Terouane.* The town of *Terouane* was anciently the capital of the *Morini,* and afterwards an epifcopal fee. It ftands on the river *Lis,* containing feveral churches and convents, but being taken, in the year 1553, by the Emperor *Charles* V. he entirely demolifhed it. The diftrict belonging to it, however, was ceded by *Spain* to *France* at the treaties of 1559 and 1659.

13. The bailiwick of St. *Omer,* the capital of which is

St. *Omer,* in Latin *Audomaropolis,* and formerly called *Sithiu,* a confiderable city feated on the river *Aa,* being one of the beft fortifications in the *Netherlands* and the fecond city of the county. It lies partly on an eminence and partly in a morafs, is the fee of a bifhop fuffragan to the archbifhop of *Cambray,* containing a diocefe of 110 parifhes, fome chapters, ten abbies and a yearly revenue of 40000 livres, its taxation to the court of *Rome* being

1000

1000 florins. It has a particular governor, a foreft-court, a cathedral, and fix parifh churches, exclufive of St. *Bertin,* a celebrated abbey of *Benedic-tines,* to which order the town owes its exiftence, two colleges of Jefuits, one hofpital and feveral convents. In the year 1677, this place was taken by the *French.*

Renty, a town and marquifate feated on the river *Aa,* where, in the year 1554, a battle was fought betwixt the *French* and *Spaniards.*

Arques, a little town and earldom, which for above 1000 years paft has belonged to the opulent *Benedictine* abbey of *S. Berthin* at *S. Omer.*

Lifbourg, a town and marquifate.

The GOVERNMENT *of*
C H A M P A G N E and B R I E.

THE province of *Champagne* terminates weftward on the *ifle de France* and *Picardy,* fouthward on *Burgundy,* eaftward on *Luxemburg* and *Lorrain,* and northward on the *Hennegau* and part of the bifhopric of *Liege,* being one of the moft confiderable provinces in the whole Kingdom, and extending from weft to fouth-eaft, or from *Lagny* to *Bourbon,* forty-fix *French* leagues in length ; and from fouth to north, or from *Ravieres* to *Rocrois,* about fifty-four. It derives its name from the large plains in its centre. Its borders are full of forefts, mountains and eminences. This country produces plenty of grain, but is particularly famous for its wine, which is exported in great quantities but with no great profit to the inhabitants, as may be eafily imagined from the fcarcity of good vintages and the charge of vineyards. The natural commodities in which it trades are grain, wine, iron, wood, cattle, hay, woollen and half-filk ftuffs, linen, &c. Its principal rivers are, the *Meufe,* or *Maas,* which receives its fource near the village of *Meufe,* or *Montigny le Roi,* and becomes navigable at St. *Thibaud.* It traverfes the diocefes of *Toul* and *Verdun,* and afterwards *Champagne, Luxemburg, Namur,* the *Auftrian* and *United Netherlands,* and having received the *Wahal* below the ifland of *Bommel,* is called the *Meruve;* falling laftly into the north fea. The *Seine,* mentioned in the Introduction. The *Marne,* which rifes in *Baffigny,* becomes navigable at *Vitry* and joins the *Seine* at *Charenton* above *Paris.* The *Aube,* the fource of which is on the frontiers betwixt *Burgundy* and *Champagne,* falls into the *Seine* near *Conflans.* Attempts have been made to render this river navigable, but hitherto without fuccefs. The *Aifne,* or *Aine,* which receives its fource above *Menehout,* on the borders of *Champagne* and *Lorrain,* joins the *Oife* half a *French* league above *Compeigne.* At *Chateau Porcien* it begins to bear veffels. Near *Bourbonne* and *Attencourt*, two *French* leagues diftant from *Vaffy*, are fome celebrated mineral waters.

In the divifion of the Kingdom betwixt the fons of *Clovis*, *Champagne* made a great part of *Auftrafia*, having *Metz* for its capital. After this it was governed by dukes, and after them by counts, who held the power till the thirteenth century, when *Count* *Theobold* V. who at the fame time was alfo King of *Navarre*, died, in the year 1270, without iffue. He was fucceeded by his brother, *Henry* III. who dying alfo in the year 1274, left a daughter named *Johanna*, who was married to *Philip the Handfome*, King of *France*; after whofe death, her fon, *Lewis* X. fucceeded to the Kingdom of *Navarre* and the county of *Champagne*, on his father's demife becoming alfo King of *France*. His brother and fucceffor affumed all thefe countries to himfelf, though *Johanna*, daughter to *Lewis*, and at that time married to *Philip*, count of *Evereux*, put in her claim to the county of *Champagne* as the property of the Queen of *Philip the Handfome*; but fhe could obtain nothing till after the death of King *Charles*, when the Kingdom of *Navarre* was given to her and her fpoufe. In the year 1335, both of them, by virtue of an agreement with *Philip de Valois*, renounced their right to *Champagne* and *Brie*, which accordingly, in the year 1361, were folemnly re-united to the crown by King *John*.

In this country are two archbifhoprics and four bifhoprics. It is governed by the parliament, the chamber of accounts and exchequer of *Paris*, the territory of *Sedan* excepted, which belongs to the parliament of *Metz*. In it are ten provincial and diftrict courts, one great forefter, feveral foreft-courts, two courts of mintage and a *generalité*, refiding at *Chalons*, which is divided into twelve elections.

The military government of this province is lodged in a governor, four general-lieutenants, one over the diftrict of *Reims*, the fecond over the diftricts of *Vitry* and *Chaumont*, the third over the diftricts of *Troyes*, *Langres* and *Sens*, and the fourth over *Brie Champenoife*. Inferior to thefe are four hereditary fub-governors, one for each department, not to mention the other officers. This country is alfo divided into the following fmaller divifions, *viz.*

I. *Champagne*, in its proper meaning, which is again fubdivided into

1. *Lower Champagne*, to which belongs

Troyes, anciently *Auguftomana*, or *Auguftobona*, the capital of the province and feated on the river *Seine*. It is the refidence of an election, a royal *prevoté*, a diftrict, a provincial and foreft-court, a falt-office, a marechauffée, a court of mintage and a particular governor; and belongs to the royal demefnes, being pretty large, but not, by a great deal, fo flourifhing and populous as formerly. It wants not, however, for fine churches, particularly the cathedral, and in it are fourteen in all, including the cathedral and two collegiate churches, with four abbies, ten convents, one college,

one

one feminary and one hofpital. The bifhop of this place is fuffragan to the archbifhop of *Sens*, has a diocefe containing 372 parifhes, ninety-eight chapels of cafe and feventeen abbies. His yearly revenue is 14000 livies, and his taxation at the court of *Rome* 2000 florins. This city carries on ftill a tolerable trade, efpecially in linen, flax, hemp and cotton, fuftians, canvas, wax and tallow candles, fewing needles, ferges and tapeftry. Befides the diftinguifhed excellence of the neighbouring paftures, the foil alfo produces plenty of grain, wine, fruits and all kinds of efculent plants.

Ifle-Aumont, a dukedom and peerage.

Nogent-fur-Seine, a tolerable town, the capital of an election and the feat of a diftrict-court, a falt-office and a *marechauffée*.

Pont-fur-Seine, a little town, receiving its name from a bridge of ftone built here over the *Seine*. In it is a royal manor and a moft magnificent feat.

Arcis, a little town, ftanding on an eminence near the *Aube*, and having a falt-office.

Mery-fur-feine, a little town, feated in a particular territory of its own, has a royal *prevoté* and a priory.

Plancy, a little town and marquifate having a chapter.

Rameru, a fmall place having a *Ceftercian* abbey, is a barony belonging to the houfe of *Luxemburg*.

S. Liebault, a little place.

Piney, or *Pigney*, a little town, raifed to a dukedom in the year 1577, under the title of *Luxemburg*; and, in 1581, to a peerage.

Lufigny, a little place, having a royal tribunal fubject to that of *Troyes*.

Beaufort Montmorency, a dukedom.

2. Into *Upper Champagne*, to which belongs

Chatillon-fur-Marne, a little town raifed, in the year 1736, to a dukedom and peerage.

Efpernay, or *Epernay*, in Latin *Sparnacum*, fituate on the *Marne*, which divides it into two equal parts, is the capital of an election and the refidence of a royal *prevoté*, a diftrict, a foreft-court and a falt-office. It has alfo an abbey and it is the election of *Efpernay*, which produces the beft *champagne*.

Ay, a town feated on the *Marne*, having a royal mayoralty, but fubject to the court of *Efpernay*, being famous alfo for the delicioufnefs of its wine.

Avenay, a little town, feated on the *Marne*, having a chapter and an abbey.

Vertus, a little town, earldom and peerage, lying at the foot of a mountain, being alfo of fome reputation for its wine, and containing a collegiate church and two abbies.

Fere Champenoife, a little town.

Dormans, a little town, feated on the *Marne*, and lately erected into an earldom.

Hautvillers,

Hautvillers, or, as it is generally pronounced, *Hauvile*, is the spot on which grows the very best of the so-much-admired wine of this province.

Pierry, a place of no consideration, except on account of its delicious wine.

II. *Chalonois*, a small territory included by some in *Proper Champagne*, contains

Chalons, a name derived from the ancient *Catalaunum*, being a large city seated on the *Marne*, the capital of a *generalité*, or intendency, and an election, the residence of an *Intendant* and *prevôt-general*, and of the *marechauffée* of the province, as also of a country and district court. The bishop of this place, who is also a count and peer, is suffragan to the archbishop of *Reims*. In his diocese are comprehended 304 parishes, with ninety-three chapels of ease and nineteen abbies, and his yearly revenue is not less than 24,000 livres, his taxation at *Rome* being 3000 florins. In this city, exclusive of the cathedral, are two chapters, eleven parish churches, one seminary, one college of Jesuits, three abbies, nine convents and two alms-houses. It carries on a good trade in the woollen stuffs which are called by its name. In the year 1592, the parliament was removed hither from *Paris*; and, in commemoration of the loyal attachment of this city to *Henry* IV. that Prince caused a medal to be struck with this inscription on it, *Catalaunenfis fidei monumentum.*

La Croisette, a little place near *Chalons*, where the inhabitants of that city, assisted by *Charles* of *Anjou*, King of *Naples*, killed 8000 *English*.

Sainte Menehoud, a little town, seated in a morass betwixt two rocks, on the highest of which stands a castle. It was once fortified, and, in the year 1590, held out a siege against the duke of *Lorrain*; but, in 1652, was taken by the *Spaniards*.

III. *Remois*, a country producing excellent wine and abounding in fine pasturages. In it are

Reims, anciently *Durocortorum* and *Civitas Remorum*, one of the most ancient and celebrated cities in the whole Kingdom, standing on the river *Vefle*. It is the largest city in all *Champagne*, being handsome and populous and the capital of an election, the residence of a district, a forest and a mintage court, and a salt-office. The archbishop of this place is the first duke and peer of *France*, perpetual legate of the see of *Rome* and primate of all *Gallia Belgica*. He also crowns the King. His suffragans are the bishops of *Soiffons*, *Chalons-fur-Marne*, *Laon*, *Senlis*, *Beauvais*, *Amiens*, *Noyon* and *Boulogne*. He has a yearly revenue of 50,000 livres, and his taxation at the court of *Rome* is 4750 florins. The churches in it, in general, are very fine, particularly the cathedral of our Lady, though of *Gothic* architecture. The principal door of this building is remarkable for its workmanship; and the great altar, at which the coronation and unction of the Kings

of

of *France* is performed, is plated with gold. The treafury of this cathedral muft be concluded to be very great, every King, at his coronation, making an offering here. The book of the Gofpel, upon which the Kings take the oath, and the cover of which is of gold fet with rough gems, is faid to be written in the *Sclavonic* language. Exclufive of this church, here are three collegiates, five abbies, a large feminary, a fine college of Jefuits, three fpacious hofpitals, nine convents, a commanderie belonging to the order of St. *Antony*, the greateft part of the revenues of which are appropriated to the invalids at *Paris*, and a commanderie of the knights of *Malta*. The abbey of *Benedictines* of St. *Remy* here is one of the nobleft belonging to that order in all *France*; and on the altar of its church, under which St. *Remigius* lies buried, is kept the *Sainte ampoulle*, or holy phial, which, according to the ftory, in the year 496, at the baptifm of *Clovis*, by Bifhop *Remigius*, was brought from Heaven, by a dove, at the prayer of that faint, the crowd hindering him from being able to come to the font with the ufual oil. It confifts of a dark red glafs about the length of one's little finger, and nearly refembling in fhape the fmall *Hungary* water bottles, being ftopped with a gold ftopple. It lies in a perforated fquare cafket, faftened on a filver falver, being feen only through a cryftal cover placed over the cafket. The ointment in it is faid to be grown dry, but, on every unction of a King of *France*, a fmall matter of it is taken out and mixed with the oil. prepared for the unction. This whole narrative is fupported only by the modern and fufpicious account of *Hincmars*, formerly bifhop of *Reims*, and is not only rejected by the late *French* hiftorians, *Chiflet* and *Bafnage*, but even acknowledged to be groundlefs by all ingenious fenfible perfons who are in the leaft converfant with the biftory of *France*. The univerfity here was founded in the years 1547 and 1549, authorized by the parliament of *Paris*. This city carries on alfo a confiderable trade in wine, woollen and filk ftuffs, and gingerbread. In it too are feveral remarkable remains of *Roman* antiquities, particularly three gates of the city, which, to this day, bear the names of fo many pagan deities, *viz.* of the *Sun*, of *Mars* and *Ceres*.

Fifmes, formerly *Fines*, a little but very ancient town, having a particular governor of its own, and a diftrict fubject to the provincial court of *Reims*. In this place, in the year 881, were held two provincial councils.

Cormicy, a little town belonging to the archbifhop of *Reims*.

· *Rocroy*, in Latin *Rupes Regia*, a ftrong town fituate in a plain on the borders of *Hennegau*, has a particular governor of its own, a royal *prevoté* dependent on the court of *S. Menehoud*, and a falt-office. In the year 1643, the *Spaniards* received a very terrible defeat here.

Avaux la Ville, a town and county lying on the river *Aifne*.

Avaux le Chateau, a little place belonging alfo to this county and ftanding on the *Aifne*.

Chateau

Chateau Porcien, a little town, feated on the *Aïne,* which divides it fiom the caftle, which ftands on a rock. This town has a falt-office, and, in the year 1561, was, with its diftrict, erected into a principality in favour of the houfe of *Mazarin*; ferges its principal commodity.

Sillery, a marquifate.

IV. *Retelois* belongs, at prefent, to the houfe of *Mazarin,* as a dutchy and peerage. Part of it confifts of woods with great numbers of iron forges in it, owing to the neighbouring mines. The other part of it confifts of meadow ground. In it are four towns, three of which belong to the dutchy of *Retel, viz.*

Retel, Regitefte, or *Reitefte,* the capital of the dutchy, lying on the river *Aifne.* It is the principal place of an election and the feat of a diftrict court, a falt-office and a *marechauffée.* In it are three convents, and *Cæfar* himfelf built a fortrefs here.

Attigny, in Latin *Attiniacum,* an old little town, feated in a fine country along the *Aine,* called *la Vallé de Bourg.* During the fpace of fome centuries, in this town ftood a royal palace. Several public meetings were alfo held here.

Mefieres, in Latin *Maceriæ,* a ftrong little town, feated on an ifland formed by the river *Maefe,* over which it has two bridges with a citadel. This place was befieged, in the year 1521, by the Emperor, *Charles* the fifth's, troops.

Charleville, a fmall but well built town, lying on the *Maefe,* and, till the year 1609, only a village called *Arches,* when it was erected into a town by *Charles* of *Gonzagues,* duke of *Nevers* and afterwards of *Mantua.*

Clofe by *Mefieres* ftands the citadel of *Mont-Olympe,* built on the other fide of the river by *Lewis* XIII. but demolifhed, in the year 1687, by *Lewis* XIV. together with the bridge of ftone leading to it.

Donchery, a little town, feated on the *Maefe,* being walled and fortified with half baftions by *Lewis* XIV. It has a *prevoté,* a falt-office and a particular governor.

Le Chatelet, Bourg and *Brieulle,* two inconfiderable places.

V. *Argonne,* a territory extending from *Champagne* to the dutchy of *Bar,* being near twenty *French* leagues in length. Some make *Sainte Menehoud,* mentioned above, under the *Chalonois,* to be the capital of this country. To it alfo belong the following places, *viz.*

Clermont, the capital of an ancient earldom, maftered formerly by the duke of *Lorrain* and united to the dutchy of *Bar. Lewis* XIII. and XIV. took it from him again, the laft of whom, in the year 1648, gave it to the Prince of *Condé,* but with the refervation of the jurifdiction and a right of appeal to the parliament of *Paris.*

Beaumont, a little town, having a royal tribunal and *prevoté,* with a royal mayoralty alfo fubject to the jurifdiction of *Reims.*

Villefranche, a fmall town on the *Meufe,* formerly fortified.

Varennes

Varennes, a little place.

Grand Pré, a little town and an ancient earldom, deriving its name from the meadows among which it lies.

Montfauçon, an old little town, having a fecularifed abbey belonging to the bifhop of *Verdun*.

VI. *Pertois*, contains only two towns, namely,

Vitry le François, a town lying on the *Marne*, being the capital of an election and royal *prevoté*, the refidence alfo of a bailiwick, a provincial and foreft-court, and a falt-office. It belongs to the royal demefnes, has a particular governor, and is ruled by its own laws. In it is a collegiate church, one college, three convents and two hofpitals, one of which is general. This town carries on a trade in corn, which is attended with confiderable advantage to it, and takes its name from its founder *Francis* I. by way of diftinction from

Vitry le brulé, a fmall place feated about one *French* league from it on the little river *Sault*, and having. been formerly a town, which being demolifhed by the Emperor *Charles* V. *Francis* I. built the new town juft mentioned. Near it, however, is an abbey, and in it a convent. The neighbouring country is the pleafanteft in all *France*.

Saint Difier, in Latin *Fanum Sancti Defiderii*, a town feated on the *Marne*, and the refidence of a royal bailiwick fubordinate to that of *Vitry le François*, having alfo a foreft-court, a falt-office and a *marechauffée*. It belongs to the royal demefnes, and has a particular governor of its own. In it are two convents and one hofpital, and near the town is an abbey. In its neighbourhood are fome iron forges. In the year 1544, this place was befieged by the Emperor *Charles* V.

VII. *Vallage*, fo called from its vallies, which are excellent paftures, abounding in fine cattle. Great quantities of military implements are alfo made here. The following places in it are the moft remarkable.

Vaffy, an old town feated on the little river *Blaife*, being the refidence of a royal *prevoté* and *caftelany*, as alfo of a falt-office and a foreft-court. It belongs to the royal demefnes, has a particular government, with a convent, an hofpital, and a manufactory of druggets. Here it was, where in the year 1562, began the bloody perfecution of the Proteftants.

Attencourt, a village about two *French* miles diftant from *Vaffy*, noted for its mineral fpring.

Joinville, a town feated on the *Marne*, at the foot of a high. mountain, on which ftands a caftle, is the chief place of a principality, which comprehends eighty-two villages, and belongs at prefent to the duke of *Orleans*; being the refidence of a Diftrict-court, of a falt-office, of an election and a *marechauffée*, with a collegiate church in the caftle, in which are to be feen feveral fine monuments of the dukes of *Guife*, and divers other lords of the

place ;

place; has also a convent, and a manufactory of druggets. The neighbouring country is hilly, and produces great quantities of wine.

Rofnay, an ancient peerage.

Brienne; in Latin *Brena*, a little town, and one of the old earldoms and
peerages of *Champagne*; at present it belongs to the house of *Lomeny*. This
town consists of two places, *viz. Brienne la ville*, lying on the *Aube*, and
Brienne le chateau, standing about 1000 paces from the other.

Bar-fur-Aube, a very old town, having the title of an earldom; is a re-
ceipt, or place of payment of taxes, and the residence of a royal *prevoté*, a
chapter, and a particular governor.

Clairvaux, a celebrated abbey of regular *Ciſtercians*.

Vignory, a little town and an earldom, seated on the *Marne*.

Chateauvilain, a little town lying on the river *Aujon*, having a collegiate
church and a caſtle.

Grancey, seated on the *Once*, a little place, belonging to the election of
Bar on the *Aube*.

VIII. *Bafzigny* belongs partly to the dutchy of *Bar*, and partly to *Champagne*. In the latter lie

Langres, anciently *Andemandunum*, the capital of this country, being
seated on a hill on the borders of *Lorrain* and *Franche comté*. This town is
the chief place of an election and the residence of a bailiwick, of a provincial court, of a salt-office, and a *marechauſſee*. Its bishop is suffragan to
the archbishop of *Lyons*, is a duke and peer of *France*, and his Diocese extends beyond the *generalité* and government of *Champagne*, containing in it
seven hundred and sixty-four pariſhes, eleven chapters, twenty-eight abbies, and a great number of priories. His yearly revenue amounts to
36,000 livres, with a taxation of 9000 florins at the court of *Rome*. Excluſive of the cathedral, here are alſo three pariſh-churches, one large and
beautiful seminary, one college of *Jeſuits*, seven convents, and two hoſpitals. In this place are made several kinds of tools, and a great quantity
of knives. The city is very ancient, and derives its name from the *Lingones*.

Aigremont, a barony, belonging to the dukedom of *Langres*.

Bourbonne, a little town, well known for its mineral water. In the year
1719, this place was deſtroyed by fire.

Chaumont, a town standing on a hill, at the foot of which runs the
Marne, being the principal place of an election and the residence of a provincial court, a royal *caſtelany* or *prevoté*, with a large Diſtrict and foreſt-
court. Its collegiate church is the only one here; but in the town is alſo
an abbey, with a college of Jeſuits and one convent. This place belongs to
the royal demeſnes.

Vignory, ſmall, but an earldom, and ſtands on the *Marne*.

Val des ecoliers, a celebrated convent.

Montigny le Roi, a little town, seated on the *Meuse*.

4

Ville

Ville neuve au Roi, a fmall place.

Vaucouleurs, a town feated on the *Maefe,* having a royal *prevoté,* a col-
·legiate church and two convents; belonged formerly to *Lorrain.*

Dom Remy, furnamed *La pucelle,* the native place of the celebrated *Joan
d'Arc,* better known by the name of the *Maid of Orleans,* who performed
fuch fignal fervices to King *Charles* againft the *Englifh,* that for her fake the
prevoté of *Vaucouleurs* has been declared ever fince exempt from all taxes.
In this country ftands the limit-ftone which the Emperor *Henry* II. and
King *Robert* fet up as a mark to diftinguifh the boundaries of their States.

Honbervaux, is a feat and barony.

Drancey le Chatel, a little town and barony, having alfo a chapter.

IX. The *Senonois* contains the following places, *viz.*

Sens, the *Agendicum* of the ancients, a city feated on the *Yonne,* and the
chief place of an election. It contains a *prevoté,* a provincial Diftrict and
a foreft-court, together with a falt-office and a *marechauffée.* This place is
an archbifhopric, having the Diocefes of *Troyes,* *Auxerre,* and *Nevers*
dependent on it, and contains feven hundred and fixty-five parifh-churches,
fix chapters, twenty-nine abbies and fix convents, communities and colleges. The yearly revenue amounts to 50,000 livres, and its taxation at
the court of *Rome* is 6166 florins. The cathedral here is very large. In it
alfo are fixteen parifh-churches, five abbies, one college of Jefuits, one
feminary and nine convents. Among the councils held here, that of the
year 1140 is the moft famous.

Pont fur Yonne, a little town, having a royal *prevoté.*

Joigny, in·Latin *Joviniacum,* a fmall town feated on the *Yonne,* having
a *prevoté,* a Diftrict-court, a falt-office, a *marechauffée,* and an election. It
confers the title of earl, which belongs at prefent to the houfe of *Villeroi.*
In it are three ·parifh-churches. The neighbouring country grows a great
deal of wine ; and in it alfo are good corn and pafture lands.

Ville neuve l'Archeveque, a little town on the *Vanne.*

Champignelles, a fmall place.

Saint Florentin, a little town feated on the *Armanfon,* and the capital of
an election. It is alfo the refidence of a bailiwick and a falt-office, and is
the principal place of an earldom, which belongs at prefent to the marquis
of *Brilliere.* In the year 1722, this place fuffered confiderably by fire.

Tonnere, in Latin *Tornodorus,* a fmall town lying on the *Armanfon,* and
the principal place of an earldom. It is the refidence of an election, a bailiwick, a falt-office, an inferior foreft-court, and a *marechauffée.* Befides its
collegiate church, in it alfo are fome other churches, with one abbey, two
convents, and one hofpital. Its neighbourhood produces good wine.

Pontigny, a fmall place, having a celebrated abbey.

Chablis, a town, its neighbourhood noted for excellent wine, and near
which, in the year 841, a bloody battle was fought.

Ancy le Franc, a ſmall town, having a fine ſeat.

Bray-ſur-Seine, a little town and barony, containing a priory and chapter.

Nogent-ſur-Seine, the principal place of an election, containing a Diſtrict_ court, a ſalt-office, and a *marechauſſée.*

Brie Champenoiſe forms a part of the province of *Brie,* in Latin *Pagus Brigenſis,* being formerly a large foreſt, part of which belongs to the government of the *iſle de France.* This part called *Brie Champenoiſe,* contains the following places, *viz.*

Meaux, in Latin *Meldi,* the capital, lying on the river *Marne,* by which it is divided into two parts. It is the principal town of an election, containing a *prevoté,* a provincial and Diſtrict-court, a ſalt-office and a *marechauſſée.* It has alſo a general lieutenant and a particular governor. The biſhop of this place is ſuffragan to the archbiſhop of *Paris.* Its Dioceſe is divided by the river *Marne* into two large archdeaconries; to each of which belong three rural deaneries: And to both, united by two hundred and twenty-ſeven pariſhes, ſeven chapters, and nine abbies. The annual revenue of its biſhop amounts to 22,000 livres, his taxation at the court of *Rome* being 2000 florins. Excluſive of the cathedral, in it are one chapter, five abbies, ſeveral convents, one alms-houſe, and one hoſpital, which is general. This city was erected into an earldom by *Henry* II. Its traffic conſiſts principally in grain, wool, and cheeſe. In this town, in the time of *Francis* I. the reformation firſt ſhewed itſelf.

Germini, a noble ſeat belonging to the biſhops, and ſeated on the *Marne.*

Saint Fiacre, a celebrated church and priory of *Benedictines,* belonging to the congregation of *S. Maur,* much reſorted to by pilgrims.

Treſmes, a dukedom and peerage.

Coulomiers, a little town lying on the *Morin,* having a ſeat, which is ſaid to have coſt two millions of livres. It is an election town, and ſtands in a fruitful country.

Roſſoy, alſo an election town, and ſtanding in a fine corn-country.

La Fortelle, an elegant ſeat, near *Roſſoy.*

Provins, in Latin *Pruvinum,* a handſome town, ſeated on the little river *Vouſie,* is the reſidence of a provincial court, a *prevoté,* an election, a. Diſtrict and foreſt-court, a ſalt-office and a *marechauſſée.* It has alſo a particular governor of its own, with three chapters, two abbies, and one college.

Monterau Faut-Yonne, a little town, ſeated at the conflux of the *Seine* and the *Yonne,* from the latter of which it receives its name. It is an election town, and contains a collegiate church. On the bridge here, which runs over both rivers, *John,* duke of *Burgundy,* was murdered in the preſence of *Charles* VII.

Sezanne, an election town and the reſidence of a *prevoté,* has a royal juriſdiction, a ſalt-office, a Diſtrict and foreſt-court, a *marechauſſée,* and a particular governor. In it is a collegiate church, an abbey, a priory, a ſmall hofpital, and two convents. In the year 1423, it was plundered and burnt by the *Englifh*; but now gives the title of earl to the youngeft fon of *Harcourt Beuvron*.

Montmirail, a little town feated on a hill, near the larger *Morin*.

Chateau Thierry, a town lying on the *Marne*, is the capital of *Brie Pouilleufe*, the feat of an election, of a provincial Diftrict, and an inferior foreft-court, and alfo of a *prevoté*. In it are three parifh-churches, one royal abbey, three hofpitals, four convents, four chapels and a fmall college. At *Valfers*, half a league diftant from the town, is an abbey. The place, as a dukedom and peerage, belongs to the duke of *Bouillon*.

Obf. The government of *Sedan* is detached from the government of *Champagne*, and in civil caufes appertains to that of *Metz*; though it lies in the *generalité* of *Champagne*. The principal places in it are

Sedan, a ftrong town fituated on the *Maefe*, and lying on the frontiers of the dutchy of *Luxemburg*, containing in it a good caftle. It is alfo the capital of an election, the refidence of a governor, of a provincial and a foreft-court, with a royal jurifdiction and *prevoté*. It confifts of the upper and lower town, together with large fuburbs : To the upper town belongs the citadel. In it is one feminary, one college of Jefuits, and two convents. Cloth and ferge are its principal manufactures. *Lewis* XIV. made a forced exchange of the dukedoms of *Albret* and *Chateau Thierry*, and the earldom of *Evereux*, with *Maurice* duke of *Bouillon*, for this town. Before the revocation of the edict of *Nantes*, in this place was an academy for the Proteftants, of very great reputation.

Monfon, in Latin *Mofomagus*, a little town feated on the *Maefe*, has a prevoté, a Diftrict-court and a *Recette*. In the year 1379, *Charles* V. compelled the archbifhop of *Reims* to exchange this place for the *caftelany* of *Velly* in the *Soiffonnois*; and in the year 1671, it was difmantled. In it is an abbey with a magnificent church. Serge is its principal commodity. And the neighbouring country, befides excellent paftures, produces great quanties of corn.

Chateau Regnaud, a town and Diftrict, having the title of a fovereign principality, to which belong twenty-feven villages. *Lewis* XIII. received it in the year 1629 from the Princefs dowager of *Conti*, by way of exchange for *Pont-fur-Seine*. *Lewis* XIV. ordered its citadel to be razed.

The GOVERNMENT *of*

B U R G U N D *Y.*

THIS government contains the dutchy of *Burgundy*, *La Breſſe*, *Le Bugy*, and the Diſtrict of *Gex*. To the north it borders on *Champagne*, eaſtward on *Franche Comté*, to the ſouth on *Lyonnois*, and weſtward on the *Bourbonnois* and *Nivernois*. From weſt to eaſt it extends above thirty *French* leagues, and from ſouth to north above forty-five. It is very fertile in corn and fruits, producing, in particular, excellent wines, among which thoſe of *Nuis*, *Chambertin*, *Beze*, *Coulange*, *Chaſſagne*, *Beaune*, and *Volenai* are much admired. The rivers here are the *Seine*, which has been already ſpoken of in the introduction to *France*; the *Dehune*, which runs into the *Soane*; and the *Brebince*, or *Bourbince*, which iſſues out of the lake of *Longpendu*; with the *Armançon*, of which it is ſaid, that is a bad river, but contains good fiſh; and the *Ouche* and *Tille*, which joins the *Soane*. But the *Soane* comes from *Lorrain* and falls into the *Rhone* near *Lyon*.

Among the four mineral waters in this dutchy, thoſe of *Apoigny* near *Seignelay*, and *Premeau* near *Nuis*, are leſs celebrated than thoſe of *Bour-bon-lancy* and *Sainte Reine*. In the Diſtrict of *Breſſe*, which is called *Montagne* and *Revermont*, are ſubterraneous lakes. The cave of *Arcy*, near *Avalon*, and the ſalt-ſpring of *Vezelay* are alſo worthy of notice. Near *Pourrain*, three *French* leagues diſtant from *Auxerre*, is a beautiful oker, much uſed in dying; and *Paily*, in the Diſtrict of *La Montagne*, produces tobacco.

The name of *Burgundy* is derived from the *Burgundians*, who, towards the beginning of the fifth century, ſettled in *Switzerland* and part of *Franche Comté*, whence ſpreading themſelves towards the rivers *Rhone* and *Soane*, they erected a kingdom of their own, which was gradually reduced by the Kings of the *Franks*. In after ages *Burgundy*, conſidered with reſpect to Mount *Jura*, now called *Mont S. Claud*, was divided into *Tranſ-Jurana*, or *Upper Burgundy*, and *Ciſ-Jurana*, or *Lower Burgundy*. The latter, at preſent ſtiled the dutchy of *Burgundy*, continued under the Kings of *France*, who governed it by dukes. But the power of theſe dukes grew to ſuch a height, that one of them named *Rodolphus*, in the time of *Charles the Simple*, was elected King of *France*: Hereupon the dutchy of *Burgundy* was given to *Hugh the Great*, duke of *France*, who proved a troubleſome neighbour to *Rudolpi*, and his ſon *Hugh Capet* ſeated himſelf and family on the throne of *France*. His ſon and ſucceſſor, *Robert*, was the founder of the firſt ducal houſe of *Burgundy*, this dutchy being conferred upon his ſecond ſon of the ſame name, with the title of the firſt duke and peer of

France.

France. This race becoming extinct, King *John,* being fon to *Johanna* a Princefs of *Burgundy,* in the year 1361, united this dutchy with the crown. In 1363, he conferred it on his fon *Philip the Bold,* in whom the fecond ducal line commenced. His great grandfon *Charles the Bold* having loft his life before *Nancy* in 1477, though *Mary* his daughter, who married *Maximilian* archduke of *Auftria,* was ftill living, and likewife *John* Prince of *Burgundy,* count of *Nevers* and *Retel,* who did not die till 1491, yet King *Lewis* XI. feized upon the dutchy and united it to his crown, which retains it to this day, notwithftanding the repeated claims and endeavours of the houfe of *Auftria* to recover it.

In this government are four bifhopricks. The diftricts and provincial courts here are fubordinate to the parliament of *Dijon,* which was erected in the year 1476. The provincial court of *Maçon* and *Auxerre,* and the Diftrict of *Bar-fur-Seine* excepted, which are fubject to the parliament of *Paris.* At *Dijon* is alfo a chamber of accounts. The ftates of the country, which confift of the reprefentatives of the clergy, nobility, and commons, meet regularly every three years by writ from the King, in order to raife the fums required of them.

The governor refides at *Dijon,* and fubordinate to him are fix general lieutenants, the firft of whom prefides over the Diftricts of *Dijon, Chatillon,* and *Bar-fur-Seine.* The fecond over the provincial court of *Chalon* ; the third over the provincial court of *Maçon* ; the fourth over *Autonois,* to which belong the tribunals of *Auxerre, Autun, Semur, Auxois,* and the county of *Charolles* ; the fifth over *Charollois* ; and the fixth over *Breffe, Beugy, Valromey,* and the country of *Gex.* It has alfo fix deputy governors : The firft for *Dijon* ; the fecond for *Chalon* ; the third for *Maçon* ; the fourth for *Autun,* the fifth for *Charolles,* and the fixth for *Breffe.* I now proceed to give a particular account of the Diftricts and territoies which compofe this government.

1. *Le Dijonois,* anciently *Pagus Ofcarenfis,* fo called from the river *Ouche,* in Latin *Ofcara,* receives its name from the capital, and abounds in wine, paftures, and woods, in the laft of which are feveral iron-forges. The places worthy of notice in it are as follow, *viz.*

Dijon, in Latin *Divio,* the capital of the country and government of *Burgundy,* is the refidence of the governor, the parliament, an intendancy, a *recette,* a taillage-office, a provincial, fupreme, and particular court ; a mintage, a falt-office, a *marechauffée,* a marble-table, a conful's court, a mayoralty, a *vicomté,* a tax-office, and other inferior courts. It is a pretty large place, the ftreets too in it are well paved, broad and ftraight, the houfes handfome, and its churches and fquares beautiful. It is alfo defended with good walls, wide moats and twelve baftions, being ftrengthened alfo with a citadel. The neighbouring country is fruitful, pleafant, and watered by the rivers *Sufon* and *Ouche.* The firft of thefe is but a

rivulet,,

rivulet, which, after running through the city moat, paſſes through the city itſelf, where it falls into the *Dijon,* which waſhes a ſuburb and a baſtion. The biſhop of this place is ſuffragan to the archbiſhop of *Lyon,* enjoys a revenue of 18,000 livres a year, out of which his taxation at the court of *Rome* is 1233 florins. In this city are ſeven pariſh-churches, four abbies, three large hoſpitals, or alms-houſes, ſeveral convents, particularly a fine *chartreuſe,* ſituate at the end of the ſuburbs of *Ouche,* in the church of which lie the laſt dukes of *Burgundy* with their ladies and children, alſo a magnificent houſe of Jeſuits, and a chapel founded in the year 1172, in which is kept a ſuppoſed miraculous hoſt. The academy of ſciences here was inſtituted by *Hector Bernard Pouſſin,* a ſenior of the parliament; and in 1723, a college of law was alſo erected here. The walks before the city are a quarter of a league in length, planted with three rows of linden-trees, and terminated by a delightful grove.

Fontaine les Dijon, a village, one *French* league diſtant from *Dijon,* is noted for being the native place of *S. Bernard;* and the ſpot on which his houſe ſtood is now converted into a monaſtery of bare-footed monks.

Selongey, a little town lying in a plain.

Saux le Duc, a ſmall place, but having a royal *caſtelany* and a ſalt-office, ſtands on an eminence in a woody country.

Auxonne, or *Auſſonne,* the capital of the earldom of this name, lies on the *Soane,* over which it has a fine bridge betwixt the two *Burgundies.* This town is fortified, has an old citadel, and is the reſidence of a *recette;* a ſalt-office, a Diſtrict-court, and particular governor. In it is one pariſh-church, three convents, and one hoſpital.

Seure, or *Bellegarde,* a town lying on the *Soane,* being the ſecond of the earldom and the Diſtrict of *Auxonne,* has a particular governor, a ſalt-office and a mayoralty. At preſent it belongs, as a marquiſate, to the houſe of *Bourbon Condé,* and was once fortified.

S. Jean de Laone, Lone, or *Laune,* a little town ſeated on the *Soane,* is the reſidence of a Diſtrict-court, of a ſalt-office, of a general *recette* for *Chalon,* and a mayoralty. In it are one college, two convents, and one hoſpital. The ſtrength of its ſituation enabled it to hold out a vigorous ſiege in the year 1636, in acknowledgment of which fidelity *Lewis* XIII exempted it for ever from the poll-tax, and empowered it to poſſeſs ennobling eſtates.

Beaune, in Latin *Belna,* a fortified town on the river *Bugeoiſe,* in the province of *Beaunois,* has a Diſtrict-court, a *recette* and a ſalt-office. In it are one collegiate and two pariſh-churches, two hoſpitals, one *chartreuſe,* one abbey, two convents, and a fine college. The *Vin de Beaune* is known all over *Europe.*

Nuits, a little town ſeated on the declivity of a mountain, lies alſo in *Beaunois;* being the reſidence of a Diſtrict-court, a *prevoté,* a ſalt-office, and

and a particular governor. Befides its collegiate church, in it is a parifh-church, with an hofpital and two convents.

Premaux, a little place, but its neighbourhood noted for excellent wine, as alfo for a lepid mineral water, which is infipid.

Cifteaux, or *Citeaux*, a little place, containing the principal abbey of the whole *Ciftercian* order, which holds immediately of the Pope. This place lies alfo in the *Beaunois*.

Fontaine Françoife, a little town, having a priory. Near this place, in the year 1595, Henry IV. defeated the *Spaniards*.

2. *La Montagne*, a territory, having a court of juftice; and fo called from its numerous mountains, contains the following places, *viz.*

Chatillon-fur-Seine, the capital of this territory, and a *recette*; being fmall, but the refidence of a provincial Diftrict and foreft-court, a mayor-alty, a royal *prevoté*, and ducal court for the jurifdiction of the bifhop of *Langres*, a *marechauffée*, a falt-office, and a particular governor. In it is a collegiate and a parifh-church, a fmall college, two abbies, two hofpitals, five convents, and a commanderie belonging to the knights of *Malta*. It is divided by the *Seine* into two parts; and in its neighbourhood are great numbers of iron-mills.

Bar-fur-Seine, the principal place of an earldom, and the refidence of a Diftrict-court, a *prevoté*, an election, a foreft-court, a falt-office, and of a particular governor; contains only one parifh-church, one chapter, with a fmall college, one hofpital, and three convents.

Aifay le Duc, a little place and *caftelany*.

Arc en Barois, a town on the little river of *Saugon*, raifed to a dukedom and peerage in the year 1703; contains a mayoralty, a falt-office, and a ducal manor-court.

Val de chaux, in Latin *Vallis caulium*, a convent, the prior of which is general to an order.

Duefme, a very ancient borough, feated in the little Diftrict of *Duef-mois*, in Latin called *Pagus Dufmifus*, to which belong feveral towns and villages.

S. Seine, an ancient abbey of *Benedictines*, belonging to the congregation of *S. Maur*, which gives name to a little place.

Chanceaux, a town ftanding on a hill, in which the *Seine* has its fource:

3. *L'Auxerrois*, an earldom with a court of juftice, extending nine leagues from north to fouth, and five from eaft to weft, abounds in vine-yards. In it are

Auxerre, in Latin *Autiffiodorum*, a town, part of which ftands on a hill near the *Yonne*, by means of which it very conveniently carries on a good trade. This place is the capital of a *recette*, has alfo a provincial court, a chamber of tythes, a royal *prevoté*, a mayoralty, a foreft-court, *&c.* toge-ther with a *marechauffée* and a falt-office. The bifhop hereof is fuffragan to the archbifhop of *Sens*, has a diocefe containing 238 parifhes, with a yearly revenue of 35,000 livres, out of which he is taxed in 4400 florins at the court of *Rome*. His palace is a moft grand ftructure. In this town too are a collegiate, eight parifh-churches, five abbies, one feminary, one college of Jefuits, fix convents, one commanderie of *Malta* and two hofpitals.

Saignelay, a little town on the *Serin*, fituate near the *Yonne*, has an old caftle and a falt-office. In it the great *Colbert* fet up two manufactures, and afterwards purchafing the place procured it to be made a marquifate.

Apoigny, a place fituate near the former, having a cold mineral water of a ferruginous quality.

Crevant, a little town on the *Yonne*, over which it has a bridge of ftone. It carries on a thriving trade in wine and other goods, for which the river is a great conveniency. In it is a falt-office, one parifh-church and one convent. In the year 1423, a battle was fought near this place.

Coulange La Vineufe, a little town, feated on the *Yonne*, and fo called from its wine which is held in great repute.

Coulange-fur-Yonne, a little town and *prevoté*, five *French* leagues diftant from the former.

4. *L'Auxois*, in Latin *Alefienfis pagus*, contains alfo a part of *Duefmois*, and was formerly an earldom. The moft remarkable places in it are

Semur furnamed *En Auxois*, a town ftanding on a high rock near the river *Armançon*, being the chief town of the diftrict and a *recette*, as alfo the refidence of an united provincial and diftrict court, of a *marechauffée* and a falt-office. It is divided into three different parts, which are walled, exclufive of fix large fuburbs. One of thefe fuburbs, which is the handfomeft and moft populous, is called *Le Bourg*; the fecond *Le Donjon*, and the third *Le Chateau*. In it is a chapter, two priories, fix convents, one college and an hofpital. Its principal trade confifts in woollen-cloth of its own manufacture.

Sainte Reine, or *Alife*, anciently *Alexia*, a town giving name to the territory. In this place are two fprings, which it is pretended are medicinal, but their virtue confifts only, or at leaft chiefly, in the artifice of the *Francifcans*, who, for their own advantage, extol the one and decry the other.

Tanlay is a little town and marquifate feated on the river *Armançon*, and having a beautiful feat.

Montbard, a little town, feated on the river *Braine*, one part of which, together with an ancient caftle, lies on a mountain, and the other by a river fide. In it is a royal *caftelany* and a falt-office.

Noyers, a little town wholly environed by the river *Serin*, having a mayoralty, a diftrict court and a falt-office, with one college, two fmall hofpitals and town convents.

Avalon,

Avalon, a little town feated on the river *Coufin,* is the principal place of a *recette* and of the little diftrict of *Avalonnois,* having alfo a particular governor, a mayoralty, which prefides over the police, a revenue-office, a provincial and foreft-court, and a falt-office. In it are one collegiate and two parifh-churches, five convents, one college and one hofpital. Befides its natural ftrength it is defended by a good citadel.

Flavigny, a little town, feated on a hill, having three churches and an abbey with a particular governor.

Arnay le Duc, a town ftanding nearly in the centre of *Burgundy,* is the capital of a *recette* and the refidence of a royal court of juftice, a diftrict-court and a falt-office. In it are one parifh-church, one priory, two convents, one college of Jefuits and one hofpital. The count of *Armagnac* is both baron and proprietor of this place.

Saulieu, in Latin *Sidoleucum,* or *Sedelaucum,* a little town, feated on a hill, containing three fuburbs which are more populous than the town itfelf. In it is a diftrict-court, with a mayoralty, a confular-court and a falt-office. It contains one collegiate church, one parifh, one college and a few convents.

5. *L' Autunois* contains the following places, *viz.*

Autun, the *Auguftodunum* and *Civitas Æduorum* of the ancients, ftanding on an eminence, near three hills, on the river *Aroux.* It confifts of the upper city, which is commanded by Mount *Cenis.* The citadel and the lower city, which is called *Marchaud,* is the refidence of a chamber of tythes, a *recette* and an upper and lower jurifdiction, together with a mayoralty, a provincial tribunal, a foreft-court, a falt-office, a *marechauffée,* &c. The bifhop of this place is fuffragan to the archbifhop of *Lyon,* but prefident of the affembly of the ftates of *Burgundy,* having a diocefe of 611 parifhes and fourteen abbies under him. His yearly revenue amounts to 17,000 livres, and his taxation at the court of *Rome* is 4080 florins. In this town is one cathedral, one collegiate, twelve parifh-churches, five abbies, two feminaries, two priories, one college of Jefuits, fix convents and two hofpitals. The infide of the city affords feveral antiquities, but more particularly the outfide ; fuch as the remains of three temples, of an amphitheatre, *&c.*

Beuveray, a little town, feated at the foot of a mountain, by fome thought to be the ancient *Bibracte.*

Bourbon l' Ancy, in Latin *Burbo Ancelli,* a little town, feated on a hill, and confifting of three parts, *viz.* the town properly fo called, together with the citadel, which ftands on a rock, a fuburb, and another fuburb called St. *Leger.* In it is a royal court of juftice, a falt-office, a chapter, three parifh-churches, two convents and two hofpitals. The fuburb of St. *Leger,* is well known for its faline, ferrugineous and fulphureous warm baths, in which the hand cannot be held much longer than in boiling water, though, notwithftanding, they burn not. Monf. *Pinot,* the fuper-intendent of this place, who, in the year 1752, publifhed an account of them, has obſerved in theſe baths the ſame phenomenon as is to be ſeen in others ; namely, that hot as they are, fire does not cauſe an ebullition in them ſooner than in common cold water. Near theſe baths are ſtill ſome remains of that magnificence which the *Romans* affected in ſtructures of this kind.

Mont Cenis, a little town, or rather borough, ſeated on an eminence betwixt two mountains, having a diſtrict-court, a *recette,* a royal *caſtelany,* a mayoralty, a ſalt-office, a pariſh-church and a convent. This place gives the title of a barony.

6. *Briennois,* a ſmall territory on the *Loire,* ſo called from the long-ſince-ruined town of *Brienne,* whence, at preſent, the only remarkable place in it is

Semur en Briennois, in Latin *Sine Murus,* a little town, ſtanding about half a *French* league from the *Loire,* yet the capital of a *recette* and the reſidence of a diſtrict-court, a particular governor, a ſalt-office, a royal *caſtelany,* a mayoralty and an inferior foreſt-court. Its pariſh-church is alſo a collegiate church.

7. The county of *Charollois,* in Latin *Pagus Quadrigellenſis,* or *Quadrelenſis,* is a fief of *Burgundy,* which, after many viciſſitudes, was ceded by *France* to *Spain,* at the treaty of the *Pyrenees,* in the year 1659.; but, by *Philip* IV. transferred to *Lewis* of *Bourbon,* Prince of *Condé.* It has its particular ſtates and contains the following places, *viz.*

Charolles, the capital, though but a ſmall place, ſituate on two little rivers. In it, however, is a diſtrict-court, a *recette,* a *prevoté,* a ſalt-office, a collegiate church, a priory, three convents and an hoſpital. It contains alſo an ancient caſtle.

Paray le Moinial, a little town, ſeated on the *Brebinche,* having a ſalt-office, a priory, a few convents and a college of Jeſuits.

Toulon, a ſmall place, ſtanding on the river *Arroux,* having a ſalt-office and a priory.

St. *Vincent,* a ſmall place.

8. *Le Chalonois,* which once had particular counts of its own, is a delightful plain, divided by the *Saone* into two almoſt equal parts: that on the right is called *Montagne,* from the mountain of *Beaune,* which extends beyond *Maçon* ; and that on the left is called *la Breſſe.* In it are

Chalon, in Latin *Cabillonum,* a city, lying on the *Saone,* with a citadel, being the capital of the country and the reſidence of a provincial court, a *recette,* a chancery, a *caſtelany,* a foreſt-court, a ſalt-office, a particular governor, and an epiſcopal-court. The biſhop of this place, who is ſuffragan to the archbiſhop of *Lyon,* enjoys a yearly revenue of 14,000 livres, his taxation at the court of *Rome* being 700 florins. Its dioceſe comprehends 186 pariſhes, or, according to others, 207. In it, beſides the cathedral, are ſeveral pariſh-churches with a few abbies and priories, ſix convents, one college of

Jeſuits

Jefuits and two hofpitals. On an ifland in the *Soane* ftands the little town of St. *Laurent*, which ferves as a fuburb to this city.

Verdun, a town, fituate at the junction of the *Saone* and *Doux*, containing the fuburb of St. *Jean*, which is both larger and handfomer than the town itfelf. This place is an earldom and carries on a good trade.

Louhans, a little town, fituate in *La Breffe Chalonoife*, being wholly environed by feveral fmall rivers, contains a falt-office, one parifh-church, one college, one convent and one hofpital, and is a ftaple for goods going from *Lyon* to *Switzerland* and *Germany*.

La Ferte fur Grofne is a borough, having an abbey.

Obf. *Seure*, or *Bellegarde*, mentioned above under *Dijonois*, is by fome placed in *Chalonois*.

9. *Le Maçonois*, or *Mafconois*, formerly enjoyed counts of its own and ftill retains its ftates. The moft remarkable places in it are as follow, *viz.*

Maçon, or *Mafcon*, in Latin *Matifco*, the capital, which ftands on an eminence near the *Soane*, having a diftrict and provincial court, a *marechauffee*, a particular and a deputy-governor. The bifhop of this place is fubject to the archbifhop of *Lyon*, his yearly revenue being 17,000 livres, and his taxation at the court of *Rome* 1000 florins. In the diocefe are 200, or, according to others, 268 parifhes. Exclufive of the cathedral, it contains one collegiate church, nine convents, one college of Jefuits and one hofpital.

Saint Gengoux le royal, a little town feated betwixt three hills on the river *Grone*, having a *caftelany*, a mayoralty, a falt-office, one parifh-church and one convent. Its neighbourhood produces the beft wine of all the *Maçonnois*.

Saint Gengoux des Seiffey, a fmall place, fituate among high mountains.

Tournus, in Latin *Tornucium*, a little town, but of great antiquity, lying on the *Soane*, contains two parifh-churches, one hofpital, one college and one abbey of *Benedictines*, which was once a caftle, but, in the year 1623, was fecularifed and converted into a collegiate church.

Cluny, a little town, feated in a valley on the *Grone*, containing a falt-office, a celebrated abbey of *Benedictines*, three parifh-churches, one convent and one hofpital.

Marcigny, fmall, but having a priory and a falt-office, ftands near the *Loire*.

10. *La Breffe*, an appellation it receives from a foreft called *Brexia*, was for fome time an earldom fubject to the dukes of *Savoy*, who, by the compact of *Lyon*, in the year 1601, ceded it to *France*, in lieu of the marquifate of *Saluzzo*. The nobility meet every three years on their public affairs. It contains the following places, *viz.*

Bourg, the capital, lying on the river *Reffouffe*, being the refidence of a governor a deputy-governor, a mayor, a provincial an inferior foreft-court, a *recette,* a *caflelany,* a falt-office, a *marcchauffée, &c.* In it is a collegiate and parifh-church, with a college of Jefuits and feven convents.

Beauge, or *Bauge,* a town, feated on an eminence, once the capital of *Breffe* and now a marquifate.

Coligny, a town and county giving name to the noble family of *Coligny-Chatillon.*

Villars, a little town on the *Chalaronne,* having the title of a marquifate.

Chatillon les Dombes, a town, feated on the *Chalaronne,* contains one parifh-church with a fmall chapter, a college, an hofpital, two convents and a falt-office.

Montuel, in Latin *Mons Lupelli,* a little town, feated on the river *Sereine,* being the capital of the diftrict of *Valbonne,* contains one collegiate and two other parifh-churches, with two convents, one college, one hofpital, a mayoralty and a falt-office.

Pont d'Ayn, a little town and marquifate feated on the *Ain.*

Loye, a town and barony on the fame river.

Pont de Vaux, a little town and dukedom, feated on the river *Refouffe,* having a falt-office, and a parifh-church, with a chapter, a college and two convents.

Pont de Vefle, a little town fo named from a bridge built at this place over the *Vefle,* and giving the title of earl. It has a particular governor, with a falt-office, one parifh-church, one hofpital and one college.

Montrevel, a fmall place, having the title of an earldom.

11. *Le Bugey,* from *Pont d'Ayn* to *Seiffel,* is about fixteen *French* leagues in extent, and from *Dortans* to *Port de Loyette* ten. It not only comprehends *Bugey,* properly fo called; but likewife *Valromey* and *La Michaille,* being anciently a part of the Kingdom of *Burgundy,* which the counts of *Savoy* gradually maftered, and kept till the year 1601, when, by the treaty of *Lyon* it was given up to *France,* that part of it excepted which lies on the other fide of the *Rhone* and conftitutes the country of *S. Cenis, Jenne* and *Loyffey,* or *Lucey.* Exclufive of four diftricts, which I fhall mention in the fequel, it has alfo its particular provincial ftates and affemblies of the towns and the nobility, and contains the following places, *viz.*

Belley, in Latin *Belica,* the capital of the country, feated among hills and fmall eminences, and the refidence of a governor, an election, a royal court fubject to the provincial tribunal of *Bourg* in *Breffe,* a marechauffée and a falt-office. The bifhop of this place is fuffragan to the bifhop of *Befançon* and ftyles himfelf a Prince of the *Roman* empire, having under him a diocefe of 221 parifhes, together with a revenue of 10,000 livres a year. His taxation at the court of *Rome* is 330 florins. In it is only one parifh-church, exclufive of the cathedral, together with four convents and one abbey.

Nantua, a little town, fituate among high mountains at the end of a lake, having in it an abbey of *Benedictines* belonging to the congregation of *Cluny*, one convent, one collegiate, one parifh-church and a college.

Seiffel, a little town, feated on the *Rhone*, being the principal place in *Valromay* and the refidence of a royal *caftelany* and a falt-office. It contains only one parifh-church and four convents. The *Rhone* divides it into two parts, and as that river here becomes navigable, furnifhes it with the conveniency of carrying on a large trade in falt to *Geneva*, *Switzerland* and *Savoy*.

S. Rambert de Joux, in Latin *Jurenfis*, fituate near an arm of Mount *Jura*, or *Joux*, is fmall, and ftands in a bottom betwixt two high mountains, owing its name and origin to the *Benedictine* abbey here of *S. Rambert*. Befides this abbey it contains alfo a parifh-church, a college and a fmall hofpital. The jurifdiction and lordfhip of this town is divided betwixt the abbot and the duke of *Savoy*, that part belonging to the latter, though a *French* fief, being raifed to a marquifate by Duke *Philibert*. The police here is adminiftered by the King's officers, and the feveral courts of the town are fubject to the parliament of *Dijon*.

Ambronay, an abbey of *Benedictines* which holds immediately of the fee of *Rome*.

S. Sorlin, a borough and marquifate belonging to the duke of *Savoy*.

Langnieu, a little town, feated on the *Rhone*, having a falt-office, but belonging to the above marquifate of *S. Sorlin*.

Pontayn and *Cerdon*, two baronies belonging alfo to the duke of *Savoy*.

Chatillon, a borough fituate in the diftrict of *Michaille*.

12. The country of *Gex* extends from *Fort d' Eclufe* to the village of *Croffay*, which is near feven leagues ; and from the town of *Gex* to *Geneva*, about three. Weftward it is bounded by Mount *Jura*, which, though it makes a very barren appearance, contains fine paftures on its fummit. The graziers of the neighbouring country truft the herdfmen, who live on this mountain with fome thoufands of heads of cattle every year, who take good care of them and return them with the moft punctual fidelity. The traffic of this country confifts in cheefe, wine and coal. The *Rhone* runs through it. It is watered alfo by the *Verfoye*, which runs into the lake of *Geneva*; and two other rivulets, which fall into the *Rhone*. By the treaty of *Lyon*, in the year 1601, the duke of *Savoy* ceded this country to *France*, and it belongs, at prefent, to the Prince of *Condé*. The *Genevans* have alfo feveral villages in it. The moft remarkable places here are

Gex, the capital, being a little place feated near the lake of *Geneva* at the foot of Mount *S. Claude*. It is divided into three parts and is the feat of a particular governor, a diftrict-court, a *caftelany*, a *marechauffée*, a mayoralty and a falt-office. It has only one parifh-church, but contains five convents, one hofpital and one fmall college.

Verfoy, a little place and marquifate lying on the lake of *Geneva*.

Le

Le fort de l' Eclufe commands the entrance into *Bugey* and *Breſſe*, being
hewn out of a rock, and lies on the *Rhone*.

Next to theſe are the priories of *Aſſerois, Dironne, Prevoiſin* and *S. Jean*.

The SOVEREIGN PRINCIPALITY *of*
D   O   M   B   E   S.

THE ſovereign principality of *Dombes*, though independent of the
government of *Burgundy*, yet lying within its circuit, comes in moſt
properly here. This country borders, eaſtward, on *Breſſe* ; northward, on
Maçonnois, and weſtward on *Beaujolois*. It is nine *French* leagues in length
and nearly as many broad, being very fruitful and lying along the *Soane*.
Formerly it made a part of the Kingdom of *Burgundy*, but, about the cloſe
of the tenth century, or towards the beginning of the eleventh, detatched
itſelf, and became an independent lordſhip, which was governed ſucceſſively
by the houſes of *Beaujé, Beaujeu*, the counts of *Forêts* and the dukes of
Bourbon ; and, laſtly, after the death of *Mary de Bourbon Montpenſier*,
dutcheſs of *Orleans*, deſcended to her daughter *Anna Maria Louiſa* of
Orleans, who, in the year 1681, left it to the duke *de Maine*. This
country was, by *Lewis* XIV. declared an abſolute ſovereign principality. The Prince *de Dombes* has power of life and death here, can
confer nobility, coin money and impoſe what taxes he pleaſes on his
ſubjects ; he alſo ſtyles himſelf, *by the grace of God, ſovereign Prince of
Dombes*, and holds his own parliament. His certain revenue is about
150,000 livres. The country is governed, in the Prince's name, by a
governor-general, contains about 230 places and is divided into twelve
caſtelanies, viz.

1. The *caſtelany de Trevoux*, the principal place of which is

Trevoux, anciently *Tivurtium*, being the capital of the principality and
the reſidence of a governor, a parliament, a diſtrict, a mintage and a
caſtelany. It lies on a hill on the *Saone* and is but a ſmall place, but was
formerly more conſiderable, having ſuffered much in the wars betwixt
the dukes of *Bourbon* and *Savoy*, particularly in the year 1431. In it is
one chapter, three convents and an hoſpital. The well-known *Memoires
de Trevoux* were from the beginning, that is, from the year 1701, not
written here, but by the Jeſuits of *Paris*, by the duke of *Maine*'s direction,
though conveyed hither to be printed.

2. The *caſtelany* of *Toiſſey*, in which are the following places, *viz.*

Toiſſey, a little town, ſituate in a pleaſant country near the rivers *Chalaronne*
and *Saone*, having one pariſh-church, one convent and a fine college.

Le

Le Port de Toiſſey, a little place about a quarter of a *French* league diſtant from *Toiſſey,* where the *Chalaronne* falls into the *Soane.* In this place the inhabitants of *Toiſſey* croſs the *Soane* and embark the goods in which they trade on that river. The old *Port de Toiſſey* is an inconſiderable place, and the only one in this principality on the other ſide of the *Soane.*

Garneraus, an earldom.

Mognenigns, a little borough ſituate near the *Soane,* and having a ſeat.

S. Didier, a borough, and one of the largeſt pariſhes of the countiy.

3. The *caſtelany* of *Montmerle,* to which belongs

Montmerle, a borough ſituate on the *Soane,* having a convent.

Amareins, a little place, having a ſeat.

Batie, an earldom.

Guerreins, a borough.

Lurcy, ſmall, but ſtiled a barony.

4. The *caſtelany* of *Beauregard,* containing

Beauregard, a little place ſituate on the *Soane,* anciently the capital of the country, and the reſidence of the parliament, having alſo a ſtrong caſtle, but has never been able thoroughly to recover the devaſtations committed there in the year 1377, by the *Savoyards.*

Flechere, a barony.

5. The *caſtelany* of *Villeneuve,* in which are

Villeneuve, a little town, having once a caſtle.

Agnereins, a ſmall place, yet once the reſidence of a *caſtelany.*

6. The *caſtelany* of *Ligneu,* being the place of that name.

7. The *caſtelany* of *Amberieu,* containing

Amberieu, a borough, and the ſeat of a *caſtelany.*

Mont Bertoud, a deanery.

Montlieu, a little place, having a ſeat.

S. Olive, another ſmall place, bearing the title of a barony.

8. The *caſtelany* of *S. Trivier,* conſiſting of the little town of the ſame name, with the title of a barony, and lying betwixt three ſmall woods and a lake. The pariſh-church here has a priory.

9. The *caſtelany* of *Chalamont,* in which are

Chalamont, a little town, ſeated on a hill near two lakes, and having formerly a caſtle.

Montfavrey, a priory.

10. The *caſtelany* of *Lent,* which has its reſidence in the little town of that name, lying on the *Veille.*

11. The *caſtelany* of *Chatelar,* containing

Chatelar, a village, and the reſidence of the *caſtelany.* This place was formerly a town, but was ruined in the *Savoy* wars.

Marlieu, a little town, ſituate betwixt two lakes on the river *Renon.*

Ville,

Ville, a fmall feat near the foregoing town.

Montrofar, a little place, having a feat.

12. The *caftelany* of *Baneins*, but not including the caftle of *Baneins*, which lies in *Breffe*.

The GOVERNMENT *of*

D A U P H I N E'.

THE province of *Dauphiné*, or the *Delphinat*, is bounded towards the north by the country of *Breffe* and the river *Rhone*; to the eaft is feparated from *Piedmont* by the *Alps* and *Savoy*; to the fouth borders on *Provence*, and weftward again on the *Rhone*. *Lower Dauphiné* produces grain, wine, olives, filk, hemp, falt, wood, vitriol, varnifh, cryftal, iron, copper and lead; but *Upper Dauphiné*, and above two thirds of the whole province, are fo barren, being very mountainous, that great numbers of the natives feck a fubfiftence in other parts; yet are thefe mountains not without ufeful produéts, containing, in particular, many curiofities. In thofe of *Ambrun* and *Die* are found *Marcafites*. Mount *Brefier*, not far from the little town of *S. Genis*, has been known to ejeét fire. The golden mountain, as it is called, yields a fpecies of diamonds. Thofe betwixt *Briançon*, *Pragelas*, and *Pignerol* are covered with larch-trees, on the rind of which is gathered a manna, of which more fhall be faid in the fequel; benjamin, of a delightful fragrancy; and agaric, which is ufed in phyfic and fcarlet dyes. Thefe mountains are alfo haunted by beafts not feen in any other part of *France*, as the *bouquettin*, or *chevrel*, of which I fhall give a more particular account when I come to the vallies of *Piedmont*, in *Italy*; the chamois, bear, and marmot, a defcription of which I alfo referve for the fame part. In it alfo are great numbers of white hares, white partridges, pheafants, eagles, hawks, &c. Its principal rivers are the *Rhone*, defcribed in the introduétion; the *Durance*, which iffues from Mount *Genevre*, being navigable at *Cavaillon*, but rapid, and often occafions great inundations. This river joins the *Rhone*. The *Ifere*, the fource of which is in Mount *Ifcran* in *Savoy*, and receives the *Acre*, the *Drai*, the *Vence*, and the *Gie* in its paffage, is fhallow at *Montmelian*, but at *Grenoble* becomes navigable for large veffels, and lofe itfelf at laft in the *Rhone*. The *Drome*, which rifes in the valley of that name near the village of *Baftie des Fonts*, is increafed by the *Meyroce*, the *Sure*, the *Roone*, the *Geroane*, the *Veoure*, and the *Befe*, and joins the *Rhone*.

Both

Both the ancient hiftorians and modern geographers give very pompous defcriptions of the wonders of *Dauphiné*, though fome make them feven, and others more; but it is only in four that they univerfally agree, and thefe are, 1. The burning-fpring, lying on a hill three *French* leagues from *Grenoble*, and half a league from *Vif*. It is a fmall brook iffuing from a foil emitting fmoak, even fmall flames being fometimes perceived in it, which communicated a heat to it; but for thefe two centuries paft this rivulet has altered its courfe to the diftance of twelve feet from that igneous foil, and is now of the coldnefs natural to water. 2. The Tower without venom, about a league from *Grenoble*, near *Seyffins* on the *Drac*, and called *Parifet*. It is faid that no venomous creature will ftay in it; but this is a notorious miftake, it being known to fwarm with adders and fpiders. 3. The inacceffible mountain, is a very fteep craggy rock, fituate on a very high mountain in the little diftrict of *Treves*, about two leagues from the town of *Die*. It is extremely difficult to climb up to the top of it, though this has been frequently done. 4. The caves of *Saffenage*, which are two excavated ftones, lying in a grotto above a village of that name, a league from *Grenoble*. The country people hereabouts relate that every year, on the fixth of *January*, they become full of water, and that the quantity of it in one of thefe ftones foretels good or bad vintage; and in another prognofticates the harveft. This ftory is of very ancient date; and for feveral centuries paft, has been kept up by the fallacious artifices of fome of the inhabitants, who themfelves fill the ftones with water. But a real curiofity here, is a water-fall in a grotto hard by thofe caves, where are ftill fhewn the chamber and table of the famous *Fee* or *Melufine*, from whom the ancient houfe of *Saffenage* claims its defcent. The other three fuppofed wonders are arbitrary; and the competition lies betwixt the following: 1. The eye-ftones of *Saffenage*, generally called the *precious ftones*, and by fome fuppofed to be the *Chelidonius lapis*; thefe are faid to be very efficacious againft the epilepfy, and therefore not to be confounded with the gems fo called. The latter, which are found among the gravel in the fprings mentioned N° 4, are very fmooth, and have a fine luftre, being like marble, exceeding foft to the touch. The beft are thofe of the bignefs and fhape of a lin-feed, which are tranfparent, and without angles: Being dropt into the eye they expel all heterogeneous impurities, which may have been bred therein, and their perfect fmoothnefs prevents them from doing the leaft damage to that orb. 2. The manna of *Briançon*, by the commonalty held to be a dew which hardens every morning on the larch-tree, but which is in fact no other than the fap of the tree exhaled by the heat. 3. The floating meadow here lies in a lake or pond in the diftrict of *Gap*, half a league from the town of *Gap*, at prefent called the *lake of Pelhotiers*. But this fuppofed meadow confifts merely of grafs and rufhes, cemented as it were together by flime and the froth of the water, which has gradually infinuated itfelf into it; by which means it floats up and down. 4. The grotto of our Lady *de la Beaume in Viennois*, in which formerly was a lake with fo violent a vortex in it that it almoft inftantaneoufly abforbed a large plank with feveral lighted torches fixed upon it, which was left behind when the curiofity of *Francis* I. prompted him to venture on the lake. This vortex has fince difappeared, and left only a fmall rivulet behind it, which is frequently dry. Some petrefactions are fome times found here. 5. *L'Oinorboe*, or ' the wine-fpring,' which has the colour and tafte of wine, is the common well of *S. Peter d'Argenfon*, a village in the diftrict of *Gap*. It is a mineral water which has been proved to be a good febrifuge, but the vinous tafte of it is wholly imaginary, though it is pretty well faturated with ferrugineous particles. 6. The rivulet near *Barberon*, in *La Valoire*, which is faid to foretel the fertility of the feafon by the quantity of its water: But this is an honour which may be attributed to many other ftreams, whofe courfe is periodical. More of thefe fuppofed miracles may be feen in *Lancelot*'s differtation in the *Hamburg* Magazine, vol. iii. p. 219.

Dauphiné, yields alfo another particular which no body has yet thought proper to range among the wonders of this country, though it merits it beyond any thing elfe: This is the waters of *La Mothe*, which are highly efteemed as a certain remedy againft all diforders of the ftomach, fluxes, and even lamenefs; being much hotter than the waters at *Aix* in *Savoy*, and fet even in competition with thofe of *Bourbon*. *La Mothe* lies betwixt *Treves* and *La Mathefine*, in the jurifdiction of *Greifivodan*, about five leagues from *Grenoble*, being a valley running betwixt two high mountains and enjoying no other profpect than that of bare and fteep rocks. The only dwellings here are wretched huts of ftraw, infomuch that the country is in every refpect difagreeable. The *Drac*, a very rapid river, proceeding from the high part of the diftrict of *Gap*, is as it were fqueezed in at *La Mothe*, betwixt two high rocks. On its fhore, at the foot of a very fteep rock is the mineral fpring, which, if the river rifes but half a foot, is covered with its turbid water, through the furface of which the fpring, however, forces its way unmixed. A little below, this river, which traverfes the whole diftrict of *La Mothe*, after receiving all the water in the valley, precipitates itfelf from the height of about thirty toifes into the *Drac*, to the great detriment of the roads thereabouts. To come at the fpring, a perfon muft clamber half a *French* league over fteep rocks and dreadful precipices which feem to threaten the paffenger with their fall; whence the approach to this fountain being fo difficult and dangerous, it is no wonder that thefe excellent waters of *La Mothe* fhould be fo little frequented.

Furthermore, thefe are not the only mineral waters in the country; for at *Pont de Baret*, not far from *Die*, betwixt *Creft* and *Montelimart*, are fprings of great efficacy againft tertian fevers. The fprings at *Bordoire*, at the foot of the mountain of *Diois*, thofe of *S. Chef*, near the abbey of *S. Antoine de Vien-*

de Viennois, thofe at *Cremieu, Lannay, Sanfon* and *Navoz*, are alfo mineral. On the borders alfo of the county of *Venaiffin*, at the foot of *Ventoux-hill*, is a fpring as cold as ice in the middle of fummer, even when the fun-beams play on it; which likewife gives rife to a river. Of the fame degree of coldnefs too is the fpring on Mount *Genevre*, in *Upper Dauphiné*, from whence iffue the *Durance* and the *Po*.

The province of *Dauphiné* confifts of tracts of land which were formerly feparate petty ftates, conftituting a part of the Kingdom of *Burgundy*. They fell afterwards under the Kings of *Arles*, till the counts of *Albon* in *Viennois* rofe to power and made themfelves mafters of the countries of *Graifivaudan, Ambrunois, Gapençois,* and *Briançonnois*. It was not till long after that the others were annexed to *Dauphiné*. Count *Guigues* IV. who lived towards the beginning of the eleventh century, was baptized by the name of *Dauphin*; and his fucceffors not only bore the fame name, but alfo gave it their country. *Guigues* IX. left one only daughter behind him, by name *Beatrix*, who being married to *Hugh* the Third, duke of *Burgundy*, brought him this country as a dowry. Their fon *Andrew*, with refpect to his grandfather by the mother's fide, affumed the title of *Dauphin* and *Guigues*: And his fon *Dauphin John*, who died in the year 1282, left his dominions to his fifter *Anne* wife to *Humbert* baron *de Tour du Pin*. *Humbert* III. their great grandfon, in confideration of 120,000 gold guilders left his dominions to *Philip* youngeft fon to *Philip de Valois*; the future Lords of the country being obliged to bear his name and arms. In the year 1349, this agreement fully took place; for the King having elected his grandfon *Charles*, eldeft fon to *John* duke of *Normandy*, Dauphin, *Humbert* refigned *Dauphiné*, embraced a monaftic life, and became afterwards patriarch of *Alexandria*, and adminiftrator of the archbifhopric of *Reims*. From the time of *Charles* V. the King's eldeft fon and prefumptive heir to the crown has been always ftiled *Dauphin*. The province of *Dauphiné* is rather a particular ftate than incorporated with the Kingdom; and the King himfelf in all inftruments relating to this province ftiles himfelf *Dauphiné le Viennois*.

In this province are two archbifhoprics and five bifhoprics. This country is one of thofe which are ftiled *Païs de droit ecrit*, that is, where the *Roman* law takes place. It has alfo particular cuftoms of its own, and alfo a parliament, which is likewife a court of taxes, one provincial court of juftice, feven prefecturates, three bailiwicks, four royal jurifdictions, and as many private lordfhips. The governor and general lieutenant of the province fit in the parliament, taking place above the chief prefident. The jurifdiction of the parliament is divided into two upper prefecturates and two bailiwicks. The upper prefecturate of *Viennois* contains the diftricts of *Vienne, Grenoble,* and *S. Marcellin*, with the royal jurifdiction of *Romans*; the prefecturate *des Montagnes* contains the diftricts of *Briançon, Ambrun, Gap* and *Buys*;

the

the diſtrict of *Die* depends immediately on the parliament. Of the two bailiffs, one preſides over the bailiwicks of *Valence, Creſſe,* and *Montelimart*; and the ſecond over the provincial court of *Valence.* With reſpect to its military government, excluſive of the governor and general lieutenant, here are alſo four deputy-governors, *viz.* one for *Grenoble* and *Briançon,* another for *Ambrun* and *Gap,* a third for *Vienne* and *S. Marcellin,* aud a fourth for *Valentinois, Diois, Tricaſtinois,* and the baronies. This government is divided into

I. *Upper Dauphiné,* which includes the following territories, namely,

1. *Graiſivaudan,* in Latin *Gratianopolitanum,* or *pagus Gratianopolitanus,* given by the Kings of *Burgundy* to the biſhops of *Grenoble,* but it devolved afterwards to the count of *Albon.* A great part of the country is over-run with large deſert mountains. The moſt remarkable places in it are

Grenoble, in Latin *Gratianopolis,* the capital of this country and of the whole government, lying on the *Iſere,* into which the river *Drac* empties itſelf in this neighbourhood. It is the reſidence of a parliament erected in the year 1453, of an intendency, of an election, of a chamber of accounts, of a court of taxes, of a court of mintage, of a *marechauſſée,* and of a court of the lord-treaſurer of *France.* It is a pleaſant, ſightly and populous city, and, excluſive of other fortifications, is defended by a citadel called the Baſtile. The biſhop of this place is ſuffragan to the archbiſhop of *Vienne,* has a dioceſe of 304 pariſhes, 240 of which are in *Dauphiné* and ſixty-four in *Savoy.* His income amounts annually to 28,000 livres, and his taxation at the court of *Rome* is 1000 florins. The river *Iſere* divides this city into two unequal parts, the ſmalleſt of which is called *La Perriere,* and contains one pariſh-church and one convent. The name of the largeſt is *Bonne,* and is alſo the beſt, being built with ſtraight handſome ſtreets. In it too are the biſhop's palace, the palace in which are held the courts of juſtice, the cathedral, a chapter, the arſenal and a general hoſpital. In the firſt religious wars, in the year 1562, this place was the ſcene of moſt horrible cruelties.

La grande chartreuſe, the moſt celebrated chartreuſe belonging to the order of *Carthuſians,* lies on the *Alps,* three leagues diſtant from *Grenoble.* Leading to it are two ways, one of which is called *Sapey,* and the other *S. Laurent du Pont*; but both are carried over ſteep rocks and precipices of a dreadful height, the terror of which is increaſed by the roarings of the river *Guyer le Mort,* during its precipitate courſe among them. The convent itſelf, which ſtands on an eminence reſembling a meadow, is environed with rocks and woody mountains of a much greater height ſtill. It is a ſpacious oblong ſquare walled in, and without are a few inns and ſtables. In the large hall of this convent is annually held the general chapter of the order of the *Carthuſians,* where the *German* priors take the precedence of all others, and at their arrival may ride into the court of the convent, which none of the reſt are permitted to do, but like ſtrangers, muſt alight at the

gate.

gate. This privilege they hold not as being the countrymen of *S. Bruno,* founder of the order, but by reason of the great share the *German Carthu-sians* had in accommodating the schisms of the papal fee. In this hall are fine paintings representing the life of *S. Bruno,* and likewise the pictures of the generals of the order. The former are copies of the admirable paintings in the chartreuse at *Paris.* In a large gallery may be seen views of all the *Carthusian* convents in Christendom, mostly executed by eminent hands. The church here is not large but very magnificent, and, among other re-liques, contains the arm of *S. Bruno,* his body being not interred here but in *Calabria,* in a *Carthusian* monastery founded by him there, this convent not having been built till after his decease. The first settlement of *Bruno* and his disciples was half a league from this place, and having no particular name was so called from *Carthreuse,* a village seated near it in a valley which is still in being. Such was the original of the general appellation of this order and its convents. From the kitchen of this monastery above 100 persons are daily fed, and among them thirty monks and forty lay-brothers. The wine-cellar in it consists of two lanes lined with casks of an extraordi-nary capaciousness, and which being immoveable, are filled through the roof by means of leather pipes. All strangers are entertained here *gratis,* setting down their names in a particular book. This convent, as the head of the order, chuses the general, who is obliged to reside here during life. In descending from this frightful eminence, we come first to a house se-cured by double doors, which are shut every evening and watched by a lay-brother who lives here. A little farther is an iron forge, and beyond it a small farm-house and a saw-mill.

Le Fort de Barreaux, or simply *Barreaux,* is a town and fort seated on the river *Isere* at the entrance of the valley of *Graisivaudan,* being the resi-dence of a particular-governor, a deputy-governor and a mayor.

Saffenage is a little place and a barony.

Corps, a small place on a hill about two leagues from *Lesdiguieres.*

Mens, a little place.

Uzille, or *Vizille,* a fine seat on the *Romanche* and the favourite residence of the brave constable *de Lesdiguieres.*

Oisan is a borough.

2. The little country of *Champsaur* lies south of *Graisivaudan,* near *Ambrunois* and *Gapençois,* being all over mountainous. It was possessed, for some hundreds of years, by the *Dauphin*'s counts of *Albon* and *Graisivaudan.* *Humbert* the last of the house of *Tour du Pin,* in the year 1336, first assumed the title of duke of *Champsaur.* *Lewis* XIV. made over a grant of this dukedom to *Francis de Bonne,* duke of *Lesdiguieres,* which belongs, at present, to the family of *Villeroy.* The principal place in it is St. *Bonnet.*

3. *Briançonnois,*

3. *Brianconnois* ſtands wholly in the *Alps*, and, through it lies one of the moſt frequented roads leading out of *France* into *Italy*. Great quantities of manna are gathered here. This country long preſerved its liberty and ſubmitted, at laſt, to the *Dauphins* of *Viennois*, only on advantageous terms. It contains

Briançon, in Latin *Brigantio*, an old little town, and the capital of this country and an election, having a caſtle ſeated on a ſteep rock. Near it unite the little rivers of *Dure* and *Ance;* which thus form the *Durance*.

Monetier is a town, the inhabitants of which drive on a great trade in all kinds of haberdaſhery-wares and prints.

Queyras is a ſmall caſtle.

Obſ. The other places here, as *Cezane, Exilles, Oux* and *Chateau Dau-phin*, belong, ſince the treaty of *Utrecht*, to the duke of *Savoy*.

4. *Ambrunois*, or *Embrunois*, lies alſo wholly among the *Alps*, and, in *Cæſar's* time, was inhabited by the *Caturigæ*. To it belong

Ambrun, or *Embrun*, in Latin *Eburodunum*, or *Ebredunum*, the principal place of the country, and ſtanding on a ſteep rock, at the foot of which runs the *Durance*. It is the reſidence of a court of juſtice and an archbiſhop who ſtyles himſelf Prince and Count of *Ambrun*, and Baron of *Guilleſtre* and *Beaufort*. To him alſo belongs one half of the lordſhip and juriſdiction of the city. His ſuffragans are the biſhops of *Digne, Graſſe, Vence, Glandeve* and *Senez*, with the biſhop of *Nice* in *Piedmont*. His yearly revenue amounts to 22,000 livres and his taxation at the court of *Rome* is 2400 florins. The epiſcopal palace here is the fineſt building in the whole city. Excluſive of the cathedral it contains five pariſh-churches, one college of Jeſuits and a Capuchin convent.

Guilleſtre is a little town, having the title of a barony of the Empire and belongs to the archbiſhop.

Beaufort is an archiepiſcopal barony, formerly alſo imperial.

S. Creſpin and *S. Clement* are two little places.

Mont Dauphin is a fortreſs near the *Durance*.

Chorges, a very old but ſmall town, a memorial of the ancient nation of the *Caturigæ*, of which word *Chorges* is manifeſtly a corruption.

Savine is a little place.

Gapençois had formerly counts of its own, and with *Provence* came at laſt under the crown of *France*.

Gap, in Latin *Vapincum*, the capital of the country, lies at the foot of a mountain near the little river *Benne*. The repeated damages it has ſuffered both in the inteſtine and foreign wars, and more particularly its laſt calamity in the year 1692, when it was burnt down by a ſudden irruption of the enemy, have ſo reduced it that it makes but a melancholy appearance at preſent. It is the principal place, however, of an election, and the ſeat of a diſtrict a diftrict and *marechauffee*. It ftill retains alfo a bifhop of its own, fubject to the archbifhop of *Aix*. His revenue is 11,000 livres a year, and his taxation at the court of *Rome* 1400 florins. The diocefe belonging to it contains 299 parifhes. In it too are four convents, one of which lies without the city.

Tharence is a bifhop's fee, but has nothing worth notice.

. *Notre Dame du Lait* is an elegant church, one *French* league and a half from *Gap*. To this place is always a great refort of pilgrims.

. *Serres* is a little town on a mountain.

Orpiere, a fmall place.

Lefdiguieres, a little place and a dutchy, belonging, at prefent, to the houfe of *Villeroy*.

Tallard, a town, feated on the *Durance*, gives the title of count.

Afpres Les Vaynes, is a little town lying among mountains.

6. *Le Royanez* is a fmall territory, about fix *French* leagues in length and four broad, the inhabitants of which were by the *Dauphins* declared free from the poll-tax. In it are

Pont de Royan, a fmall town but the capital of a marquifate.

Baume, alfo a marquifate:

Beauvoir, a place of no confideration.

7. *Les Baronies* is a territory fo called as confifting of two large baronies, formerly imperial.

1. The barony of *Meuoillon*, in Latin *Medullio*, was purchafed, in the year 1300, of its laft proprietor, by *Dauphin John*, fon to *Humbert* I. who conferred it on his brother *Henry*, after whofe death it was annexed to *Dauphiné*. In it are

Buy, the principal town and the refidence of a royal jurifdiction, on the river *Oreze*.

Meuoillon, a ruined fort.

. 2. The barony of *Montauban*, which the *Dauphin Humbert* I. feized for himfelf, whofe two fons, *Guy* and *Henry*, poffeffed it fucceffively, but after their death it became united to *Dauphiné*.

Nions, the capital, is a very fmall town, feated in a valley on the river *Aigues*, over which it has a bridge which is faid to be a piece of *Roman* architecture.

Montauban is a fmall place.

The little places of *Merindol*, *Rofans* and *Condourcet*, belonging alfo to thefe baronies, contain nothing remarkable.

II. To *Lower Danphiné* belong the following diftricts, *viz.*

1. *Viennois*, having formerly counts of its own, who, in the eleventh century, affumed the title of the *Dauphins* of *Viennois*. To it belong the following places; namely,

Vienne,

Vienne, in Latin *Vienna Allobrogum*, a very ancient town feated on the *Rhone*, being of pretty large extent, but very narrow and the ftreets ill paved. It is the refidence of a diftrict, an election, a *marechauffee* and a royal tribunal. The archbifhop of this place enjoys the title of Upper primate of *Gaul*, and within the Kingdom his fuffragans are the bifhops of *Grenoble, Viviers, Valence* and *Die*; and out of the Kingdom, the bifhops of *S. Jean de Maurienne* and *Geneva*. To his diocefe belong 440 parifhes, and his annual revenue amounts to 22,000 livres, out of which he is taxed, at the court of *Rome*, in 1854 florins. The cathedral here is a beautiful ftructure. In it alfo are three chapters, two of which are abbies, one abbey, one priory, nine convents, one feminary and one college of Jefuits. The church of *Notre dame de la vie* here is faid to have been a *Roman prætorium*. In this town are made anchors and fword-blades with other manufactures in iron and fteel, and alfo paper. *Ado* is the firft who mentions that *Pontius Pilate*, the *Roman* governor of *Judæa*, died here. In the year 1311, the fifteenth general council was held at this place.

About 100 paces diftant from the city is a remarkable *Roman* monument ftanding in the fields, which, agreeably to the general opinion, is a fepulchre. The building of it is a fquare at the bottom, and on each fide has a round vaulted aperture, through which a perfon may pafs, but on this quadrangle ftands a ftrong and high pyramid. The whole confifts of free-ftone.

S. Saphorin is a borough lying about half a *French* league diftant from the *Rhone*, in the road from *Vienne* to *Lyon*.

La Tour du Pin is a fmall town, which was formerly a free barony, but became afterwards united to *Dauphiné*.

Quirieu is a little town, ftanding high, near the *Rhone*.

Cremieu, a fmall place, feated at the foot of a mountain, one *French* league diftant from the *Rhone*.

Pont Beauvoifin, is a fmall open town divided by the river *Guier* into two parts, of which that to the eaft belongs to *Savoy* and the other to *Dauphiné*. This river, which is the boundary betwixt *France* and *Savoy*, is, from its rapid courfe, firnamed *Le vif*. The ftone bridge over it is both on the *French* and *Savoy* fide defended with a baluftrade and a guard of invalids.

Vierieu is a little place.

S. Vallier, a fmall town.

La Vulpiliere and *S. Rambert* are two little places.

Tain is a fmall town, feated on the *Rhone*, in the neighbourhood of which is produced that excellent wine called *Hermitage*.

Mantaille is a little village, in which, in the year 979, *Bofon* was elected King of *Burgundy*.

Rouffillon is a little village and earldom feated on the *Rhone*.

Albon

Albon is a little place formerly enjoying the title of an earldom, one of the counts of which, who were poffeffors of the prefent *Dauphiné*, was at his baptifm named *Dauphin*.

Beaurepaire is a little village.

Tuylins, or *Tulins*, is a fmall village, containing a priory.

Romans is a little town feated on the *Ifere*, being the principal place of an election and the refidence of a royal tribunal, but entirely demolifhed in the civil wars. It carries on a good trade, and contains two abbies, one of which being fecularifed, gave rife to the building of the place, a collegiate-church and a convent.

Triord is a delightful feat near this town.

St. *Antoine* is a fmall village lying betwixt two mountains, and owing its original to the famous abbey above-mentioned, which is as it were the capital of the order of St. *Antony* and the only abbey belonging to it. In the year 1561, it was demolifhed by the *Huguenots*, but afterwards rebuilt.

St. *Marcelin* is a well built little town feated on the *Ifere* and lying in a pleafant country, being formerly the refidence of the council of *Dauphiné*, which was erected by *Humbert* in the year 1337, but afterwards removed to *Grenoble* and changed to a parliament. In it alfo is a royal tribunal.

Pufignan is a fmall town.

S. *Jean de Bournay* is a little village.

Hoftun is a fmall place raifed, in the year 1712, to a dukedom, and, in 1715, to a peerage.

2. *Valentinois* is looked upon as a particular and diftinct ftate from *Dauphiné*, for when the King writes to *Dauphiné*, he not only ftyles himfelf *Dauphin* of *Viennois* but likewife count of *Valentinois* and *Diois*. This ancient earldom was four times raifed to a dutchy and peerage; namely, in the years 1499, 1548, 1642 and 1716, in which laft year it fell into the poffeffion of the fon of Marfhal *Matignon*, who married the eldeft daughter of the Prince of *Monaco*, formerly duke and peer of *Valentinois*. In it are the following places of note; *viz.*

Valence, in Latin *Valentia*, the capital of this diftrict, and lying on the *Rhone*, is pretty well built, of tolerable largenefs and one of the moft ancient cities in *France*, having been formerly a *Roman* colony. It is the principal place of an election, the refidence of a prefecturate, a bailiwick, a country-court and a *marechauffée*. The bifhop of this place is fubject to the archbifhop of *Vienne*, and his diocefe confifts of 105 parifhes, with a yearly revenue of 16,000 livres, out of which his taxation at the court of *Rome* is 2389 florins. The epifcopal palace here is a beautiful building. Exclufive of the cathedral, it contains alfo one chapter and three abbies, among which, that of St. *Ruf* is the principal of an order, with fix convents and one univerfity, which was founded originally at *Grenoble*, in the year 1339, but removed hither by *Lewis* XI. In this town were held three councils; namely, in the years 374, 584 and 855.

Le Valentin is a fine seat near this city having a delightful grove.

Etoile is a borough.

Crest, in Latin *Crista*, is a little town, being the capital of the dutchy of *Valentinois* and the residence of a bailiwick. In it are one chapter, one castle and one tower, which has a garrison and serves sometimes for stateprisoners.

Montelimart, properly *Monteil-Aimar*, or *Ademar*, so called from the ancient lord of that place, is a town seated on the river *Robiou*, being the capital of an election, the residence of a provincial government and a chapter, with a citadel lying on an eminence. It is neither large nor well built, but is remarkable for being the first place in all *France* which embraced the protestant religion; and among the nobility hereabouts are still many of the Reformed.

Livron is a little town, standing on an eminence, at the foot of which runs the river *Drome*. This place is but one *French* league from the *Rhone*.

Pierre Late is a small place, seated on the *Berre*, having a castle standing on a rock. This town belongs to the Prince of *Conty*.

Taulignan and *Dieu le Fit*, are two small towns.

3. *Le Diois* was formerly an earldom, which the last count thereof sold, in the year 1404, to *Charles* VI. It has been already observed, in *Valentinois*, that the Kings of *France*, in their letters to *Dauphiné*, use it as one of their titles. To this territory belong

Die, in Latin *Dea Vocontiorum*, the capital of the country and lying on the river *Drome*. It had once a citadel, is the residence of a tribunal and of a bishop, who is the lord of the city, which consists of ninety-five parishes and twenty-four castles. This bishop is suffragan to the archbishop of *Vienne*, and has a diocese of 200 parishes, with an annual revenue of 15,000 livres. His taxation at the court of *Rome* is 2126 florins. Before the revocation of the edict of *Nantes*, the Protestants had an university here.

Aoste is a little place seated on the *Drome*.

Bourdeaux is a small place, in which *Isaac Casaubon* was educated, but his native place was *Geneva*.

Luc is a little village, seated on two small lakes formed by the river *Drome*.

Satillans is a little town lying on the *Drome*.

Chatillon and *Valdrome* are two little places.

4. *Tricastin*, or *Tricastinois*, derives its name from its ancient inhabitants the *Tricastini*, and contains the following places, *viz*.

S. Paul trois chateaux, small and the only town in this little territory, lying on an eminence. The bishop of this place is subject to the archbishop of *Arles*, having a diocese consisting of thirty-four parishes, with a yearly revenue

venue of 10,000 livres. · His taxation at the court of *Rome* is 400 florins. This place affords nothing remarkable, except the cathedral, but without the town are two convents.

Suze la Rouſſe is a little place.

Donſere, a ſmall village ſeated on the *Rhone*, gives the title of a principality and belongs to the bishop of *Viviers*.

The PRINCIPALITY *of O R A N G E.*

THE principality of *Orange*, in Latin *Oranien*, is four *French* leagues in length and three broad. · The revenue of this diſtrict may be about 50,000 livres. In the ninth century this territory was under the dominion of the Kings of *Burgundy* and *Arles*. In the eleventh it had counts of its own, and ſoon acquired the title of a principality. *William*, ſon to Bertrand des Baux, firſt ſtyled himſelf, *by the grace of God, Prince of Orange.* Prince *Lewis* bought the ſovereignty of this principality, for 15,000 franks, of *Renatus*, King of *Sicily*. His ſon *William* was taken captive by *Lewis* XI. King of *France*, and compelled to diſpóſe to him the ſovereignty of his principality for 40,000 dollars. In 1500, however, his ſon, *John de Challon*, obtained the redemption of this contract of *Lewis* XII. His ſon, *Philibert de Challon*, died without iſſue, in the year 1531, but left his ſiſter's ſon, *Renatus* of *Naſſau*, heir; who having alſo no children, for that reaſon appointed his relation, *William* of *Naſſau*, his ſucceſſor in the principality of *Orange* and all his other eſtates; but he ſucceeded not to the peaceable poſſeſſion of his country till the year 1570. The ſovereignty of this principality was, in the year 1598, confirmed to the houſe of *Naſſau* by the treaty of *Vervins*, and, in 1678, by that of *Nimuegen*, and, in the year 1697, by that of *Ryſwick*. *William Henry*, Prince of *Orange*, was Statholder of *Holland*, in the year 1672, and created, in 1689, King of *England*. After his death, many heirs ſtarted up, among whom the King of *Pruſſia* was the moſt powerful; but, in the year 1713, King *Frederick William*, by the treaty of *Utrecht*, ceded this principality to the royal houſe of *Bourbon*. At the cloſe of the year 1714, it was annexed to the government of *Dauphiné*, the generality and intendency of *Grenoble*, and the election of *Montelimart*. In the year 1722, *Lewis* XV. gave it to the Prince of *Conty*. It contains one city, two ſmall towns, and about nine villages, and is exempt from all the uſual taxes in *France*. The places worthy of notice in it are

Orange, in Latin *Arauſio*, a very ancient city and the capital of the principality, lying in a ſpacious and delightful plain, which is pleaſantly watered by ſome ſmall rivers, the *Maine* waſhing the walls of the city, and the

Eigues also comes very near it. The bishop of this place is suffragan to the archbishop of *Arles*, and enjoys a yearly revenue of 10,000 livres, out of which his taxation at the court of *Rome* is 408 florins. It once had an university, founded, in the year 1365, by the Emperor *Charles* IV. In 1622, *Maurice* of *Naffau*, Prince of *Orange*, caused the castle here, which stood on an eminence, to be excellently fortified; but, in the year 1660, the fortifications, and, in 1673, the castle itself, was razed by order of *Lewis* XIV. Some councils were held here in the years 441, 529 and 1228. Among other remains of *Roman* antiquities to be seen in this place, a fine triumphal arch and an amphitheatre are both very remarkable.

Courtefon, Gigondas and *Jonquieres,* are all market-towns.

The GOVERNMENT *of*

P R O V E N C E.

THIS county, the name of which is of the same import, and derived from the Latin *Provincia,* is bounded to the north, by *Dauphiné;* to the east, by the *Alps* and the river *Var,* which divide it from the territories belonging to the King of *Sardinia;* to the south, by the *Mediterranean,* and, to the west, is separated from *Languedoc* by the *Rhone.* Its length, from *Varo* to the end of *Camargue,* is betwixt forty and forty-four *French* leagues, and its breadth, from the isles of *Hieres* to the village of *Sauze,* about thirty. *Upper Provence* enjoys a pretty temperate air, with a great deal of meadow-ground well stocked with cattle. It produces also grain, apples and pears, but little wine; yet what it does at some places, such as *Riez,* is the best in the whole province: on the contrary, *Lower Provence* is extremely hot, and near the sea-coast would be much more so were not the air somewhat refreshed by a wind called the *Bife,* lasting usually from nine or ten in the morning till evening, and of which more shall be said in *Languedoc.* The north-west wind here also cools the country, and sometimes, particularly after rains, blows fresh. *Lower Provence* produces not half corn enough for the subsistence of its inhabitants; but the soil being dry and sandy, pomegranates, oranges, lemons and olives, with mastic, cypress, palm and the *African* acacia trees thrive exceeding well here; as also shrubs of all kinds, such as bruc, which is something like box, only that the leaves are longer and more pointed, bearing also a red fruit, which continues on all the year round, and, by a singularity peculiar to itself, proceeds out of the middle of the leaf. The sea-cherry tree has a leaf like the alkermes, or scarlet oak-berry, and its fruit, both in size and
colour,

colour, refembles a cherry. The medlar-tree bears a fmall red fruit of an agreeable tartnefs. Here are alfo feveral trees of great benefit to the country. The wine in *Lower Provence* is thick, lufcious and heady, but the fort moft admired is the mufcadel. It produces vaft quantities of truffles. The beccafigo is a fmall delicious bird, feeding only on grapes and figs, and being much in requeft. Among the fea-fifh, berc is a very remarkable kind called the date, which is fo named from the refemblance of its fhape to that fruit, and is found within hollow ftones in the road and harbour of *Toulon.* In order to get them out, the ftone muft be broke; they abound alfo along the coaft of the marquifate of *Ancona* on the *Adriatic.* Iron is faid to be found near *Barles* and *Trans*; and near the town of *Hieres* and the village of *Garde Frennet*, even gold. Near *S. Baume*, and other places, are quarries of a black agate, and in the neighbourhood of *Marfeilles* are made tar and pitch.

The rivers here are but inconfiderable. The *Durance* has been already mentioned in *Dauphiné.* The *Sourge* rifes in the county of *Venaiffin*, to which alfo its fhort courfe is confined, lofing itfelf in the *Rhone.* The *Largens*, fo called from the tranfparency of its waters, has three fources, one at *S. Maximin*, the fecond on the diftrict of *S. Martin*, and the third in that of *Barjol*, falling below *Frejus* into the fea. The *Var* divides *France* from *Italy.* Along the coafts of the *Mediterranean* are feveral capes, bays, and harbours, the principal of which is *Toulon.* *Digne* and *Aix* have hot baths; and *Tortona* and the diftrict of *Morier* contain falt-fprings.

The traffic of *France* with *Spain* and *Italy*, which is almoft wholly carried on by *Marfeilles*, the centre alfo of the *Levant* trade, deferves fome notice. To *Italy, France* exports cloths, ferges, honey, prunes, figs, frefh eels, capers, olives, a fpecies of fmall fardines, oil, fail-cloth, *aqua vitæ*, cotton waiftcoats, ftockings, &c. to about three millions of livres, and the returns are made in hemp, rice, corn, fulphur, anife, manna and filk, with a balance of 200,000 livres in favour of *Italy.* To *Spain* are exported all kinds of linen, filk, wool, gold and filver ftuffs, gold, filver and thread-lace, with feveral other home and *Levant* commodities, to the amount of above nine millions of livres; but its imports from *Spain* are lefs by two millions. Their trade in filk to the *Levant* or *Conftantinople, Smyrna, Salonichi, Athens, Candia*, the iflands of the *Archipelago, Aleppo, Tripoli, Cairo* and *Alexandria*, is very confiderable both in its imports and exports.

Provence is a county, which by virtue of a marriage, devolved in the 13th century to the houfe of *Anjou*, and on the death of *Charles* of *Anjou*, king of *Jerufalem, Naples* and *Sicily*, and count of *Provence*, who was the laft of the male line of his family, his dominions, in the year 1481, defcended to *Lewis* XI. ever fince which, *Provence* has been annexed to the crown of *France.*

The

The government of *France* confifts of the united counties of *Provence*
and *Forcalquier*, and the *Terres adjacentes*, as they are called, which laft
lie difperfed all over *Provence*, being formerly poffeffed by diftinct lords,
who inftead of ferving as vaffals to the counts of *Provence*, efteemed it more
honourable to hold of the empire; and to this day conftitute not a part of
the States of a county. The King in his letters ftiles himfelf *Count* of
Provence, Forcalquier, and *des Terres adjacentes*. The high court of juftice
here, is the parliament inftituted in the year 1501 at *Aix*, and fubordinate
to it are the twelve bailiwicks into which the province is divided, with the
feven *vigueries*, or vicarfhips in the chief towns, and almoft all the other
petty jufticiaries. With refpect to the finances, *Provence* has its own ftates,
which are compofed of the clergy and the nobility, who are very numerous,
together with the reprefentations of the communities, who in their meetings
fettle the manner of levying the fums required by the crown. In this affembly the archbifhop of *Aix* prefides, and in his abfence the oldeft prelate.
In its military government the country it is fubject to has a governor, a general lieutenant, and four fub-governors. The department of the firft
fub-governor confifts of *Arles, Tarafcon, Forcalquier, Apt, Sifteron, Salon,
les Beaux, Mondragon, Alan, Grignan* and *Sault*. That of the fecond comprehends *Marfeilles, Toulon, Yeres*, part of the *viguerie* of *Aix*, and *Draguignan*. The department of the third contains *Graffe, S. Paul, Digne, Colmar, Annot, Guilleaume, Lorgues, Caftelane, Seine*, part of *Draguignan,
Val de Varenne* and *Entrevaux*; and that of the fourth *Aix, S. Maximin,
Brignolles*, and *Barjoux*.

French geographers divide *Provence* differently, one into two parts, a
fecond into three, and a third into as many Diocefes: The following appears to me the moft proper divifion of this province.

I. *Lower Provence* confifts of thefe bailiwicks, *viz.*

1. The bailiwick of *Aix*, containing only its large *viguerie*, in which
are,

Aix, in Latin *Aquæ Sextiæ*, fo called from the *Roman* commander *Sextus
Calvinus*, who built a caftle there, which he called after the warm baths
he found on the fpot, and his own name. This place is the capital of
Provence, and one of the fineft cities in the Kingdom. It ftands in a valley
of confiderable extent, planted chiefly with olives, and is rather populous
than large; the ftreets in it are well paved, and the houfes handfome,
with a moft beautiful fpacious walk called *Orbitelle*, which confifts of three
rows of trees interfperfed with fountains, and lying within the city. It is
an archbifhopric, the feat of a parliament, a *generalité*, a *recette*, a chamber of accounts, a court of taxes, a mint, an office of the treafurer of *France*,
a confulate, a royal jurifdiction and *prevoté*; and alfo of the general lieutenant's-court, the city magiftracy, the chief-bailiff of *Provence*, the chief
Prevôt de Marechauffee of the whole country, and a *Viguier*, or King's judge,
&c.

&c. Under its archbifhop are the bifhops of *Apt, Riez, Frejus, Gap,* and *Sifteron,* yet his province confifts only of eighty parifhes, but he enjoys a revenue of 32,000 livres, out of which his taxation at the court of *Rome* amounts to 2400 florins. In the cathedral here lie four counts of *Provence,* and in the church-treafury is a confecrated rofe of gold, the prefent of Pope *Innocent* IV. to Count *Raymond Berenger.* In this city alfo are two colleges, with an univerfity founded in the year 1409, an academy of the beaux arts, erected in 1668; a public library in the town-houfe, eighteen convents, a palace, formerly the refidence of the counts, now occupied by the high-bailiff, the parliament and the chamber of accounts; and feveral fightly buildings which lie particularly about its charming walk. The nobility of *Provence* refide here, efpecially in winter, fo that this city, for company, is reckoned the moft agreeable in the whole Kingdom, next to *Paris.* In its fuburbs, the warm mineral fpring, once fo famous, was found a fecond time in the year 1704, but with great decay of its ancient virtues. Its neighbourhood here produces excellent wine; but its principal trade confifts in its incomparable oil. Some ftuffs are alfo made here.

Rians, is a borough and marquifate, in a vale.

Roquevaire, a borough, feated on the *Vienne.*

Pertuis is a little town, lying in the moft fertile part of the province; being the feat of a rural-court, and belonging formerly to the county of *Forcalquier.* The Abbot *de Montmajour* is partly lord of this city. It contains alfo one parifh-church and five convents.

Lambefc is a pretty town, and principality belonging to the houfe of *Lorrain Armagne.* In this place are held the public meetings. It contains alfo two convents.

Salon, a little town ftanding high, contains one collegiate church, and belongs to the archbifhop of *Arles,* who has a fine feat here. This place is reckoned among the *Terres adjacentes.* In the *Francifcan* church here is interred the pretended prophet *Noftradamus,* with a tomb-ftone as fingular as himfelf, which is placed obliquely in the wall, and has an epitaph infcribed on it, above which is a painted bufto of the prophet taken in his 59th year. In this town, indeed, he lived, and in the year 1566 alfo died, but the place of his birth was *St. Remy.*

Iftres is a little place, near which are the falt-works of *Valduché.*

Berre, a town feated on a lake of falt-water which communicates with the fea, being five miles long, three broad, and from four to fourteen fathom deep, and every where navigable; makes great quantities of a very fine falt, but the air is unwholfome.

Martigues is a little town, ftanding on an ifland at the mouth of the above-mentioned lake towards the fea, being erected out of the ruins of the old town of St. *Genes,* and enjoying the title of a principality. It was anciently

ciently reckoned very strong, and was first taken, in the year 1591, by *Charles Emanuel,* duke of *Savoy,* after a long defence.

Jonquieres is a little town, and, together with

Ferrieres, likewise a small town, may be looked on as the suburbs of *Martigues,* both lying on each side of it, on the continent.

Before *Martigues* lies another small island, which is defended by a fort called *Tour de Bouc.*

Aubagne is a town enjoying the title of a barony.

La Ciotat, a sea-port, erected out of the ruins of *Ceireste,* is famed for its muscadel wine.

Caffis is a little place, having a deep but narrow harbour called *Mion.*

2. The bailiwick of *Arles* contains the *vigueries* of *Arles* and *Tarascon,* in which are the following places, namely,

Arles, anciently *Arelas, Arelate* and *Arelatum,* a large and well-built city on the *Rhone,* but lying in a marshy soil, and consequently enjoying not the most wholesome air. It is an archbishopric and the seat of a bailiwick, a *viguerie,* a court of admiralty and an office of the five great farms. Subject to its archbishop are the bishops of *Marseille,* St. *Paul trois chateaux, Toulon* and *Orange.* He styles himself Prince of *Montdragon,* has a province of fifty-one parishes and a yearly revenue of 33,000 livres, out of which his taxation at the court of *Rome* is 2008 florins. Exclusive of its cathedral, it contains one collegiate and seven parish churches, with an abbey, fourteen convents, one college of Jesuits, one hospital and a royal academy of sciences founded there in the year 1688. *Arles* was anciently the capital of the Kingdom of *Burgundy,* which, on its erection by *Boson,* was also called the Kingdom of *Arles.* The district belonging to this city is large, and ranked among the *Terres adjacentes.* Several councils have been held here, the most ancient and distinguished of which was that in the year 314. Among its numerous *Roman* antiquities, the most remarkable are, 1. A large collection of *Roman* sepulchral monuments, urns and the like, which is kept in the archbishop's palace; and among these is an urn of very extraordinary dimensions, containing between twenty and twenty-two gallons. 2. A large obelisk of porphyry, which was dug up here in the year 1675, and in 1676, set up again and dedicated by *Lewis* XIV. The new plain pedestal of sand-stone, belonging to it, ill agrees with the valuable monument above. 3. The upper part of an Æsculapius, with a serpent twining round it. 4. An amphitheatre; but almost wholly hid among private houses and which seems to have been compleated. 5. A fragment of a circus, consisting of a door with two capital *Corinthian* pillars of marble. 6. The remains of the *Roman* capitol which was erected here. 7. The *Elysian* fields, as they are called, which stand near the city, and are the place where the *Romans* used to bury the ashes of their dead. Accordingly we see here an incredible number of stone and marble *sarcophagi,* some stand-

2

ing

ing half, fome entirely out of the ground, with and without infcriptions. Thefe fields became afterwards a Chriftian cemetery. The fpot on which they ftand is as it were a mine of antiquities, in which a virtuofo, if not very profitably, at leaft might very laborioufly, fpend his whole life in diging and rummaging.

Near the city the *Rhone* divides itfelf into two large branches, forming the ifland of *Camargue*, called, in Latin, *Camaria*, which is one of the beft and moft fruitful tracts in this country, being interfected with feveral canals. The rapidity of the current here carries fuch a quantity of fand with it that its outlets are fo fhallow as to be fcarce practicable to fhipping; at leaft, they are very dangerous. This critical fpot is called *Tampan.* The eaft branch of the river is the broadeft, and at its mouth is called *gras de paffon:* the other mouths, or outlets, are called *gras* or *grous.* On this ifland, at the mouth of the *Leffer Rhone,* lies

Les Trois Maries, a little town, by the inhabitants of *Provence* accounted a facred place, from an opinion, that the three *Maries,* mentioned in the hiftory of our Saviour, together with St. *James* the younger, landed here.

Betwixt the *Rhone* and the lake of *Berré,* but fomewhat more northward, is a ftony plain called *Crau,* in Latin *Crava,* producing very fine herbage, which the numerous flocks of fheep feeding here feek for beneath the ftones. Here alfo grows a fine wine, and in the moraffes adjacent to the plain are falt-works.

Les Baux, or *Baulx,* in Latin *Baltium,* and corruptly *Baucium,* is a town feated on a hill and having an old caftle which was once very ftrong. The ancient lords of this place were of great power and confideration, poffeffing all that tract called at prefent, from this place, *Terres Bauffenques,* and comprehending feventy-nine places. *Bertrand,* lord of *Baux,* married *Tiburgo,* heirefs of *Orange.* This houfe became extinct in *Raymund de Baux,* Prince of *Orange,* who left two daughters behind without any male iffue. The eldeft of thefe, by name *Mary,* and married to *John de Chalon,* was Princefs of *Orange.* The youngeft, named *Elis,* enjoyed the lordfhip of *Beaux,* which, after her death, became united to the county of *Provence,* and is reckoned among the *terres adjacentes.* *Lewis* XIII. raifed it to a marquifate; and, in the year 1642, conferred it on the Prince of *Monaco.*

The following places belong to the *viguerie* of *Tarafcon;* viz.

Tarafcon, a very ancient town, feated on the *Rhone,* being large, well built, and the refidence of a *viguerie* and royal tribunal. In it is to be feen a handfome caftle fortified in the antique tafte, with one collegiate-church, one college, feven convents and one abbey.

S. Remy, is a pretty, ancient, large and populous town, feated in a fruitful country, near a lake called *La Glaciere.* In it is a collegiate-church and without the town are two convents. *Lewis* XIII. conferred this place on the Prince of *Monaco* as a dependency of *Baux.*

Chateau Regnard is a market-town.

Orgon is a fmall town feated near the *Durance.*

:. The government of *Marfeille,* to which belongs only one *viguerie.* In it are the following places of note, *viz.*

Marfeille, in Latin *Maffalia,* and *Maffilia,* the fecond city in *Provence,* but the firft in point of trade. It is large, rich, and the moft ancient place in all this province; and, as a town, is faid to have exifted 500 years before the birth of Chrift. It is feated, at the foot of a rocky mountain, near the fea, being divided into the old and new town. The former lies on an eminence, confifting of narrow crooked ftreets with mean houfes; whereas, in the latter, the ftreets are ftraight and broad, and adorned with handfome edifices. The Walk here is a very long ftreet, decorated on each fide with beautiful houfes, and, in the middle, betwixt two rows of lofty trees, it is beaten hard and contains a proper number of trenches for refting places. This city is the refidence of a bailiwick and the principal place of a *recette,* having a bifhop fuffragan to the archbifhop of *Arles,* who has a diocefe containing thirty-fix parifhes with an annual revenue of 30,000 livres, his taxation at *Rome* being 700 florins. In it is alfo a mintage, together with four parifh-churches, including the cathedral and two collegiate ones, with two abbies, an academy of the *beaux arts* and an obfervatory. It contains alfo a large arfenal well ftored with all the implements for fitting out the King's gallies. The large armory here, which confifts of four walks crofswife and is accounted the fineft in the whole Kingdom, contains arms for 40,000 men. In the arfenal is a dock for building the gallies. This is roofed over and communicates with the harbour, which is a parallelogram, having publick and private buildings on the two long fides and one of the fhorter. The other fide is the iffue into the *Mediterranean,* which is defended, on each point, by a fine ftrong fort. The entrance into the harbour, on account of a rocky cape near it, is difficult and has not a depth of water enough for men of war. Gold and filver ftuffs are made here, but of its great trade I have fpoken in the general commerce of *Provence.* In the year 1720, this city was vifited by a moft deplorable peftilence, which raged till 1722.

Oppofite to the harbour lie three fmall iflands called

1. *If,* on which *Francis* I. built a ftrong caftle, which ferves at prefent only as a place of confinement for the delinquents who are natives of *Marfeille.*

2. *Ratonneau,* anciently St. *Etienne,* on which the duke of *Guife,* towards the end of the fixteenth century, built a fortrefs.

4. The bailick of *Brignoles* confifts of the following *vigueries;* viz.

1. To the *viguerie* of *Brignoles* belong

Brignoles, in Latin *Brinonia,* a pretty large town, feated in a fine country, and deriving its name from its excellent plumbs, by the ancients called
                                                                    Brinones.

Brinones. . Exclusive of its parish-church, this place contains five convents and one hospital.

Vins is a little place erected into a marquisate about the year 1641.

Le Val, also a little place.

Carces is a small place and an earldom.

2. To the *viguerie* of St. *Maximin* belongs

St. *Maximin,* a little town, owing its original to a convent of *Benedictines* founded there, but possessed, at present, by *Dominicans.* This place has a manour-court and a *viguerie,*

La Sainte Beaume, or *Baume,* is a celebrated place, seated on a high rocky mountain, and having a cave in which, according to the real belief of the inhabitants of *Provence,* who cannot bear to hear it disputed, St. *Magdalen,* sister to *Lazarus,* spent thirty years in a rigid penance. This place is now changed into a church with a convent near it. A little way higher up, lies

St. *Pilon,* having a chapel.

3. To the *viguerie* of *Barjols* belong

Barjols, or *Barjoux,* a pretty large, handsome town, and the capital of a *recette* and a *viguerie,* containing also a royal court of justice with a collegiate-church and two convents.

5. The district of *Toulon* consists only of one *viguerie,* to which belongs

Toulon, in Latin *Tolo Martius,* being a pretty large city and the principal place of a *recette,* the residence also of a district and a manor-court, of a *marechaussée* and an admiralty. It is defended by very strong fortifications towards the land side, and the new part of the town is very handsome. The bishop of this place is subject to the archbishop of *Arles,* having under him a diocese of twenty-five parishes with a yearly revenue of 15,000 livres, out of which his taxation at the court of *Rome* is 400 florins. Exclusive of the cathedral, in it are nine convents with a fine house of Jesuits, a seminary and a college under the direction of the fathers of the oratory. The old and new harbour lie contiguous, and, by means of a canal, we pass from one to the other, both having an outlet into the spacious outer harbour, which is naturally almost of a circular figure, being surrounded with hills. Its circuit is not less than three hours, and on the entrance, on both sides, it is defended by a fort with strong batteries. The new harbour, which is a work of *Lewis* XIV. is also well defended by batteries; and round it stands the arsenal, where every man of war has its own particular store-house, but the guns and cordage are laid up separate. In it are spacious working houses for black-smiths, joiners, carpenters, lock-smiths, carvers, &c. The rope-house is built wholly of free-stone, being 320 toises in length, with three arched walks, in which as many parties of rope-makers may work at the same time. The general magazine here, which supplies whatever may be wanting in the particular store-houses for

single

fingle fhips, contains an immenfe quantity of all kinds of ftores difpofed in the greateft order and conveniency. The cannon-foundery here is alfo worth viewing. In the year 1707, the duke of *Savoy* laid fiege to this city but without fuccefs. In the year 1721, it fuffered greatly by the plague. It makes a fort of coarfe woollen cloth and its neighbourhood abounds in capers.

Sixfcurs is a town, feated on a hill, and having a harbour called St. *Cenary.* The inhabitants here of both fexes are faid to be larger than the other *provençals.*

Ollicules, properly *Olieulles,* is a town which feems to owe its name to its plenty of fine olives. The nunnery here, in the year 1730, became known all over *Europe* by the infamous practices of the Jefuit *Girard* and *Mademoifelle Martha Catherine Cadiere,* concerning which many pamphlets have been publifhed in all countries.

Seyne is a town and harbour in *Toulon* road.

6. The diftrict of *Hieres* confifts of the *viguerie* of the fame name, and contains the following places, *viz.*

Hieres, in Latin *Areæ,* a little town, formerly much more confiderable than at prefent, and having once a harbour where pilgrims bound for the Holy Land ufed to embark; but the fea here is now retired above 2000 paces from the place. This town is the feat of a diftrict-court, of a *recette* and *viguerie,* contains one collegiate with two other parifh-churches, five convents and nunneries. The foil here is excellent, and being well improved with orchards produces the nobleft fruit in *France.* A great deal of falt is made here both from the fea-water and that of a large falt lake, but the air was unwholefome till a canal was carried from the lake to the fea, whereby it is confiderably mended.

This town gives name to

The iflands of *Hieres,* in Latin *Stoëchates,* which lie in the neighbouring fea, and in conjunction with the continent form a fine road. Thefe iflands are three, each having its particular name. The principal, and that lying neareft the town, is *Porqueroles,* or *Porqueyroles,* fo called on account of the great number of wild hogs which fwim over hither from the continent to come at the acorns, abundance of oaks growing here. It is four *French* leagues in length and one broad, containing an old caftle for its defence. The fecond is *Portecroz,* that being the name of the harbour where a fort is built. The third is called *Titan,* or *Levant,* from its fituation eaftward of the two former, being four *French* leagues in length and one broad, but uninhabited. It was formerly, according to *Pliny,* called *Cavaros.* The *Marfilians* gave particular *Greek* names to thefe iflands from their fituation; namely, *Prote,* *i. e.* ' the firft;' *Mefe,* ' the middle,' afterwards *Mediana;* and *hypaèa,* ' the lower.' In 1655, thefe iflands were raifed to a marquifate.

quifate. On them are found the moft curious medicinal plants of all fpecies growing in *Italy, Spain, Greece* and *Egypt.*

Bregancon, a caftle on an ifland, was raifed to a marquifate in the year 1674.

Couloubrieres is a little place.

Souliers is a market-town, having a convent, the country around it fruitful.

Le Puget is a market-town.

Cuers is a market-town containing a collegiate-church.

7. The diftrict of *Draguignan* confifts of three *vigueries;* namely,

1. The *viguerie* of *Draguignan,* containing

Draguignan, one of the largeft towns of the whole province, and the country around it very fruitful. It is alfo the feat of a bailiwick, a *viguerie* and a tax-office; has a collegiate-church, fix convents and one college.

Frejus, or *Frejuls,* in Latin *Forum Julii* and *Colonia Pacenfis,* or *Colonia Octavanorum* and *Claffenfis,* is a fortified city, feated on the river argent, half a *French* league diftant from the fea. In it is a court of admiralty and an office of the five great farms. Its bifhop is fuffragan to the archbifhop of *Aix,* having under him a diocefe containing eighty-eight parifhes with a yearly revenue of 28,000 livres, out of which he is taxed 1000 florins at the court of *Rome.* The *regale* does not take place in this church, but on the vacancy of the fee, the vicar-general difpofes of all livings in the bifhop's gift. This city contains four convents. Among the remains of *Roman* works to be feen here, is an amphitheatre and aqueduct of confiderable length.

Grimaud, a little town, formerly fo much nearer the fea, that St. *Tropez* bay was frequently called from it. This place is a marquifate, on which *La garde du Frainet, La Molle, Cogolin, Gaffin, Ramatnel* and St. *Tropez* are dependencies.

St. *Tropez,* or *Torpez,* is a little fea-port, feated on a bay about four *French* leagues in length, to which it gives name. In it is a priory from which the town itfelf had its rife, with a convent and a good citadel.

Pignans is a market-town, containing one collegiate-church, two convents and four chapels, which lie without the town.

Le Puget is a little place.

Villecroffe is alfo fmall but has fine grottoes in its neighbourhood.

Bargemon ftands on a hill which is covered with vines and olive-trees, but furrounded with mountains.

Favas is a little place, formerly much more confiderable than at prefent, being deftroyed, in the eighth century, by the *Saracens.*

Fayence is a very ancient town, feated on the river *Benfon.*

2. The *viguerie of Aups,* fo called from

Aups,

Aups, or *Aulpes,* in Latin *Alpes,* a little town, but the feat of a royal court of juftice, a fmall manor, and a *recette;* containing alfo a collegiate church and two convents.

3. The *viguerie* of *Lorgues,* fo called from

Lorgues, a little town lying on the river *Argent,* is the feat of a royal court of juftice, of a *viguerie* and a tax-office, and contains one collegiate church. To the *viguerie* belongs the two villages of *Salegos* and *La Baftide.*

8. The diftrict of *Graffe,* in which are the following *vigueries,* viz.

1. The *viguerie* of *Graffe,* to which belong

Graffe, one of the fineft cities of the province, and the principal place of the diftrict, with a *viguerie* and a *recette,* being alfo a bifhop's fee, fubject to the archbifhop of *Ambrun.* Its diocefe confifts of twenty-two parifhes, with an annual revenue of 22,000 livres, taxed in 424 florins at the court of *Rome.* The city ftands on an eminence in a very fruitful country, is populous, and contains in it feven convents.

Antibes, in Latin *Antipolis,* one of the moft ancient cities of the province, is feated in a rich country, having a harbour for fhips of middling burden ; and a good citadel. In it are two convents, and many remains of *Roman* antiquities. This place is particularly celebrated for its delicate manner of preparing the fmall kind of fardines.

Cape *Garaup,* or *Garcupe,* projects a great way into the fea and forms the bay of *Cannes.*

Cannes is a little town, having a fmall harbour, and giving name to the above-mentioned bay.

5. The *ifles de Lerins* lie oppofite to *Cannes,* being very fertile and pleafant, and in conjunction with the continent, form a good road called *Gourjean.* In the year 1635 thefe iflands were taken by the *Spaniards;* but recovered again in 1637 by the *French :* They are, 1. *Sainte Marguerite,* formerly called *Lero,* being the largeft, though only one *French* league in length and one fourth of a league broad. It lies three leagues fouth-eaft of *Cannes,* having formerly a convent dedicated to St. *Margret,* and being defended, at prefent, by three forts called *Le fortin, Le fort Arragon,* and *Le fort Royal;* the laft of which is the principal. 2. St. *Honorat,* anciently *Lerina,* is feparated from the former by a narrow ftrait, being fo called from an archbifhop of *Arles,* who founded a celebrated abbey here which is ftill in being. For its defence it has a ftrong tower, garrifoned by a party from *S. Margueriie.* Here are alfo two very fmall iflands called *La Formique,* and *La Grenlie.*

La Napoule is a little place feated on the bay of *Cannes.*

Vence, in Latin *Vincium,* an ancient city, which was formerly an earldom, but is only a barony at prefent. The bifhop of this place, who is fubject to the archbifhop of *Ambrun,* has a Diocefe confifting of twenty-three

three parishes, with a yearly revenue of 7000 livres, being taxed at the court of *Rome* in 200 florins. He is lord also of one part of the city, but the other part belongs to the house of *Villeneuve*, with the title of a barony.

2. The *viguerie* of *S. Paul de Vence* comprises *S. Paul*, a fortified little town, the principal place of the *viguerie*, and a *recette*.

S. Laurent, a market-town near the mouth of the *Var*, producing excellent wine.

Le Puget les treizes dames, is a small place.

II. *Upper Provence* consists of the following bailiwicks, *viz.*

1. The bailiwick of *Castelane*, comprehending three *vigueries*.

1. To the *viguerie* of *Castelane* belongs

Castelane, a little town, which is the seat of the bailiwick, a *viguerie* and a *recette*. It stood formerly on a hill, but in the year 1260, the inhabitants settled at the foot of it on the river *Verdon*. In it are two convents; and it is generally the residence of the bishop of *Senez*. About a quarter of a league distant from the city is a salt-spring, the water of which issues in such abundance as at a small distance to drive a mill, afterwards joining the river *Verdon*.

Senez, in Latin *Sanitium*, is a mean place, the country around it wild, barren and mountainous, yet is it a bishop's see, suffragan to the archbishop of *Ambrun*. It has a diocese of about forty parishes, with a revenue of 10,000 livres *per annum*, out of which it is taxed at the court of *Rome* in 300 florins. The jurisdiction of the city is divided betwixt the bishop, the chapter, and the count of *Carces*.

2. To the *viguerie* of *Montiers* belong

Montiers, in Latin *Monasterium*, a pretty good town, and the seat of a *viguerie*, of a *recette*, and of a district-court. In it is a convent of *servites* *, and a manufacture of mock-china.

Riez, anciently *Albece*, a small city seated on the *Auvestre*, in a fine plain, abounding with excellent wine and fruits; gives the title of count. The bishop, who is lord of the city, is a suffragan to the archbishop of *Aix*. His diocese consists of sixty-one parishes, and his yearly revenue is 15,000 livres, out of which he is taxed at the court of *Rome* in 850 florins. In this city are three convents.

Valensole is a market-town, having two convents.

La Palud and *Montpezat* are the next places.

3. To the *viguerie* of *Aunot* belongs

---

* The *Servites* are a religious order in the church of *Rome*, founded about the year 1233, by seven *Florentine* merchants, who, with the approbation of the bishop of *Florence*, renounced the world, and lived together in a religious community on Mount *Senar*, two leagues from that city. It is pretended, that when they first appeared in the black habit given them by the bishop, the very children at the breast cried out, *See the Servants of the Virgin*, which miracle determined them to assume the name of *Servites*, or Servants.

          Aunot

Aunot or *Annet,* a little town, and the capital of the *viguerie.*

4. To the *viguerie* of *Guilleaumes* belongs

Guilleaumes, a town lying among mountains, and the principal place of the *viguerie,* of a *recette,* and a jurifdiction.

Le Puget Figette and *Le Puget Rouftan* are two little places.

Glandeves, in Latin *Glanata,* a city, now deftroyed by the inundation of the *Var,* formerly gave the title of count and was a bifhop's fee. The epif. copal palace is ftill ftanding, but the bifhop of this place refides at *Entre_vaux,* whither moft of the inhabitants have alfo retreated. This diocefe is fubject to the archbifhop of *Ambrun,* and confifts of fifty-fix parifhes. The revenue belonging to it is 10,000 livres, and its taxation at the court of *Rome* 400 florins.

Entrevaux is a town fituate near *Glandeves,* in which the bifhop of the above-mentioned place has for fome time paft refided, and in which is alfo the cathedral.

To the diftrict of *Digne* belong four *vigueries.*

1. To the *viguerie* of *Digne* belongs

Digne, in Latin *Dinia,* a fmall but very ancient city, feated at the foot of the mountains near the little river *Bleone,* being the capital of the *viguerie* and diftrict, and having alfo a *recette.* The bifhop of this place is baron of *Lauzieres,* and fubject to the archbifhop of *Ambrun,* having under him a diocefe containing fifty-fix parifhes, with an annual revenue of 10,000 livres, out of which his taxation at the court of *Rome* is 400 florins. In this city are five convents. The hot mineral waters here, faturated with fulphur and alcaline falt, are ufed both in drinking and bathing. The country around it is fine, and abounds in fruit.

Oraifon is a town, raifed in the year 1588 to a marquifate.

Champtercier, anciently *Oyfe,* is a barony, which in the year 1627 was united to the marquifate of *Villars-Brancas.*

Les Mees is a market-town.

2. To the *Viguerie* of *Seyne* belong

Seyne, in Latin *Sedena,* and *Sezena,* a town lying among the mountains, being the principal place of the *viguerie,* a *recette,* a royal *prevoté,* and country-court. In it are two convents.

3. To the *viguerie* of *Colmars* belongs

Colmars, a little town, which is the feat of the *viguerie,* of a *recette,* and a county-court. Near it is a fpring which ebbs and flows, though not periodically.

4. In the valley of *Bareme* is

Bereme, a town lying on the river *Affe*; the principal place of the valley, and the feat of a *recette.*

3. The diftrict of *Sifteron* confifts of the following places, *viz.*

1. To the *viguerie* of *Sifteron* belongs

Sifteron,

Sifteron, in Latin *Secuftero*, and *Segefterica*, a city, having a fmall citadel
feated at the foot of a rock on the *Durance*. It is the principal place of the
diftrict, of a *viguerie*, of a *recette*, and alfo of a manor-court; and the feat
likewife of a bifhop, fuffragan to the archbifhop of *Aix*. His revenue is
15000 livres *per annum*, and his taxation at the court of *Rome* 800 florins.
The diocefe confifts of fixty-four parifhes. Belonging to it are two cathedrals, *viz.* one here, and the other at *Forcalquier*. In the city too are five
convents , and on the other fide of the river is a pretty large fuburb called
La Baume, containing one parifh-church and one convent.

2. To the *viguerie* of *Cournillon* belongs
Cournillon, a market-town and a barony.

3. The earldom of *Grignan* receives its name from
Grignan, a little town, which is the capital of the earldom.

4. The diftrict of *Forcalquier* contains
1. The *viguerie* of *Forcalquier*, to which belongs
Forcalquier, in Latin *Forum Calcarium*, and corruptly, *Forcalquerium*.
This place is the feat of a diftrict-court, of a *viguerie* and a *recette*; and
the capital of an ancient county, which on the deceafe of *William*, the laft
count thereof, devolved, in 1208, to his grand-daughter and heirefs *Gaf-
fenda*, fpoufe to *Alphonfo* II. earl of *Provence*; by which means it became
united to *Provence*. The King ftill ftiles himfelf count of *Forcalquier*. The
collegiate church here is called a joint cathedral of the bifhop of *Sifteron*,
this place having been for a fhort time in the 11th century an epifcopal fee.
In it are four convents.

Lurs, Mane, and *Royallane*, are all little places.

Manofque, in Latin *Mannefca*, one of the moft populous towns in the
whole country, ftands on the *Durance*, in a very fertile and delightful valley, having in it two parifh-churches and feven convents. It is alfo a commandery of the knights of *Malta*, the commander of which ftiles himfelf
prefect and *great-crofs* of the order of St. *John* of *Jerufalem*. Near the town
is a mineral fpring.

Peyruis, is a little place, having a royal court of juftice, not a few affirm
its ancient *Latin* name to have been *vicus Petronii*, and the town itfelf the
birth-place of the celebrated *Petronius*.

2. The *viguerie* of *Apt*, to which belongs
Apt, in Latin *Apta Julia*, a city lying on the little river *Calavon*, being
the feat of a *viguerie*, of a *recette*, and alfo of a bifhop, who has the title
of *Prince of Apt*, though not lord of the city. He is a fuffragan to the
archbifhop of *Aix*, has a diocefe containing thirty-three parifhes, and enjoys
a yearly revenue of 9000 livres, out of which his taxation at the court of
Rome is 250 florins. Exclufive of the cathedral, this city contains two abbies and eight convents. This place boafts of having the body of *S. Anne*;
but it may with better reafon value itfelf on the remains of *Roman* antiqui-

Vol. II. N n n ties

ties found here. In it, in the year 1362, was held a council. The plumb of this town are in great repute.

Villard, is a little place, being a dutchy and peerage.

Cucuron, a small place, containing a convent of *Servites*.

Cadenet is a little place, which was formerly a *vicomté*.

The valley of *Sault*, in Latin *Saltus*, was erected into an earldom in the year 1562, and belongs to the neighbouring country. It receives its name from the great quantity of wood in its district, in which several glafs-houfes have been erected. In it is one borough of the fame name, and three villages.

5. Laft is the valley and diftrict of *Barcelonetta*, lying in the *Alps*, on the border of *Dauphiné*, and in the county of *Nice* in *Italy*. *Francis* I. dif-poffeffed the duke of *Savoy* of this place, adding it to *Provence*; but *Henry* II. reftored it again to *Savoy*. Laftly, at the peace of *Utrecht*, in the year 1713, it was ceded again to *France*, and in 1714, incorporated in the go-vernment of *Provence*. The principal place in it is

The little town of *Barcelonette*, founded in the year 1230, by *Raymund Berenger*, count of *Provence*, who gave it that name, in memory of the origin of his family, which he derived from *Barcelona* in *Catalonia*.

*   *   *   *   *   *

The State of *Avignon* and the County of *Venaiffin* may not improperly be introduced in this place; for though this country belongs to the Pope, and not to *France*, yet the greateft part of it being furrounded by *Provence*, it may be defcribed here by way of appendix. This moft delightful country, which particularly abounds in corn, wine, and faffron, contains the diftrict of *Avignon* and the county of *Venaiffin*, which are not to be confounded.

The city and diftrict of *Avignon* belonged formerly both to the counts of *Toulouse* and *Provence*, but continual quarrels arifing betwixt thefe lords, the inhabitants placed themfelves in a ftate of independency, which lafted till the death of the laft count of *Toulouse*. *Johanna*, his heirefs, and fpoufe to *Alphonfo*, count of *Poitiers*, brought him all her poffeffions and claims, whence uniting with his brother *Charles* of *Anjou*, who had married the heirefs of *Provence*, in the year 1261, they reduced the inhabitants of *Avignon* to their former fubjection. On the death of *Alphonfo*, his part of *Avignon* fell to *Philip the Bold*, who left it to his fon *Philip the Handfome*, and he, in 1290, conferred it on *Charles*, King of *Sicily* and count of *Pro-vence*. In the year 1348, *Johanna*, Queen of *Sicily* and countefs of *Pro-vence*, difpofed of the town and its diftrict to Pope *Clement* VI. for 80,000 florins.

Le Comtat Venaiffin, or the county of *Venaiffin*, anciently *Venaiffe*, fo called from the town of *Venafque*, the *Vendaufca*, or *Vendafca* of the an-cients

cients, was poffeffed, after the 11th century, by the counts of *Touloufe,* but reaffumed again in the 13th century, and held by Count *Raymond the Elder.* The Popes pretended to the fovereignty of this country from the time of Count *Raymond de S. Gilles,* although it is certain that the Emperors, as Kings of *Arles,* had juftly exercifed that power. In the year 1234, the Emperor *Frederick* II. transferred the Imperial rights in the towns of *Ifle* and *Carpentras,* and the other parts of the county of *Venaiffin* to *Raymond the Younger:* And the Pope found himfelf compelled to relinquifh them to him. *Raymond* left them to his daughter *Johanna,* and her hufband *Alphonfo.* From them it defcended by inheritance to *Philip the Bold,* King of *France,* who, in the year 1273, reftored it to Pope *Gregory* X. as a fief of the fee of *Rome;* and from that time it has been governed under the Popes, by officers called *Rectores.*

This country is fubject to few taxes, paying only a fmall matter, as the tenths of its products, which may make about the 60th part of its revenue, infomuch that the yearly charge of the vice-legate and the militia, with other expences, exceed what income arifes from it to the court of *Rome.* The *French* farmers-general, have by means of a fum of money, obtained the monopoly of falt and tobacco in this country, though that trade is ftill carried on in the Pope's name.

To the State of *Avignon* belongs

Avignon, in Latin *Avenio,* a city lying on the *Rhone,* which here is joined by the *Sorgne,* being a very beautiful place. It is furrounded with a wall of fine freeftone; but carries on little traffic, efpecially fince the manufactures of printed linen, which afforded a comfortable fubfiftence to great numbers of the inhabitants, were fuppreffed here in favour of the *Eaft-India* company, who purchafed this prohibition by a round fum paid into the Pope's hands. The vice-legate, as governor, ufually refides in the papal palace, which is a large ftructure of freeftone environed with moats, but far from being a regular building. The arfenal is a long lofty place, but without any other fire-arms in it than fuch as are juft neceffary to be difcharged on public rejoicings. The upper court of the vice-legate is called the *Rota,* with a right of appeal to *Rome.* In this town is alfo a *viguerie.* Its ancient diocefe was raifed in the year 1475, to an archbifhoprick, with the bifhops of *Carpentras, Cavaillon* and *Vaifon,* for its fuffragans. The cathedral here, which ftands on the fame eminence with the palace, is not very large. In two chapels in it are to be feen the monuments of Pope *Benedict* XII. and *John* XXII. The moft remarkable thing in the former is the chair, or throne, of the ancient Popes during their refidence here. In the church of the *Celeftines* is the tomb of *Clement* VII. and the patronal faint of the place is Cardinal *Peter,* of *Luxemburg,* who in his 18th year was a Cardinal, died in his 19th, and fince his death has performed feveral miracles, which, by way of proof, are finely painted in his

chapel,

chapel. In this church has alfo been erected a fplendid chapel, with a monument to it in honour of the fhepherds, who are faid to have built the ftone bridge here leading over the *Rhone.* In the church of the *Francifcans* is the tomb of the beautiful and learned *Laura,* immortalized by the poetry and paffion of *Petrarch,* but the grave belonging to it is covered only with a flat ftone. In the time of *Francis* I. when the church was repairing, the King's curiofity led him to have the grave opened, and in it was found a leaden box with a medal and fome *Italian* verfes by *Petrarch,* written on parchment; and his Majefty in return for the difturbance thus offered to her afhes honoured *Laura* with a copy of verfes, which lies alfo preferved in the leaden box. In this town are two focieties of Jefuits, to one of which belongs a fine college, with an univerfity founded in the year 1303. Betwixt the years 1305 and 1377, the Popes refided here. The *Jews* enjoy a free exercife of their religion in this city.

In the county of *Venaiffin* are the following places, namely,

1. In the jurifdiction of *Carpentras,* in Latin *Carpentoracte Meminorum,* and as the *Abbé Longuerve* fuppofes, called alfo *Forum Neronis,* the capital of the county and the feat of a court of juftice, the rector of which, is, as it were, prefect of the county. It ftands on the river *Roufe,* has a bifhop and an abbey. In the year 527 a council was held here.

Vaifon, a little town and caftle feated on a mountain, at the foot of which runs the river *Ouvefe.* The old town, which has lain in ruins for feveral centuries, is feated in the plain. Belonging to it is a bifhop who is lord alfo of the town, but has a very fmall revenue.

Venafque is a little town, ftanding on a mountain, near the river *Nafque* or *Venafque.* It was formerly the capital of the county, which derives its name from it; and likewife the feat of a bifhop, afterwards removed to *Carpentras.*

Bedouin and *Bedarides,* are market-towns, feated on the river *Ouvefe.*

Le pont de Sorgue is a town lying near the conflux of the *Sorgue* and *Ouvefe,* being the feat of a provincial judge.

Malaucene is a town, having a provincial judge.

Pernes is a fmall place on the *Sorgue,* with a provincial judge.

Savournin is a borough.

Boulene is a little town, with a provincial judge.

Caderouffe is a little place lying on the *Rhone,* being alfo the feat of a provincial judge.

2. In the diftrict of *L'ifle,* in Latin *Infula,* the beft and handfomeft town in the whole country, on an ifland in the river *Sorgue,* and the feat of the jurifdiction: The country around it extremely delightful.

Cavaillon, a mean little place, feated on an eminence in an ifland of the river *Durance,* a branch of which falls into the river *Calavon,* is a bifhop's fee, having a provincial judge, a particular judge of its own, and an abbey.

4

Menerbe

Menerbe is a little place, and the refidence of a provincial judge.

Vauclufe is alfo a fmall place, where *Petrarch* and his *Laura* refided. Here the *Sorgue* rifes, which clofe to its fource precipitates itfelf from a high rock like a ftream and foon becomes navigable. *Petrarch* has given us feveral pompous defcriptions of this fpring.

3. In the jurifdiction of *Vaulreas* is

Vaulreas, Valreas, or *Vareas,* a little town, the capital and feat of this jurifdiction, lying among the mountains of *Dauphiné.*

Vizan, a little place having a *caftelany.*

Sainte Cecile and *La Palud,* fmall places, yet having provincial judges.

Boufchet is a little place having an abbey.

Boulene, a fmall town, having a provincial judge.

Mornas, a little town, feated on the *Rhone,* and the refidence alfo of a provincial judge.

The GOVERNMENT *of*
L A N G U E D O C.

THIS province, to the eaft, is divided by the *Rhone* from *Provence,* *Avignon* and *Dauphiné;* to the north, is bounded by *Lyonnois, Auvergne* and *Guyenne;* to the weft, the *Garonne* divides a great part of it from *Gafcony;* and, fouthward, it terminates on the counties of *Rouffillon* and *Foix* and the *Mediterranean.* The eaftern coaft, from *Agde* as far as the *Rhone,* is remarkably increafed, the fea having retired from it confiderably, as appears evident on confulting ancient maps and the accounts of ancient geographers, with refpect to the fituation of feveral of its maritime places when compared with the prefent. It is a difficult matter to determine the largenefs of this country, the figure being fo irregular. Its extent, however, from weft to eaft, may be computed at about feventy *French* leagues; but from north to fouth, in the narroweft part, it is betwixt ten and twelve; and in the broadeft, weftward, almoft thirty, being, eaftward, almoft thirty-two. *Languedoc* comprifes nearly all that tract of land anciently called *Narbonenfis Prima,* with a confiderable part alfo of *Aquitania Prima,* and a part of *Viennois* and *Novempopulania.* Thefe feveral countries were united about the beginning of the thirteenth century, and are under the immediate dominion of the *French* King; whence alfo arofe the name of *Languedoc.* This newly conquered country, in order to diftinguifh it from the ancient royal dominions, being divided, according to its two general languages, or dialects, into the language of *Oc (Langue d' Oc) d' Oc)* and the language of *Oui (Langue d' Oui)* ; the former of which was the *Provinçal* and the latter proper *French*, the *Provinçals*, inftead of *Oui* faying *Oc.* The county of the language of *Oc* contained all fouth of the *Loire*, being dependent on the parliament of *Touloufe.* The country of the *Oui* tongue included thofe parts north of the *Loire*, and belonged to the parliament of *Paris* In time the pronunciation came to be fhortened from *Laugue d' Oc* to *Languedoc*, denoting the country of the language of *Oc*, which alfo was the etymology of the *Latin* word *Provincia*, or *Patria Oc̄citana*, or *Ouitania*, the prefent *Latin* for *Languedoc;*

This country is very mountainous, particularly the *Sevennes* or *Cevennes*, in Latin *Cabennæ*, which are of a great height and very fteep, yet being, as it were, the head quarters of the Proteftants, abound in people. It produces plenty of grain, fruits and fine wines, with large plantations of olives and mulberry-trees ; the latter for filk-worms and the former for oil, though its oil does not come up to that of *Provence.* The manner of threfhing * here is by caufing a number of mules and horfes coupled together to run about in a circle upon the grain which is ftrewed under them, and afterwards they alfo tread the ftraw into fuch fmall pieces as faves the trouble of chopping it. Among the products of its foil may be reckoned the paftel, or woad, here, in Latin *glaftum*, which is preferable to that of other countries. It is found particularly in *Upper Languedoc*, where the foil is rich, and there, in particular, in *Lauraguez* ; being ftill the fource of a pretty good trade to the inhabitants, though the difcovery of indigo has been no fmall detriment to it. Woad is a plant with a root ufually an inch thick, and a foot, or a foot and a half, in length, with five or fix leaves growing from it above the ground, which are a foot long and fix inches broad. The flower reprefents a crofs and bears an oblong feed, which is fometimes of a violet and fometimes of a yellow colour : of thefe the former is the beft. It is ufually fown in *February* and requires a good foil with great care after it begins to fhoot. When the herb is ripe, it is cut off clofe to the ground and immediately pounded to a pafte, which is made up into cakes, or balls, weighing a pound and a quarter ; thefe, when thoroughly dried in the fun, are reduced to a powder which gives a very beautiful and lafting blue. For above a hundred years paft a hundred thoufand cakes of paftel have been fent down the *Garonne* from *Touloufe* to *Bourdeaux*, which, at a moderate computation of thirteen livres per cake, amounts to one million and a half of livres ; but its vent, at prefent, is very fmall in comparifon of what it was formerly, indigo, as a finer powder, having fuperfeded the ufe of paftel, though this, with the preparation beftowed on indigo, would afford as good a die in every refpect. The fork-tree, in Latin *lotus arbor*

---

* We find frequent allufion to this manner of threfhing in the Scriptures, particularly in the firft book of *Corinthians*, *Job*, *Ifaiah* and the *Revelations* ; in the three latter of which it is applied to wine-preffes, and in *Moab* alfo to olives.

fructus

fruētus ceraſi celtis fruētu negricante, Aliſier, Micacoulier, Fanabrague, is alſo one of the curioſities of this country and grows particularly on Mount *Coutach,* on which the town of *Sauve* ſtands. It is alſo found in *Lower Languedoc* and *Provence,* and likewiſe in *Spain* and *Italy,* but is beſt cultivated at *Sauve.* The ſtem of this tree is from two to four feet high, when its further growth is generally checked. At the top of this ſtem grow a conſiderable number of ſtraight ſhoots, which are ſuffered to grow five or ſix feet; and, about the third year, are cut into the ſhape of three pronged forks; and, in the ſixth, ſeventh, and frequently not till the ninth, when they have compleated their growth, are carefully cut off cloſe to the ſtem, undergoing a ſecond formation in a hot oven.

The winds in this country are alſo worthy notice. Along the ſouthern coaſt of the province, which is properly a long valley extending from *Toulo.uſe* to the very ſea, frequently blows a wind which is generally weſt, but ſometimes north and ſometimes ſouth-weſt. This wind increaſes gradually, and being cool refreſhes the country in ſummer. The inhabitants call it *Cers,* in conformity to its ancient name, which was *Circius,* or *Cercius.* Oppoſite to this blows another from the eaſt, or ſouth-eaſt, and even ſouth, which is called *Autan,* and, in Latin, *Altanus.* This wind is firſt perceived near *Narbonne* and *Agde,* and at *Caſtelnaudari* is very violent. It is of a hot nature, cauſes head-aches, with loſs of appetite, and ſeems to ſwell the whole body. In the eaſtern part of the country is frequently felt a cold and very ſtrong north wind, which follows the courſe of the *Rhone* in the valley through which it runs from north to ſouth, being called *biſe,* or the *black,* and having in *Strabo* the ſynonimous name of *Melamboreas.* Sometimes too, in direct oppoſition to the former, blows a ſouth wind, called a ſea wind, which is uſually accompanied with a drizzling rain; but, when dry, has the ſame morbid effects as the *Autun* in *Upper Languedoc.* Laſtly, in the heat of ſummer, from the coaſts of *Leucate* to the *Rhone,* ſea-breezes con-ſtantly ſet in at nine or ten o'clock and laſt till about five in the evening, to the great refreſhment of the air. Theſe are called *garbin.* It is alſo to be obſerved here, that in the dioceſe of *Mirepoix,* at the foot of the *Pyrenees,* near the villages of *Blaud,* or *Eſcale,* in a narrow valley wholly environed with mountains, excepting towards the north-weſt, through certain aper-tures in, or betwixt, the mountains and two or three hundred paces wide, blows a very cool weſt, or north-weſt gale, which prevails chiefly in ſummer and then only in the night. In clear and warm weather this gale is much ſtronger than in a thick and cold air, and is called *le vent de pas.* In ſummer-time it cools the whole valley, and in winter prevents white froſts; but blowing only the night, the inhabitants of the village of *Blaud* are able to winnow their corn only at that time.

The marble dug near *Coſne* is very beautiful, and, at ſeveral places in *Lower Languedoc,* as *Laymont, Caſtres,* &c. are found turquoiſes. The

lead

lead mines too, near *Durfort,* in the diocefe of *Alais* and the iron mines here afford nothing remarkable, but here are certain fprings which are worthy of notice. The *Fonteft-orbe, i.e.* interrupted fpring,' rifing near *Belleftat,* in the diocefe of *Mirepoix,* on the river *Lers,* has its periodical ftoppages. That of *Fontanch,* betwixt *Sauve* and *Quiffac,* in the diocefe of *Nifmes,* is both periodical and mineral. *Balaruc,* near *Montpelier,* has warm baths. In this country are alfo mineral fprings at *Valhs,. Lodeve, Camares, Gabiar, Olarques, Baftide, Romeiroufe, Vendres, Guillaret, Campagne, Rennes, Maillat, S. Laurens, Toufet* and *Peyret.*

The principal rivers in the province are, the *Rhone*; the *Garonne,* which iffues from the valley of *Aran,* in the *Pyrenean* mountains; the *Aude,* which runs from *Cerdagne*; the *Tarne,* which proceeds from *Gevaudan*; the *Allier,* receiving its fource in the fame country, and the *Loire,* which iffues from the *Vivares* and falls into the *Mediterranean.*

The royal canal here is a work highly deferving a more ample defcription. The intent of it was fo to join the ocean with the *Mediterranean,* that fhips might pafs from one fea into the other without going round by *Spain.* The *Romans* are faid to have formed fuch a projeft and, under *Charles the Great, Francis* I.. and *Henry* IV.. it was again revived, but nothing done in it. Under *Lewis* XIV. *Riquet,* after employing twenty years in a minute confideration of every particular, having his gardener for his only counfellor, compleated the plan, which he executed betwixt the years 1666 and 1680. This canal begins at the harbour of *Cette,* or *Sette,* on the *Mediterranean,* and traverfes the lake of *Thau,* paffing by the towns of *Eraut* and *Orb,* and a quarter of a mile below *Touloufe,* is conveyed by three fluices into the *Garonne.* It is forty *French* leagues in length, has every where a depth of water of fix feet, fo that a cargo of 1800 quintals may be forwarded to any place on it; and the breadth, both fhores included, as far as they belong to the count of *Caraman,* is twenty-four toifes, reckoning fix feet to each toife. In the whole canal are fix *corps d'eclufes,* many of which confift of two, three and four fluices, by means of which the water is confined in the hilly parts.

For the commodioufnefs of this canal, at St. *Ferreol,* a quarter of a mile below *Revel,* betwixt two rocky mountains reprefenting a half-moon, is a large refervoir 1200 toifes in length, 500 in breadth and twenty deep, with a fuperficies 114,510 toifes, which, after receiving the rivulet of *Laudot,* iffues down from the mountain. This refervoir is further inclofed by a wall 400 toifes in length, twenty-two toifes high and twenty-nine feet thick, having a ftrong mole before it, which is again environed and further fecured by a wall of free-ftone. Under this dam runs a vaulted paffage reaching to the main wall, where three large cocks of caft brafs are turned and fhut by means of iron-bars, which cocks difcharge the water, through mouths as large as the trunk of a man's body, into a vaulted aqueduct,

duct, where it runs through the outward wall under the name of the river *Laudot,* continuing its courfe to the canal called *Rigele de la plaine,* and along this to another fine bafon near *Nauroufe,* not far from *Caftelnaudari,* which is 200 toifes in length and 150 in breadth, with a depth of feven feet. Though the above-mentioned cocks remain open for fome months fucceffively, there is, notwithftanding, no vifible diminution of the water in the great refervoir. Not far from *Beziers* are eight fluices, which form a regular and grand cafcade 160 toifes long; and, by means of thefe, the fhips crofs the river *Orb,* which runs near the city, and continue their voyage on the canal. Above it, betwixt *Beziers* and *Gapeftan,* is the *Mal pas,* where the canal is conveyed, for the length of 120 toifes, under a mountain cut into a very lofty arcade, the greateft part of which is lined with free-ftone, excepting towards the end where it is only hewn through the rock, which is of a foft fulphureous fubftance. At *Agde* is a round canal fluice with three openings, three different depths of the water meeting there; and the gates are of fuch ingenious conftruction, that veffels may pafs through by opening either the mafter pleafes; a contrivance which raifed admiration in the great *Vauban* himfelf. The leffer rivers and ftreams which might have prejudiced the canal have been carried under it by water-courfes, of which there are forty-four and eight bridges. This canal coft thirteen millions; part of which money was furnifhed by the King and part by the ftates of *Languedoc.* The King granted to *Riquet,* the inventor, and his male-heirs, all the jurifdiction and revenues belonging to it, fo that the crown does not come into poffeffion till the extinction of that line. Ships paffing on it, for every hundred weight, pay twenty fols; and the King himfelf, for military ftores, *&c.* fent by way of this canal, pays the fame impoft; fo that the revenue, efpecially in time of a brifk trade, is very confiderable: but, on the other hand, the charges attending it are alfo very great. The falaries of the feveral directors, receivers, furveyors, lieutenants, clerks and watchmen alone amount to 100,000 livres, befides the great expences of repairs. The counts of *Caraman,* defcendants of the above-mentioned *Riquet,* are alfo obliged to keep paffage-boats, which are drawn by-mules or horfes; and, though there are no paffengers, they muft go and come at ftated times.

The coaft of *Languedoc* is in itfelf dangerous and wants, befides, good and fecure harbours. The city of *Montpelier,* by means of the river *Lez* and the lakes near it, carries on, at the harbour of *Sette,* the greateft maritime trade in all the province, though the whole is of no great confideration. With refpect to the commerce of this province, it confifts principally of manufactures and other goods. Wine is exported hence to the coafts of *Italy,* oil to *Switzerland* and *Germany,* and, in a good year, corn alfo to *Italy* and *Spain*; dried cheftnuts and raifins to *Tunis* and *Algiers,* and cloth to *Switzerland, Germany* and the *Levant.* The cloth-trade to the *Levant* is the

moſt conſiderable, and three ſorts of it are ſent thither ; the beſt of which
is called *Mahon*, the ſecond *Londrins* and the worſt *Londres*.   The flight
woollen-ſtuffs exported from hence I omit, but ſilk is one of the moſt be-
neficial articles.   The traffic of particular cities ſhall be taken notice of
hereafter.

In the fifth century this country was over-run by the *Viſigoths*, who ſettled
here.   About the beginning of the eighth century theſe were driven out by
the *Saracens*, who ſoon began to encroach on their neighbours, but received
a dreadful overthrow from *Charles Martel*, and, being afterwards diſpoſſeſſed
by his ſon *Pepin*, the country fell to the crown of *France*.   At the cloſe of
the ninth century, the counts of *Toulouſe*, the margraves of *Gothien* and the
dukes or counts of *Provence* ſhared the ſovereignty of this province.   In the
tenth century, the counts of *Toulouſe*, by the junction of the marquiſates
of *Gothien* and *Provence*, with their own territories, became poſſeſſed of far
the greateſt part of *Languedoc*, being alſo maſters of many diſtricts in *Aquitaine*
of which they ſtyled themſelves Dukes or Princes.   As their families divided,
ſo they alſo divided theſe provinces; but, at the cloſe of the eleventh century,
their ſeveral territories were again united under the celebrated *Raymund de
S. Giles*, who firſt ſtyled himſelf duke of *Narbonne*, count of *Toulouſe* and
marquis of *Provence*, tranſmitting theſe honours to his deſcendants.   All
the parts of what is at preſent called *Languedoc*, were either by force or
treaty ſucceſſively brought under the immediate ſovereignty of the Kings of
France, particularly after the deceaſe of *Raymund* VIII. and his daughter
and ſon-in-law, when *Philip the Bold* took poſſeſſion of the country; but
it was the year 1361 before *Provence* was formally united to the crown.
Charles VII. ſeparated from *Languedoc* that part of *Aquitaine* which had formerly belonged to it; and *Lewis* XI. detached from it the whole part of
the government of *Toulouſe* on the left of the *Garonne*.   Since this no changes
have been wrought in the limits of *Languedoc*.   Towards the middle of
the ſixteenth century the reformation made a ſingular progreſs in this
country ; but was attended with a civil war, which was carried on with
greater rage and obſtinacy here than in the other parts of *France*.   *Languedoc* was the ſcene of continual troubles and cruelties for above one hundred
years, and more blood was ſhed here than in the whole Kingdom beſides;
yet is it ſtill thought to have a great number of ſecret Proteſtants in it.

In no one province of *France* are the clergy more numerous and wealthy
than in *Languedoc*, there being three archbiſhops and twenty biſhops here; and
the whole eccleſiaſtical revenues amount to 25,041,852 livres *per annum*.   It
has alſo a great number of inferior courts ſubject to the two ſupreme ones,
which are the parliament of *Toulouſe*, and the chamber of accounts and taxes
at *Montpelier*.   The governors of diſtricts here reſemble the bailifs in other
provinces, and are eight in number, every diſtrict having a court of juſtice, to
which lie appeals from the royal courts, or *vigueries*.   In it is alſo a ſuperior

foreſt-court with ſeven dependent on it. *Languedoc* is one of the provinces where the ſtates meet to deliberate on the concerns of the country. Theſe conſiſt of the clergy, the nobility, (namely, a count, a viſcount and twenty-one barons) and the commons, who are called together by the King every *October*. *Toulouſe* and *Montpelier* have mints of their own.

Under the governor of *Languedoc* are three general-lieutenants , namely, one over *Upper Languedoc*, or the dioceſes of *Montauban*, *Alby*, *Caſtres*, *Lavaux*, *Carcaſſonne*, *S. Papoul*, *Mirepoix*, *Rieux* and *Toulouſe* ; the ſecond over *Lower Languedoc*, or the dioceſes of *Alet*, *Limoux*, *Narbonne*, *S. Pons*, *Bezier*, *Agde*, *Montpelier* and *Lodeve* ; and the third over the dioceſes of *Nimes*, *Alais*, *Mende*, *Puy*, *Viviers* and *Uſez*. It has alſo nine inferior governors, each of which has reſpective departments. The firſt, *Toulouſe*, *Montauban*, *Rieux* and a part of *Upper Cemengues* ; the ſecond, *Caſtres* and *S. Pons* ; the third, *Papoul*, *Carcaſſonne*, *Lavaux* and *Alby* ; the fourth, *Gevaudan* ; the fifth, *Montpelier*, *Nimes*, *Alais*, *Lodeve* and the harbour of *Cette* ; the ſixth, *Beziers*, *Narbonne* and *Agde* ; the ſeventh, *Mirepoix*, *Alet* and *Limoux* ; the eighth, *Velay* and *Upper Vivarez* ; the ninth, *Uſez* and *Lower Vivarez*. The moſt general diviſion of this country, among the *French* geographers, is the following, *viz.*

I. *Upper Languedoc*, containing nine dioceſes, namely,

1. The dioceſe of *Toulouſe*, in the *Touloufain*, in which is

Toulouſe, the capital of *Languedoc*, and one of the moſt ancient cities in the Kingdom, being ſituate in a moſt delightful plain on the *Garonne*, and the reſidence of an archbiſhop and the ſecond parliament of the Kingdom. It is the ſeat alſo of an intendency and a receivers-office, of a provincial-court, of a governor, of a *viguerie*, of a royal tribunal, an admiralty, a *marechauſſee*, a forcſt-court, a mint, *&c.* Next to *Paris* it is ſaid to be the largeſt city in all *France*. The ſtreets in it are moſtly broad and the houſes built of brick. It is not, however, very populous ; and, though ſo finely ſituated, its trade is inconſiderable, its principal article that way being *Spaniſh* wool. This is chiefly imputed to their vanity, for a merchant here, after acquiring any conſiderable foitune, makes it his chief aim to be a *capitoul*, or to have a ſeat in the city council, and then gives over trade. The inhabitants too, in general, have no great inclination for commerce. The dioceſe of its archbiſhop contains in it 250 pariſhes, and his ſuffragans are the biſhops of *Montauban*, *Mirepoix*, *Lavaux*, *Rieux*, *Lombez*, St. *Popoul* and *Pamiers*. His revenue amounts to 60,000 livres a year, and his taxation at the court of *Rome* is 5000 florins. The cathedral here is very ancient, but contains nothing remarkable. In the church of *S. Sernin*, or *S. Saturnin*, are ſaid to lie thirty bodies of ſaints and among them ſeven of the apoſtles, one of whom is affirmed to be St. *James the Younger*, though St. *Jago de Compoſtella*, in *Spain*, boaſts alſo of being poſſeſſed of the body of that apoſtle. Near this church ſtands a ſecularized abbey. The church of *La Dorade* is ſo called

4

from a gilded, but now a coal-black, image of the Virgin *Mary*, which stands
on the great altar, and, in any time of calamity, is the conftant refuge of
the inhabitants. The name of the architect was *Luke*, which being marked
on the bafe, has given rife to an imagination among the commonalty, that
St. *Luke* carved this image. The church of the *Carmelites* here has a
magnificent chapel belonging to it. The *Dominican* church is large and
fplendid. Under the choir of the *Francifcan* church is a vault where bodies dry without mouldering. The fociety of the Blue Penitents is the
moft renowned fraternity in the whole Kingdom, having on its regifters
Kings, Princes of the blood and even eminent ecclefiaftics. The univerfity
here was founded in 1228, and in this city is alfo a noble college of Jefuits,
with a *collegium patrum doctrinæ Chriftianæ*, but particularly the *jeux floreaux*,
which was raifed to an academy in the year 1694. This confifts of forty-
two members, who employ their talents chiefly in poetry, and every year,
on the third of *May*, give away four prizes; *viz.* one golden and three
filver flowers. They meet in the town-houfe, which is a fpacious building
called the *capitolium*, and hence it is that the aldermen are termed *capitouls*.
The moft remarkable particulars in it are, 1. The window in the inner
court, through which the brave duke of *Montmorency* was led to the fcaffold
where he was beheaded. 2. The white marble ftatue of *Clementia*, who
is faid to have inftituted the above-mentioned floral prizes in the year 1540.
3. The chamber where the academy meets. 4. The ball, which is
adorned with the huftos of illuftrious and celebrated natives of *Touloufe*, as
alfo of two *Gothick* Kings who refided here, of fome counts of *Touloufe*
and feveral diftinguifhed civilians. 5. Fifteen large parchment folios, deli-
cately written and illuminated, being annals of the city, commencing from
the year 1288, and annually continued by the eldeft *capitoul*. Thefe
were drawn originally up in *Latin*, but, under *Francis* I. were altered to
French. The three firft volumes are principally filled with the portraits and
names of the *capitouls*, and the following contain all the memorable events
and tranfactions throughout the whole Kingdom. Here is alfo an academy
of the fciences and liberal arts. On the *Garonne*, near the city, is a large
mill confifting of fixteen wheels with horizontal water-wheels. The ftone-
bridge over the *Garonne*, which is 135 toifes long and twelve in breadth,
refting on feven arches of different magnitude, is a coftly piece of workmanfhip, and, except the cantons, which are of ftone, is wholly built of
brick. At half an hour's diftance below the city, the celebrated canal joins the *Garonne*, which here becomes navigable. In this city are
made *Bergamo* carpets, but of little value, together with fome flight filk
and woollen ftuffs. The latitude of *Touloufe*, *Caripuy* has found to be 43°
35′ 40″ and $\frac{1}{4}$.

    Grouille is a feat near *Touloufe* having fine gardens belonging to it and a
delightful grove.

Caftel-

Caſtelnau de Strettefons is a little place and a barony belonging to the ſtates.

Hauterive is a ſmall town, ſeated on the river *Ariege.*

Verfeuil too a little town.

Ville Franche de Lauragais is a little town in the country of *Lauragais.*

Mongiſcar, alſo a ſmall town.

Carming, or *S. Felix de Carming,* is a little town, having a collegiate-church.

2. To the dioceſe of *Montauban,* which city lies in *Quercy,* belong the following places ; *viz.*

Caſtel Sarazin, a little town in the *Touloufain,* ſeated on the *Garonne.* Its name is not derived from the *Saracens,* but probably from the little river *Azin,* which here falls into the *Garonne,* and thus ſignifies *Caſtel Sur Azin.* Near it formerly muſt anciently have ſtood a place called *Ville Longue,* the royal court of juſtice here ſtill bearing that name.

Montech is a little town.

Griſol, or *Grizoles,* a ſmall town, famous for good knives and ſciſſars.

Villemur is a little town ſeated on the river *Tarn.*

Roquemaux, a little town, having a fort.

3. The dioceſe of *Alby* conſtitutes the northern part of the diſtrict of *Albigeois,* from whence the *Albigenſes* take their name, being an upright people who ſo early as the eleventh century oppoſed popery, and adhered to the truth amidſt the violent perſecutions which their integrity drew upon them. But it muſt not be concealed that great numbers of worthleſs diſſolute people have mingled with them, ſpreading a lamentable corruption and licentiouſneſs among them. The country produces plenty of corn, wine, fruits and ſaffron ; is alſo populous but poor ; which is attributed to the ſickneſs of the year 1693 and their heavy impoſts. To the dioceſe of *Alby* belongs

Alby, in Latin *Albiga,* the capital of the country of the *Albigeois,* ſtanding on an eminence near the river *Tarn.* It is the reſidence of an archbiſhop, of a *recette,* of a *viguerie,* of a royal tribunal, of a *marechauſſée* and a foreſt-court. The ſuffragans of its archbiſhop are the biſhops of *Rhodez, Caſtrez, Cahors, Valres* and *Mende.* His dioceſe contains 327 pariſh-churches, and his yearly revenue amounts to 95,000 livres, out of which his tax at the court of *Rome* is 2000 florins. He is lord alſo of the city, though the high and low juriſdiction, as likewiſe the high demeſnes here, are in the King's hands. The cathedral belonging to it is one of the richeſt and fineſt in the Kingdom : the archiepiſcopal palace is alſo very noble, and the little town of *Chateauvieux* forms a handſome ſuburb. The walk called *La Lice,* a little above the city, is exceeding pleaſant. At a council held here in the year 1176, the doctrine of the *Albigenſes* was anathematiſed.

Realmont

Realmont is a little town, having a royal *prevoté*.

L'ifle d'Albigeois is a fmall town feated on the *Tarn*.

Carlus, a little town and county.

Gaillac, a town lying on the *Tarn*, which is here navigable, carries on a great trade, particularly in the excellent wine of its neighbourhood. It is alfo populous, and has large fuburbs; with a collegiate and parifh-church, two hofpitals and two convents.

Rabafteins, a town feated on the river *Tarn*, and carrying on a pretty trade, particularly in a rich good wine, has fome convents and a college.

Caftelnau de Levis is a little town feated on a hill, having the title of a barony.

Caftelnau de Bonnefons is a town and barony belonging to the States.

Caftelnau de Montmiral, a little town ftands high, and has a court of juftice.

Carmoux les Cordes is a fmall place:

4. The diocefe of *Caftres*, which makes the fouth part of *Albigeois*, contains

Caftres, the fecond city in *Albigeois*, being divided into two parts by the river *Agout*. It is the refidence of a receiver, an intendant, and a bifhop fubjeƈt to the archbifhop of *Alby*. In his diocefe are féventy-nine parifhes, and his revenue is 30,000 livres a year, out of which his tax at the court of *Rome* is 2500 florins. It formerly enjoyed the title of an earldom. On the commencement of the religious difturbances at the death of *Henry* II. the inhabitants declared for the reformation; and in the year 1567, pulled down all the catholic churches, and fortified the town; but in 1629 they were obliged to fubmit again to *Lewis* XIII. and the town was difmantled.

Villemur is a little place feated on the river *Agout*.

Caftelnau de Braffac, alfo a little town.

Mondredon is a fmall place.

Roquecuorbe, another; but having a handfome feat on the river *Agout*.

Ambres is a little town and marquifate, the owner of which votes with the barons in the affembly of the States.

Lautrec is a fmall town and *vicomté*.

La Caune, a little place, but having a royal court of juftice.

5. The diocefe of *Lavaux* conftitutes the lower part of the country of *Lauragais*, and contains

La Vaux, or *Lavaux*, in Latin *Vaurum*, a town lying on the river *Agout*, on the borders of *Lauragais* and *Albigeois*. It is the refidence of a receiver, of a royal court of juftice; and of a bifhop, fuffragan to the archbifhop of *Touloufe*, with a diocefe of eighty-eight parifhes: His income is 35,000 livres a year, and his tax at the court of *Rome* 2500 florins. This place was formerly fo ftrongly attached to the *Albigenfes*, that here it was where the religious war in the year 1561 began.

Puy-laurens

Puy-laurens is a fmall town, formerly governed by its particular lords. The Proteftants had an academy of fciences here; and fortified this place on the breaking out of the religious wars.

Revel is a little town, fortified in the religious wars by the Proteftants; but in the year 1629 the fortifications were razed.

Soreze, a fmall town, containing an abbey of *Benedictines,* was alfo fortified in the religious wars.

La Gardiolle is a little lown and barony belonging to the States.

6. The diocefe of *S. Papoul,* making the upper part of the territory of *Lauragais,* contains

S. Papoul, fmall, but the principal place of a *recette* and a bifhop's fee, fubject to the archbifhop of *Touloufe.* His revenue is 20,000 livres *per annum,* and his taxation at the court of *Rome* 2500 florins. This diocefe contains fifty-fix parifhes.

Caftelnaudari is the capital of *Lauragais,* and the feat of an *intendancy,* as alfo of a provincial, of a royal and a foreft-court. Near this place the duke of *Montmorency* was taken prifoner in the year 1652. The bafon made in thefe parts for the fupply of the royal canal, has been taken notice of above.

Laurac is a village, which gave name to the country of *Lauragais,* being formerly in much better circumftances than at prefent.

Villepinde and *la Becede* are two fmall towns.

7. The diocefe of *Rieux,* to which belong

Rieux, a little town feated on the river *Rife,* which here joins the *Garonne*; has a treafurer's office, and is a bifhop's fee fubject to *Touloufe,* containing ninety parifhes. The annual revenue of its bifhop is 18,000 livres, and his taxation at the court of *Rome* 2500 florins. In this town is made cloth.

S. Sulpice is a little place lying on the river *Lefe,* and annexed to the *Maltefe* commanderie of *Renneville.*

Montefquiou is a little town which fuffered extremely, during the religious war in the year 1586.

Cazeres is a fmall town feated on the *Garonne.*

Carmain, alfo a little town and earldom.

8. The diocefe of *Mirepoix,* containing

Mirepoix, in Latin *Mirapicæ,* and *Mirapicum,* is a little town feated on the *Lers,* containing a treafurer's office and a bifhop fuffragan to *Touloufe,* with an annual revenue of 24,000 livres, out of which his taxation at the court of *Rome* is 2500 florins. This diocefe confifts of fixty-two parifhes. It is alfo a marquifate belonging to the houfe of *Levis,* the defcendants of which have been in continual poffeffion of it above five hundred years, and fit as barons in the affemblies of the States.

Carlat de Roquefort is a little place feated on the *Befegue,* and remarkable only for being the birth-place of the celebrated *Peter Bayle.*

Fanjaux

Fanjaux is a fmall town.

Belleflat, a little place, in which, not far from the river *Lers*, is the cele-brated fpring called *Fonteft-orbe*, *i. e.* ' the interrupted fpring,' which iffues from a cavity under a rocky mountain; and in hot dry weather ceafes to run. This interruption generally falls out in the month of *June*, *July*, *Au-guft* and *September*, but a long rain fets it a running again for ten or twelve days; and in a rainy fummer its courfe is inceffant. The water of this fpring difcharges itfelf into the river *Lers*. This fountain is known by two refervoirs of water in the mountain, one higher than the other, and com-municating by a little channel, and likewife fome apertures through the lower refervoir in which the water runs off.

9. The diocefe of *Cominges*. The country of *Cominges*, or *Comenges*, be-longs properly to the government of *Guienne*; but eleven parifhes lying in *Languedoc* are called *Leffer Cominges*. Of thefe the principal are

Valentine, a little town, through which lies one of the roads to *Spain*, having a royal court of juftice.

S. Beat, a ftrong town feated betwixt two mountains, at the conflux of the *Garonne* and *Pique*, by the former of which it is divided into two parts: All the houfes here are of marble, it being the only kind of ftone here-abouts. In it is a priory; and it carries on a very profitable trade in cattle, horfes, and mules.

II. *Lower Languedoc* confifts of the following diocefes, namely,

1. The diocefe of *Alet*, being part of the county of *Razez*, in Latin *Pagus* or *comitatus Reddenfis*, which is fo called from the town of *Redda*, long fince deftroyed. In it are

Alet, in Latin *Alecta*; or *Electa*, a town feated at the foot of the *Pyrennes* on the river *Aude*, being the principal place of a collection, and owing its original to the *Benedictine* abbey founded here, which was afterwards changed into a bifhopric fubject to the archbifhop of *Narbonne*, and confifting of eighty parifhes, with an annual revenue of 18,000 livres, out of which its taxation at the court of *Rome* is 1500 florins.

Limoux, the capital of the earldom of *Rezez*, lies on the river *Aude*, in a country abounding with good white wine. This place is the refidence of a provincial governor. In it are made linen and woollen ftuffs; and the ftore-houfes alfo of all the iron-works in this neighbourhood are here.

Arques is a barony belonging to the States.

Rennes, a little place, containing fome warm baths.

Quilla, a fmall town lying on the river *Aude*, the feat of a foreft-court for the diftrict of *Sault*.

The diftrict of *Sault*, in Latin *pagus de Saltu*, has a royal court of juftice, fubject to the intendancy of *Limoux*. The principal place in it is

Efcouloubre, which, though fmall, was of confiderable importance to the *Spaniards* whilft in poffeffion of *Roufzillon*.

The

The little diftrict of *Fenouilledes*, held a confiderable time by the counts of *Roufzillon* as a fief of *France*. The only places worth notice in it are St. *Paul*, a little town lying betwixt mountains on the river *Aigli* or *Egli*. *Caudies*, a fmall place feated at the foot of the *Pyrennes*.

2. The diocefe of *Carcaffonne*, containing the ancient county of this name, in which are the following places, *viz*.

Carcaffonne, a very ancient city, the principal place of a collection and the feat of a provincial court, of an intendancy and a *marechauffee*, being divided by the river *Aude* into the upper and lower. The upper, which is old and ill built, confifts of a caftle and the palace of the bifhop, who is a fuffragan to *Narbonne*, with a diocefe of one hundred and fourteen parifhes under him, and a yearly revenue of 35000 livres, out of which his taxation at the court of *Rome* is 6000 florins. The lower city is new, and fo handfomely built that it is the moft regular town in all *Languedoc*. The palace here, in which the provincial court holds its feffions, the townhoufe, the churches, the convents and chapels are all elegant ftructures. Some beautiful fine cloths are made here.

Caunettes en Val, and *Caunettes les Mouffoulins*, are two market-towns.

La Grace, or *Graffe*, is a little town feated on the river *Orbicu*, among the mountains of *Corbiere*, containing an ancient abbey of *Benedictines*.

Mas de Cabarde is a little place, having a ftrong caftle.

Caunes, a fmall town, having an abbey.

3. The diocefe of *Narbonne* contains

Narbonne, anciently *Narbo*, with the addition of *Martius*, and *Decumanorum colonia*, being a fortified town feated on a canal which paffes through it, and unites with the river *Aude* and the large canal with lake *Robine*, in Latin *Rubrefus* or *Rubrenfis*, and thus alfo with the fea: In this lake was anciently an harbour, but this has long fince been choked up. This city is the refidence of an archbifhop, of a receiver's office, of a *viguerie*, an admiralty, a *marechauffee*, and a court of mintage. The fuffragans of its archbifhop are the bifhops of *Beziers*, *Agde*, *Carcaffonne*, *Nimes*, *Montpelier*, *Lodeve*, *Uzes*, *S. Pons de tomieres*, *Alet*, *Alais*, and *Perpignan*. Its province contains one hundred and forty parifh churches, and its revenue amounts to 90,000 livres *per annum*, which is taxed at *Rome* in 9000 florins. The moft remarkable thing in the cathedral is the marble monument of *Philip the Bold*. In this town are alfo two collegiate churches, one abbey, one feminary, &c. The remains of *Roman* antiquities ftill in being here confift of ftones and infcriptions placed for the moft part in the court walls of the archbifhop's palace, in the garden belonging to which we alfo fee a beautiful *Roman* fepulchre of white marble in the form of an altar. As this city lies in a deep bottom betwixt mountains, the water, in long rains gather here to fuch a degree that it is dangerous going out, ever fo little, of the city. This place was formerly the capital of a *vicomte*.

Sigean is a little town lying on the *Berre*, and giving name to the neighbouring lake, which communicates with the sea. In this place, in the year 737, *Charles Martel* gained a signal victory over the *Saracens.*

Capestan is a small town seated on a lake on the river *Aude.*

Bisan, a market-town.

Les Caunes, also a market-town, having an abbey.

Peyrac de Mer is a small place.

Peyrac de Minervoix, a little town.

Burban and *Tuchan*, two small places.

Rieux is a little town and county, which votes among the barons in the provincial assemblies.

10. The little district of *Corbieres*, with a valley of the same name, is famous for the terrible defeat of the *Saracens* there in the year 737.

11. *Leucate*, an old little town, and formerly fortified, stands betwixt the sea and a lake; being sometimes called *Leucate* and *Salses.*

12. The little territory of *Termenez*, so called from the ancient castle of *Termes*, which stood on a steep rock, and was one of the strongest places in this country; once had its particular lords.

4. The diocese of *S. Pons*, in which are

S. Pons de Tomieres, a town seated among mountains on the river *Jaur*, being the principal place of a collection, and the residence of a forest-court. Its bishop is subject to the archbishop of *Narbonne*, having under him a diocese of forty parishes, with a revenue of 30,000 livres, out of which he is taxed at the court of *Rome* in 3400 florins.

S. Chignan is a little town, having a manufactory of cloth.

Cessenon, a small town.

Crusy, a little town and barony.

Olargues and *Olonzac*, two towns.

5. The diocese of *Beziers* contains one of the most fruitful parts of the country. To it belongs

Beziers, anciently *Bliterræ, Biterræ, Bæterræ*, and *Beterræ*, being a pretty large city seated on an eminence near the river *Orbe*, and the great canal. It is the residence of a bishop, of a collector, and also of an intendancy and a provincial court. The bishop of this place is suffragan to *Narbonne*, with a diocese of one hundred and six parishes, and a revenue of 30,000 livres; out of which his tax at the *Roman* court is 208 florins. Exclusive of its cathedral in this city are three abbies; one secularized and one college of Jesuits. The adjacent country is scarce to be equalled in all *France* for its pleasantness.

Gignac is a little town, containing a *viguerie.*

Cellie, Bec de Rioux, Colombiere La Galliarde, and *Marviel* are all market-towns.

Gabian.

Gabian is a village, having a fpring of a mineral water; and near it is a rock, out of which runs black petroleum, famous for the relief it affords in feveral diftempers, fuch as hyfteric fits, cholics, &c. and alfo in gangrenes, tumours proceeding from colds, wounds, worms in children, and even the iliac paffion. This *petroleum* was firft difcovered in the year 1608. In this country are frequently found lumps of bitumen, which fupply all the ufes of foap.

Vendres is a market-town feated at the mouth of the river *Aude* near the lake of *Vendres.* Here alfo is a mineral water.

6. The diocefe of *Agde* is one of the richeft parts of the country, and contains

Agde, in Latin *Agatha,* a fmall but populous city on the river *Eraut,* which at half a league's diftance empties itfelf into the fea. It has an harbour for fmall veffels, defended by a little fort at the mouth of the river. Moft of its inhabitants are tradefmen, or fea-faring people. The bifhop of this place is lord of the city, and ftiled Count of *Agde;* having an annual revenue of 30,000 livres, out of which he is taxed at the court of *Rome* in 1500 florins. The diocefe is within the province of *Narbonne,* and contains eighteen parifhes. Clofe by the city is a *Capuchin* convent, much reforted to on account of an image of the Virgin *Mary* there, which ftands in a particular chapel.

Brefcou is a fort ftanding on a rock in the fea at the mouth of the river *Eraut,* near Cape *Agde.*

Pefcenas, in Latin *Pifcenæ,* is a little town feated on the river *Pein,* in a very fine country; and having a collegiate church, a college, and fome convents: It gives alfo the title of count to the Prince of *Conty.*

Florenfac, a little town feated betwixt *Agde* and *Penefas* on a branch of the *Eraut,* has the title of a barony, by virtue of which its owner, the duke of *Uzez,* enjoys a feat in the provincial affemblies.

S. Tiberi, or as it is corruptly called St. *Tuberi,* is a little town on the river *Eraut,* formerly confiderable, as a royal court of juftice.

The harbour of *Cette, Sette,* or *S. Louis,* near the cape of this name, was erected at a vaft expence there by *Lewis* XIV. yet admits only of gallies and fmall craft, which here lie covered. The agitations of the fea throwing up continually great heaps of fand, the province have eftablifhed a fund for the conftant clearing it, fo as to have always a depth of feventeen or eighteen feet. Here it is where the ftupendous work of the royal canal begins.

Meze is a little town, one third of which belongs to the jurifdiction of the bifhop of *Agde.*

7. The diocefe of *Lodéve* is barren in grain, but the trade in cattle, which are fed on the mountains, and the manufactures of hats and cloths here, enrich the inhabitants. In it are

　　　　　　　　　　　　Lodéve,

Lodéve, in Latin *Leuteva*, and *Luteva*, an old city, which is the principal place of a collection, and an intendancy. Its bifhop is lord of the place, and ftiles himfelf count of *Lodéve*. The diocefe belonging to it, which confifts of fifty-one parifhes, is fubject to the archbifhop of *Narbonne*, and its yearly revenue is 22,000 livres, out of which his taxation at the court of *Rome* is 1060 florins. There is alfo an abbey here.

Clermont, a town ftanding on a rock in the *Lergue*, makes fine cloth and hats ; being a barony with a vote in the affembly of the States.

Villeneuve les Clermont, half a league diftant from *Clermont*, has a fine woollen manufacture, fupplying moft of the cloths which are fent to the *Levant*.

Canet is a market-town.

8. The diocefe of *Montpellier* is a delightful country, producing all kinds of grain, and ftanding amidft vineyards and plantations of olives. In it are

Montpellier, in Latin *Mons Peffalanus*, a city feated on a hill, on the river *Lez*, which owes its original to the deftruction of the city of *Magnelone*, which ftood in an ifland on the lake of *Thau*. Next to *Touloufe*, it is the moft confiderable city in all *Languedoc*; and contains a chamber of taxes and accounts, with an intendancy, a treafurer's office, a mint, a provincial court of juftice, a fheriff's court, an independent court, a foreft-court, a *marechauffée*, a confulate, &c. It has a great many fine houfes, but the ftreets are very narrow; yet with this conveniency, that in hot weather they have linen awnings drawn over them, under which a perfon may walk without being incommoded by the fun. The diocefe of this place is fubject to the archbifhop of *Narbonne*, and contains in it 107 parifhes. The revenue of its bifhop is 32,000 livres; and his tax at the court of *Rome* 4000 florins. The univerfity here is particularly famous for phyfic, being founded by phyficians who were expelled *Spain* in the year 1180. The celebrated fcarlet gown of *Rabelais*, with which all doctors of phyfic are invefted at taking their degrees, has long fince ceafed to exift as an original, the ftudents having from time to time cut off little fnips ; and thus reduced this venerable relic ufed in the ceremony fo much that the robe now ufed is at leaft the third or fourth fubftitute. In the year 1706, an academy of fciences was erected here. In it is an excellent phyfic-garden, in which public lectures on botany are read. On the *place de Louis le Grand*, near the moft delightful walk called *Peyrou*, is an equeftrian ftatue of brafs of that monarch ftanding on a marble pedeftal, erected by the States of *Languedoc*; and the city gate leading to it is in form of a triumphal arch. Near this place lies the citadel, which commands the city and the neighbouring country. The principal commodity here is verdigreafe, in which it carries on a very confiderable trade, as alfo in wool which is brought from the *Mediterranean*, in wine, *aqua vitæ*, *Hungary* water, cinnamon-water, &c. It makes alfo great quantities of woollen carpets and fuftian.

4 *Perault*

Perault is a village one *French* league diftant from *Montpellier.* Near it, clofe by the lake of *Thau,* is a pit where the rain-water which gathers in it boils continually, yet retains its ufual coldnefs. The inhabitants bathe in it for diarrhæas.

Balaruc is a town lying on the lake of *Thau,* with warm baths which, according to *Aftruc,* contain little or no fulphur but fome falt. The water of thefe baths is lighter than the ufual drinking-water of the place, and, though not of fufficient heat to boil eggs, yet thefe being hung in it in a dry veffel, it hatches them as foon as a hen. The water after being taken out retains its heat at leaft eight hours, is flower in boiling than common water and the ebullition lefs vehement. This manifefts that its heat is not de-rived from any fubterraneous fire, but from an inward fermentation of its parts. It has alfo a very good effect when ufed for drinking, bathing and fomentation.

Frontignan is a fmall town, feated on the lake of *Thau,* famed for its ex-cellent *mufcadine* wine. From hence alfo come thofe excellent *mufcadine* grapes called *pafferillis.*

Obf. The lake of *Thau* is twelve *French* leagues in length, being feparated from the fea by a very narrow tract of land, but, at a place called *Grau de Pelavas,* communicates with the gulph of *Lion,* which takes not its name from the city of *Lion,* this lying at a very great diftance from the fea, but rather from the violent ftorms fo frequent in this fhallow part of the fea, and which deftroy the fhips as a furious lion does its prey.

Lunel is a little town, anciently fortified.

Ganges, a little town and marquifate on the river *Aude,* is one of the ba-ronies which vote in the Diet.

Barave, a fmall town, feated on the river *Vene,* and belonging to the bifhop of *Montpellier,* lies in the marquifate of *Marquerofe.*

Caftries is a town and barony belonging to the States,

Anianne, or *S. Benoit,* is a fmall town, feated at the foot of a mountain near the river *Arre,* having an old abbey of *Benedictines.*

9. The diocefe of *Nimes* contains a level country, fruitful in grain, wine and oil ; and where alfo great quantities of filk are made. In it are

Nimes, or *Nifmes,* in Latin *Nemaufus,* a large city, very pleafantly fituated, having hills covered with vines on one fide and on the other a very fertile plain. In it is a collector's office, an intendancy and a provincial court of juftice. The ftreets are in general narrow, but kept pretty clean, and the houfes of ftone and handfomely built. One third part of the inhabitants are fecret Pro-teftants. Its bifhop is fubject to the archbifhop of *Narbonne,* and has a dio-cefe of 215 parifhes, with an annual revenue of 26,000 livres, out of which his taxation at the court of *Rome* is 1200 florins. In it is a Jefuits college and other convents, with an academy of *belles-lettres,* founded here in the year 1682. The citadel confifts of four baftions. In this town is fuch a

number

number of manufactures that its cloth and silk trade alone overbalances that of the whole province. The antiquities in and about this city are, 1. The celebrated amphitheatre called *Les Arenes*, which, though built on the inside with houses, is, of all the amphitheatres now exifting, the leaft damaged. The free-ftone in many places is of an amazing length and bignefs. 2. The quadrangular houfe, being an oblong ftructure embellifhed with thirty-two exquifite pillars of the *Corinthian* order, and not improbably a temple built by the Emperor *Adrian* in honour of *Plotina* confort to *Trojan*. 3. A temple, confifting of moft beautiful free-ftone, above one half of which is ftill ftanding, and is fuppofed to have been dedicated to *Diana*. 4. The fountain, with a very full and fine fpring, at the foot of a high rocky mountain, and lately difcovered on the cleanfing of a marfhy pond. From feveral remains it appears that this place was a publick bath, and lower, where the fpring widens, a fquare infular edifice for pleafure. 5. The octangular tower, which is fifteen toifes in height and of a proportionate bignefs, the whole being a folid ftructure from the bottom to the top. It is remarkable, that the heads of all the *Roman* eagles found here are ftruck off; which is imputed to the *Goths*, who, on their reduction of this country, did this by way of infult to that haughty enemy whom they had fubdued. This city is of great antiquity. The *L'efplanade*, or ' walking place,' without the city, is extremely delightful.

Caverac is a little place, having a fine feat.

Le Pont du Guard is a moft valuable piece of *Roman* antiquity, lying half a *French* league from the market-town of *Remoulins*, and being a part of the great aqueduct by which the fpring of *Eure*, near the little town of *Uzez*, was carried the length of nine hours diftance to *Nimes*, and there diftributed in the amphitheatre, publick fountains and private houfes. The bridge in queftion confis of three ranges of arches built one upon another and traverfing the river *Gardon*, and thus connecting the high rocks on both fides the river fo as to afford a free paffage to the water of the fpring above-mentioned from one mountain to the other. The lower bridge confifts of fix arches, is 438 feet in length and eighty-three high. The fecond is fupported by eleven arches, each fifty-fix feet in diameter and fixty-feven high ; and of fo extraordinary a width, that, befides the pillars on which the third bridge, or range of arches refts, there is a horfe and foot way leading through it which is alfo fecured by a breaft-work. Laftly, the third range, which refts on the fecond, and is 585 feet and a half in length, confifts of thirty-five arches, each of which is feventeen feet in diameter ; and over this uppermoft range runs the aqueduct, which is of fuch a breadth and height, that, with a little inclination of the body, a perfon may walk in it. The whole is of very hard and durable free-ftone. From an eminence towards *Uzez* are here and there feen entire ranges of arches, all which belonged to this vaft aqueduct.

Beaucaire,

Beaucaire, in Latin *Belliquadrum*, is a town, lying on the *Rhone*, directly facing *Tarafcon*, and faid to derive its name from a fquare caftle deftroyed in the year 1632. In it is a collegiate-church, and its fair, which begins on the twenty-fecond of *July* and lafts ten days, is particularly famous for the vaft quantities of raw filk fold at it.

S. Gilles is a little town, owing its origin to a convent.

Sommieres, in Latin *Sumerium*, a town, feated on the *Vidourle*, is the refidence of a *viguerie* and a royal jurifdiction, having alfo a caftle.

Aimargues, in Latin *Arniafanicæ*, is a little town, fituate among fens, on the river *Viftre*, having the title of a barony.

Aiguefmortes is alfo a little town, ftrong by reafon of its marfhy fituation. It ftood formerly on the fea and had a harbour, but is now above two *French* leagues up the country and the harbour quite filled up. It is the refidence of a court of admiralty, of a *viguerie*, of a royal court of juftice and an office of the five large farms. In the neighbouring parts too are feveral lakes.

Peccais is a market-town with a fort built for the defence of the falt-works here.

Quiffac is a little place, feated on the *Vidourle*, betwixt which and the town of *Sauve*, at the foot of a mountain near the river, is a periodical fpring, running and intermitting twice in twenty-four hours. The flow lafts feven hours twenty-five minutes, and the intermiffion five hours. This muft be occafioned by a cave, or refervoir, in the mountain, which being filled every five hours, difcharges itfelf through a canal in form of a fyphon. This water is drank as a mineral, and when heated is ufed for bathing. It contains a confiderable quantity of fulphur in it and is reputed very good for all diforders of the eyes.

Le Grand Galargues is a place where a fine blue and red die is made out of the nightfhade (in *French Morelle)* or funflower, which they call the *Tournefol*.

Calviffon is a little town in the diftrict of *Vonage*, and one of the voting baronies.

10. The diocefe of *Uzez*, commonly called *l' Ufege*, or *Ufegais*, is one of the largeft diocefes, extending from the mountains of the *Cevennes*, in which are feveral of its parifhes, as far as the river *Rhone*. It abounds in grain, oil, filk, good wine, and particularly in numerous breeds of fheep. In it are

Uzez, in Latin *Ucetia*, a little town lying among mountains on the river *Eyfent* and bearing the title of a dukedom. The bifhop of this place is fuffragan to the archbifhop of *Narbonne*, with a revenue of 25,000 livres, his tax to the court of *Rome* being 1000 florins. This diocefe contains 181 parifhes in it, and the duke's palace here is a large building. Below that of the bifhop rifes the fpring of *Aure*, which formerly fupplied the aqueduct

of

of *Pont du Guard* with water. On the reformation in *France, Jean de S. Gelais*, bifhop of the city, together with his whole chapter, relinquifhed popery and became Proteftants; which was the religion of the whole place.

Peyret is a mineral fpring ufed both for drinking and bathing.

Aramon, a little town and barony fituate on the *Rhone*.

Villeneuve d'Avignon, is a little town, ftanding high by the fide of the *Rhone*, directly oppofite to *Avignon*, but lying indeed in the diftrict of *Uzez*, and being alfo under its collection. It belongs to the diocefe of *Avignon*. In it is an abbey and a fine monaftery of *Carthufians*.

Roquemaure, a town, in the diocefe of *Avignon* and feated on a fieep rock by the *Rhone*, is a barony and the refidence of a *viguerie*, in the diocefe of *Avignon*.

Bagnols, or *Baignols*, a little town ftanding on a rock near the river *Cefe*, belongs to the Prince of *Conti*. In it is a priory and one convent.

Pont. S. Efprit is a little ill built town, feated on the *Rhone*, but having a ftately bridge of ftone over it with nineteen large and feven fmall arches. This bridge is 420 rods long and two rods four feet in breadth. In it is a priory and a hofpital, with a citadel for the defence of the place.

Barjac, or *Bargeac*, is a little town, bearing the title of barony, near which is a fine feat called *Banos*.

S. Ambroife is a fmall town lying on the river *Cefe*.

Youfet, a village, much celebrated for its excellent mineral fprings of fulphur.

11. The diocefe of *Alais* lies among the mountains of the *Cevennes*, and makes a part of the province of the *Cevennes*. In it are

Alais, or *Alez*, a pretty large populous city, feated on the river *Gardon*, being the principal place of a collection, with the title both of a county and barony; the former belonging to the Prince of *Conti*, and the latter fhared betwixt two proprietors. The bifhop of this place is fuffragan to the archbifhop of *Narbonne* with an annual revenue of 16,000 livres, and his tax at *Rome* is 500 florins. The diocefe confifts of eighty parifhes. This bifhopric was founded in the year 1691, for forwarding the converfion of the great numbers of Proteftants hereabouts; and the Jefuits college was added as an auxiliary to it; but, in the year 1689, a citadel had been before built to awe the religionifts. The quantity of unwrought filk carried out of this city in a year amounts at leaft to 1,200,000 pounds weight.

Anduze, a town, bearing the title of a barony and feated on the river *Gardon*, carries on a confiderable trade in ferges and other woollen ftuffs.

S. Hupolite is a new town on the river *Vidourle*, the greater part of its inhabitants being alfo new converts to popery.

12. The diocefe of *Mende* contains the county of *Gevaudan*, formerly *Gavauldan*, or *Gabauldan*, a name derived from its ancient inhabitants the *Gabali*.

Gabali. Upper Gevaudan lies almoft wholly in the mountains of *La Marguerite* and *Aubrac* ; but *Lower Gevaudan* conftitutes a part of the *Cevennes.* In it are

Mende, in Latin *Memmate*, a populous town, feated on a mountain on the river *Lot;* being a county and the refidence of a collection and intendency. Its diocefe contains 208 parifhes, and the revenue of its bifhop is 40,000 livres, out of which his taxation at the court of *Rome* is 3500 florins. In it is one college and four convents.

Javoux is a very ancient market-town in the barony of *Peyre*, and formerly a refidence of the bifhop.

Marvejols, or *Marjejols*, *Marvege*, in Latin *Marilogium*, is a handfome, populous, trading town, fituate in a charming vale watered by the river *Colange*. In it is a collegiate-church and a convent. Under *Henry* III. the inhabitants of this place embraced Proteftantifm; but, in the year 1586, were obliged to furrender at difcretion to the King's forces, when the Papift general, after many cruelties exercifed on the inhabitants, deftroyed the whole town.

Florac is a little town and barony.

Efpagnac, a fmall place on the river *Tarne*, containing a priory and a collegiate-church.

Canourgue is a priory of *Benedictines.*

13. The diocefe of *Viviers* contains the county of *Vivarais*, which is divided by the river *Erieu* into the *Upper* and *Lower*. A part of the former belongs to the archbifhop of *Vienne*. This country has its particular ftates, in the meetings of which the bifhop of *Viviers* prefides; and thefe fend deputies to the general diet of *Languedoc.*

1. *Lower Vivarais* is, by the induftry of its inhabitants, fo well cultivated, as to fupply itfelf with wine, and makes alfo great quantities of filk. It contains

Viviers, in Latin *Vivarium*, the capital of all the *Vivarais*, and the refidence of a collector's office and a *marechauffee*. It is but fmall and mean, and lies on the *Rhone*. Its diocefe contains, however, 314 parifhes in it, and its bifhop, exclufive of the title of Prince of *Donzere*, which is a village in *Dauphine*, enjoys an annual revenue of 30,000 livres, his taxation at the court of *Rome* being 4400 florins. In it is but one convent.

S. Andiol, a large market-town, feated on the *Rhone* at its junction with the *Ardeche*, is the ufual refidence of the bifhop of *Viviers*, and contains two convents.

Villeneuve de Berg, a little town, feated on the river *Ibbie*, is the refidence of a governor and foreft-court.

Aps, or *Alps*, in Latin *Albe*, is a little market-town, formerly the capital of the *Helvi.*

Aubenas is a little town and barony containing a college of Jefuits.

Joyeuze, a fmall town, feated on the river *Beaune,* which, at a fmall diftance from this place, joins the *Ardèche*; together with *Balbiac* and *Rezieres,* forms a dutchy and peerage.

Argentiere is a fmall town.

Vals, a walled town near the little river *Volane,* ftands in a valley, having five celebrated cold mineral fprings clofe by it. That of *La Marie,* being acid and diuretic, is ufed for nephritic diforders. *La Marquife* is rather faline than acid. *La S. Jean* is lefs acid than the other. *La Camufe* abounds more in falt than acidity; and *La Dominique* is of a moft difagreeable tafte and offenfive to the ftomach.

S. Laurent is a fmall place famous for its baths.

Boulogne, a barony, belonging to the ftates.

Privat, a market-town.

Pouzin, a little place.

2. *Upper Vivarais,* though every-where mountainous, is well cultivated. In it are

Annonay, a fmall town, feated on the little river *Deume,* being the capital of *Upper Vivarais,* a marquifate and an intendancy, belongs to the family of *Rohan-Soubife.*

Andance is a little town, lying at the foot of a mountain, near the conflux of the rivers *Dome* and *Rhone.*

S. Agreve is a fmall place at the foot of the mountains.

Tournon, in Latin *Taurodunum,* is a little town, ftanding on a hill near the *Rhone,* with a caftle, a celebrated college of Jefuits and a convent. In this town was born the famous geographer *Pierre d'Avity.*

14. The diocefe of *Pui* contains the country of *Velay,* which, though its mountains are of fo great a height as to be half the year covered with fnow, yet does it not want for a fufficiency of grain. It has alfo particular ftates of its own. In it is

Pui, the capital of the country, ftanding on the ruins of the ancient city of *Anis* and receiving its name from its fituation on a mountain, *puich,* or *puech,* in the *Aquitain* tongue fignifying a hill. It is pretty large and populous, has a collector's office, an intendancy and a provincial court of juftice. Its bifhop is lord of the city and holds immediately of the Pope; but, in ex- ternals, his diocefe, which confifts of 239 parifhes, belongs to the arch- bifhop of *Bourges.* His revenue is 25,000 livres, and the *Roman* taxation 2650 florins. In the cathedral here is an image of the Virgin *Mary,* which is worfhipped with particular devotion. In it alfo are kept a great num- ber of relicks. Among the numerous churches and convents here the moft remarkable are the Jefuits college and the abbey of St. *Clair.* Not far alfo from this city is the fource of the *Loire.*

Polignac is a country town and marquifate giving title to a celebrated fa- mily of that name.

Montfalcon,

Montfalcon is a little place, having a royal court of juſtice.

Moniſtrol is a ſmall town lying betwixt two rocks, in whic h the biſhop of *Pui* has a moſt charming ſeat.

Solignac, *Tanſe*, *Crapone* and St. *Diſier*, are all ſmall places.

Obſ. To the dioceſe of the biſhop of *Pui* belong alſo certain places in *Auvergne*, as *S. Paulien*, &c.

The GOVERNMENT *of* FOIX.

THE government of *Foix* contains the diſtrict and county of *Foix*, with the territories of *Douneſan* and *Andorre*. In the ſouth it borders on the *Pyrenean* mountains and *Rouſzillon*; weſtward on *Gaſcogne*, and eaſtward and northward on *Languedoc*. The principal rivers here are the *Auriege*, which riſes on the borders of *Foix* and *Cerdagne*, becoming navigable at *Hauterive* in *Languedoc*, and a mile above *Touloiſe* falls into the *Garonne*; and the *Rize*, which iſſues from a mountain near *Maz d'Azil*. This country is a dependency of the parliament of *Touloiſe* and conſtitutes part of the lands of the ſtates, who are every year called together by the King, and conſiſt of the clergy, the nobility, the commons and the peaſants. The trade of it conſiſts of cattle, reſin, turpentine, pitch, cork, jaſper and, particularly, iron. Excluſive of the governor, it has alſo a general-lieutenant.

1. The county of *Foix*, ſo called from its capital, had formerly counts of its own, deſcendants from *Roger* II. count of *Carcaſſonne*, whoſe ſon *Bernard*, in the eleventh century, was the firſt count of *Foix*. *Henry* IV. in the year 1607, re-united this county to the crown. It is divided into *Upper* and *Lower Foix*.

1. *Upper Foix* lies among the mountains, whence all its products are wood, iron and mineral waters, with ſome paſtures. In this county are ſeveral caverns with very ſingular figures in them formed by the petrifying waters, and it contains the following places, *viz.*

Foix, the capital of the government, lying on the river *Auriege*, has a caſtle and abbey, the abbot of which takes place in the Diet next to the biſhop.

Taraſcon is a little town on the river *Auriege*, and one of the four principal in the country, but, in the firſt year of this century, was almoſt totally conſumed by fire. In it are ſeveral iron manufactures.

Acqs is a little town at the foot of the *Pyrenees*, which are hereabouts of a ſtupendous height, and is ſo called from its hot waters.

La Baſtide de Seron is a little town ſeated on an eminence.

2. *Lower*

2. *Lower Foix* produces fome grain and wine. In it are the following places ; namely,

Pamiers, or *Pamiés*, formerly *Apamiés*, *Apamiæ* or *Apamia*, and the ancient *Fredelas*, a little town, feated on the river *Auriege*, fuffered extremely in the religious broils. It is the refidence of a bifhop under the archbifhop of *Touloufe*. Its diocefe contains 103 parifhes, with a revenue of 25,000 livres, and its taxation at the court of *Rome* is 2500 florins. In it is alfo a college of Jefuits. The town does not belong to the county, but pays its taxes feparately. It is a part of the government.

Mazeres, one of the four principal towns of the country, formerly the refidence of the counts of *Foix*, is now but a fmall place. The *Huguenots* fortified it in the fixteenth century and maintained it till the year 1629, when they were obliged to fubmit, and the fortifications were razed.

Barilles is a town and intendancy on the *Auriege*.

Saverdun, alfo one of the four principal towns of this country, though in itfelf but fmall, lies on the river *Auriege*, being formerly accounted the ftrongeft place of the whole country. The lower town, which has alfo fuburbs, is handfomer and more populous than the upper.

Maz d'Azil is a little town, feated on the *Rize*. Before the repeal of the edict of *Nantes*, the inhabitants of this place were all Proteftants. They fortified and kept it till the year 1629. Both its origin and name are owing to the *Benedictine* abbey here called *Manfum Afyli*.

Lezat is a little town on the river *Leze*. Its taxes make the twenty-fourth part of thofe of the country, and it pays them feparately. In it is one abbey.

S. Ibars and *Montaut* are two fmall towns.

2. The little diftrict and fovereignty of *Donnezan*, three *French* leagues in length and of the fame breadth, being feparated from the county of *Foix* by a chain of mountains, belonged formerly to the counts of *Foix*, on whom it was conferred, as a fief, by *Peter* II. King of *Arragon* and Count of *Roufzillon*; but, in the fourteenth century, rendered itfelf independent. *Henry* IV. annexed it to the crown. It contains nine market-towns, or villages; among which the moft remarkable is

Guerigu, or *Querigu*, a town, having a fort, formerly looked upon as the barrier of *Upper Languedoc*, is the principal place in the country.

Son is a fort fituate near the foregoing, both commanding the road to *Roufzillon* over the *Pyrenees*.

III. The territory, or valley, of *Andorre* contains feveral villages, of which *Ourdines* is the principal.

The GOVERNMENT *of*
R O U S Z I L L O N.

THE county of *Roufzillon* is feparated to the north from *Languedoc* by the *Leffer Pyrenees*; to the caft it borders on the *Mediterranean*, and to the fouth-weft is divided by the large *Pyrenean* mountains from *Catalonia* and *Cerdagne*. Its length, from caft to weft, is eighteen *Spanifh* leagues. The name of *Roufzillon* is derived from its ancient capital *Rufcino*, which was once a *Roman* colony and the capital of the *Cardoni*. This name, by degrees, became altered to *Rofzilio*, or *Rufzilio*. Among the feveral mountains here the higheft are thofe of *Maffane* and *Canigou*, the latter of which is faid to be 1440 toifes in height. This country being environed with mountains on all fides, the fummer heats are very intenfe, infomuch that the inhabitants are almoft univerfally fwarthy and meagre. The foil is very fruitful in corn, wine and forage, and, in fome parts, of fuch uncommon fertility, that after the corn-harveft is got in, they fow millet, and fuch kinds of feed, thus procuring two and fometimes three harvefts in a year. Mules * are here ufed for the plough. The great wealth of the country confifts in olives and oranges, which are almoft as common in this country as pears and apples in *Normandy*. Very little wood grows here, and that only of fhrubs ; and, by reafon of their want of navigable rivers, their only way of being fupplied with goods from other countries, is by means of mules. It carries on a confiderable trade in fheep, which are much efteemed for the uncommon delicacy of their flefh. Oxen too are fattened here for the gratification of the wealthy, but the breeding of cows is neglected, their milk being bad. Their pigeons, quails and partridges are excellent. The chief branch of trade in this province is oil, which they traffic in to the annual amount of 200,000 livres. The other articles it exports are corn, millet and wool: The *Tet*, *Tec* and *Agly*, its only rivers, are but large rapid brooks, which difcharge themfelves into the fea. Here are alfo hot baths; and, at *Cannet* and the lake of *S. Nazaire*, the fea-water is conveyed into canals and there by the fun prepared into falt.

This country, from the dominion of the *Romans*, came under the *Vifi-goths*. Its next mafters were the *Saracens*, who, about the year 796, were difpoffeffed by *Charles the Great* and his fon *Lewis*, who inftituted counts here as governors; but thefe found means, by degrees, to render themfelves independent lords proprietors of the country. *Guinard*, or *Gui-*

---

* As alfo in *Normandy* and moft other parts of *France*.

rad, the laft of them, in the year 1173, transferred this country, by will, to *Alphonfo,* King of *Arragon* and Count of *Barcelona.* *John* II. King of *Arragon,* in the year 1462, mortgaged the county to *Lewis* XI. King of *France;* but *Charles* the eighth's confeffor having perfuaded him, that he could not confcientioufly keep the country, in the year 1493, he reftored it *gratis* to *Ferdinand the Catholic;* but *Lewis* XIII. took it again, and, by the peace of the *Pyrenees,* in the year 1659, it was entirely ceded to *France.* It contains but one bifhopric. At *Perpignan* is a fuperiour court, or coun_cil, to which lie appeals from all the inferior courts. In it is alfo a mint. The contributions payable by the inhabitants confift in a poll-tax, which produces about 40,000 livres. Exclufive of the governor, here is alfo a general-lieutenant and deputy-governor.

The government confifts of the county of *Roufzillon* and part of *Cer_dagne.*

The county of *Roufzillon* is divided into two *vigueries, viz.*

1. The *viguerie* of *Perpignan,* containing the following places ; namely,

Perpignan, in Latin *Perpeniacum,* at prefent the capital of the country, lies on the river *Tet,* ftanding partly in a plain and partly on a hill, being fortified with a high and thick wall and baftions. Its citadel is a fine work, and the fuburbs alfo are defended by a fort. It is a place of no great extent but is very populous, being the refidence of a fuperior royal council, or high court of juftice, an intendancy, a collection, a chancery, a mint and a falt-office. Its bifhop ftyles himfelf bifhop of *Elne,* the fee having been for_merly at that place. He is fuffragan to the archbifhop of *Narbonne* and has a diocefe under him of 180 parifhes. His revenue is 18,000 livres, out of which his taxation at the court of *Rome* is 1500 florins. Exclufive of its cathedral, it contains alfo four parifh-churches, two colleges of Jefuits, one feminary, twelve convents, and feveral alms-houfes and hofpitals, with an univerfity founded there in the year 1349. In it is a noble cannon-foundery. The only thing wanting here is good drinking-water.

The town of *Roufzillon* is a remnant of the old town of *Rufcino,* which gave name to the country, and is faid to have been deftroyed about the year 828. It ftands on the river *Tet,* not far from *Perpignan.*

Salfes, in Latin *Salfulæ,* is an old fort, on the borders of *Languedoc,* com_manding the great road from *Perpignan* to *Narbonne,* fituate betwixt the mountains and the lake of *Salfes,* or *Leucate.* It was built by the Emperor *Charles* V. is of a fquare form with very thick walls and towers at the angles, and has good mines. In the year 1639, this place was taken by the *French;* in 1640, by the *Spaniards;* and, in the year 1642, again by the *French.* Near it is a village which was formerly a town.

La Franquin is a road in *Cape Leucate.*

Canet, a market-town and *vicomté.*

S. Nazaire,

S. Nazaire, a village, giving name to a neighbouring lake, betwixt which and *Canet* are moraffes, where falt is prepared by the heat of the fun.

Rivefaltes is a large market-town on the *Agly*, producing a fine foit of mufcadine wine.

Opouls, or *Apouls*, is a little town.

Millas is a country town and marquifate.

Vernet, a village famous for the virtues of its baths.

Elne, in Latin *Helena*, is an old town, ftanding on an eminence by the river *Tec*, being the fecond town in the country, but of little traffic, having been deftroyed, in the years 1285, 1474 and 1642. It was formerly a bifhop's fee, which in 1602 was removed to *Perpignan*. In it anciently ftood the city of *Illiberis*.

Collioure, in Latin *Caucoliberis*, is a little fea-port, having two forts and a harbour for barks.

Port Vendres, in Latin *Portus Veneris*, is a little narrow harbour, defended by two forts.

The two following places lie in *Val-Spir*, called in Latin *Vallis Afperia*, which was formerly a county, but at prefent only an inferiour *viguerie* belonging to *Perpignan*, viz.

Bellegarde, a ftrong place feated on a mountain, and confifting of five regular baftions, befides a fort lying a little below it. It is inhabited only by military men, the fole intent of this place being to defend the difficult pafs from *Roufzillon* into *Catalonia*, called *Col de Pertuis*.

Prats de Molo, or *Mouliou*, a fortified little town feated on the river *Tec*; the fortification quite irregular.

Arles is a little town lying at the foot of Mount *Canigou*, by the river *Tec*, and containing two parifh-churches, with an abbey of *Benedictines*, which is the largeft in the country.

Le Fort des Bains ftands on a hill, at the foot of which runs the river *Tec*. Its principal fortifications are four regular baftions. This place was built in the year 1670, by *Lewis* XIV. under it is the little village of *Bains*, fo called from its warm baths.

Ceret is a little town, feated at the foot of the *Pyrenees* near the river *Tec*, and containing a fuburb much larger than itfelf, with two convents.

Ille is a pretty little town, feated in a plain on the river *Tet*.

Boulou, a market-town.

2. The *viguerie* formerly the county of *Conflans*, is environed by the *Pyrenean* mountains, and watered by the *Tec*. It was united with *Roufzillon* in the year 1659. and contains the following places, namely,

Ville Franche, a little town ftanding amidft mountains on the *Tet*, but the principal place in *Conflans*. It contains in it a handfome parifh-church, without a convent. Under *Lewis* XIV. a ftrong fort was built near this place. *Prades,*

Prades is a handfome little town lying in a charming plain near the river *Tet*, the lordfhip of this place belongs to the *Abbé de Graffe*. Without It lies a convent of *Capuchins*; and in a narrow valley about a quarter of a league diftant, ftands an abbey.

Vinca is a fmall town.

The little diftrict of *Capfir* belonged formerly to *Cerdagne*, but now is only an inferior *viguerie* of *Conflans*. The principal place in it is

Pu-Valedor, or *Valadier*, feated at the entrance of *Languedoc*.

II. *La Cerdagne* is a little province, one part of which belongs to *Spain*, the other to *France*. The latter, which was ceded to *France* in the treaty of 1660, is about one *French* league and a half broad, containing only one place. in it worthy notice, *viz.*

Mont-Louis, a well fortified regular little town, founded in the year 1681, and having a good citadel; it ftands on an eminence among the *Pyrenees*, to the right of *Col de la Perche*, which is the name of the road from *Roufzillon* to *Cerdagne*.

The GOVERNMENT of
NAVARRE and BEARN.

NAVARRE and *Bearn* borders northward on *Gafcogne*; and eaftward on *Bigorre*; being feparated from *Spain* to the fouth by the *Pyrennean* mountains; and weftward bordering on *Labourd*. Lower *Navarre* is one of the fix *Merindades* or bailiwicks, which formerly compofed the kingdom of *Navarre*, and by the *Spaniards* was called *Merindada de Ultra Puertos*, it being to them beyond the *Pyrenees*, and the road which leads over them, in their language called *puertos*, *i. e.* ' the gates.' Ferdinand King of *Arragon* and *Caftile*, having under a frivolous pretence in 1512, poffeffed himfelf of the Kingdom of *Navarre*, all that *Catherine* the lawful heirefs and her hufband *John* of *Albert* could procure to be reftored to them was this little fpot. *John*, their fon, had no better fuccefs; his fovereignty being confined to this fcanty remnant of *Navarre*, though with the title of a Kingdom. *Johanna*, his daughter by his fpoufe *Margaret*, in 1548, was married to *Antony* of *Bourbon*, to whom fhe brought the above-mentioned remnant as a dowery, together with the jufteft pretenfions againft the King of *Caftile*. Their fon *Henry* IV. arrived to be King of *France*, and his fon *Lewis* XIII. in 1620, annexed Lower *Navarre* and *Bearn* to the crown of *France*.

The country or *vicomté* of *Bearn*, has paffed through a feries of tranfitions from the dukes of *Gafcogne* to the dukes of *Aquitaine*, and fince the

twelfth

twelfth century, by feveral marriages to the vifcounts of *Gavar* and the
houfes of *Moncade, Foix, Grailly,* and *Albret. Lewis* XII. declared the
above-mentioned *Catherine,* Queen of *Navarre,* and her hufband *John
d'Albret,* independent poffeffors of *Bearn*; and not long after this country
was united to the crown of *France,* at the fame time with *Navarre.*

· Part of *Lower Navarre* belongs to the diocefe of *Acqs,* and the reft to
that of *Bayonne.* In the whole country there is neither abbey, chapter,
nor convent; the only ecclefiaftical buildings remaining here being four
parochial priories. Under *Henry* II. King of *Navarre,* the reformation
took place here; but in the year 1694, there was, at leaft in appearance,
a general return to the *Romifh* church. In *Bearn* are two bifhoprics. The
whole government is fubjeét to the jurifdiétion of the parliament of *Pau.*
Not only *Lower Navarre,* but *Bearn* alfo has States of its own, each paying
the King annually, as an ordinary contingent 4860 livres, with 2000
more for the maintenance of the troops. To the governor they allow 7740,
and the deputy-governor 2740.

I. The Kingdom of *Lower Navarre* is feparated to the fouth from
Spanifh Upper Navarre by the *Pyrenean* mountains; weftward it borders
on *Labour,* eaftward on the country of *Soule* and *Bearn,* and northward
is bounded by part of *Gafcogne,* being but eight *French* leagues in length
and five broad; mountainous alfo, and producing little. The principal
rivers in it, and thofe too very fmall ones, are the *Nive,* which iffues out
of the *Spanifh* mountains, falling near *Bayonne* into the *Adour,* and the
Bidoufe, which has its fource here, and falls alfo into the *Adour.* The inhabitants are all *Bafques,* and fpeak the *Bafque* dialeét, which is the fame
with that of *Bifcay.* This country confifts of the five diftriéts of *Amix,
Cize, Baigorri, Arberou* and *Oftabaret,* which contain the following places;
namely,

Saint Palais, in Latin *oppidum Sanéti Pelagii,* a fmall town in the diftriét of *Amix,* lying on an eminence on the river *Bidoufe.*

Garrix, a large village, formerly a town in the diftriét of *Amix,* but
now known only as a bailiwick.

La Baftide de Clarence, a little town in the diftriét of *Amix,* built in the
year 1306.

S. Jean Pie de Port, the capital of the country, ftanding on the *Nieve*
in the diftriét of *Cize*; and containing a citadel feated on an eminence which
commands the road into *Spain.*

In the three other quarters or diftriéts of this country there are no towns.
But the dutchy of *Grammont* and the county of *Luffe* are not to be paffed
over in filence.

The COUNTRY *and* PRINCIPALITY *of BEARN.*

THE country and principality of *Bearn* is fo called from *Beneharnum,* a town which formerly had exiftence, but the fight of which is now uncertain. This principality is fixteen *French* leagues in length, and twelve broad; being mountainous, and, excepting in the plains, barren. Very little wheat or rye is produced here; but great quantities of *Mailloc,* which is a kind of *Indian* corn, and flax: The very rocks are planted with vines, and fome places make an excellent wine. The mountains of *Monein* contain mines of iron, copper and lead, and are alfo covered with pine-trees, which ferve for mafts and planks. In the vallies of *Offau* and *Afpe,* and alfo at *Oleron,* are mineral waters. The principal rivers in it are the *Gave Bearnois,* and *Gave d'Oleron,* both which are very rapid, and fall into the *Adour.* At *Saillies* is a falt-fpring, which fupplies the whole government with falt. In the year 1695, the inhabitants of *Bearn* were calculated at 198,000.

In it are the following places, *viz.*

1. In the bailiwick of *Pau* is the town of

Pau, in Latin *Palum,* which is the capital, and lies on an eminence on the *Gave Bearnois,* being indeed fmall, but well built, and the feat of a parliament, a bailiwick, and a collection. In the palace here was born *Henry* IV. Exclufive of an academy of fciences and liberal arts, in it is a college of Jefuits, with five convents and two hofpitals. It contains alfo manufactures of cloth.

Lefear, which ftands on an eminence, is a bifhop's fee fubject to the archbifhop of *Auch,* with a diocefe of one hundred and feventy-eight parifhes, and a revenue of 15,000 livres. It contains nothing remarkable but one college.

Nay, a town, or rather a large village, feated on the *Gave Bearnois,* was burnt in the year 1545.

Pontac is a little town.

2. In the bailiwick of *Ortez,* the only place worth notice is

Ortez, a fmall town feated on the *Gave Bearnois,* and containing a bailiwick-court. In it formerly was an univerfity for the Proteftants.

3. In the bailiwick of *Sauveterre* lies

Sauveterre, a little town, which ftands high on the *Gave d'Oleron,* and contains a bailiwick-court.

Navarreins, or *Navarrinx,* a fortified town on the *Gave d'Oleron.*

Saillies, a little town, noted for its fine falt-fpring.

4. In the bailiwick of *Oleron* is

Oleron,

Oleron, a fmall but populous town lying on the river *Gave*, to which it gives name, and the feat of a bailiwick-court. The trade of this place was formerly more confiderable than at prefent. Its bifhop, who is fuffragan to the archbifhop of *Auch*, has a diocefe containing two hundred and feventy-three parifhes, with a yearly revenue of 13,000 livres, out of which his taxation at the court of *Rome* is 600 florins.

S. Marie, a little town feparated from *Oleron* only by the river, over which it has a bridge of ftone. In this town is the cathedral and bifhop's palace.

Ogeu, or *Ogen*, a village, having fome cold mineral waters.

Monein, a fmall but populous town fituate in a wine country.

At *Aiguefcaudes*, in the valley of *Offau*, which is one of the fineft in the whole country, is a hot fpring containing a water of a foapy, fpirituous, oily nature, refembling the fmell of a bad egg. This water is reckoned an excellent vulnerary, and is alfo ufed with fuccefs for inward diforders.

6. In the valley of *Afpe*, the capital of which is the town of *Acous*, are feveral cold fprings, particularly thofe of *Efcot*.

In the bailiwick of *Morlas*, lies

Morlas, a poor town, though formerly the refidence of the vifcounts and Princes of *Bearn*, and alfo of a mint.

Lembeye, a little town ftanding high on the borders of *Armagnac*.

The GOVERNMENT *of*
G U Y E N N E and G A S C O G N E.

THIS borders to the fouth on the *Pyrenean* mountains, to the weft on the ocean, to the north on *Saintonge*, *Angoumois*, *Limofin*, and *Auvergne*; and to the eaft on *Auvergne* and *Languedoc*. Its extent from fouth to north, or from *Vio de Sos* in the *Pyrenean* mountains to *Niort* in *Poitou*, is eighty *French* leagues; and from *S. Jean de Luz* to *S. Geniez* in *Rouergue*, about ninety. The name of *Guyenne*, firft ufed about the year 1360, is derived by corruption from *Aquitaine*; this country, according to *Pliny*, being fo called on account of the mineral waters in it. The government of *Guyenne* does not contain the whole ancient Kingdom, and afterwards dutchy of *Aquitaine*, which was of much greater extent. It abounds, however, in corn, wine, fruits, hemp, and tobacco, whence arifes a very confiderable trade in wine (of which *Bourdeaux* alone annually exports 100,000 tuns) brandy, prunes, and many other commodities. In it are alfo feveral fprings of medicinal waters, together with copper, coal and other mines, and fine

　　　　　quarries

quarries of marble of all colours. In the election of *Figeac*, grows a plant called *radoul*, which is ufed in dying and tanning. The principal rivers in this government are, 1. The *Garonne*, mentioned in the introduction to *France*. This receives into it the little river of *Tarn*, which at *Gailliac*, becomes navigable, together with the *Baife* and the *Lot*, in Latin *Olitis*, *Olde*, and *Oulde*, which by means of fluices are alfo rendered navigable; the *Drot*, and *Dordogne*. Its tide flows up as far as *Langon* and *Macaire*; and therefore by confequence thirty *French* leagues from its mouth. 2. The *Adour*, which receives its fource in the mountains of *Bigorre*, at a place called *Tremoula*, being navigable at *Grenade* in *Marfan*; and falling into the ocean through an outlet called *Boucault*. This river has fome harbours on the ocean.

Charles the Great created his fon *Lewis* King of *Aquitaine*; but this Kingdom was foon divided into the two dutchies of *Aquitaine* and *Gafcogne*, which by means of the marriage of *William* IV. duke of *Aquitaine*, with *Brifce* heirefs to *Gafcogne* were united and continued in *William's* family till the year 1150; when by virtue of the marriage of *Eleanor* with *Henry* II. King of *England*, this dutchy fell to that crown, under which it continued near three hundred years. In the year 1453, *Charles* VII. difpoffeffed the *Englifh* of it. *Lewis* XI. in 1469, conferred it on his brother *Charles*, who was the laft duke of *Guyenne*, it being again, on his deceafe, re-annexed to the crown. In 1753, the title of Duke of *Aquitaine* was again revived in the perfon of the Dauphin's fecond fon. In this government is an archbifhop with nine fuffragans. It contains two generalities; namely, that of *Bourdeaux*, which is under the parliament of *Bourdeaux*; and that of *Montauban*, under the parliament of *Touloufe*. The former is compofed of nine and this of four large *fenechauffées*, or bailiwicks. In each is alfo a court of aids, under the direction of the governor, whofe income is about 10,000 livres; with two general-lieutenants, *viz.* one for *Guyenne* and the other for *Gafcogne* and *Bigorre*: And befides thefe thirteen other deputy-governors, that is to fay, one in each election. The government confifts of two main parts.

1. *Proper Guyenne* makes the northern part, and confifts of the following countries, namely,

1. *Bourdelois*, being the moft confiderable among them, and properly called the bailiwick of *Guyenne*. Wine is its principal commodity, yet is it in general fruitful, though the foil is fomething fandy. It abounds particularly in cheftnut and fig-trees of an extraordinary fize; and alfo propped vines, which ftand almoft like trees, in the open field. In it are the following places, *viz.*

Bourdeaux, or *Bordeaux*, in Latin *Burdigala*, the capital of *Guyenne* and the refidence of an archbifhop, of a parliament, an intendancy, a treafurer's office, a court of aids, a provincial tribunal, a *fenechauffée*, an admiralty, an election,

election, a *marechauffée, &c.* lies on the *Garonne* in form of a half-moon, being pretty large and populous ; and having a great number of ſtately houſes built of ſtone, but old, and the ſtreets very narrow. The neweſt and handſomeſt part in the whole town is the *place royale*, near the harbour, in which ſtands a grand magazine, with the exchange and a ſtatue of *Lewis* XIV. in braſs. The ſuburb of *Chartron*, or *Chartreaux*, alſo makes a fine appearance. The city and harbour are defended by three forts. The *Chateau trompette* is a citadel owing its preſent beauty and ſtrength to *Lewis* XIV. and ſerving both to cover the harbour and as a bridle to the city. It is a noble work, and built entirely of free-ſtone. The ramparts are all ruined, and a perſon may walk quite round them. The two other forts of *Le chateau de Haa*, and *S. Louis*, or *S. Croix*, are of no great importance. The archbiſhop of this place has nine ſuffragans under him, with a province of four hundred and fifty pariſhes, excluſive of about fifty chapels of caſe. His revenue is 55,000 livres *per annum*, and his taxation at the court of *Rome* 4000 florins. To him belong alſo the diſtricts of *Montravel, Belvez, Bigaroque, &c.* The cathedral is a *Gothick* building, large, bare, and de-ſolate. The ſacriſty treaſury contains nothing extraordinary in it ; and the large ſilver ſhrine of relicks on the great altar is the only thing worth ſee-ing. The church and convent of the *Dominicans* are new and elegant; but thoſe belonging to the *Carthuſian* monks of a ſplendor ſeldom ſeen in any convents of that order. In this city is alſo an abbey of *Benedictines*, of the order of *S. Maur.* The Jeſuits have likewiſe a fine college here ; beſides which there is alſo another college. The univerſity was founded in the year 1441, and in 1712 the King erected here an academy of the ſciences and *beaux arts*, the library belonging to which, though not very numerous, conſiſts entirely of ſelect pieces, placed in a moſt elegant hall. In *Bourdeaux* are alſo three ſeminaries. The trade of this city is very conſiderable, and for its improvement, a toleration is granted to the *Engliſh, Dutch, Danes, Hamburghers* and *Lubeckers*, and even to the *Portugueſe Jews*, though the laſt have no ſynagogue here, nor are Proteſtants indulged in the public ex-erciſe of their worſhip; only the *Engliſh* are connived at, and have a mi-niſter in a lay-habit. Other Proteſtants are not prohibited from reading a ſermon privately to their family. The rich traders here of foreign countries call themſelves *Negotians*. In this place are ſtill viſible ſome remains of *Roman* antiquities ; namely, an amphitheatre built by the Emperor *Gal-lienus*, whoſe palace it is called, and of which, beſides ſome parts of its ſide-walls are ſtill ſtanding the two main portals : *La porte baſſe*, an ancient gate of very large freeſtone, neither cemented with mortar nor vaulted, but built horizontally, with houſes on it : part of the walls belonging to the palace of the *Dei tutelares* and of the fountains. In the new hoſpital with-out the city is a fine manufactory of lace.

Bourg,

Bourg, a little town feated on the *Dordogne*, having a little harbour, whence wine is fhipped to different parts.

Blaye, in Latin *Blavia*, or *Blavutum*, a fmall town lying on the *Garonne*, having a citadel feated on a high rock. In this town is an abbey of *Bene-dictines*. All fhips going up to *Bourdeaux* muft leave their cannon and arms here, which they take in again at their return. The *Garonne* being very wide near this town, a battery is erected on an ifland in it to keep off an enemy's fhips. Directly oppofite to *Blaye*, on the other fide of the river in the country of *Medoc*, is alfo another fort.

The country of *Medoc* conftitutes a great part of *Bourdelois*, and lies in form of a peninfula betwixt the *Garonne* and the fea, but is neither populous nor fruitful, the high tides overflowing the north part of it. On a rock at the mouth of the *Garonne*, ftands a fine light-houfe called *le tour de Cordoue*. In *Medoc* are no towns, the little places of *Souillac, Caftillon, Efpare* and *Chauneuf* excepted.

The country of *Buch*, originally *Buyes*, the former owners of which, namely, the barons of *Captal*, or *Capoudal*, were famous in the biftory of *Guyenne*. Since the year 1715, this country belonged to the marquifes of *Contaut*.

La Tête de Buch is a market-town on the harbour of *Arcachon*, the entrance to which is dangerous on account of the many fand-banks in it. The inhabitants of this place are moftly fifhermen.

Caftres is a market-town.

Rions, a little place.

Creon, a market-town.

Fronfac is a fmall town, dutchy and peerage.

Coutras is a large market-town, feated near the conflux of the *Ifle* and the *Droume*, being a marquifate belonging to the dutchy of *Fronfac*. In this neighbourhood, in the year 1587, the army of the *Ligue* was defeated by *Henry* IV.

Libourne is a fmall, but well inhabited town, ftanding in a good fituation on the river *Dordogne*, into which the *Ifle* falls at this place. In this town is a bailiwick, a provincial court of juftice and fome convents. Its principal trade confifts in falt. The country betwixt *Libourne* and *Bourdeaux* is called the land betwixt two feas, the two rivers of the *Garonne* and *Dordogne*, into which the tide runs, being here of fuch a breadth as to be called feas.

Cadillac is a little town on the banks of the *Garonne*, and is the capital of the earldom of *Benagues*, containing a fine caftle, with a collegiate-church and a convent.

2. *Perigord*, which derives its name from its ancient inhabitants the *Petricorii*, is thirty-three *French* leagues in length and twenty-four broad,

being

being a ftony country and not very fruitful, but containing fome good mines of iron and mineral waters.   It is divided into two parts.

1. *Upper Perigord,* called alfo the *White,* contains the following places; namely,

Perigueux, in Latin *Petricordium,* or *Petricorium,* the capital of the country and the refidence of a bifhop, an election, a bailiwick and a country court of juftice, being feated on the river *Ifle,* in a fine and pretty well inhabited country.   The bifhop of this place is fubordinate to the archbifhop of *Bourdeaux,* and has a diocefe containing 450 parifhes with a revenue of 24,000 livres, out of which the taxation he pays the court of *Rome* is 2590 florins.   In the town is a college of Jefuits, together with four convents, one hofpital and fome remains of *Roman* antiquities; as, namely, an amphitheatre, a temple of *Venus,* &c.   The old town is called *La Cité,* and the new town, which is about a hundred paces diftant from the former, is named *la Ville.*

Bourdeilles, Brantôlme or *Brantôme,* containing an abbey of *Benedictines* of the congregation of *S. Maur, Ribeyrat,* or *S. Martin de Riberat,* both fmall market towns.

Mucidan, or *Mufzidan* in Latin *Mulcedonum,* is a little town, which the Proteftants formerly fo fortified that it ftood out more than one fiege, particularly in 1579.

Bergerac, anciently *Braierac* is a town feated in a delightful plain on the river *Dordogne,* confifting of two fmall towns, namely *S. Martin de Bergerac* and *Madelaine.*   In this place is a bailiwick.   The town was formerly ftrongly fortified by the Proteftants, but *Lewis* XIII. on its reduction, in the year 1621, caufed the works to be razed.

Limeuil is a fmall town fituate near the conflux of the *Vezerre* and *Dordogne.*

La Force is a dutchy and peerage.

2. *Lower Perigord,* which on account of its many woods is alfo called the *Black,* comprehends the following places; *viz.*

Sarlat, a poor town, difadvantageoufly fituated on a fmall river in a valley betwixt hills, the refidence, however, of a bifhop, a court of juftice, a royal jurifdiction, an election, a *marechauffée* and a falt-magazine, being alfo reckoned the fecond city of the country of *Perigord.*   Its bifhop, who is fuffragan to the archbifhop of *Bourdeaux,* enjoys a diocefe of about 250 parifhes with a revenue of 12,000 livres a year, out of which he pays to the court of *Rome* 742 florins.   This was one of the fortified towns belonging to the Proteftants, and for that reafon twice befieged in the fixteenth century, and the third time in the year 1652.

Teraffon is a little town, feated on the *Vezere,* having an abbey of *Benedictines.*

Domme is a fmall town, but ftrong from its fituation and defended alfo by a caftle.   *Beaumont,*

Beaumont is a little market-town on the river *Coufe.*

Caftillon is a fmall town, near which, in the year 1453, the *French* defeated the *Englifh.*

3. *Agenois* takes its name from the town of *Agen,* being the moft fruitful part of this government and formerly giving the title of count. The *Nitiobrigæ* were the ancient inhabitants of this fine country. In it are the following places; *viz.*

Agen, the capital of the country, which is pretty large and well inhabited, being the feat of a bifhopric, a bailiwick and an election. The bifhop hereof ftyles himfelf count of *Agen,* though he is not lord of it, but a fuffragan to the archbifhop of *Bourdeaux,* having a diocefe under him of 373 parifhes and 191 chapels of cafe, with a yearly revenue of 35,000 livres, out of which his taxation to the court of *Rome* is 2440 florins. Exclufive of the cathedral and a collegiate-church, here are alfo two parifh-churches with feveral convents, one college of Jefuits and one feminary. In this town lived *Julius Scaliger,* and it is alfo famed for being the birth-place of his fon *Jofeph;* is finely fituated for trade but makes no advantage of it.

Valence is a little town on the *Garonne.*

Teneins is a town confifting of two large villages, one of which belongs to the duke *de la Force,* and the other to the count *de Vauguion.*

Aiguillon is a little town, feated on the *Garonne,* at the influx of the *Lot,* and having a caftle. This place is a dutchy and peerage, and carries on a confiderable trade in hemp, tobacco, grain, wine and brandy. In the year 1346, it held out a fiege againft *John* duke of *Normandy.*

Monheurt is a fmall town, lying on the *Garonne* and formerly fortified.

Clerac, or *Clairac,* a little town on the river *Lot,* drives a good trade in tobacco, wine and brandy. In it is an abbey, a convent and a fine college of Jefuits. This place fuffered much in the religious war.

Marmande is a pretty large town, feated on the *Garonne* and carrying on a confiderable trade in corn, wine and brandy.

Duras is a fmall town but a dutchy and peerage.

Sainte Foi, a town, lying on the *Dordogne,* was ftrongly fortified by the Proteftants and firft taken from them by the King in the year 1622.

Villeneuve d'Agenois is a little town, ftanding in a fruitful plain on the river *Lot,* having a royal court of juftice. Over the river it has a bridge.

Salvetat, or *Sauvetat,* is a fmall town.

Mas d'Agenois is a market-town.

Monflanquin is a little town, ftanding on the *Lez,* and containing a royal court.

4. *Quercy,* formerly *Cahourcin,* was anciently inhabited by the *Cadurci.* This country abounds greatly in grain and wine, and is divided into two parts; *viz.*

1. *Upper*

1. *Upper Quercy*, which forms the north part and contains the following places; namely,

Souillac, a little town, feated in a rich fertile valley on the little river of *Borefe*, near the *Dordogne*, and containing an abbey of *Benedictines* of the congregation of *S. Maur.*

Lauzerte, a fmall town ftanding on a rock.

Martel, a little town, feated on a rock near the *Doraogne*, and containing a royal court of juftice with a particular bailiwick, belongs to the vifcounts of *Turenne.*

Uffedun, a little place on a hill on the *Dordogne*, in all probability, the *Uxellodunum* of *Cæfar.*

S. Cere, a fmall town belonging to the *vicomté* of *Turenne.*

Gourdon, a little town having an abbey.

Roquemadour, alfo a fmall town with an abbey.

Fons, a little place containing a royal tribunal.

Figeac, a town lying on the *Sele*, the feat of an election, an abbey and a chapter. This was a place of fome ftrength in the time of the religious war; but falling again, in 1622, into the King's hands, he demolifhed the fortifications and citadel.

Cadenac, a very ancient little town on a fteep rock, almoft entirely furrounded by the *Lot.* This place never fubmitted to the *Englifh.*

Cajarc, a market-town lying on the *Lot.*

Mier, a little place noted for its mineral waters.

Cahors, anciently *Devona*, or *Dibona*, and in Latin *Cadurcum*, the capital of *Quercy*, lying partly on the *Lot* and partly on a rock, is the feat of a bifhopric, an election and a provincial court, but meanly built, and an inconfiderable place. Its bifhop ftyles himfelf count of *Cahors* and is a fuffragan to the archbifhop of *Albi*, with a diocefe containing 800 parifhes. His revenue is 45,000 livres, and his taxation at the court of *Rome* 1000 florins. The vifcount of *Ceffac*, or *Seffac*, is his vaffal. The univerfity here, which was founded in the year 1332, was entirely fuppreffed by the King, and of its three colleges one was turned into the town-houfe and the other two given to the Jefuits.

Puy l' Eveque, a little town.

Caftelnau de Bretenons, a fmall town and a barony, with a chapter.

2. *Lower Quercy* conftitutes the fouth part, to which belong the following places, *viz.*

Albenque, *Moncucq* and *Montpezat*, all little towns.

Caftelnau de Montatier, a fmall place containing a chapter.

Lauzerte, a little town ftanding on a rock.

Caylus, a fmall town.

Moiffac, a little town, feated at the foot of a mountain near the river *Tarn*, which falls into the *Garonne* near this place. In it is a chapter.

Negrepeliſſe, a ſmall town, ſeated on the *Aveyrou,* formerly fortified by the Proteſtants, but, in the year 1621, the works were demoliſhed.

Real ville la Françoiſe, a little town ſtanding on the river *Aveyrou.*

Couſſade, a ſmall town, fortified during the time of the religious diſturbances.

Montauban, in Latin *Mons Albanus,* a well built handſome city on the river *Tarn,* being the ſeat of a biſhopric, a generality, an election, a court of aids, a provincial court, a bailiwick and a *vicomté.* It conſiſts properly of three parts; namely, the old and new towns with the *Ville Bourbon,* which laſt lies on the other ſide of the river and is a *fauxbourg.* The biſhop of this place is ſuffragan to the archbiſhop of *Touloufe* and has a dioceſe of ninety-ſix pariſhes, excluſive of chapels of eaſe, with a revenue of 25,000 livres *per annum,* out of which he pays a tax of 2500 florins to the court of *Rome.* In this town are two chapters, an academy of *belles lettres,* one college of Jeſuits, one ſeminary, eight convents and a general hoſpital. Its principal trade conſiſts in woollen ſtuffs. In the year 1562, the inhabitants became Proteſtants and fortified the city ſo ſtrongly that *Lewis* XIII. beſieged it without ſucceſs in the year 1621, and was not able to reduce it till 1629, when its fortifications were razed.

Bourniquel, a market-town.

5. *Rouergue* anciently inhabited by the *Rutheni,* produces little or no grain, but abounds in cattle, together with iron, copper, alum, vitriol and ſulphur. The extent of the country, from *S. Jean de Breuil* to *S. Antonin,* is about thirty *French* leagues; and, from *S. Pierre d'Yſſis* to *Mur de Barres,* twenty. It conſiſted formerly of the earldoms of *Rodez* and *Milhaud,* but at preſent is divided into three parts; namely,

1. An earldom; to which belongs

. *Rodez,* anciently *Segodunum,* the capital of *Rouergue* and the reſidence of a biſhopric, an election, a bailiwick, a country-court and a *marechauſſee,* ſtanding on a hill on the river *Aveyrou.* The biſhop of this place is a ſuffragan to the archbiſhop of *Albi,* ſtyles himſelf count of *Rodez,* being alſo lord of the city, and has a dioceſe containing about 450 pariſhes. His yearly revenue is 40,000 livres, and to the court of *Rome* he pays a taxation of 2326 florins. In it the Jeſuits have a fine college, and, beſides this, in this city are eight convents with a ſeminary and a hoſpital. This place never took part with the Proteſtants.

S. Geniez de Rivedolt, a town, containing a royal court of juſtice and a convent.

Entraigues, a little town and an earldom ſituate near the conflux of the *Lot* and *Truyere.*

Guiolle, a ſmall town.

Le Mur de Barez, a little town containing a collegiate-church and two convents.

Eſtain, Albin and *Caſſagnettes,* all ſmall towns.　　　　　2. In

2. In the *Upper marché* of *Rouergue*, in which are

Millan, or *Milhaud*, in Latin *Æmilianum*, a town feated on the river *Tarn* and the capital of this part of *Rouergue*, containing an election. In it are five convents and a commanderie of the knights of *Malta*. The Proteftants formerly fortified it, but *Lewis* XIII. difmantled it in the year 1629. In 1744, the Proteftant inhabitants had three troops of dragoons quartered upon them, the maintenance of whom for three months coft them 30,000 livres ; by which means the town was totally ruined.

Nant, a fmall town, containing an abbey of *Benedictines.*

Pont de Camarares, a place celebrated for its mineral waters.

S. Rome de Tarn, a little town lying on the *Tarn.*

S. Sernin, a fmall place containing a collegiate-church.

Belmont, a little town.

Sainte Frique, or *Sainte Afrique,* a fmall town, fortified by its inhabitants after they became Proteftants; but reduced, in the year 1629, by *Lewis* XIII.

Vabres, a fmall city on the river *Dourdan,* and the refidence of a bifhop fubordinate to the archbifhop of *Albi,* with a diocefe of 150 parifhes and a revenue of 20,000 livres *per annum,* out of which his tax to the court of *Rome* is 1000 florins.

Severac le Chateau, a little town and dutchy.

Severac l' Eglife, a market-town and earldom.

3. The *Lower marché* of *Rouergue* contains

Ville Franche, in Latin *Francopolis,* the principal town of that part of *Rouergue,* and next to *Rodez,* the moft confiderable in the whole country, being feated on the river *Aveyrou,* and containing one chapter, one college, three convents and a *chartereufe* without the walls.

S. Antonin, a little town on the river *Aveyrou,* with a chapter and three convents. The principal trade of this place confifts in faffron and fine large plumbs.

Najac, a little town, fituate on the river *Aveyrou,* and the feat of a bailiwick. Hereabouts, in the years 1672 and 1673, was difcovered a mine of copper.

Sauveterre, a fmall town, lying on the *Aveyrou,* and containing a royal court of juftice.

Conques, a market-town having a chapter.

Peyruffe, anciently *Petrucia,* a little town, ftanding on a mountain and the refidence of a bailiwick.

Villeneuve, a fmall town.

S. Juft, a market-town.

Cranfac, a little place noted for its excellent fulphureous mineral waters. In the neighbourhood alfo are coal-pits.

G A S C O G N E.

GASCOGNE conftitutes the fouth part of the government of *Guy-enne*, and includes the country lying between the *Garonne*, the ocean and the *Pyrenean* mountains. It receives its name from the *Gafcones* and *Vafcones*, by the moderns called *Bafques* or *Vafques*, a people who lived on the *Pyrenean* mountains in *Spain*, and towards the clofe of the fixth century fettled on the north fide of the *Pyrenees*, where they defended themfelves againft the *Franks* and extended their conquefts in *Novempopulania*, but were obliged at laft to fubmit to the King of the *Franks*. Under the *Carlovingian* line they chofe a duke of their own, and after that' family became extinct, fell, in the eleventh century, under the dominion of the dukes of *Aquitaine*. To it belong the following countries ; namely,

1. *Bafadois*, taking its name from the people called *Vafatæ*. The fouth part of this province is fandy, and in it are the following places ; *viz.*

Bafas, anciently *Coffio*, and in Latin *Civitas Vafatica*, the capital of the country, receiving alfo its name from the *Vafatæ*. This town ftands on a mountain and is the refidence of a *prevoté*, a bailiwick, a country and a royal court of juftice. Its bifhop is fuffragan to the archbifhop of *Auch*, has a diocefe of 234 parifhes and thirty-feven chapels of cafe; with a revenue of 18,000 livres, out of which his taxation at the court of *Rome* is 600 florins. Exclufive of the cathedral in this town are three parifh-churches and one college.

Langin, a little town and marquifate, feated on the *Garonne* in a good wine country, and having the tide reaching up to it.

Reole, in Latin *Regula*, a little town on the *Garonne*, deriving its name from an old abbey of *Benedictines*. The inhabitants of this place deal in grain, wine and brandy. In 1676, the parliament of *Bourdeaux* was removed hither till the year 1689.

Captieux, a market-town and barony.

2. *Condomois* is a fruitful country containing the following places ; *viz.*

Condom, the capital, which lies on the *Baïfe* and is the feat of a provincial-court, a bailiwick and a bifhop fubject to the archbifhop of *Bourdeaux*, with a diocefe of 140 parifhes and eight chapels of cafe. His revenue is 50,000 livres, and his taxation at *Rome* 2500 florins. Exclufive of the cathedral, in it are two other churches and five convents. This place fuffered much in the religious wars.

Mezin, a little town containing a royal court of juftice.

3. The dutchy of *Albret* contains the following places ; namely,

Albret,

Albret, or. *Lé Bret,* a little town, raifed to a dutchy in the year 1556.

Nerac, the capital of the dutchy, lying on the river *Baife,* which here becomes navigable. It is divided into *Great* and *Little Nerac.* In it is a caftle with a country and a ducal court of juftice and four convents. The Kings of *Navarre,* as dukes of *Albret,* had once a palace here. In the fixteenth century, moft of the inhabitants became Proteftants; but, in the year 1621, they were forced to furrender to *Lewis* XIII.

Cdftel-Jaloux, a fmall town, feated on the little river *Avence,* containing a chapter and a pretty trade in wine, cattle and honey.

Caftel Moron, a market town.

Tartas, a little town; feated on the *Midore* near the river *Adour,* being handfome and containing a bailiwick-court, with a church and two convents.

4. The little diftrict of *Gabardan,* or *Gavardan,* which enjoyed formerly vifcounts of its own, who became afterwards vifcounts of *Bearn,* contains

Gabaret, a fmall town on the river *Genife,* and the capital of this country.

5. The little diftrict of *Marfan,* once a *vicomté,* containing

Mont de Marfan, a fmall town, feated on the river *Medouze* and the principal place of the country.

Roquefort de Marfan, a little town lying on the fame river.

S. Juftin, a market-town.

6. The little diftrict of *Turfan* always had the fame vifcount as *Marfan.* In it is

Aire, in Latin *Atura,* or *Adura,* and *Vicus Julii,* a town, feated on the river *Adour,* but having more the appearance of a large village, though, notwithftanding, a bifhop's fee with a diocefe of 241 parifhes belonging to it, fubject to the archbifhop of *Auch.* His revenue is 30,000 livres, and his taxation at the court of *Rome* 1200 florins. This place fuffered greatly in the religious wars.

Mas, a town having a collegiate church.

Grenade, a fmall place.

7. The little territory of *Chaloffe,* contains

S. Sever, commonly called *Cap de Gafcogne,* one of the prettieft towns in all *Gafcogne,* and feated on the river *Adour,* owing both its name and origin to an abbey of *Benedictines.* This town contains a bailiwick-court.

Mugron, a fmall town.

Tolofette, a place of no confideration.

8. *Les Landes,* or *Lannes,* is a narrow level flip of land, not very fruitful, and ftill worfe inhabited; in it, however, are great numbers of bees, and confequently large quantities of honey and wax. In a more extenfive fenfe,

it includes the whole fea-coaft fouth of *Bourdeaux* to the mouth of the river *Adour*, together with a part of *Bourdelois* and *Bofadois*. But in its moft confined and ufual import, it comprehends only the following diftricts, namely,

1. The *vicomté* of *Acqs*, to which belongs

Acqs, or *Dâx*, in Latin *Aquæ Tarbellicæ*, and *Aquæ Augufiæ*, lying on the river *Adour* in the diftrict of *Auribat*. In it are a provincial and baili-wick-court, with an election ; and it is alfo a bifhop's fee fubject to the archbifhop of *Auch*, with a diocefe of 243 parifhes. His income is 14,000 livres, and the *Roman* tax upon it 500 florins. In this place are fix con-vents, one college, and an hofpital. The fortifications, and even the caftle, are of no great importance. The town and the neighbouring country boaft feveral warm baths.

Port des Lannes, a little place feated on the *Adour*.

2. The country of *Marancin* borders on the fea-coaft, on which are *Marennes*, the choaked up harbour called *le Vieux Boucault*; and *Cape Breton*, a village noted for the goodnefs of its wine.

3. The *vicomté* of *Aorte*, or *Urt*, fo called from *Urt*, now but a mid-dling village, though the principal place of the country.

Peire Hourade, in Latin *Petra forata*, a fmall place lying on the river *Adour*, which is here joined by the *Gave*, was anciently the refidence of the vifcount, who had a caftle here called *Afpremont*.

Obf. *Albret* and *Tartas*, once *vicomtés*, and lying in this territory, be-long to the dutchy of *Albret*, N° 3.

9. *Labour*, or *Labourd*, fo called from the ancient city of *Lapurdum*, now *Bayonne*, and inhabited formerly by the *Tarbellii*, is part of the country of the *Bafques*, extending anciently almoft as far as *S. Sebaftian*, in the *Spanifh* province of *Guipufcoa*; but all beyond the river *Bidaffoa* belongs to the King of *Spain*. It produces little corn, and lefs wine ; but enjoys great plenty of fruit. The tribute its inhabitants pay to the King is very inconfi-derable. In this diftrict are the following places ; namely,

Bayonne, in Latin *Lapurdum*, the capital of the country, lying at the junction of the *Adour* and *Nive*, at no great diftance from the mouth of the former. It is of pretty confiderable extent, and the refidence of a bi-fhop, a collection, a bailiwick court, an admiralty and a mint. The name of *Bayonne* is compounded of the two *Bafque* words *Baia* and *Ona*, figni-fying, a good harbour : It well deferves that title, and is accordingly much frequented and of great advantage to the country, though the fhallows render the entrance to it fomewhat difficult. The river *Nive* runs through the city ; and the *Adour* clofe by its walls ; foon after which they unite, and divide the city into three parts; the large town as it is called, lies on this fide the *Nive* ; the fmall town is betwixt the *Nive*, and the *Adour* ; and the fuburb of *S. Efprit*, in which great numbers of *Jews* refide, lies on the

other

other fide of the river. The two firft parts, befides their other works, have each a fmall fort; but the fuburb, exclufive of its good fortifications, has a regular fquare citadel ftanding on an eminence, which thus commands the three feveral parts of the city, the harbour, and the adjacent country. Its bifhop is a fuffragan to *Auch*, has a diocefe under him containing feventy-two parifhes, with a revenue of 19,000 livres, out of which his taxation at the court of *Rome* is only 100 florins. Exclufive of its cathedral and chapter, the fuburb alfo contains another fmall chapter. In it is likewife one college and five convents.

S. Jean de Luz, in the *Bafque* tongue *Luis*, or *Loitzun, i. e.* ' a muddy place,' and *Sibour*, two large villages, feparated only by a rivulet called *Ninette*, over which they have a communication by means of a fmall bridge. Both contributed to form the little harbour of *Socoa* for the fecurity of their fifhing barks. In the firft of thefe villages Cardinal *Mazarine* refided during the congrefs for the treaty of peace with *Spain*, which was held in *Pheafant Ifland* in the river of *Bidaffoa*; and there alfo *Lewis* XIV. married *Maria Terefa* the Infanta of *Spain*.

Andaye, a fort and large village on the river *Bidaffoa*, oppofite to *Fontarabia*, makes a fine brandy.

Uftariz, a market-town.

Bidache, a principality belonging to the family of *Grammont*.

Guiche, in Latin *Guiffunum*, a county.

10. The country or vale of *Soule*, properly *Subola*, is included in *Lower Navarre* and *Bearn*, lies in the *Pyrenean* mountains, and belongs to the country of the *Bafques*, having States of its own, and containing fixty-nine parifhes. A confiderable part of the inhabitants feck for work in *Spain*. The mountains here produce good fhip-timber, but with no conveniency for fpeedy carriage. The principal place of this vale, formerly a *vicomté*, is *Mauleon*, a town and caftle on the river *Gave*.

11. *Armagnac*, with the adjacent lordfhips, was formerly a *comté*; is twenty-two *French* leagues in length and about fixteen in breadth, producing both grain and wine; and being divided into *Upper* and *Lower Armagnac*.

1. *Upper Armagnac* lies among the *Pyrenean* mountains, and contains four vallies.

1. The vale of *Magnoac*, the principal place of which is *Caftelnau dé Magnoac*, a little town on the river *Gers*, the feat of the bailiwick of the four vales, and of a collegiate church.

2. The vale of *Neftez*, in which is *Barte*, a large village, feated on the river *Neftez*.

3. The vale of *Barouffe*, in which is *Mauleon*, a little town.

4. The vale of *Aure*, in it are

Arreu,

Arreu, a ſmall town.

Sarrancolin, a little town containing a priory. A very fine kind of marble is dug in its neighbourhood. It has alſo a thriving glaſs-houſe.

2. *Lower Armagnac,* which is larger and more fruitful than the former, conſiſts of

1. *Proper Armagnac,* in which is

Auch, anciently *Eluſaberris,* or as it is alſo found written *Climberris, Ellimberris,* and afterwards *Auguſta,* the capital of *Armagnac* and all *Gaſcogne.* This place is the ſeat of an archbiſhopric, an intendancy, and an election; contains a treaſurer's office, a bailiwick, a provincial and royal court, with an office alſo of the finances. The lordſhip of the town is divided betwixt the archbiſhop and the count of *Armagnac.* The dioceſe conſiſts of 372 pariſh-churches and 277 chapels of caſe. The annual revenue of its archbiſhop is 90,000 livres, out of which his taxation at the court of *Rome* is 10,000 florins. Excluſive of the cathedral, in this city is alſo a chapter and priory.

Nogaro, a little town on the river *Douſe,* having a collegiate church, is one of the five towns given to the duke of *Bouillon* in exchange for his principality of *Sedan.*

2. The earldom of *Fezenzac,* in Latin *Comitatus Fidentiacus,* in which is

Vic, with the addition of *de Fezenzac,* formerly *Fidentia,* a little town ſeated on the river *Douſe.* It contains one collegiate church, and is the principal place of the earldom.

3. The little country of *Eauſan* contains

Eauſe, in Latin *Eluſa,* a ſmall town lying on the *Geliſe* near the town of *Eauſe* or *Eluſa;* which, for a conſiderable time, was the capital of *Novempopulania,* and gave name to the *Eluſatæ.* This ancient place, which was afterwards called *Civtat, i. e.* ' the city,' is in a ruinous condition, and near it is built the preſent new town.

4. The county of *Gaure.* In it is

Fleurence, a little town, but the capital.

5. The ſmall tract and *vicomté* of *Brullois.* In it is

Leyrac, a little town.

6. *Lomagne,* formerly a *vicomté.* In it are

Lectoure, in Latin *Lactura* or *Lactora,* a ſtrong town, having a caſtle on a declivity near, the river *Gers.* It contains a bailiwick and a provincial court; is alſo the place of an election and a biſhop's ſee, under the archbiſhop of *Auch,* with a dioceſe of ſeventy-three pariſhes. His revenue is 18,000 livres, and the *Roman* tax 1600 florins.

Vic, now a ſmall place, but formerly the reſidence of the viſcounts.

Beaumont, a little town lying on the *Gimone.*

7. *Fezenzaguel,* formerly a *vicomté,* containing,

Mauveſin,

Mauvefin, fmall, but makes great quantities of falt-petre. This place was formerly one of the ftrongeft holds of the *Huguenots*; but in the year 1621, furrendered to *Lewis* XIII.

8. The country of *Riviere* contains

Verdun, a little town feated on the *Garonne*, and the principal place of a particular lordfhip, to which alfo belong the following places, *viz*.

Grenade, a little town lying on the *Garonne*.

Le Mas-Garnier, a fmall place, feated on the *Garonne* and having an ab-bey of *Benedictines*.

L'Ifle Jourdain, anciently *Caftellum Ictium*, a little town lying on the *Save*, formerly the principal place of an earldom, has a collegiate-church.

Sainte Foy de Peyroheres, a fmall town.

9. The *comté* of *Aftarac*, belonging to the ducal houfe of *Roquelaure*, being a fruitful populous country, with the following places in it; namely,

Mirande, the capital and feat of an election but fmall.

Caftelnau de Barbarens, a little town.

Berdoues, an abbey.

Simorre, a town, containing an abbey, and the refidence of the duke *de Roquelaure's* court.

Maffe Oube, a market-town.

Roquelaure, giving name to the dukedom.

10. *Comenges*, or *Comminges*, in Latin *Convenæ*, eighteen *French* leagues in length, and about fix broad; being remarkable only for fome marble quarries; is divided into the upper and lower.

1. *Upper Comenges*, the air of which is cold by reafon of its fituation on the mountains. The principal places in it are as follow, *viz*.

S. Bertrand, the capital of the *comté* fmall, and ftanding on an eminence near the *Garonne*; is a bifhop's fee fubject to the archbifhop of *Auch*, with a diocefe containing two hundred parifhes, fome of which lying in the pro-vince of *Languedoc*, the bifhop of this place is one of the States. His re-venue amounts to 28,000 livres *per annum*, and his taxation at the court of *Rome* is 5000 florins. Near this place formerly ftood the *Lugdunum Con-venarum* of the ancients.

S. Beat, a ftrong town feated betwixt two mountains on the *Garonne*, which is here joined by the *Pique*, contains a priory. All its houfes are of marble, that being the only ftone hereabouts. The fmall town of *Valentine*, though belonging to the diocefe of the bifhop of *S. Bertrand*, is reckoned in *Upper Languedoc*.

The little country of *Nebouzan*, was formerly a vifcounty, and ftill votes in the affembly of the States. It contains the following places, *viz*.

S. Gaudens, the capital, which is well inhabited, and the feat of the country court. It contains alfo a collegiate church and three convents.

Barbazan, a little place, having mineral waters.

Caffagnabere, fmall, but noted for being the birth-place of Cardinal *Offat*.
Nifos, an abbey.

La Roque, a little place and barony.

Capver, a village.

Mont-regeau, or *Monrejau*, a little town ftanding high on the *Garonne*, not far from the place where it is joined by the *Nette*, belongs to the duke. of *Antin*.

S. Martory, a large market-town lying on the *Garonne*, is famous for the faint of that name.

2. *Lower Comenges*, is feated in a plain, and· enjoys a warmer air than the upper part of the country.   In it are

1. The lordfhip of *Sammatan*, containing

Sammatan, a little town, which is the feat of a *caftelany*.

Lombez, a little town feated on the *Save*, and a bifhop's fee, fubject to the archbifhop of *Touloufe*. Its diocefe confifts of ninety parifhes; the bifhop's revenue is 20,000 livres, out of which his taxation at the court of *Rome* is 2500 florins.

2. The lordfhip of *Muret*, containing

Muret, a fmall town lying on the *Garonne*, and the feat of a royal and foreft-court. Near this place *Peter* of *Arragon*, who befieged it in 1213, was defeated and flain in the field.

L'Ifle Dodon, fmall, and ftanding high on the river *Save*, contains a King's court and *caftelany*.

Montpezat, a market-town.

11. The country of *Conferans*, or *Couferans*, formerly a vifcounty, lies among the *Pyrenean* mountains, and contains

S. Lizier, a little town feated on the *Salat*, and the principal place of this country being a bifhop's fee fubject to the archbifhop of *Auch*. Its diocefe confifts of eighty-two parifhes, and the bifhop of this place enjoys a revenue of 24,000 livres, out of which he is taxed at the court of *Rome* in 1000 florins. This place was anciently called *Auftria*.

Conferans, a market-town ftanding on an eminence, and formerly the capital of the country.

S. Girons, a little town on the river *Salat*.

Maffat, a fmall town, containing a collegiate church.

Bigorre, once a county, and anciently inhabited by the *Bigerreri* or *Begerroni*, has States of its own, which are compofed of the bifhop of *Tarbc*, four abbots, two priories, a commander of the order of *Malta*, eleven barons, and the citizens and peafants. It is divided into three parts, *viz.*

1. The levels, containing

Tarbe, the principal place, which ftands on the fide of the ancient city of *Begorra*, is fmall, but has a bailiwick and foreft-court; being alfo a bifhop's fee fubject to *Auch*, with a diocefe of three hundred and eighty-four
parifhes.

parifhes and chapels. His revenue is 22,000 livres, arid his taxation at the court of *Rome* 1200 florins. In it, befides the cathedral, are one parifh-church and two convents. The town is alfo defended by a fort.

Vic de Bigorre, a little town, which its firft counts frequently made their place of refidence.

Antin confers the title of duke and peer.

Bagneres, a town fituate in the vale of *Campan* on the river *Adour*, and celebrated for the virtues of its mineral fprings, which were known fo early as the times of the *Romans*, and of which M. *Salignac* has publifhed a particular account. Here are both cold and hot fprings.

Campan, a market-town, having quarries of marble in its neighbour-hood.

Lourde, a little town fituate in the vale of *Lavedan* on the river *Gave*; has a fort ftanding on a rock.

2. The mountainous part contains

S. Savin, a very ancient abbey of *Benedictines*, fituate in the vale of *Lavedan*.

Baredge, a little place lying in the valley of *Lavedan*, at the foot of Mount *Tormalet*, celebrated for its warm baths.

3. *Ruftan* contains

S. Sever, a little town feated on the river *Rouffe*, which takes its name from the ancient abbey of *Benedictines* feated there.

Jornac, a fmall place.

Obf. The valley of *Lavedan*, which lies among the *Pyrenean* mountains, is about ten or twelve *French* leagues in length and in fome places feven or eight in breadth, being a vifcounty, but not comprehending at prefent the whole vale; *Lourde*, *S. Savin*, and *Baredge*, belonging to the county of *Bigorre*, in which I have alfo included them. Among the places pertaining to this vifcounty is *Caftellobon* and *Beaufent*.

The G O V E R N M E N T *of*

S A I N T O N G E *and* A N G O U M O I S,

Contains the greateft part of S A I N T O N G E and A N G O U M O I S.

1. S *A I N T O N G E* terminates eaftward on *Angoumois* and *Perigord*; fouthward on *Bourdelois* and the *Garonne*; weftward is bounded by the ocean, and northward by the country of *Aunis* and *Poitou*. This diftrict is about twenty-five *French* leagues in length and twelve broad. Its

name

name is derived from the *Santoni*, the ancient inhabitants thereof. This country abounds in grain, wine, and all kinds of fruit; and particularly near the sea makes great quantities of excellent salt. Its horses also are greatly esteemed. In it too are some mineral springs. Its principal rivers are the *Charente*, which abounds in fish, and receives its source near *Charennac*, running into the sea; and the *Butonne*, the source of which is at *Chef-Boutonne* in *Poitou*, being navigable at *S. Jean d'Angely*, and falling into the *Charente*. This country was formerly an earldom, and in the middle-ages frequently changed masters betwixt the *French* and the *English*; but *Charles* V. fixed it to the crown of *France*. It is subject to the parliament of *Bourdeaux*, a few parishes only excepted, which are within the jurisdiction of *Angoumois*. The governor-general is generally also deputy-governor of this country, which the *Charente* divides into south and north.

1. In the south part or *Upper Saintonge* lies .

Saintes, the capital of the country being seated on the *Charente*, and a bishop's see. It contains an election, a provincial and a bailiwick court, with a *marechaussée*. Its bishop is suffragan to the archbishop of *Bourdeaux*, having a diocese under him containing five hundred and sixty-five parishes and chapels, with a revenue of 20,000 livres, out of which his *Roman* tax is 2000 florins. It is small, and has narrow mean streets; but in the suburbs is an abbey and college of Jesuits. Some councils were held here in the years 563, 1075, 1080, 1088 and 1096.

Pons, a handsome town, stands high on the little river *Seigne*, over which it has several bridges, whence probably it received its name: It is divided into the upper and lower. In it are three parish-churches, three convents, three alms-houses, and a commanderie of the knights of *Malta*. It has also a mineral spring, is an ancient *Sirauté*, the proprietors of which stile themselves *Sires de Pons*, with fifty-three parishes and two hundred and fifty noble fiefs. At present it belongs to the house of *Lothring-Marsan*.

Jonsac, a little town.

Barbesieux, a small town and marquisate belonging to the house of *Louvois*. In it are two parish-churches and one convent, with a mineral spring in its neighbourhood.

Montausier, a dukedom and peerage, having seven parishes belonging to it.

Mortagne, a town seated on the *Garonne*, and bearing the title of a principality.

Tallemond, a market-town standing on an eminence on the *Garonne*, with the title also of a principality.

Pont l'Abbé, a market-town.

2. In the north part, or *Lower Saintonge* lies

Saint Jean d'Angely, in Latin *Angeriacum*, a town lying on the *Butonne*, and the capital of an election, with a royal court of justice. Whilst this place continued in the hands of the Protestants it was populous and well-fortified;

5

fortified; but being reduced in the year 1621 by *Lewis* XIII. the fortifications were razed, and the city deprived of its privileges. In it, however, is an abbey and three convents. Its brandy too is much efteemed; and it alfo makes woollen ftuffs.

Taillebourg, a little place lying on the *Charente,* having a caftle feated on a rock. It has a chapter, and gives title of count to the houfe of *Tremouille.*

Tonnay Charente, a town and principality on the *Charente,* with a caftle, an abbey, and an harbour; has belonged for fome time to the houfe of *Rochechouart.*

Tonnay-Boutonne, a little town feated on the *Boutonne.*

Fontenay la Battu, a market-town, which in the year 1714, was raifed to a dukedom and peerage under the title of *Rohan.*

A N G O U M O I S,

WHICH derives its name from its capital, is bounded to the weft by *Saintonge,* and to the fouth by *Perigord.* Eaftward it borders on *Limofin,* and northward on *Poitou;* being betwixt fifteen and eighteen *French* leagues in length and about fixteen broad. The country, which is full of hills, but none of any confiderable bignefs, produces wheat, rye, barley, oats, *Spanifh* corn, faffron, wine, and all kinds of fruits. It yields alfo fome excellent mines of iron; but is particularly famous for paper. Its principal rivers are the *Charente* and *Touvre,* the latter of which rifes here and runs into the former. This country was anciently an earldom; but count *Francis* coming to be King of *France,* raifed it to a dutchy. It is fubject to the parliament of *Paris,* and contains a bailiwick and territorial jurifdiction. In it are the following places, *viz.*

Angoulefme, in Latin *Inculifma,* the capital of the country, feated on a hill on the river *Charente.* It confers the title of duke, and is the refidence of a bifhopric, an election, a *prevoté,* a bailiwick, a country and foreft-court, an office of the five great farms, and a *marechauffie.* Its bifhop is fuffragan to the archbifhop of *Bourdeaux,* with a diocefe of two hundred parifhes, and a revenue of 20,000 livres, out of which his tax at the court of *Rome* is 1000 florins. In it alfo is a college of Jefuits, with twelve parifh-churches, and the abbey of *S. Cibard,* in which lie buried the old counts of *Angoumois,* together with ten convents and a general hofpital.

Coignac, or *Cognac,* the fecond town of the country, pleafantly fituated on the *Charente,* has a caftle, in which *Francis* I. was born, together with three convents. In the year 1238, a council was held here.

Jarnac, a market-town feated on the *Charente,* near which, in the year 1569, the *Huguenots* were defeated by *Henry* III.

Chateau-

Chateau-neuf, a town and county, containing a royal *prevoté* and *cha-tellany.*

La Valette, a little town, having a diftrict of thirteen parifhes and forty fiefs, was formerly a dutchy and peerage.

Aubeterre, a fmall town and a marquifate with an abbey and one colle-giate-church.

Rochefoucaut, a little town, dukedom and peerage, lying on the *Tardouere,* having a collegiate-church and a convent.

Blanzac, a fmall town, containing a chapter.

Verteuil, a litte town and barony, feated on the *Charente* and belonging to the ducal houfe of *Rochefoucaut.*

Ruffec, a little town and marquifate.

Chabanois, or *S Quentin de Chabanois,* a fmall town and principality.

Montbrun, a little town and the capital of a county, to which belong eigh-teen parifhes and forty fiefs.

The GOVERNMENT *of* AUNIS.

THE province of *Aunis,* or *Aunix,* in Latin *Alnifium,* is bounded to the north by *Saintonge,* to the weft by the ocean, and northward and eaftward by *Poitou.* It is watered by the river *Seure,* which rifes in *Poitou,* becomes navigable at *Niort,* after which it receives the *Vendie,* which is alfo navigable, and by means of the *Charente.* It has good harbours along the coaft. The country is rather too barren, but produces good grain and a great deal of wine, the fwampy parts alfo affording good pafturage. An excellent falt is made in the falt-marfhes hereabouts. The province depends on the parliament of *Paris,* but is partly governed by its own common law, founded on cuftom. Under the governor is a general-lieutenant and a deputy-governor.

1. The country of *Brouageais,* which forms a part of *Saintonge* and makes very good fea-falt, contains

Brouage, a little ftrong fea-port fituate among marfhes.

Fort Chapus lies on a rock at the mouth of the *Seudre,* which renders it a tolerable good harbour.

Marennes, a fea-port and alfo an election with a court of admiralty. The adjacent country is very fruitful and its wine good. A great deal of falt is alfo made here. The *Sires de Pons* are ftyled counts of *Marennes.*

Arvert, a town feated on a peninfula.

La Temblade, a handfome, populous and thriving town.

Sauion, a town, having a caftle, feated on the river *Seudre,* was once fortified.   *Royan,*

Royan, a little town, lying on the *Garonne*, with an harbour in which are caught great numbers of fardines. In it are two convents. It also confers the title of marquis. When under the *Huguenots* it was well fortified, and so vigorously defended against *Lewis* XIII. in the year 1621, that he was obliged to draw off his troops; but he afterwards severely avenged this difgrace, demolishing it so entirely, that the prefent place is only the fuburbs of the former.

Soubife, a town, lying on the *Charente* and having a chapter, is the capital of a principality, to which belong feven parifhes. Its annual produce is about 12,000 livres. Dependent alfo on this town is

L'ifle de Madame, lying at the mouth of the *Charente*, and *Fort Lupin*. In its neighbourhood are the mineral waters of *Roufxililaffe*.

Oleron, in Latin *Uliarus*, an ifland about three *French* leagues diftant from the main land, being five in length and two broad and extremely fruitful. The inhabitants are good feamen. It belongs to the government of *Aunix*, but is fubject to the judge of *Saintonge*, appeals from whence are carried to the parliament of *Bourdeaux*. On the eaft of it ftands a ftrong caftle, near which a town has been erected in which are two hofpitals and one convent. The light-houfe of *Chauffiron* ftands on a point of the ifland.

3. The ifle of *Re*, in Latin *Radis*, lies betwixt two and three *French* leagues from the continent, being four miles in length and two broad. It produces plenty of wine, of which is made a very fine fort of brandy. It abounds alfo in falt and is populous. In it are

S. Martin, a fortified little town, having a citadel and harbour.

Fort La Prée defends the entrance into *Pertius Breton*.

Fort Samblanceau commands the ftraits of *Pertius d'Antioche*.

Fort Martray lies on one fide of it.

4. The country of *Aunis* Proper contains

Rochefort, a new regular built town, feated on the *Charente*, having an excellent dock and magazines well replenifhed with naval ftores. In it is alfo a royal marine academy with an hofpital for feamen, a feminary and a convent; but the neighbouring falt-marfhes give an unwholefomenefs to the air. The approach to the town up the river is well defended by feveral forts; as, namely, that on the ifle of *Aix**, the redoubt facing it called *Aiguille*, Forts *Fourax*, *de la Pointe* and *Vergeron*; befides which there is a pallifado carried along the river.

Surgeres, a neat market-town.

Rochelle, in Latin *Rupella*, the capital of the province and government, lies on the fea, having a good harbour and being rather handfome than large.

---

* On the twenty-firft of *September*, 1757, this fort, together with the ifland on which it ftands, was taken by the brave Captain *Howe*, in the *Magnanime*, after about half an hour's refiftance; the whole garrifon, which confifted of near fix hundred men, being made prifoners of war.

large. It is the refidence of an intendency, an election, a provincial and
bailiwick court, an admiralty, a chamber of commerce and a marechauſſée.
It has alfo a mint and an academy of *belles lettres*, with a college of Jefuits,
a medicinal, botanical and anatomical fchool and a fugar manufactory. It
is likewife a bifhop's fee, fuffragan to *Bourdeaux*, with a diocefe of one
hundred and eight parifhes, and a revenue of 17,000 livres, out of which
he is taxed at the court of *Rome* in feven hundred and forty-two florins.
Its maritime trade in general, and efpecially to the *French* colonies in *Africa*
and *America*, is very confiderable, the harbour being fafe, though the
entrance to it is narrow and none of the deepeft. In the fixteenth century, the inhabitants joined in the reformation, fortified the town and held
out a fiege. In the year 1622, *Lewis* XIII. in order to compel them to a
furrender, ordered *Fort Louis* to be erected at the entrance of their harbour;
and, in the year 1628, to prevent their receiving any fuccours by fea a mole
was raifed which furrounded the haven. Famine, at length, obliged them
the fame year to capitulate; in confequence of which their privileges were
taken from them and the fortifications demolifhed; but, in the reign of
Lewis XIV. thefe were again repaired. The air here is no better than at
Rochefort and from the fame caufe.

Marans, a large market-town, lying among the falt marfhes, carries on
a good trade in falt, malt and meal.

Charon, a market-town, fituated on the fea and having an abbey.

Marfilly, alfo a market-town.

The GOVERNMENT *of*
P   O   I   T   O   U.

THE province of *Poitou* borders, to the fouth, on *Saintonge*, *Angoumois* and *Aunix*; to the eaft on *La Marche* and part of *Berry*; to the
north on *Touraine* and *Anjou*, and weftward on the ocean. From weft to caft
it is forty-eight *French* leagues, and from north to fouth twenty-two. It
receives its name from the ancient *Pictoni*, or *Pictavi*. The foil is various,
according to the different parts of the country, but it abounds, in general,
with grain and cattle; and the principal trade of the inhabitants confifts in
oxen, mules, horfes and woollen ftuffs. The largeft rivers in it are, the
Vienne, which rifes on the borders of *Limoufin*, becomes navigable fome
miles above *Chatelleraud*, and being joined by the *Creufe* falls into the
Loire. The *Sevre Niortoife*, the fource of which is fome leagues above
S. Maixent, becomes navigable at *Niort* and, after receiving the *Vendée*, difcharges itfelf into the fea. The *Clain*, which rifes in the frontiers of *Angoumois* and mingles with the *Vienne*.   This

This province was erected into an earldom by *Charlemagne. Eleanor*, daughter to the laſt duke of *Aquitaine*, brought it to her ſpouſe King *Henry* of *England. Philip Auguſtus* made himſelf maſter of it, and *Henry* III. King of *England*, formally ceded it to *France*; but afterwards thoſe two crowns had frequent diſputes concerning it, during which it was given as an appenage to the Princes of the blood. Since the year 1436, this province has been perpetually annexed to the crown. It is ſubject to the parliament of *Paris* and has but one provincial court. Under the governor is a general-lieutenant and two deputy-governors.

It is divided into two parts.

1. *Upper Poitou*, which conſtitutes the eaſtern part of the country, is larger, more fruitful, pleaſant and healthy than the *Lower*. The principal places in it are

Poitiers, in Latin *Auguſtoritum*, the capital of the country, lying on the river *Clain*. It is of conſiderable extent, and perhaps, next to *Paris*, the largeſt town in the whole Kingdom, but is mean, ſmoky and thinly inhabited. It is the ſeat of a biſhopric, an intendancy, an election, a country court and royal juriſdiction, a *marechauſſee* and a mintage. The biſhop hereof is ſuffragan to the archbiſhop of *Bourdeaux*, with a dioceſe of ſeven hundred and twenty-two pariſhes and a revenue of 22,000 livres, out of which he is taxed at the court of *Rome* in 2800 florins. The cathedral here is of uncommon bigneſs and built in the *Gothick* ſtyle. In this city are alſo four collegiate-churches, ſeventeen other pariſh-churches, twenty-one convents, four abbies, with an univerſity founded in the year 1431, one college of Jeſuits, two ſeminaries and three hoſpitals. On the *place royale*, as it is called, the inhabitants, in the year 1687, erected a pedeſtrian ſtatue to *Lewis* XIV. The handicraftmen in this town are princpally glovers and comb-makers. It exports alſo round woollen caps and ſtockings to the *American* colonies. It is not likewiſe without ſome remains of *Roman* antiquities; and, though the amphitheatre here lies greatly buried among gardens and ſmall houſes, yet is it wholly diſtinguiſhable. Of a triumphal arch, or rather gate, erected at the beginning of a military way, all that remains at preſent is the Arc with the two pillars which ſupport it. The palace and thick round tower cloſe by it are indeed ſaid to be *Roman*, but they carry evident traces along with them of *Gothick* ſtructure, not to mention others of more modern date. In the neighbourhood here, in the year 1356, was fought a battle betwixt the *French* and the *Engliſh* to the diſadvantage of the former, and in which *John*, King of *France*, was taken priſoner.

Luſignan, an old town ſeated on a hill.

Vivonne, a little town, lying on the river *Clain*, gives the title of count.

Niort, the beſt trading town in this province, lies on the river *Seure*, being the ſeat of an election, a bailiwick, a royal juriſdiction, a foreſt-court and a *marechauſſée*. It contains a caſtle, two pariſh-churches, nine convents, a general hoſpital, and has ſome manufactures of wool.

S. Maixant, a little town, ſeated on an eminence on the river *Seure*, being the reſidence of an election and a royal court of juſtice. In it are three pariſh-churches, one abbey of *Benedictines*, four convents, one college and one hoſpital. It belongs to the duke of *Mazarine*.

Melle, a little town, the ſeat of a royal *prevoté*, contains three churches, one ſmall college and a manufactory of ſerge.

Chizay, a market-town, having a royal *prevoté*.

Aunay, a little place, conferring the title of count.

Civray, a ſmall town, lying on the *Charente*, in which is a royal juriſdiction, a royal bailiwick, a provincial *marechauſſee*, one pariſh-church and two convents.

Charroux, or *Chairoux*, a ſmall town, ſeated on the *Charente*, in the little territory of *Briou*, contains an abbey.

Marſillac, a principality.

Rochechouart, a little town, lying on a mountain and having a caſtle, is the principal place of a viſcounty.

L' iſle Jourdain and *Luſſac* are two ſmall towns, and the latter alſo a marquiſate.

Mortemar is a dukedom and peerage belonging to the houſe of *Rochechouart*.

Montmorillon, a town, which is the ſeat of a royal juriſdiction, a bailiwick and a *marechauſſee*, contains one pariſh, one collegiate-church and three convents.

Tremouille, or *Trimouille*, a little town, lying on the *Venaiſe*, is a dukedom and peerage.

S. Savin, a ſmall place, contains a celebrated abbey of *Benedictines*.

Chauvigny is a little town lying on the *Vienne*.

Chatellerand, in Latin *Caſtellum Eraldi*, or *Caſtrum Airaudi*, a town lying on the *Vienne*, in the country and dukedom of *Chatellerandois*, belongs, at preſent, to the duke of *Tremouille*. It is the ſeat of an election, a royal bailiwick, a *marechauſſee* and a foreſt-court. In it alſo is a collegiate-church and four convents with a fine bridge of ſtone leading over the river to the ſuburbs.

Richelieu, a regular built town, founded by Cardinal *Richelieu*, together with a ſtately ſeat, lying on the little rivers *Amable* and *Vide*, gives the title of duke and peer, and is the ſeat of an election and a ſalt magazine.

The little diſtrict of *Mirebalais* contains

Mirebau, a ſmall town and barony, which is the principal place of the diſtrict. In it are five pariſh-churches, one chapter and ſeveral convents.

Moncontour, a little town, ſeated on the *Dive*, near which, in the year 1567, the *Huguenots* were defeated.

The

The little country of *Gaſtine*, its principal place

Partenay, a town, the ſeat of a *prevoté*, a royal juriſdiction and a *marechauſſée*. In it alſo are one chapter and three convents.

The little country of *Loudunois*, the principal place of which is

Loudun, a town and the reſidence of an election, a royal *prevoté*, a juriſdiction and a *marechauſſée*. In it is alſo one chapter, two pariſh-churches and eight convents. Much labour and art has been uſed to bring over the Proteſtants of this town to the *Romiſh* church.

Fontevraut, a market-town, having an abbey, lies on the borders of *Anjou*.

Thours, a town ſtanding on a hill on the river *Toue*, is the principal place of an election and the ſeat of a *marechauſſée*. It contains two chapters, three pariſh-churches, one abbey, five convents, two hoſpitals and a ſmall college. It confers alſo the title of a dukedom and peerage, to which belong 1700 fiefs.

II. *Lower Poitou* conſtitutes the weſtern part, and contains the following places ; *viz.*

Argenton, a little town.

Mauleon, a ſmall town and the principal place of an election.

Mortaigne, a little town and a dukedom.

Montaigu, a market-town and marquiſate.

Garnache, a market-town and barony.

S. Gille, a ſmall ſea-port.

La Roche ſur Yon, a market-town and principality lying on the river *Yon* and belonging to the houſe of *Bourbon-Conti*.

Mareuil, a market-town, ſituate on the river *Lay*.

Les Sables d' Olonne, a town, lying on the ſea, the principal place of an election, containing alſo a ſmall harbour and being the ſeat of an admiralty: In it are two pariſh-churches and four convents. The inhabitants of this place are reckoned good ſailors.

Talmont, a little town, bearing the title of a principality. In it are two pariſh-churches and one abbey.

Fontenay la Comte, a town, lying on the *Vendée*, having three pariſh-churches, two hoſpitals, four convents and a college of Jeſuits. It contains a royal bailiwick, a *marechauſſée* and a foreſt-court.

Meilleraye, a dukedom and peerage.

Luçon, a town, lying in a moraſs and enjoying the title of a barony, is a biſhop's ſee and the reſidence of a bailiwick and *marechauſſée*. Its biſhop is lord and baron of the town, being ſuffragan to the archbiſhop of *Bourdeaux*, with a dioceſe of two hundred and thirty pariſhes, and a revenue of 20,000 livres, out of which his taxation at the court of *Rome* is 1000 florins. Excluſive of the cathedral it contains one pariſh-church, one ſeminary and two convents. The air here is unwholeſome.

Maillezais,

Maillezais, a little town, lying on an Island formed by the rivers *Seure* and *Antize.* The bishop's see, which was formerly here, is removed to *Rochelle.*

The Island of *Noirmoutier,* in Latin *Nigrum Monasterium* and *insula Dei,* is three *French* leagues in length and populous, confers the title of marquis, belongs to the younger branch of *Tremouille,* and contains the following places; *viz.*

Noirmoutier, a little town having a priory.

Barbastre, a market-town.

The island of *Yeu,* in Latin *Oya,* is three *French* leagues in length.

The GOVERNMENT *of*

B R E T A G N E.

THIS province is a peninsula surrounded on all sides by the sea, excepting towards the east, where it joins *Normandy,* *Maine,* *Anjou* and *Poitou.* Its greatest length, from west to east, is sixty *French* leagues, and its greatest breadth forty-five; but in many places it is very narrow. Its name it receives from the old *Britons,* or *Britts,* who being driven out of *Britain* about the middle of the fifth century by the *Angles* and *Saxons,* crossed the sea into *Gaul,* and, after wandering about for some time, settled in the country of the *Curiosolitæ* and *Ofismi,* who were branches of the *Armorici,* and possessing themselves almost of the whole country of the *Vanni,* gave name at length to this province. *Gregory* of *Tours,* however, is the first who mentions it under this title. In succeeding times the *Britons* were obliged to submit to the Kings of the *Franks.* *Charles the Great* had a fleet here to act against the *Normans.* Under his successors, *Numenoius,* a chief of the *Bretons* in that part of the country which at present bears the name of *Upper Bretagne,* and whose inhabitants were originally *Gauls,* created himself King. His second successor was taken off by some conspirators who made themselves masters of the country, but only under the title of counts. The *Normans* indeed reduced it but could not keep quiet possession of it. These disturbances were brought, however, to a period under *Conan,* count of *Bretagne,* who married his only daughter, *Constantia,* to *Gottfried,* count of *Anjou,* son to *Henry* II. of *England* and duke of *Normandy.* The daughter and heiress of this *Constantia* marrying with *Peter* of *Dreux,* *Bretagne* fell to the royal family of *France,* *Peter* being a Prince of the blood. His grandson, *John* II. was a duke and peer of *France.* After the death of *Francis* II. duke of *Bretagne,* *Anne,* his heiress, was successively married to the two Kings *Charles* VIII. and *Lewis* XII. and by the latter left issue

two

two daughters, the eldeſt of whom, called *Claudia*, was married to *Francis* I. who, at the deſire of the ſtates, in the year 1532, united *Bretagne* to the crown. His ſucceſſor, *Henry* II. aboliſhed the title of duke of *Bretagne*.

This province contains ſome very 'good harbours but few navigable rivers except the *Loire*, which terminates its courſe here, and the *Vilaine*, in Latin *Vicinonia*. The country is in ſome parts level, in others hilly, particularly in *Upper Bretagne*, through which runs a whole chain of mountains called *Le Mont Arré*. It produces no great quantity of grain or wine, but its extenſive and fruitful paſtures yield a very profitable trade in butter. It produces alſo hemp and flax, of which are made great quantities of linen and ſail-cloth. At *Carnot*, in the dioceſe of *Quimpir*, is a lead mine, but the lead found in it is not a near ſo good as that of *England*. The dioceſe of *Nantes*, in ſome places, yields alſo pit-coal, but that too much inferior to the *Engliſh*. On the coaſts are taken great numbers of ſardines and other fiſh. The earldom of *Nantois* contains a fair manufactory. Horſes are alſo another branch of its trade. The inhabitants on the ſea-coaſt are excellent ſea-men. In *Upper Bretagne* they ſpeak *French*, but *Lower Brittany* has a particular language, which is ſuppoſed to be the ſame with the *Celtick* *.

This country has its own parliament, which is held at *Rennes*, as alſo its own laws and particular ſtates; the latter of which conſiſt of the clergy, nobility, burghers and peaſants. They are ſummoned by the King every two years. The governor is alſo admiral of *Bretagne*, and under him are two general-lieutenants, one of whom has the ſuperintendency of eight dioceſes and the other only of the earldom and biſhopric of *Nantes*. Beſides theſe there are three other ſub-governors; *viz.* the firſt for *Rennes, Dol, S. Malo* and *Vannes*; the ſecond for *S. Brieu, Treguier, S. Paul de Leon* and *Quimper*; and the third for *Nantois*. In the meeting of the ſtates and aſſeſſment of taxes, the country is divided according to the nine biſhopricks, of which the firſt five belong to *Upper Bretagne*.

1. The biſhopric of *Rennes* contains the following places; namely,

Rennes, anciently *Condate*, the capital of the whole country, deriving its name from the *Rhedones*, the moſt famous people among the *Armorici*. It is divided by the *Vilaine* into two parts, is pretty large, populous, and the reſidence of a biſhop, the parliament of the whole country, an intendancy, a collection, a chamber of aids, a country-court, a foreſt-court and a conſulate. It contains many well-built houſes, but the ſtreets in it are narrow, dark and dirty. Its biſhop is ſubordinate to the archbiſhop of *Tours*, has a dioceſe conſiſting of two hundred and ſixty-three pariſhes, with an income of 14,000 livres, out of which he is taxed to the court of *Rome* in 1000 florins. Ex-

---

* The author ought rather to have ſaid the *Welch*, which, making ſome allowances for a *French* accent, it certainly is in greater purity than what is ſpoke in many parts of that country.

cluſive

clufive of the cathedral, in it are eight parifh-churches, feventeen convents and a fine college of Jefuits, with a nòble ftatue of *Lewis* XV. erected
by the ftates of the province. The vifcounty of *Rennes* belongs to the duke
of *Tremouille.*

Chateaubourg, an earldom.

Vitrei, a fmall town, lying on the *Vilaine,* having a chapter and priory.
It is the feat of the firft barony in the country and belongs to the duke of
Tremouille.

S. Aubin du Cormier, a little town, near which the *Bretons* and their confederates defeated the army of *Charles* VIII.

Fougeres, in Latin *Filiceriæ,* a town, feated on the river *Cuefnon,* containing a caftle, two parifh-churches and an abbey.

2. The bifhopric of *Nantes* includes the earldom of *Nantois,* which is
divided by the *Loire* into two parts, being fruitful in wine and corn, good
paftures and confequently alfo in good cattle. It yields likewife falt and
pit-coal, and enjoys in all parts a flourifhing trade. In it are the following
places; *viz.*

Nantes, anciently *Condivincum,* or *Condivicnum,* which is in rank the
fecond city in *Brittany,* and receives its name from the *Namnetes,* a people
of *Gallia Armorica.* It lies on the *Loire,* is one of the greateft trading cities
in the Kingdom, being pretty large, populous, well fituated, and containing four fuburbs. It is alfo a bifhop's fee, the feat of an intendancy, a
collection, a chamber of accounts, a board of finances, a mint-office, a
country-court, a bailiwick, a *prevoté,* an admiralty, a foreft-court, and a
confulate. Its bifhop is fuffragan to the archbifhop of *Tours,* with a diocefe
of two hundred and twelve parifhes exclufive of chapels of cafe, and enjoys
a revenue of 30,000 livres, out of which his taxation at the court of *Rome*
is 2000 florins. Exclufive of the cathedral and a collegiate church, in it
are eleven parifh-churches, fourteen convents, two hofpitals, one college,
and an univerfity, founded about the year 1460. This city carries on a very
large trade to the *French* colonies in *America,* as alfo to *Spain* and *Portugal,* nòt to mention other *European* fhipping which frequent its port,
though fhips of burden can come no farther up the *Loire* than to *Paim-
bocuf,* a market-town, where the cargoes are put into fmaller veffels and
carried to *Nantes.* The city is famous in biftory for the edict iffued here
in the year 1598, by *Henry the Great,* for granting to the Proteftants the
public exercife of their religion, but this edict *Lewis the Great* revoked in
1685.

Ancenis, a little town and marquifate lying on the *Loire,* and belonging
to the houfe of *Bethune-Caroft.*

Chauteau-Briand, a fmall town, containing one parifh-church and two
convents, belongs at prefent, as a barony, to the houfe of *Bourbon Condé.*

Derval, a barony.

Coiflin,

Coiflin, a little place, raifed to a dukedom in the year 1603.

La Roche-Bernard, a market-town and barony lying on the *Vilaine,* and belonging to the dukedom of *Coiflin.*

Pont-chateau, a barony, belongs alfo to the duke *de Coiflin.*

Guerande, a little town ftanding clofe by the fea, amidft marfhes, in which a good falt is made. Its collegiate church is alfo the parifh-church, befides which it contains likewife two convents.

Croifie, a fmall town and fea-port, in the neighbourhood of which are falt-marfhes.

Paimboeuf, a market-town, containing an harbour on the river *Loire,* which is properly that of *Nantes.* Vide *Nantes.*

The country of *Retz* confifts of that part of the bifhopric of *Nantes* which lies to the fouth fide of the *Loire,* and belongs at prefent, as a dukedom and peerage, to the houfe of *Villeroi.* In it are the following places, viz.

Bernerie, a fmall market-town and fea-port, its inhabitants fubfifted chiefly by fifhing.

Bourgneuf, a little town having a fea-port. In the neighbouring marfhes are made great quantities of falt.

Machecou, Machecol, or *Machequoleu,* the principal place of this country, is a market-town containing two parifh-churches, and ftanding on the ruins of the town of *Ratiate,* which has been long fince demolifhed, but from which the country of *Retz* receives its name.

Cliffon, a little town and barony, having a collegiate church.

The greateft part of the ifland of *Bouin* belongs alfo to this divifion.

3. The bifhopric of *Dol* is the leaft, being only five miles in circuit.

Dol, the only town in it, is fmall, thinly inhabited, and its fituation unhealthy, this place ftanding on marfhes. It contains a treafurer's office and a court of admiralty, and is a bifhopric fubject to the metropolitan of *Tours,* with eighty parifhes, and a revenue of 22,000 livres, out of which the *Roman* tax is 4000 florins. The bifhop of this place, as lord of the city, bears the title of count of it, and formerly affumed the ftate of an archbifhop. Even to this day, within his own diocefe the crofs is carried before him, and at a meeting of the States, he takes place of all other bifhops.

4. The bifhopric of *S. Malo* is of pretty large extent, and its territory produces corn and fruits. In it are

S. Malo, in Latin *Maclovium,* and *Maclopolis,* a fmall but populous city, lying on a little ifland in the fea, formerly called *S. Aaron,* and joined to the continent by a mole, at the head of which is a ftrong fort. It contains a treafurer's office and a court of admiralty, and is alfo a bifhop's fee fubject to *Tours,* with a diocefe of one hundred and fixty parifhes. The bifhop of this place is lord of the city; and his income amounts to 35,000 livres *per annum,* out of which his tax to the court of *Rome* is 1000 florins. The harbour

2   bour

bour is large, but the entrance very rocky, and at ebb almost dry. On the adjacent rocks have been erected several forts, the principal of which are *Sezembre*, *la Conchée*, *le Fort Royal*, *le grand Bay*, *le petit Bay*, *L'iſle à Rebours*, *le Fort du Cap*, *Roteneuf*, and *le Chateau de Latte*. It carries on a conſiderable trade with *England*, *Holland*, and *Spain*, and in time of war ſtands very convenient for privateering.

At *Solidor* harbour, one *French* league from *S. Malo*, formerly ſtood the town of *Aleth*, which was alſo a biſhop's ſee, but in the year 1149 removed to *S. Malo*. The place where the ruins are ſtill viſible is called *Quidaleth*, or *Guichaleth*, *i. e.* ' *Aleth* town.'

Cancale, a port, giving name to a neighbouring cape, noted for fine oyſters.

Chateauneuf, a market-town and marquiſate.

Dinan, a town ſtanding on a hill, having a good caſtle, two convents, and an hoſpital. The States have ſometimes met here.

Combourg, an earldom.

Montfort, ſurnamed *la Cane*, *Ploermel*, and *Joſſelin*, containing an abbey, together with *Comper*, which lies on the river *Men*, are all little places.

5. The biſhopric of *S. Brieu*, a good corn and fruit country, with three iron-works at *Loudeac*, *Hardouinaye*, and *Vaublanc*. The moſt remarkable places in it are

Saint Brieu, in Latin *Oppidum S. Brioci*, or *Briocenſe*, lying among mountains which intercept the ſight of the ſea, though but half a league off, and on which it has a ſmall harbour, contains a receiver's office and a court of admiralty, being alſo a biſhopric ſubject to *Tours*, and conſiſting of two hundred pariſhes. Its revenue is 22,000 livres, and its taxation to *Rome* 800 florins. Excluſive of the cathedral, it contains alſo one chapter and one college.

Lambale, a little town, having ſome manufactures of linen, belongs to the dukedom of *Penthievre*.

Montconthour, alſo ſmall, and belonging to the ſame dukedom.

Quintin, or *Lorge*, a town and dukedom, containing a collegiate church, is alſo one of the ſtate baronies.

Jugon, another little place, conſtituting a part of the dukedom of *Penthievre*.

Loudeac, ſmall, but makes a great deal of thread; contains an iron-mill.

II. To *Lower Bretagne* belong four biſhoprics.

1. The biſhopric of *Treguier*; its territory produces corn and hemp, and is noted alſo for its breed of horſes. In it are

Treguier, in Latin *Trecorium*, a ſea-port, ſtanding on a peninſula formerly named *Trecor*. The *Bretons*, in their tongue, call this place *Lantriguier*, *i. e.* ' the town of *Treguier*.' It contains a collector's office, and is a biſhopric ſubject to the archbiſhop of *Tours*, with a dioceſe of ſeventy pariſhes.

parifhes. Its bifhop is lord and count of *Treguier*, and his revenue amounts to 20,000 livres, out of which his tax to the court of *Rome* is only 460 florins.

Lanmur, a little place, but having a royal court of juftice.

Morlaix, properly *Monrelais*, and in Latin *Mons relaxus*, a fmall town lying on a navigable river, with the tide coming up to it, which has not a little contributed to its trade. Its harbour is defended by fort *Taureau*, which ftands on an ifland. It contains two parifh and one collegiate church, and in its large fuburb of *Venice*, two convents and an hofpital.

The dukedom and peerage of *Penthievre*, belonging to the count *de Touloufe*, contains the following places, *viz.*

Guingamp, a fmall town, but the principal place of the dukedom, and having an abbey.

Lanion, alfo fmall, but extremely well fituated for trade.

Lambale, *Moncantour*, and *Jugon*, three little places in the bifhopric of *S. Brien*, alfo belong to it.

2. The bifhopric of *Vannes* contains the following places, *viz.*

Vannes, or *Vennes*, the *Dariorigum* of the ancients, a fea-port receiving its name from the *Veneti.* It is the feat of a provincial, an admiralty, and a foreft-court, a collection and a confulate; with a bifhopric alfo fubject to the archbifhop of *Tours*. Its diocefe confifts of one hundred and fixty parifhes, and its bifhop, who is lord of part of the city, enjoys a revenue of 24,000 livres, out of which his taxation at *Rome* is 350 florins. One of its two fuburbs called *le grand Marche*, is larger than the city itfelf; the name of the other is *S. Paterne*. In it are feveral churches and convents, together with a college of Jefuits. The harbour in *Morbian* bay is one of the fafeft and moft fpacious in the whole Kingdom.

The peninfula of *Ruys*, in Latin *Reuvifium*, or *S. Gildas*, containing an abbey of *Benedictines*, and a fort.

Auray, a little town and harbour feated on *Morbian* bay, and carrying on a good trade. In the year 1364, a battle was fought here betwixt *John* Count *de Montfort* and *Charles de Blois*.

Port Louis, a town having a very good road and harbour; its principal trade eels and pilchards. A citadel and other fortifications were erected here by order of *Lewis* XIV. who alfo gave his name to the place, which was before called *Blavet*.

L'Orient, well known for being the particular port of the *French Eaft-India* company.

Hennebond, a little town feated on the river *Blavet*, and once fortified, confifts of three parts, *viz.* the new town, the walled town, and the old town. In it are three parifh-churches and a fmall harbour, but it enjoys a good trade.

Redon, a little town ftanding on the *Vilaine*, and having an abbey.

Guemene, a little town containing a collegiate church, enjoys the title of a principality, and belongs to the house of *Rohan-Soubife*.

Rohan, a town lying on the *Aoufle*, and bearing the title of a dutchy and peerage; gives name to a very diftinguifhed family.

Pontivi, the principal place of the dukedom of *Rohan*.

Maleftroit, a little town and barony, feated on the river *Louft*.

Belle-Ifle, anciently *Colonefus*, about fix *French* leagues from the continent; being alfo fix long and two broad; confers the title of marquis. It is environed fo on all fides with rocks that it has only three landing-places; at one of thefe lies *Palais*, a little fortified place, having a citadel and a good road.

Bangor, a market-town.

Of the other places on this ifland, the principal are *Sauzon* and *Lomaria*.

Grouaix, or *Groays*, a little town lying oppofite the mouth of the river *Blavet*. Its principal employment catching of eels.

3. The bifhopric of *Quimper*, or *Cornouaille*, i. e. ' *Cornu Galliæ*,' this part of *Gaul* ftretching itfelf like a horn into the fea. It includes the whole earldom of *Cornouaille*, and in it are

Quimper, or *Quimper-Corentin*, a town feated on the river *Oder*, where it is joined by the *Bedet*. It is both large and populous, and contains a provincial court, a bailiwick, an admiralty, and a treafurer's office, being likewife a bifhop's fee fubject to the archbifhop of *Tours*, with a diocefe of above two hundred parifhes. The bifhop is lord of the city, and his revenue is 22,000 livres, out of which his taxation to the *Roman* court is 1000 florins. Exclufive of a fine college of Jefuits, in it are alfo two convents and an abbey.

Douarnenes, a little town, having an harbour feated on a bay of the fame name; in which are caught vaft quantities of pilchards.

Audierne, a little maritime town.

Concarnean, or *Coneq*, a little place, having an harbour lying on the fea.

Quimperle, a fmall town feated on the little river *Iffotte*, and containing an abbey and two parifh-churches.

Carhaix, a little town containing a foreft-court.

Chateaulin, a market-town feated on the river *Aufon*, and carrying on a great trade in falmon and flates. In its neighbourhood are found both copper and iron mines.

4. The bifhopric of *S. Pol de Leon*, contains the following places, *viz.*

S. Pol de Leon, or fimply *Leon*, and in Latin *Legio*, flanding on the fea, and receiving the name of *S. Pol* from a bifhop of that place; but *Leon* is the proper name of the diftrict. The bifhop hereof, who ftiles himfelf count of *Leon*, is fuffragan to the archbifhop of *Tours*, with a diocefe of one hundred and twenty parifhes, and a revenue of 15,000 livres, out of which he is taxed at the court of *Rome* in 800 florins.

Rofcof,

Rofcef, a town, having a harbour about a league diftant from *S. Pol.*
The ifle of *Bafs,* lying oppofite to it, forms a fine road.

Lefneven, and *S. Renand,* two towns belonging to the crown.

Landernau, a little town, but the principal place of the barony of *Leon,* belonging to the houfe of *Rohan.* It lies on the river *Elhorne,* and contains three parifhes.

Le Chatel, a fine lordfhip.

Le Conquet, a little town, ftanding on the weft point of *Britany.*

Breft, a fmall but ftrong town, having a fpacious and fine road and harbour, which is reckoned the heft and fafeft in the whole Kingdom. The entrance to it, however, is difficult on account of the many rocks in it, which lie under water. The harbour ftands betwixt the town and the fuburb called *Recouvrance,* being defended by a ftrong fort and a tower. In the town is a court of admiralty and a bailiwick, with two parifh-churches, a feminary under the direction of the Jefuits, and a convent. On the fouth fide of the harbour, in the year 1750, a foundation was laid for a building which is to ferve as a receptacle in winter to the flaves; and likewife for magazines of all kinds of naval ftores. Here is alfo an academy for fea-officers, infomuch that this place may be called the capital of the *French* marine.

Oueffant, in Latin *Uxantis,* a fmall ifland lying in the fea oppofite *Conquet,* and being about eight *French* leagues in circumference, gives the title or marquis to the family of *Rieux.* Befides the light-houfe here for the conveniency of fhips going into *Breft,* in it is alfo a fort. Some leffer iflands lying hereabouts likewife receive their name from it.

Sayn, in Latin *Sena,* a fmall ifland ftanding oppofite to the bay of *Douaruenes,* from which it is feparated by the *paffage du Ras.* The rocks and fhallows about it are very dangerous.

The G O V E R N M E N T of
N O R M A N D T.

NORMANDT terminates weftward on *Britany,* fouthward on *Beauffe, Perche,* and *Maine;* eaftward on *Picardy* and *L'ifle de France,* and northward on the *Brittifh* channel. Its extent from weft to eaft is upwards of fixty *French* leagues, and from fouth to north thirty. Its name, as will appear from the fequel, it derives from the *Normans.* It is one of the moft fruitful provinces in the whole Kingdom, as alfo one of the moft profitable to the King. It abounds in grain, flax, hemp, and vegetables for

dying,

dying, but what little wine it has is none of the heſt. It yields, however, great quantities of apples and pears, of which the natives make cyder and perry for their uſual drink. It is alſo a fine country for cattle, being full of excellent paſtures. The ſea ſupplies it with plenty of fiſh, and of its water are made great quantities of ſalt. The many iron-works in it are alſo of no ſmall advantage to the country. It has likewiſe ſome mines of copper, whence of courſe it is not without its mineral waters; and thoſe of *Forges* are in great repute.

The principal rivers in it are the *Seine*, mentioned in the introduction to *France*; the *Eure*, which riſes in *Grande-Perche*, and becomes navigable at *Maintenon*, receiving into it the rivers *Aure* and *Iton*, after which it falls into the *Seine* at *Pont d'Arche*; the *Andelle*, which takes its ſource in the pariſh of *Forges*, and falling into the *Seine* is very ſerviceable in conveying to *Paris* the fuel for firing which is cut in the foreſts of *Lions* and *Pitre*; the *Rille*, or *Riſle*, which receives its origin in the pariſh of *S. Vandrille*, and at *Roque* mingles with the *Seine*; the *Dive*, riſing in the pariſh of *Cour-Menil*, and after receiving the *Vie* becomes navigable, diſcharging itſelf at laſt into the ſea; the *Lezon*, which begins in the *Liévin* and, being joined by the *Orbiquet*, is afterwards called the *Tonques*, and becoming navigable loſes itſelf in the ſea; the *Carentone*, the ſource of which lies alſo in *Liévin*, and, after receiving the *Cernant*, falls into the *Rille*; the *Ante*, which iſſues beyond *Falaiſe* and, after a courſe of four *French* leagues, joins the *Vire*; the *Orne*, which riſes not far from *Seez*, receives the *Noirau*, *Guigne*, *Laize* and *Oudon*, and empties itſelf into the ſea. This river is navigable only from *Caen* to its mouth. The *Aure*, which riſes in the pariſh of *Parfouru*, ſix *French* leagues from the ſea and, in the pariſh of *Maiſons*, joins the *Drome*, which takes its ſource in the pariſh of the ſame name, but afterwards gradually loſes itſelf, and, as is thought, makes its appearance again at *Port en Beſzin*.

This country, under the *Roman* Emperors, was the ſecond *provincia Lugdunenſis*, and under the Kings of the *Franks* conſtituted a part of the Kingdom of *Neuſtria*. Under *Charles the Bold* it was over-run by ſwarms of pyratical *Normans*, who ſeated themſelves in *Neuſtria*, and, in the year 912, obliged *Charles the Simple* to cede it to them as a fief of *France*. Their duke and leader, *Rollo*, having been baptized, the laſt-mentioned *Charles* gave him his daughter *Giſle* in marriage. The ſucceeding dukes roſe to great power both here and beyond ſea; and *William*, in the year 1066, became King of *England*. In the year 1135, the male line of this King and duke became extinct in the perſon of *Henry* I. and his daughter *Matilda* married *Godfrey*, count of *Anjou*. The fruit of this marriage was *Henry* II. King of *England*, duke of *Normandy*, lord of *Guyenne*, *Poitou* and *Saintonge*. He left three ſons, *Richard*, *Godfrey* and *John*; the laſt of whom ſeized on the dominions of both his brothers, and even had *Arthur*,

the

the fon of *Godfrey*, taken off; for which, in the year 1202, *Philip Auguſtus*, with the confent of the peers, deprived him of moſt of his territories in *France*; and, in the year 1203, *Normandy* was annexed to the crown. *Henry* III. of *England* ceded to *Lewis the Pious*, and his fucceſſors, all his claim to this province, which, afterwards, to the end of the fourteenth century, fome Kings beſtowed on their eldeſt fons, with the title of duke of *Normandy*, till that of *Dauphin* was inſtituted. The animoſities betwixt the houſes of *Orleans* and *Burgundy* gave the *Engliſh* an opportunity of overrunning not only *Normandy* but a great part of *France*. This province they held about thirty years, when they were driven out by *Charles* VII.

Subjeƈt to the archbiſhop of *Rouen* are the ſix biſhoprics of *Normandy*, and theſe feven dioceſes contain eighty abbies and 4299 pariſhes. The country is governed by its own law, which is called ' the Wife;' and it is for this reaſon *Normandy* is ſtyled, *La pais de la ſapience*, *i. e.* ' the land of wiſdom.' *Rouen* has a parliament, on which all the other courts of the province are dependent. With refpeƈt to its finances, in it are three *generalities*; namely, at *Rouen*, *Caen* and *Alençon* ; from which the King is faid to have drawn 20,000000 of livres a year. The government of *Normandy* is one of the moſt conſiderable in the whole Kingdom. Under the governor are two general-lieutenants, one for *Upper* and the other for *Lower Normandy* ; and each of the feven great diſtriƈts has a deputy-governor of its own. Theſe diſtriƈts are *Rouen, Ceaux, Giſors, Evreux, Caen, Coutances* and *Alençon.* The country is divided into two grand parts.

I. *Upper Normandy* conſiſts of four great diſtriƈts and the following countries ; *viz.*

1. The country of *Ceaux*, excluſive of that part which conſtitutes the government of *Havre de Grace*, being fo called from its ancient inhabitants the *Caleti*, is a high level, containing little good water in it, but being very fertile in all kinds of grain. It belongs to the diſtriƈt of *Ceaux.* In it are

Caudebec, in Latin *Calidum Beccum, i. e.* ' the hot bee, or rivulet,' being a fmall but populous town, feated on the river *Seine*, into which, near this place, falls a rivulet which runs through the town and gives name to it. In it is a bailiwick, a provincial, an admiralty and a foreſt-court, as alfo a falt and treaſurer's office. It contains two convents, one pariſh-church and an hoſpital. Its manufaƈture of hats was formerly much more advantageous than at prefent, though the place ſtill preferves a pretty good trade by fea. In the year 1419, it was taken by the *Engliſh*; in 1562, by the Proteſtants; but, in the year 1592, retaken by the Royaliſts.

Lillebonne, a fmall town and principality, lying on the *Seine* and containing two churches and an old palace. In the years 1080 and 1162, the provincial fynods were held here.

· *Yvetot*, a large country town, containing a caſtle and one collegiate-church. The inhabitants pay no taxes. The lords of this place ſtyle themſelves felves Princes of *Yvetot*, but that the town has formerly been a Kingdom carries with it much the air of a fable.

Cailli, a market-town and barony, having the title of a marquifate.

Longueville, a town, conferring the title of duke.

S. Valeri, a large market-town, having a fmall harbour, and containing alfo an admiralty, a mayoralty and a falt-office.

Aumale, in Latin *Albamarla*, a little town but the principal place of a dutchy and peerage, being alfo the feat of a bailiwick, a *vifcomté* and a foreft-court, ftands on an eminence and contains two parifh-churches, one abbey and two convents; woollen manufactures its principal traffic.

Arques, a little town, but having more the appearance of a village, ftands on the river of the fame name. It contains, however, feveral courts and offices, and has alfo an abbey. Near this place, in the year 1589, *Henry* IV. defeated the forces of the league.

Dieppe, a fea-port, as well as its caftle, very irregularly fortified, contains two fuburbs, a harbour, a court of admiralty and a falt-office. In the year 1694, this place was laid in afhes by a bombardment from the *Englifh*, but afterwards rebuilt much better. It is famous for its laces and various ivory works.

Baqueville, a town and county.

Eu, a town, county and peerage, lying on the *Brefle*, contains a bailiwick, an admiralty, a foreft-court and a falt-office, together with one collegiate and three parifh-churches, one college of Jefuits, one hofpital, one priory, two convents and two forts.

Treport, a town, feated at the mouth of the *Brefle*, is the harbour to the town of *Eu*, and contains a falt-office, a mayoralty, an office of the farms and an abbey.

The marquifate of *Claire* and *Panilleufe*.

2. The little diftrict of *Bray* is very mountainous and its vallies marfhy; accordingly the country derives its name from the word *Bray*, *i. e.* ' pitch or flime.' The foil is not very fit for tillage, but makes amends for this defect by its excellent paftures and fertility in fruit. The principal places in it are the following, *viz.*

Neufchatel, a fmall town, containing three parifhes and one college.

Gournay, a little town, lying on the river *Ette*, or *Epte*, being a bailiwick, vifcounty, a mayoralty and a falt-office, contains one collegiate and one parifh-church, with four convents. This place has a manufacture for ferges and deals largely in butter and cheefe,

La Ferté, a town, containing a vifcounty-court and a *caftelany*, belongs to the houfe of *Matignon*.

Forges, a town, famous for its ferrugineous mineral-waters.

3. *Le Vexin Normand*, fo called by way of diftinction from *Vexin François*, the latter of which belongs to the government of *L' ifle de France*. In

place formerly lived the *Velocaſſi*, or *Bellocaſſi*, from whom the names of *Vulcaſſinum*, *Veulgueſzin*, or *Veulqueſzin* and *Vexin* are derived. The ſoil of this country is more fruitful than that of *Ceaux*. In it are

Rouen, the *Rothomagus* of the ancients, afterwards *Rothomum*, or *Rodomum*, the capital of *Normandy*, being environed on three ſides by mountains. The fourth ſide is watered by the *Seine*, over which it has a pontoon, or bridge of boats, which riſes and falls with the ebb and flood and is paved. It is a large populous city, and, excluſive of its great trade, is the reſidence of a parliament, an intendancy, an election, a chamber of accounts, a tax-office, a *viſcomté*, a bailiwick, a royal and foreſt-court, an admiralty, a ſalt-office, a mint, a conſulate, a *marechauſſée* and a *prevoté*. It has ſix ſuburbs, in one of which are three or four fine mineral ſprings. It contains above 7200 houſes, thirty-ſix pariſh-churches and fifty-ſix convents, among which are four abbies and a college of Jeſuits. The ſtreets are ſhort and narrow. Subject to its archbiſhop are ſix ſuffragans and a province of 1388 pariſh-churches, excluſive of chapels. His annual revenue is 80,000 livres, and his taxation to the *Roman* court 12,000 florins. He ſtyles himſelf primate of *Normandy*. In the cathedral are to be ſeen the monuments of ſeveral Kings, prelates and lords. In the butter tower, as it is called, which is one of the three towers of this church, hangs a bell ten feet in height, being the ſame in diameter, and weighing 36,000 pounds. On the market-place, *aux veaux*, ſtands the image of the celebrated maid of *Orleans*, who was burnt here as a witch by the *Engliſh*, and is repreſented kneeling before *Charles* VII. but that erected in the new market, in the year 1721, to *Lewis* XV. is a very ordinary piece of workmanſhip. In this city are ſeveral manufactures, and its trade is very conſiderable.

Jumiege, a market-town, lying on the *Seine*, containing two churches and one abbey.

Ecouis, a large market-town and barony, having a collegiate-church and one hoſpital.

Duclair and *Pavilli*, both market-towns.

La Londe, *Appeville*, *Charleval*, *Mailleraie*, *Quevilli*, *Til*, *Roſai*, *Tourni*, *Belbeuf* and *Varneville* are all marquiſates.

Mauteville, an earldom.

The four following towns and viſcounties belong to the diſtrict of *Giſors*.

Giſors, a town, ſeated on the river *Epte*, and having three ſuburbs, bears the title of a viſcounty. It is the principal place of one of the ſeven large diſtricts of *Normandy*, and contains an election, a *marechauſſée*, a mayoralty, a police, a foreſt-court and a ſalt-office. In it is only one pariſh-church, but it has ſix convents and an hoſpital.

Andeli, the *Andelaus*, or *Andelagus*, of the ancients, of which has been made *Andelium*, is the name of two towns lying near one another, being both together called *Les Andelis*.

Great Andeli, which is but a fmall place, lies in a valley on the river *Gambon,* being an election and containing a royal, a provincial, a *vicomté,* and a forcft-court with a falt-office. In it is a collegiate-church and a priory, with two convents and a fmall college.

Little Andeli, which is ftill fmaller, lies on the *Seine* and was formerly fortified. In it is one parifh-church, one hofpital and one convent.

Lions, the feat of a *vicomté,* an election and a foreft-court, with two convents, is furrounded with a foreft which is the largeft in all *Normandy.*

Vernon, a town and *vicomté,* lying in a fine valley on the river *Seine,* contains an election, a falt-office, a foreft and a diftrict-court. Under its parifh-church is alfo a collegiate one. In it are likewife fix convents and one college.

Neuf-Marche en Lions, a market-town, feated on the *Epte* and formerly fortified, contains a priory.

4. *Campagne,* belonging partly to the diftrict of *Rouen* and partly to that of *Evreux,* is divided into

1. *Campagne de Neubourg,* containing

Pont de l'Arche, in Latin *Pons Arcus,* or *Arcuatus,* or alfo *de Arcis,* a little town, having a bridge of ftone over the *Seine.* It is the feat of an election, a falt-office, a *vicomté,* a bailiwick and a foreft-court; has alfo a ftout fort, ftanding on an ifland, with a parifh-church, two convents and a very fine manufacture of cloth.

Louviers, in Latin *Lupariæ,* a fmall town and earldom, belonging to the archbifhop of *Rouen,* lies on the river *Eure* and contains a falt-office. It makes alfo a great deal of cloth.

Neubourg, a large borough, or market-town, giving name to the country and conferring the title of marquis. In it is a caftle, one parifh-church, one convent and an hofpital.

Harcourt, a market-town, dukedom and peerage, was formerly a marquifate under the name of *Tury.*

Evreux, anciently *Mediolanum,* fo called from the *Eburovici,* corruptly pronounced *Ebroicæ,* is a little town, having large fuburbs, lying on the river *Iton.* It is the principal place of an earldom belonging to the duke of *Bouillon,* with an election and a falt-office; is likewife a bifhopric fubject to *Rouen,* with a diocefe of four hundred and eighty parifhes. The revenue of its bifhop is 20,000 livres, and his taxation at the court of *Rome* 2500 florins. In it are nine parifh-churches, two abbies and feveral convents. Near it ftands the caftle of *Navarre.*

Gaillon, a market-town, lying on the *Seine,* containing a collegiate-church and a ftately feat of the archbifhop of *Rouen,* who is lord of the place. Not far from it is a fine *Carthufian* monaftery; the church of which is the burial-place of the counts of *Soiffons-Bourbon.*

2. *Campagne de S. André,* in which are

S. André,

S. André, a market-town.

Nonancourt, lying on the river *Aure*, being a small town and viscounty, containing a bailiwick and forest-court.

Verneuil, in Latin *Vernolium*, a town and marquisate, lying on the *Aure*, being also the seat of a collection, and containing a bailiwick, viscounty and forest-court, with a salt-office. It has also a college, two convents and some churches.

Breteuil, a little town and viscounty, seated on the river *Iton*, contains a forest-court.

Conches, a town and marquisate, which is the seat of a viscounty and a bailiwick-court, a *marechauſſée*, a mayoralty, a police and a salt-office, contains one abbey, three parish-churches and an hospital, and belongs to the earldom of *Evreux*.

5. *Le Roumois* lies betwixt the rivers *Seine* and *Rille*, abounding in grain and fruit and producing also good breeds of cattle, particularly sheep. In the country belonging to the district of *Rouen* are

Pont-Audemer, a town, lying on the river *Rille*, being the principal place of an election, and the seat of a viscounty, a district and a forest-court, having also a salt-office, two parish-churches and a small harbour.

Quillebeuf, a little town, lying on the *Seine*, contains a court of admiralty, and was formerly fortified.

Elbeuf, a little town on the *Seine*, being a dukedom and peerage, makes fine cloth and carpets. In it is one parish-church and one convent.

La Bouille, a market-town, containing a salt-office and some manufactures of cloth.

Boucachard, or *Bourg-achard*, having a collegiate-church; *Routot*, *Bonneville* and *Brióne*, are market-towns, seated on the *Rille*.

Annebaut, a market-town on the *Rille* enjoying the title of a marquisate.

Bec, a market-town, seated at the conflux of the *Rille* and *Bec*, contains one abbey of *Benedictines*.

Montfort, a little town lying on the *Rille*.

Bourg-Theroude, a market-town, containing a collegiate-church, and an hospital.

6. *Leiuvin*, a small tract of land, but being a good corn, pasture, and flax-country. In it are

Cormeilles, a large borough, or market-town and barony, containing three parish-churches and one abbey.

Lievray, another market-town, the principal place of this little territory.

Tiberville, a market-town.

7. The country of *Ouche*, in Latin *Pagus Uticensis*, contains the following places; *viz.*

Bernay, a town, standing on the little river *Charentonne*, being an election and the seat of a viscounty-court, and having also a salt-office, two parifh-churches, one college, a rich abbey of *Benedictines* and fome convents.

Beaumont Le Roger, a little open town and earldom, lying on the *Rille* and containing a priory.

L'Aigle, in Latin *Aquila*, a fmall town and marquifate ftanding on the *Rille*, is the feat of a vifcounty and a fuperior court of juftice. It contains alfo a falt-office, three fubuibs, three parifh-churches and two convents.

II. *Lower Normandy* confifts of three large diftricts and the following fmall territories; *viz.*

1. The country of *Auge*, fo named from its fine paftures. In it are

Lifieux, feated at the conflux of the rivers *Orbec* and *Gaffey*, fo called from the *Lexovii*. It is the feat of a collection and the fee of a bifhop who enjoys the title of count of *Lifieux*, with a diocefe of five hundred and eighty pa-rifhes. His revenue is 40,000 livres *per annum*, and his taxation at the court of *Rome* 4000 florins. In it is one abbey and feveral ftuff and linen manufactures.

Pont l' Eveque, a little town, feated on the *Touque*, containing an elec-tion, a vifcounty, a bailiwick and a foreft-court.

Beaumont en Auge, a market-town, containing a foreft-court, a college and a priory.

Honfleur, a populous town, fituate near the mouth of the *Seine*, having a vifcounty and an admiralty-court with a falt-office. In it are alfo two parifh-churches and three convents.

2. *Le Campagne de Caen* extends to *Falaife* and has only one fingle town in it ; namely,

Caen, in Latin *Cadomus*, lying at the conflux of the rivers *Orne* and *Odon*, in a valley betwixt two large meadows. It is the fecond town in *Normandy*, and contains an intendancy, an election, a provincial, a *marechauffée*, an admiralty and a foreft-court, with a falt-office and an univerfity founded in the year 1452, together with an academy of fciences inftituted in 1706. In it are two large fuburbs, twelve parifh-churches, one collegiate-church, fourteen convents, a houfe of Jefuits, a general and another hofpital, and alfo two abbies. In the *place royale*, which is both very fpacious and re-gular, ftands an equeftrian ftatue of *Lewis* XIV. The principal trade of this town and its election confifts in cloths and fine linen.

Argences, a market-town and barony on the river *Meauce*.

Tronard, a market-town containing an abbey.

3. The country of *Beffin* is a fingular inftance of the effects of perfevering induftry, being rendered very fruitful with a great variety of productions, and in feveral places covered with fine orchards. The principal places in it are

Bayeux, lying on the river *Aure*, containing an election, a foreft, an ad-miralty, a *marechauffée*, a diftrict-court and a falt-office, with feventeen parifh-churches, nine convents, two hofpitals, one college of Jefuits and

one

one feminary. The bifhop hereof is fuffragan to *Rouen*, with a diocefe of fix hundred and fifteen parifhes and a revenue of 60,000 livres, out of which his taxation to the court of *Rome* is 4433 florins.

Saint Lo, in Latin *Sanctus Landus*, a town, feated on the *Vire*, contains an election, a foreft, a bailiwick and a vifcounty-court, with a mint. It has alfo a chapter, four parifh-churches, feveral convents, two hofpitals and a college, and makes ferges and druggets.

4. The country of *Cotantin*, or *Coutantin*, contains many rocky mountains and vallies, and confequently feveral fprings and little rivers. It breeds alfo great numbers of cattle. *Cape la Hogue* in it is well known for the important victory gained there, in the year 1692, by the *Englifh* fleet over that of *France*. In it are

Coutances, or *Conflances*, in Latin *Conflantia*, a town, containing a bifhop's fee, an election, an admiralty, a *vicomté*, a mayoralty, a country, a diftrict and a foreft-court, with two parifh-churches, one abbey, five convents, one college, a feminary and two hofpitals. Its bifhop, who is fuffragan to *Rouen*, enjoys a diocefe of five hundred and fifty parifhes, with a revenue of 22,000 livres, out of which his taxation at the court of *Rome* is 2500 florins.

Carentan, a little town, lying near the fea, contains an election, an admiralty, a vifcounty and a bailiwick-court, but only one parifh-church, one convent and one hofpital. It is unhealthy on account of its fituation among marfhes.

S. Sauveur, furnamed *Le Vicomté*, a little town with a ftately abbey of *Benedictines*, on the river *Bauptois*. The diftrict here is alfo called by its name.

Valogne, in Latin *Valoniæ*, is an election and the feat of a vifcounty, a foreft and a bailiwick-court. In it are two parifh-churches and one collegiate, with three convents, one feminary and two hofpitals. The diftrict in which it ftands is called *Hague*, and is a peninfula.

Cherbourg, in Latin *Cæfaris Burgus*, a fea-port, containing an admiralty, a bailiwick and a vifcounty-court, a mayoralty, &c. It has alfo one abbey and a general hofpital. It was formerly fortified, and for that reafon befieged by the *Englifh* in the year 1418, and in 1450 by the *French*. Cloth and ferge are made here.

Barfleur, a little fea-port, contains a vifcounty and an admiralty-court. It was formerly a place of confideration, its harbour being the beft in all *Normandy*. In the year 1346, this town was laid in ruins by the *Englifh*.

Pirou, a marquifate.

Granville, a little fea-port.

Ville Dieu, a large and opulent market-town, containing a commanderie of *Malta*.

　　　　　5. The

5. The country of *Avranchin* abounds in grain, fruits, flax and hemp, but little pasture-ground. Salt is made along the coasts. In it are the following places ; *viz.*

Avranches, in Latin *Abricantæ* and *Abrincæ*, a town, standing high on the river *See*. In it is a viscounty and a bailiwick-court, with an election, *&c.* It is also a bishopric subject to the archbishop of *Rouen*, with a diocese of one hundred and eighty parishes. Its revenue is 15,000 livres, and its taxation at the *Roman* court 2500 florins. Exclusive of the cathedral, in it are three parish-churches, one convent, one hospital, one college and a seminary.

Mont Saint Michel, a little town, abbey and castle, standing in a bay on a rock called *Tumba*, being about the eighth part of a *French* league in circumference. At low water a person may go on foot to it from the continent. This town, on account of its dangerous situation, is called *Mons S. Michaelis in periculo maris.* In the eighth century a small chapel was founded here, which, in process of time, gave rise to an abbey ; and near that a little town was afterwards erected. The latter lies lower than the former, and, though strong by situation, is nevertheless fortified. It is also much frequented by pilgrims. At a small distance from it is the rock of *Tumbella*, or *Tombellaine*, on which formerly stood a castle.

Pont Orson, a small town, formerly fortified.

Mortain, in Latin *Moritolium*, a little town, seated on the river *Lances*, being the principal place of an earldom and the seat of an election, a district, a viscounty, a *marechaussée* and a forest-court, belonging at present to the house of *Orleans*. In it is one collegiate-church.

6. The small territory of *Bocage* contains the following places ; *viz*,

Torigni, a large market-town, standing near the river *Vire*, and the principal place of an earldom. In it is a bailiwick, a viscounty and a superior court. It contains also a very fine seat, two parish-churches, one abbey, one priory and one hospital.

Vire, the principal place of this little country, lies on a river of the same name, and contains an election, a viscounty, a bailiwick, a forest-court and five convents. It has also a manufactory of fine linen.

Condé, a little town, seated on the *Nereau*, or *Noireau*, and containing a mayoralty, two parish-churches and one hospital, belongs to the house of *Matignon*.

7. *Les Marches*, a small country, so called as lying on the frontiers of *Perche* and *Maine* ; in which is

Alençon, a town, seated on the *Sarte*, which rose gradually to be a marquisate, an earldom and a peerage ; together with a dukedom and a peerage. It is the seat of a generality and an election ; contains a provincial, a royal, a forest and a viscount's court, with a salt-office, two parish-churches, one college of Jesuits, five convents and two hospitals.

Seez,

Seez, in Latin *Saii*, or *Sagii*, a town, fituate on the river *Orne*, contains an election and falt-office. Exclufive of its cathedral, in it are five parifh-churches, one abbey, one convent, two feminaries, one college and an hofpital. Its bifhop is fuffragan to the archbifhop of *Rouen*, with a revenue of 16,000 livres *per annum* and a taxation to the court of *Rome* of 3000 florins. The diocefe confifts of five hundred parifhes.

Effay, a little town, contains a bailiwick and a vifcounty-court, with one parifh-church and one hofpital.

Argentan, a town, feated on the *Orne*, is a marquifate and a vifcounty, with an election, diftrict and foreft-court. It contains alfo a falt-office, three parifh-churches, four convents and two hofpitals. In the town and neighbourhood are feveral manufactures of fine linen, eftamine and other flight ftuffs.

Falaife, a town and marquifate, lying on the river *Ante*, containing an election, a diftrict, an under-foreft-court and falt-office; has alfo two parifh-churches, two convents, one abbey and two hofpitals.

Domfront, in the little country of *Paffais*, the feat of an election, a vifcounty, a bailiwick, a foreft-court, *&c.* contains alfo a church and one or two convents.

The little territory of *Houlme*, alfo belonging to this diftrict, is mountainous, and confequently little fit for tillage ; but has fine apple-orchards and iron-works. In it are the following places ; *viz.*

Briouze, a market-town and barony.

Carouges, a market-town and earldom.

The GOVERNMENT *of*
H A V R E   D E   G R A C E.

THIS diftrict conftitutes the weftern part of the country of *Caux*, in *Upper Normandy*, and, though a particular military government, is fubject notwithftanding to the civil and ecclefiaftical government of *Normandy*. To it belong the following places ; *viz.*

Le Havre de Grace, a ftrong town, feated at the mouth of the *Seine* and built by *Francis* I. who called it alfo, after his own name, *Ville Francois*; whence it is accordingly, by fome, called *Francifcopolis* ; but this name is now fuperfeded by the other. *Havre* is the capital and feat of the government, and contains a naval intendant, a vifcounty, an admiralty and a royal-court of juftice, together with a falt-office, one parifh-church; one feminary and two convents. It has a good harbour lying betwixt the town and the citadel, which is fmall but regularly fortified. In the year 1562,

this

this place was furprifed by the *Huguenots*, but in the following year again recovered.

Harfleur, formerly *Hareflot*, a little town ftanding on the *Lezarde*, and containing a vifcounty and an admiralty court, together with a police, a falt-office, a mayoralty, &c. It has only one parifh-church, and one convent. The profperity of *Havre de Grace* has occafioned the decline of this town; and its harbour is fit now only to receive barks. It was twice taken by the *Englifh*, namely, in the years 1415 and 1440.

Montivilliers, a little town lying on the *Lezarde*, contains a bailiwick and a vifcounty-court, with an election, three parifh-churches, and one abbey.

Fefcan, or *Fefcamp*, in Latin, *Fifcamnum*, a town and barony lying on the fea, and containing a fmall harbour, a bailiwick and an admiralty-court, with a falt-office. The buildings here dedicated to facred ufes are ten parifh-churches, three of which lie without the walls, one abbey, two convents, one college, one hofpital, and a commanderie of *Malta*.

The GOVERNMENT *of*
M A I N E and P E R C H E.

THIS government comprehends the province of *Maine*, together with the country and earldom of *Laval*, and the greateft part of the country and earldom of *Perche*.

1. The county of *Maine* borders to the north on *Normandy*, eaftward on *Perche*, and to the fouth is bounded by *Touraine* and *Vendomois*, weftward by *Anjou* and *Brittany*. Both the government and its capital derive their name from the *Cenomani*. Its length from weft to caft is thirty-five *French* leagues, and its breadth from fouth to north twenty. It is pretty fruitful, and contains fome iron-works, mineral waters, and two quarries of marble. The principal rivers in it are the *Maienne*, which rifes at *Linicres*, on the frontiers of *Normandy*, and after receiving the *Sarte*, falls into the *Loire*, being partly of itfelf, and partly by means of fluices navigable from *Laval* to its mouth; the *Huine*, which receives its fource in *Perche*, and runs into the *Sarte*; the *Sarte*, which rifes in *Perche*, receives the *Orne*, *Huine*, *Enferne*, and *Loire* in its paffage, and is navigable as far as *Mans*, after which it falls into the *Maienne*. This country was formerly an earldom; but has been united to the crown ever fince the year 1584. It is governed by laws of its own, but is fubject to the parliament of *Paris*, having a particular fub-governor. It confifts of three parts.

1. *Upper*

1. *Upper Maine,* containing

Maienne, or *Main la Juhele,* or *la Juée,* in Latin *Meduana,* a town, duke-dom, and peerage. It ſtands on a river of the ſame name, is the principal place of an election, a *marechauſſée* and a foreſt-court, with three pariſh-churches, and one convent.

Ernee, a little town, in which is a ſalt-office, one convent, and one hoſ-pital.

Sorron, a large maiket-town, ſeated on the *Coëſnon.*

Ambrieres, a little town and barony.

Laſſai, a ſmall town and a marquiſate.

Evron, a ſmall town, containing an abbey.

Vilaine la Juel, a market-town and a marquiſate.

Sille le Guillaume, a ſmall town and barony, formerly fortified.

Beaumont le Vicomte, a little town lying on the *Sarte,* and containing a royal diſtrict-court, a ſalt-office, a *marechauſſée,* one pariſh-church, and one convent.

Memers, or *Mamers,* in Latin *Mamercia,* a ſmall town ſeated on the *Dive,* being the principal place of the *Sonnois,* and the ſeat of a royal di-ſtrict, a *prevoté,* a ſalt-office, and a foreſt-court.

Freſnay, a ſmall town and barony, ſituate on the *Sarte.*

Ballon, a little town and marquiſate, ſtanding on the *Orne.*

Bonneſtable, a little town, which formerly bore the oppoſite name of *Maleſtable.*

2. *Lower Maine,* in which are

Mans, in Latin *Cenomanum,* the capital of the province of *Maine,* ſtands on a hill on the river *Sarte,* being the ſeat of a biſhopric, an election, a di-ſtrict, a bailiwick, a provincial-court, a ſalt-office, a foreſt-court, a *mare-chauſſée, &c.* In it are ſixteen pariſh-churches, the cathedral and two col-legiate churches being included, together with four abbies, eight convents, one college, and one ſeminary. Its biſhop is ſubordinate to the archbiſhop of *Tours,* has a dioceſe of ſix hundred and ninety-ſix pariſhes, and a re-venue of 35,000 livres, out of which he pays a taxation to the court of *Rome* of 2216 florins.

Montfort, a little town and marquiſate, lying on the river *Huiſne.*

La Ferte-Bernard, a ſmall town and a barony, ſtanding on the river *Huiſne,* being the ſeat of a *marechauſſée,* a foreſt-court, and a ſalt-office; and containing a caſtle, one pariſh-church, two convents, and one hoſpital. This place belongs to the duke of *Richelieu.*

Vibrais, a large market-town and marquiſate, ſeated on the river *Brais.*

Chateau du Loir, a ſmall town ſtanding on an eminence on the river *Loir,* and containing a royal court of juſtice, an election, a *marechauſſée,* a foreſt-court, a ſalt-office, and two pariſh-churches. It is the capital of the

the fmall territory of *Vaux du Loir*, and well known in the county, having held out a feven years fiege againft *Herbert* Count of *Mans*.

Saint Calais, a little town and barony, containing an abbey and one convent.

Montdoubleau, a fmall town, barony, and peerage, containing a falt-office.

La Sutz, an earldom, fituate on the *Sarte*.

S. Suzanne, a fmall town and a barony, lying on the *Erve*.

Sable, a town feated on the *Sarte*, and containing two parifh-churches, one convent, and a falt-magazine. This place was formerly fortified. It confers the title of marquis, dependent on which are the barony of *S. Germain*, together with the *caftelanies* of *Malicorne*, *Garlande*, *Viré*, and about fifty other fiefs.

Entrafme, a barony.

3. The earldom of *Laval*, which has enjoyed counts of its own ever fince the eleventh century, and belongs at prefent to the houfe of *Tremouille*.

Laval, in Latin *Vallis Widonis*, or *Guidonis*, a town feated on the *Maine*, being the principal place of an election, a country and a foreft-court, a falt-office, &c. In it are two parifh and two collegiate churches, one priory, and eight convents.

La Gravelle, a market-town fituate on the frontiers of *Brittany*, and containing a falt-office. In the year 1424, the *Englifh* were defeated near this town.

P E R C H E.

THE country of *Perche* is fifteen *French* leagues in length and twelve broad, being bounded to the fouth by the river *Maine*, to the weft and north by *Normandy*, and bordering to the eaft on *Timerais* and *Chartrain*. The eminences here produce only grafs for cattle; the vallies and plains abound in all kinds of corn, hemp, hay, and particularly in apples, of the juice of which their ufual liquor is made. It affords, however, but little wine, and that too very indifferent. In it are fome iron-mines. In the middle of the foreft of *Bellefme*, at *Mortagne*, is a mineral fpring called *la Herfe*, the water of which is of a ferruginous quality, and very falutary in many cafes. The water of the fpring of *Chefnegallon* is alfo of the fame nature, but not fo ftrong.

This country had formerly counts of its own, the laft of whom, *Stephen*, left behind him a daughter, who bequeathed it to *S. Lewis*, and he annexed it to the crown. It enjoys alfo its own laws, is fubject to the parliament of *Paris*, and governed by a particular fub-governor. The whole country

does

does not belong to this government, *Perche Gouet* being subject to the parliament of *Orleanois*, and *Timerais* to that of *l'ifle de France*, whence, consequently only two parts of *Perche* are to be described here, namely,

1. Great *Perche*, containing

Mortagne, in Latin *Moritonia*, or *Moritania*, the capital of the country, and the principal place of an election, the feat alfo of a district-court, a vifcounty, a *marechauffee*, a falt-office, and a foreft-court. In it is one collegiate and three parifh-churches, four convents, one hofpital, and fome confiderable manufactures of coarfe linen.

Bellefme, a little town, difputing the rank of capital with the preceding place, is the refidence of a royal vifcounty, a foreft-court, a falt-office and a district belonging to the country-court of *Chartres*. In the neighbouring wood is the mineral fpring of *Herfe*, of which I have fpoken above.

Nogent le Rotrou, a populous market-town fituate on the river *Huifne*, and receiving the addition to its name from Count *Rotrou*.

Saint Denis, a market-town belonging to the convent of the fame name.

La Trappe, a *Cifercian* abbey, noted for the aufterity of its monks.

2. The *French Country*, as it is called, is a fmall tract, and *La Tour grife* is the principal, being the feat of a royal judge or lieutenant, with a jurifdiction extending over twenty-two parifhes.

The G O V E R N M E N T *of*

O R L E A N O I S.

CONSISTS of feveral fmall countries, namely, *Orleanois Proper*, *Sologne*, *Beauffe Proper*, or *Chartrain*, *Dunois*, *Vendomois*, *Blaifois*, the greateft part of *Gatinois* and *Perche Gouet*. To the north it terminates on *Normandy* and *l'ifle de France*, to the eaft on *l'ifle de France*, *Champagne*, and *Burgundy*; to the fouth it is bounded by *Nivernois* and *Berry*; and weftward by *Touraine* and *Maine*. The rivers which run through this government, or have their fource in it are the *Loire*, of which mention has been made in the introduction to *France*, the *Loiret*, which rifes a league from *Orleans*, and after a courfe of two leagues, in which it receives the rivers *S. Cyre* and *Cobray*, falls into the *Loire*; the *Cher*, which has its fource in *Cumbraille* belonging to the government of *Auvergne*, is navigable above *Vierzon* in *Berry*, and runs into the *Loire*; the *Laconie*, which rifes in the wood of *Oleans*, and lofes itfelf in the *Loire*; the *Aigle*, the fource of which lies near *Mée* in *Beauſſe*, and mingles alſo with the *Loir*; and the *Hyere*, which loſes itſelf under ground, and when it appears again afterwards near *Montigny* is called *le Ganelon*; and joins the *Loire*. Mention has already been made of the rivers *Yonne*, *Eure*, and *Loire*.

In this diſtrict are alſo ſome remarkable canals. That of *Briare*, which receives its name from a little town near it, was begun in the time of *Henry the Great*, and finiſhed under *Lewis* XIII. being the firſt conſiderable work of this kind in *France*. It joins the river *Loire* with the *Loing*, which falls into the *Seine*, and conſequently opens a communication betwixt the coun-tries lying on the *Loire* and the city of *Paris*. Near *Briare* it falls into the *Loire*, and near *Montargis* into the *Loing*. Since the making of the canal of *Orleans*, its revenues are much diminiſhed: The latter alſo joins the above-mentioned rivers, begins about two miles above the city of *Orleans*, in the diſtrict of *Portmorant*, is near eighteen *French* leagues in length, contains thirty ſluices, and ends in the *Loing* near the village of *Cepoy*. This canal was begun in the year 1682, and finiſhed in 1692.

The whole government is ſubject to the parliament of *Paris*, and con-tains four large and three ſmall juriſdictions. The trade carried on here by means of the *Loire* is the moſt extenſive in the whole Kingdom; compre-hending not only all that which comes from the ſouthern and weſtern parts, but likewiſe from foreign countries. The principal ſtaple is at *Orleans*. Sub-ject to the governor are three general-lieutenants, and three ſub-governors: The firſt general-lieutenant and ſub-governor have under their juriſdiction the country and dutchy of *Orleans*, *Dunois*, and *Vendomois*; the ſecond the country of *Chartrain*, and under it *Gatinois Orleanois*; the third is for *Blaiſois*.

1. *Proper Orleanois* is one of the fineſt countries in *France*, being fertile in grain, wine, and excellent fruit; and abounding in cattle, game, and fiſh. The foreſt of *Orleans* is the largeſt in the whole Kingdom. The coun-try is divided into

1. *Upper Orleanois*, to which belongs

Orleans, anciently *Genabum*, or *Cenabum*, afterwards *Aurelianum*, and *Aureliana civitas*, the capital of the government, and lying on the *Loire*, over which it has a fine bridge of ſtone, on which is to be ſeen a beautiful monument of caſt braſs, but the pedeſtal is only of a ſandy ſtone, with ſome ornaments in the *Gothic* taſte. This monument conſiſts of the fol-lowing particulars: In the centre is a crucifix, on the top of which ſtands a pelican with its brood, pecking its own breaſt; before the crucifix is the Virgin *Mary* ſitting with the dead body of *Chriſt*, which is reclined on her lap. On the right ſide is King *Charles* VII. kneeling, and ſtretching out both his hands towards the croſs; and on the left ſide the famous maid of *Orleans* or *Joanne de Arc*, alſo repreſented kneeling, who in 1429, relieved the city when inveſted by the *Engliſh*. Both are armed *Cap a pie*, with
ſwords

fwords by their fides; the King has the *French* coat of arms, with the crown
and a helmet on, but the helmet of the maid ftands by her. Her hair is tied
clofe at her neck, the reft of it hangs loofe on her back. Thefe ftatues are
not as big as the life. A folemn proceffion is obferved here annually on the
12th of *May*, in commemoration of the deliverance of the place.

This city is one of the largeft in the Kingdom, but meanly built, and a
few trading people excepted, the inhabitants are poor. It is a bifhop's fee,
and contains an intendancy, an election, a *caftelany*, and a diftrict which extends over the whole dutchy, together with a country-court, a *prevoté*, a
falt-office, a foreft-court, a hunting jurifdiction, and a *marechauffée*. It has,
befides a fine cathedral built in the *Gothic* ftile, three chapters, twenty-two
parifh-churches, one abbey, an univerfity confifting only of the fingle faculty
of civilians, which was formerly famous, but at prefent in no great repute,
one college of Jefuits, one feminary, in which divinity is taught, and a
public library. Its bifhop is fubordinate to the archbifhop of *Paris*, has a
diocefe of two hundred and feventy-two parifhes, and a revenue of 24,000
livres, out of which his taxation to the court of *Rome* is 2000 florins. The
fuburb on the farther fide of the river makes a tolerable appearance; in it is
a *Carthufian* monaftery. The public walk is properly a part of the city rampart, levelled and planted with beautiful rows of trees. The city, on account
of its fituation in the middle of the *Loire*, is the magazine of the whole trade
of the Kingdom, efpecially in grain, wine, brandy, and fpices. It carries on
alfo a confiderable trade in ftockings, of which great quantities are made
here, and likewife in fheep-fkins. In it are alfo fome fugar-bakers. And in
the years 511, 533, or 536, 538, 541, 549, *&c.* councils were held here.
During the time of the *Merovingian* line this city was the royal refidence
for upwards of one hundred years, till *Clotharius* II. united this State to his
crown. It was afterwards an earldom. In the year 1344, it was raifed to a
dukedom and peerage, and is ufually an appenage to Princes of the blood.
Lewis XIV. gave it his brother *Philip*, in which houfe it continues.

Beaugency, in Latin *Balgentiacum*, a town and earldom feated on the
Loire, over which it has a bridge of ftone. It is the feat of an election, a
royal *prevoté*, a diftrict, a falt-office, a foreft-court, a *caftelany* belonging to
the jurifdiction of *Orleans*, and a hunting diftrict, together with a chapter.
Some councils were held here in the years 1104 and 1157.

Meun, *Meuing*, or *Mehun*, on the *Loire*, a little town ftanding on an
eminence, and containing one collegiate church.

Boigny, the principal commanderie and meeting-place of the knights of
St. *Lazarus*.

Chateauneuf, a market-town, having a fine caftle feated on the *Loire*.

Pithiviers, *Piviers*, or *Pluviers*, a little town feated on the river *Deuf*,
near the foreft of *Orleans*, belongs to the bifhop of *Orleans*, being the feat
of an election, a *caftelany*, and a chapter.

Pithiviers le Vieul is a village lying about one *French* league from the preceding place, belonging alſo to the biſhop of *Orleans.*

2. *Lower Orleanois,* containing

Clery, a market-town, having a collegiate church, · built by *Lewis* XI. who alſo lies buried there.

Jurgeau, or *Gergeau,* in Latin *Gurgorilum,* a little town ſeated on the *Loire,* over which it has a bridge of ſtone. In it, excluſive of the pariſh-church,· is a collegiate church, and it belongs to the biſhop of *Orleans.* In the year 1428, this place was taken by the *Engliſh,* but they loſt it again the following year.

Obſ. Under the name *Beauſſe,* or *Beauce, Belſia,* or *Belſa,* are uſually comprehended the countries of *Chartrain, Dunois, Vendomois, Mantois,* and *Hurepois,* though it never was, a proper country and lordſhip. The two laſt countries belong to the government of *l'iſle de France.* In this place too the three firſt are to be deſcribed.

II. The country of *Chartrain,* called alſo *Proper Beauce,* being a fine corn country. In it is

Chartres, the *Autricum* of the ancients, being divided into two parts by the river *Eure,* the largeſt of which ſtands on an eminence, and has very narrow ſtreets. This town is the ſeat of an election, a provincial and a baili-wick-court, together with a ſalt-office, *&c.* It is alſo a biſhopric ſubject to the archbiſhop of *Paris,* with a dioceſe of eight hundred and ten pariſhes ; the annual revenue of which is 25,000 livres, and its taxation at *Rome* 4000 florins. Excluſive of the cathedral, which is very fine, in it are ſix pariſh-churches, not including thoſe in the ſuburbs, three abbies, one priory,˙ nine convents, one ſeminary, and two hoſpitals. It enjoys the title of dukedom, and belongs to the houſe of *Orleans.*

Gallardon, a little town ſtanding on .the *Voile,* and being a *chatelany.*

Nogent le Roi, a ſmall town, lying in a valley on the river *Eure,* con-taining a *chatelany* and a royal court of juſtice.

Epernon, the principal place of a dukedom.

Maintenon, a market-town ſeated on the *Eure,* and having the title of a marquiſate conferred on the celebrated female favourite of *Lewis* XIV. In it is a chapter ; and on the *Eure* ſtands an unfiniſhed aqueduct which was intended to convey water to *Verſailles.*

Bonneval, a little town ſtanding on the *Loire,* and ſo called from its ſitua-tion in a fruitful valley; is the ſeat of a royal *prevoté* and a mayoralty, con-taining one abbey, three pariſh-churches, and one hoſpital.

III. The country of *Dunois* is an earldom and very fruitful in grain. It contains

Chateaudun, in Latin *Dunum,* an old town and viſcounty, ſeated on an eminence on the river *Loir.* This is the principal town of the country, and
contains

contains an election and a bailiwick-court, with two collegiate and six parish-churches, one abbey, three convents, two hospitals and an old castle.

Patay, near which, in the year 1429, the *English* were defeated. *Pui-feaux*, *Marchenois* and *Freteval*, all little towns.

IV. *Vendemois*, a county and afterwards a dukedom and peerage. This country is also very fertile in corn and is divided into two parts.

1. *Upper Vendomois* contains

Vendome, in Latin *Vindocinum*, which stands on the *Loire*, and contains an election, a *marechauffée* and a bailiwick-court, together with a salt-office and a collegiate-church, in which lie the counts and Princes of *Vendome*, one abbey, one college, five convents and an hospital. In it are also forty-five parishes.

2. *Lower Vendomois* contains

Montoire, a little town, seated on the *Loire*, containing a *chatelany* and a salt-office, with forty-two parishes.

Obf. Some also include here the towns of *Montdoubleau* and *S. Calais*, which have been mentioned in *Maine*.

V. *Le Perche Gouet*, or *Leffer Perche*, forms a part of the province of *Perche*, which bears the name of its ancient lords the *Gouets*, and now confifts of five baronies; *viz. Halluye*, or *Alluye*, the principal, *Auton*, *La Bafoche*, *Montmirail* and *Brou*.

VI. *Le Blaifois*, anciently an earldom, is a fine country and divided into the *Upper* and *Lower*. The principal places of note in it are

Blois, in Latin *Blefæ*, the capital of the country, standing partly on an eminence and partly on a plain on the river *Loire*, over which it has a well built bridge of stone. It had formerly the title of an earldom, is a bishop's fee, the feat of an election, a district, a chamber of accounts, a *marechauffée* and a salt-office, contains one college of Jefuits with a fine church, in which lies interred the mother of King *Stanislaus*, and several other churches and convents. Its bishop is fuffragan to the archbishop of *Paris*, enjoys a diocefe of about two hundred parishes and a revenue of 24,000 livres, out of which he is taxed at the court of *Rome* in 2533 florins. Its celebrated castle, or palace, stands on a rocky eminence. Many Princes have made additions and alterations to it, and, among others, *Lewis* XII. who was born here. An equestrian statue of that Monarch is to be seen over one of the gates. In this palace is the *black chamber*, where the duke of *Guife* was murdered, and also that where his brother, the cardinal, fuffered the fame fate; also the public hall, containing the chimney in which both their bodies were burnt to ashes. The castle is now inhabited by several noble families.

Chambord, a royal palace, feated in a wood on the river *Coffon*, being a magnificent *Gothick* edifice of free-stone, built by *Francis* I. The tower, in particular over the centre, makes a grand appearance. Its principal grand winding stair-cafe is of fo singular a construction that two perfons can at the

2　　　　　　　　　　　　　　　　　　　　　　　　　　　　　　　fame

fame time go up it in different places and he always parallel to each other. In this palace _Staniſlaus_, the dethroned King of _Poland_, on his firſt reception in _France_, reſided near nine years. It was afterwards conferred on the celebrated Count _Maurice_ of _Saxe_, who died here in the year 1750, when the King gave it his heir Count _de Frieſe_, who died here in the year 1755.

Montfrault, _Les Montils_ and _Herbault_, three royal palaces.

Ville Savin, _Chiverny_, _Beuregard_, _Menards_, _Nozieux_, _Chaumont_, _Unzain_, _Bury_, &c. feats belonging to particular noblemen.

Mer, a little town, belongs to the marquifate of _Menards_, and contains a falt-office.

Saint Die, a large market-town, fituate on the _Loire_ and having a convent.

Millarey, a little town, containing a royal _caſtelany_.

Coutres, a market-town.

Pont le Voi, a little place, containing a celebrated abbey of _Benedictines_, in which is a college ; but the revenue of the abbot has been annexed to the fee of _Blois_.

La Ferte Aurain, a dukedom and peerage.

VII. _Sologne_, in Latin _Secalaunia_, or _Segalonia_, a country, the boundaries and appertenances of which are not eafily determined. In it are reckoned the following places :

Romorentin, in Latin _Rivus Morentini_, the principal town, and containing an election, a _caſtelany_, a _marechauſſée_, a bailiwick, a royal and a foreſt-court, with a falt-office and a collegiate-church ; trafficks in ferges and other woollen ſtuffs.

La Ferte Imbaut and _La Ferte Senneterre_, two little places.

La Chapelle d'Angillon, a fmall town, _caſtelany_ and barony on the _Leſſer Sandre_.

Aubigny, a little town, lying on the _Nerre_, is a dukedom and peerage containing three convents.

Sully, a little town, ftanding on the _Loire_, contains a collegiate-church and falt-office, being alfo a dukedom and peerage.

VIII. _Le Gatinois Orleannois_, fo called by way of diſtinction from the _Gatinois François_, belongs to the government of _l'iſle de France_. In it are

Montargis, in Latin _Mons Agiſus_, or _Mons Argi_, the capital, being alfo handfome and populous. It ftands on the _Loing_, is a dukedom and peerage, with an election, a _prevoté_, a foreſt, a bailiwick, a provincial, a _marechauſſée_-court and a falt-office. In it is an old caftle, and one parifh-church, with feveral convents and a college.

Lorris, a little town, having a _caſtelany_ for the bailiwick of _Montargis_.

Chateau-Renard, a little town, ftanding on the _Ouaine_, and once fortified, contains a woollen manufactory.

Chatillon on the _Loing_, a fmail town, containing a dukedom and a peerage.　　　　　　　　　　　　　　　　　　　　　　　　　　_Gien_,

: *Gien*, in Latin *Giemum*, a town, feated on the *Loire*, and bearing the title of an earldom, is the feat of an election, a bailiwick-court, a falt-office, and a *prevoté*, with one collegiate church and three convents.

Briare, a little town feated on the *Loire*, near which the canal, which unites the *Loire* and *Seine*, begins.

The little country of *Puifaye* contains the following places ; *viz.*

S. Fargeau, a fmall town, ftanding on the river *Loing*, being a dukedom, containing a bailiwick-court and a falt-office.

Saint Amand, a fmall town.

Bleneau, a little town, having a bailiwick-court.

Cofne, the ancient *Condate*, which name was afterwards changed into *Condida*, *Coñada*, and laftly into *Cona*, a fmall town, near the river *Loire*, containing one collegiate-church, three convents, a priory and a falt-office, and having alfo fome iron-works in its neighbourhood.

Obf. Some include here the towns of *Milly* and *Etampes*, which are placed by others in the government of *l'Ifle de France*, where I have already defcribed them.

The GOVERNMENT *of* NIVERNOIS,

T O the north, is bounded by *Gatinois* and *Auxerrois* ; to the eaft, by *Burgundy* ; to the fouth, by *Bourbonnois* ; and terminates to the weft on *Berry*. Its figure is pretty nearly circular, with a diameter of about twenty *French* leagues, *Morvant* excepted, which is mountainous and barren. It produces wine, fruit and grain. It enjoys alfo plenty of wood and pit-coal, together with fome mines of iron. Among the many rivers which water this country three of them are navigable ; *viz.* the *Loire*, which has been mentioned above ; the *Allier*, which runs into the *Loire*, and the *Yonne*, the fource of which lies in the borders of this country, two *French* leagues from *Chateauchinon*, near which it runs into the fea. The other rivers here are the *Nevre*, the *Arron*, the *Alaine*, the *Quenne*, the *Andarge*, the *Yffeure*, the *Creffonne*, the *Acolin*, the *Abron*, the *Befbre*, the *Acolaftre*, the *Aubois*, the *Narcy*, the *Guerchy*, the *Noaix*, the *Arrou*, &c. At *S. Parife* and *Bougues* are fome mineral fprings. This country had been a confiderable earldom from the ninth century, and in the year 1588, was erected, by *Francis* I. into a dukedom and peerage ; but, in the year 1707, fell again to a country. It is fubject to the parliament of *Paris*, has its own particular law, and over it is a governor, a general-lieutenant and a deputy-governor. It is divided into the following eight diftricts ; *viz.*

I. *Le vaux de Nevers*, abounding in wine, corn, wood, and paftures, in which are the following places ; namely,

Nevers,

Nevers, anciently *Noviodunum,* and afterwards *Nivernum,* the capital of the country, ſtanding in form of an amphitheatre on the river *Loire,* which is here joined by the little river *Nievre,* over which it has a handſome bridge of ſtone.   It contains an election, a foreſt, a bailiwick, and a *marechauſſée* court, with a ſalt-office ; and excluſive of its cathedral, in it are eleven pariſh-churches, two abbies, ſeveral convents, one college of Jeſuits, and a caſtle. The biſhop of this place is ſuffragan to *Sens,* with a dioceſe of two hundred and ſeventy-one pariſhes, and beſides his revenue of 20,000 livres, is lord of the *caſtelanies* of *Premery, Urſy,* and *Parſy;* being taxed at the *Romiſh* court in 1250 florins. This place is celebrated for its porcelain and glaſs-houſes ; as alſo for its enamel-works.

Pougues, a village about two *French* leagues diſtant from *Nevers,* on the road to *Paris,* ſtands at the foot of a mountain, having near it a ferrugineous ſpring of great virtue.

Chamlemy, a little town, ſtanding near one of the ſprings of the *Nievre.*

2. *Les Arnognes,* very fertile in corn and wine, together with fine paſtures, and plenty alſo of wood; but containing neither town nor village.

3. The vallies of *Montenoiſon,* which are alſo fertile, and derive their name from a caſtle on a hill, at the foot of which is the village of *Noiſon.* Its principal places are

Montenoiſon, a *caſtelany.*

Premery, a little town and *caſtelany,* containing a chapter.

Champalemond, a *caſtelany.*

4. The vallies of *Yonne* lie along the river *Yonne,* being eſteemed the moſt fertile diſtrict in the whole country. To it belongs

Clamecy, in Latin *Climiciacum,* or *Clameciacum,* a town ſeated on the *Yonne,* which here becomes navigable, and is joined by the *Buvron.*   In it is a *caſtelany* and a ſalt-office. *Pantenor,* one of its ſuburbs ſtanding on the other ſide of the *Yonne,* and called a town, was in the year 1180, made the reſidence of the biſhop of *Bethlehem,* who had been compelled to leave *Paleſtine.* This biſhop, who ſtill ſtiles himſelf biſhop of *Bethlehem,* is created by the count *de Nevers,* and enjoys all the privileges of the other *French* biſhops, though his revenue be only 1000 livres, and this town his whole dioceſe. He frequently, however, performs the epiſcopal functions of other prelates ; and is in reality *ſervus ſervorum Dei,* if the biſhops may be accounted God's true ſervants.

Vezelay, a little town ſtanding on a hill cloſe by the river *Eure,* which ſome include in the diſtrict of *Morvant.* This place is the ſeat of an election, a *marechauſſée* and a bailiwick-court, together with a ſalt-office ; and contains one abbey, one collegiate church, and one convent.

Tannay, containing a chapter, and *Varzy,* are two market-towns.

Corbigny, or *S. Leonard,* a ſmall town, having an abbey of *Benedictines.*

5. The

5. The diftrict of *Morvant*, in Latin *Morvinus pagus*, which is covered with wood and mountains, and contains few fertile fpots, ftands partly in the dutchy of *Burgundy.* In it is

Chatel, or *Chateau Chinon,* in Latin *Caftrum Caninum,* a little town feated on an eminence on the river *Yonne,* and containing an election, a *mare-chauffee,* and a falt-office. It is the principal place of a lordfhip, with the title of an earldom; has ten parifhes and five manors belonging to it, of which this place makes one.

Ouroux, or *Auroux,* and *Lorme,* two fmall towns and manors.

Braffy, and *Dun les places,* two little places having manors.

6. *Bazois,* a diftrict, confifting of vallies lying along the mountains of *Morvant,* produces little wheat or rye; but abounds in good pafturage, woods, and coal-pits. In it is

Moulins-Engilbert, fmall, but having a *caftelany,* a falt-office, a collegiate church, two convents, and one hofpital.

Montrouillon and *Cercy,* together with *Tour de Coddes,* and two *caftelanies.*

Defize, in Latin *Dececia,* a very ancient town feated on a rocky ifland in the river *Loire,* where it is joined by the *Airon.* It contains a *caftelany,* a falt-office, an old caftle, and three convents.

S. Saulge, a little town, having a *caftelany,* a falt-office, a mayoralty, one parifh-church, and a priory.

Luzy, a fmall town, containing a falt-office.

7. The little country lying betwixt the rivers *Loire* and *Allier,* begins at the conflux of thefe two rivers and runs along them as far as the *Bourbonnois.* Some parts of it abound in good pafture-grounds, others are woody, and others produce corn and a little wine. In it are

S. Pierre le Moutier, a little town feated among hills on a marfhy lake belonging to the King, and containing a provincial and a diftrict-court, with a falt-office, a chapel, a priory, and two convents.

La Ferte Chauderon, a little town and ancient barony, the proprietor of which is ftiled marfhal, and governor of *Nivernois.*

D'orne, a market-town, having a chapter.

8. *Le Donziois,* a diftrict, formerly a barony, difmembered from the county of *Nevers;* but in the year 1552 re-annexed to it: It is ftill, however, a fief in the gift of the bifhop of *Auxerre,* and contains

Donzy, the capital, fmall, and lying on the river *Nohin.* In it is a collegiate church, a priory, a convent, and an hofpital.

Antrain, or *Entrain,* in Latin *Interamnis,* a town and *caftelany* environed with lakes.

Dreve, a little town containing a *caftelany,* ftands on a mountain.

S. Sauveur, Corvol Lorgueilleux, Billy, and *Eftaiz,* caftelanies.

The G O V E R N M E N T *of*

B O U R B O N N O I S.

B OURBONNOIS terminates to the north on *Nivernois* and *Berry*, to the weſt on *Upper-Marche*, to the ſouth on *Auvergne*, and to the eaſt on *Burgundy* and *Forez*; being about thirty *French* leagues in length, and twenty broad. It is pretty fertile, particularly in corn, fruits, and forage, producing alſo good wine, though not in a ſufficient quantity for exportation. It boaſts alſo a few coal-pits, and great numbers of mineral ſprings and warm baths. Its rivers are the *Loire*, the *Allier*, the *Cher*, together with ſome other ſmaller ones. In the month of *July*, when the ſnows melt on the mountains of *Auvergne*, great damages are done by the inundations of the *Allier*. This country had formerly *Sires* of its own, who were alſo ſtyled Princes, barons, and counts. Towards the end of the year 1327, it was raiſed to a dukedom. Duke *Lewis* had two ſons, *Peter* and *James*, the latter was count of *la Marche*, and his deſcendants ſtill ſit on the *French* throne; but the former was the founder of the other dukes of *Bourbon*. On his taking up arms againſt his ſovereign *Francis* I. his dukedom was forfeited, and annexed to the crown. At the peace of the *Pyrenees*, in the year 1659, it was again ſeparated from the demeſnes and given to *Lewis de Bourbon*, Prince of *Condé*, in lieu of the dutchy of *Albret*. It is ſubject to the parliament of *Paris*; though the duke *de Bourbon* has the nomination of all civil offices, yet their proper maſter is the King. Excluſive of the governor and general-lieutenant, here are alſo two ſub-governors. In *Bourbonnois* are twenty-two towns, the moſt remarkable of theſe are,

Moulins, in Latin *Molinæ*, which is the capital of the country, lies on the *Allier*, being well built, and one of the pleaſanteſt in the Kingdom. In it is an intendancy, an election, a *caſtelany* with ſeveral courts, and a chamber of demeſnes, &c. It contains likewiſe one collegiate church, with a college of Jeſuits, five convents and an hoſpital; and cloſe by the town is a mineral ſpring.

Villeneuve, a market-town.

Bourbon l'Archeambaud, in Latin *Burbo Archembaldi*, a little town environed by four hills, on one of which ſtands an old caſtle containing three chapels in it. That called the *Holy* is very ſplendid. In the town is alſo a royal *caſtelany*, and a diſtrict-court, with one pariſh church, one chapter, a priory, a convent and two hoſpitals. This place is noted for the virtues of its hot baths and cold mineral ſprings.

Le

Le Veurdre, feated on the *Allier*. *S. Amand* lying on the river *Cher*. *Hertſſon*, cantaining a chapter. *Villefranche, Souvigny, Gouzon, Huriel* and *Le Montet aux Moines*, all little towns; the four latter of which are but mean.

Montluçon, a town ſtanding on a rock near the river *Cher*, contains an election, a *caſtelany* and a ſalt-office, with one collegiate church, two pariſh churches, four convents and one hoſpital.

Neris, a borough or market-town, feated on a rock. Its warm baths are much frequented.

Montmeraut Verneuil, Jaligny, Varennes, Prilly, and *La Palice*, all little towns.

Vichy, a ſmall town ſtanding on the *Allier*; noted for its mineral waters and baths.

Saint Pourcain, a little place lying on the river *Scioule*, owes its name and origin to an abbey of *Benedictines*, which is at preſent but a priory. In it are alſo three convents and an hoſpital.

Gannat, a little town, the feat of an election, a *caſtelany* and a diſtrict-court, with a ſalt-office, a collegiate church and three convents.

The G O V E R N M E N T *of*

L I O N N O I S,

C O N T A I N S the three ſmall provinces of *Lionnois, Forez,* and *Beau-jolois*, bordering to the north on *Maconnois* and *Burgundy*; to the eaſt it is ſeparated by the *Soane* and *Rhone* from *Breſſe* and *Dauphiny*; to the ſouth is bounded by *Vivarois* and *Velais*; and weſtward by *Auvergne* It produces a ſufficiency of grain, wine, and fruits, particularly a kind of large cheſtnuts called *Marrons*. Excluſive of the three large rivers of the *Rhone, Soane,* and *Loire*, mentioned above; in it are ſeveral ſmall ones, as the *Furan, Lignon, Rhin, Azergue, &c.* Not far from the village of *Cheſſey*, four *French* leagues from *Lyon*, is a mine of copper and vitriol. *S. Galmier, Moin, S. Alban, &c.* are mineral ſprings. The courts of this government proceed according to the civil-law, and appeals from them lie to the parliament of *Paris*. Excluſive of its governor and general-lieutenant, it has two ſub-governors, one over *Lyonnois* and *Beaujolois*, the other over *Forez*.

1. *Lionnois* is about twelve *French* leagues in length and ſeven in breadth. During the time of the *Merovingian* Kings it was governed by counts or *Stattholders*, who by degrees got the country into their own hands. Theſe

counts and the archbifhops of *Lyon* were at perpetual variance about the limits of their power, till in the year 1173, *Guy* II. count of *Forez*, and archbifhop *Guichard* came to an agreement, whereby the former relinquifhed to the church of *Lyon* all his poffeffions in the city of that name, and the *Lionnois*; the latter ceding to him moft of his poffeffions in *Forez* and *Beaujolois*, with the addition of 1100 filver marks. King *Philip the Handfome* compelled the bifhop to take an oath of allegiance to him; and in the year 1307, raifed the lordfhip of *Lyon*, which was but a barony to an earldom, conferring it, together with its jurifdiction, on the archbifhop and the chapter, and it is by virtue of this that the prebends ftile themfelves earls. At length, in the year 1563, the jurifdiction devolved to the crown. In it are

Lyon, in Latin *Lugdunum*, or *Lugdunum Segufianorum*, the capital of the province, and of the whole government, lying at the conflux of the *Rhone* and *Soane*. The latter runs through a part of the city, having two bridges of wood and one narrow one of ftone over it; whereas the ftone bridge over the *Rhone* is a noble ftructure. It is about a fourth part as large as *Paris*, and populous, though moft of the ftreets are narrow. It contains, however, two fine fquares, particularly one very large and pleafant one, in which ftands an equeftrian ftatue of brafs of *Lewis* XIV. and on the other is the town-houfe, which is a very beautiful building. Its exchange is remarkable only for the multiplicity of bufinefs tranfacted on it. Round the city lie fome mountains, which being variegated with convents, feats, gardens, and vineyards, form a very delightful appearance. This place is an archbifhopric an intendancy, and an election, with a mint, a provincial and other courts. Its archbifhop is primate over the five archbifhoprics of *Lyon*, *Rouen*, *Tours*, *Sens*, and *Paris*, fo that appeals lie from them to him. He has alfo fix bifhops for his fuffragans with a diocefe of feven hundred and fixty-four parifhes, and a revenue of 48,000 livres, out of which his taxation to the *Roman* court is 3000 florins. I have already obferved that the metropolitan of this place ftiles himfelf count of *Lyon*. Exclufive of the cathedral, in it are three chapters or collegiate churches, thirteen parifh-churches, two colleges of Jefuits, the larger of which is the moft ftately in all the Kingdom. It has alfo a very numerous and well arranged library with an obfervatory, feveral convents, and three hofpitals. The arfenal here is a fine building, and well ftored with military implements. Here are alfo three forts, namely, *Pierre en Cife* (the only one which is garrifoned, and which ferves for a flate prifon) *S. Jean*, and *S. Clair*. Moft of the inhabitants are makers of filk, gold and filver ftuffs, with gold and filver laces. Formerly, while its other manufactures were wholly fupplied from *France*, the looms in and about this city amounted to 18,000. But in the year 1698 this number was found reduced to 4000. It is ftill, however, in great repute for the above-mentioned manufactures; particularly for its bombafines, the beautiful luftre

of

of which is owing to an invention of *Octavius May*: And its trade extends not only all over *France*, but even to *Spain*, *Italy*, *Switzerland*, *Germany*, the *Netherlands*, and *England*. The remnants of antiquity here are now scarce visible. In the years 1245 and 1274 some councils were held in this city.

Ance, or *Anse*, a little town seated near the *Soane*, in which some provincial synods are held.

Tarare, a town lying on the *Tardive* in a vale at the foot of a mountain of the same name.

La Bresle, a little town seated betwixt hills on the river *Tardive*, suffered extremely by an inundation in the year 1715.

Condrieux, a small place on the *Rhone*, containing one parish-church and two convents.

S. Chaumond, a town lying on the *Gier*, having a castle and a chapter.

2. *Forez*, which is equal in extent to *Lyonnois*, and *Beaujolois* put together, had formerly counts of its own, whose male issue failing in the year 1369, the last count hereof was succeeded by his sister *Johanna* married to *Beraud the Great*, Dauphin of *Auvergne*, whose daughter in 1371, at her marriage with *Lewis* II. duke of *Bourbon*, brought him this county, which continued in their lineage till the year 1521, when *Susannah* of *Bourbon* dying, her succession occasioned great disputes betwixt her husband the constable *de Bourbon*, *Louisa* of *Savoy* mother to *Francis* I. and the Princess *de la Roche Sur Yon*. But in the year 1532, *Francis* I. annexed *Forez* to the crown. It consists of two parts, *viz.*

1. *Upper Forez*, containing

Feurs, or *Fors*, in Latin, *Forum Segusianorum*, a little town lying on the *Loire*, and giving name to the country, being formerly much more considerable than at present. In it is a royal *castelany*. One *French* league from hence at the foot of a rock called *Dinzy*, is a sulphureous spring.

Saint Galmier, a little town standing on an eminence near the river *Loire*, and having a royal *castelany*. At the end of its suburbs is a spring called *Font-Forte*, of a vinous taste, and very wholsome.

S. Etienne de Furans, a populous town seated on the river *Furans*, and next to *Lyon*, the most considerable in this government. The inhabitants of this place carry on a great trade in hard-ware, which is their principal business.

2. *Lower Forez*, in which is

S. Rambert, a little town seated on the *Loire*, and containing a chapter.

Montbrison, the capital of *Forez*, lies on the little river *Vezise*, being the principal place of an election, a *castelany*, a *prevoté*, a forest and several other courts. In it also, exclusive of its collegiate church, is one college,

with

with feveral other churches and convents. At no great diftance from it are the mineral fprings of *Moin*.

Rocheforte, a little town lying on the river *Lignon*.

S. Germain Laval, a fmall town having a *caftelany*.

Roanne or *Rouane*, in Latin *Rodumna*, a very ancient town feated on the *Loire*, which here becomes navigable. This is the ftapie for all goods fent from *Lyon* to *Paris*, *Orleans*, *Nantes*, &c. It is alfo the feat of an election and bailiwick. The little country of *Roannois*, or *Roannez*, in which this town lies, has been erected into a dutchy and peerage.

S. Alban, a village, one *French* league and a half from *Roanne*, having three mineral fprings.

3. *Beaujolois* is about ten *French* leagues in length, and eight in breadth, being a very fruitful country. It was formerly a barony, which Baron *Edward* II. in the year 1400, conferred, together with the lordfhip of *Dombes*, on *Lewis* II. duke of *Bourbon*, from which houfe it defcended by inheritance to the houfe of *Orleans*.

Beaujeu, a little town ftanding on the *Ardiere*, and containing an old caftle on a mountain. It was formerly the capital of the country, but is now only a large market-town, from which the country takes its name.

Ville Franche, the capital of the country, lies near the *Soane* on the little river *Morgon*, being the feat of an election, and a falt-office, with a collegiate church, and an academy of the *beaux arts*.

Belleville, a little town containing an abbey.

The GOVERNMENT *of*
A U V E R G N E.

THIS county, which takes its name from its ancient inhabitants the *Arverni*, is bounded to the eaft by *Forez*, to the north by *Bourbonnois*, to the weft by *Limofin*, *Quercy*, and *la Manche*, and to the fouth by *Rouergue* and *Sevennes*. Its extent from fouth to north is about forty *French* leagues, and from weft to eaft thirty. *Lower Auvergne* is a very fertile pleafant country, abounding in wine, grain, forage, fruit, and hemp, and far excelling *Upper Auvergne*, which is cold and full of mountains, being covered with fnow feven or eight months in the year, though its paftures neverthelefs are excellent; and accordingly it deals largely in cattle. The fituation of the mountains occafions fuch variablenefs and eddies in the winds, that no wind-mills are found to fucceed there. Its principal rivers are the *Allier*, which rifes at *Chabellier* in *Gevaudan*, becomes navigable at *Viale*,

not

not far from *Maringue*, and falls into the *Loire:* The *Dardogne*, which issues from the highest mountain in this country, called *Mont d'or*, and loses itself in the *Garonne*. The *Alagnon*, the source of which is at *Cantal*, being very rapid; is not·very navigable, and falls into the *Allier*.

At *Pontgibaud* is a filver mine; but the produce not anfwering the charges, it has been difcontinued : Whereas the coal-mines in the neighbourhood of *Braffac* are very profitable articles. Exclufive of its mineral fprings, in which it exceeds any province in *France*, here are alfo feveral others equally remarkable. The higheft mountains in this country are *Le Pui de Dome*, in Latin *Mons Dominans*, the perpendicular height of which is eight hundred and ten toifes; the *Cantal*, nine hundred and eighty-four, and the *Mont d'or* one thoufand and thirty: The two latter are no lefs noted for the curious plants growing on them. Exclufive of corn, wine, cattle, cheefe, coals, and other products of the earth; this country carries on alfo a great trade in manufactures, fuch as all kinds of filk ftuffs, cloths, very beautiful laces, and paper, which is accounted the beft in all *Europe*. Every year fome thoufands of labourers go from hence into *Spain*, and return with the heft part of their earnings. *Auvergne* is an ancient county raifed to a dukedom and peerage in the year 1360, but in 1531, again re-united to the crown, a fmall part of the ancient county excepted, which ftill enjoys the title of an earldom; and belongs to the houfe of *Bouillon*. The whole country is fubject to the parliament of *Paris*, but under different laws; *Lower Auvergne* having a particular code of its own, but in *Upper Auvergne* the civil-law takes place. In it are five large diftricts and two bailiwicks. Subject to its governor are two general-lieutenants, and two fub-governors.

. I. To *Upper Auvergne*, which lies among the mountains, belong the following places, *viz.*

S. Flour, the capital of *Upper Auvergne*, lying on a mountain of difficult accefs. It is a bifhop's fee, containing a diocefe of two hundred and feventy parifhes, with a revenue of 12,000 livres, and a tax to the court of *Rome* of 900 florins. Exclufive of its cathedral here is alfo a chapter, and a college of Jefuits. It carries on likewife a good trade in grain, this city being, as it were, the general magazine of the neighbouring country, which produces a great deal of rye. Its carpets, cloths, and knives are alfo greatly efteemed.

Aurillac, a town, difputing the title and rank of capital with *S. Flour*. It lies in a valley on the river *Jordane*, being pretty well built and populous, and conferring the title of count. It is the feat of an election, a diftrict, a bailiwick, and a *marechauffée*; it contains a caftle feated on a high iock, together with a collegiate church, which is properly a fecularized abbey, the abbot of which is lord of the town, and holds immediately of the Pope. In it alfo is a college of Jefuits, with one abbey and four convents. In this town are fome manufactures of tapeftry of *haute* and *baffe lieu*, as alfo of lace.

4

Murat,

Murat, a town and vifcounty feated on the river *Allangon*, is the feat of a diftrict and a foreft-court, together with a royal *prevoté*. The inhabitants principally braziers and lace-makers.

4. The vifcounty of *Carlades*, conferred in the year 1644, on the Prince of *Monaco*; contains

Vic, a large market-town, lying on the river *Eure*, and the feat of a diftrict-court, having alfo a mineral fpring, the water of which is of a vitriolic nature.

Carlat, a little town, the capital of the country, and having formerly a foitified caftle.

Maurs, Montfalvi, la Roquebrou, and *Pleaux*, all mean little towns.

Mauriac, a fmall town, ftanding not far from the *Dordogne*, and containing a college of Jefuits, and an abbey.

Salers, a little town, but the feat of a royal diftrict, the greateft part of which belongs to the baron of *Salers*; the other part is the property of the count *de Caylus*.

Chades Aigues, in Latin *Aquæ calidæ*, a little town and barony, taking its name from the warm mineral waters found there.

II. To *Lower Auvergne* belongs alfo the great valley of *Limagne*, through which runs the *Allier*. In it are

Clermont, anciently *Auguftonemetum*, afterwards *Arverna*, or *Urbs Arvernorum*, the capital of the whole county, ftanding on a fmall eminence betwixt the rivers *Artier* and *Bedat*. It is populous, but has very narrow ftreets, and its houfes are dark. This town is alfo a bifhop's fee; contains a tax-chamber, an election, a bailiwick, a country-court, &c. and was formerly the principal place belonging to the counts of *Auvergne*, who on this account ftiled themfelves alfo counts of *Clermont*. The bifhop hereof is the firft fuffragan to the archbifhop of *Bourges*, is alfo lord of the fmall towns of *Billon* and *Croupieres*; and enjoys a diocefe of eight hundred parifhes, with a revenue of 15,000 livres, out of which his taxation at the court of *Rome* is 4550 florins. In it, exclufive of its cathedral, are three collegiate churches, and three abbies. In that of *S. Allire*, the bodies of feveral faints lie depofited in its chapel of *S. Venerand*; and in that of *S. André*, the tombs of the old counts of *Clermont*, and the Dauphins of *Auvergne*; with many convents, and a college of Jefuits.

In the neighbourhood of this city are wells where any fubftances laid in them foon contract a lapideous cruft; but the moft remarkable of thefe is that in the fuburb of *S. Allire*, which has formed the famous ftone bridge mentioned by fo many hiftorians. It is a folid rock compofed of feveral ftrata, formed during the courfe of many years, by the running of the petrifying waters of this fpring over it. It has no cavity or arches till after above fixty paces in length, where the rivulet of *Tiretaine* forces its way through. This petrifying fpring, which falls on a much higher ground

than

than the bed of the rivulet, gradually leaves behind it some lapideous matter, and thus in process of time has formed an arch, through which the *Tiretaine* has a free passage. The necessity which this petrifying matter seemed to be under of forming itself into an arch, could continue no longer than the breadth of the rivulet; after which the water of the spring ran regularly again under it, and there formed a new petrification resembling a pillar. The inhabitants of these parts to lengthen this wonderful bridge, have diverted the brook out of its old channel, and made it to pass close by the pillar; whereby the spring formed a second arch, and thus as many arches and pillars as they please, might by the same means have been produced. But the great resort of people to see this natural curiosity becoming troublesome to the *Benedictines* of the abbey of *S. Allier*, within whose jurisdiction the spring lies; in order to lessen its petrifying virtue they divided the stream into several branches, which has so well answered their intent, that at present it only covers with a thin crust those bodies on which it falls perpendicularly, but in those over which it runs in its ordinary course no traces of this petrifying quality are any longer perceivable. It is the only drinking-water in this suburb, and no bad effect is felt from it.

Hard by the city are the mineral springs of *S. Pierre* and *Jaude*.

Montferrand, a little town standing on a high hill, and containing a district-court, a chapter, two commanderies, and one convent.

Riom, in Latin *Ricomagus*, a well built but ill inhabited town, containing an intendancy, an election, a country, a *marechauslée*, a mint-court, *&c.* together with three chapters, and a college.

Volvic, a village, celebrated for its stone bridge.

The dukedom and peerage of *Montpensier*, to which is united the principality of *Dauphine d'Auvergne* and the barony of *Combrailles*, belongs to the ducal house of *Orleans*, and contains the following places, *viz.*

Aigueperse, in Latin *Aqua Sparsa*, the capital of the dukedom lying in a fine plain on the river *Luzon*. It is small, but contains a royal court of justice, one abbey and two chapters. Not far from hence is a spring having a continual ebullition, but its water nevertheless is cold and has nothing remarkable in its taste. Near this place lie the ruins of the ancient castle of *Montpensier*.

Vodable, a little town, which is the residence of an extensive *castelany*, constituting the ancient *Dauphiné d'Auvergne*, and to which also belong the places of *Lestoing* and *Alt-Brionde*.

Montegu, Chambon on the *Voise* and *Evaux*, all small places, together with *Sermur* a market-town, belong to the barony of *Combrailles*.

Ebreuille, a little town, seated on the river *Sioule*, and containing an abbey.

Cuffet, a little town, but the seat of a royal district and a *prevoté*, containing also a chapter and an abbey.

Maringue, a little town lying close by the river *Allier*, where the dealers in corn have their granary.

Thiers or *Tiern*, a town and vifcounty fituate near the *Durolle*, in the country of *Limagne*, which by its trade has rendered it the moft confiderable and moft populous town in *Auvergne*. In it is a royal court of juftice, one collegiate church, and one abbey.

Vic le Comté, a little town, the refidence of the laft-counts of *Auvergne*, and containing a chapter. In its neighbourhood are four mineral fprings.

Pont du Chateau, a little thriving town, feated on the *Allier*. This place bears the title of a marquifate, and belongs to the houfe of *Canillac*.

Billon, a poor town belonging to the bifhop of *Clermont*, contains a chapter, and a college of Jefuits.

Iffoire, or *Yffoire*, in Latin *Iciodorus*, a fmall town feated on the river *Coufe*, which not far from hence falls into the *Allier*, contains an election and 'a *prevoté*. The abbot of the *Benedictine* abbey of the congregation of *S. Maur* is lord of this place.

Saucilanges, a fmall town, containing a priory of *Benedictines*.

Ambert, a town, the principal place of the little country of *Liuradois*, and pertaining to the marquis of *Roche Baron*, who is a defcendant of the houfe of *Rochefoucault*.

Uffon, a little and poorly inhabited town ftanding on a fteep mountain, and having a royal court of juftice.

Aufon, a little town and barony.

Brioude, in Latin *Brivas*, a very ancient town feated on the river *Allier*, over which it has a bridge of ftone which is looked upon as a *Roman* work, of whom it is not unworthy. This town is properly called *Old Brioude*.

Brioude Glife, lies alfo hard by the river *Allier*. In it is a collegiate church called *S. Julien*, the order of which is noble, and to it belongs the lordfhip of the town.

Saint Germain Lambron, fmall, but the principal place of the little country of *Lambron*, which is fruitful in corn and wine.

Langeac, a fmall town but the feat of a royal *prevoté*, and belonging at prefent to a branch of the houfe of *Rochefoucault*.

Ardes, a little town, the principal place of the ancient dukedom of *Mercœur*, and ftanding not a great way from the *Mercœur* palace.

S. Amant and *S. Saturnin*, two little towns belonging to the marquis *de Broglio*.

The mineral waters and baths of *Mont d'Or* receive their name from the mountain of *Mont d'Or*.

Hermant, a little town, conferring the title of baron, and having a chapter.

Artonne, a fmall town containing a chapter. Not far from this place, near the village of *Saint Myon*, are two mineral fprings.

The GOVERNMENT *of*
L　I　M　O　S　I　N.

LIMOSIN, or *Limoufin*, which derives its name from the ancient *Lemovices*, borders to the east on *Auvergne*, to the south on *Quercy*, to the west on *Perigord* and *Angoumois*, being bounded to the north by *La Marche* and *Poitou*. The extent of this country from south to north is about twenty-five *French* leagues, and from west to east somewhat less. *Upper Limofin* is very mountainous, and consequently cold: *Lower Limofin* is more temperate. The former produces but little wine and that too very indifferent, but the latter is very good. It produces such numbers of cheftnut-trees, that the inhabitants derive their principal maintenance from them. The grain growing here is rye, barley, and *Turkish* corn. Their principal trade consists in horned-cattle and horses. Its chief rivers are the *Vienne*, the source of which lies in the borders of *Lower Limofin* and *La Marche* in the parish of *Millévanches*; the *Vezere*, which rises in the same country, and becomes navigable near *Saraffon*; the *Coureze*, which has its source above *Maignac*, and falls into the *Vezere*; and the *Dordogne* which divides *Limofin* from *Auvergne* and *Quercy*. In this province have been discovered mines of lead, copper, tin, and steel; it also has some iron-works, but not so profitable as those of *Angoumois*. This country was formerly an earldom, and afterwards became a viscounty, which *Henry* IV. united to the crown. It is governed by the *Roman* law, and is subject to the parliament of *Bourdeaux*: Subordinate to its governor are one general-lieutenant, and two sub-governors. The country is divided

I. Into UPPER LIMOSIN, in which are

Limoges, in Latin *Lemovicæ*, the capital of the country, standing partly on a hill and partly in a valley on the river *Vienne*, being but an indifferent place. It is notwithstanding a bishop's see, and contains an intendancy, an election, a bailiwick, a country, a *prevoté*, a royal and a *marechauffée* court, together with a mint. Its bishop is suffragan to the archbishop of *Bourges*, with a diocese which extends itself over *Upper* and a part of *Lower Limofin*, *La Marche*, and a part also of *Angoumois*, and contains nine hundred parishes, with a revenue of 20,000 livres, out of which its taxation at the court of *Rome* is 1600 florins. Exclusive of its cathedral, in it are one collegiate church, three abbeys, one convent, one college of Jefuits, and a feminary.

Souteraine, a little town.

S. Junien, a small town seated on the *Vienne*, belongs to the bishop of *Limoges*, and contains a chapter.

S. Leonard, a little town lying on the *Vienne*, partly belonging to the King and to the biſhop of *Lemoges*. In it is a chapter, with a manufacture of cloths and paper.

Pierre Buffiere, a little town, bearing the title of the firſt barony of *Limoſin*; but this the barony of *La Tour* diſputes with it.

S. Irier de la Perche, anciently *Atanus*, a ſmall town containing a collegiate church.

Chalus, a little town bearing the title of an earldom.

II. Into Lower Limosin, containing

Tulle, properly *Tuelle*, and in Latin *Tutela*, a town ſituate near the conflux of the little rivers *Coureze* and *Solane*, has a biſhop's fee, an election, a viſcounty, a bailiwick, a country-court, &c. Its biſhop, who is lord and viſcount of the town, is ſuffragan to the archbiſhop of *Bourges*, and enjoys a dioceſe of ſeventy pariſhes, with a revenue of 12,000 livres, out of which his taxation to the court of *Rome* is 1400 florins. In it is a college of Jeſuits and ſix convents.

Brive la Gaillarde, properly *Brive*, on the *Coureze*, contains an election, a country court of juſtice, a bailiwick, a collegiate church, and a college. Its firſt name it probably derives from its pleaſant ſituation and beauty; but the ſecond ſignifies properly a bridge over the *Coureze*.

Uſerche, in Latin *Uſerca*, a ſmall town, containing three pariſh-churches and one abbey, the abbot of which is lord of the place.

The dukedom and peerage of *Ventadour* comprehends

Ventadour, an ancient caſtle.

Uſſel, a little town, the capital of the dukedom, and the ſeat of a court of juſtice.

Bord, a ſmall town lying on the *Dordonne*, and containing a convent.

III. The viſcounty of *Turenne*, which is eight *French* leagues in length and ſeven broad, was in ancient times independent; but towards the beginning of the 10th century, the viſcount thereof did homage to the King, with this proviſo that the viſcounty ſhould not be given out of the King's hands, and the viſcounts ſhould ſtill enjoy all the *regalia*. At preſent the duke of *Bouillon* is proprietor of this viſcounty. The taxes in it are impoſed by the States, and he ſummons them together. Excluſive of the poll-tax, which uſually exceeds 30,000 livres, the viſcount has introduced here almoſt all the other impoſts uſual in the Kingdom. To this country belong the following places, *viz.*

Turenne, the capital, being ſmall but containing a caſtle and chapter.

Beaulieu, a little town containing an abbey.

Argentat, a ſmall town lying on the *Dordonne*.

Saint Cere, Meſſat, Calonges, &c. all little towns; together with

Ninety other market-towns and pariſh-churches, moſt of them in *Lower Limoſin*.

The

The GOVERNMENT *of*
L A   M A R C H E.

THIS country borders to the east on *Auvergne*; to the south on *Li-mosin*; to the west on *Poitou*, and northward on *Berry*; is about twenty-two *French* leagues in length and eight in breadth, being watered by the *Vienne*; the *Great* and *Little Creuse*, which runs into the *Vienne*; and the *Cher* and the *Gartempe*, the latter of which joins the *Creuse*. About *Bellac* and *Dorat* are vineyards; and the upper parts of this district are pretty fruitful in grain. This country had formerly counts of its own. In 1316 it was erected into a peerage, and by *Charles* IV. into a dukedom and peerage. Ever since the year 1531, it has been united to the crown. It contains two bailiwicks, and is governed by its own laws. Subordinate to its governor is one general-lieutenant, and two sub-governors. It consists of the following parts; *viz.*

I. UPPER MARCHE, containing

Gueret, in Latin *Waractus*, the capital of the *Upper* and whole *Marche*. It stands on the river *Gartempe*, and has an election, a bailiwick, a provincial, a royal *castelany*, a *marechaussée*, a forest-court, *&c.* In it are one parish-church, one priory, two convents, one college, and one hospital.

Aubusson, in Latin *Albucum*, or *Albucium*, a small but pretty populous town seated on the river *Creuse*, thrives by its manufactory of tapestry.

Felletin, a little town seated on the *Creuse*, containing a *castelany* and college. This place also makes tapestry.

Chenerailles, *Jarnage*, *Ahun* on the *Creuse*, having an abbey, and *Bourganeuf*, containing an election, are small towns.

Grandmont, a small town, containing a celebrated abbey, which is the head of a religious order.

II. The LOWER MARCHE contains

Bellac, the capital of this part of the *Marche*, and the seat of a bailiwick and provincial court of justice, taking its name from an old fortified castle.

Souteraine, a market-town.

III. The little country of *Franc Allen* lies on the borders of *Auvergne*, and belongs to the bailiwick of *Upper Marche*. It contains the little towns of

Bellegarde and *Croc*, which has a collegiate church, together with the market-town of *Pont Charrod*.

The

The GOVERNMENT of BERRY.

THE county of *Berry*, which derives its name from the *Cubi Vituriges*, is bounded to the fouth by *Bourbonnois* and *Marche*; to the weft by *Touraine*; to the north by *Orleanois*; and eaftward by *Nivernois*. Its extent from weft to caft is betwixt twenty-feven and twenty-eight *French* leagues; and from fouth to north between thirty-five and thirty-fix. The air here is temperate, and the foil produces wheat, rye, and wine. In fome places, as at *Sancerre*, St. *Satur*, and *Lavernuffe* is equal to that of *Burgundy*. Its other products alfo are good, and its rich paftures feed great numbers of cattle, particlarly fheep, which are remarkable for the finenefs of their wool. It abounds likewife in hemp and flax. The parifh of *S. Hilaire*, at *Vierzon*, yields an oker which is feldom met with in *France*; and at *Bourges* is a mineral fpring. The principal rivers here are the *Loire*, the *Creufe*, the *Cher*, which have been mentioned above; the *Large* and *Leffer Saudre*; the *Nerre*, which rifes three leagues above *Aubigny*, falling into the *Larger Saudre*; the *Indre*, the fource of which is in this country, becomes navigable at *Chatillon*, and runs into the *Loire*; the *Orron*, which iffues from the only lake in *Bourbonnois*, and as well as the *Aurette* and the *Moulon*, falls into the *Evre*; which *Evre*, or *Yevre*, receives its fource near *Neronde*, and runs into the *Cher*. In the neighbourhood of the little town of *Linieres* is the lake of *Villiers*, which is between feven and eight leagues in circuit.

This country had formerly counts of its own who were ftiled counts of *Bourges*, and afterwards vifcounts, the latter of whom, in the year 1100, fold the country to *Philip* I. fince which it has been united to the crown. But in 1360, King *John* conferred it on his third fon as a dukedom; and the like grant was alfo made to Princes of the blood in fucceeding reigns. It is under the jurifdiction of the parliament of *Paris*, and is governed by laws of its own. Subordinate to the governor are one general-lieutenant and two fub-governors, and it is divided into the *Upper* and *Lower Berry*.

I. UPPER BERRY, contains the following places, *viz.*

Bourges, anciently *Bituriges*, and *Bituricæ*, as alfo *Avaricum*, the capital of the whole country. It lies on the river *Evre*, is an archbifhopric, an intendancy, and an election, with a falt-office, a *marechauffee*, an independent royal tribunal, and feveral other courts, &c. Its univerfity was either founded or revived in the year 1463, and confifts of four faculties. Here is alfo a beautiful and large college of Jefuits. Exclufive too of its cathedral, it contains four collegiate churches, befides the two annexed to the feminary; fixteen parifh-churches, four abbies, &c. fo that the clergy

and

and their dependents form the greateſt part of the inhabitants. It is alſo the reſidence of ſeveral genteel families, all living in proper elegance and ſociablenefs. The old city ſtands higher than the new, the archbiſhop, who is ſtiled patriarch and primate of *Aquitaine*, has five fuffragans, with a dioceſe of nine hundred pariſhes, and a revenue of 30,000 livres, out of which his taxation to the court of *Rome* is 4033 florins. One part of the old palace is appropriated to the reſidence of the governor, and the other to the above-mentioned courts. In it particularly is a moſt magnificent ſpacious hall where the States of the country hold their meetings. Here are alfo mineral waters.

Dun le Roi, a little town belonging to the royal demeſnes, and unalienable, contains a falt-office, a collegiate and a pariſh-church.

Chateauneuf, a ſmall town lying on the river *Cher*, is divided into the upper and lower; being an old barony, and containing one collegiate church.

Montrond, a caſtle ſtanding on a mountain, being once a conſiderable fortification.

Meun, or *Mehun*, in Latin *Magdunum*, a ſmall town ſeated on the *Evre*, contains a *prevoté*, a manor-court and a chapter.

Virzon, in Latin *Virſic*, a ſmall town lying on the rivers *Evre* and *Cher*, is an earldom containing a manor-court, an abbey, three convents, and one college.

Chatillon ſur Loire, a ſmall town.

Concorſault, Concourceaut, or *Concreſſaut*, alſo a ſmall town, little better at preſent than a large village.

La Chapelle Dam-Gilon, in Latin *Capella Domini Gilonis*, a town and barony, ſtanding high on the *Leſſer Saudre*, and containing a *caſtelany*.

Les Aix-Dam-Gilon, a market-town, having an old caſtle and a chapter.

Sancerre, in Latin *Sincerra*, called improperly *Sacrum Cæſris*, a town ſeated on the river *Loire*, and the capital of an ancient earldom, contains a falt-office, one pariſh-church, and one convent. In the years 1569 and 1572, this place was bravely defended by the Proteſtants, but in 1573, after another tireſome and bloody ſiege, they were compelled by famine to ſurrender; on which the fortifications were immediately demoliſhed.

II. Lower Berry, in which are the following places; viz.

Iſſoudun, in Latin *Exolidunum*, a large town, and the ſecond in rank in the province, lies in a beautiful plain on the river *Theols*, containing an election, a royal *prevoté*, a manor-court and a falt-office. It is divided into the upper and lower town, and has in it a caſtle, with four pariſh and two collegiate churches, one abbey, five convents, and two hoſpitals. It ſuffered extremely by fire in the years 1135, 1504, and 1651.

Charoſt, in Latin *Carophum*, a little town and dukedom ſeated on the river *Arnon*, containing a palace, a caſtle, one pariſh-church, and one priory.

Linieres, a ſmall town, containing a caſtle and one collegiate church. It had formerly lords of its own who ſtiled themſelves barons, *Sircs*, and Princes of *Lineres*.   In

In this neighbourhood is the lake of *Villiers,* which is about seven *French* leagues in circuit.

S. Chartier, a little town.

La Chatre, a small town lying on the *Indre,* being an old barony and the seat of an election, and a salt-office. In it are two churches, one of which is a collegiate; three convents, one hospital, and an old castle which serves for a prison.

Chateau Meillant, a little town and earldom, containing one chapter, one parish-church, one priory, an hospital, and an old castle.

Agurande, or *Aigurande,* a little town, having a *castelany.*

Boussac, a small town, containing a seat.

Argenton, a town seated on the *Creuse,* by which it is divided into the upper and lower. In the former is a chapel and college; and in the latter a convent and church.

Blanc, in Latin *Oblincum,* a town lying on the *Creuse,* and the seat of an election, together with a *marechaussée* and a receiver's office, is divided into the upper and lower. In the former is the castle, and in the latter a priory and one convent.

Chateau-Roux, in Latin *Castrum Radulphi,* lying on the river *Indre,* contains an election and a dukedom, together with a chapter, three convents, four parish-churches, and a considerable manufactory of cloth.

Deols, or *Bourg-Deols,* called also *Bourg-Dieux,* a little town lying on the *Indre,* and bearing the title of a principality, once contained three parish-churches and a celebrated abbey. Of the former two are still remaining, though one only is parochial, and of the abbey the chapel is still to be seen.

Levroux, a little town containing a collegiate church.

Valençay, a small town, having a castle seated on the river *Nahon.*

Saint Agnan, a little town standing on the river *Cher,* and containing a castle, with one collegiate church and two convents.

Selles, properly *Celle,* a small town, having a castle or seat on the river *Cher,* and exclusive of an hospital and convent, containing also an abbey.

Vasten, or *Vestan,* a small town, having a castle and chapter.

Graçay, a little town and an ancient barony, the proprietors of which stiled themselves barons, *Sires,* and Princes.

Lury, the smallest town in the whole province, belongs to the cathedral of *Bourges.*

*   *   *   *   *   *

The sovereign principality of *Bois-Belle,* or *Henrichemont,* lies within *Upper Berry,* and belongs, at present, to the duke of *Sully,* of the house of *Bethune;* being about twelve *French* leagues in circumference. The soil in most parts is none of the best, and the inhabitants are only six thousand and

some

odd hundreds. The prince's demefnes fcarce produce him 2000 livres; but this is made up by an annual prefent of 24,000 livres from the King's farmers-general for the monopoly of falt made here, and this is the only tax with which the fubjects are burdened. To it belong the following places, *viz.*

Henrichemont, in Latin *Henricomontanum*, the capital, ftanding on an eminence.

Boisbelle, a market-town near the former, making a part of the parifh of *Menneton-Salon*, otherwife called *Fief-pot*, and conftituting alfo a part of the parifh of *Quantilly*.

The GOVERNMENT *of*
T O U R A I N E.

THIS province and its capital derive their name from the *Turones*, be-ing bounded to the north by the river *Maine*; to the eaft by *Orlean-nois*; to the fouth by *Berry* and *Poitou*; and to the weft terminating on *Anjou*. Its greateft length from caft to weft is twenty-two *French* leagues, and its breadth from fouth to north twenty-four. The air here is temperate, and the country fo delightful, that it is called the garden of *France*; though notwithftanding it is not every where alike. *Les Varennes*, which lies along the *Loire*, is a fandy foil producing rye, barley, millet, and garden-plants, as alfo an herb which makes a good yellow die. The tract called *Le Verron* lies fomething higher, being richer, and yielding corn, wine, very fine fruit, and particularly large plumbs. *La Champagne*, a fmall ftrip of land lying be-twixt the rivers *Cher* and *Indre*, abounds in grain, particularly wheat. *La Brenne* is an uliginous fwampy country. The eminences along the *Loire* and *Cher* are covered with vineyards. *La Gafline* is a harfh land very diffi-cult to the plough. The country of *Noyers* contains fome mines of iron, and one of copper in it. At *Rochepofay* is a mineral fpring. This province had formerly counts of its own. In the year 1044, it was taken by the counts of *Anjou*; in 1202, united with the crown; and in 1356, raifed to a dukedom and peerage. It has frequently been granted to the royal chil-dren; and after the death of *Francis* duke of *Alençon*, brother to *Henry* III. was again united to the crown, fince which it has never been alienated. It is governed by laws of its own, but with a right of appeal to the parliament of *Paris*. Exclufive of its governor, here is one general-lieutenant, and one fub-governor. Of the twenty-feven towns or boroughs in this country, eight only are demefnes; the others belong to particular proprietors.

Tours, the *Cæſarodunum*, or *Turoni* of the ancients, and the capital of the country, lies in a plain running along the river *Loire*, betwixt that and the *Cher*, being an archbiſhopric, an intendancy and an election, with two *ma- rechauſſées*, a receiver's and a ſalt-office, a mint, a provincial, bailiwick, and a foreſt-court. Excluſive of its cathedral, in it are five chapters, three ab- bies, one college of Jeſuits, twelve convents, a caſtle called *Le Pleſſis les Tours*, with a ſtone bridge over the *Loire*, and a ſilk and cloth manufacture. Subject to its archbiſhop are eleven ſuffragans, three hundred pariſhes, ſe- venteen abbies, twelve chapters, ninety-eight priories, and one hundred and ninety-one chapels. His revenue is 40,000 livres, and his taxation at the court of *Rome* 9500 florins. The houſes being well built and of a very white ſtone, form a very pleaſing appearance. The ſtreets too are pretty clean.

Luynes, anciently an earldom called *Maille*, but in the year 1619, raiſed to a dukedom and peerage under its preſent name; is a little town ſeated on the *Loire*, containing a caſtle, a chapter, two pariſh-churches, and two convents.

Langeai, *Langey*, or *Langez*, a ſmall town ſeated on the *Loire*, contain- ing a caſtle and two pariſh-churches; one of which has a chapter.

Samblançay, *Villebourg*, *Bueil* and *Neufvi*, all market-towns.

Chateau Renaud, anciently *Carament* and *Villemoran*, a little town lying on the *Branſle*, and bearing the title of a marquiſate, contains one pariſh- church and one convent.

Amboiſe, in Latin *Ambaſia*, or *Ambacia*, ſituate at the conflux of the *Loire* and *Amaſſe*; is an election with a royal, a *marechauſſée*, and a foreſt- court, containing alſo a ſalt-office, &c. In it are two pariſh-churches, four convents, and one hoſpital. Near this town is a very large caſtle ſtanding on a high rock, in which are to be ſeen the ſtatues of *Charles* VIII. and his conſort *Anne*, as alſo a ſtag's head of very extraordinary dimenſions, being ten feet high, and from the extremity of one horn to the other eight feet broad: It is, however, only factitious, being made of wood. Some other curioſities are alſo to be ſeen here. It has likewiſe a chapel. In this town it was that *Charles* VIII. met with his death, either by running againſt a door, or according to others, by being ſhot through the head. This was alſo the place where the civil war in the year 1561 firſt broke out; and and where the name of *Huguenot* had its riſe.

Mont-Louis, a market-town ſeated betwixt the *Loire* and the *Cher*, where in the year 1174, a peace was concluded betwixt *Lewis* VII. and *Henry* II. of *England*.

Veret and *Chenonceau*, ſeats lying on the river *Cher*.

Mont-Treſor and *Paluau*, two little towns and earldoms ſeated on the river *Indre*.

Loches, in Latin *Luccæ*, a mean town, lying on the *Indre*, but an earl- dom, and the principal place of an election, with a bailiwick, a *caſtelany*, and
a ſalt-

a falt-office. In it is alfo one parifh-church and fix convents. It has like.
wife a caftle ftanding on a rock, formerly a very important fortification. In
this caftle are four ranges of fubterraneous paffages running over each other,
in the uppermoft of which *Lewis Sforza*, duke of *Milan*, was kept prifoner
for ten years, and where alfo he died. In a large tower in it are two cages
or moveable rooms, with very ftrong oak grates covered with iron, in one
of which Cardinal *Balve*, bifhop of *Algiers*, was confined by *Lewis* XII.
In the caftle is alfo a collegiate church. By means of a bridge over the river
Loches, this place communicates with

Beaulieu, a little town and barony.

Chatillon, a fmall town lying on the *Indre*, and containing one collegiate
church and two convents ; is the capital of the little country of *Brenne*.

Cormery, a fmall town lying on the *Indre*, and having an abbey.

Montbuzon, a little town on the *Indre*, being a dukedom and peerage, to
which alfo belongs the little town of *Sainte Maur*, which is a barony, and
the borough or market-town of *S. Catherine de Fierbois*.

Affay, or *Azay*, which bears the furname of *Rideau*, a fmall town lying
on the *Indre*.

Chinon, a town ftanding on the river *Vienne*, and having a ftrong caftle,
four parifh-churches, and fifteen convents.

Cande, a little town feated at the conflux of the *Loire* and *Vienne*.

S. Efpin, a fmall town.

L'ifle Bouchard, a little town ftanding on the *Vienne*, and containing a
feat, belongs as a barony to the dukedom of *Richelieu*. In it are two parifh-churches, and two convents.

Prefzigny, a fmall town feated on the river *Clere*, and bearing the title of
a barony, contains a feat, a chapter, and one parifh-church.

Paulmy, a feat lying on the river *Brignon*.

Pruilly, a fmall town and barony fituate on the river *Claiffe*, and containing five parifh-churches.

La Haye, a little town and barony feated on the *Creufe*, and containing
two parifh-churches.

La Guierche, a fmall town ftanding on the *Creufe*, having a pretty ftrong
caftle.

La Rochepoffay, feated on the *Creufe*, noted for its mineral fprings.

Ligueil, a fmall town and barony.

Champigny, a little town ftanding on the *Vende*, or *Vetle*, and bearing the
title of a barony, contains one parifh-church, two convents, and a fmall
college.

The

The GOVERNMENT *of* ANJOU.

THIS country, which is so called from the ancient *Andes*, or *Andegavi*, is
bounded to the east by *Touraine*; to the south by *Poitou*; to the west
by *Bretagne*; and northward by the river *Maine*. Its greatest length from
west to east is twenty-six *French* leagues; and from north to south twenty-
four. The country is a pleasant succession of hills and vallies, producing
wine, grain, peafe, beans, flax, hemp, and all kinds of fruit-trees, &c. Its
fine paftures too furnish it with rich breeds of cattle. It yields also mines of
coal, iron, and salt-petre, together with quarries of slate, stone, and marble.
It has likewife feveral mineral waters, but thefe are neglected. In it are
reckoned no lefs than forty-nine great and small rivers; but of them only
six are navigable, *viz.* the *Loire*, *Vienne*, *Toue*, *Maienne*, *Loir*, and *Sarte*.
This country formerly confifted of two counties, which towards the end
of the ninth century were united. In the year 1202, *Philip Auguftus* annexed it to the crown. In 1256, St. *Lewis* conferred it on his brother
Charles, founder of the first line of *Anjou* which afcended the throne of
Sicily. In the year 1297, *Philip the Handfome* raifed it to a dukedom and
peerage, and soon after the country devolved to the crown. King *John*
gave this dukedom to his son *Lewis* I. the founder of the fecond houfe of
Anjou, fome branches of which arrived alfo to be Kings of *Sicily* and *Na-*
ples. In 1481, it efcheated again to the crown; but *Henry* III. conferred
it on his brother *Francis*. *Philip* of *France*, duke of *Orleans*, and brother
to *Lewis* XIV. bore the title of *Anjou*, which feems properly to belong to
the third Prince of *France*. This country is alfo under the parliament of
Paris, but has a law of its own. Subordinate to the governor is one general-
lieutenant and two sub-governors; and in it are the following places, *viz.*

Angers, anciently *Juliomagus*, and in Latin *Andegavum*, the capital of
the country, divided by the river *Maienne* into two parts. It is large and
populous, and the feat of a bifhop, a manor, a bailiwick, and a provincial
court, together with a mint, a falt-office, a *marechauffée*, &c. It contains
likewife an univerfity founded here by St. *Lewis*, together with an academy
of *Belles lettres*, inftituted in the year 1685. In it alfo is a ftrong caftle
feated on a fteep rock, with a cathedral, feven other chapters, fixteen parifh-churches, four abbies, feveral other convents, and one feminary. The
bifhop of this place is suffragan to the archbifhop of *Tours*, with a diocefe
of six hundred and fixty-eight parifhes. His revenue is 26,000 livres, and
his taxation at the court of *Rome* 1700 florins. This town makes camblets,
ferges, and mixed ftuffs; and carries on alfo a confiderable trade in other
commodities.

Beauge

Beauge le Vieux, and *Beauge* on the *Coefnon,* two little towns lying near each other.

Le Verger, a regular caftle.

Jarze, a fmall town and marquifate feated on a lake, and containing a collegiate church.

Briffac, a little town lying on the *Aubance :* near it in the year 1667 was fought a battle. In 1611, this place was erected into a dutchy and peerage.

Vaujour, a dukedom and peerage.

Chateau Gontier, a pretty large and populous town feated on the *Maienne,* and bearing the title of a marquifate, contains a collegiate and three parifh-churches, with fome convents.

Lude, a little town fituate on the *Loir,* belongs to the duke of *Roquelaure.*

Durtal, a fmall town and county belonging to the duke *de Rochefoucault.*

La Fleche, a town feated on the *Loir ;* containing an election and a provincial court. In this town is the moft celebrated college of Jefuits in all Chriftendom. It contains likewife a fine feat erected by the marquis of *Varranne.*

Le Pont de See, a fmall town feated on the *Loire,* and having a caftle.

Treves, a little town and barony ftanding on the *Loire,* and containing a feat.

Poance, or *Pouance,* a fmall town and barony feated on a lake.

Chateaux-Ceaus, in Latin *Caftrum Celfum,* a fmall town ftanding on an eminence on the river *Loire,* and belonging to the duke of *Bourbon.*

Chollet, a little town and barony feated on the *Maine,* containing a fine feat, one parifh-church, and three convents.

Doe, or *Doue,* a little town, having one parifh and one collegiate church.

Ingrande, a little town and barony lying on the *Loire.*

Craon, in Latin *Credonium,* a fmall town and barony, feated on the *Udon.* The proprietor of this place ftiles himfelf premier baron of *Anjou ;* and the whole country round it for fome diftance is called *Le Craonois.*

Chantoce, a barony fituate on the *Loire.*

Chateauneuf, a fmall town and barony lying on the river *Sarte.*

Cande, which bears the furname of *Lamée,* a little town feated on the rivers *Mandie* and *Erdre,* enjoys the title of a barony.

Chemille, a fmall town and barony feated on the river *Irome,* contains one collegiate church.

Vehiers, a town and earldom feated on a lake, and containing four parifh-churches.

Montforeau, a little town and earldom ftanding on the *Loire,* has a chapter.

Paffavant, a market-town and earldom feated on the *Layon.*

Montrevau, a fmall town and an earldom fituate on the river *Ifere.*

Beaufort la Valée, a little town, but containing two parifh-churches and one convent.   *Beaupreau,*

Beaupreau, a fmall town, fituate on the *Ifere*, contains two parifh and one collegiate-church, and bears alfo the title of a dukedom and peerage. *Le Puy de la Garde*, a celebrated convent of *Auguftines*.

The G O V E R N M E N T *of*

S   A   U   M   U   R.

THE diftrict of *Saumurois* contains a part of *Anjou* and *Upper Poitou*, being under the direction of a governor, a general-lieutenant and a fubgovernor. In it is

Saumur, in Latin *Salmurus*, the capital and the feat alfo of. an election; with a *prevoté*, a *marechauffée*, a bailiwick and a falt-office. It contains alfo a fine caftle, three parifh-churches, nine convents and one royal college. It was much more opulent while in the poffeffion of the. *Huguenots*, but has ftill an univerfity. Near it is a moft delightful abbey of *Benedictines* of the congregation of *S. Maur*.

Montreuil-Beilay, or *Berlay*, a little town and barony, feated on the *Toue*, contains an election, a foreft-court and a *marechauffée*, with a fine caftle in which is a collegiate-church. It contains alfo an hofpital and one convent.

Richelieu and *Mirebau* belong likewife to this government, but of thefe an account has already been given in *Poitou*.

The G O V E R N M E N T *of*

F   L   A   N   D   R   E   S,

COMPREHENDS certain tracts in the *Netherlands*, as, namely, a part of the earldom of *Flanders*, the *Cambrefis*, the county of *Hennegau*, or *Hainau*, the bifhopric of *Liege* and the county of *Namur*. Thefe are commonly called the *French Netherlands*. This government terminates, to the fouth, on *Artois*; to the eaft, on the *Auftrian Netherlands*; to the north, on the *Seine*, and likewife on the *German* ocean; and, weftward, is bounded alfo by that fea. The quality and hiftory of thefe countries fhall be treated of in the *Netherlands*, where I fhall give a diftinct account of the feveral earldoms, of which this government includes certain portions. Far the greateft part of the government is fubject to the parliament of *Douay*; and the judicial proceedings in it are grounded on royal edicts, ufages and

the

the civil law. The taxes of the country are affeffed by the intendants; the *caftelany* of *Ifle* and *Cambrefis* excepted, in which the ftates of the country ftill retain that privilege. Subordinate to the governor is one general-lieu-tenant and three fub-governors. The feveral parts of this government are as follow, *viz.*.

1. *French Flanders, la Flandre Françoife*, being a part of the earldom of *Flanders*, which *Lewis* XIV. over-ran in the year 1667. It abounds in grain and garden vegetables of all kinds, as alfo in flax. Its paftures are ex-cellent and confequently produce fine breeds of cattle; but for want of wood turf is their general fuel here. It is divided into three *quartiers*, viz.

1. The *quartier de Terre Franche*, containing three *caftelanies*, which re-ceive their names from fo many towns. In it is

Gravelines, or *Gravelingen*, a fmall but ftrong town, lying near the fea on the river *Aa*. This place, exclufive of its fortifications, is defended on the land fide by a good citadel, and to the fea by a fort. In the year 1383, it was deftroyed by the *Englifh*. In 1528, fortified again by *Charles* V. In the year 1558, the *French* were defeated here by the *Spaniards*. In 1644, it was taken by the *French*; in 1652, by the *Auftrians*; in 1658, again by the *French*, who alfo retained it at the treaty of the *Pyrenees*; but, in the year 1694, it was entirely laid in afhes.

Bourbourg, or *Broueborg*, a fmall town, feated on the *Kolme*, but greatly reduced by the numerous calamities of war to which it has been expofed; ftill contains a nunnery of *Benedictines* and a convent of *Capuchins*.

Bergues, or *Berg*, furnamed *Saint Vinox*, or *Vynoxberge*, and feated on the river *Colme*, an ill built but well fortified town, not to mention two forts belonging to it called *Le Lapin* and *Suiffe*. All the country round this place, from *Fort Suiffe* to the canal of *Dunkirk*, may be laid under water. The town is the feat of a bailiwick, a vifcounty and a collection, together with an abbey and a college of Jefuits. One *French* league from this place is

Fort Saint François, lying on the canal of *Dunkirk* but dependent on *Bergues*.

Hondefchrote, a market-town, fituate in the *caftelany* of *Bergues*, but has belonged to the *French* ever fince the year 1667.

2. The *quartier* of *Caffel* contains

Caffel, in Latin *Caftellum Morinorum*, a fmall town, ftanding on a high hill in a very fine plain. This place is the capital of a large *caftelany*, con-taining four towns and forty-feven villages; but by frequent fires has loft a great part of its former profperity. Some battles were fought here in the years 1072, 1328 and 1677. This town was ceded to *France* at the treaty of *Nimuegen*.

Watten, Hafebroeck, Merville, or Merghem, Stegers, or *Etaires*, all little towns belonging to the *caftelany* of *Caffel*.

Bailleul,

3. *Bailleul*, or *Belle*, in Latin *Balliolum*, and anciently *Balgiolum*, a small open place, five or six times confumed by fire, yet ftill the principal town of a *caftelany*. In it is alfo a college of Jefuits and a *Capuchin* convent. *Nienkerke*, a market-town.

3. The *quartier* of *L'Ifle*, or *Lille*, enjoys ftates of its own, who are annually fummoned together by the King for the affeffment of the taxes. It confifts of three *caftelanies*; viz.

1. The *caftelany* of *L'Ifle*, or *Lille*, which takes its name from

L'Ifle, *Lille*, or *Ryffel*, the capital of *French Flanders* and of all the *French* conquefts in the *Netherlands*, being the refidence alfo of the governor-general. It is called by the three above-mentioned names, . the laft of which is *Flemifh*. Geographers are not agreed about the two firft, fome maintaining that *L'Ifle*, and the Latin *Infula*, are the right name, the city lying betwixt the rivers *Lis* and *Deule:* others deny it to be an ifland, and justify the name *Lille* from the Latin *Ifla*. It is a moft important fortification, with an admirable citadel and fort. The city too itfelf is large, handfome and populous. In it is an intendancy, a mint, a *caftelany*, a bailiwick and a foreft-court, with a collegiate-church and about thirty other churches, feven of which are parochial, a great number of convents and a noble hofpital called *l' hopital comteffe*. It carries on a large trade in its camblets, which are excellent, its cloths and other ftuffs. In the year 1667, *Lewis* XIV. made himfelf mafter of it. The Confederates took it again in 1708, after a very long and bloody fiege; but, in the year 1713, it was reftored to *France* at the treaty of *Utrecht*. The *caftelany* of *Lille* is divided into feven *quartiers*, to which belong one hundred and thirty-feven villages and fome towns.

1. The *quartier Ferain* lies north of the city along the river *Lys*. In it is *Comines*, a little town, divided by the *Lys* into two parts, of which that towards *L'Ifle*, by virtue of the peace of *Utrecht*, belongs to *France*, and that on the left to the houfe of *Auftria*. It contains a collegiate-church, but its fortifications have been demolifhed.

2. The *quartier La Wepe* lies alfo on the *Lys*, and contains *Armentieres*, a fmall town, feated on the *Lys*, but difmantled by order of *Lewis* XIV. A good kind of cloth is made here.

Baffee, a little town, ftanding on the *Deule* and once fortified.

3. The *quartier Melantois*, the name of which is derived from *Medenantum*, contains *Seclin*, in Latin *Sacilinium*, a market-town, having a chapter.

4. The *quartier Carembauld*, in which is *Phalempin*, the capital containing an abbey.

5. The *quartier la Peule*, containing *Bouvines*, a market-town, fituate on the river *Marque*, near which, in the year 1214, a great battle was fought.

6. The

6. The *quartier*, or earldom, of *Lannoy*, contains
Lannoy, a market-town, having a feat.

7. The *quartier* of *Auvede La Lefcaut*. In it are
Roubaix and *Turcoim*, two market-towns; their principal trade, the making half filk ftuffs.

2. The *caftelany* of *Orchies*, containing
Orchies, a little town and bailiwick noted for its woollen ftuffs.

Marchiennes, a little town, ftanding in a marfhy country on the *Scarpe*, but having a celebrated abbey.

S. Amand, a little town, feated on the *Scarpe*, and formerly belonging to *Tournaifis*; the *French*, however, got it by the treaty of *Utrecht*. Its abbey is very famous, and the abbot here is lord of the place. At a little diftance from it is a mineral fpring.

Mortagne, a country-town, feated on the *Scheld* near its junction with the *Scarpe*. Before the peace of *Utrecht* this place belonged to the *Tournaifis*.

3. The bailiwick of *Douay*, fo called from
Douay, a pretty large and well fortified town, having a fort. It ftands on the river *Scarpe*, being the refidence of a parliament for *French Netherlands*, and alfo of a bailiwick-court. In it is an univerfity founded in the year 1559, one collegiate-church and feven parochial ones. In the year 1667, this place was taken by the *French*. In 1710, the Confederates made themfelves mafters of it, but loft it again in the year 1712. A famous *Englifh* feminary has been erected here.

2. The *Cambrefis*, from the village of *Arleux* to *Chatillon*, on the *Sambre*, is about ten *French* leagues in length, and from five to fix broad, though in fome places not above two or three. It is fruitful, populous, and has ftates of its own. The principal places in it are
Cambray, or *Camerick*, in Latin *Cameracum*, and *Camaracum*, the capital, which lies on the *Scheld*, being a pretty, large, well fortified town, and defended moreover by a citadel and fort. This place is the feat of a collection and an archbifhopric; and, exclufive of its cathedral, contains two chapters, ten parifh-churches, two abbies and two hofpitals. The archbifhop ftiles himfelf Prince of the holy *Roman* Empire, as indeed he formerly was, and count of *Cambrefis*. He is, however, lord of the city. His diocefe confifts of near eight hundred parifhes, and his revenue is 100,000 livres, out of which his taxation at the *Roman* court is 6000 florins. This town is famous, beyond any place in *Europe*, for a kind of fine linen made here which is called after its name. Ever fince the year 1667, it has been in the hands of *France*.

Chateau, or *Cateau-Cambrefis*, the capital of the earldom of *Cambrefis*. The archbifhop, who has here a very grand palace, is lord of this place. It was once fortified, but at prefent lies open and contains one abbey. A peace was concluded here, in the year 1559, betwixt *France* and *Spain*.

Crevecœur, a market-town, fituate on the *Scheld*, makes ferge.

Valincourt, a little place, but a peerage, contains a chapter.

Vaucelles, as fmall, but having an abbey.

3. *French Hainaut*, or its part of the earldom of *Hennegau*, contains

Valenciennes, in Latin *Valentinianæ*, which is large and populous, but the buildings and fortifications very irregular. It contains a good citadel, feated on the *Scheld*, and was once, together with its diftrict, an independent country of *Hainaut*. It is the principal place of a jurifdiction called *Prevoté la Comté*. That part of the town on the right of the *Scheld* belongs to the diocefe of *Cambray*, and contains one collegiate-church and one abbey; whereas the part on the left fide belongs to the diocefe of *Arras*. In the year 1677, this place was taken by the *French*.

Obf. The country lying betwixt this city and the *Scarpe* is called *Oftre-vand*.

Famars, a village, formerly called *Fanomarte*, and having once a diftrict belonging to *Valenciennes*.

Condé, in Latin *Condate*, a fmall town, but an important fortification, fituate on the *Haine* and *Scheld*. The neighbouring country may be laid entirely under water. In it is a receiver's office and a collegiate-church. In the year 1676, this place was taken by the *French*.

Denain, a village, fituate betwixt *Valenciennes* and *Bouchain* not far from the river *Scheld*, containing a collegiate-church. In the year 1712, the Allies, after the departure of the *Englifh*, were here furprifed by the *French* and defeated with a confiderable flaughter.

Bouchain, in Latin *Buccinium*, a ftrong little town, divided by the *Scheld* into the *Upper* and *Lower*, was, in the year 1676, taken by the *French*.

Pequincourt, a ftrong place.

Quefnoy, in Latin *Quercetum*, fmall, but pretty ftrong, contains a *prevoté*, a bailiwick-court and an abbey. In the year 1711, this place was taken by the Allies, but retaken again from them in the following year.

Bavay, in Latin *Bagacum*, a very ancient little town, containing a *prevoté* and a collection, with two convents and one college.

Maubeuge, in Latin *Malbodium*, a fortified town, feated on the *Sambre*, being the principal place of an intendancy, a *prevoté* and a collection, with two chapters, one college of Jefuits and feveral convents. *Lewis* XIV. having obtained this place at the treaty of *Nimeguen*, caufed it to be ftrongly fortified.

Longueville, a peerage.

Landrecy, or *Landrechies*, fmall but ftrong, ftands on the river *Sambre*, containing a royal *prevoté* and one convent. In the year 1655, this town was taken by the *French*. In 1711, Prince *Eugene* befieged it, but was obliged to draw off his forces.

Avefne, a fortified little town, fituate on the river *Hefpres*, contains a collection, a royal bailiwick and a peerage, and alfo a chapter.

Solre,

Solre le Chateau, an earldom.

Marienbourg, a small town, containing a *prevoté* and a collection. This place was built in the year 1547, by *Maria* of *Austria,* sister to *Charles* V. the land having been procured, about a year before, by way of exchange from the bishop of *Liege.* In the year 1554, *Henry* II. made himself master of it, and, at the treaty of the *Pyrenees,* it was ceded to *France,* on which *Lewis* XIV. in the year 1674, ordered it to be dismantled, leaving only a bare rampart.

Philippeville, a well fortified little town, being the principal place of a collection and a *prevoté.* It was formerly only a market-town, called *Corbigny,* which *Mary* of *Austria* left fortified in the year 1577, calling it after *Philip* II. *Lewis* XIV. greatly augmented its fortifications.

The principality of *Chimai* belongs to the *Hennin* family, the little town of *Chimay* being its principal place.

IV. The *French* part of the earldom of *Namur* consists of the following places; *viz.*

Charlemont, a fortified little town, standing on a steep rock on the *Maas,* receives its name from its founder, *Charles* V. and its situation on a mountain. The land on which it stands, together with its territory, was purchased in the year 1555, of the bishop of *Liege,* yet was the town annexed to *Namur.* At the peace of *Nimeguen* this place was ceded to *France.* At the foot of the mountain on which the fort is built lies

Givet Saint Hilaire, new, regular and handsome, and, together with *Givet nôtre Dame,* lying on the other side of the *Maas,* at the foot of *Mont d'Or,* which is also environed with works, forms a small fortified place.

The G O V E R N M E N T *of*
D U N K I R K,

CONTAINS only the town of *Dunkirk* and some neighbouring villages; but since the peace of *Utrecht,* when the fortifications were demolished and the harbour filled up, no general-governor has been appointed over it, a particular governor only being instituted for the town. I do not find, however, that it is annexed to any other government. There is too much appearance therefore that this government will one day be revived.

I shall here give a full description of this celebrated place.

Dunkerk, Dunkirch, or *Dunkerque,* lies on the sea, or rather, as it is called, a canal, deriving its name from a church built there on the *Duns, i. e.* ' the sand-banks,' and erected, as it is said, by St. *Eloi,* who first

preached

preached *Chriſtianity* among the *Flemings*. Some houſes gradually were built near it, ſo as to form a little town, which *Baldwin*, earl of *Flanders*, ſurrounded with a wall in the tenth century, and the place enjoying a commodious harbour for trade, engaged in commerce, by which it grew large and wealthy, and even had ſome ſhips of war; inſomuch that, in the twelfth century, it fitted out a little fleet againſt the *Norman* pyrates, who committed frequent depredations on the ſea. In this expedition it performed ſuch eminent ſervices, that *Philip*, earl of *Flanders*, conferred ſome conſiderable privileges on the town. In the thirteenth century it was ſold to *Godfrey* of *Condé*, biſhop of *Cambray*, who conſiderably enlarged it and likewiſe improved the harbour. In the year 1288, his heirs again transferred the town to *Guy*, earl of *Flanders*, whoſe ſon, *Robert de Bethune*, diſmembered it from the earldom of *Flanders*, and, in the year 1320, conferred it as a particular lordſhip on his ſon *Robert de Caſſel*, whoſe daughter *Jolanda* brought it, in the year 1343, to her huſband *Henry* IV. count of *Bar*. In the year 1395, this *Jolanda* held *Dunkirk* as a fief from *Philip* duke of *Burgundy* and earl of *Flanders*, which, together with ſome other places, ſhe conferred on her couſin *Robert*, earl of *Marle*, who, in the year 1400, fortified it. In 1435, it devolved by marriage from the houſe of *Bar* to that of *Luxembourg*, and, in the year 1487, from the latter to the houſe of *Bourbon*, upon the marriage of *Mary* of *Luxembourg* with *Francis* of *Bourbon*, count of *Vendome*; but the ſovereign juriſdiction of this place belonging to the houſe of *Auſtria*, the Emperor, *Charles* V. erected a caſtle here in the year 1538. In 1558, it was taken and deſtroyed by the *French*; and, in the following year, by the peace of *Cateau-Cambreſis*, fell again under the *Spaniſh* dominion, when *Antony* of *Bourbon*, King of *Navarre*, couſin and heir to the above-mentioned *Mary* of *Luxembourg*, received this town and ſome other places which he held as a fief from *Philip* II. King of *Spain*, as earl of *Flanders*. The town recovered itſelf again, but had a great ſhare in the diſturbances which aroſe in ſucceeding times in the *Netherlands*. In the years 1646 and 1658, it was taken by the *French*, and, in the laſt mentioned year, ceded to the *Engliſh*, for aſſiſting *France* againſt *Spain*. In the year 1662, *Charles* II. of *England* ſold the town to *France* for five millions of livres; by which means *Lewis* XIV. acquired alſo the villages which the *Engliſh* had built round *Dunkirk*, as, namely, the village and fort of *Mardik*, *Great* and *Little Sainte*, *Arenbouts-Capel-Capelle*, *Coudekerke*, *Teteghem*, *Uxem* and *Ghyvelde*, with *Lefferinchouke* and *Zuytcote*. Upon this *Lewis* ordered the town to be well fortified, erected a fine citadel, and built *Fort Louis*, which ſtands to the ſouth, on the canal of *Bergen*, about half a *French* league from the town. The harbour was alſo put into a moſt excellent condition, two moles of piles, forming a canal in the ſea, of 1000 toiſes in length and about forty in breadth, inſomuch that a man of war of ſeventy guns might at any time paſs through it; and at each

end

end of the moles, two batteries were erected, one of which is called *Chateau verd*, the other *Chateau de bonne esperance.* Exclusive of these, on each side of the dyke, was a fort, together with the battery of *Revers* on the west, the castle of *Gaillard* on the cast, and a little farther within *Fort Blanc.* Betwixt all these forts ships were to pass in their way to the harbour, contiguous to which also was a large bason. This town enjoyed a flourishing trade, and, in the year 1706, contained 1639 houses, in which were 14,274 inhabitants. The *English*, for the safety of their commerce, which had sustained immense damage from the *Dunkirk* privateers, at the peace of *Utrecht*, in the year 1713, compelled *France* to destroy all the fortifications of the town, at her own charge to fill up the harbour and demolish the dams and sluices, and even never after, in the least, to repair these works. Upon this a beginning was actually made, but it was found that, by filling up the harbour, the country round, for about ten *French* leagues, would be in danger of being overflown. To remedy this inconveniency several fruitless negociations passed betwixt *France* and *England*, and, in the mean time, a new canal was made at *Mardyk.* In the year 1717, in the treaty concluded at the *Hague* betwixt *France*, *England* and *Holland*, it was determined, that the large entrance of the new sluice at *Mardyk*, which was forty-four feet broad, should be entirely demolished, and that no harbour, sluice, or bason, should ever be made either at *Mardyk* or *Dunkirk*, or within two *French* leagues round, and what remained to be demolished of the works at *Dunkirk*, should be entirely accomplished: but *France* making, no great haste in the demolition, it was, in the year 1748, again made an article of the peace at *Aix la Chapelle.* *France*, however, not only never completed this demolition, but rather, under-hand, begun some new works, which gave fresh umbrage to the *English* court, who received certain intelligence that the town was again fortified on the land side, the bason widened, and by that means rendered as commodious for the reception of ships as ever the harbour had been. The *French* court replied, that these works had no other view than to free the inhabitants from the exhalations of the stagnating water. Sir *Richard Steele*, an *English* gentleman, in an express treatise, shews the great advantage which would accrue to *England* by the demolition of *Dunkirk*, as, by that means, seven ninths of the *English* trade would be secured, the *French* having no other harbour in the Channel than that of St. *Malo*, which is capable only of receiving ships of between thirty and forty guns. Since the demolition of its harbour, this town has greatly declined.

Mardyk, a village lying on the sea, about a league and a half west to *Dunkirk*, was noted only for a fort on the *Duns*, about one mile from hence towards *Dunkirk*. Over against this fort of *Mardyk*, in the sea, stood *Fort de Bois.* Fort *Mardyk* has been frequently besieged and taken, namely, in the years 1645, 1646, 1652, and 1657. But in the years 1664 and

1665, it was demolifhed, and *Fort de Bois* alfo fuffered to fall to ruin. In later times *Mardyk* became celebrated for the noble canal there, which, after the peace of *Utrecht*, *Lewis* XIV. caufed to be erected under the direction of *Le Blanc*, being the length of 3338 toifes and two feet. This canal began at the canal of *Bergen* near *Dunkirk*, extending itfelf with a breadth of between twenty-five and thirty toifes, no lefs than 1500 from eaft to weft in length, at which place it winded from fouth to north, and 300 toifes farther had an incomparable fluice with two bafons on it, one of which was forty-four feet broad, being contrived for the reception of large veffels, the other twenty-fix feet in breadth, and intended for fmall. After this it extended ftill farther through the *Duns* to the main fea. *England* refenting this new canal, infifted at the above-mentioned treaty in the year 1717, that *France* fhould engage to demolifh the large paffage of the new fluice, the fmall fluice to be continued of the fame depth but with a reduction of its breadth to fixteen feet; and all the other works, together with the dams and fluices of the new canal to be deftroyed and levelled, and none erected any more on that coaft, within two leagues of *Dunkirk* and *Mardyk*.

The GOVERNMENT of *METZ*,

CONSISTS of the countries of *Mefzin*, *French*, *Barrois*, *la Saare*, and French *Luxembourg*. Its principal rivers are the *Maas*, of which mention has been made in *Champagne*; the *Mofel*, which takes its fource in the *Faucilles*, one of the *Wafgau* mountains, and is navigable all the year from *Metz*, difcharging itfelf into the *Rhine*; the *Meurte*, which rifes alfo in the *Wafgau* chain, being navigable for two or three miles above *Nancy*, and runs into the *Mofel*; the *Orney*, or *Ornain*, having its fpring at *Meureaux* in *Champagne*, and lofing itfelf in the *Marne*; the *Saare*, which has its fource near *Salme*, becomes navigable at *Saralbe*, and runs into the *Mofel*; the *Seille*, which iffues from the lake of *Lindre*, and falls likewife into the *Mofel*; the *Saone*, which receives its fource in the *Wafgau* mountains, becomes navigable near *Traves* above *Auffonne*, and receiving the *Crone* with fome other fmall rivers, falls into the *Rhone*.

1. The *Mefzin*, or the country round the city of *Metz*, is of tolerable fertility, and produces a little wheat. It was anciently a part of the Kingdom of *Auftrafia*, of which *Metz* was for a long time the capital, and the ufual refidence of the King. When the children of *Charles the Great* and *Lewis the Pious*, divided the dominions of that crown, the Kingdom of *Lorrain* rofe out of the remains of that of *Auftrafia*, and towards the end of the fecond royal line of *France*, *Metz*, *Toul*, and *Verdun*, three principal towns, fhook off the yoke, and put themfelves

2   as

as free cities under the protection of the Emperor. The power was divided betwixt the bishop and the magistrates; but the latter, with the assistance of the people, got the superiority over the former, insomuch that the bishop had no other power remaining to him in the city of *Metz* and country of *Mefzin* than a share in the election of the magistrates, and in the administration of the oath of office. The magistrates enjoyed an unlimited authority in the city and in the country of *Mefzin*, and the bishop in the demesnes of his bishopric on this side the *Vis*; though in certain cases appeals lay to the Imperial chamber, which they acknowledged as the supreme power of the empire. In the reign of the Emperor *Charles* V. the members of the *Smalcaldean* union applied to King *Henry* II. for assistance; upon which it was stipulated that the cities of *Metz*, *Verdun*, and *Toul*, should be delivered up by way of security into the King's hands; but upon his auxiliaries getting possession of *Metz* in the year 1552, the city was obliged to put itself under the protection of the *French*, which example was followed by the two others. Thus *France* kept possession of these three cities under the name of the protected-district till the year 1648, when at the peace of *Weftphalia*, the three bishoprics were absolutely transferred to *France*. In the year 1556, the bishop was obliged to cede to the King the right he had in the election of the magistrates, and the administration of the oath to them. In the *Mefzin* are the following remarkable places, *viz.*

Metz, in Latin *Metæ*, and *Metis*, the capital of the country and the seat of a bishopric, a parliament, an intendancy, a collection-office, a chamber of accounts and imposts, a country, mint, a forest-court, *&c.* It lies betwixt the *Mofel* and *Seille*, which unite their streams in this place. The old town is large, but the streets are narrow. The houses, however, though old fashioned are handsome. The new town is likewise large, and of much more beauty than the former. Exclusive of its fortifications it has three citadels. The bishop, who stiles himself a Prince of the *Roman* empire, is suffragan to the archbishop of *Treves*, with a diocese of six hundred and thirteen parishes, and a revenue of 120,000 livres, out of which his taxation at the court of *Rome* is 6000 florins. Besides the cathedral, it contains three chapters, sixteen parish-churches, six abbies, and a college of Jesuits. In this city are also a great number of *Jews*, who have a synagogue here. It was formerly a free Imperial city.

Montigny, a castle.

Ennery, *Ury*, *Bionville*, and *Borlife*, all little places.

Obf. The bishopric of *Metz* and its district must be distinguished from the city of *Metz*, and its district the *Mefzin*. It is a long but irregular tract of land, the lord of which is the bishop, but he has been deprived of this temporal jurisdiction, which at present belongs chiefly to the duke of *Lorrain*, particularly the salt-works, in lieu of which he only receives 30,000 livres and 400 bushels of salt, which the duke, or at present the King, is

obliged

obliged to deliver annually into the bifhop's magazine. What little of it belongs yet to the bifhopric confifts of the following *caftelanies*, lordfhips and
places, *viz.*

Helfedange, Hauboudange, and *Inqueffange,* lordfhips, and held as fiefs
from the bifhop.

Remilli, a *caftelany,* never difmembered from the bifhop's board-land.

Vic, a town lying on the *Seille,* the feat of the bifhop's chancery, of a
fupreme-diftrict and a *caftelany.* Salt was formerly made here.

La Garde, a lordfhip, the caftle of which lies on a lake, out of which
iffues the river *Saulnon.*

Fribourg, a *caftelany.*

Rechicourt, or *Richecourt,* an earldom, which, together with the lordfhip of *Marimont,* belongs to the counts of *Leiningen ;* but is a fief fubject
to the bifhop.

Turqueftain and *Chatillon,* feated on the *Vezouze,* are lordfhips.

Bacara, a *caftelany,* and a lordfhip of confiderable produce, belonging to
the bifhop, but the jurifdiction to the duke of *Lorrain.*

Rambervillers, a little town ftanding on the river *Mortagne,* and the feat
of one of the fineft *caftelanies* belonging to the bifhopric.

II. FRENCH BARROIS forms a part of the dutchy of *Bar,* and contains

1. The *prevoté* of *Longvic,* or *Longwi,* formerly an earldom, ceded at
the peace of *Nimeguen* to *France.* This *prevoté* derives its name from

Longwi, a little town, which is the feat of the *prevoté* and a diftrict.
The upper or new town which ftands high is regularly fortified ; the lower
or old town lies in a valley, being furrounded only with an ancient wall.
To this *prevoté* belong ten villages.

2. The *prevoté* of *Jametz,* its principal place

Jametz, a little town, formerly fortified. In the year 1641, this place
was ceded to *France* by *Lorrain,* and conferred by *Lewis* XIV. on the
houfe of *Condé.*

Invigny, an abbey of *Benedictines.*

3. The *prevoté* of *Dun,* belonging formerly to the country of *Dormois,*
or *Doulmois,* and the property of duke *Godfrey the Crooked,* who, in the year
1066, beftowed it on the bifhopric of *Verdun.* But in the following century it was fold to the duke of *Bar,* and afterwards united to that dutchy.

Obf. The *prevoté* of *Stenay,* which had its feat in the fmall and formerly
fortified town of *Stenai,* anciently *Sathenai,* on the *Maas,* was fold by *Godfrey*
of *Bouillan* to the bifhop of *Verdun,* from whom it defcended to the duke
of *Bar,* and in the year 1641, was ceded by the duke of *Lorrain* to *France.*
This *prevoté* was conferred by *Lewis* XIV. on the houfe of *Condé,* and
though in *Barrois,* belongs to the government of *Champagne.*

III. FRENCH LUXEMBOURG has been difmembered from the dutchy
of the fame name, being ceded to *France* by the peace of the *Pyrenees* in
the year 1659. To it belong   1. The

1. The *prevoté* of *Thionville*, containing

Thionville, or *Diedenhofen*, in Latin *Theodonis villa*, a fortified little town, feated on the *Mofel*, over which is a beautiful bridge of ftone. This place is the feat of a diftrict and a *prevoté*, the inhabitants being principally *German*. In the years 1558 and 1643, it was taken by the *French*.

Budingen, or *Budange*, a lordfhip.

2. The *prevoté* of *Damvilliers* lies in the circuit of *Verdunois*, and exclu‑ five of feven villages, contains

Damvilliers, a little town fituate in a marfhy country, fortified in the year 1528, by *Charles* V. but taken by the *French* both in the reign of *Henry* II. and *Lewis* XIII. and by the peace of the *Pyrenees* ceded to *Lewis* XIV. who in 1673, demolifhed its fortifications.

3. The *prevoté* of *Marville* and *Arancey*, the principal place of which is *Marville*, a little town feated on the river *Ottin*, being defended only by an old wall and a few towers.

4. The *prevoté* of *Montmedy* has its feat in *Montmedy*, a fortified little town on the river *Chier*, which divides it into the upper and lower. This place was taken by *France* in the year 1657.

5. The *prevoté* of *Carignan*, formerly *Yvoy*; erected into a dukedom in the year 1662, has its feat in

Carignan, a little town lying on the river *Chier*, and formerly called *Yvoy*, *Ipfch*, *Epufus*, or *Epoifus*. This place being conferred by *Lewis* XIV. on the count of *Soiffons* of the houfe of *Savoy*, he altered it from that count to its prefent name.

6. The dutchy of *Bouillon* belongs alfo to this place, and is an ancient lordfhip which was difmembered from the earldom of *Ardenne*. The dukes of *Bouillon* had long difputes about it with the bifhop of *Liege*, who alfo poffeffed it for a confiderable time; but *Lewis* XIV. having taken the town of *Bouillon* in the year 1676, conferred it two years after on the duke of *Bouillon*, then firft lord of the bed-chamber.

The town of *Bouillon*, in Latin *Bullio*, which ftands on a rock on the river *Semois*, is fortified, and on the fteepeft part of the rock, contains a ftrong caftle.

IV. The town and territory of SAAR-LOUIS.

Saar-Louis is a new town and fortification, begun by *Lewis* XIV. on the river *Saar* in the year 1680, and finifhed about four or five years after. Its ftreets are regular, and the fortifications a regular hexagon. It is the feat of a country-court, contains one parifh-church and two convents. By the peace of *Ryfwick*, *France* continued in poffeffion of this place, and in the year 1718, the duke of *Lorrain* ceded alfo to *France* the fite of the town of *Valderfange*, *Vaudrefange*, or *Valderfingen*, now in ruins, with the vil- lages of *Liftorf*, *Emftorf*, *Fraloutre*, *Roden*, and *Beaumaris*, which confti- tute the diftrict of this town.

Valdegaze, or *Valdegafk*, an abbey of *Premonftratenfes*.

The G O V E R N M E N T *of*

L O R R A I N *and* B A R.

THE dutchy of *Lorrain,* called alfo *Lothier,* deiives its name from *Lotharius* II. grandfon to the Emperor *Lewis* I. to whom it was affigned by his father *Lotharius* in the partition made betwixt him and his brothers, and is by the *Netherlanders* called *Lothar's Ryk,* which in time was changed into *Lothring.* This Kingdom was formerly of very confiderable extent, comprehending in it *Germania prima* and *fecunda,* as alfo *Belgica prima* and a part of *Belgica fecunda,* and being likewife looked upon as a part of the Kingdom of *Auftrafia.* It was afterwards divided, all that part of it lying between the *Rhine, Maas,* and *Scheld* as far as the fea, being included in *Lower Lorrain, Upper Lorrain* comprifing the countries betwixt the *Rhine* and the *Mofel* as far as the *Maas,* and which form the prefent *Lorrain.* To the eaft this country is bounded by the *Lower Palatinate* and *Alface,* from the latter of which it is feparated by the *Vogean* mountains; to the north it terminates on the provinces of the *Netherlands;* weftward on *Champagne;* and fouthward on *Franche Comté.* Its greateft breadth from fouth to north is about twenty-fix common *German* miles, and its greateft length from weft to caft nearly the fame. It is a very temperate climate and a fruitful foil, abounding particularly in grain, though with many woods and mountains, in which, however, we meet with good game, and yielding alfo excellent pafture. This country has likewife its falt-fprings, together with fome mines of iron, copper, tin, and filver. Its lakes too abound in fifh, particularly the lake of *Lindre,* the profits of which are faid to amount to 16,000 livres *per annum;* and in the mountains of *Wafgau* are found agates, granates, chalcedonies, and other gems, as alfo a particular fubftance for making cups and other veffels. Concerning the above-mentioned lofty mountain of *Wafgau,* which lies between the country of *Lorrain,* the *Sundgau, Alface,* and *Burgundy;* a further account fhall be given in *Alface.* The moft noted rivers in this dutchy are the *Maas,* the *Mofel* and the *Saar;* the *Saone,* indeed, has its fource likewife in the *Wafgau* mountains betwixt *Burgundy* and *Lorrain,* but fcarce touches the borders of the latter: In this province are alfo the little rivers of *Voloy, Mortaire,* and *Meurte.* The inhabitants have in times of old been reputed brave foldiers, and its modern dukes have always kept on foot a particular army, which performed fignal fervice, and *France* has not been wanting in policy to turn the power of *Lorrain* frequently to its own great advantage, but to the unfpeakable prejudice of the duke. The language ufed here is *French,* the *German* diftricts only excepted, where *German* is fpoken, and the *Romifh* religion

gion prevails over all the country. The revenue of both dutchies is said to be about two millions of *Rhenish* guilders.

Lotharius, younger son to the Emperor of the same name, may be looked upon as the first founder of the State of *Lorrain*, its name being derived from him. After his decease the country was divided among his relations, *Lewis* King of *Germany*, and *Charles* King of *Neustria*, the former of whom obtained all that part which we here include in *Upper Lorrain*. Soon after also the remaining part of the State of *Lorrain* was given by the sons of *Lewis the Stammerer*, to the younger *Lewis*, son to *Lewis* the *German*, by which means it came entirely into the *German* family. *Arnolph* the Emperor conferred it on his natural son *Zwentipold*, under the title of a Kingdom, but he held it only five years, being slain in the year 900, in a battle. Upon this *Lorrain* devolved to *Arnolph's* son, Prince *Lewis*, afterwards Emperor, and then was laid the foundation of the continual disputes which afterwards arose betwixt the Emperor's of *Germany* and the Kings of *France* concerning the country of *Lorrain*. The Emperor *Henry the Fowler* created his sister's husband *Giselbert* (who was a powerful lord of *Lorrain*) duke of the country and the Emperor *Otho* I. conferred this dutchy on his son-in-law *Conrad the Wise*, of *Franconia*. The Emperor *Otho* II. made *Charles*, brother to *Lewis* V. King of *France*, heir to the dutchy of *Lorrain*, investing him with it as a fief; but he having no issue, nominated for his successor *Godfrey the Younger*, count of *Verdun* in *Lorrain*, after adopting him with the consent of the Emperor, as lord paramount of the fief. The Emperor *Henry* III. in the year 1048, granted the investiture of this dutchy to *Gerhard* of *Alsace*, who is said to be the original founder both of the present house of *Lorrain* and that of *Austria*. His descendants are interwoven in the history of the *German* Emperors according to their inclinations towards the empire. But a later, and indeed a more remarkable period in the history of *Lorrain* is that of *Charles the Bold*, who dying in the year 1430, without heirs male, his daughter *Isabella* married *Renat* of *Anjou*, titular King of *Naples* and *Sicily*, in whose time the dutchy of *Bar* having been conferred upon him as a present by Cardinal *Lewis*, the last duke of that country, it became united with *Lorrain*. *Renat* II. grandson to count *Antony* of *Vaudemont*, whose son *Frederick* had married *Jolantha* daughter to the above-mentioned *Isabella*, succeeded to the government, and in 1477 defeated his enemy *Charles the Bold* of *Burgundy*, who had formed a design upon *Lorrain*, and from him are descended various families, his sons *Antony* being the founder of the chief line and *Claudius* of the collateral, which spread greatly in *France*. *Antony* died in 1544, and his eldest son *Francis* succeeded him in the government, but *Nicolas* the other obtained the dutchy of *Mercœur*. *Charles*, son to the former, in 1558, married *Claudia* daughter to *Henry* II. King of *France*, but lost the bishoprics of *Metz*, *Tull*, and *Verdun*, these being taken from him in 1552, by King

Henry II. and afterwards by the peace of *Weftphalia*, fettled for ever on the crown of *France.* Of his three fons, *Henry* the eldeft, fucceeded him, and left two princeffes behind him who were married to *Charles* and *Nicholas Francis*, fons to Prince *Charles*, upon which the government came into the hands of *Charles*, who diftinguifhed himfelf greatly both by his intrepidity and amours; but fiding in the thirty years war with the houfe of *Auftria*, *France* levelled its whole malice againft him, till he was driven àt length out of his country. In the year 1659, he was reftored to it upon very hard conditions. In 1662, he entered into a treaty with *France*, in which it was agreed that *Lorrain* fhould defcend, after his death, to *France*; and, on the other hand, that the whole houfe of *Lorrain* fhould be reckoned among the Princes of the blood; but new differences arifing betwixt him and *France*, he was driven a fecond time out of his country in the year 1670, and died in the Emperor's fervice in 1675. *Charles Leopold* IV. fon to his brother *Nicolas Francis*, fucceeded indeed to his dominions, but never enjoyed the fovereignty, as he could not brook the conditions on which he was to be reftored to them by the peace of *Nimeguen.* His eldeft fon, *Leopold Jofeph*, was, by the peace of *Ryfwick*, put again in poffeffion of his father's dutchy, after it had been twenty-feven years in the hands of the *French. Francis Stephen*, his fon, after his father's deceafe, in the year 1729, entered indeed upon the government, but, in 1733, *France* took poffeffion of his dutchy; and, in the preliminaries of the peace, in 1735, it was concluded, that not only the dutchy of *Bar*, but likewife that of *Lorrain*, the earldom of *Falkenftein* alone excepted, fhould be ceded to *Staniflaus*, King of *Poland*, and father-in-law to *Lewis* XV. and after his demife, both dutchies, together with the abfolute fovereignty of them, for ever to the crown of *France.* On the other hand, the Emperor *Charles* VI. engaged, in return for this, to cede the great dutchy of *Tufcany* to his fon-in-law Duke *Francis Stephen.* All this was fettled in 1736, and the following year not only *Staniflaus* arrived to the actual poffeffion of the dutchies of *Bar* and *Lorrain*, but likewife the duke of *Lorrain* to that of *Tufcany.*

The dukes of *Lorrain* bear the following title, N. N. by the grace of God, duke of *Lorrain* and *Merccuur*, King of *Jerufalem* and *Marchis*, duke of *Calabria, Bar* and *Guelders*, margrave of *Pont à Mouffon* and *Nomeny*, and count of *Provence, Vaudemont, Blamont, Zutphen, Saarwerden* and *Salen.* From this may be feen not only the countries which the dukes actually poffefs, but likewife their pretenfions. The eldeft fon, during his father's life-time, is ftyled count of *Vaudemont*, and alfo writes himfelf, whilft unmarried, margrave of *Pont à Mouffon*, but after his marriage he gives himfelf the title of duke of *Bar.* The houfe of *Lorrain*, by virtue of a treaty in the year 1736, retains all the titles, arms and privileges with the rank and quality of fovereign Princes, which it formerly had; yet without any claim to the ceded countries.

We

We fhall give an account of each dutchy in particular.

I. The dutchy of LORRAIN confifts of

  1. Three large diftricts ; namely,

    1. The diftrict of *Nancy*, by fome called alfo the *French* diftrict. To it belongs

Nancy, the capital of the dutchy and the ancient refidence of the dukes, being fituate in a delightful plain, not far from the river *Meurte*, and divided into the old and new town. The former is both the fmalleft and makes the meaneft appearance, being built with narrow ftreets ; but in it ftands the palace, which was begun indeed by Duke *Leopold Jofeph*, but only the main body finifhed. The fpacious long fquare, or wide ftreet, before it has a fine afpect. The new town is larger, contains broad and ftraight ftreets, with feveral beautiful boufes, only that, according to the fafhion of this country, they are generally low. In this town are three parifh and three collegiate churches, with two abbies, feventeen convents, one college, one noviciate of Jefuits, one hofpital and one commandery of the knights of *Malta*. Among the collegiate-churches, here is the new ftately cathedral, or primate-church, which is immediately fubject to the Pope. In the collegiate-church of St. *George* ftands not only the monument of *Charles the Bold*, duke of *Burgundy*, who was flain during the fiege of this city in the year 1476, but alfo the feat of the ancient dukes of *Lorrain*. The latter dukes are interred in the church of the *Capuchins*. It contains alfo an academy of fciences. The city was formerly fortified ; but, by the peace of *Ryfwick*, difmantled ; the works of the old town excepted, which were left untouched.

Malgrange, a ducal feat, ftanding on a little eminence about half a mile diftant from the city, and very magnificently begun, but never finifhed.

Perni, or *Prenei*, a market-town, the feat of a *prevoté*, was formerly fortified.

Fruart, or *Frouart*, a little place, fituate on the *Mofel*, belonging to which is a *prevoté* and a *caftelany*.

Condé, a market-town, feated on the *Mofel*.

Amance, in Latin *Efmantia*, an open place, formerly fortified, and the feat of a *prevoté* or *caftelany*.

Gondreville, a fmall town, fituate on the *Mofel*, the feat alfo of a *prevoté*.

S. Nicolas, a market-town, formerly a village, called *Port*, and owing its reputation to the reliques of St. *Nicolas*, which are kept here in the fine church dedicated to him. This town is the feat of a *prevoté*, has a houfe of Jefuits, four convents and one hofpital.

Rofieres aux Salines, the principal place of a *prevoté*, lies on the *Meurte*, having fome very profitable falt-fprings.

Einville, alfo the capital of a *prevoté*.

Luneville,

Luneville, in Latin *Lunaris villa*, a fmall, very ancient, and once fortified town, on the river *Vejouze*, ftands in a low marfhy country. This place was formerly a county, and is, at prefent, the capital of a *prevoté*, having a fine palace, in which not only the two laft dukes of *Lorrain* chiefly lived, but which was likewife the refidence of *Staniflaus*, till, in the year 1755, the right wing of this palace was entirely burnt down. In it is alfo an abbey, with a commandery of *Malta* and three convents. The gymnaftick academy here is altered to a foundation for Cadets, one half *Lorrainers* and the other *Poles.*

Beaupré, an abbey of Proteftant *Bernardines*, but following the *Ciftercian* rule of St. *Benedict*, feated in a delightful country, about an hour's diftance from *Luneville*, has an abbot, nine priefts, nineteen religious and twenty-three lay-brothers, with a yearly revenue of 80,000 *Lorrain* livres.

The *prevoté* of *Raon* and *S. Diei* lies among the mountains of *Wafgau*, being fo called from the little towns of *Raon*, furnamed *L' Etape*, fituate on the *Meurte*, and St. *Diei*, which is feated in a valley, on the fame river, called *Val de Galilee*, and contains a chapter.

Eftival, or *S. Eftival*, an abbey, feated on the river *Meurte.*

2. The diftrict of *Vofge, Voge*, or *Vauge*, fo called from the *Wafgau* mountains, is of confiderable extent, comprehending in it all the fouth part of *Lorrain*. This diftrict confifts of feveral *prevotés* and *caftelanies.* In it is

Mirecourt, in Latin *Mercurii Curtis*, a little town, feated on the river *Madon*, containing a bailiwick and a *prevoté* court, with four convents.

Chatenoi, a market-town, giving name to a *caftelany.*

Neuf-Chateau, a little town, feated on the *Maas*, but the principal place of the *caftelany* of *Chatenoi*, contains one abbey, one priory, one houfe of knights of *Malta*, one hofpital and five convents.

Darney, a market-town, lying on the *Saone*, and the principal place of a *prevoté.*

Charmes, a little town, ftanding on the *Mofel*, and containing one *prevoté* with two convents.

Dompaire, a little town, alfo a *prevoté.*

Arches lies on the *Mofel*, and is the principal place of a *prevoté*, or *caftelany*, which extends to the frontiers of *Alface*, and contains the whole country of *Havend*, lying in the *Wafgau* mountains. Part of the lordfhip of this town belongs to the chapter of *Remiremont.*

Plombieres, a fmall place, but its mineral waters of great reputation.

Remiremont, a market-town, fituate on the left fhore of the *Mofel*, and containing a celebrated chapter of ladies of noble defcent, as alfo a convent and an hofpital. The lordfhip of *Remiremont* was once an imperial fief.

Brujeres, the principal place of a *prevoté.*

3. The *German* diftrict, which lies along the *Saar*, or *Sarre*, being fo called from the inhabitants fpeaking *German.* In it is

Chateau Salins, the capital of a *prevoté*, feated on the river *Seille*, and con_ taining good falt fprings.

Guemunde, or *Saarguemines,* a little town, lying on the *Saar*, and the feat of a *prevoté*, was once fortified.

Boulai, Boula, or *Bolſhen,* a little place, which is the feat of a *câſtelⁿny.*

Beltain, or *Beaurain,* a market-town, the capital of a lordſhip.

Bouzonville, a lordſhip.

Feiſtorff, a *prevoté.*

Sirque, or *Sirck,* a fmall town, ftanding on the *Moſel,* and the piincipal place of a *prevoté*, was once fortified.

Schauenburg, the feat of a confiderable *prevoté*, in which lies the *Bene-dictine*-abbey of *Tholei,* in Latin *Theologia.*

2. Several other lordſhips and diftricts, acquired from time to time by the dukes of *Lorrain,* particulaily the diocefe of *Metz.* Among them is

Nomeni, a little town and marquifate, ftanding on the *Seille,* belonged formerly to the bifhopric of *Metz.* In the feventeenth century the dukes received this marquifate as a fief of the Empire.

Delme, a fmall place and lordſhip belonging to the marquifate of *Nomeni.*

The county of *Vaudemont* was erected in the eleventh century, and after-wards became a fief of the county and dutchy of *Bar.* *Renatus* united it with the dukedom of *Lorrain* and *Bar,* and the dukes afterwards made it the title of their youngeſt fon. To it belong

Vaudemont, a little town, or market-town, once the capital place.

Vezeliſe, a little town, feated on the river *Brenon,* at prefent the principal place of an earldom and the feat of a *prevoté*, with two convents.

Chaſtel, or *Chatte,* a lordſhip, which has always been feparated from *Lor-rain,* being a fief of the dutchy of *Bar.* From the counts of *Vaudemont* it defcended to the houfe of *Neufchatel,* and from thence to the counts of *Iſen-burg,* of whom *Antony,* duke of *Lorrain,* procured ∴, by way of exchange, in the year 1543. The little town of this name lies on the *Moſel,* and was once fortified.

The town and territory of *Epinal,* or *Eſpinal,* lies on the *Moſelle,* and was one of the moſt ancient demefnes belonging to the church of *Metz.* In the year 1444, the inhabitants of this place revolted from the epifcopal au-thority and put themfelves under *Charles* VII. King of *France*; but, in the prefent century, this place devolved to the houfe of *Lorrain,* which had alfo been put in poffeſſion of it at the treaty of the *Pyrenees,* in the year 1659, and at that of *Vincennes* in 1661. The town is but fmall, though once fortified. It is divided by the *Moſel* into two parts, is the feat of a *prevoté*, and contains one fecular abbey, four convents, one college of Je-fuits and two hofpitals.

The *prevoté* of *Deneuvre,* or *Denevre,* was alfo formerly a demefne of the bifhopric of *Metz,* and firſt efcheated to the lords of *Blamont,* pertaining

afterwards

afterwards as an epifcopal fief to the dukes of *Lorrain*, who, in the year 1561, obtained by compact the full fovereignty of it.

Deneuvre is a market-town, feated on the *Meurte*, and containing one collegiate and parifh-church.

Blamont, once a lordfhip, at prefent a county, was formerly an epifcopal fief; but, in the year 1542, became a fief of the Empire, and was bequeathed as a legacy by bifhop *Oulry* to *Renat*, duke of *Lorrain*. The little town of *Blamont*, or *Blankenberg*, lies on the river *Vezouze*, containing one collegiate-church and two convents.

The lordfhip of *Chatillon* borders on the *Vezouze* and is a dependency of *Blamont*.

The *caftelany* and lordfhip of *Marfal* belonged formerly to the bifhop of *Metz*. The town of *Marfal*, which lies in a marfhy country, but has good falt fprings, is the fear of this *caftelany*.

The little town of *Moyenvic*, containing alfo fome falt fprings, was once fortified, and belonged to the diocefe of *Metz*; but at the peace of *Munfter*, in the year 1648, was ceded to *France*.

The *caftelany* of *Dieufe* was firft held by the dukes of *Lorrain* as a fief of the diocefe of *Metz*; but ever fince the year 1347, the fief duties have been difcontinued.

The little town of *Dieufe*, in Latin *Decempagi*, is very old and contains fine falt fprings. The village of *Affurange*, belonging to this *caftelany*, was ceded to *France* at the treaty of *Vincennes*, in the year 1661. The lordfhip of *Sarbourg*, or *Sarbruck*, in which lies the town of *Kaufmanns-Sarbourg*, or *Sarbruck*, belonged formerly to the church of *Metz*; but, in the year 1475, the duke of *Lorrain* made himfelf mafter of it; and, in 1561, the bifhop was obliged to give it up entirely; as, in the year 1661, the duke was alfo compelled to give up *Sarbourg* and *Niederfwiller* to *France*, retaining only *Sareck* and fome villages.

The lordfhip of *Fauquemont*, or *Falkenburg*, belonged formerly to the diocefe of *Metz*, but the dukes of *Lorrain* afterwards made themfelves mafters of it, and even fo long ago as the beginning of the fifteenth century were in poffeffion of a part of it.

The town of *Fauquemont*, or *Falkenburg*, is but fmall.

The *caftelany* of *S. Avod*, or *Avau* and *Homburg*, belonged once to the diocefe of *Metz*, but was frequently alienated, and fometimes poffeffed by the dukes of *Lorrain*, who, in the year 1582, purchafed it for ever.

S. Avod is a little town and abbey, formerly called *S. Nabor*; but this name was afterwards changed into that of *S. Navau*, *S. Avau*, *S. Avold* and *S. Avod*.

Hombourg, a little town, about two miles diftant from the former.

The lordfhip of *Albe*, or *Aube*, is alfo an old fief of the bifhopric of *Metz*, which the duke of *Lorrain* procured, together with the full fovereignty of it, in the year 1561.

Sar

Sar-Albe, the principal place of this lordſhip, lies on the river *Saar*.

The earldom of *Sarwerden*, part of which was formerly a fief of the biſhopric of *Metz*, and the ſubject of a long diſpute betwixt the duke of *Lorrain* and the houſe of *Naſſau Sarbruck*. In the Dyet of the year 1669, it was determined by an agreement, that *Lorrain* ſhould keep the towns of *Sarwerden* and *Bockenkeim*, or *Bouquenon*, on the *Saar*, and deliver up the reſt to its competitor.

The earldom, or lordſhip, of *Bitſch*, which lies on the other ſide of the *Saar*, on the borders of the dutchy of *Zweybruck*, or *Deuxpont*, and *Lower Alſace*, belongs alſo to *German Lorrain*, and is an old inheritance appertaining to the ducal houſe, which beſtowed it as a fief on the counts of *Deuxpont*. Count *James* dying in 1570, this earldom deſcended to the counts of *Hanau*, Duke *Charles* beſtowing the fief on them, but ſoon after reaſſumed the earldom, which from that time has been conſtantly annexed to *Lorrain*.

The little town of *Bitſch* being taken by *Lewis* XIV. he fortified it; but, on his reſtoration of it at the peace of *Ryſwick*, his firſt care was to demoliſh the fortifications.

The earldom of *Falkenſtein*, anciently an imperial fief; but the Emperor, *Frederick* of *Auſtria*, in 1458, inveſted the houſe of *Lorrain* with the feoffment; and, in 1667, the property of it was alſo ſold to the ſame houſe by *William*, count of *Falkenſtein*. On the ſurrender of *Lorrain* to *France*, in the year 1736, this dutchy was excepted as appertaining to the duke.

3. The following lordſhips, lying in *German Lorrain* and belonging to the *German* Empire, *viz.*

1. *Crichingen*, in *French Creange*, ſituate near *Falkenburg*, a ſmall town, though once a barony. In the year 1617, it was erected into an earldom by the Emperor *Matthias*. The ancient barons of *Crichingen* had by marriage and inheritance obtained

The lordſhip of *Pitlingen*, in *French Putelange*, which was an old fief of the biſhopric of *Metz*.

The lordſhip of *Morchingen*, in *French Morhange*, has for ſome centuries belonged to the foreſt-counts; and by the daughter and heireſs of the laſt of them, *John Simon*, devolved to the *Rhine*-count *John*, her ſpouſe.

The lordſhip of *Finſtringen*, in French *Feneſtrange*, formerly belonged to a houſe which bore the ſame name. On the death of its laſt lord, *Nicolas*, his two daughters divided the lordſhip. *Barbara*, the eldeſt, brought her ſhare to *Nicolas*, count of *Sarverden*, and, by their daughter *Johanna*, it devolved to *John*, count of *Salm*. *Margaret*, the youngeſt, carried hers to her huſband *Ferdinand* of *Neufchatel*, by whoſe daughter it deſcended to the Baron *de Fontenoy*, of the houſe of *Dommartin* in *Lorrain*, and from his family devolved by marriage to *Charles Philip de Croi*, marquis of *Havrec* in *Hanau*; and the male iſſue of the *Havrec* family failing, the daughter of the laſt brought it to *Philip Francis de Croui*.

VOL. II.   4 F   *Finſtringen*,

Finſtringen, the capital, is but ſmall and ſtands on the *Saar*.

4. The earldom of *Salm*, in the *Waſgau* mountains, formerly a fief of the biſhops of *Metz*, but belonging at preſent to the Empire. Count *John* of *Salm* had two ſons ; namely, *John* and *Simon*, who divided this earldom into two equal parts. *Paul*, the laſt male deſcendant of Count *John*, left a daughter, by name *Chriſtina*, who married to *Francis* of *Lorrain*, count of *Vaudemont*, bringing him half of the earldom of *Salm* as a portion, to-gether with other eſtates pertaining to his father. The only daughter and heireſs of Count *Simon* brought the ſecond half of the earldom to her huſband *John*, *Foreſt* and *Rhine*-count, from whom deſcended the *Rhine*-count *Philip Otho*, lord of half the earldom of *Salm*, who, in the year 1623, was raiſed to be a Prince of the Empire. The caſt part of this earldom, which lies near *Alſace*, belongs to the diocese of *Straſburg* and the weſt, which borders on *Lorrain*, belongs to the diocese of *Toul*.

II. The dutchy of BAR, or BARROIS, was formerly an earldom, ſo called from the caſtle of *Bar*, built in the year 964, by *Frederick* I. whom the Emperor *Otho* created earl of *Bar*. It has been a long diſpute among the *German* and *French* hiſtorians, when, and by whom, the county of *Bar* was raiſed to a dukedom. The *Germans* affirm, that the Emperor *Charles* IV. being at *Metz* in the year 1354, raiſed this earldom to a duke-dom at the ſame time with the earldom of *Luxembourg*, and in the ſame year erected the earldom of *Pont à Mouſſon* into a marquiſate. The latter is certain, but vouchers are wanting to the former. Some *French* hiſtorians are very fond of attributing the erection of this dutchy to *John* II. King of *France* : but the weakneſs of this pretence is acknowledged even by ſome experienced *French* writers ; yet even theſe, on the other hand, aſcribe the erection to *Charles*, ſon of the above-mentioned King : This is unqueſti-onable, that, as early as the year 1355, *Robert de Bar* ſtyled himſelf duke of *Bar*. No leſs true is it, that the lordſhip of *Bar*, or the country on the other ſide of the *Maas*, was, in the year 1354, a fief of *France*, and conti-nued as ſuch ; but, in more ancient times, it was a fief of the Empire as well as the country on this ſide the *Maas*, for which homage had always been done to the Empire as a fief of it. How the dutchy of *Bar* came to be annexed to *Lorrain* has been ſhewn above in the article of this dutchy, and ſince that time it has ſhared the fate of *Lorrain*.

The *French* divide this dutchy into

1. *Barrois Mouvant*, which was formerly detached from *France* as a fief, and is compoſed of. two large diſtricts, *viz.*

1. The diſtrict of *Bar*, conſiſting of two *prevotés*.

1. The *prevoté* of *Bar le duc*, containing the following places ; namely, *Bar le duc*, the capital of the dutchy of *Bar*, which conſiſts of the upper and lower town with ſome ſuburbs, and lies on the river *Ornei*. In the upper town once ſtood the caſtle ; and to this day the ducal palace. In

it

it are alſo two chapters, and in the whole town are one pariſh-church, one priory, ſeven convents, one college of Jeſuits and one hoſpital.

Loupi le chateau, a ſmall town, bearing the title of a lordſhip.

2. The *prevoté* of *Souilliers* contains only ſome mean country towns and villages.

3. The earldom of *Ligni,* in which is

Ligni, the capital, and the ſecond town of the dutchy, lying on the river *Ornei,* and containing a pariſh and collegiate-chuich, with one college and five convents.

Dammarie, a little place, having a priory. The ſupreme juriſdiction of this town belongs to the counts, the middle and lower to the prior.

2. The diſtrict of *Baſzigni* includes only one part of the lands, or heaths, of *Baſzigni,* the other belonging to the government of *Champagne.* It conſiſts of ſix diſtricts or juriſdictions.

1. The *prevoté* of *Gondrecourt,* the feat of which is in the market-town of that name.

2. The bailiwick of *La Motte* and *Bourmont* contains

La Motte, or *La Mothe,* once an important caſtle, ſituate among mountains, but taken, in the years 1634 and 1648, by the *French;* and the laſt time totally deſtroyed.

Bourmont, a little town, the feat of a manor and a bailiwick-court, containing a pariſh-church, two chapters and two convents.

3. The *prevoté* of *La Marche,* the feat of which is

La Marche, a ſmall place, containing a convent; and a quarter of a *French* league diſtance from it a priory.

4. The *prevoté* of *Chatillon,* in the little place of the ſame name.

5. The *prevoté* of *Conflans.*

6. The lordſhip of *S. Thieboud,* its principal place, but inconſiderable.

2. *Barrois non mouvant, i. e.* ' not ſeparated from *France*' as a fief, and comprehended in the bailiwick of *Michel,* being ſo large that it extends betwixt the *Maas* and *Moſel* as far as the frontiers of *Luxemburg.* The feat of it is

S. Mihel, or *S. Michel,* a little town, lying on the *Maas,* and owing its origin to an old *Benedictine* abbey there. To this manor belong the following *caſtelanies* and lordſhips; *viz.*

1. The *caſtelany* of *Sanci.*

2. The *caſtelany* of *Foug,* or *Fau;* its feat

Foug, in Latin *Fagus,* a country-town.

3. The *caſtelany* of *Bouconville,* its feat in the little place of the ſame name, ſituate on the river *Maid.* To it belong the lordſhips of *Trognon* and *Thiaucourt.*

4. The lordſhips of *Mandre aux quatre tours ſur Amermont,* formerly fiefs of the church of *Metz,* but ſince the ſixteenth century the dukes have poſſeſſed it independently.

5. The

5. The *caftelany* of *Chauffée*, feated on the river *Iron.*

6. The *caftelany* of *Conflans*, formerly a demefne belonging to the dio-cefe of *Metz*; and in the year 1561, ceded by the bifhop to the duke of *Lorrain.* Its feat is *Conflans,* furnamed *Jernifi,* which lies betwixt two fmall rivers.

7. The *prevoté* of *Eflain,* or *Etain,* belonged formerly to the chapter of *Verdun,* which in the year 1224, was compelled to cede it to *Henry* count of *Bar.* In it is

Eflain, an old little town.

8. The *prevoté* of *Briei,* once an earldom, in the year 1225 conferred as a fief by the bifhop of *Metz* to *Henry* count of *Bar,* the grant being afterwards made abfolute.

9. *Mufzi,* or *Mufzei,* a town formerly having a caftle, and once the prin-cipal place of a *caftelany,* which the counts of *Bar* held as a fief of the bifhop of *Verdun;* but they afterwards cancelled that dependance.

10. The *prevoté* of *Longuion,* an ancient demefne of the counts of *Bar.*

11. The marquifate of *Pont à Mouffon* confifts of the *caftelany* of *Mouffon,* and the *prevoté* of *Pont à Mouffon.* In it is

Pont à Monffon, or *Mouffon.* This town is divided by the *Mofel* into two parts, containing two parifh-churches, one chapter, an univerfity, founded here in the year 1573, one college of Jefuits, one feminary, one abbey, one hofpital, and nine convents. In the year 1354, *Charles* IV. raifed it to an Imperial city and a marquifate.

12. The *caftelany* of *Condé* lies on the *Mofel,* and formerly belonged to the demefnes of the bifhop of *Metz,* but in the year 1561 was wholly ceded to the dukes of *Lorrain.*

13. The lordfhips of *Avantgarde* and *Pierre-forte,* old fiefs of *Bârrois.* *Avantgarde* lies on the *Mofel.*

3. The earldom of *Clermont* in *Argonne* belonged anciently, under *cafte-lanies,* to the bifhopric of *Verdun,* who raifed themfelves to independency. About the beginning of the 13th century, *Thibaud,* count of *Bar,* made himfelf mafter of it, and not only his defcendants but the dukes of *Lorrain* alfo held it as a fief of the bifhop of *Verdun.* The lords of *Clermont,* on ac-count of feveral fiefs belonging to it lying in *Champagne,* did homage for it to the counts of *Champagne,* and afterwards to the Kings of *France.* In the year 1564, the bifhop of *Verdun,* for a fmall fum of money purchafed the feudal jurifdiction, but the earldom remained a fief to the empire. Duke *Charles* III. transferred it to *France,* and *Lewis* XIV. beftowed it on *Lewis* of *Bourbon* Prince of *Condé.*

The town of *Clermont* lies on the river *Air;* the lordfhips of *Varennes* and *Vienne* belong alfo to this earldom.

4. Betwixt the *Maas* and *Mofel* lie feveral lordfhips, dependent neither on *Lorrain* nor *Bar;* but which may moft properly be treated of here.

1. The

1. The lordſhip of *Apremont* and its barony borders on the manor of *S. Michel*, and is one of the oldeſt fiefs of the biſhopric of *Metz*, but was divided among ſeveral proprietors. In the 16th century it deſcended to the houſe of *Lorrain*.

2. The lordſhip of *Commerci* is an ancient fief of the biſhopric of *Metz*, which has been in ſeveral hands, and by degrees has ſhaken off the biſhop's feudal power. The property to this lordſhip was purchaſed of the *Gondi* family by the houſe of *Lorrain*. The town of *Commerci* lies on the *Maas*, and contains a pariſh and collegiate church, with two convents and one hoſpital.

3. The marquiſate of *Hatton-chaſtel* lies in the country of *Vaivre*, and belónged formerly as a lordſhip to the church of *Verdun*, which in the year 1564, ceded it to *Lorrain*, with its full juriſdiction and ſovereignty. In 1567, duke *Charles* II. was inveſted with it as a fief of the empire; and the Emperor *Maximilian* II. raiſed it to a marquiſate. The town of this name had formerly a caſtle noted for its ſtrength.

4. The lordſhip of *Dieulonard* lies on the *Moſel* not far from *Pont à Mouſſon*; being one of the moſt ancient demeſnes of the church of *Verdun*, but came into the poſſeſſion of the houſe of *Lorrain*.

Dieulonard, in Latin, *Deſlonardum*, is ſmall, and was formerly fortified.

5. The lordſhip of *Gorze*, lying betwixt the *Meſzin* and *Verdunois*, belongs to the ſecularized abbey of *Gorze*. The dukes of *Lorrain*, in the year 1621, annexed the lands belonging to it to the cathedral of *Nancy*, in which it continued till the year 1661. At the peace of *Vincennes* duke *Charles* III. ceded the lordſhip of *Gorze* to *France*, when the abbey was again ſeparated from the cathedral; to which the King has the nomination.

6. The little town and diſtrict of *Malatour*, anciently *Mars la Tour*, has paſſed through ſeveral proprietors; but all ſubordinate to the archbiſhop of *Metz*. The duke of *Lorrain* relinquiſhed the independent poſſeſſion of it: But at the peace of *Vincennes* was compelled to cede it to *France*.

The GOVERNMENT *of*

VERDUN and the VERDUNOIS.

THE country of *Verdunois* extends itſelf along the *Maas*, and abounds in large boroughs and villages, which are divided into ſeveral *prevotés*; but its only city is *Verdun*, from which it receives its name. At the peace of *Munſter*, in the year 1648, the empire yielded up to *France* the full ſovereignty of the biſhopric, the city, and its diſtrict. In it is

1. The

1. The city of *Verdun* and its district.

Verdun, in Latin *Verodunum*, or *Veredunum*, the capital of the country and the feat of a bishop, a provincial and a manor-court, together with a collection, is large and well peopled, and consists of three parts, *viz.* the upper, lower, and new town. Exclusive of its fortifications, this place is further defended by a fine citadel. The bishop, before the city and district was annexed to the crown of *France*, was a prince of the empire; and still stiles himself such, as also earl of *Verdun*. The archbishop of *Triers* is his metropolitan. This diocese consists of one hundred and ninety-two parishes. The bishop's revenue is 50,000 livres, and his taxation to the court of *Rome* 4466 florins. Exclusive of the cathedral, in this city is one collegiate and nine parish-churches, six abbies and one college of Jesuits. *Verdun* was formerly an Imperial city. In the year 1552 it put itself under the protection of *France*, and in 1648 fell absolutely under its power. To the district of this city belong several hamlets.

2. The proper bishopric of *Verdun*, or the district over which the bishop has a jurisdiction subordinate to *France*; is composed of one hundred and six parishes. The marquisate of *Hatton-chastel*, the lordship of *Sampigny* on the *Maas*, and the feudal jurisdiction of the lordship of *Clermont*, *Vienne*, and *Varenne* belonged formerly to the bishops of *Verdun*, but do not at present.

The GOVERNMENT *of* TOUL *and* TOULOIS.

THE earldom of *Toulois*, being small, is quite hemmed in by *Lorrain*, and together with the city of *Toul*, and the two other bishoprics of *Metz* and *Verdun*, put itself, in the year 1552, under the protection of *France*, which, in 1648, got the absolute sovereignty over them. In it are

1. The city of *Toul*.

Toul, in Latin *Tullum*, the principal town of the government, and the feat of a bishop, a provincial and district-court, together with a collection, lies on the *Mosel*, over which it has a fine bridge of stone, with a regular fortification. It was formerly an Imperial city, and its bishop a Prince of the empire; which title, together with that of earl of *Toulois*, he still assumes. He is a suffragan to the archbishop of *Triers*, with a diocese of 1400 parishes; yet his income little exceeds 17000 livres, and at the court of *Rome* he is taxed in 2500 florins. Exclusive of its fine cathedral, in this town is one collegiate and four parish-churches, three abbies, two priories, seven convents, two hospitals, one commandery of *Malta*, and one seminary.

The

The city has its particular diſtrict.

The biſhopric or diſtrict of *Toul*, the temporal lord of which is the bi-ſhop, is ſubordinate to *France*, conſiſts of ſix *prevotés*, including only market-towns and villages, the two principal of which are *Luverdon* and *Vicheri*.

The G O V E R N M E N T *of*
A L S A C E.

ALSACE terminates weſtward on *Lorrain* and *Burgundy*, ſouthward on *Swiſſerland* and *Elſgau*; eaſtward on *Ortenau* and *Briſgau*; and northward on the *Palatinate*. The beſt map of it is that by *Homann*'s heirs, in two ſheets, called *Alſatia* ―― *Unâ cum Sumgoviâ.* Its extent from ſouth to north is about eighteen common *German* miles. It was anciently inha-bited by the *Rauraci, Sequani,* and *Mediomatrici.* Its name firſt occurs in the hiſtory of *France* under the *Merovingian* Kings, being of *Frank* origi-nal; and the moſt juſt derivation of it deduced from the river *Ell,* or *Ill,* the inhabitants of whoſe borders were called *Elſaſſen,* from whom after-wards the country itſelf came to be called *Elſas,* in Latin *Eliſatia, Aliſatia, Alſatia,* &c. The country is in general very pleaſant, and abounds in all kinds of grain, fruits, eſculent vegetables, flax, tobacco, wood, &c. Its paſtures alſo are rich; and its wine very palatable, and of a good body. The coun-try betwixt the *Ill, Haardt,* and *Rhine* is narrow, and but of indifferent fertility; having wine, and few paſtures; rye, barley, and oats being its only grain. But that part lying betwixt the mountains, the *Ill,* and the plain of *Soults* in *Upper Alſace,* to the diſtance of two *French* leagues be-yond *Hagenau,* yields an exuberance of grain, wine, and paſtures. The country beyond *Soults* and *Befort,* running along the mountains in a breadth from two to three *French* leagues, abounds in wood, but little corn land; which is in ſome meaſure made up by its good paſtures and large breeds of cattle. The country towards *Switzerland,* or about *Altkirchen, Baſel,* and *Muhlhauſen* is very fertile. The diſtrict about *Hagenau,* which is called the plain of *Marienthal,* is a ſandy heath, yielding only *Turkiſh* corn. The country from mount *Saverne,* and the levels about *Strasburg* to the *Rhine,* is incomparably fertile and delightful, being rich in all kinds of grain, tobacco, culinary vegetables, ſafron and hemp. The tracts betwixt the *Hagenau* mountains and the *Rhine,* to *Landau* and *Ger-merſheim,* is moſtly a woody and uncultivated country; affording more forage than other conveniences. But the fine plain about *Landau* pro-duces a great deal of corn. From *Landau* to *Weiſſenburg* is a wine country. The chief chain of mountains in this country is the *Waſgau,*

in

in *French* called *La Voſge, Les Voſges,* and *Vauges,* in Latin *Vogeſus.* This chain begins in the neighbourhood of the town of *Langres*; and ſtretching itſelf at firſt from weſt to eaſt towards the country of *Befort,* ſeparates the country of *Burgundy* from *Lorrain,* being called *Montagne de Bourgogne,* as it is alſo on account of its good paſtures, named *Mont de Faucilles.* After this it winds towards the north, ſeparates *Lorrain* from *Alſace,* and forms another curve towards the Electorate of *Triers.* Its length is from ſouth to north, as its breadth from weſt to eaſt; and the narroweſt part of the latter is at *Zaberner Steeg*; ſo that here is the ſhorteſt and eaſieſt road from *Alſace* into *Lorrain.* It contains the ſources of great numbers of rivers and ſtreams. Its higheſt pikes are the *Belg* or the *Balon,* which is the higheſt of all, the mountain of *S. Odilia,* and the pike of *Frankenberg* or *Framont.* Both the ſummits and the vallies of the *Waſgau* chain, as well as the plains of *Alſace,* produce one hundred and fifty kinds of trees and ſhrubs, and fifteen hundred and fifty ſpecies of herbs, which all grow wild. They have alſo ſome excellent paſtures, and the eminences where the ſoil is favoured by a ſunny expoſure, produce a very agreeable red and white wine, of which is made a great deal of brandy, vinegar, and tartar. The large foreſts here are the *Haardt,* or *Hart,* which lies betwixt the *Ill* and the *Rhine*; extending from *Sundgau* to *Upper Alſace,* a length of about eight miles and a breadth of two, and belonging to the King. The *Hagenau* foreſt, which is five miles in length and four broad, belonging half of it to the King, and half to the town of *Hagenau*; and the *Bewald,* properly called the *Bienwald,* or *Foreſt of Bees,* which is of the ſame largeneſs with the preceding, and lies on the frontiers of *Lower Alſace,* belonging to the biſhop of *Spires.* Theſe foreſts abound in deer and game of all kinds.

The *Waſgau* chain contains alſo its inward treaſures; and for ſeveral centuries has been famous for its ſilver, copper, and lead. The mine-works of ſilver in *Leber* or *Hagenthal,* at *Furtelbach,* were richer in the 16th century than at preſent, the neat produce of the *Markirch* mines being at preſent only 1500 marks of metal. In *Weiler-vale* is alſo a ſilver ore. In *Upper Alſace,* towards *Burgundy,* in the vale of *Roſenberg,* which belongs to the diſtrict of *Befort* at *Giromany,* and *Upper Auxelles,* are alſo profitable mines of ſilver. Several parts alſo abound in iron mines, yielding a very good metal. Near *Dambach* is a mine of ſtcel, and alſo of copper and lead; not to mention antimony, cobalt, ſulphur, and many other minerals which are found here. In *Weiler-vale* is dug a reſinous coal, and at *Nieder-Ehenheim* turf. From the *Waſgau* chain alſo iſſue mineral waters. Among the moſt noted baths are thoſe of *Niederbronn, Watweiler,* and others ſituate not far from *Benfeld,* as thoſe alſo at *Sulz,* and near *Molzheim.* Laſtly, of theſe mountains it muſt be further obſerved, that they are thick ſet with a very uncommon number of churches, convents, and chapels.

The principal rivers in *Alſace* are the following, *viz.*

5

1. The *Rhine*, of the source and course of which an account shall be given in the third volume. This river serves as a security to the country; but frequently causes terrible devastations and those not only in winter, but even in the middle of summer, when the snow melts on the *Alps*. Its inundations ruin the fields by covering them with sand. The violent torrents of the *Rhine*, which happen generally every year, frequently alter the situation of the islands in it, and the banks on the *Alsace* side, which *Old Brisac*, *Rheinau*, and the convent of *Honau* have particularly experienced. One singularity of this river is, that in its sand are found particles of gold which the torrents in their fall wash from the *Alps*, and bring into the *Rhine*. Accordingly it is only below *Basel* that the sand contains this precious mixture, which, in some depths called the *gold-grounds*, run together, and in autumn and winter, when the river is at the lowest, are drawn out among the sand. After this, passing through several waters, they are worked by means of quicksilver into lumps or *laminæ*. The golden particles are much more scarce betwixt *Brisac* and *Strasburg* than betwixt *Strasburg* and *Philipsburg*; and betwixt *Fort Lewis* and *Germersheim* are in much greater quantity; the stream here abating of its rapidity. It is very seldom that they are equal to a grain of millet. The gold is indeed very fine and beautiful; but at present so scarce that the city of *Strasburg*, which has a right of gathering gold for the length of 4000 paces, scarce collects five ounces in a year. The proprietors of the village of *Blohsheim*, which is three *French* leagues from *Strasburg*, make but three guilders a year of the gold-wash there, though they sweep the river for near a *French* league. The *Rhine* also contains many crystals, and particularly pebbles, which receive a polish like diamonds, with other gems. They are much used in *France* under the name of *Rhine-pebbles*. The following rivers issue from the *Wasgau* chain, and run into *Lower Alsace*.

2. The *Leber* in *Leberthale*, which runs into the *Scher*.

3. The *Cher*, in Latin *Scara*, in *Weilerthale*, running into the *Andlau*.

4. The *Andlau*, which joins the *Ill*.

5. The *Ergers*, in Latin *Ergitia*, at its beginning called the *Ehn*, mingling also with the *Ill*.

6. The *Breusch*, in Latin *Brusca*, and *Bruscha*, issuing from the earldom of *Salm*, and dividing itself in the district of *Dachstein* into two branches; one of which receives the *Mosly*, and through the canal cut by *Lewis* XIV. to the length of four *French* leagues with a breadth of twenty-four feet and a depth of eight for the more commodious transportation of all materials for building, falls into the *Ill* above *Strasburg*, whilst the other runs through that city, and below it joins the *Ill*. The *Breusch* receives the lesser rivers of *Sauvel*, *Moszig*, or *Mosig*, *Hasel*, &c.

7. The *Sorr*, in Latin *Sorna*, discharging itself into the *Rhine*.

8. The *Motter*, in Latin *Matra*, which receives the rivers *Zinfel* and *Sauer*, or *Sur*, and falls into the *Rhine*.

9. The *Seltzbach* and *Lauter*, which run into the *Rhine*.

10. The *Queich*, which traverses *Landcu*, and near *Germerfheim* falls into the *Rhine*. From this river, before it comes into the plain, not only a canal has been cut as far as *Landau*; but likewife by means of it and fome leffer ftreams has been made a confiderable fortification of moats and ramparts, which, extends from *Landau* towards the market-town of *Herte* as far as the *Rhine*.

In *Upper Alface* are the following rivers, *viz.*

11. The *Ber*, or *Berre*, which runs into the *Rhine*.

12. The *Ill*, anciently the *Ell*, which has its fource at the market-town of *Winkel* in *Sundgau*, receives the rivers *Larg*, *Tolder*, *Thur*, *Lauch*, *Fech*, *Zembs*, *Scheer*, *Andiau*, *Ergers*, and *Breufch* in its paffage, and run-ing through *Strasburg* falls below *Wanzenaw* into the *Rhine*.

In *Alface* are alfo feveral lakes, the moft noted of which are the *Schwarze*, *Weiffe*, and the *Daren-See*, in *Upper Alface*, and the *Wafgau* mountains.

The number of inhabitants in *Alface* is computed to be about half a million; in *Upper Alface* and *Sundgau*, are thirty-two large and fmall towns, and in *Lower Alface* thirty-nine, and in both upwards of 1000 market-towns and villages. The common language of the inhabitants is the *German*, and they are partly *Lutheraus* and partly *Roman* catholics.

This country from the *Celtæ* fell under the dominion of the *Romans*, be-coming fubject next to the *Franks*. *Lewis the Pious* conferred it on his fon *Lotharius*, after whofe deceafe it fell to *Lewis the German*, and in 870 was a province of *Germany*. It bore anciently the title of an earldom, and for a long time belonged to the counts of *Egenfheim*; after the extinction of which family, the Emperor divided this country as a fief betwixt the counts of *Oettingen* and the counts of *Habspurg*. In the year 1359, the former fold their fhare to the bifhop of *Strasburg*, who ftiled himfelf landgrave of *Alface*, which title the counts of *Habspurg* alfo affumed. The government of *Al-face* was afterwards conferred by the Emperors on feveral houfes, till *Fer-dinand* I. gave it to the *German* line of his own houfe; and accordingly it continued in the houfe of *Auftria*. At the peace of *Munfter*, in the year 1648, the Emperor, not only in behalf of himfelf but alfo of the houfe of *Auftria* and the empire, ceded for ever to the crown of *France* all right to the town of *Brifac*, the landgravate of *Upper* and *Lower Alface*, *Sundgau*, and the diftrict of the ten united Imperial cities in *Alface*, with the whole fovereignty belonging to them. On the other hand *France* engaged that not only the bifhops of *Strasburg* and *Bafel*, but· alfo all the immediate States of the holy *Roman* empire throughout *Aface*, together with the ab-bies of *Murbach* and *Luders*, the convent of *S. Georgenthal*, the *Pfalzgraves*

of

of *Lutzelſtein*. The counts and barons of *Hanau*, *Fleckenſtein* and *Oberſtein* ; the nobility in *Lower Alſace* and the above-mentioned ten imperial cities of the diſtrict of *Hagenau* ſhould continue unmoleſted in the enjoyment of their freedom, and to be contented with the power which the houſe of *Auſtria* had over them, and ceded to it by that treaty, yet without any prejudice to the ſovereignty which at the ſame time was transferred to *France*. In the ſucceeding wars *France* took the ten imperial cities in *Alſace*; and, as by the peace of *Nimeguen* no expreſs ſtipulation was made for their reſtitution, the King looked upon himſelf as entitled to treat them as part of his acquiſitions. He alſo erected at *Briſac* a chamber of appeal, to which not only theſe cities but alſo the immediate nobility of the Empire, and all ſubjects holding imperial lands in *Alſace*, were to bring their proceſſes, and no longer to the imperial chamber or tribunal of the Empire. At the peace of *Ryſwick*, in the year 1697, the Emperor and the Empire ceded to *France* the perpetual ſovereignty of the city of *Strasburg* and of all its dependencies on the left ſide of the *Rhine*.

To the government of *Alſace* belongs

1. *Lower Alſace*, containing

Straſburg, in Latin *Argentoratum*, the capital of the whole country, being a royal, free, imperial city, and the ſeat of a mint-court. It lies a good quarter of a *French* league from the *Rhine*, and is ſurrounded by the rivers *Ill* and *Breuſch*, which alſo run through it, beſides an arm of the *Rhine* which is conveyed to the city. It is not only very well fortified but may be laid under water; and eaſtward, or towards the *Rhine*, has a regular citadel, the canon of which reaches *Kehl*. It is likewiſe large and populous, and contains about 32,000 houſes, 4300 families and 40,000 inhabitants. The ſtreets are in general narrow, two only excepted, and adorned with few handſome houſes. The public buildings in it are the town houſe, the epiſcopal palace, the intendant's reſidence, the arſenal and the play-houſe. The city hoſpital is a building in which is alſo kept grain of a great age, and wine which is ſome hundred years old. The *French* hoſpital is likewiſe a fine building, being erected by *Lewis* XIV. for the reception of ſoldiers. The cathedral here is indeed an old edifice, but contains ſomething remarkable in it, its foundation being laid in water and a clayey ſoil. A boat might formerly paſs under the loweſt vaults of it, but the entrance has been now for ſome years walled up. In the church is not only a very curious clock and organ, but more particularly a very ſplendid altar-cloth, which was a preſent from *Lewis* XIV. and ſaid to have coſt 600,000 dollars. To it, excluſive of a triple ſet of miſſal veſtments and altar furniture, belong ſix large ſilver chandeliers, each of which requires a ſtrong man to carry it, and a crucifix of double that weight. All theſe ſeven pieces of plate put together weigh 1600 marks, or 1066 pounds eight ounces. The church-tower is of a pyramidal figure, being five hundred and ſeventy-four feet in height. Excluſive of theſe, in it are three chapters, or collegiate-churches, two of which are alſo parochial, together with four others belonging to the *Romiſh* community ; but the *Lutherans* are in poſſeſſion of St. *Thomas*'s church, in which, in the year 1751, the illuſtrious Count *Maurice* of *Saxony* was buried.   To the *pediger*, or ' new church,' belong the churches of St. *Ni-colas* and St. *Aurelia.* The Proteſtants perform their public worſhip at *Wolfiſheim,* and not in the city.   The univerſity and gymnaſium here is *Lutheran,* and it has alſo an anatomical theatre, a phyſic garden, a college of Jeſuits, and a royal ſociety founded in the year 1752, and particularly inſtituted for the natural hiſtory of *Alſace*; as alſo ſix convents.   The city-council is half *Lutheran* and half *Roman* Catholic, but moſt of the burghers are *Lutherans.* The college of magiſtrates conſiſts of three orders, or cham-bers.   To the firſt belong the *Dreyzehner,* or thirteen, who preſide over matters of juſtice ; to the ſecond, the *Funfzehner,* or fifteen, who look to the rights and privileges of the city, the hoſpitals, police and finances; and, in the *Gin* and *Zwanziger,* or twenty-one, is lodged the ordinary govern-ment of the city.   Over theſe is the great council, which conſiſts of thirty members, *viz.* ten noblemen and twenty burghers.   It has alſo an infe-rior council for the determination of leſs important affairs.   The ſupreme magiſtrate here, ſince the city has fallen under the dominion of *France,* is the royal *prætour,* who takes care of the King's rights, and ſees that nothing be done in the college of magiſtrates repugnant to his Majeſty's pleaſure.   The inhabitants pay nothing to the King, but all the impoſts levied here are ex-pended in the ſupport of the city.   The nobleſſe of *Lower Alſace* ſtill hold their monthly meetings in it.   It was formerly an imperial free city.   In the year 1681, it ſubmitted to *Lewis* XIV. who confirmed to them all their rights and privileges.

The long wooden bridge over the *Rhine* belongs to the city, as alſo the five following diſtricts or lordſhips ; namely, thoſe of *Ilkirch, Werlenheim, Marlenheim, Waſlenheim, Herrenſtein* and *Bar.*

2. The diſtrict, or government, of *Hagenau* includes the ten cities of *Alſace,* formerly imperial, which are now entirely diſmembered from the Empire and reduced to the obedience of *France.* Their deſtiny has already been ſpoken of. *Lewis* XIV. gave this diſtrict to Cardinal *Mazarin,* with full power of transferring it by inheritance to his ſiſter's ſon and his de-ſcendants.   The ſix following towns lie in *Lower Alſace.*

Hagenau, a little town, ſeated on the *Motter,* and formerly the reſidence of the imperial bailiwick, but at preſent is the ſeat of a royal *prevoté* and a foreſt-court.   In it is alſo a college of Jeſuits.   It was anciently better for-tified than at preſent, being environed only with a rampart and moat.   In the year 1673, this place was taken by the *French,* and, in 1675, diſ-mantled by the Imperialiſts; but, in the year 1706, it fell again into the hands of the *French.* Its preſent condition is very indifferent.   To the di-ſtrict of *Hagenau* belong upwards of ſixty good villages.

Weiſſenburg,

Weiſſenburg, or *Kron-Weiſſenburg*, a little town, lying on the *Lauter*, is the ſeat of a diſtrict, and contains a chapter, the probſly of which was annexed to the biſhoprick of *Spires*, together with an abbey of *Benedictines*. In the year 1673, its fortifications were demoliſhed. *Staniſlaus*, king of *Poland*, reſided here a conſiderable time, and near it, in 1744, the *Auſtrians* were defeated by the *French*.

Landau, an important fortification, ſituate on the river *Queich*, and, though betwixt five and ſix *French* leagues from *Alſace*, yet, being ſurrounded by the Palatinate, belongs to *Lower Alſace*. The town, in general, is well built and regular, and contains four churches and a chapter. As it is almoſt commanded on every ſide by the eminences ſurrounding it, *Vauban* exerted all his ſkill in fortifying this place, and near it ſtands a fort on a hill. In the year 1702, it was taken by the *Auſtrians*; and, in 1703, recovered again by the *French*. In the year 1704, the Imperialiſts made themſelves maſters of it a ſecond time; and, in 1713, it was again taken from them. In 1714, by the treaty of *Baden*, this town, together with the three villages and fortifications belonging to them, was ceded to *France*. Its canal, which is chiefly ſupplied with water from the river *Queich*, comes from *Anweil*.

Roſheim, a little town, ſeated on the river *Magel*.

Ehenheim, or *Upper-Ehenheim*, ſo called with reſpect to the village of *Nieder-Ehenheim*, is a little town, ſtanding on the river *Ergers*, and containing a chapter.

Schletſtat, a fortified town, ſeated in a marſhy country on the *Ill*. The Jeſuits have a college here. In the year 1673, *Lewis* XIV. ordered it to be diſmantled, but, in 1679, its fortifications were again repaired.

3. The diſtrict of *Fort Louis* receives its name from, and has its ſeat at

Fort Louis, a very handſome fortification, ſituate on an Iſland in the river *Rhine*, being partly within the territory of the Empire, and partly in that of the baron of *Flectenſtein*. It was erected in the year 1686, and conſiſts of a ſpacious and regular quadrangle, with four large baſtions and the ſame number of half-moons. Near it is a little town with ſtraight ſtreets, and the whole iſland is fortified round. At the peace of *Ryſwick*, and alſo at that of *Baden*, *France* retained this fortification.

4. Eccleſiaſtical foundations; namely,

1. The biſhopric of *Strasburg*, which comprehends a conſiderable tract of territory ſubject to the biſhop as temporal lord; but this tract lies ſcattered in *Upper* and *Lower Alſace*, and on the other ſide of the *Rhine*. The biſhop has alſo two diſtricts. I ſhall here give an account of the whole biſhopric. He ſtyles himſelf a Prince of the Empire and Landgrave of *Alſace*; and, though at preſent ſubject to the dominion of *France*, yet, by virtue of the two diſtricts on the other ſide of the *Rhine*, is actually a ſtate of the Empire with a ſeat and vote in the Dyet; and his ſovereignty over

both

both theſe diſtricts is the ſame with that of other Princes of the Empire over their territories. He has alſo a great many vaſſals under him, and among them belong almoſt the whole nobility in the *Lower Alſace*, together with a great part of thoſe in the *Upper*. His revenue is ſaid to amount annually to 250,000 livres, and it was formerly much more conſiderable. As a biſhop he is ſubordinate to the archbiſhop of *Mentz*.

Belonging to him in *Lower Alſace* is

1. The diſtrict of *Zabern*, containing
Zabern, or *Alſace-Zabern*, in Latin *Tabernæ Alſatiæ*, a little town, lying on the river *Sor*, which was the uſual reſidence of the biſhop of *Strasburg*, till this city embraced the *Lutheran* religion. His palace and gardens are delightful; but what is more particularly remarkable, is the beaten broad road carried behind the city up to the top of a high mountain, which contains ſo many windings in it that a perſon may either deſcend or aſcend with the greateſt caſe. In it is a collegiate-church one hoſpital and two convents.

2. The diſtrict of *Rochersberg* derives its name from an old ruined caſtle and near, it is covered with fine villages.

3. The diſtrict of *Wanzenau* contains
Wanzenau, a market-town, near the conflux of the *Ill* and the *Rhine*.

4. The half of the market-town of *Marlenheun*, noted for its excellent red wine; the other half belonging to the city of *Strasburg*; and a part of the little town of *Wangen*.

5. The diſtrict of *Dachſtein*, in which is
Dachſtein, a little town, ſeated on the *Breuſch*.
Berghietheim, a large market-town.
Sulz, a village, formerly a town, and famous for its medicinal waters.
Molzheim, a little town, containing a college of Jeſuits.
Biſchoffhelm, a large market-town.

6. The diſtrict of *Schirmek* contains
Mutzig, a little town.
Hermolſheim, a village and convent of *Franciſcans*, near which begins the canal leading to *Strasburg*.
Haſlach, a collegiate-church and probſtey.
Schirmek, a village and ancient caſtle.

7. The diſtrict of *Benfeld* contains
Benfeld, a little town, ſtanding on the *Ill* and anciently fortified.
Rheinau, formerly a large town, but more than one half of it at preſent carried away by the *Rhine*. There being ſeveral iſlands in that river, the *French* in their wars have generally made uſe of them, for throwing bridges over the *Rhine* into *Breiſgau*.
Epfig, a large market-town.
Dambach, a little town, ſeated on the *Scheer*.

Eberſheim,

Eberſheim Munſter, a ſmall town, having an abbey, lies betwixt two branches of the *Ill*.

In *Upper Alſace* is

8. The diſtrict of *Markolſheim*, or *Margelſheim*, having its ſeat in a little town of the ſame name, and comprehending alſo ten large villages.

9. The *Upper-Mundat*, as it is called, conferred by King *Dagobert* on the biſhopric, includes

Egiſheim, a little town, having a very ancient caſtle.

Geberſweiher, a handſome market-town, ſurrounded with vineyards.

Pfaffenheim, a large market-town, containing an old caſtle.

Ruffach, a little town, the ſeat of the upper diſtrict of *Mundats*, containing one college of Jeſuits, one convent of *Franciſcans*, a commandery of the *Teutonick* order and one hoſpital.

Iſenburg, a caſtle, ſtanding on a hill which is covered with vineyards.

Weſthalten, a handſome market-town.

Sulzmath, a market-town, containing a mineral ſpring.

Sulz, a little town and the ſeat of a juſticiary. In it is alſo a commandery of the knights of *Malta*.

On the oppoſite bank of the *Rhine*, on the *Breiſgau* ſide, lies

10. The diſtrict of *Oberkirch*, in which is

Oberkirch, a little town, ſtanding on the river *Renich*.

Oppenau, a ſmall town, ſituate on the ſame river.

Wald-Ulm, a market-town.

11. The diſtrict of *Ettenheim*, containing

Ettenheim, a little town.

The chapter of the cathedral of *Strasburg* conſiſts of twelve *Capitularians* and the ſame number of *Domicellarians*, among whom the *Germans* muſt all be Princes or counts of the Empire, and the *French*, Princes, dukes, peers or marſhals of *France*. To the cathedral belong the following places ; *viz.*

Lampertheim, a large village.

Geiſpolzheim, commonly called *Geiſpitzen*, a large market-town.

Erſtein, one of the largeſt market-towns in *Alſace*, containing a convent. To it alſo belongs the inn * and brick-kiln of *Kraft*.

Borſch, a ruinous little town.

Upper and *Lower Munſterhorf*, in the town of *Upper-Ehenheim*.

Eberſheim, a large village.

Scheerweiler, a large market-town.

---

* We do not know what uſe it may be of for the ſettling of the boundaries in a conteſted country which is divided among ſo many potentates, to be preciſe in mentioning circumſtances of this nature ; but, in *England*, we ſhould hardly think a brick-kiln or inn worthy notice, even though the one ſupplied us with the beſt of materials for building, and the other with the beſt of liquor.

Roſtenholz,

Roftenholz, alfo a large market-town, with a manfion-houfe for the ufe of the fteward of the chapter.

3. The probftey of *Kron-Weiffenburg* has its feat in the once imperial city of *Kron-Weiffenburg,* and was once an abbey. The Emperor *Charles* V. and Pope *Paul* III. added it to the board-lands of the bifhop of *Spires.* To it belong a great many fiefs and the following places ; *viz.*

Lauterburg, a little town, fituate not far from the *Rhine,* and the feat of a diftrict.

Salmbach, a market-town.

The villages of *S. Remich,* part of *Altftadt, Babenthal, S. German, S. Paul Lchn, Vierthurnen, S. Walpurg* and *Weiller.*

4. The *Ciftercian* abbey of *Neuenburg* lies near the *Motter.* To it belongs the convent of *Baumgarten,* with two villages.

5. The nunnery of *Andlau,* inftituted for ladies of noble extraction, ftands in the little town of the fame name, which was formerly an imperial free city. To it alfo belong the caftle of *Freudeneck,* together with the convent of *Hugfhofen.*

5. The lorfhip of *Lichtenberg* and *Ochfenftein,* after the failure of the lords of *Lichtenberg,* defcended by marriage to the counts of *Hanau,* the laft of whom ceded it, during his life-time, to his fon-in-law the landgrave of *Heffe-Darmftadt,* &c. It confifts of the following fcattered diftricts ; *viz.*

1. The diftrict of *Hatten,* or the little country of *Hatgau,* the capital of which is the market-town of *Hatten,* lying on the *Rhine.*

2. The diftrict of *Word,* containing

Word, a little town, feated on the river *Sor.*

Gerfdorf, or *Gerlingfdorf,* a fmall town.

3. The diftrict of *Niederbrun,* in which is

Niederbrun, a little place, having a medicinal fpring containing fulphur and copper.

Waffenburg and *Arnfperg,* two old caftles feated on hills.

Reichfhofen, a fmall town.

4. The diftrict of *Jugweiler,* to which belongs

Lichtenberg, a ftrong caftle ftanding on a hill.

Jugweiler, a little town, lying on the river *Motter.*

5. The diftrict of *Pfaffenhofen,* containing

Pfaffenhofen, a fmall town, feated on the river *Sor.*

6. The diftrict of *Bufchweiler,* in which is

Bufchweiler, the beft town in this lordfhip, with a feat.

Neuweiler, a little town, having a collegiate probftey.

7. The diftrict of *Brumath,* containing

Brumath, or *Brumpt,* a market-town, having a feat.

8. The diftrict of *Offendorf,* in which is

Drufenheim, a market-town, fituate at the conflux of the *Motter* and *Rhine.*

I   *Offendorf,*

Offendorff, a large market-town, ſeated near the conflux of the rivers *Sor* and *Rhine*.

9. The diſtrict of *Lichtenau* lies on the other ſide of the *Rhine*, but be-longs not properly to *Alſace.* In it is

Lichtenau, formerly a town, but now only a large village.

10. The diſtrict of *Wilſtadt* lies on the other ſide of the *Rhine*, containing *Wilſtadt*, a good town, and ſeveral villages.

11. The diſtrict of *Wolfiſheim*, comprehending

Wolfiſheim, a little place, in which the Proteſtants of *Strasburg* and its neighbourhood are allowed the public exerciſe of their religion.

12. The diſtrict of *Weſthofen* contains

Weſthofen, a little town and the ſeat of the diſtrict.

Balbron, a ſmall town, through which runs the river *Moſig*.

13. The lordſhip of *Ochſenſtein* lies on the borders of *Lorrain*, near the *Sor*, below *Zabern*.

6. To the counts of *Leiningen* belong

Dagsburg, a ruinous caſtle, ſeated on a mountain on the borders of *Lorrain*.

Glaſhut, ſo called from its making a fine kind of glaſs.

Morzweiler, Neiſſeren, Oberbrun, Zinſweiler, Weyherſheim:

Rauſchenburg, containing a fine caſtle, and being in itſelf alſo a very pretty place.

Obſ. Of the lordſhip of *Kleeburg*, and the other places belonging to *Deux-ponts*, together with *Kopenum*, a diſtrict of *Baden*, an account ſhall be given in *Germany*, Vol. 4.

7. To the dutchy of *Birkenfeld* belongs

Biſchweiler, a handſome market-town, lying cloſe by the river *Motter*, with a fine caſtle, which was formerly the reſidence of the duke of *Birkenfeld.*

The barony of *Fleckenſtein* comprehends ſeveral places and little territories.

9. The ſmall town and mark of *Maurſmunſter* have different lords, who are called *Mark-lords.* In it is an abbey.

10. The lordſhip of *Thanweiler*, or *Thalweiler*, in which are the little towns of *Weiller*, or *Wiler*, together with the convent of *Hugſhofen*, and ſome other little places.

Obſ. The eſtates and diſtricts of the nobleſſe, which, though once be-longing to the Empire and conſiderable, yet our brevity will not permit us to enumerate them here.

II. Upper Alsace contains

1. Four cities, once belonging to the Empire, and ſubordinate to the im-perial bailiwick of *Hagenau*; viz.

Colmar, in Latin *Columbaria*, the capital of *Upper Alſace*, and the ſeat of the ſovereign council and intendancy of the country. It is the principal place of a collection-office, lies on the river *Lauch*, and was formerly fortified, but in the year 1673 diſmantled. In it is a collegiate-church. This town belongs to the little town of *Heiligkreutz*, with a caſtle in which the chief magiſtrate of the bailiwick reſides.

Turkheim, a little town, near which, in the year 1675, Marſhal *Turenne* defeated the Imperialiſts.

Kaiſersbeg, a ſmall town, not far from which is produced an excellent wine. This place ſuffered much in the wars of 1652, 1674 and 1675.

Munſter, in the valley of *Gregorien*, a little town, containing a very rich abbey of *Benedictines*, and being formerly in immediate ſubjection to the Empire. To this town and caſtle belong ſeveral villages.

2. Eccleſiaſtical foundations and lands.

The diſtricts belonging to the biſhoprick of *Strasburg*, and lying here, have been already ſufficiently deſcribed. The cathedral of *Strasburg* has ſeveral incomes and contingencies ariſing to it from thoſe foundations, particularly at *Zellenburg*.

1. The princely imperial foundation of *Murbach*, inſtead of being transferred to *France*, by the peace of *Munſter* continued an immediate ſtate of the Empire; but, in the year 1680, fell under the dominion of that rapacious crown. To it belongs

Murbach, the *Benedictine* abbey itſelf, the abbot of which is a Prince. It was founded in the year 724. Not far from hence is the high mountain of *Belch*.

Gebweiler, a little town, ſtanding on the river *Lauch*, and containing a court of juſtice, a mint and a receiver's-office. *Sering*, a mountain contiguous to it, produces a delicious kind of wine. In this place alſo the *Murbach* and *Lauchbach* unite their ſtreams, by means of which, *Lewis* XIV. for the more ſpeedy conveyance of materials for building, cauſed a canal to be made which reaches to *New Briſach* by the way of *Ruffach* and *Herliſheim*, where it receives another canal from the river *Ill* in its way from *Enſiſheim*.

Angretſtein and *Hungerſtein*, ſeats and fiefs, ſituate on the river *Murbach*. *Lautenbach*, a large market-town, containing an opulent canonſhip.

Watweiler, a little town, having a manour-court and medicinal bath.

S. Amarin, or *Dantarin*, a ſmall town and the reſidence of a juſticiary. Of the old caſtle of *Friederichsburg*, near it, only one tower now remains. At the village of *Urbis* is a paſs leading towards *Lorrain*. Omitting many other villages and places belonging to it, its chief appendix is

The princely foundation of *Luders* in *Upper Burgundy*. The abbey and little town belonging to it, lie on an iſland in a lake. To this city alſo appertain the bailiwicks of *Blanchier* and *Paſſavant*; namely, the ſeat of *Paſſavant*, in the town of that name, with its dependencies and other lands.

2. The free Imperial foundation in the valley of *Gregorien*, which has already been mentioned in the town of *Munſter*. The juriſdiction of this valley

valley, in which are several villages, is lodged both in the abbot and the town.

3. The earldom of *Rappolstein,* in *French Ribaupierre,* belonging to the duke of *Deux-ponts,* contains

1. The bailiwick of *Goemar,* or *Gemar,* being a little town, having a seat near the *Ill.*

2. The bailiwick of *Berkheim.* In it

Berkheim, a little town, from which to *Gemar* runs a dry ditch, as the boundary betwixt *Lower* and *Upper Alface.*

3. The bailiwick of *Rappolsweiler* ; in it

Rappolsweiler, a town, lying lengthwise betwixt two mountains.

Rappolstein, the name of three ruined castles which stood on a hill not far from the town, and are still fiefs of the bishop of *Baffel.*

Dreykirchen, or *Dufenbach,* a convent with a church, much frequented by pilgrims.

4. The bailiwick of *Markirch* contains

Markirch, in French *S. Marie aux Mines,* a little town.

In *Leber-Thale* are some mine-works of copper and silver. In it also is *Leberau,* a market-town, and *Furtelbach.* The mine-works here belong partly to *Rappolstein.*

5. The bailiwick of *Hohenack;* in it

Hohenack, a ruined castle, seated on a hill. *Urbis,* a market-town, with some other places.

6. The bailiwick of *Zellenberg* contains

Zellenberg, a little town and castle, standing on a mountain, which, together with the whole country, is almost one continued vineyard. The chapter of *Strasburg* have a seat here.

7. The bailiwick of *Weyer.* In it

Weyr, or *Wihr,* a castle, seated on a hill.

8. The bailiwick of *Heydern,* consisting only of some villages, lies near the *Rhine.*

4. The lordship of *S. Hippolite* belonged formerly to the bishop of *Strasburg,* who was compelled, in the year 1372, to cede it to the duke of *Lorrain,* which cession was confirmed to it, in 1718, by *France.*

S. Pild, or *Bild,* instead of *Hippolite,* a little town, containing a seat.

5. To the duke of *Wurtemburg* belongs

1. The lordship of *Reichenweyer,* containing

Biblstein, a castle in ruins, lying not far from the town of *Rapolsweiler.*

Reichenweyer, a pretty town and seat, its neighbourhood noted for the heft wine in all *Alface.* The castle of *Reichenstein,* which lies at no great distance from it, is now in ruins.

Nunnenweyer, Mittelweyer, and *Munchweyer,* all market-towns.

2. The earldom of *Harburg,* to which, among other places, belongs

Often, a market-town, seated on the *Ill.*

4 H 2

Bifchwihr,

Bifchwihr and *Forfchwihr*, two little places.

Harburg, or *Horburg*, once a very strong and beautiful castle, seated on the *Ill* and lying in ruins.

Obf. In this earldom, and likewise in the *Wurtemburg* territories, stands

New Breifach, not far from the *Rhine* and opposite to *Old Breifach*, being a regular octogon most admirably fortified, and of such exact symmetry, that from the market-place a person may see the four gates of the place. It was built by *Lewis* XIV. after the peace of *Ryfwick*. Near it, on the arm of the *Rhine*, stands *Fort Mortier*.

Formerly, at a small distance from this important fortress, stood

Jacobs Schanz, or *James-fort*, on an island, near which, by order of *Lewis* XIV. was built a town with the name of *S. Lewis*, but since the peace of *Ryfwick* it is fallen to decay.

Here also mention may properly be made of *Fort Sponck*, which lies on the other side of the *Rhine*.

6. The barony of *High Landfperg*, so called from a ruined castle, belongs at present to Baron *Leyhen*. In it is

Ammerfweyer, a little town.

Kuhnfheim, or *Konigfheim*, another small town.

Winzenen, or *Winzenheim*, a large market-town, seated among vineyards.

Wettelfheim, or *Wedelfheim*, also a large market-town.

7. The barons of *Schauenburg* possess

Schauenburg, a pilgrimage, standing on an eminence.

Sulzbach, a little town, having a mineral spring of great repute.

Herlifheim, a little place, containing a seat, stands on the *Lauch*.

Hatfiat, a large market-town, but the castle called *Hohen-Hatfiat* lies in ruins.

8. The lordship of *Enfifheim*. In it

Enfifheim, a pretty town, seated on the *Ill*, and containing a college of Jesuits:

9. The lordship of *Eifenheim* contains

Eifenheim, a little place, seated on the river *Lauch*.

10. The lordship of *Sennen* contains

Sennen, a small town.

III. SUNDGAU, i. e. *Pagus Meridionalis*, in opposition to *Nordgau*, borders, to the north, on *Upper Alface*; to the eastward, on the *Rhine* and the canton of *Bafel*; to the south, on the bishopric of *Bafel*, the earldom of *Mumpelgarde* and *Franche comté*; and, westward, on *Lorrain*. Its length, from west to east, is twelve *French* leagues, and its breadth nearly the same. The whole country speaks *German*, except the borders of *Elgrau* and *Burgundy*, where a corrupt *French* prevails. The inhabitants are principally Papists. This place was formerly a fief of the bishopric of *Bafel*, and had counts of its own, the last of whom died in the year 1324, leaving only one daughter, by name *Johanna*; who marrying with *Albert*, duke of *Austria*,

brought

brought him this earldom. At the peace of *Munſter*, In the year 1648, it was ceded by the Emperor and *Empire* to *France*. The King firſt granted it to the marquis *de Suſe*, afterwards Cardinal *Mazarin* procured it for himſelf and left it to the duke of *Mazarin*, whoſe deſcendants ſtill poſſeſs it under the King. This country conſiſts of the following bailiwicks; *viz.*

1. The bailiwick of *Landſer*, which extends itſelf along the *Rhine*, and includes a large foreſt called the *Hart*, being about five *French* leagues in length and, in ſome places, one broad. It contains

The market-towns of *Lanſeren*, *Habſen* and *Ottmarſen*, in the laſt of which is a foundation for ladies.

The fort of *Funingen* lies on the *Rhine* oppoſite to the city of *Baſel*, and ſo near that both towns are within gun-ſhot of each other. *Lewis* XIV. ordered this fort to be built in the year 1679.

3. The lordſhip of *Landſkron* lies in the *Blue* mountains called, in *French*, *Blaumont*, or *Laumont*, which is a branch of the *Jura* or *Jurten* chain, and contains not one remarkable place in it except

Landſkron, a ſmall fortification, ſtanding on a hill.

2. The bailiwick of *Pfird*, in French *Ferrette*, had formerly counts of its own, who were lords of the whole *Sundgau*. The places of note in it are

Pfird, the capital of the earldom of this name, being a ſmall town, containing a caſtle. *Alt-fird* is the ſuburb, and *Hohenpfird* a ruined caſtle, ſeated on a high rock.

Feldbach, a village and convent.

Liebenſtein, a village and caſtle, ennobling its proprietors.

Lupach, a convent and village.

Bietherthal, a caſtle and village, nobility annexed to it.

St. *Brix* and *S. Blaſi*, two pilgrimages.

Munchenſtein, a market-town and caſtle; its proprietor a noble.

The ancient lordſhip of *Morſperg* had formerly lords of its own, but at preſent forms a part of a bailiwick.

Morſperg is a market-town.

3 The bailiwick of *Altkkirch*, containing

Altkirch, a little town and caſtle, which, with ſome other villages, conſtitutes a lordſhip.

The noble villages and caſtles of *Lummeſweiler*, *Freningen* and *Blotzen*.

Our Lady zum grunen wald, a pilgrimage.

4. The bailiwick of *Thann*, in which is

Thann, or *Dann*, a town, not far from the river *Thur*, containing a foundation removed hither from St. *Amarin*. The church here is remarkable for the height of its tower, and without the town are two convents. A very ſtrong wine is produced here.

Rothenburg, *Brun*, *Morzweiler*, *Amerſweiler* and *Giltweiler*, villages and caſtles conferring nobility.

Dammerkirch, a large market-town.

5. The

5. The diſtrict of *Befort*, or *Betfort*, containing

Befort, a fortified town, ſeated on the *Hall*. In its neighbourhood are ſome excellent mines of iron. Cloſe by lies an old caſtle, which is ſeated on a rock, and oppoſite to it is a fort built in another hewn rock between which two places the road goes towards the town, and near them terminates *Alſace*.

Roſenberg, a little town, having an under diſtrict.

Dattenried, or *Dell*, a ſmall town, alſo containing an under diſtrict.

Grandvillar, a little town, belonging to the under diſtrict of *Dattenried*.

Laſtly, in *Sundgau* alſo lie

1. The lordſhip of *Blumberg*, in French *Florimont*, conſiſting of the little town and caſtle of *Blumberg*, with five villages, and belonging to the lords of *Pfird*.

2. The lordſhip of *Maſmunſter*, lying in a valley. *Lewis* XIV. made a preſent of this lordſhip to Baron *Ratſki*. To it belong

Maſmunſter, a little town, ſeated on the river *Tolder*, and containing a *Benedictine* foundation for ladies of noble birth.

Seeben, a large market-town, ſituate on a lake through which runs the river *Tolder*.

3. The town of *Muhlhauſen*, with its diſtrict, was formerly an imperial town, but, in the year 1506, entered into the union of the *Swiſs* cantons, in which it ſtill continues. In the year 1532, this town embraced the Proteſtant religion. It lies on the river *Ill*, and to its diſtrict belongs

Ilzach, a large market-town, which, together with *Montenheim* and the conſiſtory of *Sauſheim*, with all its dependences and the high and low juriſdiction, were, in the year 1437, bought by the town of *Mulhauſen* of the Counts *Ulrick* and *Lewis* of *Wurtemberg*.

4. The lordſhip of *Lutſzelſtein*, now belonging to the *Pfalſgrave*, of *Deuxponts*, comprizes,

Lutzelſtein, a little town, having a ſtrong caſtle, in which is the burial-place of the old counts of *Veldenz*.

Borſpach, *Rugelſtein*, *Birſingen*, *Inſtall*, *Eſperance*, *Matten*, *Briſlo*, *Little Biſch*, *Neupalberg*, *Dorſchweiler*, *Chamberic*, *Nettenweiler* and *Hanſbach*.

5. The principality of *Phalzburg* conſiſts of the villages and caſtles which belonged formerly to the lordſhip of *Lutzelberg*, and were diſmembered from the demeſnes of the biſhoprick of *Metz*. In the ſixteenth century, the ſovereignty of *Lutzelburg* fell to the duke of *Lorrain*, who built the caſtle of *Phalzburg*, which he ceded, in the year 1661, to *France*. *Lewis* XIV. cauſed this caſtle to be erected into an admirable fortreſs, and, in the year 1718, by the treaty of *Paris*, obtained from the duke the caſtle of *Lutzelburg* and the whole diſtrict of *Phalzburg*, which bore the title of a principality.

The

The G O V E R N M E N T *of*

F R A N C H E   C O M T E.

T H E earldom of *Burgundy,* or *Franche Comté,* is bounded to the north by *Lorrain*; to the eaft by the earldom of *Mumpelgard* and *Switzerland*; and to the fouth and weft borders on the government of *Burgundy* and *Champagne..* Its extent from fouth to north is thirty *French* leagues, and from fouth-eaft to north-weft twenty. Almoft one half of it is a level country, abounding in grain, wine, paftures, hemp, &c. The other half is mountainous, but produces a good breed of cattle, and alfo fome corn and wine. Its principal rivers are the *Saone,* the *Ougnon,* the *Doux,* the *Louve,* and the *Dain.* This country contains mines of copper, iron, lead and filver, and at *Luxeuil* and *Repes,* are mineral fprings. Near the village of *Touillon* is a fpring which runs and ceafes at ftated times, and the town of *Salins* has very profitable falt-fprings and marfhes. In this diftrict are alfo quarries of alabafter and marble. One *French* league from *Quingey,* about fifty paces from the river *Doux* is a large grotto, in which nature has formed pillars, monuments, and a furprifing variety of figures. Near *Leugne* is a natural cavern of ice, which thaws in winter and freezes in fummer.

This country was anciently a part of the Kingdom of *Burgundy,* and conferred by *Lewis the Pious* on his fon *Lotharius,* to whom fucceeded *Charles the Bald,* in whofe time it was called *Upper Burgundy.* It had afterwards counts of its own, and was a fief of the *German* empire. Count *Rheinold* III. withdrew himfelf from his allegiance to the Emperor *Lotharius* II. and from this defection the country is faid to have acquired the name of *Franche Comté.* Count *Otho* affumed the title of Palfgrave. The country defcended by marriage to *Philip* of *France,* furnamed the *Bold,* who was the founder of the fecond ducal line of *Burgundy,* from whence the laft duke, *Charles the Bold,* derived his pedigree. *Mary,* his daughter and heirefs, brought the country by marriage to *Maximilian* of *Auftria:* *Charles* V. united the dutchy and earldom of *Burgundy* with the *Netherlands,* by which means it belonged as a part of the circle of *Burgundy,* to the *German* empire. In the years 1668 and 1674, *France* took and retained it by virtue of the peace of *Nimeguen.* *Franche Comté* has a parliament of its own; is fubject to a governor, general-lieutenant, and a fub-governor; and confifts of four large diftricts; *viz.*

I. The diftrict of BESANÇON, containing

Befançon, in Latin *Vefontio,* and *Befontium,* the capital of the country, and the feat of an archbifhopric, a parliament, an intendancy, a collection, a bailiwick, a country and a mint-court, as alfo of a marble-table and a foreft-tribunal, &c. being divided by the river *Doux* into the *upper* or *old town,* or *lower* or *new town.* Till the peace of *Weftphalia* it was a free imperial

perial city, but at that time was transferred to *Spain*. *Lewis* XIV. made it an important fortification, which was afterwards ſtrengthened by the addition of two citadels. It contains two chapters, eight pariſh-churches, four abbies, one ſeminary, one univerſity, one college of Jeſuits, twelve convents, and three hoſpitals. The archbiſhop of this place ſtiles himſelf a Prince of the empire, has three ſuffragans under him, and a diocefe of eight hundred and thirty-eight pariſhes, with a revenue of 36,000 livres, out of which his taxation at the court of *Rome* is 1023 florins.

II. The diſtrict of DOLE or MILIEU comprehends the three under diſtricts of DOLE, QUINGEY, and ORNANS.

Dole, a town on the river *Doux*, in a diſtrict, which for its beauty and fertility is ſtiled *Val d'Amours*, was formerly, while *Beſançon* continued an imperial free city, the capital of the country, the ſeat of a parliament, a chamber of accounts and an univerſity, being alſo fortified; but on its reduction by *Lewis* XIV. in the year 1668, he cauſed the fortifications to be demoliſhed; and the *Spaniards* having fortified it again, and *France* again in the year 1674, making itſelf maſter of the place, it was a ſecond time diſmantled, and the parliament and univerſity afterwards removed to *Beſançon*, but the chamber of accounts ſtill remain here. In it alſo is one chapter, one college of Jeſuits, eleven convents, and one hoſpital.

Samuans, a village noted for its quarry of marble.

Quingey, a little town ſeated on the *Louve*, and the ſeat of a bailiwick, one pariſh-church, and two convents.

Ornans, a ſmall town on the *Louve*, containing a bailiwick. This place, together with *Villafans* and *Bracons*, anciently formed a ſeparate lordſhip.

III. The diſtrict of AMONT or GRAY includes the three under diſtricts of VESOUL, GRAY, and BEAUME.

Gray, in Latin, *Gradicum*, is a little town on the *Saone*, the ancient works of which *Lewis* XIV. cauſed to be demoliſhed in the year 1668. In it is a collegiate and one pariſh-church, five convents, and one college of Jeſuits. In this place is ſhipped the grain and iron conſigned to *Lyon*.

Veſoul, in Latin *Veſolum*, or *Veſullum*, a little town ſtanding on a hill, at the foot of which runs the river *Durgeon*. In it is one collegiate-church, three convents, and one college of Jeſuits. This place ſuffered greatly by the wars.

Beaume les Nonnes, a little town lying on the *Doux*, almoſt entirely deſtroyed in the wars.' In it is one pariſh-church and two convents.

Peſmes, *Marnay*, *Gy*, *Villers* on the *Scey*, *St. Hipolyte*, *Jonvelle*, *Dampierre*, *Hericourt*, the iſland in the *Doux*, *Clairevaux*, *Champlitte*, *Amance*, *Belvoir*, *Bouclans*, *Faucogné*, *Charié*, &c. are ſmall towns and larger villages.

In this diſtrict are included the three following places and tribunals, which are not ſubject to the bailiwick of *Amont*, but depend immediately on the parliament of *Beſançon*.

Luxueil, or *Luxeu*, a little town at the foot of the *Waſgau* mountains, owes its origin to a celebrated abbey founded there by St. *Colomban*, in the

year

year 602, which embraced the rule of St. *Benedict.* The abbey was for.
merly an immediate ſtate of the empire. Near this town is a mineral ſpring.

Lure, an abbey of *Benedictines,* which was united to the abbey of *Murbach* in *Alſace.*

Vauvillers, a little place ſeated on the borders of *Lorrain.*

IV. The bailiwick of Aval compriſes the under bailiwicks of Poligny,
Salins, Arbois, Pontarlier and Orgelet.

Salins, in Latin *Salinæ,* a pretty large town in a valley on the little river of
Furieuſe, is the ſeat of a country and a diſtrict-court, with a collection-office.
In it are four chapters, four pariſh-churches, ten convents, one college, and
one hoſpital. Its ſalt-ſprings and marſhes are of great advantage to it.   In
the neighbourhood of the town are quarries of jaſper, beautiful alabaſter,
and black marble. Near it ſtands *Fort Belin,* which is ſeated on a hill, and
on another the *Redoubte* of *Fort Bracon,* together with the caſtle of *Fort
S. Andre.* This town had anciently lords of its own.

Arbois, a little town containing a chapter, a priory, and three convents.

Pontarlier, formerly *Pont Elie,* lies near Mount *Jurten,* on the borders of
Switzerland, which forms here a commodious road, defended by a caſtle
ſeated on a mountain at about half a mile's diſtance from the town, and
called the caſtle of *Joux.* In it is one pariſh-church, four convents, and a
houſe of Jeſuits.

Poligny, in Latin *Polemniacum,* a ſmall town in the country and earldom of *Woraſch.* In it is a chapter and five convents.

Lons, or *Lions le Saunier,* in Latin *Leodo,* a ſmall town, containing one
pariſh-church and five convents. Its ſalt-ſpring is not uſed for want of wood.

Orgelet, a little town containing one convent.

*Monmoret, Chatel-Chalon, Clereval, Nozeroy, Jouge, Montfleur, Saint-Amour,
Chavannes, Selieres, Bleterans, S. Julien, &c.* are all ſmall towns.

Immediately ſubject to the parliament of *Beſançon* is

S. Claude, or *S. Oyen de Joux,* a little town ſeated on a hill, and having
a noble abbey belonging to it. Excluſive of the caſtle, in it are alſo three
convents.

End of the Second Volume.

9 781333 725757